The Dictionary of the Etymology of Dinosaur Names

Written by

MAYUMI MATSUDA

Supervised by

Dr. **SHIN-ICHI FUJIWARA**
The Nagoya University Museum

HOKURYUKAN

For those who love dinosaurs
and are interested in the meaning of their names.

The Dictionary of the Etymology of Dinosaur Names

December 20, 2021, publication of the first edition

published by The HOKURYUKAN Co., ltd.
3-17-8, Kamimeguro, Meguro-ku
Tokyo, 153-0051, Japan

ISBN978-4-8326-1011-8

Printed in Japan

This book is an etymological dictionary of dinosaur scientific names, newly written in English based on the Japanese edition of the *Scientific names of Dinosaurs and their Etymologies* published in 2017.

Contents

Explanatory Notes

This book lists etymologies of type species name of non-avialan dinosaurs, as well as of some early avialans and non-dinosaurs. Taxonomy of extinct taxa listed on this book is based on *The Dinosauria 2nd edition* (2004), edited by D. B. Weishampel, P. Dodson and H. Osmólska. For the new species named after that, I referred to each paper. Among the taxa listed in this book, the taxonomy of avialan and non-dinosaurian taxa are square-bracketed after the type-species name. Scansoriopterygids and anchiornithids have been classified in the clade Avialae after Cau *et al.* (2017). The family names are specified for the reference, though the exact taxonomic positions are controversial in some taxa.

Heading: type species of the genus

> senior synonym of the species that have been included in the genus, including *nomina dubia* and reclassified species.
⇒ objective synonym or alteration
= subjective synonym

Signs

-> see
: explanation
---* a supplementary explanation, if necessary
/ = (equal) or "or"
>>> alteration. The generic name in parentheses is omitted from the description of the scientific name. Please refer to the renamed generic name after >>>.

Abbrevations

acc. = accusative
acc. = according to
adj. = adjective
adv. = adverb
Chin. = Chinese
cf. = confer, compare
dat. = dative
dim. = diminutive
e.g. (*exempli gratia*) = for example
Eng. = English
etc. = *et cetera* (= and so on)
f. = feminine ← ἡ (Gr.)
Fr. = French
gen. = genitive
Gr. = Greek
i.e. = *id est* (= that is)
indecl. = indeclinable
Jpn. = Japanese
Lat. = Latin
m. = masculine ← ὁ (Gr.)
Mod. Gr. = Modern Greek
Mong. = Mongolia, Mongolian
myth. = mythological or mythology
n. or *neut.* = neuter ← το (Gr.)
n.b. (nota bene) = note well
neg. = negative
nom. = nominative
pl. = plural
praep. or *prep.* = preposition
pron. = pronoun
Port. = Portuguese
sing. = singular
Sp. = Spanish
subst. = substantive

Technical terms

[anat.] anatomical, anatomy
[geol.] geological, geology
[med.] medical, medicine

Attribute of the proprietary nouns

(acronym): a word formed from the first letters of the words
(demonym): tribal names
(institution name): name of university, name of company, *etc.*
(person's name)
(place-name)
(stratal name)

Pronunciation of Latin, Greek, Japanese

Pronunciation of Latin

Vowel:

Y was Greek Υ, being equivalent to French U.

Consonants:

C was always pronounced as K.

G was always pronounced as in 'got' or 'give'.

X was always pronounced as KS.

TH, PH, and CH were probably sounded as T, P, and K followed by puff of air.

These letters (TH, PH, CH) were only in the transliterations of Greek words.

Pronunciation of Greek

alphabets	sound	examples
Α, α	[ā, ă]	ἀήϱ/āēr
Β, β	[b]	βάκτϱον/baktron
Γ, γ	[g]	γέϱᾰνος/geranos
Δ, δ	[d]	δάκτῠλος/daktylos
Ε, ε	[ĕ]	ἐχῖνος/echīnos
Ζ, ζ	[zd, dz]	ζῶστϱον/zōstron
Η, η	[ē, e]	ἡμέϱα/hēmerā
Θ, θ	[th]	θήϱ/thēr
Ι, ι	[ī, ĭ]	ἰχθύς/ichthys
Κ, κ	[k]	κόσμος/kosmos
Λ, λ	[l]	λᾰγός/lagos
Μ, μ	[m]	μάϱμᾰϱος/marmaros
Ν,ν	[n]	νᾶνος/nānos
Ξ, ξ	[ks]	ξένος/xenos
Ο, ο	[ŏ]	ὄϱος/oros
Π, π	[p]	πηλός/pēlos
Ρ, ϱ	[rh]	ῥάβδος/rhabdos
Σ, σ, ς	[s]	σαῦϱος/sauros
Τ, τ	[t]	τϱάχηλος/trachēlos
Υ, υ	[y]	ὕλη/hylē
Φ, φ	[ph]	φίλος/philos* *generally, 'fi'
Χ, χ	[kh]	χείϱ/cheir
Ψ, ψ	[ps]	ψιττακός/psittakos
Ω, ω	[ō, o]	ὤψ/ōps

Transliteration from Greek into the Roman alphabet

ζ = z, κ = c, ξ = x, ϱ (initial position) = rh, χ = ch, ψ = ps, γγ = ng, αι = ae (ai), οι = oe (oi), ει = ei, i, e, ου = u, υ = y, -ος (ending position) = -us (-os), -ον (ending position) = -um (-on)

Pronunciation of Romanized Japanese

Cha is pronounced as in 'challenge'.
e.g.: cha 茶 (= tea)

Chi is pronounced as in 'chicken', never in 'Chicago'.
e.g.: Chiba-ken 千葉県 (= Chiba Prefecture)

Chu is pronounced as in 'chew'.
e.g.: Chūō-ku 中央区 (= Chuo Ward)

Cho is pronounced as in 'chocolate' or 'chalk'.
e.g.: Chōshi 銚子 (=Choshi City)

G is pronounced as in 'go'.
e.g.: Ginza 銀座 (= a district of Chūō, Tokyo)

J is pronounced as in 'jinx'.
e.g.: jinja 神社 (= shrine)

Lexicon

— word list used in the generic names of dinosaurs and so on —

A

a-, an- (Gr.) ἀ-, ἀν-: alpha privative, expressing want or absence.
e.g.: *Abrictosaurus*; *Acristavus*; *Adelolophus*; *Adynomosaurus*; *Agnosphitys* [Silesauridae]; *Alectrosaurus*; *Anodontosaurus*; *Anoplosaurus*; *Anurognathus* [Pterosauria]; *Arrhinoceratops* ; *Astrophocaudia*

-a-, -an-
e.g.: *Megapnosaurus*; *Pteranodon* [Pterosauria]

aard- (Afrikaans) aarde: Earth, earth, ground. | aardvark: earth pig.
e.g.: *Aardonyx* -> -onyx

Abdarain- (Rus. place-name) Abdarain Nuru: Abdrant Nuru locality.
e.g.: *Abdarainurus* -> -ur(o)-

Abel (person's name) Roberto Abel: the former director of the Provincial Museum of Cipolletti in Argentina.
e.g.: *Abelisaurus*

abricto- (Gr.) ἄ-βρικτος/a-briktos: wakeful, awake.
e.g.: *Abrictosaurus*

abro- (Gr.) ἁβρός/habros: pretty, graceful, dainty, tender. (Mod. Gr.) αβρός/abros: gracious, delicate.
e.g.: *Abrosaurus*

Abydo- (Gr. place-name) Ἄβῡδος/Abydos: the oldest city of ancient Egypt.
e.g.: *Abydosaurus*

acantho- (Gr.) ἀκανθ(ο)-/akanth(o)-: spined. ἄκανθα/akantha, *f.*: a thorn, prickle, the vertebrae (backbone, spine).
e.g.: *Acantholipan* -> -lipan; *Acanthopholis* -> -pholis

-acantho-, acanthus
e.g.: *Acrocanthosaurus*; *Haplocanthosaurus*; *Metriacanthosaurus*; *Polacanthus* -> poly-; (*Polacanthoides*)

Achelou- (Gr. myth.) Ἀχελῷος/Achelōos, *m.*: shape-shifting Greek river god.
e.g.: *Achelousaurus*

Achero- (Lat.) Acherōn (*nom.*), Acherontis (*gen.*), *m.* = (Gr.) Ἀχέρων/Acherōn: Acheron, River of Woe, one of the rivers of the world below.
e.g.: *Acheroraptor* -> -raptor

Achille-, Achillo- (Lat.) Achillēs (*nom.*), Achillis (*gen.*), *m.* = (Gr. myth.) Ἀχιλλεύς/Achilleus: a Grecian hero.
e.g.: *Achillesaurus*; *Achillobator* -> -bator

acrist- {(Gr.) a- [ἀ-/a-: un-] + (Lat.) crist- [crista, *f.*: a tuft, comb, crest, plume]} non-crested, crestless.
e.g.: *Acristavus* -> -avus

acr(o)- (Gr.) ἄκρος/akros: at the furthest point, highest, topmost.
e.g.: *Acrocanthosaurus* -> -acantho-; *Acrotholus* -> -tholus

actio- (Gr.) ἄκτιος/actios: haunting the shore.
e.g.: (*Actiosaurus* Sauvage, 1883) [Choristodera?]

Ada (Mong. myth.) ад: an evil spirit in the mythology of Mongolia.
e.g.: *Adasaurus*

Adamanti- (stratal name) Adamantina Formation (Late Cretaceous), São Paulo state, Brazil.
e.g.: *Adamantisaurus*

adelo- (Gr.) ἄ-δηλος/a-dēlos: unknown.
e.g.: *Adelolophus* -> -lophus

adeo- (Lat.) adeo, *adv.*: to designate a limit, so far, so much, so very.
e.g.: *Adeopapposaurus* -> -pappo-

adratiklit {(Berber) adras (= adrar): mountain + tiklit: lizard}
e.g.: *Adratiklit*

adyn-
cf. (Gr.) ἀ-δύνᾰμος/a-dynamos: weak.
e.g.: *Adynomosaurus* -> -omo-

Aegypto- (Gr. place-name) Αἴγυπτος/Aigyptos, *f.*: Egypt. *cf. m.*: the river Nile.
e.g.: *Aegyptosaurus*

Aeolo- (Gr. myth.) Αἴολος/Aiolos, *m.*: Aeolus, lord of the winds. As *adj.*, αἰόλος/aiolos: quick-moving.
e.g.: *Aeolosaurus*

aepi- (Gr.) αἰπεινός/aipeinós: high, lofty.
e.g.: *Aepisaurus*

aepy- (Gr.) αἰπύς/aipys: high and steep, lofty.
e.g.: *Aepyornithomimus* -> -ornitho-, -mimus; (*Aepysaurus*) >>> *Aepisaurus*

aer(o)- (Gr.) ἀήρ/aēr (*nom.*), ἀέρος/aeros (*gen.*), *m./f.*: air. | (Lat.) āēr (*nom.*), āeris (*gen.*), *m.*: the air, atmosphere, sky.
e.g.: *Aerosteon* -> -osteon

-aero-
e.g.: *Vectaerovenator*

aet(o)- (Gr.) ἀετός/aetos, *m.*: an eagle.
e.g.: (*Aetonyx* Broom, 1911) -> -onyx

Afro- (Lat.) Āfer, (*m.*), Āfra (*f.*), Āfrum (*n.*), *adj.*: African.
e.g.: *Afromimus* -> -mimus; *Afrovenator* -> -venator

aga- (Gr.) ἄγαν/agan: very, very much, too, too much.
e.g.: *Agathaumas* -> -thaumas

agili- (Lat.) agilis (*m./f.*), agile (*n.*), *adj.*: nimble, quick, agile.
e.g.: *Agilisaurus*

agno- (Gr.) α-γνώς/a-gnōs: unknown.
e.g.: *Agnosphitys* [Silesauridae]

-agreus (Gr.) ἀγρεύς/agreus, *m.*: a hunter.
e.g.: *Tanycolagreus*

agro- (Gr.) ἀγρός/agros, *m.*: a field, a farm, the country.
e.g.: *Agrosaurus*

Aguja (stratal name) Aguja Formation (Late Cretaceous), Texas, USA. | (Sp.) aguja: needle.
e.g.: *Agujaceratops* -> -ceratops

Agustin (person's name) Agustin Martinelli: Senior Researcher at the CONICET (Consejo Nacional de Investigaciones Cientificas y Técnicas [Argentinian National Scientific and Technical Research Council]), the discoverer of fossils.
e.g.: *Agustinia*

Ahshisle (place-name) Ah-shi-sle-pah Wash locality (formerly Meyer's Creek) in New Mexico, USA.
e.g.: *Ahshislepelta* -> -pelta

ajanc (Mong.) аян: travel. аянч: familiar with travel, (noun) traveller.
e.g.: *Ajancingenia* -> Ingen-

Ajka (place-name) a city in Hungary.
e.g.: *Ajkaceratops* -> -ceratops

akaina (Gr.) ἄκαινα/akaina, *f.*: thorn, goad.
e.g.: *Akainacephalus* -> -cephalus

Alamo (stratal name) Ojo Alamo Formation (Late Cretaceous), New Mexico, USA. | (Sp.) álamo: poplar.
e.g.: *Alamosaurus*

Al Anqa (Arab.) a phoenix-like mythological flying creature.
e.g.: *Alanqa* [Pterosauria]

Alashan (Chin. place-name) Ālāshàn 阿拉善: a league of Inner Mongolia, China.
e.g.: (*"Alashansaurus"* Chure, 2000) >>> *Shaochilong*

Alaska (place-name) a northwesternmost state, USA.
e.g.: *Alaskacephale* -> -cephale

alba- (Lat.) albus (*m.*), alba (*f.*), album (*n.*), *adj.*: white, bright.
e.g.: *Albalophosaurus* -> -lopho-

Alberta-, Albert(o)- (place-name) a western province of Canada.
e.g.: *Albertaceratops* -> -ceratops; *Albertadromeus* -> -dromeus; *Albertavenator* -> -venator; *Albertonykus* -> -onykus; *Albertosaurus*

Albi- (Lat. place-name) the Albis = the Elbe: a river that arises in northwestern Czech Republic.
e.g.: *Albisaurus* [Archosauria]

albin (legend) wandering lights: evil influences.
e.g.: *Albinykus* -> -onykus

Alcmona (place-name) Alcmona: ancient name for a river Altmühl in Bavaria, Germany.
e.g.: *Alcmonavis* -> -avis [Avialae]

Alcova (place-name) Alcova Quarry in Natrona County, Wyoming, USA.
e.g.: *Alcovasaurus*

alectro- (Gr.) ἄλεκτρος/alectros: unbedded, unmarried.
e.g.: *Alectrosaurus*

aleto- (Gr.) ἀλήτης/alētēs (*nom.*), ἀλήτου/alētou (*gen.*), *m.*: a wanderer. As *adj.*, vagrant, roving.
e.g.: *Aletopelta* -> -pelta

Algoa (place-name) Algoa Bay: a bay in the Eastern Cape, South Africa.
e.g.: *Algoasaurus*

alio- (Lat.) alius (*m.*), alia (*f.*), aliud (*n.*), *adj.*, *pron.*: other, another, different.
e.g.: *Alioramus* -> -ramus

Aliwal (place-name) Aliwal-Noord = Aliwal North: a town in central South Africa.
e.g.: *Aliwalia*

allo- (Gr.) ἄλλος/allos = (Lat.) alius: another, other, different.
e.g.: *Allosaurus*

Almas (Mong. folklore) wild man or snowman: a man-like creature.
e.g.: *Almas*

alnashetri (Tehuelche*) 'slender thighs'.
*one of the languages spoken in Patagonia, South America.
e.g.: *Alnashetri*

aloc(o)- (Gr.) ἄλοξ/alox (*nom.*), ἄλοκος/alokos (*gen.*), *f.*: a furrow.
e.g.: *Alocodon* -> -odon

alti- (Lat.) altus (*m.*), alta (*f.*), altum (*n.*), *adj.*: high, deep.
e.g.: *Altirhinus* -> -rhinus; *Altispinax*

-alul- (Lat.) alula (*dim.* of āla, *f.*: wing): bastard wing.
e.g.: *Eoalulavis* [Avialae]

Alvarez (person's name) Gregorio Álvarez: an Argentine historian.
e.g.: *Alvarezsaurus*

Al Walker (person's name) Alick Walker (1925–1999): a British paleontologist.
e.g.: *Alwalkeria*

Alxa (place-name) Alxa League or Ālāshàn League 阿拉善盟, in Inner Mongolia, China. | Alxa (= Alashan) Desert.
e.g.: *Alxasaurus*

Amanz (person's name) Amanz Gressly: a Swiss geologist and paleontologist.
e.g.: *Amanzia*

Amarga (place-name) La Amarga. | (type locality) La Amarga Arroyo. | (stratal name) La Amarga Formation (Early Cretaceous), Neuquén Basin, Argentina.
e.g.: *Amargasaurus*s; (*Amargastegos* Ulansky, 2014) -> -stegos *(nomen dubium)*; *Amargatitanis* -> -titanis

Amazon (place-name) the Amazon Basin: the drainage basin where the Amazon River flows.
e.g.: *Amazonsaurus*

ambi- (Lat.) ambi-: around. | ambiguus, *adj.*: going two ways, uncertain.
e.g.: *Ambiortus* -> -ortus [Avialae]

ambo- (Lat.) ambō: both.
e.g.: *Ambopteryx* -> pteryx [Scansoriopterygidae]

ammo- (Gr.) ἄμμος/ammos, *f*.: sand.
e.g.: (*Ammosaurus* Marsh, 1891) >>> *Anchisaurus*

ampelo- (Gr.) ἄμπελος/ampelos, *f*.: a vine. ἀμπελών/ampelōn, *m*.: a vineyard.
e.g.: *Ampelosaurus*

amphi- (Gr.) ἀμφί/amphi: around, on both side.
e.g.: *Amphicoelias* -> -coeli-. (*Amphisaurus* Marsh, 1877) (preoccupied) >>> *Anchisaurus*

Amto- (place-name) Amtgai (= Amtgay): the Amtgai locality in Mongolia.
e.g.: *Amtocephale* -> -cephale; *Amtosaurus*

Amuro- (Russian place-name) Амур/Amur: the Amur River, forming the border between the Russian Far East and Northeast China. | Amur Oblast.
e.g.: *Amurosaurus*

amygdal(o)- (Lat.) amygdalum (*nom.*), amygdalī (*gen.*), *n.*: almond = (Gr.) ἀμύγδαλον/amygdalon: an almond. | (Gr.) ἀμύγδαλος/amygdalos, *f*.: an almond-tree.
e.g.: *Amygdalodon* -> -odon

Ana (Chin. place-name) 阿納 [Pinyin: Ēnà]: a village in Yunnan Province (雲南省), China.
e.g.: *Analong* -> -long

Ana Biset (person's name) Ana María Biset: an archaeologist from Neuquén Province in Argentina.
e.g.: *Anabisetia*

Anasazi* (demonym) the ancestors of the modern Pueblo peoples.
*Contemporary Puebloans do not want this term to be used because it was used to mean 'ancient enemies'.
e.g.: *Anasazisaurus*

anato- (Lat.) anas (*nom.*), anatis (*gen.*), *f*.: a duck.
e.g.: (*Anatosaurus* Lull & Wright, 1942) >>> *Edmontosaurus*; *Anatotitan*

-anax (Gr.) ἄναξ/anax (*nom.*), ἄνακτος/anaktos (*gen.*), *m*.: a lord, master, king.
e.g.: *Lythronax*; (*Polyonax*); *Saurophaganax*

anchi- (Gr.) ἄγχῐ/anchi: near, nigh, close.
e.g.: *Anchiceratops* -> -ceratops; *Anchiornis* -> -ornis [Anchiornithidae]; *Anchisaurus*

Andes (place-name) the Andes Mountains: a continuous highland along the western edge of South America.
e.g.: *Andesaurus*

Andhra (place-name) Andhra Pradesh: a state of India.
e.g.: (*Andhrasaurus* Ulansky, 2014) *(nomen dubium)*

angaturama (Tupi) 'noble, brave': a protective spirit.
e.g.: *Angaturama*

Angola (place-name) the Republic of Angola: a country in Southern Africa.
e.g.: *Angolatitan* -> -titan

angulo-, anguli- (Lat.) angulus (*nom.*), angulī (*gen.*), *m*.: an angle, corner.
e.g.: *Angulomastacator* -> -mastax

Anhui (Chin. place-name) Ānhuī 安徽: a province in China.
e.g.: *Anhuilong* -> -long

anikso- (Mod. Gr.) ἀνοιξη/aniksē, *f*.: spring.
e.g.: *Aniksosaurus*

animant- (Lat.) animāns (*nom.*), animantis (*gen.*), *adj*.: animate, living. As *subst*., a living being, animal.
e.g.: *Animantarx* -> -arx

ankylo- (Gr.) ἀγκύλος/ankylos: crooked, bent. | fused, stiff ← [med.] αγκυλωσις/ankylōsis: ankylosis.
e.g.: *Ankylosaurus*

anodonto- (Gr.) {ἀν-/an: -less + odont(o)- [ὀδών/odōn: tooth]} toothless.
e.g.: *Anodontosaurus*

anomali- (Late Lat.) anomalus (*nom.*), anomalī (*gen.*), *adj.* = (Gr.) ἀν-ώμᾰλος/an-ōmalos: uneven, irregular.

e.g.: *Anomalipes* -> -pes

anoplo- (Gr.) ἄνοπλος/anoplos {ἀν-/an-: un- + ὅπλον/hoplon: large shield} unarmed.
e.g.: *Anoplosaurus*

anseri- (Lat.) ānser (*nom.*), anseris (*gen.*), *m.*: a goose.
e.g.: *Anserimimus* -> -mimus

antarcto- {(Gr.) ἀντ-/ant-: opposite + ἄρκτος/arktos, *f.*: a bear, the constellation Ursa Major, the North} Antarctic, southern.
e.g.: *Antarctopelta* -> -pelta; *Antarctosaurus*

ante- (Lat.) *adv.* and *praep.*: before, in front.
e.g.: *Antetonitrus* -> -tonitrus

-anth(o)- (Gr.) ἄνθος/anthos, *n.*: a blossom, flower.
e.g.: *Paranthodon*

antro- (Lat.) antrum (*nom.*), antrī (*gen.*), *n.* = (Gr.) ἄντρον/antron: a cave, grotto, cavern.
e.g.: *Antrodemus* -> -demus

anuro- {(Gr.) ἀν-/an-: -less + οὐρά/oura [ūrā], *f.*: the tail} tailless. *cf.* Anura: taxonomic name of frogs.
e.g.: *Anurognathus* -> -gnathus [Pterosauria]

Anzu (Sumerian myth.) a large bird (eagle or vulture) that lived in the realm of the gods.
e.g.: *Anzu*

aoni (Tehuelche) south.
e.g.: *Aoniraptor* -> -raptor

Aorun (Chin. novel) Āorùn 敖閏: a Chinese mythical deity, the Dragon (龍) King (王) of the West (西) Sea (海) in the epic *Journey to the West* (西遊記).
e.g.: *Aorun*

apat(o)- (Gr.) ἀπάτη/apatē, *f.*: a trick, fraud, deceit. | (Gr. myth.) Ἀπάτη/Apatē: an evil spirit released from Pandora's Box.
e.g.: (*Apatodon* Marsh, 1877) *(nomen dubium)*; *Apatoraptor* -> -raptor; *Apatornis* -> -ornis [Avialae]; *Apatosaurus*

-apno- (Gr.) ἄ-πνοος/a-pnoos: without breath, lifeless.
e.g.: *Megapnosaurus*

Appalachio- (place-name) Appalachia. | Appalachian Mountains: a system of mountains in eastern North America.
e.g.: *Appalachiosaurus*

Apsara (Hindu myth.) a female spirit of the clouds and waters prominent.
e.g.: (*Apsaravis* Norell & Clarke, 2001) -> -avis [Avialae]

aquil-, aquila- (Lat.) Aquila (*nom.*), aquilae (*gen.*), *f.*: an eagle.
e.g.: *Aquilarhinus* -> rhin(o)-; *Aquilops* -> -ops.

-araatan (Mong.) араатан/araatan: beast.
e.g.: *Bagaraatan*

Arago- (Sp. place-name) Aragón: an autonomous community, in northeastern Spain.
e.g.: *Aragosaurus*

Aralo- (place-name) the Aral Sea: a saltwater lake of Central Asia. It lies between Kazakhstan in the north and Uzbekistan in the south.
e.g.: *Aralosaurus*

arata- (Tupi) arata {ara: born + atá: fire} born of fire.
e.g.: *Aratasaurus*

-Arce (Gr. myth.) Ἄρκη/Arkē: the winged messenger for Titans. She is a fraternal twin sister of Iris. Arke is associated with the faded second rainbow.
e.g.: *Mirarce* [Avialae]

archae(o)- (Gr.) ἀρχαῖος/archaios: ancient, primeval, olden. | ἀρχή/archē, *f.*: a beginning, origin, first cause.
e.g.: *Archaeoceratops* -> -ceratops; *Archaeodontosaurus*; *Archaeopteryx* -> -pteryx [Avialae]; (*Archaeoraptor* Sloan, 1999) >>> *Microraptor*; *Archaeorhynchus* -> -rhynchus [Avialae]; *Archaeornithoides* -> -ornitho-; *Archaeornithomimus* -> -ornitho-, -mimus; *Archaeornithura* -> -ornitho-, -ura [Avialae]

-archae(o)-
e.g.: *Protarchaeopteryx*

-archon (Gr.) ἄρχων/archōn, *m.*: a ruler, commander, chief, captain.
e.g.: (*Thalattoarchon*) [Ichthyosauria]

Arco- (place-name) l'Arc: a long river in le Midi (the South of France).
e.g.: *Arcovenator* -> -venator

arcto- (Gr.) ἀρκτος/arctos, *f.*: a bear, the Wain, the region of the bear, the North.
e.g.: (*Arctosaurus* Adams, 1875) [Archosauromorpha]

arcu- (Lat.) arcus (*nom.*), arcūs or arcī (*gen.*), *m.*: a bow, the rainbow.
e.g.: *Arcusaurus*

Areny (Catalan, place-name) Areny de Noguera = Arén: a municipality located in Aragón, Spain.
e.g.: *Arenysaurus*

Argentino- (place-name) Argentina: Argentine Republic.
e.g.: *Argentinosaurus*

argyro- (Gr.) ἄργῠρος/argyros, *m.*: white metal, *i.e.* silver.
e.g.: *Argyrosaurus*

aristo- (Gr.) ἄριστος/aristos: best, noblest, bravest.
e.g.: (*Aristosaurus* van Hoepen, 1920) >>> *Massospondylus*; *Aristosuchus* -> -suchus

Arkan- (institution name) University of Arkansas. | (place-name) Arkansas: a state of USA.
e.g.: ("Arkansaurus" Sattler, 1983) >>> *Arkansaurus*

Arkhara (Russian, place-name) Архара/Arkhara, in Amur Oblast, Russia.
e.g.: *Arkharavia* -> -via

arrhinocerat- (Gr.) ἄρ-ῥῖνο-κερατ-/ar-rhino-kerat-: without nose horn.
e.g.: *Arrhinoceratops*

Arstano- (place-name) Arstan: an ancient region of Kazakhstan.
e.g.: *Arstanosaurus*

-arx (Lat.) arx (*nom.*), arcis (*gen.*), *f.*: a castle, citadel, fortress, stronghold.
e.g.: *Animantarx*; *Invictarx*

Asfalto (stratal name) Cañadón Asfalto Formation (Middle Jurassic), Chubut, Argentina. | (Sp.) asfalto: asphalt.
e.g.: *Asfaltovenator* -> -venator

Asia-, Asiato- (Gr. place-name) Ἀσιᾱ/Asia, *f.*: Asia. *cf.* (Gr. myth.) one of Oceanids.
e.g.: *Asiaceratops* -> -ceratops; (*Asiahesperornis* Nesov & Prizemlin, 1991) -> -hesper(o)- [Avialae]; (*Asiamericana* Nesov, 1995) [Actinopterygii] >>> *Richardoestesia*; *Asiatosaurus*

-Asia (Lat. place-name) Asia (*nom.*), Asiae (*gen.*), *f.*: Asia.
e.g.: *Caenagnathasia*

asili- (Swahili) ancestor, foundation, nature.
e.g.: *Asilisaurus* [Silesauridae]

-aste- (Gr.) ἄστυ/asty (*nom.*), ἄστεος/asteos (*gen.*), *n.*: a city, town.
e.g.: *Heptasteornis*

astr- 《root》 aster- 《comb. form》 (Gr.) ἀστήρ/astēr, *m.*: a star, a flame, light, fire.
e.g.: *Astrodon* -> -odon

astropho- (Gr.) ἄ-στροφος/a-strophos: without turning away, unturning.
e.g.: *Astrophocaudia* -> -caudia

asylo- (Gr.) ἄ-σῦλος/asylos: safe from violence, inviolate.
e.g.: *Asylosaurus*

Atacama (place-name) the Atacama Desert: a plateau in South America.
e.g.: *Atacamatitan* -> -titan

Atlanto- (Gr. myth.) Ἄτλας/Atlas (*nom.*), Ἄτλαντος/Atlantos (*gen.*), *m.*: Atlas, one of the elder gods, who bore up the pillars of heaven. Later, one of the Titans.
e.g.: *Atlantosaurus*

Atlas (place-name) Mount Atlas in Africa, regarded as the pillar of heaven.
e.g.: *Atlasaurus*

Atlas Copco (institution name) Atlas Copco (Compagnie Pneumatique Commerciale).
e.g.: *Atlascopcosaurus*

atroci- (Lat.) atrōx (*m./f./n., nom.*), atrōcis (*gen.*), *adj.*: savage, fierce, wild, cruel, harsh, severe.
e.g.: *Atrociraptor* -> -raptor

atsingano- (Byzantine Gr.) αθίγγανος: gypsy.
e.g.: *Atsinganosaurus*

au- (Gr.) αὖ/au: again, anew, afresh, once more.
e.g.: *Aublysodon* -> -odon

Auca (Mapuche, place-name) Auca Mahuevo: the fossil locality, a Cretaceous Lagerstätte (deposit places), Argentina.
e.g.: *Aucasaurus*

Augustyno- (person's name) Augustyn family, who helped support the Los Angeles County Museum.
e.g.: *Augustynolophus* -> -lophus

aurora, aur- (Lat.) aurōra (*nom.*), aurōrae (*gen.*), *f.*: the morning, dawn, daybreak. | (Roman myth.) the goddess of dawn: the East, Orient.
e.g.: *Auroraceratops* -> -ceratops; *Aurornis* -> -ornis [Anchiornithidae]

Aussie (place-name) a shortened form of Australian.
e.g.: *Aussiedraco* -> -draco [Pterosauria]

australo- (Lat.) austrālis (*m./f.*), austrāle (*n.*), *adj.*: southern. *cf.* auster, *m.*: the south, southwind.
e.g.: *Australodocus* -> -docus; *Australovenator* -> -venator

Austria (Eng. / Medieval Lat. place-name) Austria = (Ger.) Republik Österreich: the German name for Austria, Österreich, meant 'eastern realm' in Old High German.
e.g.: *Austriadactylus* -> -dactylus [Pterosauria]

austro- (Lat.) auster (*nom.*), austrī (*gen.*), *m.*: the south wind, the south.
e.g.: *Austrocheirus* -> -cheirus; *Austroposeidon* -> -poseidon; *Austroraptor* -> -raptor; *Austrosaurus*

Ava (person's name) Ava Cole: the wife of Eddie Cole who found the first remains.
e.g.: *Avaceratops* -> -ceratops

Avalon (Matter of Britain) 'Island of Apples': a legendary island in Arthurian legend.
e.g.: (*Avalonia* Seeley, 1898) (preoccupied) >>> (*Avalonianus*? Kuhn, 1961) (*partim*) >>> *Camelotia*

avi- (Lat.) avis (*nom.*), avis (*gen.*), avī or ave (*abl.*), *f.*: a bird. *pl.* avēs.
e.g.: *Avimaia* -> -maia; *Avimimus* -> -mimus; (*Avipes* von Huene, 1932) (*nomen dubium*); *Avisaurus* [Avialae]

-avi-
e.g.: *Soroavisaurus*

-avis: the common suffix used to denote members of the Avialae and so on.
e.g.: *Alcmonavis* [Avialae]; (*Apsaravis*) [Avialae]; *Brodavis* [Avialae]; *Cerebavis* [Avialae]; *Cratoavis* [Avialae]; *Gargantuavis* [Avialae]; (*Horezmavis*) [Avialae]; *Iteravis* [Avialae]; (*Kizylkumavis*) [Avialae]; *Lectavis* [Avialae]; *etc.*

avia (Lat.) avia (*nom.*), aviae (*gen.*), *f.*: a grandmother.
e.g.: *Aviatyrannis* -> -tyrannis

-avus (Lat.) avus (*nom.*), avī (*gen.*), *m.*: a grandfather, an ancestor.
e.g.: *Acristavus*; (*Plateosauravus*)

Azendoh (place-name) a village in Atlas Mountains, Morocco.
e.g.: *Azendohsaurus* [Archosauromorpha]

Azhdarcho (myth.) the name of a dragon in Persian mythology.
e.g.: *Azhdarcho* [Pterosauria]

B

Baal (place-name) Baal: the name of the dinosaur site. | (Phoenician myth.) Baal: the fertility god.
e.g.: *Baalsaurus*

bactro- (Gr.) βάκτρον/baktron, *n.* = (Lat.) baculus, *m.*, baculum, *n.*: a staff, stick, cudgel.
e.g.: *Bactrosaurus*

baga (Mong.) бага/baga: small.
e.g.: *Bagaceratops* -> -ceratops; *Bagaraatan* -> -araatan

Bagual (place-name) Bagual Canyon, Chubut, Argentina.
e.g.: (*Bagualia* Pol *et al.*, 2020)

bagualo- (dialect of Rio Grande do Sul) bagual: 'strongly built fellow'.
e.g.: *Bagualosaurus*

Baharia (stratal name) Baharija Formation (Late Cretaceous), el-Bahariya, Egypt.
e.g.: *Bahariasaurus*

Baino- (place-name) Bain: Bain-Dzak / Bayn Dzak. | (Mong.) баян: rich.
e.g.: *Bainoceratops* -> -ceratops

Bajada (Sp. place-name) Bajada Colorada: the locality. | (Sp.) bajada: downhill.
e.g.: *Bajadasaurus*

Balaur (Romanian folklore) a creature, similar to an European dragon.
e.g.: *Balaur* [Avialae?]

Balochi (demonym) Balochi: Baloch tribes of Pakistan.
e.g.: *Balochisaurus*

Bambi (novel) *Bambi, Eine Lebensgeschichte aus dem Walde*: an Austria novel written by Felix Salten. | (Ital.) bambino: child, baby.
e.g.: *Bambiraptor* -> -raptor

ban- (Chin.) bàn 半: half.
e.g.: *Bannykus* -> -onykus

banji (Chin.) bānjí: striped crest.
e.g.: *Banji*

Baotianman (Chin. place-name) Bǎotiānmàn 宝天曼, located in Henan Province (河南省), China. | Baotianman National Nature Reserve.
e.g.: *Baotianmansaurus*

bapt(o)- (Gr.) bapt- [βάπτω/baptō: to dip in water, dive].
e.g.: *Baptornis* -> -ornis [Avialae]

barapa (Indian languages) big-legged, big leg.
e.g.: *Barapasaurus*

bar(o)-, bary (Gr.) βᾰρυς/barys: heavy. *cf.* βάρος/baros, *n.*: a weight, burden, load.
e.g.: *Barilium* -> -ilium; *Barosaurus*; *Baryonyx* -> -onyx

Barrosa (place-name) Sierra Barrosa: a mountain range in Neuquén, Argentina.
e.g.: *Barrosasaurus*

Barsbold (Mong. person's name) Барсболд: Rinchen Barsbold: a Mongolian paleontologist and geologist.
e.g.: *Barsboldia*

-bator (Mongol.) баатар/baatar: a hero.
e.g.: *Achillobator*; *Imperobator*

batracho- (Gr.) βάτραχος/batrachos, *m.*: a frog.
e.g.: *Batrachognathus* -> -gnathus [Pterosauria]

batyro- (Kazakh / Qazaq) батыр/batyr: a hero.
e.g.: *Batyrosaurus*

Bauru (stratal name) Bauru Group (Late Cretaceous), Bauru Sub-basin and Paraná Basin, Brasil.
e.g.: *Baurutitan* -> -titan

bauxit- (mineral) bauxite: a sedimentary rock containing aluminium. The name *bauxite* is derived from Les Baux-de-Provence in France.
e.g.: *Bauxitornis* -> -ornis [Avialae]

Bayannuro- (Mong. place-name) Bayannur League: a prefecture-level city, Inner Mongolia, China. | (Mong.) баян нуур/bayan nuur: 'rich lake'.
e.g.: *Bayannurosaurus*

Beckles (person's name) Samuel H. Beckles (1814–1890): a fossil collector.
e.g.: *Becklespinax* -> -spinax

Beg (Tibetan Buddhism) Beg-tse: a Himalayan-deity.
e.g.: *Beg*

-bei- (Chin.) běi 北: north.
e.g.: *Jinbeisaurus*

beibei (Chin.) bèibèi 貝貝. *cf.* 貝: treasure. | (transliteration) baby.

e.g.: *Beibeilong* -> -long

Beipiao (Chin. place-name) Běipiào 北票: a city in Chaoyang Prefecture, Liaoning Province (遼寧省), China.
e.g.: *Beipiaognathus* -> -gnathus; *Beipiaosaurus*

Beishan (Chin. place-name) Běishān 北山: a typically arid area in Gansu Province (甘肅省).
e.g.: *Beishanlong* -> -long

-bellator (Lat.) bellātor (*nom.*), bellātōris (*gen.*), *m.*: a warrior.
e.g.: *Dineobellator*

bellu- (Lat.) bellus (*m.*), bella (*f.*), bellum (*n.*), *adj.*: pretty, handsome, pleasant, nice.
e.g.: *Bellubrunnus* -> -Brunn [Pterosauria]; *Bellusaurus*

-bema (Gr.) βῆμα/bēma, *n.*: a step, pace, stride.
e.g.: *Hypsibema*

Berber (demonym) the Berbers: an ethnic group indigenous to North Africa.
e.g.: *Berberosaurus*

beta (Gr.) "β"
e.g.: *Betasuchus* -> -suchus

bicentenaria (Sp.) bicentenario = (Lat.) bicentenarius {bi-*: 2 + centēnārius: centennial} bicentennial.
*bi- (Lat. prefix) bis, *adv. num.*: twice, double. (comb. form): two, twice, double.
e.g.: *Bicentenaria*

Bien (Chin. person's name) 卞美年 [Pinyin: Biàn Měinián]: China's geologist.
e.g.: *Bienosaurus*

Bissekti- (stratal name) Bissekty Formation (Late Cretaceous), situated in the Kyzyl Kum Desert of Uzbekistan.
e.g.: *Bissektipelta* -> -pelta

Bistahi [Bis-tah-he] (Navajo) Bistahí (=Bisti/De-Na-Zin Wilderness): a wilderness area located in San Juan County, New Mexico, USA.
e.g.: *Bistahieversor* -> -eversor

Blasi (place-name) Blasi-1 site: the Blasi 1 locality of Arén, Huesca, south-central Pyrenees of Spain.
e.g.: *Blasisaurus*

Blikana (place-name) Blikana Mountain in the Herschel District, Cape Province, South Africa.
e.g.: *Blikanasaurus*

Bo (Chin. person's name) Bó 薄: Bo Hai-chen and Bo Xue.
e.g.: *Bolong* -> -long

Bohai (Chin. place-name) the Bóhăi Sea 渤海: an inland sea in China.
e.g.: *Bohaiornis* -> -ornis [Avialae]

-Bohai-
e.g.: *Parabohaiornis* [Avialae]

Boluochi (Chin. place-name) Bōluóchì 波羅赤: a village located in Chaoyang County, Liaoning Province, China.
e.g.: *Boluochia* [Avialae]

Bonaparte (person's name) José Fernando Bonaparte: an Argentine paleontologist.
e.g.: *Bonapartenykus* -> -onykus; *Bonapartesaurus*; *Bonatitan* -> -titan

Bonita (place-name) La Bonita Hill: the fossil quarry, Río Negro Province, Northwestern Patagonia, Argentina.
e.g.: *Bonitasaura* -> -saura

boreal(o)-, bore- (Lat.) boreālis (*m./f.*), boreāle (*n.*) *adj.*: of north wind, northern. *cf.* Boreās (*nom.*), Boreae (*gen.*), *m.*: the north wind, the north. | (Gr. myth.) βορέας/boreas, *m.*: the Greek god of the cold north wind.
e.g.: *Borealopelta* -> -pelta; *Borealosaurus*; *Boreonykus* -> -onykus

Borogov- (novel) Borogove: something like a live mop, in the nonsense poem Jabberwocky, coined by Lewis Carroll.
e.g.: *Borogovia*

bothrio- (Gr.) βόθριον/bothrion {βόθρος/bothros, *m.*: any hole or pit dug in the ground + diminutive, -ion} small trench.
e.g.: *Bothriospondylus* -> -spondylus

brachio- (Gr.) βρᾰχίων/brachiōn, *m.*: the arm = (Lat.) brachium/bracchium (brāch-), *n.*: the forearm, lower arm.
e.g.: *Brachiosaurus*

brachy- (Gr.) βρᾰχύς/brachys: short.
e.g.: *Brachyceratops* -> -ceratops; *Brachylophosaurus* -> -lopho-; (*Brachyrophus* Cope, 1878) >>> *Camptosaurus*; *Brachypodosaurus* -> -podo-; *Brachytrachelopan* -> -trachelo-, -pan

brady- (Gr.) βρᾰδύς/bradys: slow.
e.g.: *Bradycneme* -> -cneme

Brasil (Port. place-name) Brazil.
e.g.: *Blasilotitan* -> -titan

Brava (place-name) Laguna Brava National Park in Argentina.
e.g.: (*Bravasaurus* Hechenleitner *et al.*, 2020)

Bravo (Sp. place-name) Río Bravo del Norte: 'wild river of the north'. | (Sp.) bravo: brave, rough, wild.
e.g.: *Bravoceratops* -> -ceratops

brevi- (Lat.) brevis (*m./f.*), breve (*n.*), *adj.*: short, small, shallow, brief, short-lived, concise.
e.g.: *Breviceratops* -> -ceratops

Brod- (person's name) William Pierce Brodkorb: an American ornithologist and paleontologist.
e.g.: *Brodavis* -> -avis [Avialae]

Brohi (demonym) Brohi tribe (=Brahui): 'mountain dwellers' living in Pakistan.
e.g.: *Brohisaurus*

bronto- (Gr.) βροντή/brontē, *f.*: thunder.
e.g.: *Brontomerus* -> -merus; *Brontosaurus*

bruhathkaya (Sanskrit) huge body, heavy-bodied.
e.g.: *Bruhathkayosaurus*

-Brunn- (place-name) Brunn: a municipality, Upper Palatinate, Bavaria, Germany.
e.g.: *Bellubrunnus* [Pterosauria]

bugena {(Gr.) βου-/bū-: huge [← βους/bous [būs], *m./f.*: bull] + (Lat.) gena, *f.*: cheek}
e.g.: *Bugenasaura* -> -saura

Buitre- (place-name) La Buitrera: a famous fossil site and a village and municipality in Neuquén Province in southwestern Argentina. | (Sp.) buitre: vulture.
e.g.: *Buitreraptor* -> -raptor

Burian (person's name) Zdeněk Burian: Czech palaeoartist.
e.g.: *Burianosaurus*

Buriol (person's name) Buriol family. | (place-name) Buriol ravine in São João do Polêsine, Brazil.
e.g.: *Buriolestes* -> -lestes

Byron (person's name) Byron Jaffe.
e.g.: *Byronosaurus*

C

caen(o)- (Gr.) καινός/kainos: new, fresh.
e.g.: *Caenagnathasia* -> -gnath-, -asia; *Caenagnathus* -> -gnathus

caihong (Chin) cǎihóng 彩虹: rainbow.
e.g.: *Caihong* [Anchiornithidae]

calamo- (Gr.) κάλᾰμος/kalamos, *m.*: a reed.
e.g.: *Calamosaurus*; *Calamospondylus* -> -spondylus

-callio- (Gr.) κάλλος/kallos (*nom.*), κάλλεος/kalleos (*gen.*), *n.*: beauty. καλός/kalos: beautiful. 《comb. form》 kal-, kalo-, kall-, kallo- [k ↔ c]
e.g.: *Sinocalliopteryx*

Callovo-, Callovi- (Lat. geologic time scale) Callovium. | [geol.] Callovian: Middle Jurassic.
e.g.: *Callovosaurus*

camara (Gr.) κᾰμάρα/kamara, *f.*: camera, anything with an arched cover, a covered carriage.
e.g.: *Camarasaurus*

-camerotus
e.g.: *Eucamerotus*

Camarillas (stratal name) Camarillas Formation (Early Cretaceous), Teruel and La Rioja, Spain.
e.g.: *Camarillasaurus*

Camelot (legend) the legendary center of King Arthur's realm.
e.g.: *Camelotia*

Camp (person's name) Charles Lewis Camp (1893–1975): an American paleontologist and zoologist.
e.g.: *Camposaurus*

campt(o)- (Gr.) καμπτός/kamptos: curved. κάμπτω/kamptō: to bend, curve.
e.g.: *Camptodontornis* -> -ornis [Avialae]; (*Camptonotus* Marsh, 1879) (preoccupied) >>> *Camptosaurus*

campyl(o)- (Gr.) καμπύλος/kampylos: bent, crooked, curved.
e.g.: (*Campylodon* von Huene, 1929) (preoccupied) >>> *Campylodoniscus* -> -odon

Canada (place-name) Canada.
e.g.: (*Canadaga* Hou, 1999) [Avialae]

canard (Fr.) canard, *m.*: duck.
e.g.: *Canardia*

carchar(o)- (Gr.) κάρχᾰρος/charcharos: sharp-pointed, jagged, with sharp or jagged teeth.
e.g.: *Carcharodontosaurus* -> -odont(o)-

-carchar-
e.g.: *Eocarcharia*

cardi(o)- (Gr.) καρδία/kardia, *f.*: the heart.
e.g.: *Cardiodon* -> -odon

carno- (Lat.) carō (*nom.*), carnis (*gen.*), *f.*: flesh.
e.g.: *Carnotaurus* -> -taurus

Case (person's name) Ermine Cowles Case (1907–1941): a distinguished vertebrate paleontologist.
e.g.: *Caseosaurus*

catharte- (Gr.) κᾰθαρτής/kathartēs, *m.*: a purifier. | (Sp.) buitre: vulture.
e.g.: *Cathartesaura* -> -saura

Cathay (place-name) the old poetic name of China.
e.g.: *Cathayornis* -> -ornis [Avialae]

catheto- (Gr.) κάθετος/kathetos: let down, of a fishing-line. | = (Lat.) perpendiculum: perpendicular.
e.g.: *Cathetosaurus*

caudi- (Lat.) cauda (*nom.*), caudae (*gen.*), *f.*: a tail.
e.g.: *Caudipteryx* -> -pteryx

-caudi-
e.g.: *Similicaudipteryx*

-caudia
e.g.: *Astrophocaudia*; *Opistocoelicaudia*; *Rugocaudia*; *Wamweracaudia*

caudo-
e.g.: (*Caudocoelus* von Huene, 1932) -> -coel- >>> *Teinurosaurus*

caul(o)- (Gr.) καυλός/kaulos, *m.*: the shaft of a spear, the hilt of a sword, the stalk of a plant.
e.g.: (*Caulodon* Cope, 1877) -> -odon >>> *Camarasaurus*

-cavus (Lat.) cavus: hollow, excavated. As *subst., n.*, hole, opening.
e.g.: *Terminocavus*

cedar(o)- (Eng. stratal name) Cedar Mountain Formation (Early Cretaceous), Utah, USA.
e.g.: *Cedarosaurus*; *Cedarpelta* -> -pelta

cedr(o)- (Gr.) κέδρος/kedros, *f.*: the cedar-tree. | (Lat.) cedrus (*nom.*), cedrī (*gen.*), *f.*: the cedar.
e.g.: *Cedrorestes* -> -Orestes

centro- (Gr.) κέντρον/kentron, *n.*: any sharp point. = (Lat.) stimulus: a prick, goad. This word (centro-) is derived from the same Greek word κέντρον/kentron.
e.g.: *Centrosaurus*

-centr-
e.g.: *Dacentrurus*

-cephale (Gr.) κεφᾰλή/kephalē, *f.*: the head of man or beast. This word (cephale) is the common suffix used to denote members of the Pachycephalosauridae.
e.g.: *Alaskacephale*; *Amtocephale*; *Colepiocephale*; *Ferganocephale*;

Foraminacephale; *Goyocephale*; *Homalocephale*; (*Microcephale*); *Prenocephale*; *Texacephale*; *Tylocephale*

-cephalo-
e.g.: *Dinocephalosaurus* [Archosauromorpha]; *Micropachycephalosaurus*; *Nodocephalosaurus*; *Pachycephalosaurus*

-cephalus
e.g.: *Akainacephalus*; *Euoplocephalus*; (*Saurocephalus*) [Actinopterygii]; (*Stereocephalus*); *Tatankacephalus*

-ceras (Gr.) κέρας/keras (*nom.*), κέρᾱτος/keratos (*gen.*), *n.*: the horn of an animal, anything made of horn.
e.g.: *Stegoceras*

cerasin- (Lat.) cerasinus (*m.*), cerasina (*f.*), cerasinum (*n.*), *adj.*: cherry red.
e.g.: *Cerasinops* -> -ops

cerat(o)- (Gr.) κέρας/keras, *n.*: the horn of an animal.
e.g.: *Ceratonykus* -> -onykus; *Ceratops* -> -ops; *Ceratosaurus*

-ceratops (Gr.) {κέρατ-/kerat- [κέρας/keras: horn] + ὤψ/ōps: the eye, face}
e.g.: *Agujaceratops*; *Ajkaceratops*; *Albertaceratops*; *Anchiceratops*; *Archaeoceratops*; *Arrhinoceratops*; *Asiaceratops*; *Avaceratops*; *Bagaceratops*; *etc.*

cereb- (Lat.) cerebrum (*nom.*), cerebrī (*gen.*), *n.*: the brain.
e.g.: *Cerebavis* -> -avis [Avialae]

-cervix (Lat.) cervix (*nom.*), cervīcis (*gen.*), *f.*: a head-joint, neck, nape, throat.
e.g.: ("Megacervixosaurus")

cetio- (Gr.) κῆτος/kētos, *n.*: any sea-monster or huge fish (whale), an abyss. κήτειος/kēteios: of sea monsters. *cf.* Cetacea: a clade of aquatic mammals that consists of whales, dolphins and porpoises.
e.g.: *Cetiosauriscus* -> -sauriscus; *Cetiosaurus*

chang (Chin.) cháng 長: long.
e.g.: *Changyuraptor* -> -yu, -raptor; *Changzuiornis* -> -zui-, -ornis [Avialae]

Changcheng (Chin. place-name) Chángchéng 長城: the Great Wall of China.
e.g.: *Changchengornis* -> -ornis [Avialae]

Changchun (Chin. place-name) Chángchūn 長春: the capital city of Jilin Province (吉林省), China.
e.g.: *Changchunsaurus*

Changma (Chin. place-name) Chāngmǎ 昌馬 | Changma Basin of Gansu Province, China.
e.g.: *Changmaornis* -> -ornis [Avialae]

changmian (Chin.) chángmián 長眠: long sleep.
e.g.: *Changmiania*

Chaoyang (Chin. place-name) Cháoyáng 朝陽: a prefecture-level city of Liaoning Province (遼寧省), China.
e.g.: *Chaoyangia* [Avialae]; *Chaoyangsaurus*

Charon (Gr. myth.) Χάρων/Charōn: the ferryman of the Styx, from his bright fierce eyes.
e.g.: *Charonosaurus*

chasmo- (Gr.) χάσμα/chasma (= khasma) (*nom.*), χάσματος/chasmatos (*gen.*), *n.*: a yawning hollow, chasm, gulf. Generally, any wide expanse.
e.g.: *Chasmosaurus*

Chas Sternberg (person's name) Charles Mortram Sternberg: an American-Canadian fossil collector and paleontologist, son of Charles Hazelius Sternberg.
e.g.: (*Chassternbergia* Bakker, 1988) >>> *Edmontonia*

cheb (colloquial Arabic) young man, teenager.
e.g.: *Chebsaurus*

-cheirus (Gr.) χείρ/cheir (*nom.*), χειρός/cheiros (*gen.*), *f.*: the hand, the hand and arm, the arm.
e.g.: *Austrocheirus*; *Cruxicheiros*; *Deinocheirus*; *Haplocheirus*; *Ornithocheirus* [Pterosauria]

Chenani- (place-name) Sidi Chennane: the Sidi Chennane phosphate mines in Morocco.
e.g.: *Chenanisaurus*

cheneo- (Gr.) χήνεος/chēneos: of or belonging to a goose. | (Gr.) χήν/chēn (*nom.*), χηνός/chēnos (*gen.*), *m./f.* = (Lat.) ānser (*nom.*), anseris (*gen.*), *m.*: a goose, the wild goose.
e.g.: (*Cheneosaurus* Lambe, 1917) >>> *Hypacrosaurus*

Chialing (Chin. place-name) 嘉陵 [Pinyin: Jiālíng]: the Chialing River in southern China.
e.g.: *Chialingosaurus*

Chiayu (Chin. place-name) 嘉峪 [Pinyin: Jiāyù] 'Excellent Valley' (Pass): the pass located at the west end of the Great Wall of China.
e.g.: (*Chiayusaurus* Bohlin, 1953) *(nomen dubium)*

Chilantai (Chin. place-name) 吉蘭泰 [Pinyin: Jílántài], Inner Mongolia, China.
e.g.: *Chilantaisaurus*

Chile (Sp. place-name) Chile: Chili.
e.g.: *Chilesaurus*

Chinde- (Navajo, place-name) Chinde Point: derived from Chinde Point of the Petrified Forest National Park. | (Navajo) chindi/chii(n)dii: ghost, evil spirit.
e.g.: *Chindesaurus*

Chingkankou (Chin. place-name) 金剛口 [Pinyin: Jīngāngkǒu]: a village in Shandong Province (山東省), China.
e.g.: *Chingkankousaurus*

Chinshakiang (Chin. place-name) 金沙江 [Pinyin: Jīnshājiāng] 'Jinsha River': the name for the upper stretches of the Yangtze River / Chang Jiang 長江.
e.g.: *Chinshakiangosaurus*

chiro- (Gr.) χείρ/cheir (*nom.*), χειρός/cheiros (*gen.*), *f.*: the hand. 《comb. form》 cheir-, chir-, chiro-.
e.g.: *Chirostenotes* -> -stenotes

-chiro-
e.g.: *Epichirostenotes*

Chocon (place-name) Villa El Chocón, Neuquén Province, Argentina.
e.g.: *Choconsaurus*

chondr(o)- (Gr.) χονδρός/chondros, *m.*: granular. | [anat.] cartilage.
e.g.: *Chondrosteosaurus* -> -osteo-

Chongming (Chin. myth.) Chóngmíngniǎo 重明鳥 (double-pupil bird): a Chinese mythological bird, which is shaped like a chicken and sing like a phoenix. It can dispel evils.
e.g.: *Chongmingia* [Avialae]

Choyr (place-name) a city in Mongolia.
e.g.: *Choyrodon*

chromogi- {(Gr.) χρῶμα/chrōma, *n.*: the surface. Generally, colour + γῆ/gē, *f.*: earth}: referring to Valle Pintado (=Painted Valley).
e.g.: *Chromogisaurus*

Chuandong (Chin. place-name) Chuāndōng 川東: the east part of Sichuan Province (四川省), China.
e.g.: *Chuandongocoelurus* -> -coel-, -ur(o)-

Chuanjie (Chin. place-name) Chuānjiē 川街: the township in Lufeng County (禄豊県), Yunnan Province (雲南省), China.
e.g.: *Chuanjiesaurus*

chuanqi (Chin.) chuánqí 傳奇: legendary.
e.g.: *Chuanqilong* -> -long

Chubuti- (place-name) Chubut: a province in southern Argentina.
e.g.: *Chubutisaurus*

Chungking (Chin. place-name) 重慶 [Pinyin: Chóngqìng]: a major city in Southwest China.
e.g.: *Chungkingosaurus*

chupka (Ainu*) cupka: eastern.
*a language spoken by members of the Ainu ethnic group, Hokkaido, Japan.
e.g.: *Chupkaornis* -> -ornis [Avialae]

Chuxiong (Chin. place-name) Chǔxióng 楚雄: a county-level city and the seat of the Chuxiong Yi Autonomous Prefecture, Yunnan Province, China.
e.g.: *Chuxiongosaurus*

Cimolo-
e.g.: (*Cimolopteryx* Marsh, 1892) -> -pteryx [Avialae]

cion(o)- (Gr.) κίων/kiōn (*nom.*), κίονος/kionos (*gen.*), *m./f.*: a pillar.
e.g.: *Cionodon* -> -odon

citi- (Lat.) citus (*m.*), cita (*f.*), citum (*n.*) *adj.*: quick, swift, rapid.
e.g.: *Citipes*

Citipati {(Sanskrit) citi: pyre + pāti: lord} funeral pyre lord.
e.g.: *Citipati*

clao- (Gr.) κλάω-/klaō-: broken.
e.g.: *Claorhynchus* -> -rhynchus; *Claosaurus*

clasm- (Gr.) κλάσμα/klasma (*nom.*), κλάσματος/klasmatos (*gen.*), *n.*: a fragment.
e.g.: (*Clasmodosaurus* Ameghino, 1898) -> -odon *(nomen dubium)*

-clidus [anat.] cleido- [(Gr.) κλείς/kleis (*nom.*), κλειδός/kleidos (*gen.*), *f.*: a bar or bolt, a key] the collar-bone, clavicle.
e.g.: (*Cryptoclidus*) [Plesiosauria]

-clon- (Gr.) κλών/klōn, *m.*: a twig, spray.
e.g.: *Monoclonius*; (*Diclonius*)

-cneme (Gr.) κνήμη/knēmē, *f.*: the part between the knee and ankle, the leg.
e.g.: *Bradycneme*; *Eucnemesaurus*

Coahuila (place-name) the state located in Northeastern Mexico on the US border.
e.g.: *Coahuilaceratops* -> -ceratops

coel(o)- (Gr.) κοῖλος/koilos, *adj.*: hollow, hollowed. As *subst.*, κοῖλον/koilon, *n.*: a hollow, cavity.
e.g.: *Coelophysis* -> -physis; *Coelurus* -> -ur(o)-; (*Coeluroides* von Huene & Matley, 1933) *(nomen dubium)*

-coel-
e.g.: (*Caudocoelus*); *Chuandongocoelurus*; *Microcoelus*; *Pleurocoelus*; (*Sinocoelurus*); (*Thecocoelurus*)

-coeli-
e.g.: *Amphicoelias*; *Opisthocoelicaudia*

-col- (Gr.) κῶλον/kōlon, *n.*: a limb, the leg.
e.g.: *Tanycolagreus*

colepio- (Lat.) colepium: knuckle. | (Gr.) κωλήπιον/kōlēpion: (Fr.) jarret.
e.g.: *Colepiocephale* -> -cephale

-collis (Lat.) collis (*nom.*), collis (*gen.*), *m.*: hill.
e.g.: *Fumicollis*

-collum (Lat.) collum (*nom.*), collī (*gen.*), *n.*: the neck.
e.g.: *Macrocollum*

Coloradi- (place-name) Los Colorados Formation (Late Triassic), La Rioja Province, Argentina.
e.g.: (*Coloradia* Bonaparte, 1978) (preoccupied) >>> *Coloradisaurus*

-colossus (Lat.) colossus (*m.*), colossa (*f.*), colossum (*n.*), *adj.*: gigantic. As *subst.*, colossus (*nom.*), colossī (*gen.*), *m.*: a gigantic statue.
e.g.: *Iguanacolossus*; *Notocolossus*

Comahue (place-name) a sub-region of Argentina.
e.g.: *Comahuesaurus*

compso- (Gr.) κομψός/kompsos: well-dressed, elegant, exquisite.
e.g.: *Compsognathus* -> -gnathus; *Compsosuchus* -> -suchus

-con- (Gr.) κῶνος/kōnos *m.*: a pine-cone, a cone.
e.g.: *Priconodon*

Conca, Conc- (place-name) Cuenca: a province of central Spain.
e.g.: *Concavenator* -> -venator; *Concornis* -> -ornis [Avialae]

concho- (Gr.) κόγχη/konchē, *f.* = (Lat.) concha (*nom.*), conchae (*gen.*), *f.*: a bivalve, shell-fish, mussel.
e.g.: *Conchoraptor* -> -raptor

Condor (place-name) Cerro Cóndor: a village and municipality in Chubut Province, Argentina.
e.g.: *Condorraptor* -> -raptor

Confucius (Lat. person's name) Cōnfūcius = (Chin.) 孔夫子 [Pinyin: Kong fūzǐ] 'Master Kǒng': a Chinese philosopher.

e.g.: *Confuciusornis* -> -ornis [Avialae]

coni(o)- (Gr.) κόνις/konis (*nom.*), κόνιος/konios (*gen.*), *f.*: dust, ashes.
e.g.: (*Coniornis* Marsh, 1893) [Avialae] >>> *Hesperornis*

convolo- (Lat.) convolo- [convolō: fly together]
e.g.: *Convolosaurus*

corono- (Lat.) corōna (*nom.*), corōnae (*gen.*), *f.*: a garland, chaplet, wreath, a crown.
e.g.: *Coronosaurus*

corytho- (Gr.) κόρῠς/corys (*nom.*), κόρῠθος/corythos (*gen.*), *f.*: a helmet, helm, casque, the head.
e.g.: *Corythoraptor* -> -raptor; *Corythosaurus*

-coxa (Lat.) coxa (*nom.*), coxae (*gen.*), *f.*: the hip.
e.g.: *Planicoxa*; *Sellacoxa*

crasped(o)- (Gr.) κράσπεδον/kraspedon, *n.*: the edge, border, skirt or hem of a thing.
e.g.: *Craspedodon* -> -odon

cratae- (Gr.) κρᾰταιός/krataios: strong, mighty. = καρτερός/karteros.
e.g.: (*Crataeomus* Seeley, 1881) >>> *Struthiosaurus*

cratero- (Gr.) κρᾱτήρ/krātēr (*nom.*), κρᾱτῆρος/krātēros (*gen.*), *m.*: a mixing vessel, esp. a large bowl, any cup-shaped hollow, a basin in a rock.
e.g.: *Craterosaurus*

Crato- (stratal name) Crato Formation (Early Cretaceous), Araripe Basin, Brazil.
e.g.: *Cratoavis* -> -avis [Avialae]

creo- (Gr.) κρέας/kreas (*nom.*), κρέως/kreōs (*gen.*), *n.*: flesh, meat.
e.g.: (*Creosaurus* Marsh, 1878) >>> *Allosaurus*

Crichton (person's name) Michael Crichton: the American best-selling auther of *Jurassic Park* (1990).
e.g.: *Crichtonpelta* -> -pelta; *Crichtonsaurus*

-crist- (Lat.) crista (*nom.*), cristae (*gen.*), *f.*: a tuft, comb, crest.
e.g.: *Acristavus*

cristatu- (Lat.) cristātus (*m.*), cristāta (*f.*), cristātum (*n.*), *adj.*: crested, plumed.
e.g.: *Cristatusaurus*

Crittenden (stratal name) Fort Crittenden Formations (Late Cretaceous), Arizona, USA.
e.g.: *Crittendenceratops*

cruci- (Lat.) crux (*nom.*), crucis (*gen.*), *f.*: a gallows, cross.
e.g.: *Cruxicheiros* -> -cheirus

-crus- (Lat.) crūs (*nom.*), crūris (*gen.*), *n.*: the leg, shin.
e.g.: *Longicrusavis* [Avialae]

cryo- (Gr.) κρύος/kryos, *n.*: icy cold, chill, frost.
e.g.: (*Cryodrakon* Hone *et al.*, 2019) -> -drakon [Pterosauria]; *Cryolophosaurus* -> -lopho-

crypto- (Gr.) κρυπτός/kryptos: hidden, secret.
e.g.: (*Cryptoclidus* Seeley, 1892) -> -clidus [Plesiosauria]; (*Cryptodraco* Lydekker, 1889) >>> *Cryptosaurus*; (*Cryptovolans* Czerkas *et al.*, 2002) >>> *Microraptor*

Cumnor (place-name) Cumnor Hurst: a wooded hill, Cumnor, Oxfordshire, UK.
e.g.: *Cumnoria*

-cursor (Lat.) cursor (*nom.*), cursōris (*gen.*), *m.*: a runner, racer, courier.
e.g.: *Diluvicursor*; *Mahuidacursor*; *Parvicursor*

cuspi- (Lat.) cuspis (*nom.*), cuspidis (*gen.*), *f.*: a point, pointed end, blade, head.
e.g.: *Cuspirostrisornis* -> -rostris, -ornis [Avialae]

cymbo- (Lat.) cumba or cymba, *f.* = (Gr.) κύμβη/kymbē, *f.*: a boat, skiff.
e.g.: (*Cymbospondylus* Leidy, 1868) -> -spondylus [Ichthyosauria]

cyno- (Gr.) κυνώ/kynō, *f.*: a female-dog.
e.g.: (*Cynognathus* Seeley, 1895) -> -gnathus [Mammaliaformes]

D

da- (Gr.) δᾰ-/da-: intensive Prefix.
e.g.: *Dacentrurus* -> -centr-, -ur(o)-

Daan (Chin. place-name) Dàān 大安: a district of the city of Zigong, Sichuan Province (四川省), China.
e.g.: *Daanosaurus*

-dactylus (Gr.) δάκτῠλος/daktylos, *m.*: a finger.
e.g.: *Austriadactylus* [Pterosauria]; (*Megadactylus*); *Preondactylus* [Pterosauria]; *Pterodactylus* [Pterosauria]; (*Tupandactylus*) [Pterosauria]

daemono- (Lat.) daemon (*nom.*), daemōnis (*gen.*), *m.* = (Gr.) δάιμων/daimōn (*nom.*), δαίμονος/daimonos (*gen.*), *m./f.*: a god, goddess, an evil spirit, a demon, devil.
e.g.: *Daemonosaurus*

dahalokely [dah-HAH-loo-KAY-lee] {(Malagasy) dahalo: bandit, rustler + kely: small} small bandit.
e.g.: *Dahalokely*

dako- (Gr.) δάκος/dakos, *n.*: an animal of which the bite is dangerous, a noxious beast.
e.g.: (*Dakosaurus* von Quenstedt, 1856) [Metriorhynchidae]

Dakota (place-name) North Dakota and South Dakota.
e.g.: *Dakotadon* -> -odon; *Dakotaraptor* -> -raptor

Dalian (Chin. place-name) Dàlián 大連: the southernmost city of Northeast China.
e.g.: (*Dalianraptor* Gao & Liu, 2005) -> -raptor [Avialae]; *Daliansaurus*

Dandako- (Hindu epic) Dandaka-aranya = Dandak Forest: the abode of the demon Dandak.
e.g.: *Dandakosaurus*

Danubio- (Lat. place-name) Dānubius: the Danube River.
e.g.: *Danubiosaurus*

Darwin(o)- (person's name) Charles Robert Darwin (1809–1882): an English naturalist and geologist.
e.g.: *Darwinopterus* -> -pterus [Pterosauria]; *Darwinsaurus*

Dashanpu (Chin. place-name) Dàshānpū 大山鋪: the area in Zigong, in Da'an District, Sichuan Province (四川省), China.
e.g.: *Dashanpusaurus*

daspleto- (Gr.) δασπλής/dasplēs (*nom.*), δασπλῆτος/dasplētos (*gen.*), *m./f.* | δασπλῆτις/dasplētis: horrid, frightful.
e.g.: *Daspletosaurus*

Datang (Chin.place-name) Dàtáng 大塘: the town near Nanning, Guangxi, China.
e.g.: *Datanglong* -> -long

Datong (Chin. place-name) Dàtóng 大同: a city in northern Shanxi Province (山西省), China.
e.g.: *Datonglong* -> -long

datou (Chin.) dàtóu 大頭: big head. | (Malay) 首領: chieftain.
e.g.: *Datousaurus*

Dauro- (place-name) Dauro-: "Daria, ancient Russian name of western areas of Transbaikalia and eastern areas of the Amur Region". (acc. Alifanov & Saveliev, 2014)
e.g.: (*Daurosaurus* Alifanov & Saveliev, 2014) >>> *Kulindadromeus*

-daustro (Lat.) -d'austro [de + auster]: from the south.
e.g.: *Pterodaustro* [Pterosauria]

Daxia (Chin. place-name) Dàxià 大夏: the Daxia River (大夏河), a tributary of the Yellow River in southern Gansu Province (甘粛省), China.
e.g.: *Daxiatitan* -> -titan

-dectes (Gr.) δήκτης/dēktēs, *m.*: a biter. δάκνω/daknō: to bite.
e.g.: *Genyodectes*

dein(o)- (Gr.) δεινός/deinos: fearful, terrible.
e.g.: *Deinocheirus* -> -cheir-; *Deinodon* -> -odon; *Deinonychus* -> -onychus

de Lapparent (person's name) Albert-Félix de Lapparent: a French paleontologist.
e.g.: *Delapparentia*

delta (Gr.) δέλτα/delta, *n. indecl.*: delta.
e.g.: *Deltadromeus* -> -dromeus

Demanda (place-name) Sierra de la Demanda: a mountain chain situated in the northern Iberian Peninsula.
e.g.: *Demandasaurus*

-demus (Gr.) δέμας/demas, *n.*: body.
e.g.: *Antrodemus*

-dendro- (Gr.) δένδρον/dendron, *n.*: a tree.
e.g.: *Epidendrosaurus*

-dens (Lat.) dēns (*nom.*), dentis (*gen.*), *m.*: a tooth, ivory, point, spike.
e.g.: *Fruitadens*; *Manidens*

-dent(i)-
e.g.: *Oculudentavis* [Squamata]

Denver (place-name) the capital of the U. S. state of Colorado. | the Denver Museum of Natural History.
e.g.: *Denversaurus*

desmato- (Gr.) δέσμα/desma (*nom.*), δέσματος/desmatos (*gen.*), *n.* poetically for δέσμός/desmos, *m.pl.*: anything for binding, a band, bond. δεω/deō: to bind.
e.g.: (*Desmatosuchus* Case, 1920) -> suchus [Aetosauria]

di (Chin.) dì 帝: emperor.
e.g.: *Dilong* -> -long

di- (Gr.) δι-/di, δίς/dis: twice.
e.g.: (*Diceratops* Lull vide Hatcher, 1905) (preoccupied) >>> (*Diceratus* Mateus, 2008) >>> *Nedoceratops*; *Diclonius* -> -clon-; (*Didanodon* Osborn, 1902) >>> *Lambeosaurus* 1923; *Dilophosaurus* -> -lopho-; (*Dimodosaurus* Pidancet & Chopard, 1862) >>> *Plateosaurus*; *Dimorphodon* -> -morph(o)-, -odon [Pterosauria]

-di-
e.g.: *Eudimorphodon* [Pterosauria]

diablo (Sp.) devil.
e.g.: *Diabloceratops* -> -ceratops

Diamantina (place-name) Diamantina River: a river in Central West Queensland and far north of South Australia.
e.g.: *Diamantinasaurus*

dicraeo- (Gr.) δικραιος/dikraios {δίς/dis: twice, doubly + κάρā/karā: the head + -ios} bifurcated.
e.g.: *Dicraeosaurus*

diluvi- (Lat.) dīluvium (*nom.*), dīluviī (*gen.*), *n.*: flood, deluge.
e.g.: *Diluvicursor* -> -cursor

Dineo- (Navajo, demonym) Diné, Navajo: a native American people.
e.g.: *Dineobellator* -> -bellator

Ding (Chin. person's name) Ding Wenjiang丁文江, "the father of Chinese geology".
e.g.: *Dingavis* [Avialae]

Dinheiro (place-name) Porto Dinheiro: a locality in central-west Portugal.
e.g.: *Dinheirosaurus*

dino- (Gr.) δεινός/deinos: fearful, terrible.
e.g.: *Dinocephalosaurus* -> cephalo- [Archosauromorpha]; *Dinodocus* -> -docus

Diodorus (person's name) Diodorus: legendary king of the Berber people. | Diodorus Siculus: an ancient Greek historian.
e.g.: *Diodorus* [Silesauridae]

diplo- (Gr.) διπλόος/diploos: twofold, double. {δίς/dis: twice + ἁπλόος/aplóos: single, simple, natural, plain}
e.g.: *Diplodocus* -> -docus; *Diplotomodon* -> -tom(o)-, -odon

dirac-
e.g.: (*Diracodon* Marsh, 1881) *(nomen dubium)* >>> *Stegosaurus*

-docus (Gr.) δοκός/dokos, *f.* later, *m.*: a bearing-beam, in the roof or floor of a house. Generally, a balk or beam. The term *docus* is the common suffix to denote members of the diplodocid sauropod.
e.g.: *Australodocus*; *Dinodocus*; *Diplodocus*; *Kaatedocus*

dolicho- (Gr.) δολῐχός/dolichos: long.
e.g.: *Dolichosuchus* -> -suchus

Dollo (person's name) Louis Antoine Marie Joseph Dollo (1857–1931): a French-born Belgian paleontologist.
e.g.: *Dollodon* -> -odon

Dongbei (Chin. place-name) Dōngběi 東北: a geographical region of China which consists of the three provinces of Liaoning (遼寧), Jilin (吉林) and Heilongjiang (黒龍江).
e.g.: *Dongbeititan* -> -titan

Dongyang (Chin. place-name) Dōngyáng 東陽: a city, Zhejiang Province (浙江省), China.
e.g.: *Dongyangopelta* -> -pelta; *Dongyangosaurus*

doryphoro- (Gr.) δορῠ-φόρος/dory-phoros: spear-bearing.
e.g.: (*Doryphorosaurus* Nopcsa 1916) >>> *Kentrosaurus*

draco- (Lat.) dracō (*nom.*), dracōnis (*gen.*), *m.* = (Gr.) δράκων/drakōn, *m.*: a dragon, or serpent of huge size, a python.
e.g.: *Draconyx* -> -onyx; *Dracopelta* -> -pelta; *Dracoraptor* -> -raptor; *Dracorex* -> -rex; *Dracovenator* -> -venator

-draco
e.g.: *Aussiedraco*; (*Cryptodraco*); *Hippodraco*; (*Lonchodraco*) [Pterosauria]; *Pantydraco*; *Vectidraco* [Pterosauria]

-drakon (Gr.) δράκων/drakon, *m.*: dragon.
e.g.: (*Cryodrakon*) [Pterosauria]

Dravido- (place-name) Dravida Nadu: a region in the southern part of India.
e.g.: *Dravidosaurus*

Dreadnought (Eng.) Dreadnought: the predominant type of battleship. | (Old English) dreadnought: fearing nothing.
e.g.: *Dreadnoughtus*

-drinda (Urdu / Seraiki) drinda: beast.
e.g.: *Vitakridrinda*

Drinker (person's name) Edward Drinker Cope (1840–1897): an American paleontologist and comparative anatomist.
e.g.: *Drinker*

dromaeo- (Gr.) δρομαῖος/dromaios: runnning at full speed, swift, fleet.
e.g.: *Dromaeosauroides* -> -saur(o)-; *Dromaeosaurus*

-dromeus (Gr.) δρομεύς/dromeus, *m.*: a runner.
e.g.: *Albertadromeus*; *Deltadromeus*; *Kulindadromeus*; *Orodromeus*; *Oryctodromeus*; *Pampadromaeus*

dromiceio- (—) Dromiceius: a word coind by the author for the difference from the emu (*Dromaius*).
e.g.: *Dromiceiomimus* -> -mimus

Drusila (person's name) Drusila Ortiz de Zárate, a young member of the family who owns the María Aika Ranch, Santa Cruz Province, Argentina.
e.g.: *Drusilasaura* -> -saura

dryo- (Gr.) δρῦς/drys (*nom.*), δρυός/dryos (*gen.*), *f.*: originally a tree, commonly the oak. | (Lat.) quercus (*nom.*), quercūs (*gen.*), *f.*: an oak, oak-tree.
e.g.: *Dryosaurus*

-dryo-
e.g.: *Eousdryosaurus*

drypto- (Gr.) drypto- [δρύπτω/dryptō: to tear].
e.g.: (*Dryptosauroides* von Huene, 1932 *vide* von Huene & Matley, 1933) *(nomen dubium)* -> -saur(o)-; *Dryptosaurus*

Dsungari- (place-name) Dsungar (= Junggar) Basin: a basin in northern Xinjiang, China.
e.g.: *Dsungaripterus* -> -pterus [Pterosauria]

Dubreuillo- (person's name) André Dubreuil: in 1944, the mayor of Conteville in Normandy.
e.g.: *Dubreuillosaurus*

Duria (Lat. place-name) = Dorset: a county of southwestern England.
e.g.: *Duriatitan* -> -titan; *Duriavenator* -> -venator

dy- (Gr.) δύο/dyo: 2, two. *cf.* δίς/ dis, *adv.*: twice, doubly. 《comb. form》 di- -> dipl(o)-.
e.g.: *Dyoplosaurus* -> -oplo-; *Dystylosaurus*

dy-, dys- (Gr.) δῦσ-/dys-, insepar. prefix, like un- or mis-: destroying the good sense of a word, or increasing its bad sense.
e.g.: *Dysalotosaurus*; (*Dysganus* Cope, 1876) *(nomen dubium)*; (*Dyslocosaurus* McIntosh, *et al.*, 1992); *Dystrophaeus* -> -strophaeus

dynamo- (Gr.) δύνᾰμις/dynamis, *f.*: power.
e.g.: *Dynamosaurus*; *Dynamoterror* -> -terror

-dyn-
e.g.: *Adynomosaurus*

E

echin(o)- (Gr.) ἐχῖνος/echinos, *m.*: the urchin, hedgehog.
e.g.: *Echinodon* -> -odon

Edmark (person's name) Bill Edmark: a scientist and paleontologist.
e.g.: (*Edmarka* Bakker *et al.*, 1992) >>> *Torvosaurus*

Edmonton-, Edmonto- (place-name) Edmonton: the capital of Alberta, Canada. | (stratal name) Edmonton Formation (= Horseshoe Canyon Formation of the Edmonton Group) (Late Cretaceous), Alberta, Canada.
e.g.: *Edmontonia*; *Edmontosaurus*

E Fraas (person's name) Eberhard Fraas (1862–1915): a German paleontologist.
e.g.: *Efraasia*

einio- (Blackfoot) eini [eye-knee]: a buffalo or American bison.
e.g.: *Einiosaurus*

ekrixinato- {(Gr.) ekrixi-: explosion + nātus: born} explosion-born. *cf.* ἐκρήξω: to break out.
e.g.: *Ekrixinatosaurus*

Elal [ee-lal] (myth.) the god of the Tehuelche people of Chubut Province, Argentina.
e.g.: *Elaltitan* -> -titan

elaphro- (Gr.) ἐλαφρός/elaphros: light in weight.
e.g.: *Elaphrosaurus*

elasmo- (Mod. Gr.) ἔλασμα/elasma, *n.*: thin plate.
e.g.: *Elasmosaurus* [Plesiosauria]

elector(o)- (Gr.) ἤλεκτρον/ēlectron, *n.*: amber.
e.g.: (*Elektorornis* Xing *et al.*, 2019) [Avialae]

elmi (Mong.) elmi: pes.
e.g.: *Elmisaurus*

elo- (Gr.) ἕλος/helos, *n.*: low ground by rivers, a marsh-meadow.
e.g.: *Elopteryx* -> -pteryx; (*Elosaurus* Peterson & Gilmore, 1902) >>> *Brontosaurus*

Elrhaz (stratal name) Elrhaz Formation (Early Cretaceous), Gadoufaoua, Ténéré Desert, Agadez, Niger.
e.g.: *Elrhazosaurus*

els- (Mong.) элс/els: sand.
e.g.: *Elsornis* -> -ornis [Avialae]

EMAU (acronym) Ernst Moritz Arndt Universität Greifswald.
e.g.: *Emausaurus*

Emba (place-name) the Emba River: a river in west Kazakhstan.
e.g.: *Embasaurus*

enali(o)- (Gr.) ἐνάλιος/enálios: belong to the sea, of the sea.
e.g.: *Enaliornis* -> ornis [Avialae]

enanti(o)- (Gr.) ἐν-αντίος/en-antios: opposite, fronting, face to face.
e.g.: *Enantiophoenix* -> -phoenix [Avialae]; *Enantiornis* -> -ornis [Avialae]

-enantius
e.g.: *Nanantius*

enigma (Gr.) αἴνιγμα/ainigma (*nom.*), ἀινιγματος/ ainigmatos (*gen.*), *n.*: a dark saying, riddle.
e.g.: *Enigmosaurus*

eo- (Gr.) ἠώς/ēōs, *f.*: the morning-red, daybreak, dawn, the East. | (Gr. myth.) Ἠώς/Ēōs: Aurora, the goddess of morn, who rises out of her ocean-bed.
e.g.: *Eoabelisaurus*; *Eoalulavis* [Avialae]; *Eobrontosaurus*; *Eocarcharia*; *Eocathayornis* [Avialae]; (*Eoceratops* Lambe, 1915) >>> *Chasmosaurus*; *Eoconfuciusornis* [Avialae]; *Eocursor*; *Eodromaeus*; *Eoenantiornis*

[Avialae]; *Eogranivora* [Avialae]; *Eolambia*; *Eomamenchisaurus*; *Eoplophysis*; *Eoraptor* ; *Eosinopteryx* [Anchiornithidae]; *Eotrachodon*; *Eotriceratops*; *Eotyrannus* -> -tyrannus; (*Eozostrodon* Parrington, 1941) [Mammaliaformes]

Eous (Lat.) Ēōus (*m.*), Ēōa (*f.*), Ēōum (*n.*), *adj.*: of dawn, eastern. | Ēōs (only *nom.*), *f.* = (Gr. myth.) Ἠώς/Ēōs: the dawn, Aurora, the goddess of morn.
e.g.: *Eousdryosaurus* -> -dryo-

epachtho- (Gr.) ἐπαχθής/epachthēs: heavy, ponderous.
e.g.: *Epachthosaurus*

epi- (Gr.) ἐπί/epi, *prep.*: upon, over, after.
e.g.: *Epichirostenotes* -> -chiro-, -stenotes; *Epidendrosaurus* -> -dendro- [Scansoriopterygidae]

epidexi- (Gr.) ἐπίδειξις/epideixis, *f.*: a shewing forth, making known, an exhibition, display.
e.g.: *Epidexipteryx* -> -pteryx [Scansoriopterygidae]

equijubus {(Lat.) equi- [equus (*nom.*), equī (*gen.*), *m.*: horse] + juba [= iuba, *f.*]: mane, crest} 'horse mane'.
e.g.: *Equijubus*

erecto- (Lat.) ērēctus (*m.*), ērēcta (*f.*), ērēctum (*n.*), *adj.*: upright, lofty, noble, haughty.
e.g.: *Erectopus* -> -pus

Erketu (Mong. legend) Erketü Tengri = 'Mighty Heaven': a deity.
e.g.: *Erketu*

Erlian (place-name) = Erenhot: a city in the Inner Mongolian Autonomous Region, China.
e.g.: *Erliansaurus*

Erlik (myth.) the Demon king Erlik from Turk-Mongolian mythology.
e.g.: *Erlikosaurus*

erythro- (Gr.) ἐρυθρός/erythrós: red.
e.g.: (*Erythrovenator* Müller, 2020)

Eshan (Chin, place-name) Éshān 峨山: Eshan Yi Autonomous County, Yuxi (玉溪), Yunnan Province (雲南省), China.
e.g.: *Eshanosaurus*

-etes (Gr.) ἔτης/etēs, *m.* mostly in *pl.*: the kinsmen of a great house, cousins. Later, = δημότης/dēmotēs: a townsman, neibour.
e.g.: *Texasetes*

eu- (Gr.) εὖς/eus: good, brave, noble.
e.g.: (*Euacanthus* Tennyson, 1897) >>> *Polacanthus*; *Eucamerotus* -> -camerotus; (*Eucentrosaurus* Chure & McIntosh, 1989) >>> *Centrosaurus*; (*Eucercosaurus* Seeley,1879) *(nomen dubium)*; *Eucnemesaurus* -> -cneme; (*Eucoelophysis* Sullivan & Lucas, 1999) [Silesauridae]; *Eudimorphodon* -> -di-, -morph(o)- [Pterosauria]; *Euhelopus* -> -helo-, -pus; *Euoplocephalus* -> -oplo-, -cephalus; *Euparkeria* [Archosauromorpha]; *Eurhinosaurus* -> -rhino- [Ichthyosauria]; *Euskelosaurus* -> -skelo-; *Eustreptospondylus* -> -strepto-, -spondylus

Eur(o)- (New Lat.) (comb. form) European. | (Lat.) Eurōpa (*nom.*), Europae (*gen.*), *f.*: mythical princess of Tyre. | (Gr.) Εὐρώπη/Eurōpē, *f.*: Europa, Europe.
e.g.: *Euronychodon* -> -onycho-, -odon; *Europejara* [Pterosauria]; *Europelta* -> -pelta

Europa-
e.g.: *Europasaurus*; *Europatitan* -> -titan

-eversor (Lat.) ēversor (*nom.*), ēversōris (*gen.*), *m.*: a subverter, destroyer.
e.g.: *Bistahieversor*

Excalibo- (legend) Excalibur: the legendary sword of King Arthur.
e.g.: *Excalibosaurus* [Ichthyosauria]

F

Fabro- (person's name) Jean Fabre: a French geologist.
e.g.: *Fabrosaurus*

falcarius (Lat.) falcārius (*nom.*), falcariī (*gen.*), *m.* [falx: a curved blade]: a sickle-maker, scythe-maker, a sickle-cutter.
e.g.: *Falcarius*

Fergana-, Fergano- (place-name) the Fergana Valley: a valley in Central Asia.
e.g.: *Ferganasaurus*; (*Ferganastegos*

Ulansky, 2014) *(nomen dubium)* -> -stegos; *Ferganocephale* -> -cephale

ferri-, ferro- (Lat.) ferrum (*nom.*), ferri (*gen.*), *n.*: iron.
e.g.: *Ferrisaurus*; (*Ferrodraco* Pentland *et al.*, 2019) [Pterosauria]

-fiss- (Lat.) fissum (*nom.*), fissī (*gen.*), *n.*: a cleft, fissure.
e.g.: (*Multifissoolithus*) 〚Eggshell〛

flex- (Lat.) flexus (*nom.*), flexūs (*gen.*), *m.*: bent, winding.
e.g.: *Flexomornis* -> -omo-, -ornis [Avialae]

foramina (Lat.) foramen (*nom.*), forāminis (*gen.*), *n.*: an opening, aperture, hole.
e.g.: *Foraminacephale* -> -cephale

-fortis (Lat.) fortis (*m./f.*), forte (*n.*): strong, powerful, brave.
e.g.: *Jinguofortis*

Foster (person's name) John Russell Foster: an American paleontologist.
e.g.: *Fosterovenator* -> -venator

Fostor (person's name) Robert Foster: an opal miner, who discovered the bone bed.
e.g.: *Fostoria*

Frenguelli (person's name) Joaquín Frenguelli: a paleontologist, physician, naturalist.
e.g.: *Frenguellisaurus*

-frons (Lat.) frōns (*nom.*), frontis (*gen.*), *f.*: the forehead, brow.
e.g.: *Velafrons*

Fruita (place-name) Fruita: a Home Rule Municipality located in western Mesa County, Colorado, USA.
e.g.: *Fruitadens* -> -dēns

Fukui (Jpn. place-name) Fukui 福井: a prefecture located in the Chūbu region (中部地方) on Honshū (本州), Japan.
e.g.: (*Fukuipteryx* Imai *et al.*, 2019) [Avialae]; *Fukuiraptor* -> -raptor; *Fukuisaurus*; *Fukuititan* -> -titan; *Fukuivenator* -> -venator

Fuleng- (anagram) of Lùfēng (禄豊): a county, Chuxiong Yi Autonomous Prefecture, Yunnan Province (雲南省), China.
e.g.: (*Fulengia* Carroll & Galton, 1977) *(nomen dubium)*

fulguro- (Lat.) fulgur (*nom.*), fulguris (*gen.*), *n.*: lightning, flash, brightness.
e.g.: *Fulgurotherium* -> -therium

fumi- (Lat.) fūmus (*nom.*), fūmī (*gen.*), *m.*: smoke.
e.g.: *Fumicollis* -> -collis [Avialae]

Fushan (Chin.) Fùshān 傅山.
e.g.: (*Fushanosaurus* Wang *et al.*, 2019)

Fusui (Chin. place-name) Fúsuí 扶綏: a county in southern Guangxi 広西, China.
e.g.: *Fusuisaurus*

Futaba (Jpn. place-name) Futaba 双葉 | (stratal name) the Futaba Group (Late Cretaceous) of Fukushima Prefecture (福島県), Japan.
e.g.: *Futabasaurus* [Plesiosauria]

futalognko (Mapudungun) giant chief.
e.g.: *Futalognkosaurus*

G

galeam- (Lat.) galea (*nom.*), galeae (*gen.*), *f.*: a helmet.
e.g.: *Galeamopus* -> -opus

galleono- (Eng.) galleon: a large sailing ship.
e.g.: *Galleonosaurus*

galli- (Lat.) gallus (*nom.*), gallī (*gen.*), *m.*: a cock, a domestic cock.
e.g.: *Gallimimus* -> -mimus

Galton (person's name) P. M. Galton: an American vertebrate paleontologist.
e.g.: (*Galtonia* Hunt & Lucas, 1994) [Crurotarsi] >>> *Thecodontosaurus*

Galve-, Galveo- (place-name) Galve: a municipality located in the province of Teruel, Aragón, Spain.
e.g.: *Galveosaurus*; *Galvesaurus*

Gannan (Chin. place-name) Gànnán 贛南: the district area located in Ganzhou, Jiangxi Province (江西省), China.
e.g.: *Gannansaurus*

Gansu- (Chin. place-name) Gānsù 甘肃: a province located in the northwest of China.
e.g.: *Gansus* [Avialae]

Ganzhou (Chin. place-name) Gànzhōu 贛州: a city of Jianxi Province (江西省), southern China.
e.g.: *Ganzhousaurus*

Gargantua (Fr. folklore) the giant of French folklore made by François Rabelais.
e.g.: *Gargantuavis* -> -avis [Avialae]

gargoyleo- (Eng.) gargoyle: a carved or formed grotesque with a spout. | (Fr. legend) Gargouille: a dragon from the legend of Saint Romanus of Rouen.
e.g.: *Gargoyleosaurus*

garriga (Occitan) garriga: 'dry thicket'.
e.g.: (*Garrigatitan* Díaz et al., 2020)

Garudi- (Sanskrit, myth.) Garuda: a large mythical bird in both Hindu and Buddhist mythology.
e.g.: *Garudimimus* -> -mimus

gaso- (Eng.) gas.
e.g.: *Gasosaurus*

Gasparini (person's name) Zulma Brandoni de Gasparini: an Argentinian paleontologist and zoologist.
e.g.: *Gasparinisaura* -> -saura

Gaston (person's name) Robert Gaston: discoverer of the fossil site.
e.g.: *Gastonia*

gemini- (Lat.) geminus (*nom.*), geminī (*gen.*), *adj.*: born together, twin-born, twin-. *cf.* geminī (*nom.*), geminōrum (*gen.*), *m. pl.*: twins.
e.g.: *Geminiraptor* -> -raptor

genu- (Lat.) genū (*nom.*), genūs or genū (*gen.*), *n.*: a knee.
e.g.: *Genusaurus*

genyo- (Gr.) γένῠς/genys, *f.*: the under jaw. *pl.* the jaws, the mouth.
e.g.: *Genyodectes* -> -dectes

geo- (Gr.) γῆ/gē, *f.*: earth, a land, country.
e.g.: (*Geosaurus* Cuvier, 1824) [Crocodylomorpha]

gerano- (Gr.) γέρᾰνος/geranos, *f./m.*: a crane.
e.g.: *Geranosaurus*

Getty (person's name) Mike Getty: a field paleontologist.
e.g.: *Gettyia* [Avialae]

Gideon Mantell (person's name) Gideon Algernon Mantell (1790–1852): an English obstetrician, geologist, paleontologist.
e.g.: *Gideonmantellia*

Giga- (Gr. myth.) Γίγας/Gigas (*nom.*), Γίγαντος/Gigantos (*gen.*), *m.* [mostly in *pl.* Γίγαντες/Gigantes: the Giants]: a savage race destroyed by the gods, the sons of Gaia. As *adj.*, mighty.
e.g.: *Giganotosaurus* -> -noto-

Gigant(o)-
e.g.: *Gigantoraptor* -> -raptor; *Gigantosaurus*; (*Gigantoscelus* van Hoepsen, 1916) *(nomen dubium)*; *Gigantspinosaurus* -> -spino-

Gilmore (person's name) Charles Whitney Gilmore (1874–1945): an American paleontologist.
e.g.: *Gilmoreosaurus*

giraffa (New Latin) ← (Arab.) zarāfa(h): giraffe.
e.g.: *Giraffatitan* -> -titan

glaciali- (Lat.) glaciālis, *m./f.*, glaciāle, *n.*, *adj.*: icy, frozen.
e.g.: *Glacialisaurus*

glishades {(Lat.) glis: mud + (Gr. myth.) Ἅιδης/Haidēs: Hades| (Gr.) ἀ-ϊδής/aidēs: unseen}: 'consealed in mud'.
e.g.: *Glishades*

glypt(o)- (Gr.) γλυπτός/glyptos: carved.
e.g.: *Glyptodontopelta* -> -odont- , -pelta.

-gnathus (Gr.) γνάθος/gnathos, *f.*: the jaw.
e.g.: *Anurognathus* [Pterosauria]; *Batrachognathus* [Pterosauria]; *Beipiaognathus*; *Caenagnathus*;

Compsognathus; *Cynognathus* [Mammaliaformes]; *Huaxiagnathus*; *Penelopognathus*; *Priodontognathus*; *Scaphognathus* [Pterosauria]

gnatho-
e.g.: *Gnathovorax* -> -vorax

-gnath(o)-
e.g.: *Caenagnathasia*; *Protognathosaurus*

Gobi (Mong.) Говь: the Gobi Desert.
e.g.: *Gobiceratops* -> -ceratops; *Gobihadros*; *Gobipteryx* -> -pteryx [Avialae]; *Gobiraptor*; *Gobisaurus*; *Gobititan* -> -titan; *Gobivenator* -> -venator

Gojira (Jpn. Roman alphabet) Gojira: a fictional giant monster. *cf.* (Eng.) Godzilla.
e.g.: *Gojirasaurus*

Gondwana (Sanskrit) gondavana: Forest of Gondi, a region in north central India. | [geol.] Gondwana: the ancient southern supercontinent.
e.g.: *Gondwanatitan* -> -titan

Gongbu (Chin.) Gōngbù 工部: Ministry of Works (imperial China). The poet and politician Dù Fǔ 杜甫 was called Gongbu (工部) due to his position in the ministry.
e.g.: *Gongbusaurus*

Gongpoquan (Chin. place-name) Gōngpóquán 公婆泉: Gongpoquan Basin in Jiuquan Area, Gansu Province, China.
e.g.: *Gongpoquansaurus*

Gongxian (Chin. place-name) Gǒngxiàn 珙縣: a county located in Sichuan Province (四川省), China.
e.g.: *Gongxianosaurus*

-gonio- (Gr.) γωνία/gōnia, *f.*: a corner, angle.
e.g.: (*Orthogoniosaurus*)

-gono- (Gr.) γόνος/gonos, *m./f.*: offspring.
e.g.: *Syngonosaurus*

gorgo- (Gr.) γοργός/gorgos: grim, fierce, terrible. | (Gr. myth.) Γοργών/Gorgōn: a Gorgon.
e.g.: *Gorgosaurus*

goyo- (Mong.) гоё: decorated, elegant.
e.g.: *Goyocephale* -> -cephale

gracili- (Lat.) gracilis (*m./f.*), gracile (*n.*), *adj.*: thin, slender, slight, meager, poor.
e.g.: *Graciliceratops* -> -ceratops; *Graciliraptor* -> -raptor

-grani- (Lat.) grānum (*nom.*), grānī (*gen.*), *n.*: a grain, seed.
e.g.: *Eogranivora* [Avialae]

gravi- (Lat.) gravis (*m./f.*), grave (*n.*), *adj.*: heavy, weighty.
e.g.: ("Gravisaurus" Chabli, 1988) >>> *Lurdusaurus*; *Gravitholus* -> -tholus

Gressly (person's name) Amanz Gressly: a Swiss geologist and paleontologist.
e.g.: (*Gresslyosaurus* Rütimeyer, 1856) >>> *Plateosaurus*

grypho- (Eng.) gryphon (= griffin) ← (Lat.) gryps, *m.* = (Gr.) γρύψ/gryps.
e.g.: *Gryphoceratops* -> -ceratops

-gryphus, -grypho-
e.g.: *Hagryphus*; *Macrogryphosaurus*

grypo- (Gr.) γρυπός/grypos: hook-nosed, with aquiline nose, curved.
e.g.: *Gryponyx* -> -onyx; *Gryposaurus*

Guaiba (place-name) Rio Guaíba Hydrographic Basin. | the Guaíba River: a waterway in Rio Grande do Sul, Brazil.
e.g.: *Guaibasaurus*

Gualicho (Mapuche, myth.) Gualicho = Gualichu: an evil spirit or demon.
e.g.: *Gualicho*

guan (Chin.) guān 冠: crown.
e.g.: *Guanlong* -> -long

Gurilyn (Mong. place-name) Gurilyn Tsav locality, Ömnögovi, Mongolia.
e.g.: *Gurilynia* [Avialae]

gypo- (Gr.) γύψ/gyps (*nom.*), γῡπός/gypos (*gen.*), *m.*: a vulture.
e.g.: *Gyposaurus*

H

Ha (myth.) a god of the deserts to the west of Egypt.
e.g.: *Hagryphus* -> gryphus

hadro- (Gr.) ἁδρός/hadros: thick, strong, great, large, fine, well-grown.
e.g.: *Hadrosaurus*; ("Hadrosauravus" Lambert, 1990) >>> *Gryposaurus*

-hadro-, -hadros
e.g.: *Microhadrosaurus*; *Plesiohadros*; *Protohadros*; *Riabininohadros*; *Tethyshadros*

Haesta (person's name) Haesta would have been the chieftain or ruler. | (demonym) Haestingas: one of the tribes of Anglo-Saxon Britain.
e.g.: *Haestasaurus*

halim(o)- (Gr.) ἅλιμος/halimos: belong to the sea. ἅλς/hals: the sea.
e.g.: *Halimornis* -> -ornis [Avialae]

Halszka (person's name) Halszka Osmólska: a polish paleontologist.
e.g.: *Halszkaraptor* -> -raptor

haltico- (Gr.) ἁλτικός/haltikos: good at leaping.
e.g.: *Halticosaurus*

Hans Sues (person's name) Hans-Dieter Sues: a German-born paleontologist.
e.g.: *Hanssuesia*

haplo- (Gr.) ἁπλόος/haploos: single, simple, natural, plain.
e.g.: (*Haplocanthus* Hatcher, 1903) *(nomen rejectum)* >>> *Haplocanthosaurus (nomen conservandum)* -> -acantho-; *Haplocheirus* -> -cheir-

Harger (person's name) Oskar Harger: Marsh's assistant.
e.g.: (*Hargeria* Lucas, 1903) [Avialae] >>> *Hesperornis*

Harpy (Gr.) ἅρπυια/harpyia: a snatcher. | (Gr. myth.) Ἅρπυιαι/Harpyiai: the Harpies. (Eng.) Harpy: a female monster in the form of a bird with a human face.
e.g.: *Harpymimus* -> -mimus

Haya (Sanskrit) haya-griva: having the mane/neck of a horse. | (Mong. legend) Hayagrīva: an important deity in Tibetan and Japanese Buddhism.
e.g.: *Haya*

Hebei (place-name) Héběi 河北: a province of China. Its abbreviation is 冀 Jì.
e.g.: ("Hebeiornis" Xu *et al.*, 1999) [Avialae] >>> *Vescornis* [Avialae]

hecata- (Gr. myth.) Ἑκάτη/Hekatē, *f.*: goddess of magic and enchantment in the classical mythology.
e.g.: (*Hecatasaurus* Brown, 1910) >>> *Telmatosaurus*

Heishan (Chin. place-name) Hēishān 黒山: 'Black Mountain', in Gansu Province, China.
e.g.: (*Heishansaurus* Bohlin, 1953) *(nomen dubium)*

helio- (Gr. myth) Ἥλιος/Helios, *m.*: the sun, the sun-god.
e.g.: *Helioceratops* -> -ceratops

helo-, -helo- (Gr.) ἕλος/helos: a marsh-meadow.
e.g.: (*Helopus* Wiman, 1929) (preoccupied) >>> *Euhelopus*

hept(a)- (Gr.) ἑπτά/hepta, *indecl.*: seven.
e.g.: *Heptasteornis* -> -aste, -ornis

Herrera (person's name) Victorino Herrera: the rancher who discovered the first specimen.
e.g.: *Herrerasaurus*

hesper(o)- (Gr.) ἕσπερος/hesperos: of or at evening, the evening-star, western.
e.g.: *Hesperonychus* -> -onychus; *Hesperornis* -> -ornis [Avialae]; *Hesperornithoides*; *Hesperosaurus*

-hesper(o)-
e.g.: (*Asiahesperornis*) [Avialae]; *Parahesperornis* [Avialae]

heter(o)- (Gr.) ἕτερος/heteros = (Lat.) alter: the other, one of two. *cf.* alius: another, different.
e.g.: *Heterodontosaurus* -> -odont(o)-; (*Heterosaurus* Cornuel, 1850) >>> *Mantellisaurus*

hexing (Chin.) hèxíng 鶴形: like a crane.
e.g.: *Hexing*

He Xinlu (Chin. person's name) Hé Xīn-Lù 何信禄: a professor of Chengdu University of Technology 成都理工大学.
e.g.: *Hexinlusaurus*

Heyuan (Chin. place-name) Héyuán河源: a prefecture-level city of Guangdong Province (広東省), China.
e.g.: *Heyuannia*

hiero- (Gr.) ἱερός/hieros: super-human, mighty, divine, wonderful, holy, hallowed.
e.g.: *Hierosaurus*

hippo- (Gr.) ἵππος/hippos, *m./f.*: a horse, mare. = (Lat.) equus.
e.g.: *Hippodraco* -> -draco

Histria (Lat. place-name) an old name of Istria.
e.g.: *Histriasaurus*

Holland (person's name) the Holland family.
e.g.: *Hollanda* [Avialae]

homalo- (Gr.) ὁμαλός/homalos: even, level, equal.
e.g.: *Homalocephale* -> -cephale

Hongshan (Chin.) Hóngshān 紅山: the Hongshan culture, a Neolithic culture in northeastern China. | (place-name) Hongshan: a district of the city of Chìfēng (赤峰市), Inner Mongolia.
e.g.: *Hongshanornis* -> -ornis [Avialae]; *Hongshanosaurus*

-Hongshan-
e.g.: *Parahongshanornis* [Avialae]

hoplito- (Gr.) ὁπλίτης/hoplitēs: heavy-armed, armed. As *subst.*, a heavy-armed foot-soldier.
e.g.: *Hoplitosaurus*

Horezm (place-name) Khorezm or Xorazm: a region of Uzbekistan.
e.g.: (*Horezmavis* Nessov & Borkin, 1983) -> -avis [Avialae]

Horsham (place-name) a town in West Sussex, South East England. | (institution name) Horsham Museum.
e.g.: *Horshamosaurus*

hortalo-
e.g.: *Hortalotarsus*

Hou (Chin. person's name) Dr. Hóu Liánhǎi 侯連海: Chinese ornithologist and paleontologist.
e.g.: *Houornis* -> -ornis [Avialae]

Huabei (Chin. place-name) Huáběi 華北 'North China': a geographical region of China.
e.g.: *Huabeisaurus*

hualian (Chin.) huāliǎn 花臉: ornamental face, ugly face*.
*{hua 花: painted, ornamented + liǎn 臉: face}. Jìng 浄 is one of the roles on Peking Opera and known as Hualian with a painted-face.
e.g.: *Hualianceratops* -> -ceratops

huanan (Chin.) huánán 華南 'southern China'.
e.g.: *Huanansaurus*

Huang He (Chin. place-name) Huáng Hé 黄河 'the Yellow River'.
e.g.: *Huanghetitan* -> -titan

Huangshan (Chin. place-name) Huángshàn 黄山: a mountain range in southern Anhui Province (安徽省) in eastern China.
e.g.: *Huangshanlong* -> -long

Huaxia-, Huaxiao- (Chin.) Huáxià 華夏: the ancient word for China.
e.g.: *Huaxiagnathus* -> -gnathus; *Huaxiaosaurus*

Huayang (Chin. place-name) Huáyáng 華陽: an alternate name for Sichuan 四川, China.
e.g.: *Huayangosaurus*

hudie (Chin.) húdié 蝴蝶 / 胡蝶: butterfly.
e.g.: *Hudiesaurus*

Huehuecanauhtlus [UEUE-CANA-UHh-TLUS] (Náhuatl) ancient duck.
e.g.: *Huehuecanauhtlus*

Huincul (stratal name) Huincul Formation (Late Cretaceous), Patagonia, Argentina.
e.g.: *Huinculsaurus*

Hulsan (place-name) Hulsan = Khulsan: the type locality in Ömnögovi Province, Mongolia.
e.g.: *Hulsanpes* -> -pes

Hungaro- (place-name) a combining form of 'Hungary' and 'Hungarian'.
e.g.: *Hungarosaurus*

Hupeh (Chin. place-name) 湖北 [Pinyin: Húběi]: a province located in the Central China region.
e.g.: (*Hupehsuchus* Young, 1972) -> -suchus [Ichthyosauromorpha]

Huxley (person's name) Thomas Henry Huxley: an English biologist.
e.g.: *Huxleysaurus*

hylaeo-, hylo- (Gr.) ὑλαῖός/hylaios: of the forest, savage. ὕλη/hyle, *f.*: a wood, forest, woodland.
e.g.: (*Hylosaurus* Fitzinger, 1843) >>> *Hylaeosaurus*

hypacro- (Gr.) ὕπακρος/hypakros: neary the highest. hyp(o)- (Gr.) ὑπό-/hypo-: under, beneath.
e.g.: *Hypacrosaurus*

hypselo- (Gr.) ὑψηλός/hypsēlos: high, lofty, high-raised.
e.g.: *Hypselosaurus*; *Hypselospinus* -> -spinus

hypsi-, hypso- (Gr.) ὑψί-/hypsi-: on high, aloft.
e.g.: *Hypsibema* -> -bema

hypsiloph(o)- (Gr.) ὑψί-λοφος/hypsi-lophos: high-crested.
e.g.: *Hypsilophodon* -> -odon; (*Hypsirhophus* [*sic*] Cope, 1878) *(nomen dubium)*

-hypsiloph(o)-
e.g.: *Notohypsilophodon*

I

iace- (Lat.) iace- [iaceō: to lie, to be neglected].
e.g.: *Iaceornis* -> -ornis [Avialae]

ichthy(o)- (Gr.) ἰχθύς/ichthys, *m.*: fish.
e.g.: *Ichthyornis* -> -ornis [Avialae]; (*Ichthyosaurus* De la Beche & Conybeare, 1822) [Ichthyosauria]; *Ichthyovenator* -> -venator

ignavu- (Lat.) īgnāvus (*m.*), īgnāva (*f.*), īgunāvum (*n.*), *adj.*: idle, lazy, listless, cowardly.
e.g.: *Ignavusaurus*

ignoto- (Lat.) īgnōtus (*m.*), īgnōta (*f.*), īgnōtum (*n.*), *adj.*: unknown, strange.
e.g.: *Ignotosaurus* [Silesauridae]

iguana-, iguan- (Arawak / Taino / Carib) iwana: a large lizard. | (generic name) *Iguana*.
e.g.: *Iguanacolossus* -> -colossus; *Iguanodon* -> -odon

ilio- [anat.] īlium (*nom.*), īliī (*gen.*), *n.* ilium.
e.g.: *Iliosuchus* -> -suchus

-ilium
e.g.: *Barilium* -> bary-

ilokelesio {(Mapuche) ilo: flesh + kelesio: lizard}
e.g.: *Ilokelesia*

impero- (Lat.) impero- [impero: to command] *cf.* imperiōsus: powerful.
e.g.: *Imperobator* -> -bator

incisivo- (Lat.) incisivus (*m.*), incisiva (*f.*), incisivum (*n.*), *adj.*: incisive. incīdō: to cut into, to cut through.
e.g.: *Incisivosaurus*

Indo- (Gr. place-name) Ἰνδός/indos, *m.*: an Indian, the river Indus.
e.g.: *Indosaurus*; *Indosuchus* -> -suchus

Indra (myth.) an ancient Vedic deity.
e.g.: *Indrasaurus* [Squamata]

Ingen (place-name) Ingen Khoboor Depression of Bayankhongor Province, Mongolia.
e.g.: *Ingenia*

-Ingen-
e.g.: *Ajancingenia*

ingent- (Lat.) ingēns (*nom.*), ingentis (*gen.*), *adj.*: huge.
e.g.: *Ingentia*

Ingrid (person's name) Ingrid Wellnhofer, wife of Peter Wellnhofer.
e.g.: (*Ingridia* Unwin & Martill, 2007) >>>

Tupandactylus

Ino- (place-name) In Tedreft, in Agadez desert in central Niger.
e.g.: *Inosaurus*

Inti- (Quechua) the sun god.
e.g.: *Intiornis* -> -ornis [Avialae]

invict- (Lat.) invictus (*m.*), invicta (*f.*), invictum (*n.*), *adj.*: unconquerable.
e.g.: *Invictarx* > -arx

Iriso- (Gr. myth.) Ἶρις/Īris (*nom.*), Ἴριδος/Īridos (*gen.*) *f.*: the goddess of the rainbow, messenger of the gods.
e.g.: *Irisosaurus*

irritator (Lat.) irrītātor (*nom.*), irrītātōris (*gen.*), *m.*: irritator.
e.g.: *Irritator*

Isa Berry- (person's name) Isabel Valdivia Berry, who discovered the holotype material.
e.g.: *Isaberrysaura* -> -saura

Isan (place-name) northeastern region of Thailand.
e.g.: *Isanosaurus*

Isasi (person's name) Marcelo Pablo Isasi: a technician at Museo Argentino de Ciencias Naturales.
e.g.: *Isasicursor*

Ischi- (stratal name) Ischigualasto Formation (Late Triassic), Argentina.
e.g.: *Ischisaurus*

-ischia (Lat.) ischia: *pl.* of ischium, *n.* = (Gr.) ισχίον/ischion, *n.*: the hip-joint.
e.g.: *Mirischia*

ischio- (Gr.) ισχίον/ischion, *n.*: the hip-joint, the ischium.
e.g.: *Ischioceratops* -> -ceratops

ischyro- (Gr.) ἰσχῡρός/ischyros: strong, mighty.
e.g.: *Ischyrosaurus*

ISI (acronym) Indian Statistical Institute.
e.g.: *Isisaurus*

Itapeua (place-name) Itapeua, in northern Brazil.
e.g.: *Itapeuasaurus*

Itemir (place-name) the village in the Kyzylkum Desert in Uzbekistan. Kyzylkum means 'Red Sand'.
e.g.: *Itemirus*

iter (Lat.) iter (*nom*)., itineris (*gen*), *n.*: journey.
e.g.: *Iteravis*

Iutico- (demonym) Iuti: the Jutes.
e.g.: *Iuticosaurus*

J

Jain (person's name) Sohan Lal Jain: an Indian paleontologist.
e.g.: *Jainosaurus*

Jaklapalli (place-name) the town in Andhra Pradesh, central India.
e.g.: *Jaklapallisaurus*

Janensch (person's name) Welner Ernst Martin Janensch: a German paleontologist and geologist.
e.g.: *Janenschia*

Jaxarto- (Gr. place-name) Ἰαξάρτης/Iaxartēs: Jaxartes river (= the Syr Darya): a river in Central Asia.
e.g.: *Jaxartosaurus*

Jehol (place-name) Jehol = Rèhé 熱河: a former Chinese special administrative region and province. | Jehol Biota: the Early Cretaceous ecosystem.
e.g.: *Jeholornis* -> -ornis [Avialae]; *Jeholosaurus*

Jenghiz Khan (person's name) Jenghiz Khan = Genghis Khan: the founder of the Mongol Empire.
e.g.: (*Jenghizkhan* Olshevsky, 1995) >>> *Tarbosaurus*

jeyawati [HEY-a-WHAT-ee] {(Zuni) jeya: grind + awati: mouth} grinding mouth.
e.g.: *Jeyawati*

Jianchang (Chin. place-name) Jiànchāng 建昌: a county of Liaoning Province (遼寧省), China.
e.g.: *Jianchangornis* -> -ornis [Avialae]; *Jianchangosaurus*

Jiangjun (Chin. place-name) Jiàngjūnmiao 将軍廟, Junggar Basin, Xinjiang, China.
e.g.: *Jiangjunosaurus*

Jiangshan (Chin. place-name) Jiāngshān 江山: a county-level city, Zhejiang Province (浙江省), China.
e.g.: *Jiangshanosaurus*

Jiangxi (Chin. place-name) Jiāngxī 江西: Jianxi Province of China.
e.g.: *Jiangxisaurus*

Jianianhua (Chin. institution name) Jiāniánhua 嘉年華: a Chinese company.
e.g.: *Jianianhualong* -> -long

Jibei (Chin. place-name) Jìběi 冀北.
e.g.: (*Jibeinia* Hou, 1997) [Avialae]

Jinbei (Chin. place-name) {Jìn 晋: abbreviation of Shanxi Province 山西省 + běi 北: north}
e.g.: *Jinbeisaurus*

Jinfeng (Chin.) Jīnfèng 金鳳 {jīn金 + (Chin. myth) Fènghuáng 鳳凰} 'golden phoenix'.
e.g.: *Jinfengopteryx* -> -pteryx

Jingshan (*sic*) (Chin. place-name) 金山 [Pinyin: Jīnshān]: a town of Jinshan, Lufeng County (禄豊県), Yunnan Province (雲南省), China.
e.g.: *Jingshanosaurus*

jinguo (Chin.) jīnguó 巾幗 'woman's headdress': woman.
e.g.: *Jinguofortis* -> -fortis [Avialae]

Jinta (Chin. place-name) Jīntǎ 金塔: a county of Jiuquan City (酒泉市), Gansu Province (甘粛省), China.
e.g.: *Jintasaurus*

Jinyun (Chin. place-name) Jìnyún 縉雲: a county of south-central Zhejiang Province (浙江省), China.
e.g.: *Jinyunpelta* -> -pelta

Jinzhou (Chin. place-name) Jǐnzhōu (formerly Chinchow) 錦州: a prefecture-level city of Liaoning Province (遼寧省), China.
e.g.: (*Jinzhouornis* Hou *et al.*, 2002) [Avialae] >>> *Confuciusornis*; *Jinzhousaurus*

Jiuquan (Chin. place-name) Jiǔquán 酒泉: a prefecture-level city in Gansu Province (甘粛省), China.
e.g.: *Jiuquanornis* -> -ornis [Avialae]

Jiutai (Chin. place-name) Jiǔtái 九台: a district in Jilin Province (吉林省), Northeast China.
e.g.: *Jiutaisaurus*

jixiang (Chin.) jíxiáng 吉祥: happy.
e.g.: *Jixiangornis* -> -ornis [Avialae]

Jobar (Tuareg myth.) Jobar: a local giant beast.
e.g.: *Jobaria*

Jubbulpur (place-name) = Jabalpur / Jubbulpore: the second-tier city in the state of Madhya Pradesh, India.
e.g.: *Jubbulpuria*

-jubus (Lat.) iuba (*nom.*), iubae (*gen.*), *f.*: a mane.
e.g.: *Equijubus*

Judi- (stratal name) Judith River Formation (Late Cretaceous), Montana, USA.
e.g.: *Judiceratops* -> -ceratops

Judin (person's name) Konstantin Alekseyevich Yudin (= Judin).
e.g.: *Judinornis* -> -ornis [Avialae]

Juehua (Chin. place-name) Juéhuā 覚華: the island called the 'Chrysanthemum island'.
e.g.: *Juehuaornis* -> -ornis [Avialae]

jun (Chin.) jùn 俊: beautiful.
e.g.: *Junornis* -> -ornis [Avialae]

Jura (Lat.) Jūra| (Ger.) (Fr.) (Eng.) the Jura Mountains.
cf. Jurassic: of the Jurassic period, the geological period between the Triassic and the Cretaceous. | (place-name) of the Jura Mountains.
e.g.: *Juratyrant* -> -tyrant; *Juravenator* -> -venator

K

kaate (Crow / Absaroka*) káata: small, little.
*Crow (Absaroka) language: "one of the Native American Tribes of northern Wyoming". (acc. Tschopp & Mateus, 2012)
e.g.: *Kaatedocus* -> -docus

Kaijiang (Chin. place-name) Kāijiāng 開江: a county of Sichuan Province (四川省), China.
e.g.: *Kaijiangosaurus*

kaiju (Jpn.) kaiju 怪獣: monster.
e.g.: *Kaijutitan* -> -titan

Kakuru (Aboriginal myth.) one of the names for the Rainbow Serpent of Australian Aboriginal mythology.
e.g.: *Kakuru*

Kamuy (Ainu*) god.
*Ainu: a language spoken by members of the Ainu ethnic group, Hokkaido, Japan.
e.g.: *Kamuysaurus*

Kangnas (place-name) Farm Kangnas: a farm in the Orange River Valley of northern Cape Province, South Africa.
e.g.: *Kangnasaurus*

Karonga (place-name) Karonga: a district located on the western shore of Lake Nyasa, Northern Region, Malawi.
e.g.: *Karongasaurus*

katepen- (Tehuelche) katepenk: hole.
e.g.: *Katepensaurus*

Katsuyama (place-name) 勝山: a city in Fukui Prefecture, Japan.
e.g.: ("Katsuyamasaurus"Lambert, 1990) >>> *Fukuiraptor*

Kayenta (stratal name) Kayenta Formation, Glen Canyon Group (Late Triassic–Early Jurassic), Colorado Plateau, USA.
e.g.: *Kayentavenator* -> -venator

Kazak(h) (place-name) the ancient/old name of Kazakhstan. | (demonym) a Turkic ethnic group inhabiting mainly Kazakhstan.
e.g.: *Kazaklambia* -> -lamb-

Keichou (Chin. place-name) Keichou = Kweichou 貴州 [Pinyin: Guìzhōu]: now Guizhou Province (貴州省), China.
e.g.: *Keichousaurus* [Sauropterygia]

Kelmayi- (place-name) 克拉瑪衣 [Pinyin: Kèlāmǎyi]: a city of the Xinjiang Uygur Autonomous Region, China.
e.g.: *Kelmayisaurus*

Kem Kem (stratal name) Kem Kem Beds (Late Cretaceous), the border between Morocco and Algeria.
e.g.: (*Kemkemia* Cau & Maganuco, 2009) [Crocodylomorpha]

kentr(o)- (Gr.) κέντρον/kentron, *n.*: any sharp point, a horse goad, a goad, spur.
e.g.: (*Kentrurosaurus* Hennig, 1916) -> -uro-, >>> *Kentrosaurus*

Kerbero- (Gr. myth.) Κέρβερος/Kerberos, *m.* = (Lat.) Cerberus: the fifty-headed dog of Hades, which guarded the gate of the nether world, later, with three heads or bodies.
e.g.: *Kerberosaurus*

khaan (Mong.) хаан: lord.
e.g.: *Khaan*

Khetran (demonym) the Khetran tribe of Barkhan district in Pakistan.
e.g.: *Khetranisaurus*

Kholumolumo (Sotho folklore) a reptilian monster.
e.g.: *Kholumolumo*

kileskus (Khakas) lizard.
e.g.: *kileskus*

Kinnaree (Sanskrit, Thai myth.) Kinnarī: Kinnaree 緊那羅: graceful beings of Thai mythology, with the body of a woman and the legs of a bird.
e.g.: *Kinnareemimus* -> -mimus

Kizylkum (place-name) the Kyzylkum (= Qizilqum) Dessert. | (Turkic) 'Red Sand'.
e.g.: *Kizylkumavis* -> -avis [Avialae]

klameli- (Chin. place-name) 克拉美麗 [Pinyin: Kèlāměilì]: the fossil locality near the Kelameilishan Mts.
e.g.: *Klamelisaurus*

kol (Mong.) foot.
e.g.: *Kol*

komps(o)- (Gr.) κομψός/kompsos: elegant. *cf.* compso-.
e.g.: (*Kompsornis* Wang *et al.*, 2020) [Avialae]

Kongonaphon {(Malagasy) kongona: bug + (Gr.) φονεύς/phoneus: slayer}
e.g.: (*Kongonaphon* Kammerer *et al.*, 2020) [Dinosauromorpha?]

koparion (Gr.) κοπάριον/koparion: a small surgical knife.
e.g.: *Koparion*

Korea-, Koreano- (Eng. place-name) Korea.
e.g.: *Koreaceratops* -> -ceratops; *Koreanosaurus*

Koshi (Jpn. place-name) 越: an old Japanese regional name including Fukui Prefecture (福井県), Japan.
e.g.: *Koshisaurus*

kosmo- (Gr.) κόσμος/kosmos, *m.*: good order, an ornament, the world of universe.
e.g.: *Kosmoceratops* -> -ceratops

Kota (stratal name) Kota Formation (Early Jurassic–Early Cretaceous), Telangana, India.
e.g.: *Kotasaurus*

koutali (Mod. Gr.) κουτάλι/koutali [← (Gr.) κώταλις/kōtalis]: spoon.
e.g.: *Koutalisaurus*

krito- (Gr.) κρῐτός/kritos: picked out. κρίνω/krinō: to separate, part, put asunder, to pick out.
e.g.: *Kritosaurus*

Krono- (Gr. myth.) Κρόνος/Kronos, *m.*: Cronus, son of Uranos and Gaia.
e.g.: *Kronosaurus* [Plesiosauria]

krypt(o)- (Gr.) κρυπτός/kryptos: hidden, secret, covered. κρύπτω/kryptō: to cover, to hide.
e.g.: *Kryptops* -> -ops

-ktenos (Gr.) κτῆνος/ktēnos, *n.*: a beast.
e.g.: *Meroktenos*

Kukufeld (place-name) Cuckfield (= Cuckoo's field): a village, Mid Sussex District of West Sussex, England.
e.g.: *Kukufeldia*

kul (place-name) Khodzhakul Locality. | (stratal name) Khodzhakul Formation (mid Cretaceous), Khodzhakul Locality, Uzbekistan. | (Uzbek) ko`l: lake.
e.g.: *Kulceratops* -> -ceratops

Kulinda (place-name) a site on the banks of the Olov River, in Zabaikal District, Siberia.
e.g.: *Kulindadromeus* -> -dromeus; *Kulindapteryx* -> -pteryx

kunbarra (Mayi*) shield.
*a small family of extinct Australian languages of Queensland.
e.g.: *Kunbarrasaurus*

Kunduro- (place-name) Kundur: a locality of Amur region, Far Eastern Russia.
e.g.: *Kundurosaurus*

-kupal (Mapuche) family.
e.g.: *Leinkupal*

kuszholi (Kazakh) kus zholi / Kus zholy: the Milky Way.
"According to the Kazakh beliefs the Milky Way showed the direction of the birds' migration, that's where it got its name 'Bird's path' (Kus zholy)." (quot. Dzhelbuldin, Y. & Jeteyeva, D. (2014). Traditions and Customs of Kazakhs.)
e.g.: *Kuszholia* [Avialae]

kwana- (Ute) eagle.
e.g.: *Kwanasaurus* [Silesauridae]

L

La Bocana (stratal name) La Bocana Roja Formation (Late Cretaceous), Baja California, Mexico. Roja means 'red'.
e.g.: *Labocania*

labro- (Gr.) λαβρός/labros: greedy, furious.
e.g.: (*Labrosaurus* Marsh, 1878) >>> *Allosaurus*, *Ceratosaurus*

Laelaps (Gr. myth.) Λαῖλαψ/Lailaps: a mythological dog.
e.g.: (*Laelaps* Cope, 1866) (preoccupied) >>> *Dryptosaurus*

laevis (Lat.) levis (*m./f.*), leve (*n.*), *adj*.: light, light-armed.
e.g.: *Laevisuchus* -> -suchus

lago- (Gr.) λἄγός/lagos, *m*.: a hare.
e.g.: *Lagosuchus* -> -suchus [Dinosauriformes]

Laiyang (Chin, place-name) Láiyáng 萊陽: a city of Shandong Province (山東省), China. | the Laiyang Basin.
e.g.: *Laiyangosaurus*

Lajas (place-name) Las Lajas: the city of Las Lajas in Neuquén, Argentina.
e.g.: *Lajasvenator*

lama (Tibetan) blama 喇嘛 (=Lama): a title for a teacher of the Dharma in Tibetan Buddhism.
e.g.: *Lamaceratops* -> -ceratops

Lambeo- (person's name) Lawrence Moris Lambe (1863–1919): a Canadian geologist and paleontologist.
e.g.: *Lambeosaurus*

-lamb-
e.g.: *Kazaklambia*

Lameta (stratal name) Lameta Formation (Late Cretaceous), Jabalpur, India.
e.g.: *Lametasaurus*

Lamplugh (person's name) Pamela Lamplugh Robinson: founder of the Indian Statistical Institute.
e.g.: *Lamplughsaura* -> -saura

lana (Lat.) lāna (*nom.*), lānae (*gen.*), *f.* : wool.
e.g.: *Lanasaurus*

Lanzhou (Chin. place-name) Lánzhōu 蘭州: the capital city of Gansu Province (甘粛省), China.
e.g.: *Lanzhousaurus*

lao- (Gr.) λᾶας/laas (*nom.*), λᾶος/laos (*gen.*), *m.*: a stone.
e.g.: *Laosaurus*

La Pampa (place-name) La Pampa Province, Argentina.
e.g.: *Lapampasaurus*

La Plata (place-name) the capital city of Buenos Aires, Argentina. | (place-name) Río de la Plata. | (Sp.) plata: silver.
e.g.: *Laplatasaurus*

Lapparent (person's name) Albert-Félix de Lapparent (1905–1975): a French paleontologist.
e.g.: *Lapparentosaurus*

La Quinta (stratal name) La Quinta Formation (Early Jurassic), Serranía del Perijá, northwestern Venezuela.
e.g.: *Laquintasaura* -> -saura

largi- (Lat.) largus (*m.*), larga (*f.*), largum (*n.*) *adj.*: large.
e.g.: *Largirostrornis* -> -rostr-, -ornis [Avialae]

lateni- (Lat.) latēns (*nom.*), latentis (*gen.*) *adj.*: lying, hid, hidden, concealed, secret, unknown.
e.g.: *Latenivenatrix* -> -venatrix

lati- (Lat.) lātus (*m.*), lāta (*f.*), lātum (*n.*), *adj.*: broad, wide, extensive.
e.g.: *Latirhinus* -> -rhinus

Lavocati- (person's name) René Lavocat: a French researcher.
e.g.: *Lavocatisaurus*

Leaellyn (person's name) Leaellyn Rich: the daughter of the Australian paleontologist couple Tom Rich and Patricia Vickers-Rich who discovered.
e.g.: *Leaellynasaura* -> -saura

lect- (Lat.) lectus (*nom.*), lectī (*gen.*), *m.* = (Sp.) lecho 'bed': the Lecho Formation (Late Cretaceous), Argentina.
e.g.: *Lectavis* -> -avis [Avialae]

-lectro- (Gr.) λέκτρον/lektron = (Lat.) lectus: a couch, bed.
e.g.: *Alectrosaurus*

Ledumahadi (Sesotho) a giant thunderclap.
e.g.: *Ledumahadi*

Leinkupal {(Mapuche*) lein: vanishing + kupal: family}
*= Mapudungun: the language of the Mapuche people that inhabit northern Patagonia.
e.g.: *Leinkupal* -> -kupal

Le Nes- (person's name) Lev A. Nesov: Russian ornithologist and paleontologist.
e.g.: *Lenesornis* -> -ornis [Avialae]

Leonera- (place-name) Cañadón Las Leoneras, south of Cañadón del Zaino, Chubut Province, Argentina. | (stratal name) Las Leoneras Formation (Early Jurassic) in Argentina.
e.g.: *Leonerasaurus*

lepido- (Gr.) λεπίς/lepis (*nom.*), λεπίδος/lepidos (*gen.*), *f.*: scale.
e.g.: (*Lepidocheirosaurus* Alifanov & Saveliev, 2015) >>> *Kulindadromeus*

lepidus (Lat.) lepidus (*m.*), lepida (*f.*), lepidum (*n.*), *adj.*: pleasant, charming, neat, witty, fascinating.
e.g.: *Lepidus*

lepto- (Gr.) λεπτός/leptos: peeled, husked, threshed out, fine, small, thin. Generally, small, weak, impotent.
e.g.: *Leptoceratops* -> -ceratops; *Leptorhynchos* -> -rhynchus

Leshan (Chin. place-name) Lèshān 楽山: a prefecture-level city in Sichuan Province (四川省), China.
e.g.: *Leshansaurus*

Lesotho (place-name) Lesotho: an enclaved, landlocked country surrounded by South Africa.
e.g.: *Lesothosaurus*

Lessem (person's name) Don Lessem: the founder of The Dinosaur Society and the Jurassic Foundation, and a popular science writer.
e.g.: *Lessemsaurus*

-lestes (Gr.) ληστής/lēstēs, *m.*: a robber, plunderer.
e.g.: *Buriolestes*; *Ornitholestes*; *Sarcolestes*; *Saurornitholestes*

Lev Nesov- (person's name) Lev A. Nesov (1947–1995): a Russian ornithologist and paleontologist.
e.g.: *Levnesovia*

Lewis (person's name) Arnold D. Lewis, chief preparator.
e.g.: (*Lewisuchus* Romer, 1972) [Silesauridae]

Lexovi- (Lat. demonym) Lexoviī (*nom.*), Lexoviōrum (*gen.*). *m. pl.*: an ancient Gallic tribe. | (Fr.) lexoviens: people of Armorica, established in Lisieux.
e.g.: *Lexovisaurus*

Leyes (person's name) Leyes family: inhabitants of a small town Balde de Leyes, San Juan Province, Argentina, who discovered the holotype.
e.g.: *Leyesaurus*

Liao (Chin. abbr. place-name) Liáo 遼: Liaoning Province (遼寧省) of China.
e.g.: *Liaoceratops* -> -ceratops

Liaoning(o)- (Chin. place-name) Liáoníng 遼寧: a province in the Northeast China.
e.g.: *Liaoningornis* -> -ornis [Avialae] ; *Liaoningosaurus*; *Liaoningotitan* > -titan; *Liaoningvenator* -> -venator

Liaoxi (Chin, place-name) Liáoxī遼西: part of Liaoning and Jilin provinces.
e.g.: *Liaoxiornis* -> ornis [Avialae]

Lias [geol.] the earliest epoch of the Jurassic period.
e.g.: ("Liassaurus")

Ligabue (person's name) Giancarlo Ligabue (1931–2015): an Italian paleontologist.
e.g.: *Ligabueino*; *Ligabuesaurus*

Lilienstern (person's name) Hugo Rühle von Lilienstern: a German doctor and amateur paleontologist.
e.g.: *Liliensternus*

Limay (place-name) Río Limay: the Limay River

in the northwestern Patagonia, Argentina.
e.g.: *Limaysaurus*

limen (Lat.) līmen (*nom.*), līminis (*gen.*) *n.* : a cross-piece, threshold.
e.g.: *Limenavis* -> -avis [Avialae]

limno- (Gr.) λίμνη/limnē, *f.*: a pool of standing water left by the sea or a river, a marshy lake, mere.
e.g.: (*Limnosaurus* Nopcsa, 1899) (preoccupied) >>> *Telmatosaurus*

limus (Lat.) līmus (*nom.*), līmī (*gen.*), *m.*: mud, slime, dirt.
e.g.: *Limusaurus*

Lingwu (Chin. place-name) Língwǔ 靈武: a city of Ningxia Hui Autonomous Region, China.
e.g.: *Lingwulong* -> -long

Lingyuan (Chin. place-name) Língyuán 凌源: a city of Liaoning Province, China.
e.g.: *Lingyuanosaurus* (*Lingyuanornis* Ji & Ji, 1999) [Avialae] >>> *Liaoxiornis*

Linhe (Chin. place-name) Línhé 臨河: a district in Bayannur, Inner Mongolia.
e.g.: *Linhenykus* -> -onykus; *Linheraptor* -> -raptor; *Linhevenator* -> -venator

lio- (Gr.) λεῖος/leios: smooth.
e.g.: *Liopleurodon* -> -pleur(o)- [Plesiosauria]

-Lipan (demonym) Lipan (= Lápai-Ndé): the "Gray* People", a tribe of the Apache.
*The ancient Lipan Apaches migrated from the north (white) and moved east (black) into Texas. When you mix these two colors, you get the color gray. They represent the direction by a color, such as, west (yellow) and south (blue). (acc. The official website of the Lipan Apache Tribe of Texas.) <http://www.lipanapache.org/Museum/museum_lipanname.html>
e.g.: *Acantholipan*

liraino- (Basque) lirain: slender.
e.g.: *Lirainosaurus*

-lithus (Gr.) λίθος/lithos, *m.*: a stone, a precious stone. 《comb. form》 lith-, lithi-, litho-.
e.g.: (*Multifissoolithus*) 〚Eggshell〛; (*Nipponoolithus*) 〚Eggshell〛

Liubang (Chin. place-name) Liùbàng 六榜: a village in Fusui County (扶綏県), the Guangxi Zhuang Autonomous Region, China.
e.g.: *Liubangosaurus*

Lo Hueco (place-name) a paleontologic site in Fuentes, Cuenca, Castile-La Mancha, Spain.
e.g.: *Lohuecotitan* -> -titan

loncho- (Gr.) λόγχη/lonchē, *f.*: lance.
e.g.: (*Lonchodraco* Rodrigues & Kellner, 2013) -> -draco [Pterosauria]

lonco-
e.g.: (*Loncosaurus* Ameghino, 1898) *(nomen dubium)*

Long (person's name) Robert A. Long: an American paleontologist.
e.g.: (*Longosaurus* Welles, 1984) >>> *Coelophysis*

-long (Chin.) lóng 龍: dragon.
e.g.: *Anhuilong*; *Beibeilong*; *Beishanlong*; *Bolong*; *Chuanqilong*; *Datanglong*; *Datonglong*; *Dilong*; *Guanlong*; *Huangshanlong*; *Jianianhualong*; *Lingwulong*; *Qiaowanlong*; *Qijianglong*; *Qiupalong*; *Shaochilong*; *Taohelong*; *Tianyulong*; *etc.*

longi- (Lat.) longus (*m.*), longa (*f.*), longum (*n.*), *adj.*: long, extended. *cf.* longī (*gen.* of longus)
e.g.: *Longicrusavis* -> -crus, -avis [Avialae]; *Longipteryx* -> -pteryx [Avialae]; *Longirostravis* -> -rostr-, -avis [Avialae]

longus-
e.g.: *Longusunguis* -> -unguis [Avialae]

lopho- (Gr.) λόφος/lophos, *m.*: the crest, a ridge. = (Lat.) crista, *f.*: a tuft, comb, crest.
e.g.: *Lophorhothon* -> -rhothon; *Lophostropheus* -> -stropheus

-lopho-
e.g.: *Albalophosaurus*; *Brachylophosaurus*; *Cryolophosaurus*; *Dilophosaurus*; *Hypsilophodon*; *Monolophosaurus*

-lophus
e.g.: *Adelolophus*; *Augustynolophus*; *Parasaurolophus* ;*Saurolophus*

loricato- (Lat.) lōrīcātus (*m.*), lōrīcāta (*f.*), lōrīcātum (*n.*), *adj*.: clothed in mail, harnessed, mailed.
e.g.: *Loricatosaurus*

lorico- (Lat.) lōrīca (*nom.*), lōrīcae (*gen.*), *f.*: leather cuirass.
e.g.: *Loricosaurus*

Losilla (Sp. place-name) a village in the district of Serranos (Valencia, Spain).
e.g.: *Losillasaurus*

Lourinha (Port. stratal name) Laurinhã Formation (Late Jurassic), western Portugal.
e.g.: *Lourinhanosaurus*; *Lourinhasaurus*

Luanchuan (Chin. place-name) Luánchuān 欒川: a county in Luoyang (洛陽), Henan Province (河南省), China.
e.g.: *Luanchuanraptor* -> -raptor

Luanping (Chin. place-name) Luánpíng 灤平: a county of Hebei Province, China.
e.g.: ("Luanpingosaurus" Cheng *vide* Chen, 1996) >>> *Psittacosaurus*

Luciano (person's name) Luciano Leyes, who first reported the remains.
e.g.: *Lucianovenator* -> -venator

Lufeng (Chin. place-name) Lùfēng 禄豊: a county located in Chuxiong Yi Autonomous Prefecture, Yunnan Province, China.
e.g.: *Lufengosaurus*

Lukou (Chin. place-name) 盧溝 [Pinyin: Lúgōu] | 盧溝橋 [Pinyin: Lúgōuqiáo] 'Lugou Bridge': a stone bridge located 15 km southwest of the downtown of Beijing where Sino-Japanese War started. It is also known by the name of Marco Polo Bridge.
e.g.: (*Lukousaurus* Young, 1940) [Crochodylomorpha]

Luoyang (Chin. place-name) Luòyáng 洛陽: a city located in western Henan Province (河南省), China.
e.g.: *Luoyanggia*

lurdu- (Lat.?)
cf. (Fr.) lourd: heavy.
e.g.: *Lurdusaurus*

Lusitano- (place-name) Lūsītānia (*nom.*), Lūsītāniae (*gen.*), *f.*: the ancient Latin name for Portuguese.
e.g.: *Lusitanosaurus*

Luso- (place-name) Luso-: of Lusitania, Portuguese. | (demonym) Luso: the inhabitants of ancient Lusitania.
e.g.: *Lusotitan* -> -titan; *Lusovenator*

lutungutali (Bemba*) high hip.
*a language spoken in Zambia by the Bemba people.
e.g.: *Lutungutali* [Silesauridae]

lyco- (Gr.) λύκος/lykos, *m.*: a wolf.
e.g.: *Lycorhinus* -> -rhinus

lystro- [*sic*], **listro-** (Gr.) λίστρον/listron, *n.*: a kind of shovel.
e.g.: (*Lystrosaurus* Huxley, 1859) [Therapsida]

lythro- (Gr.) λύθρον/lythron, *n.*: defilement from blood, gore.
e.g.: *Lythronax* -> -anax

M

maaqwi (Coast Salish) ma'aqwi: water bird.
e.g.: *Maaqwi* [Avialae]

machaira (Gr.) μάχαιρα/machaira, *f.*: a large knife or dirk, a short sword or dagger.
e.g.: *Machairasaurus*; *Machairoceratops* -> -ceratops

macr(o)- (Gr.) μακρός/makros: long, large.
e.g.: *Macrocollum* -> collum; *Macrogryphosaurus* -> grypho-; (*Macrophalangia* Sternberg, 1932) >>> *Chirostenotes*; *Macrurosaurus* -> -ur(o)-

Madsen (person's name) James Henry Madsen: an American paleontologist.
e.g.: ("Madsenius" Bakker, 1990) >>> *Allosaurus*

magna-, magni-, magno- (Lat.) māgnus (*m.*), māgna (*f.*), māgnum (*n.*), *adj.*: great, large, big, tall. *cf.* magnī (*gen.* of magnus).
e.g.: *Magnamanus* -> -manus; *Magnapaulia* -> -Paul-; *Magnirostris* -> -rostris; *Magnosaurus*

Magyaro- (demonym) Magyar = Hungarian.
e.g.: *Magyarosaurus*

Mahakala (Sanskrit) Mahākāla {mahā: grand + kāla: *adj.* black, *noun*, time}: a deity common to Hinduism, Buddhism and Sikhism. Daikokuten (大黒天) in Japanese.
e.g.: *Mahakala*

mahuida- (Mapudungun) mountain.
e.g.: *Mahuidacursor* -> cursor

maia (Gr.) μαῖα/maia, *f.*: good mother, dame. | (Gr. myth.) Μαῖα/Maia: daughter of Atlas, mother of Hermes.
e.g.: *Maiasaura* -> -saura

-maia
e.g.: *Avimaia*; *Nemegtomaia*

Majunga (Fr. place-name) = Mahajanga: a city, northwestern Madagascar.
e.g.: *Majungasaurus*; *Majungatholus* -> -tholus

Malargue (place-name) Malargüe: a city of Mendoza, Argentina.
e.g.: *Malarguesaurus*

Malawi (place-name) the Republic of Malawi: a landlocked country in southeast Africa.
e.g.: *Malawisaurus*

Maleev (person's name) Evgeny Aleksandrovich Maleev: a Russian (Soviet) paleontologist.
e.g.: (*Maleevosaurus* Carpenter, 1992) >>> *Tarvosaurus*; *Maleevus*

Mamenchi (Chin. place-name) 馬門溪*: 馬鳴溪 Mămíngxī situated in Sichuan, China.
*due to an accentual mix-up.
e.g.: *Mamenchisaurus*

Mandschuro- (Eng. place-name) Manchuria = (Chin. place-name) 満州 [Pinyin: Mǎnzhōu]: now Northeast China [Dōngběi 東北].
e.g.: *Mandschurosaurus*

mani-, manu- (Lat.) manus (*nom.*), manūs (*gen.*), *f.*: a hand.
e.g.: *Manidens* -> -dens

mano- (Gr.) μανός/manos: porous.
e.g.: (*Manospondylus* Cope, 1892) *(nomen dubium)* >>> *Tyrannosaurus*

Mansoura (place-name) a city in Egypt. | (institution name) Mansoura University, in Mansoura. | (Egypt.) 'victorious'.
e.g.: *Mansourasaurus*

Mantell (person's name) Gideon Algernon Mantell (1790–1852): an English obstetrician, geologist and paleontologist.
e.g.: *Mantellisaurus*; *Mantellodon* -> -odon

-manus (Lat.) manus (*nom.*), manūs (*gen.*), *f.*: a hand.
e.g.: *Magnamanus*

mapu (Mapuche) earth, land.
e.g.: *Mapusaurus*

mara (animal) the mara = Patagonian cavy: a rodent.
e.g.: *Marasuchus* [Dinosauriformes]

maraapuni (Southern Utah) Ma-ra-pu-ni (pron. mah-rah-poo-nee): huge.
e.g.: *Maraapunisaurus*

Mari (demonym) the Mari tribe of Pakistan.
e.g.: *Marisaurus*

marmaros (Gr.) μάρμᾰρος/marmaros, *m.*: marble. | (stratal name) Forest Marble Formation (middle Jurassic), Dorset, UK.
e.g.: (*Marmarospondylus* Owen, 1875) *(nomen dubium)* -> -spondylus >>> *Bothriospondylus*

Marsh (person's name) Othniel Charles Marsh (1831–1899): an American paleontologist.
e.g.: *Marshosaurus*

Martha (person's name) Martha Hayden: She co-discovered the site and has served as the assistant to paleontologists of Utah.
e.g.: *Martharaptor* -> -raptor

Martin (person's name) Larry D. Martin: an American paleontologist and curator.
e.g.: *Martinavis* -> -avis [Avialae]

masiaka (Malagasy) fierce.
e.g.: *Masiakasaurus*

masso- (Gr.) μάσσων/massōn [comp. of μακρός/makros or μέγας/megas] longer, greater.
e.g.: *Massospondylus* -> -spondylus

-mastac-, **-mastax** (Gr.) μάσταξ/mastax (*nom.*), μάστᾰκος/mastakos (*gen.*) *f.*: that with which one chews, the mouth.
e.g.: *Angulomastacator*; *Pegomastax*

Matheron (person's name) Philippe Matheron: a French palaeontologist and geologist.
e.g.: *Matheronodon* -> -odon

Maxakali (demonym) one of the tribes living in Brazil.
e.g.: *Maxakalisaurus*

Medusa (Gr. myth.) Μέδουσα/Medūsa: a Gorgon.
e.g.: *Medusaceratops* -> -ceratops

meg(a)- (Gr.) μέγας/megas: big, great, strong. 《comb. form》 meg-, mega-, megalo-.
e.g.: ("Megacervixosaurus" Zhao, 1985) -> -cervix; (*Megadactylus* Hitchcock, 1865) (preoccupied) >>> *Anchisaurus*; *Megapnosaurus*-> -apno-; *Megaraptor* -> -raptor

megalo- (Gr.) μέγας/megas (*nom.*), μεγάλος/megalos (*gen.*), *adj.*: big, great, mighty.
e.g.: *Megalosaurus*

mei (Chin.) mèi 寐: to sleep soundly.
e.g.: *Mei*

melan(o)- (Gr.) μέλᾱς/melās (*nom.*), μέλᾰνος/melanos (*gen.*), *adj.*: black, dark.
e.g.: *Melanorosaurus* -> -oro-

Mendoza (place-name) a province of Argentina.
e.g.: *Mendozasaurus*

Mercuri- (Lat. myth.) Mercurius (*nom.*), Mercuriī (*gen.*), *m.*: Mercury, the messenger of the gods, the god of trade, thieves, speech and the lyre.
e.g.: *Mercuriceratops* -> -ceratops

mero- (Gr.) μηρός/mēros, *m.*: the thigh, femur.
e.g.: *Meroktenos* -> -ktenos

-merus (Gr.) μηρός/mēros, *m.* = (Lat.) femur, *n.*: the thigh, thigh bone.
e.g.: *Brontomerus*; *Ornithomerus*; *Orthomerus*

metri(o)- (Gr.) μέτριος/metrios: moderate.
e.g.: *Metriacanthosaurus* -> -acantho-; (*Metriorhynchus* von Meyer, 1830) -> -rhynchus [Crocodylomorpha]

micr(o)- (Gr.) μῑκρός/mīkros: small, little.
e.g.: *Microceratops*; *Microceratus*; *Microcoelus* -> -coel-; ("Microdontosaurus" Zhao, 1983); *Microhadrosaurus* -> -hadros; *Micropachycephalosaurus* -> -pachy-, -cephalo-; *Microraptor* -> -raptor; *Microvenator* -> -venator

Miera (person's name) Bernardo de Miera y Pacheco: a cartographer of New Spain.
e.g.: *Mierasaurus*

Mifune (Jpn. place-name) 御船町Mifune-machi: a town located in Kamimashiki District (上益城郡), Kumamoto Prefecture (熊本県), Japan.
e.g.: ("Mifunesaurus")

-mimus (Lat.) mīmus (*nom.*), mīmī (*gen.*), *m.* = (Gr.) μῖμος/mīmos, *m.*: a mimic actor, mime, an imitator, mimic.
e.g.: *Aepyornithomimus*; *Anserimimus*; *Archaeornithomimus*; *Avimimus*; *Dromiceiomimus*; *Gallimimus*; *Garudimimus*; *Harpymimus*; *Kinnareemimus*; *Ornithomimus*; *Pelecanimimus*; (*Saltillomimus*); *Sciurmimus*; *Sinornithomimus*; *Struthiomimus*; *Suchomimus*; *Timimus*; *Tototlmimus*; *etc.*

-mimoides
e.g.: *Ornithomimoides*

Minmi (place-name) Minmi Crossing in Queensland, Australia.
e.g.: *Minmi*

Minotaura- {(Gr. myth.) Minotauros [Μίνως/Minōs, *m.*: son of Zeus and Europa, king of

Creta + ταῦρος/tauros, *m.*: a bull]} Minotaur.
e.g.: *Minotaurasaurus* -> -taurus

Miragaia (Locality) Miragaia: a village in Lourinhã, Portugal. | (geological unit) Miragaia unit of Sobral Formation (Late Jurassic), Lourinhã Group. | mira Gaea {(Lat.) mirus: wonderful + (Gr. myth.) Γαῖα/Gaia: Gaia, Earth} 'wonderful goddess of the Earth'.
e.g.: *Miragaia*

mir(i)- (Lat.) mīrus (*m.*), mīra (*f.*), mīrum (*n.*), *adj.*: wonderful, marvellous, astonishing, strange. *cf.* mīrī (*gen.* of mīrus).
e.g.: *Mirarce* -> Arce [Avialae]; *Mirischia* -> -ischia

mixo- (Gr.) μιξο-/mixo-: half-, mixed.
e.g.: (*Mixosaurus* Baur, 1887) [Ichthyosauria]

mnyamawamtuka {(Kiswahili) mnyama: beast + wa: of + Mtuka: the river drainage} beast of the Mtuka.
e.g.: *Mnyamawamtuka*

Moab (place-name) Moab: a city in Grand County, Utah, USA.
e.g.: *Moabosaurus*

mochl(o)- (Gr.) μοχλός/mochlos, *m.*: a bar used as a lever, a crowbar, hand-spike.
e.g.: *Mochlodon* -> -odon

Mogan (Chin. person's name / sword name) Mò Yé 莫邪 & Gān Jiāng 干将: Gan Jiang and Mo Ye, a swordsmith couple who lived during the Spring and Autumn period of Chinese history.
e.g.: *Moganopterus* -> -pterus [Pterosauria]

mojo (African-American term) a magic charm or talisman.
e.g.: *Mojoceratops*

-moloch (legend) Moloch: putative god of fire, associated with child sacrifice, demon, horrid king.
e.g.: *Stygimoloch*

Mongolo- (Mong. place-name) Монгол: Mongolia. | (demonym) Mongolian.
e.g.: *Mongolosaurus*; *Mongolostegus* -> -stegus

Monkono- (place-name) Monko: Markam County = (Chin.) 芒康 [Pinyin: Mángkāng]: a county of the Chamdo Prefecture in the Tibet Autonomous Region.
e.g.: *Monkonosaurus*

mon(o)- (Gr.) μόνος/monos: alone, only.
e.g.: *Monoclonius* -> -clon-; *Monolophosaurus* -> -lopho-; (*Mononychus* Perle *et al.*, 1993) (preoccupied) >>> *Mononykus* -> -onykus

Montano- (place-name) Montana: a state in northwestern United States. | (Sp.) montaña: mountain.
e.g.: *Montanoceratops* -> -ceratops

Morella (stratal name) Arcillas de Morella Formation (Early Cretaceous), Morella locality, Spain.
e.g.: *Morelladon* -> -odon

Morino- (demonym) Morini: a Belgic tribe of northern Gaul.
e.g.: (*Morinosaurus* Sauvage, 1874) *(nomen dubium)*

moro- (Gr.) μωρός/mōros: dull, sluggish, stupid.
e.g.: *Morosaurus*

moros (Gr.) μόρος/moros, *m.*: fate, doom.
e.g.: *Moros*

-morph(o)- (Gr.) μορφή/morphē, *f.*: form, shape. = (Lat.) forma.
e.g.: *Dimorphodon*; *Eudimorphodon*

Morro (place-name) El Morro: the site of James Ross Island, Antarctica.
e.g.: *Morrosaurus*

Mosa (Lat. place-name) Mosa = (Dutch) Maas = (Eng.) the Meus: a major European river.
e.g.: (*Mosasaurus* Conybeare, 1822) [Squamata]

mosaic- (Mod. Lat.) mosaicus: mosaic.
e.g.: *Mosaiceratops* -> -ceratops

Moshi (Jpn. place-name) Moshi 茂師, Iwaizumi (岩泉), Iwate Prefecture (岩手県), Japan.
e.g.: ("Moshisaurus" or "Moshiryu")

multi- (Lat.) multus (*m.*), multa (*f.*), multum (*n.*), *adj.*: many. *cf.* multī (*gen.* of multus)
e.g.: (*Multifissoolithus* Imai *et al.*, 2020) -> fiss, -oo-, lithus 〖Eggshell〗

murus (Lat.) mūrus (*nom.*), mūrī (*gen.*), *m.*: wall.
e.g.: *Murusraptor* -> -raptor

mus (Lat.) mūs (*nom.*), mūris (*gen.*), *m./f.* : a mouse. rat. | (Gr.) μῦς/mys, *m.*: a mouse, a muscle of the body.
e.g.: *Mussaurus*

Muttaburra (place-name) a small grazing town, located in Queensland, Australia.
e.g.: *Muttaburrasaurus*

Muyelen (Mapuche, place-name) one of the names of the Colorado River, Argentina.
e.g.: *Muyelensaurus*

-mylos (Gr.) μύλος/mylos, *f.*: a millstone.
e.g.: (*Sasayamamylos*) [Mammalia]

Mymoora- (person's name) Mygatt-Moore: Peter and Marilyn Mygatt, and John D. and Vanetta Moore.
e.g.: *Mymoorapelta* -> -pelta

N

Naashoibito (place-name) Naashoibito Member of the Kirkland Formation. | (Navajo) 'lizard creek'.
e.g.: *Naashoibitosaurus*

Nambal (place-name) Nambal: a village in Andhra Pradesh, India.
e.g.: *Nambalia*

Nankang (Chin. place-name) Nánkāng 南康: a district of Ganzhou City, Jianxi Province (江西省), China.
e.g.: *Nankangia*

Nanning (Chin. place-name) Nánníng 南寧: a city of the Guăngxī 広西, Zhuang Autonomous Region, China.
e.g.: *Nanningosaurus*

nan(o)- (Gr.) νᾶνος/nānos, *m.*: a dwarf.
e.g.: *Nanantius* -> -enantius [Avialae]; *Nanosaurus*; *Nanotyrannus* -> -tyrannus

Nanshiung (Chin. place-name) 南雄 [Pinyin: Nánxióng]: a county-level city, Guandong Province (広東省), China.
e.g.: *Nanshiungosaurus*

nanuq (Iñupiaq*) polar bear.
*the Inuit language spoken by the Iñupiat, a group of Alaska Natives.
e.g.: *Nanuqsaurus*

Nanyang (Chin. place-name) Nányáng 南陽: a prefecture-level city, Henan Province (河南省), China.
e.g.: *Nanyangosaurus*

Narambuena (place-name) Puesto Narambuena, located in the Neuquén Basin, Argentina.
e.g.: *Narambuenatitan* -> -titan

Narinda (place-name) Narinda Bay: a bay located in Madagascar.
e.g.: *Narindasaurus*

nasuto- (Lat.) nāsūtus (*m.*), nāsūta (*f.*), nāsūtum (*n.*), *adj.*: with a large nose, large-nosed.
e.g.: *Nasutoceratops* -> -ceratops

-natator (Lat.) natātor (*nom.*), natātōris (*gen.*), *m.*: a swimmer.
e.g.: *Terminonatator*

Natrona (place-name) Natrona: a county of Wyoming, USA.
e.g.: "Natronasaurus"

Navajo (demonym) a native American people.
e.g.: *Navajoceratops*

nebula (Lat.) nebula (*nom.*), nebulae (*gen.*), *f.*: mist, vapour, fog, cloud.
e.g.: *Nebulasaurus*

-nectes (Gr.) νήκτης/nēctes: swimmer. νήχω/nēchō: to swim. νηκτός/nēktos: swimming.
e.g.: *Opallionectes* [Plesiosauria]

Ned Colbert (person's name) Edwin Harris Colbert (1905–2001): known as Ned. A distinguished American vertebrate paleontologist.
e.g.: *Nedcolbertia*

nedo- (Russian) недо-: insufficient.
e.g.: *Nedoceratops* -> -ceratops

Neimongo- (place-name) Nei Monggol: Inner Mongolia.
e.g.: *Neimongosaurus*

Nemegt (stratal name) Nemegt Formation (Late Cretaceous). | Nemegt Basin, northwestern Gobi Desert, Ömnögovi Province, southern Mongolia.
e.g.: *Nemegtia*; *Nemegtomaia* -> -maia; *Nemegtonykus* -> -onykus; *Nemegtosaurus*

-nemo- (Gr.) νέμος/nemos, *n.*: a wooded pasture, glade. = (Lat.) nemus (*nom.*), nemōris (*gen.*), *n.*: a wood, grove, forest.
e.g.: *Pycnemosaurus*

neo- (Gr.) νέos/neos: young, new, fresh.
e.g.: *Neosaurus*; (*Neosodon* Moussaye, 1885) -> odon; *Neovenator* -> -venator

Neuquen (place-name) Neuquén: a province located at the northern end of Patagonia, Argentina.
e.g.: *Neuquenornis* -> -ornis [Avialae]; *Neuquenraptor* -> -raptor; *Neuquensaurus*

Ngexi (place-name) Ngêxi Township 埃西郷: a village, Qamdo (昌都), Xizang (= Tibet), China.
e.g.: ("Ngexisaurus" Zhao, 1983)

ngwevu (Xhosa*) grey.
*one of the official languages, spoken in South Africa.
e.g.: *Ngwevu*

nhandumirim {(Tupi·Guarani) nhandu*: running bird, rhea + mirim: small}
*nhanda: austrich (acc. Kowyama R (1951). *Vocabulario Tupy Português-Japonês*.)
e.g.: *Nhandumirim*

niebla (Sp.) mist.
e.g.: (*Niebla* Aranciaga Rolando *et al.*, 2020)

niger (Lat.) niger (*m.*), nigra (*f.*), nigrum (*n.*), *adj.*: black, dark. | (Fr. place-name) Niger. | (Tuareg) n'egiren.
e.g.: *Nigersaurus*

Ningyuan (Chin. place-name) Níngyuǎn 寧遠: an old name of Xīngchéng City (興城市), Liaoning Province (遼寧省), China.
e.g.: *Ningyuansaurus*

Niobrara (stratal name) Niobrara Formation / Niobrara Chalk (Late Cretaceous), middle of North America.
e.g.: *Niobrarasaurus*

Nippon (Jpn. place-name) 日本: Japan.
e.g.: (*Nipponoolithus* Tanaka *et al.*, 2016) -> -oo-, -lithus 〖Eggshell〗; *Nipponosaurus*

NOA (acronym) Northwestern Argentina.
e.g.: *Noasaurus*

nodo- (Lat.) nōdus (*nom.*), nōdī (*gen.*), *m.*: a knot, knob.
e.g.: *Nodocephalosaurus* -> -cephalo-; *Nodosaurus*

Nomingi- (place-name) Nomingiin Gobi: a nearby part of the Gobi Desert.
e.g.: *Nomingia*

Nopcsa (person's name) Franz Nopcsa (1877–1933): a Hungarian-born aristocrat.
e.g.: *Nopcsaspondylus* -> -spondylus

Normannia (place-name) an ancient name of Normandy. | (Fr.) Normandie.
e.g.: *Normanniasaurus*

nota (Lat.) nota (*nom.*), notae (*gen.*) *f.*: feature.
e.g.: *Notatesseraeraptor* -> -tesserae, -raptor

notho- (Gr.) νόθος/nothos: a bastard, baseborn child. Generally, spurious, counterfeit, supposititious, illegitimate.
e.g.: (*Nothosaurus* Müster, 1834) [Sauropterygia]

nothr(o)- (Gr.) νωθρός/nōthros: sluggish, slothful, torpid.
e.g.: *Nothronychus* -> -onychus

noto- (Gr.) νότος/notos, *m.*: the south or south-west wind.
e.g.: *Notoceratops* -> -ceratops; *Notocolossus* -> -colossus; *Notohypsilophodon* -> -hypsiloph(o)-

-noto-
e.g.: *Giganotosaurus*

Nqweba [n-KWE-bah] (Xhosa, place-name) the local name of Kirkwood Formation (Early Cretaceous) in South Africa.
e.g.: *Nqwebasaurus*

Nullo (person's name) Francisco E. Nullo, geologist of Argentine Geological Survey.
e.g.: *Nullotitan*

nur (Mong.) нуур: lake.
e.g.: "Nurosaurus"

nuthetes (Gr.) νουθέτης/nūthetēs (an abbreviation of nouthetetes): one who admonishes or a monitor. νουθετέω: to put in mind, to admonish, warn, advise.
e.g.: *Nuthetes*

Nyasa (place-name) Lake Nyasa = Lake Malawi.
e.g.: *Nyasasaurus* [Dinosauromorpha]

nycto- (Gr.) νύξ/nyx (*nom.*), νυκτός/nyktos (*gen.*) *f.*: night. | (Gr. myth.) Νύξ/Nyx: the goddess of Night.
e.g.: (*Nyctosaurus* Marsh, 1876) [Pterosauria]

-nykus -> -onykus
e.g.: *Shishugounykus*; *Xiyunykus*

O

oceano- (Lat.) Ōceanus, *m.* = (Gr. myth.) Ὠκεανός/Ōkeanos: a Greek water god. (Eng.) Oceanus, ocean, the great sea.
e.g.: *Oceanotitan* -> -titan

oculu- (Lat.) oculus (*nom.*), oculī (*gen.*) *m.*: eye.
e.g.: *Oculudentavis* -> -dent(i)- [Squamata]

-odon (Gr.) ὀδών/odōn (*nom.*), ὀδόντος/odontos (*gen.*), *m.*, Ion.* for ὀδούς/odūs: a tooth.
*Ion. = in the Ionic dialect.
e.g.: *Amygdalodon*; (*Apatodon*); *Astrodon*; *Aublysodon*; (*Campylodon*); *Cardiodon*; *Craspedodon*; *Dakotadon*; *Deinodon*; *Dimorphodon*; *Dollodon*; *Echinodon*; *Iguanodon*; *Mantellodon*; *Mochlodon*; (*Neosodon*); *Owenodon*; *etc.*

-odon-
e.g.: *Campylodoniscus*

-odonto-
e.g.: *Anodontosaurus*; *Archaeodontosaurus*; *Carcharodontosaurus*; *Glyptodontopelta*; *Heterodontosaurus*; (*Polyodontosaurus*); *Priodontognathus*

Ohmdeno- (place-name) Ohmden: a municipality in the district of Esslingen, Baden-Württemberg, Germany.
e.g.: *Ohmdenosaurus*

Ojo (stratal name) Ojo Alamo Formation (Mesozoic / Cenozoic boundary), New Mexico, USA.
e.g.: *Ojoceratops* -> -ceratops; *Ojoraptorsaurus* -> -raptor-

Oksoko [oak-soak-oh'] (Altaic myth.) Öksökö: the three-headed eagle.
e.g.: (*Oksoko* Funston et al., 2020)

oligo- (Gr.) ὀλίγος/oligos: few, little, scanty, opp. to πολύς/polys.
e.g.: *Oligosaurus*

olor (Lat.) olor (*nom.*), olōris (*gen.*), *m.*: a swan.
e.g.: *Olorotitan* -> -titan

Omei (Chin. place-name) 峨眉山 [Pinyin: Éméi Shān]: a mountain in Sichuan Province (四川省), China.
e.g.: *Omeisaurus*

omnivoro- {(Lat.) omni- [omnis: all, every, any] + voro- [vorō: to swallow whole, devour]} omnivorous.
e.g.: *Omnivoropteryx* -> -pteryx [Avialae]

om(o)- (Gr.) ὦμος/ōmos, *m.* = (Lat.) umerus = (Eng.) humerus: the shoulder with the upper arm.
e.g.: *Omosaurus*

-om(o)-
e.g.: *Adynomosaurus*; (*Crataeomus*); *Flexomornis*

omphalo- (Gr) ὀμφᾰλός/omphalos, *m.*: the navel, anything central.
e.g.: *Omphalosaurus* [Ichthyosauria]

onycho- (Gr.) ὄνυξ/onyx (*nom.*), ὄνυχος/onychos (*gen.*), *m.* = (Lat.) unguis: a claw, talon, hoof.
e.g.: (*Onychosaurus* Nopcsa, 1902) >>> *Zalmoxes*

-onych(o)-
e.g.: *Euronychodon*; *Paronychodon*; *Stenonychosaurus*

-onychus (Gr.) ὄνυξ/onyx (*nom.*), ὄνῠχος/onychos (*gen.*), *m.*: claw, talon. = (Lat.) unguis.
e.g.: *Deinonychus*; *Hesperonychus*; *Nothronychus*

-onykus, -nykus (Gr.) ὄνυξ/onyx, *m.*: claw, talon.
e.g.: *Albertonykus*; *Albinykus*; *Bannykus*; *Bonapartenykus*; *Boreonykus*; *Ceratonykus*; *Linhenykus*; *Mononykus*; *Nemegtonykus*; *Patagonykus*; *Qiupanykus*; *Shishugounykus*; *Xixianykus*

-onyx (Gr.) ὄνυξ/onyx, *m.*: claw, talon.
e.g.: *Aardonyx*; (*Aetonyx*); *Baryonyx*; *Draconyx*; *Gryponyx*; *Scipionyx*

-oo- (Gr.) ᾠόν/ōion, *n.* = (Lat.) ōvum, *n.*: an egg.
e.g.: (*Multifissoolithus*) 〚Eggshell〛; (*Nipponoolithus*) 〚Eggshell〛

oohkotok- (Blackfoot) animate noun,*ooh'kotoka* : large stone or rock.
e.g.: *Oohkotokia*

opallio- (Gr.) ὀπάλλιος/opallios, *m.*: opal.
e.g.: *Opallionectes* [Plesiosauria] -> -nectes

ophthalmo- (Gr.) ὀφθαλμός/ophthalmos, *m.*: the eye.
e.g.: (*Ophthalmosaurus* Seeley, 1874) [Ichthyosauria]

opistho- (Gr.) ὀπισθο-/opistho- [ὄπισθεν/opisthen: behind, at the back]
e.g.: *Opisthocoelicaudia* -> -coeli, -caudia

-oplites (Gr.) ὁπλίτης/hoplitēs: heavy-armed, armed. As *subst.*, a heavy-armed foot-soldier.
e.g.: *Peloroplites*; *Sauroplites*

oplo- (Gr.) ὅπλον/hoplon, *n.*: a tool, implement: in *pl.*, also, implements of war, arms. Rarely in *sing.*, a weapon.
e.g.: *Oplosaurus*

-oplo-
e.g.: *Anoplosaurus*; *Dyoplosaurus*; *Euoplocephalus*

-ops (Gr.) ὤψ/ōps, *f.*: the eye, face, countenance.
e.g.: excepted "-ceratops": *Aquilops*; *Cerasinops*; *Ceratops*; *Kryptops*; (*Pachysaurops*); *Rugops*; *Sauroniops*; *Spinops*

-opsis
e.g.: *Ornithopsis*

-opus (Lat.) opus (*nom.*), operis (*gen.*), *n.*: work, need, want.
e.g.: *Galeamopus*

-Orestes (Lat.) Orestēs: 'mountain dweller'. = (Gr. myth.) Ὀρέστης {ὄρος/oros: mountain + ἵστημι/hístēmi: to stand} 'stands on a mountain'.
e.g.: *Cedrorestes*

orient- (Lat.) oriēns (*nom.*), orientis (*gen.*), *m.*: the rising sun, morning sun.
e.g.: *Orienantius* [Avialae]

Orko- (Tehuelche) Orr-Korr 'Toothed River': La Leona River of Patagonia, Argentina.
e.g.: *Orkoraptor* -> -raptor

ornato- (Lat.) ōrnātus (*m.*), ōrnāta (*f.*), ōrnātum (*n.*), *adj.*: adorned, decorated. ōrnō: to fit out, decorate. ōrnātī (*gen.* of ōrnātus)
e.g.: *Ornatotholus* -> -tholus

-ornis (Gr.) ὄρνις/ornis (*nom.*), ὄρνῑθος/ornithos (*gen.*), *m.* / *f.*: a bird. 《comb. form》 ornitho-.
e.g.: *Anchiornis* [Avialae?]; *Aurornis* [Avialae?]; *Camptodontornis* [Avialae]; *Cathayornis* [Avialae]; *Changchengornis* [Avialae]; *Conficiusornis* [Avialae]; *Enantiornis* [Avialae]; *Heptasteornis*; *Hesperornis* [Avialae]; *Houornis* [Avialae]; *etc.*

ornith(o)-
e.g.: *Ornithocheirus* -> -cheir- [Pterosauria]; (*Ornithodesmus* Seeley, 1887); *Ornitholestes* -> -lestes; *Ornithomerus* ->

-merus ; *Ornithomimoides* -> -mimoides; *Ornithomimus* -> -mimus; *Ornithopsis* -> -opsis; (*Ornithotarsus* Cope, 1869) *(nomen dubium)*

-ornith(o)-
e.g.: *Aepyornithomimus*; *Archaeornithoides*; *Archaeornithomimus*; *Archaeornithura*; *Saurornithoides*; *Saurornitholestes*; *Sinornithoides*; *Sinornithomimus*; *Sinornithosaurus*

oro- (Gr.) ὄρος/oros, *n.*: a mountain, hill.
e.g.: *Orodromeus* -> -dromeus; (*Orosaurus* Huxley, 1867) >>> *Euskelosaurus*

-or(o)-
e.g.: *Cedrorestes*; *Melanorosaurus*

ortho- (Gr.) ὀρθός/orthos: straight, right. = (Lat.) rēctus.
e.g.: (*Orthogoniosaurus* Das-Gupta, 1931) *(nomen dubium)*; *Orthomerus* -> -merus

-ortus (Lat.) ortus (*nom.*), ortūs (*gen.*), *m.*: a rising, a rise, beginning, origin. As *adj.*, sprung, descended, born.
e.g.: *Ambiortus* [Avialae]

orycto- (Gr.) ὀρυκτός/oryctos: formed by digging, opp. to a natural channel.
e.g.: *Oryctodromeus* -> -dromeus

osmaka (Lakota) ósmaka 'a valley, canyon': Calico Canyon in western South Dakota, USA.
e.g.: *Osmakasaurus*

Ostafrika (place-name) Deutsch-Ostafrika: a German colony in East Africa.
e.g.: *Ostafrikasaurus*

-osteo- (Gr.) ὀστέον/osteon, *n.*: a bone.
e.g.: *Aerosteon*; *Chondrosteosaurus*; (*Ponerosteus*) [Archosauromorpha]; (*Tichosteus*)

Ostrom (person's name) John Ostrom: a palaeontologist, who identified the Haarlem specimen.
e.g.: *Ostromia* [Anchiornithidae]

Othniel (person's name) Othniel Charles Marsh (1831–1899): an American paleontologist.
e.g.: *Othnielia*; *Othnielosaurus*

Otog (place-name) Otog Banner 鄂托克旗: a banner of southwestern Inner Mongolia, China.
e.g.: *Otogosaurus*

-oura (Gr.) οὐρά/oura [ūrā], *f.*: tail.
e.g.: *Schizooura* [Avialae]

ourano- (Arab.) ---/ourane: valour, courage, recklessness. | (Tuareg) ourane: the sand monitor.
e.g.: *Ouranosaurus*

Overo (place-name) Cerro Overo: the locality in the Anacleto Formation (Late Cretaceous), Patagonia, Argentina.
e.g.: *Overosaurus*

overo (Sp.) piebald.
e.g.: *Overoraptor*

ovi- (Lat.) ōvum (*nom.*), ovī (*gen.*), *n.*: an egg.
e.g.: *Oviraptor* -> -raptor

Owen (place-name) Richard Owen (1804–1892): an English biologist, comparative anatomist and paleontologist.
e.g.: *Owenodon* -> -odon

Oxala (myth.) Oxalá: the African deity.
e.g.: *Oxalaia*

Oz (place-name) Oz (= Ozzie, Aussie): Australia, Australian.
e.g.: *Ozraptor* -> -raptor

P

pachy- (Gr.) πᾰχύς/pachys: thick, stout.
e.g.: *Pachycephalosaurus* -> cephalo-; *Pachyrhinosaurus* -> -rhino-; (*Pachysauriscus* Kuhn, 1959) >>> *Plateosaurus*; (*Pachysaurops* von Huene, 1961) >>> *Plateosaurus*; (*Pachysaurus* von Huene, 1907–1908) >>> *Pachysauriscus*; (*Pachyspondylus* Owen, 1854) >>>?*Massospondylus*; *Pachysuchus* -> -suchus *(nomen dubium)*

-pachy-
e.g.: *Micropachycephalosaurus*

Padilla (person's name) Carlos Bernardo Padilla: the founder of el Centro de Investigaciones Paleontológicas, Colombia.
e.g.: *Padillasaurus*

Paki- (place-name) Pakistan.
e.g.: *Pakisaurus*

palaeo- (Gr.) πᾰλαιός/palaios: old, aged, ancient.
e.g.: *Palaeopteryx* -> -pteryx; (*Palaeosaurus* Riley & Stutchbury, 1840) [Archosauria]; (*Palaeoscincus* Leidy, 1856) -> -scincus

paludi- (Lat.) palūs (*nom.*), palūdis (*gen.*), *f.*: a swamp, marsh, pool.
e.g.: *Paludititan* -> -titan

Paluxy (place-name) Paluxy River: a river in Texas, USA.
e.g.: *Paluxysaurus*

pampa (Quechua*) pampa: plain. | (place-name) the Pampas. | (demonym) the Indian Pampas people.
*a language spoken in Andes and highlands of South America.
e.g.: *Pampadromaeus* -> -dromaeus; *Pamparaptor* -> -raptor

pan- (Gr.) πᾶν-/pan-: all-. (Gr.) πᾶς/pās (*m.*), πᾶσα/pāsa (*f.*), πᾶν/pān (*n.*): all, the whole. *cf.* (Lat.) omnis (*m./f.*), omne (*n.*).
e.g.: *Panphagia* -> -phag(o)-

-Pan (Gr. myth.) Πάν/Pan (*nom.*), Πᾶνός/Panos (*gen.*), *m.*: Pan, god of Arcadia, son of Hermes.
e.g.: *Brachytrachelopan*

Pan American (institution name) the Pan American Energy company.
e.g.: *Panamericansaurus*

Pandora (place-name) Caja de Pandora 'Pandora's box': the type locality in Chubut Province, Argentina. | (Gr. myth.) παν-δώρα/pan-dōra: giver of all. Pandora, *i.e.*, the All-endowed.
e.g.: *Pandoravenator* -> -venator

Pangu (Chin. myth.) Pángŭ 盤古: P'an Ku, the first living being and the Creator. | [geol.] 盤古大陸: Pangaea.
e.g.: *Panguraptor* -> -raptor

panoplo- (Gr.) πάν-οπλος/pan-oplos: in full armor, full-armed.
e.g.: *Panoplosaurus*

Panty- (abbr. place-name) Pant-y-ffynnon Quarry. | (Welsh) Pant-y-ffynnon {pant: a hollow place, valley + ffynnon: a well or spring.}
e.g.: *Pantydraco* -> -draco

-pappo- (Lat.) pappo- [pāpo or pappo: to eat].
e.g.: *Adeopapposaurus*

par-, para- (Gr.) πᾰρά/para: beside. (prefix) par(a)-: similar.
e.g.: *Paranthodon* -> -anth(o)-; *Pararhabdodon* -> rhabd(o)-, -odon; *Parasaurolophus* -> -saur(o)-, -lophus; (*Parasuchus* Lydekker, 1885) -> -suchus [Phytosauria]; *Paronychodon* -> -onycho-, -odon

para- (Lat.)
e.g.: *Parabohoaiornis* -> -Bohai-, -ornis [Avialae]; *Parahesperornis* -> -hesper(o)-, -ornis [Avialae]; *Parahongshanornis* -> -Honshan, -ornis [Avialae]; *Parapengornis* [Avialae]; *Paraprotopteryx* [Avialae]

parali- (Gr.) παρ-άλιος/par-alios: by the sea.
e.g.: *Paralititan* -> -titan

paraxeni- (Gr.) παρά-ξενος/para-xenos: half-foreign, counterfeit.
e.g.: *Paraxenisaurus*

pareisactus (Mod. Gr.) παρείσακτος/pareisaktos: intruder. | (Gr.) introduced privily.
e.g.: *Pareisactus*

-Parker- (person's name) William Kitchen Parker (1823–1890): an English physician, zoologist, comparative anatomist.
e.g.: *Euparkeria* [Archosauromorpha]

Parks (person's name) William Arthur Parks: a Canadian geologist and paleontologist.
e.g.: *Parksosaurus*

Parr (person's name) Albert Eide Parr (1900–1991): a Norwegian-born, American marine biologist, zoologist and oceanographer.
e.g.: *Parrosaurus*

parvi- (Lat.) parvus(*m.*), parva (*f.*), parvum (*n.*) *adj.*: small, little, slight.
e.g.: *Parvicursor* -> -cursor

Pasquia (place-name) Pasquia Hills region, Saskatchewan, Canada.
e.g.: *Pasquiaornis* [Avialae] -> -ornis

Patag(o)- (place-name) Patagonia.
e.g.: *Patagonykus* -> -onykus; *Patagopteryx* -> -pteryx [Avialae]; *Patagosaurus*; *Patagotitan* -> -titan

-Paul- (person's name) Paul Haaga.
e.g.: *Magnapaulia*

Paw Paw (stratal name) Paw Paw Formation (Early Cretaceous), Texas, USA.
e.g.: *Pawpawsaurus*

pectin(i)- (Lat.) pectin (*nom.*), pectinis (*gen.*), *m.*: a comb.
e.g.: *Pectinodon*

pedo-, **pedi-** (Lat.) pēs (*nom.*), pedis (*gen.*), *m.*: a foot.
e.g.: *Pedopenna* -> -penna [Anchiornithidae]

pego- (Gr.) πηγός/pēgos: well put together, compact, strong.
e.g.: *Pegomastax* -> -mastax

Peishan (Chin. place-name) 北山 [Pinyin: Běishān]: Peishan, 'North Mountain', in the Xinjiang Uygur Autonomous Region, China.
e.g.: *Peishansaurus*

Pekin (stratal name) Pekin Formation (Late Triassic), North Carolina, USA.
e.g.: (*Pekinosaurus* Hunt & Lucas, 1994) [Suchia] >>> *Revueltosaurus*

pelecani- (Latin) pelecānus (*nom.*), pelecanī (*gen.*), *m.* = (Mod. Gr.) πέλεκάνος/pelekanos, *m.*: pelican. *cf.* πέλεκυς/pelekus, *n.*: axe.
e.g.: *Pelecanimimus* -> -mimus

-pelix (Gr.) πέλυξ/pelyx, *m.* = [anat.] pelvis.
e.g.: *Stenopelix*

Pellegrini (place-name) a town in Buenos Aires Province, Argentina. | Lake Pellegrini.
e.g.: *Pellegrinisaurus*

pelo- (Gr.) πηλός/pēlos, *m./f.*: clay, earth.
e.g.: (*Peloneustes* Seeley, 1889) [Plesiosauria]

peloro- (Gr.) πέλωρος/pelōros: monstrous, prodigious, huge, gigantic.
e.g.: *Peloroplites* -> -oplites; *Pelorosaurus*

-pelta (Lat.) pelta (*nom.*), peltae (*gen.*), *f.* = (Gr.) πέλτη/peltē, *f.*: a light shield, shaped like a half-moon, pelt, Thracian shield.
e.g.: *Ahshislepelta*; *Bissektipelta*; *Borealopelta*; *Cedarpelta*; *Crichtonpelta*; *Dongyangopelta*; *Dracopelta*; *Europelta*; *Mymoorapelta*; *Sauropelta*; *Stegopelta*; *Zaraapelta*; *etc.*

-pelto-
e.g.: (*Sinopeltosaurus*)

penelopo- (Gr.) πηνέλοψ/pēnelops (*nom.*), πηνέλοπος/pēnelopos (*gen.*), *m.*: a kind of *duck* with purple stripes. *cf.* (Gr. myth.) Πηνελόπη/Pēnelopē: the name of Odysseus' wife, forced to fend off suitors while her husband is away fighting at Troy. When Penelope was a baby, she was thrown into the sea, but she was saved by ducks. In one account, believing that her husband was dead, she threw herself into the sea, and she was again saved by ducks. [acc. *The Dictionary of Mythology*]
e.g.: *Penelopognathus* -> -gnathus

Peng (Chin. folklore) Péng 鵬: a mythological giant bird.
e.g.: *Pengornis* -> -ornis [Avialae]

-penna (Lat.) penna (*nom.*), pennae (*gen.*), *f.*: a feather, plume (on a bird). *pl.*, a wing.
e.g.: *Pedopenna*

penta- (Gr.) πέντε/pente: five.
e.g.: *Pentaceratops* -> -ceratops

-pes (Lat.) pēs (*nom.*), pedis (*gen.*), *m.*: a foot.
e.g.: *Anomalipes*; *Hulsanpes*; *Velocipes*

peteino- (Gr.) πετεινός/peteinos: able to fly, winged.
e.g.: *Peteinosaurus* [Pterosauria]

Petrobras (institution name) El Petróleo Brasileiro S. A.: the petroleum industry, Rio de Janeiro, Brazil.
e.g.: *Petrobrasaurus*

phaedro- (Gr.) φαιδρός/phaidros: bright, beaming with joy, joyous, jocund, elated.
e.g.: (*Phaedrolosaurus* Dong, 1973) *(nomen dubium)*

-phag(o)- (Gr.) phago- [φᾰγεῖν/phagein: to eat up, devour.]
e.g.: *Panphagia*; *Saurophaganax*

-phalang(o)- (Gr.) φᾰλᾰγξ/phalanx (*nom.*), φᾰλᾰγγος/phalangos (*gen.*), *f.*: [anat.] the phalanx.
e.g.: (*Macrophalangia*)

philo- (Gr.) φιλ(ο)-/philo-: φίλος/philos, pass. *loved, beloved, dear*. (Lat.) amicus, carus. in act. sense, like φίλιος/philios, *loving, friendly.* | (person's name) Phil (= Phillip J. Currie).
e.g.: *Philovenator* -> -venator

-phoenix (myth.) the phoenix, a mythological bird.
e.g.: *Enantiophoenix* [Avialae]

-pholis (mod. Gr.) φόλίς/pholis, *f.*: scale (of fish, *etc.*)
e.g.: *Acanthopholis*

-phoro- (Gr.) φορός/phoros: bringing on one's way. φέρω/pherō: to bear or carry.
e.g.: *Spinophorosaurus*

-phoneus (Gr.) φονεύς/phoneus, *m.*: a murderer, slayer, homicide.
e.g.: *Teratophoneus*

Phu Wiang (place-name) a district of Khon Kaen Province, northeastern Thailand.
e.g.: *Phuwiangosaurus*; *Phuwiangvenator* -> -venator

phyll(o)- (Gr.) φύλλον/phyllon, *n.*: a leaf, a petal, a medical herb.
e.g.: *Phyllodon* -> -odon

-physis (Gr.) φύσις/physis, *f.*: form, stature, nature.
e.g.: *Coelophysis*; *Eoplophysis*

-phytalia (Gr.) φῠτᾰλία/phytalia, *f.*: a planted place, an orchard or vineyard, opp. to corn-land.
e.g.: *Theiophytalia*

Piatnitzky (person's name) Alejandro Matveievich Piatnitzky (1879–1959): a Russian-born Argentine geologist.
e.g.: *Piatnitzkysaurus*

picr(o)- (Gr.) πικρός/pikros: pointed, sharp.
e.g.: (*Picrodon* Seeley, 1898) *(nomen dubium)* -> -odon [Archosauria]

Pilmatue (place-name) Pilmatué: a locality in Neuquen, Argentina.
e.g.: *Pilmatueia*

pinaco- (Gr.) πίναξ/pinax (*nom.*), πίνᾰκος/pinakos (*gen.*), *m.*: a board, plank.
e.g.: *Pinacosaurus*

Pisano (person's name) Juan Arnaldo Pisano: an Argentine paleontologist.
e.g.: *Pisanosaurus* [Silesauridae]

pitekun (Mapuche | Mapudungun) pitëkun: to discover.
e.g.: *Pitekunsaurus*

Piveteau (person's name) Jean Piveteau (1899–1991): a French vertebrate paleontologist.
e.g.: *Piveteausaurus*

plani- (Lat.) plānus (*m.*), plāna (*f.*), plānum (*n.*), *adj.*: even, level, flat, plane.
e.g.: *Planicoxa* -> -coxa

plateo-, platy- (Gr.) πλᾰτύς/platys: wide, broad, flat, level.
e.g.: (*Plateosauravus* von Huene, 1932) *(nomen dubium)* -> -avus; *Plateosaurus*; *Platyceratops* -> -ceratops; *Platypelta* -> -pelta; (*Platypterygius* von Huene, 1922) -> -pterygius [Ichthyosauria]

plesio- (Gr.) πλησίος/plēsios: near, close to.
e.g.: *Plesiohadros* -> -hadros; (*Plesiosaurus* Conybeare, 1821) [Plesiosauria]

pleuro- (Gr.) πλευρά/pleura, *f.* = πλευρόν/pleuron, *n.*: *a rib*. in *pl.* the ribs, the side.
e.g.: *Pleurocoelus* -> -coel-; (*Pleuropeltus* Seeley, 1881) >>> *Struthiosaurus*

-pleur(o)-
e.g.: *Liopleurodon*

-pleuron
e.g.: *Poekilopleuron*

plio- (Gr.) πλείων/pleiōn (=more): *more, larger*, both of number and size.
e.g.: (*Pliosaurus* Owen, 1841) [Plesiosauria]

-plio-
e.g.: (*Sinopliosaurus*) [Plesiosauria]

pneumato- (Gr.) πνεῦμα/pneuma (*nom.*), πνεύματος/pneumatos (*gen.*), *n.*: a blowing, a wind, blast, breath.
e.g.: *Pneumatoraptor* -> -raptor

-podo- (Gr.) πούς/pūs (*nom.*), ποδός/podos (*gen.*), *m.*: a foot.
e.g.: *Brachypodosaurus*

podoke- (Gr.) ποδώκης/podōkēs: swiftfooted. Generally, swift, quick.
e.g.: *Podokesaurus*

poecilo-, **poekilo-**, **poicilo-**, **poikilo-** (Gr.) ποικίλος/poikilos: many-coloured, spotted, mottled.
e.g.: *Poekilopleuron* -> -pleuron

pol-, poly- (Gr.) πολύς/polys: many, opp. to ὀλίγος/oligos: much, mighty, great, large, wide, wide-stretched.
e.g.: (*Polacanthoides* Nopcsa, 1928) >>> *Hylaeosaurus*; *Polacanthus* -> -acanthus; (*Polyodontosaurus* Gilmore, 1932) *(nomen dubium)*; (*Polyonax* Cope, 1874) -> -anax *(nomen dubium)*

poner(o)- (Gr.) πονηρός/poneros: bad, useless.
e.g.: *Ponerosteus* -> -osteo- [Archosauromorpha]

-Poseidon (Gr. myth.) Ποσειδῶν/Poseidōn, *m.*: Poseidon, god of the sea.
e.g.: *Austroposeidon*; *Sauroposeidon*; *Xenoposeidon*

potam(o)- (Gr.) ποτᾰμός/potamos: a river.
e.g.: *Potamornis* -> -ornis [Avialae]

Powell (person's name) Jaime Eduardo Powell: an Argentine paleontologist.
e.g.: *Powellvenator* -> -venator

Pradhan (person's name) Dhuiya Pradhan: an Indian fossil collector.
e.g.: *Pradhania*

preno- (Gr.) πρηνής/prēnēs: with the face downwards, inclined. (Lat.) prōnus: turned forward, prone.
e.g.: *Prenocephale* -> -cephale; *Prenoceratops* -> -ceratops

Preon- (place-name) the Preone Valley of the Italian Alps.
e.g.: *Preondactylus* -> -dactylus [Pterosauria]

pri- (Gr.) πρίων/priōn, *m.*: a sawyer, a saw.
e.g.: *Priconodon* -> -con-, -odon; *Priodontognathus* -> -odonto-, -gnathus

pro- (Gr.) πρό/pro: before.
e.g.: *Probactrosaurus*; *Probrachylophosaurus*; (*Proceratops* Lull, 1906) >>> *Ceratops*; *Proceratosaurus*; (*Procheneosaurus* Parks, 1920) >>> *Lambeosaurus*; *Procompsognathus*; *Prodeinodon*; *Propanoplosaurus*; *Proplanicoxa*; *Prosaurolophus*

proa (Sp.) a bow, prow: the shape of a bow on a boat.
e.g.: *Proa*

prot(o)- (Gr.) πρῶτος/prōtos: first.
e.g.: *Protarchaeopteryx* -> -archae(o)-, -pteryx [Avialae]; *Protoavis* -> -avis; *Protoceratops* -> -ceratops; *Protognathosaurus* -> -gnathus; *Protohadros* -> -hadros; *Protopteryx* -> -pteryx [Avialae]; (*Protosuchus* Brown, 1934) -> -suchus [Crocodylomorpha]

psittacos (Gr.) ψιττᾰκός/psittakos, *m.*: a parrot.
e.g.: *Psittacosaurus*

pter(o)- (Gr.) πτερόν/pteron, *n.*: a feather, mostly in *pl.* feathers (= πτέρυξ/pteryx, *f.*: a bird's wing.)
e.g.: *Pteranodon* -> an-, -odon [Pterosauria]; *Pterodactylus* -> -dactylus [Pterosauria]; *Pterodaustro* -> -daustro [Pterosauria]; (*Pteropelyx* Cope, 1889) *(nomen dubium)*; (*Pterospondylus* Jaekel, 1913) *(nomen dubium)*

-pterus
e.g.: *Darwinopterus* [Pterosauria]; *Dsungaripterus*

[Pterosauria]; *Moganopterus* [Pterosauria]; *Sinopterus* [Pterosauria]

-pterygi- (Gr.) πτερύξ/pteryx, *f.*: wing. *cf.* (Mod. Gr.) πτερύγιον/pterygion, *n.*: fin.
e.g.: (*Platypterygius*) [Ichthyosauria]; (*Stenopterygius*) [Ichthyosauria]

-pteryx (Gr.) πτέρυξ/pteryx, *f.*: the wing of a bird, a winged creature, a bird, anything like a wing, the flap or skirt of a coat of armour, anything that covers or protects like wings.
e.g.: *Ambopteryx*; *Archaeopteryx*; *Caudipteryx*; *Elopteryx*; *Epidexipteryx*; *Fukuipteryx*; *Jinfengopteryx*; *Kulindapteryx*; *Longipteryx*; *Omnivoropteryx*; *Palaeopteryx*; *Scansoriopteryx*; *Sharovipteryx* [Pterosauria]; *Similicaudipteryx*; *etc.*

Puerta (person's name) Pablo Puerta, a remarkable fossil-hunter.
e.g.: *Puertasaurus*

Pukyong (institution name) Pukyong 釜慶: Pukyong National University.
e.g.: *Pukyongosaurus*

pulane (Sesotho) rain-maker.
e.g.: *Pulanesaura* -> -saura

Puna (place-name) Puna: the local name, distinguishing "the oxygen-depleted atmosphere typical of the high Andes". (Hechenleitner *et al.*, 2020)
e.g.: (*Punatitan* Hechenleitner *et al.*, 2020)

-pus (Gr.) πούς/pous [pūs], *m.*: a foot.
e.g.: *Erectopus*; *Euhelopus*; *Saltopus* [Dinosauromorpha]

pycno- (Gr.) πυκνός/pyknos: close, compact, closely-packed, crowded, thick, dense.
e.g.: *Pycnonemosaurus* -> -nemo-

pyro- (Gr.) πῦρ/pyr (*nom.*), πῦρός/pyros (*gen.*), *n.*: fire.
e.g.: *Pyroraptor* -> -raptor

Q

QANTAS (acronym, institution name) Queensland and Northern Territory Aerial Services.
e.g.: *Qantassaurus*

Qianzhou (Chin. place-name) Qiánzhōu 虔州: the ancient name of the city of Ganzhou (贛州市), Jiangxi Province (江西省), China.
e.g.: *Qianzhousaurus*

Qiaowan (Chin.) Qiáowān 橋湾: a cultural relic in Gansu Province (甘粛省).
e.g.: *Qiaowanlong* -> -long

Qijiang (Chin. place-name) Qíjiāng District (綦江区), Chongqing Municipality (重慶市), Sichuan Province (四川省), China.
e.g.: *Qijianglong* -> -long

qilian (Xiongnu*) qílián 祁連 'heaven': the Qilian Mountains that lies to the south of the Changma Basin, China.
*Xiongnu: an ancient confederation of nomadic tribes that inhabited central Asia.
e.g.: *Qiliania* [Avialae]

qingxiu (Chin.) qīngxiù 清秀 = shānqīngshuǐxiù 山清水秀: a picturesque scenery of mountains and water.
e.g.: *Qingxiusaurus*

Qinling (Chin. place-name) Qínlǐng 秦嶺: a major east-west mountain range in southern Shaanxi Province (陝西省), China.
e.g.: *Qinlingosaurus*

Qiupa (Chin. stratal name) 秋扒 [Pinyin: Qiūbā]: Qiupa Formation (Late Cretaceous), Henan Province (河南省), China.
e.g.: *Qiupalong* -> -long; *Qiupanykus* -> -onykus

quaesito- (Lat.) quaesītus (*m.*), quaesīta (*f.*), quaesītum (*n.*), *adj.*: sought out, special. *cf.* quaerō: to seek, look for.
e.g.: *Quaesitosaurus*

quetec (Milcayac*) fire.
*the language used by the people who inhabited the region of Mendoza, Argentina.
e.g.: *Quetecsaurus*

Quetzalcoatl (Classical Nahuatl, Aztec myth.) Quetzalcoatl: the Mesoamerican feathered serpent god.
e.g.: *Quetzalcoatlus* [Pterosauria]

Quilmes (demonym) the Quilmes: an indigenous tribe living in Argentina.
e.g.: *Quilmesaurus*

R

Rahioli (place-name) Rahioli Village: the fossil site, Gujarat, western India.
e.g.: *Rahiolisaurus*

rahona (Malagasy*) rahona: cloud, menance.
*the official language in Madagascar.
e.g.: (*Rahona* Forster *et al.*, 1998) (preoccupied) >>> *Rahonavis* -> -avis

raja (Sanskrit) a king.
e.g.: *Rajasaurus*

-ramus (Lat.) rāmus (*nom.*), rāmī (*gen.*), *m.*: a branch.
e.g.: *Alioramus*

rapator (New Latin)
e.g.: *Rapator*

rapax (Lat.) rapāx (*nom.*), rapācis (*gen.*), *adj.*: tearing, furious, grasping.
e.g.: *Rapaxavis* -> -avis [Avialae]

Rapeto [ruh-PAY-tu] (Malagasy folklore) the giant.
e.g.: *Rapetosaurus*

raptor (Lat.) raptor (*nom.*), raptōris (*gen.*), *m.*: a plunderer, robber.
e.g.: *Raptorex* -> -rex

-raptor
e.g.: *Aoniraptor*; *Apatoraptor*; *Austroraptor*; *Bambiraptor*; *Buitreraptor*; *Changyuraptor*; *Conchoraptor*; *Condorraptor*; *Corythoraptor*; *Dakotaraptor*; (*Dalianraptor*) [Avialae]; *Dracoraptor*; *Fukuiraptor*; *Geminiraptor*; *etc.*

-raptor-
e.g.: *Ojoraptorsaurus*

Ratchasima (place-name) Nakhon Ratchasima: one of the Isan provinces of Thailand.
e.g.: *Ratchasimasaurus*

rati- (Lat.) ratis (*nom.*), ratis (*gen.*), *f.*: a raft, float.
e.g.: *Rativates* -> -vates

Rayoso (stratal name) Rayoso Formation (Early Cretaceous), Neuquén Province, Argentina.
e.g.: *Rayososaurus*

Rebbachi- (demonym) Aït Rebbach: a Berber tribe.
e.g.: *Rebbachisaurus*

regali- (Lat.) rēgālis (*m./f.*), rēgālē (*n.*), *adj.*: of a king, royal, regal.
e.g.: *Regaliceratops* -> -ceratops

Regno- (demonym) Regni: a British tribe.
e.g.: *Regnosaurus*

Revuelto (place-name) Revuelto Creek: a stream located in Quay County, New Mexico. | (Sp.) revuelta: revolution.
e.g.: *Revueltosaurus* [Suchia]; ("Revueltoraptor" Hunt, 1994) >>> *Gojirasaurus*

-rex (Lat.) rēx (*nom.*), rēgis (*gen.*), *m.*: an arbitrary ruler, absolute monarch, king.
e.g.: *Dracorex*; *Raptorex*; *Rhinorex*

rhabd(o)- (Gr.) ῥάβδος/rhabdos, *f.*: a rod, stick.
e.g.: *Rhabdodon* -> -odon

rhadino- (Gr.) ῥᾰδῐνός/rhadinos: slender.
e.g.: (*Rhadinosaurus* Seeley, 1881) *(nomen dubium)*

rhampho- (Gr.) ῥάμφος/rhamphos, *n.*: a beak, bill.
e.g.: *Rhamphorhynchus* -> -rhynchus [Pterosauria]

rhino- (Gr.) ῥίς/rhis (*nom.*), ῥῑνός/rhīnos (*gen.*), *f.*: the nose.
e.g.: *Rhinorex* -> -rex.

-rhin(o)-, -rhynus
e.g.: *Altirhinus*; *Aquilarhinus*; *Arrhinoceratops*; *Eurhinosaurus* [Ichthyosauria]; *Latirhinus*; *Lycorhinus*; *Pachyrhinosaurus*

Rhodano- (Lat. place-name) Rhodanus: Rhone River. *cf.* (Fr.) le Rhône, (Ital.) il Ròdano: the Rhone. | (Gr.) ῥοδᾰνός/rhodanos: waving, flickering.
e.g.: (*Rhodanosaurus* Nopcsa, 1929) *(nomen dubium)*

Rhoeto- (Lat.) Rhoetus (*nom.*), Rhoetī (*gen.*), *m.*: one of Giants. The character is mentioned by Ovid in Book V. 38 of *Metamorphoses*. Rhoetus was slain by Perseus.
e.g.: *Rhoetosaurus*

-rhothon (Gr.) ῥώθων/rhōthōn, *m.*: the nose.
e.g.: *Lophorhothon*

-rhynchus, -rhynchos (Gr.) ῥύγχος/rhynchos, *n.*: a snout, muzzle of swine, a beak of birds.
e.g.: *Archaeorhynchus* [Avialae]; (*Claorhynchus*); *Leptorhynchos*; (*Metriorhynchus* von Meyer, 1830)[Crocodylomorpha]; *Rhamphorhynchus* [Pterosauria]

Riabinin (person's name) Anatoly Nikolaevich Riabinin: a Russian geologist and vertebrate paleontologist.
e.g.: *Riabininohadros* -> -hadros

Richardo Estes (person's name) Richard Estes (1932–1990): an American paleoherpetologist.
e.g.: *Richardoestesia*

Rinchen (person's name) Rinchen Barsbold: a Mongolian paleontologist and geologist.
e.g.: *Rinchenia*

Rincon (Sp. place-name) Rincón de los Sauces 'place of the willows': a town in the Neuquén Province, Argentina.
e.g.: *Rinconsaurus*

Rio Arriba (place-name) Rio Arriba: a county located in New Mexico, USA.
e.g.: (*Rioarribasaurus* Hunt & Lucas, 1991) *(nomen rejectum)* >>> *Coelophysis* 1887 *(nomen conservandum)*

Rioja (place-name) La Rioja Province, Argentina.
e.g.: *Riojasaurus*

Roca (place-name) General Roca, northeast of Río Negro, northern Patagonia, Argentina.
e.g.: *Rocasaurus*

-rostris (Lat.) rōstrum (*nom.*), rōstrī (*gen.*), *n.*: a beak, bill, snout, muzzle, mouth.
e.g.: *Cuspirostrisornis* [Avialae]; *Magnirostris*

-rostr-
e.g.: *Largirostrornis* [Avialae]; *Longirostravis* [Avialae]

rubeo- (Lat.) rubeus, *adj.*: of bramble, bush. | rubus (*nom.*), rubī (*gen.*), *m.*: a bramble-bush.
e.g.: *Rubeosaurus*

Ruehle (person's name) Hugo Ruehle von Lilienstern (1882–1946): a German amateur paleontologist.
e.g.: *Ruehleia*

rug(o)- (Lat.) rūga (*nom.*), rūgae (*gen.*), *f.*: a crease in the face, wrinkle. rūgō: to become creased.
e.g.: *Rugocaudia* -> -caudia; *Rugops* -> -ops

Rukwa (place-name) Lake Rukwa: a lake in the Rukwa Valley of southwestern Tanzania.
e.g.: *Rukwatitan* -> -titan

Ruyang (Chin. place-name) Rǔyáng 汝陽: a county in Henan Province (河南省), China.
e.g.: *Ruyangosaurus*

S

Saci (Brazilian folklore) [pron.: sasi]: a one-legged creature.
e.g.: *Sacisaurus* [Silesauridae]

Sahaliyan (Manchu) black. | (place-name) Sahaliyan ula 'Black River': 黒龍江 [Pinyin: Hēilóngjiāng] .
e.g.: *Sahaliyania*

saichan (Mong.) сайхан: beautiful.
e.g.: *Saichania*

Salta (place-name) Salta Province, Argentina.
e.g.: *Saltasaurus*

Saltillo (place-name) the capital of Coahuila State, Mexico.
e.g.: ("Saltillomimus" Aguillón Martínez, 2014) -> -mimus *(nomen ex dissertatione)*

salto- (Lat.) salto- [saltō: to dance]
e.g.: *Saltopus* -> -pus [Dinosauromorpha]

Saltrio (place-name) a commune in the Province of Varese, Lombardy, Italy.
e.g.: ("Saltriosaurus" Dal Sasso, 2000) >>> *Saltriovenator* -> -venator

San Juan (place-name) San Juan Province, western Argentina.
e.g.: *Sanjuansaurus*

Sanpa (Chin. place-name) 三巴 [Pinyin: Sānbā]: three regions of eastern Sichuan.
e.g.: *Sanpasaurus*

Santana (stratal name) Santana Group (Early Cretaceous), Araripe Basin, northeastern Brazil. A geologic Lagerstätte.
e.g.: *Santanaraptor* -> -raptor

Sanxia (Chin. place-name) Sānxiá 三峡: 'Three Gorges' of Yangtze River (= Chang Jiang), China.
e.g.: *Sanxiasaurus*

SAPE (acronym) Society for Avian Paleontology and Evolution.
e.g.: *Sapeornis* -> -ornis [Avialae]

Sarah (person's name) Sarah (Mrs Ernest) Butler.
e.g.: *Sarahsaurus*

Saraiki massoom {(Saraiki) Saraiki: an Indo-Aryan language spoken in Pakistan + masoom: innocent}
e.g.: (*Saraikimasoom* Malkani, 2014)

sarco- (Gr.) σάρξ/sarx (*nom.*), σαρκός/sarkos (*gen.*), *f.*: flesh. = (Lat.) carō (*nom.*), carnis (*gen.*), *f.*
e.g.: *Sarcolestes* -> -lestes; *Sarcosaurus*; (*Sarcosuchus* Broin & Taquet, 1966) -> suchus [Crocodylomorpha]

Sarmiento (place-name) a town in Chubut Province, Argentina.
e.g.: *Sarmientosaurus*

Sasayama (Jpn. place-name) Sasayama 篠山: a city of Hyōgo Prefecture (兵庫県), Japan.
e.g.: (*Sasayamamylos* Kusuhashi *et al.*, 2013) -> -mylos [Mammalia]

Saturnalia (Lat.) Sāturnālia (*nom.*), Sāturnālium or Sāturnāliōrun (*gen.*), *n. pl.*: festival of Saturn in December.
e.g.: *Saturnalia*

-saura (Gr.) σαύρα/saura, *f.* = σαῦρος/sauros, *m.*: a lizard. *cf.* (Lat.) lacerta, *f.*: lizard.
e.g.: *Bonitasaura*; *Bugenasaura*; *Cathartesaura*; *Drusilasaura*; *Gasparinisaura*; *Isaberrysaura*; *Lamplughsaura*; *Laquintasaura*; *Leaellynasaura*; *Maiasaura*; *Pulanesaura*; *Trinisaura*

-sauriscus
e.g.: *Cetiosauriscus*; (*Pachysauriscus*)

saur(o)- (Gr.) σαῦρος/sauros, *m.*: a lizard.
e.g.: (*Saurocephalus* Harlan, 1824) -> -cephalus [Actinopterygii]; *Saurolophus* -> -lophus; *Sauropelta* -> -pelta; *Sauropaganax* -> -phag(o)-, -anax; *Sauroplites* -> -oplites; *Sauroposeidon* -> -Poseidon; *Saurornithoides* -> -ornitho-; *Saurornitholestes* -> -ornitho-, -lestes; (*Saurosuchus* Reig, 1959) -> -suchus [Paracrocodylomorpha]

-saur(o)-
e.g.: *Dromaeosauroides*; (*Dryptosauroides*); *Parasaurolophus*; *Sinosauropteryx*

-saurus (Gr.) σαῦρος/sauros, *m.*: a lizard.
e.g.: *Abelisaurus*; *Abrictosaurus*; *Abrosaurus*; *Abydosaurus*; *Achelousaurus*; *Achillesaurus*; *Acrocanthosaurus*; (*Actiosaurus*); *Adamantisaurus*; *etc.*

Sauroni- (novel) Sauron: a character of *The Lord of the Ring*.
e.g.: *Sauroniops* -> -ops

savanna (Spanish / Taino) zavana (= savanna).
e.g.: *Savannasaurus*

scansorio- (Lat.) scansōrius (*m.*), scansōria (*f.*), scansōrium (*n.*), *adj.*: adapted for climbing.
e.g.: *Scansoriopteryx* -> -pteryx [Scansoriopterygidae]

scapho- (Gr.) σκάφος/skaphos, *n.*: the hull of a ship, a ship. σκάφη/skaphe: anything dug or scooped.
e.g.: *Scaphognathus* -> -gnathus [Pterosauria]

scel(o)- (Gr.) σκέλος/skelos, *n.*: the leg. *cf.* skel(o)-
e.g.: *Scelidosaurus*

schizo (Gr.) schizo- [σχίζω/schizō: split]
e.g.: *Schizooura* -> -oura [Avialae]

Schleitheim (place-name) the type locality in Switzerland.
e.g.: *Schleitheimia*

-scincus (Gr.) σκίγγος/skinkos: skink.
e.g.: (*Palaeoscincus*)

Scipio (person's name) Scipione Breislak, the 18th century geologist, and Scipio Africanus, the famous Roman consul fighting against Hannibal.
e.g.: *Scipionyx* -> -onyx

sciuru- (Lat.) sciūrus (*nom.*), sciūrī (*gen.*), *m.* = (Gr.) σκίουρος/skiouros, *m.*: squirrel.
e.g.: *Sciurumimus* -> -mimus

scolo- (Gr.) σκῶλος/skōlos, *m.*: a pointed stake.
e.g.: *Scolosaurus*

scutello- (New Lat.) scutellum (diminutive of scūtum), *n.*: a small shield.
e.g.: *Scutellosaurus*

Sebec- (Lat.) Sebek = (Gr.) Σοῦχος/Sūchos: an ancient Egyptian deity with a complex and fluid nature.
e.g.: (*Sebecus* Simpson, 1937) [Crocodylomorpha]

secerno- (Lat.) secerno- [sēcernō: to put apart, sever, separate.]
e.g.: *Secernosaurus*

sefapano- (Sesotho) sefapano 'cross': the cross-shaped astragalus or talus bone in its ankle.
e.g.: *Sefapanosaurus*

Segi (place-name) Segi Canyon / Tsegi Canyon: a canyon in Navajo County, Arizona, USA.
e.g.: *Segisaurus*

segno- (Lat.) sēgnis (*m./f.*), sēgne (*n.*): slow, sluggish, lazy.
e.g.: *Segnosaurus*

seismo- (Gr.) σεισμός/seismos, *m.*: a shaking, shock, an earthquake.
e.g.: *Seismosaurus*

Seitaad (Navajo, myth.) a mythological sand 'monster' from the Diné folklore.
e.g.: *Seitaad*

sekten (Tehuelche) sekten: island.
e.g.: *Sektensaurus*

sella (Lat.)sella (*nom.*), sellae (*gen.*), *f.*: a seat, chair.
e.g.: *Sellacoxa* -> -coxa

sello-
e.g.: *Sellosaurus*

Serendip (place-name) an ancient name of Sri Lanka. | "serendipity" coined by Horace Walpole (in a letter written to Horace Mann, 1754) from *The Three Princes of Serendip*: "always making discoveries, by accidents and sagacity, of things which they were not in quest of."
e.g.: *Serendipaceratops* -> -ceratops

serici- (Lat.) sēricum (*nom.*), sēricī (*gen.*), *n.* : silk.
e.g.: (*Sericipterus* Andres *et al.*, 2010) [Pterosauria]

serik(o)- (Gr.) σηρῐκός/sērikos: Seric, silken. *cf.* Σήρ/Sēr, *m.*, Σῆρες/ Sēres, *pl.*: the Seres, an Indian people from whom the ancients got silk.
e.g.: *Serikornis* -> -ornis [Anchiornithidae]

shamo (Chin.) shāmò 沙漠: desert.
e.g.: *Shamosaurus*

Shanag: Black-hatted dancers in the Buddhist Tsam festival.
e.g.: *Shanag*

Shanshan (Chin. place-name) Shànshàn 鄯善: a county within the Xinjiang Uyghur Autonomous Region, China.
e.g.: *Shanshanosaurus*

Shantung (place-name) 山東 [Pinyin: Shāndóng]: a coastal province of China.
e.g.: *Shantungosaurus*

Shanweiniao (Chin.) Shànwěiniǎo 扇尾鳥: fan-tail bird.
e.g.: *Shanweiniao* [Avialae]

Shanxi (Chin. place-name) Shānxī 山西: a province of China.
e.g.: *Shanxia*

Shanyang (Chin. place-name) Shānyáng 山陽: a county of Shangluo, Shaanxi (陝西省), China.
e.g.: *Shanyangosaurus*

shaochi (Chin.) shāchǐ 鯊歯: shark toothed.
e.g.: *Shaochilong* -> -long

Sharov (person's name) Aleksandr Grigorevich Sharov: a Russian paleontologist.
e.g.: *Sharovipteryx* -> -pteryx [Pterosauria]

Shengjing (Chin. place-name) Shèngjīng 盛京: the ancient name of Shenyang (瀋陽市), Liaoning Province (遼寧省), China.
e.g.: *Shengjingornis* -> -ornis [Avialae]

Shen qi (Chin.) shén qī 神7: 神舟7号 [Shenzhou qi]: China's third human mission into space in 2008.
e.g.: *Shenqiornis* -> -ornis [Avialae]

Shenshi- (Chin. institution name) Shěnyáng Normal University 瀋陽師範大学.
e.g.: (*Shenshiornis* Hu *et al.*, 2010) >>> *Sapeornis*

Shenzhou (Chin.) Shénzhōu 神州 'divine land': an old name for China.
e.g.: (*Shenzhouraptor* Ji *et al.*, 2002) [Avialae] >>> *Jeholornis* [Avialae]; *Shenzhousaurus*

Shidai (Chin. institution name) the Jīn-Shídài Company 金時代社, China.
e.g.: *Shidaisaurus*

shingopana {(Swahili / Kiswahili) shingo: neck + pana: wide} wide neck.
e.g.: *Shingopana*

Shishugou (stratal name) Shishugou Formation (Late Jurassic), Xinjiang, China.
e.g.: *Shishugounykus* -> -unykus

Shixing (Chin. place-name) Shǐxīng 始興: a county of Shaoguan, Guangdong Province (広東省), China.
e.g.: *Shixinggia*

Shoni- (place-name) the Shoshone Mountains, Nevada, USA.
e.g.: *Shonisaurus* [Ichthyosauria]

shringa (Sanscrit) horn.
e.g.: *Shringasaurus* [Archosauromorpha]

Shuangbai (Chin. place-name) Shuāngbǎi 双柏 (雙柏): a county located in Chuxiong Yi Autonomous Prefecture, Yunnan Province (雲南省), China.
e.g.: *Shuangbaisaurus*

Shuangmiao (Chin. place-name) Shuāngmiào 雙廟 'twin temples': the village of Shuangmiao, Beipiao, Liaoning Province (遼寧省), China.
e.g.: *Shuangmiaosaurus*

Shuno- (Chin. place-name) Shǔ 蜀: an old name for the Sichuan region, China.
e.g.: *Shunosaurus*

Shuvo (person's name) Shuvo: the name of a paleontologist Sankar Chatterjee's son.
e.g.: *Shuvosaurus* [Paracrocodylomorpha]

Shuvuu (Mong.) шувуу: bird.
e.g.: *Shuvuuia*

Siam (place-name) the ancient name for Thailand.
e.g.: *Siamodon* -> -odon; *Siamosaurus*; *Siamotyrannus* -> -tyrannus; *Siamraptor*

Siats (Ute, legend) = See-atch: a predatory, man-eating monster.
e.g.: *Siats*

Sibiro- (place-name) Siberian.
e.g.: *Sibirotitan* -> -titan

Sigilmassa (place-name) the ancient city in the Sahara Desert, Morocco.
e.g.: *Sigilmassasaurus*

Sile- (place-name) Silesia: a region of Central Europe.
e.g.: *Silesaurus* [Silesauridae]

Silu (Chin. place-name) Sīlù 絲路: Silk Road.
e.g.: *Siluosaurus*

silvi- (Lat.) silva (*nom.*), silvae (*gen.*), *f.*: a wood, forest, plantation.
e.g.: *Silvisaurus*

simili- (Lat.) similis (*m./f.*), simile (*n.*), *adj.*: like, similar.
e.g.: *Similicaudipteryx* -> -caudi-, -pteryx

Sin(o)-* (Med. Lat.) China ← (Gr.) Sínai | (Arabic) Ṣīn ← (Chin.) Ch'in ←秦[Qín]
*Sino- (nationality prefix)
e.g.: *Sinankylosaurus*; *Sinocalliopteryx* -> -call-, -pteryx; *Sinoceratops* -> -ceratops; *Sinocoelurus* -> -coel, ur(o)-; (*Sinopeltosaurus* Ulansky, 2014) -> -pelto-; (*Sinopliosaurus* Young, 1942) [Plesiosauria]; *Sinopterus* -> -pterus [Pterosauria]; *Sinornis* -> -ornis [Avialae]; *Sinornithoides* -> -ornitho-; *Sinornithomimus* -> -ornitho-, -mimus; *Sinornithosaurus* -> -ornitho-; *Sinosauropteryx* -> -saur(o)-, pteryx; *Sinosaurus*; *Sinotyrannus* -> -tyrannus; *Sinraptor* -> -raptor; *Sinovenator* -> -venator; *etc.*

sinu- (Lat.) sinus (*nom.*), sinūs (*gen.*), *m.*: curve, fold.
e.g.: (*Sinucerasaurus* Xu & Norell, 2006) >>> *Sinusonasus*

Sirindhorn- (person's name) Princess Maha Chakri Sirindhorn, Thailand.
e.g.: *Sirindhorna*

-skelo- (Gr.) σκέλος/skelos, *n.*: the leg. *cf.* -scel(o)-
e.g.: *Euskelosaurus*

skorpio- (Lat.) Scorpius, *m.* = (Gr.) σκορπίος/skorpios, *m.*: a scorpion.
e.g.: *Skorpiovenator* -> -venator

Smitano- (Old Saxon) smitan: referring to J. August Smith and Smithsonian Institution.
e.g.: (*Smitanosaurus* Whitlock & Wilson, 2020)

-soma (Gr.) σῶμα/sōma (*nom.*), σῶματος/sōmatos (*gen.*), *n.*: the body.
e.g.: *Spondylosoma* [Archosauria]

Songling (Chin. place-name) Sōnglǐng 松嶺 'pine ridge': the mountain range in Chaoyang County, China.
e.g.: *Songlingornis* -> -ornis [Avialae]

Sonido- (place-name) 蘇尼特(旗) 'Sonid Left Banner and Sonid Right Banner': banners of Inner Mongolia, China.
e.g.: *Sonidosaurus*

Sonora (place-name) the Sonoran Desert in southern Arizona, USA.
e.g.: *Sonorasaurus*

sordes (Lat.) sordēs (*nom.*), sordis (*gen.*), *f.*: dirt, squalor, shabbiness. | (folklore) evil spirits in local folklore.
e.g.: (*Sordes* Sharov, 1971) [Pterosauria]

Soria (place-name) a province of central Spain.
e.g.: *Soriatitan* -> -titan

soro- (Lat.) soror (*nom.*), sorōris (*gen.*), *f.*: a sister.
e.g.: *Soroavisaurus* -> -avi- [Avialae]

Soumya (person's name) Sankar Chatterjee's son.
e.g.: *Soumyasaurus* [Silesauridae]

spectro- (Lat.) spectrum (*nom.*), spectri (*gen.*) *n.*: an appearance, form, ghost.
e.g.: (*Spectrovenator* Zaher et al., 2020)

sphaero- (Gr.) σφαῖρᾰ/sphaira, *f.*: a ball.
e.g.: *Sphaerotholus* -> -tholus

spheno- (Gr.) σφήν/sphēn, *m.*: a wedge.
e.g.: (*Sphenospondylus* Seeley, 1883) -> -spondylus

spiclypeus {(Lat.) spīca, *f.*: spike + clipeus, *m.*: shield}
e.g.: *Spiclypeus*

-spinax (Lat.) *cf.* spīna (*nom.*), spīnae (*gen.*), *f.*: a thorn, princkle, spine, backbone.
e.g.: *Altispinax*; *Becklespinax*

spin(o)- (Lat.) spīna (*nom.*), spīnae (*gen.*), *f.*: a thorn, prickle, spine, backbone.
e.g.: *Spinophorosaurus* -> -phoro-; *Spinops* -> -ops; *Spinosaurus*; *Spinostropheus* -> -stropheus

-spino-, -spinus
e.g.: *Gigantspinosaurus*; *Hypselospinus*

spondylo- (Gr.) σπόνδῠλος/spondylos = σφόνδῠλος, *m.*: a vertebra.
e.g.: *Spondylosoma* -> -soma [Archosauria]

-spondylus
e.g.: *Bothriospondylus*; *Calamospondylus*; *Cymbospondylus* [Ichthyosauria]; (*Marmarospondylus*); *Massospondylus*; *Nopcsaspondylus*; (*Pachyspondylus*); (*Pterospondylus*); (*Sphenospondylus*); *Streptospondylus*; (*Thecospondylus*)

staur(o)- (Gr.) σταυρός/stauros, *m.*: the Cross, Constellation of Southern Cross.
e.g.: *Staurikosaurus*

stego- (Gr.) στέγος/stegos, *n.*: a roof. στέγω/stegō: to cover closely, generally to keep off, fend off weapons, *etc.*
e.g.: *Stegoceras* -> -ceras; *Stegopelta* -> -pelta; *Stegosaurus*

-stegos
e.g.: (*Amargastegos*); (*Ferganastegos*)

-stegus
e.g.: *Mongolostegus*

stella (Lat.) stella (*nom.*), stellae (*gen.*), *f.*: a star.
e.g.: *Stellasaurus*

sten(o)- (Gr.) στενός/stenos: narrow, strait, close, confined.
e.g.: *Stenonychosaurus* -> -onycho-; *Stenopelix* -> -pelix; (*Stenopterygius* Jaekel, 1904) -> -pterygius [Ichthyosauria]; (*Stenotholus* Giffin *et al.*, 1988) >>> *Stygimoloch*

-stenotes (Gr.) στενότης/stenotēs, *f.*: narrowness, straitness.
e.g.: *Chirostenotes*; *Epichirostenotes*

stephano- (Gr.) στέφᾰνος/stephanos, *m.*: a crown, wreath.
e.g.: (*Stephanosaurus* Lambe, 1914)

stereo- (Gr.) στερεός/stereos: stiff, solid.
e.g.: (*Stereocephalus* Lambe, 1902) -> -cephalus >>> *Euoplocephalus*

Stokes (place-name) William Lee Stokes: a pioneer Utahan geologist.
e.g.: *Stokesosaurus*

Stormberg (stratal name) Stormberg Group (Late Triassic–Jurassic) in Karoo Basin, Southern Africa.
e.g.: *Stormbergia*

strenu- (Lat.) strēnuus (*m.*), strēnua (*f.*), strēnuum (*n.*), *adj.*: brisk, quick, active, strenuous.
e.g.: (*Strenusaurus* Bonaparte, 1969) >>> *Riojasaurus*

strepto- (Gr.) στρεπτός/streptos: flexible, pliant, to be bent or turned, curved.
e.g.: *Streptospondylus* -> -spondylus

-strepto-
e.g.: *Eustreptospondylus*

-stro-
e.g.: (*Tanystrosuchus*)

-strophaeus, -stropheus (Mod. Gr.) στροφεύς/stropheus, *m.*: hinge. [anat.] hinge joint. | (Gr.) the socket. στρόφος/strophos, *m.*: a twisted band or cord. στροφή/strophe: a turning, a twist.
e.g.: *Dystrophaeus*; *Lophostropheus*; *Spinostropheus*; (*Tanystropheus*) [Archosauromorpha]

struthio (Late Lat.) strūthiō (*nom.*), strūthiōnis (*gen.*), *m.*: ostrich | (Gr.) στρουθός/strūthos: the ostrich, Struthio.
e.g.: *Struthiomimus* -> -mimus; *Struthiosaurus*

Stygi- (Gr. myth.) Στύξ/Styx (*nom.*), Στῠγός/Stygos (*gen.*), *f.*: the river Styx in Greek mythology. *cf.* Στύγιος/Stygios: Stygian.
e.g.: *Stygimoloch* -> -moloch; (*Stygivenator* Olshevsky, 1995) >>> *Tyrannosaurus*

-stylo- (Gr.) στῦλος/stylos, *m.*: a pillar, as a support or bearing.
e.g.: *Dystylosaurus*

styraco- (Gr.) στύραξ/styrax (*nom.*), στύρᾰκος/styrakos (*gen.*), *m.*: the spike at the lower end of a spear-shaft.
e.g.: *Styracosaurus*

sucho- (Gr.) Σοῦχος/Souchos [Sūchos]: Egyptian crocodile god Sobek.
e.g.: *Suchomimus* -> -mimus; *Suchosaurus*

-suchus (Gr.) Σοῦχος/Souchos [Sūchos].
e.g.: *Betasuchus*; *Compsosuchus*;

(*Desmatosuchus*) [Aetosauria]; (*Dolichosuchus*); (*Hupehsuchus*) [Diapsida]; *Iliosuchus*; *Indosuchus*; *Laevisuchus*; *Lagosuchus* [Dinosauromorpha]; (*Pachysuchus*); *Parasuchus* [Phytosauria]; (*Protosuchus*) [Archosauriformes]; (*Pterosuchus*); *Sarcosuchus* [Crocodylomorpha]; (*Saurosuchus*) [Paracrocodylomorpha]; (*Tanystrosuchus*); (*Walgettosuchus*)

Sulaimani- (place-name) the Sulaiman Fold Belt, northwestern Pakistan.
e.g.: *Sulaimanisaurus*

sulc- (Lat.) sulcus (*nom.*), sulcī (*gen.*), *m.*: a furrow.
e.g.: *Sulcavis* -> -avis [Avialae]

super- (Lat.) super: above, on top, over. | (Eng.) super-
e.g.: *Supersaurus*

suski- (Zuni) suski: coyote.
e.g.: *Suskityrannus* -> -tyrannus

suu wassa (Crow*) ancient thunder.
*one of Native American tribes living in North America.
e.g.: *Suuwassea*

Suzhou (Chin. place-name) Sùzhōu 肃州. Jiuquan (酒泉), formerly known as Suzhou (肃州), is a prefecture-level city in Gansu Province (甘肃省), China.
e.g.: *Suzhousaurus*

syn- (Gr.) σύν/syn, *prep.* 《comb. form》 sy-, syl-, sym-, syn-, sys- : along with, in company with, together with.
e.g.: *Syntarsus* -> -tarsus; *Syngonosaurus* -> -gono-

syrmo- (Gr.) συρμός/syrmos, *m.*: any lengthened sweeping motion.
e.g.: *Syrmosaurus*

Szechuan (place-name) 四川 [Pinyin: Sìchuān]: a province in southwest China.
e.g.: ("Szechuanoraptor" Chure, 2001) -> -raptor; *Szechuanosaurus*

T

Tachira (place-name) Táchira: one of the 23 states of Venezuela.
e.g.: *Tachiraptor* -> -raptor

talar(o)- (Gr.) τάλᾰρος/talaros, *m.*: a basket.
e.g.: *Talarurus* -> -ur(o)-

talenkauen (Tehuelche / Aónikenk) talenk kauen: small skull.
e.g.: *Talenkauen*

Talos (Gr. myth.) Τάλως/Talōs: fleet-footed protector of Crete, often depicted as winged, who succumbed to a wound on the ankle.
e.g.: *Talos*

Tamba (Jpn. place-name) 丹波: a city of Hyōgo Prefecture (兵庫県), Japan.
e.g.: *Tambatitanis* -> -titanis

Tan (Chin. person's name) Tán Xīchóu 譚錫疇: a Chinese Paleontologist.
e.g.: *Tanius*

Tangvay (place-name) Tang Vay, Savannakhet Province, Laos.
e.g.: *Tangvayosaurus*

Taniwha (Māori myth.) beings that live in deep pools in rivers, dark caves, or in the sea. a supernatural, aquatic creature.
e.g.: (*Taniwhasaurus* Hector, 1874) [Mosasauridae]

tany- (Gr.) τανυ-/tany-: long. τᾰνύω/ tanyō: to extend.
e.g.: *Tanycolagreus* -> col-, -agreus; (*Tanystrophaeus* Cope, 1887) >>> Coelophysis; (*Tanystropheus* Meyer, 1852) -> -stropheus [Archosauromorpha]; (*Tanystrosuchus* Kuhn, 1963) -> -stro-, -suchus

Taohe (Chin. place-name) Táohé 洮河: a tributary of Huanghe (黄河), China.
e.g.: *Taohelong* -> -long

Tapejara (Tupi, myth.) the old being. *cf.* (place-name) Tapejara [tape: a road + jara: a lord]: a municipality, Rio Grande do Sul, Brazil.
e.g.: *Tapejara* [Pterosauria]

Tapuia (demonym) indigenous people that inhabited the inner regions of Brazil.
e.g.: *Tapuiasaurus*

Tarasco- (folklore) Tarasque: a devouring monster from Occitan and Spanish folklore.
e.g.: *Tarascosaurus*

tarbo- (Gr.) τάρβος/tarbos, *n.*: fright, alarm, terror. ταρβέω/tarbeō: to be frightened, alarmed.
e.g.: *Tarbosaurus*

tarchi- (Mong.) тархи/tarkhi: brain.
e.g.: *Tarchia*

-tarsus (Lat.) tarsus, *m.*: tarsus. | (Gr.) ταρσός/tarsos, *m.*: a stand or frame of wicker-work, a crate. τ. ποδός: the flat of the foot, the part between the toes and the heel.
According to *Kato's integrated English-Japanese medical dictionary*, tarsus includes astragalus, calcaneus, navicular, cuboides and cuneiforme. And tarsus also means tarsus palpebrarum.
e.g.: *Syntarsus*

Tastavins (place-name) Río Tastavins (river) in Spain and also the name of the village. | (Catalan) tastavin: 'wine taster'.
e.g.: *Tastavinsaurus*

tatanka (Lakota) bison.
e.g.: *Tatankacephalus* -> -cephalus; *Tatankaceratops* -> -ceratops

Tataouine (place-name) a city in southern Tunisia.
e.g.: *Tataouinea*

Tati (place-name) 大地 [Pinyin: Dàdì]: the village of Dadi, Yunnan Province (雲南省), China.
e.g.: *Tatisaurus*

tauro- (Lat.) taurus (*nom.*), taurī (*gen.*), *m.* = (Gr.) ταῦρος/tauros, *m.*: a bull.
e.g.: *Taurovenator* -> -venator

-taur-
e.g.: *Carnotaurus*; *Minotaurasaurus*

Taveiro (place-name) Taveiro: a village southwest of Coimbra, Portugal.
e.g.: *Taveirosaurus*

Tawa (Chin. place-name) 大窪 [Pinyin: Dàwā]: a village near Lufeng, China.
e.g.: (*Tawasaurus* Young, 1982) >>> *Lufengosaurus*

Tawa (Hopi, myth.) Puebloan sun god.
e.g.: *Tawa*

Tazouda (place-name) Tazouda: the village of Toundoute in the Province of Ouarzazate, High Atlas of Morocco.
e.g.: *Tazoudasaurus*

Techno- (institution name) Texas Tech University, which was called until 1969 Texas Technological College. | (Gr.) τεχνο-/techno- [τέχνη/technē: skill].
e.g.: *Technosaurus* [Silesauridae]

Tehuelche (demonym) the Tehuelche people native to the Chubut Province, Argentina.
e.g.: *Tehuelchesaurus*

teihi- (Arapaho*) teihiihan: strong.
*a member of Native people living in mainly Colorado and Wyoming.
e.g.: *Teihivenator* -> -venator

tein(o)- (Gr.) tein(o)-: extended [τείνω/teinō: to stretch out, to extend]
e.g.: (*Teinurosaurus* Nopcsa, 1928) *(nomen dubium)* -> -ur(o)-

telmato- (Gr.) τέλμα/telma (*nom.*), τέλματος/telmatos (*gen.*), *n.* : standing water, a pool, pond, marsh, swamp.
e.g.: *Telmatosaurus*

tenanto- -> tenont(o)-
e.g.: ("Tenantosaurus" Brown *vide* Chure & McIntosh, 1989) >>> *Tenontosaurus*

Tendaguru (stratal name) Tendaguru Formation (Middle Jurassic–Early Cretaceous), southeastern Tanzania. | (place-name) Tendaguru hills. | (Wamwera) Tendaguru: 'steep hill'.
e.g.: *Tendaguria*

Tengri (Mongorian-Turkish myth.) one of the names for the primary chief deity.
e.g.: *Tengrisaurus*

tenonto- (Gr.) τένων/tenōn (*nom.*), τένοντος/tenontos (*gen.*), *m.*: a sinew, tendon, the foot.

e.g.: *Tenontosaurus*

terato- (Gr.) τέρας/teras (*nom.*), τέρατος/teratos (*gen.*), *n.*: a sign, wonder, a monster.
e.g.: *Teratophoneus* -> -phoneus; (*Teratosaurus* von Meyer, 1861) [Rauisuchidae] >>> *Efraasia*

termino-: (Lat.) termino-: last, terminal. [terminus (*nom.*), terminī (*gen.*), *m.*: a boundary-line, bound, limit, end]
e.g.: *Terminocavus* -> -cavus.; (*Terminonatator* Sato, 2003) -> -natator [Plesiosauria]

-terror (Lat.) terror (*nom.*), terrōris (*gen.*), *m.*: great fear, terror.
e.g.: *Dynamoterror*

-tesserae- (Lat.) tessera (*nom.*), tesserae (*gen.*), *f.*: mosaic tiles.
e.g.: *Notatesseraeraptor*

Tethys (Gr.myth.) Τηθύς/Tēthys, *f.*: wife of Oceanus. 《geol.》 The Tethys Ocean: an ocean that separated the super-continents of Gondwana and Laurasia during much of the Mesozoic era.
e.g.: *Tethyshadros* -> -hadros

tetragono- (Gr.) τετρά-γωνος*/tetra-gōnos : with four equal angles, rectangular or square.
*γωνία/gōnia, *f.*: a corner, angle.
e.g.: (*Tetragonosaurus* Parks, 1931) *(nomen rejectum)* >>> (*Procheneosaurus*) *(nomen conservandum)* >>> *Lambeosaurus*

Texas (place-name) Texas: a south central state, USA.
e.g.: *Texacephale* -> -cephale; *Texasetes* -> -etes

Teyuwasu (Tupi*) big lizard.
*Old Tupi was spoken by the native Tupi people of Brazil.
e.g.: *Teyuwasu*

thalatto- (Gr.) θάλαττα/thalatta, *f.* = θάλασσα/thalassa, *f.*: the sea.
e.g.: (*Thalattoarchon* Fröbisch *et al.*, 2013) -> -archon [Ichthyosauria]

thanato- (Gr.) θάνᾰτος/thanatos, *m.*: death. | (Gr. myth.) Θάνᾰτος/Thanatos: Death.
e.g.: *Thanatotheristes* -> -theristes

Thanos (comic character) Marvel Comic character Thanos: a fictional supervillain.
e.g.: *Thanos*

-thaumas (Gr.) θαῦμα/ thauma (*nom.*), θαῦματος/thaumatos (*gen.*), *n.*: a wonder, marvel.
cf. In Greek mythology, Thaumas, an old sea god, can be translated as 'miracle' or 'wonder'. His name was derived from *thaumatos*.
e.g.: *Agathaumas*

theco- (Gr.) θήκη/thēkē, *f.*: a case, a box.
e.g.: *Thecocoelurus* -> -coel-, ur(o)-; *Thecodontosaurus*; *Thecospondylus* -> -spondylus

theio- (Gr.myth.)θεῖος/theios: of or from the god, sent by the god. *cf.* θεός/theos, *m.*: god.
e.g.: *Theiophytalia* -> -phytalia

-theristes (Gr.) θεριστής/theristes, *m.*: a reaper, harvester.
e.g.: *Thanatotheristes*

-therium (Gr.) θηρίον/therion, *n.*: in form a *dim*, of θήρ/thēr, *m.*: a wild animal, beast.
e.g.: *Fulgurotherium*

therizino- (Gr.) therizino- [θερίζω/therizō: to mow, reap].
e.g.: *Therizinosaurus*

thescelo- (Gr.) θέσκελος/theskelos: marvelous, wondrous.
e.g.: *Thescelosaurus*

thespesius (Gr.) θεσπέσιος/thespesios: divine, wondrous, marvelous.
e.g.: *Thespesius*

-tholus (Gr.) θόλος/tholos, *f.*: a round building with a conical roof, a vaulted chamber.
e.g.: *Acrotholus*; *Gravitholus*; *Majungatholus*; *Ornatotholus*; *Sphaerotholus*; (*Stenotholus*)

thotobolo (Sesotho*) trash heap.
*one of the languages spoken primarily in South Africa and Lesotho.
e.g.: ("Thotobolosaurus" Ellenberger, 1970) >>> *Kholumolumo*

Tianchi (Chin. place-name) Tiānchí 天池 'heavenly pond': a pond in Tiānshān (天山), Xinjian Uyghur Autonomous Region, China.
e.g.: *Tianchisaurus*

Tianyu (Chin. place-name) Tiānyŭ 天宇 'sky'. (institution name) the Shandong Tianyu Museum of Nature (山東省天宇自然博物館).
e.g.: *Tianyulong* -> -long; (*Tianyuornis* Zheng *et al.*, 2014) [Avialae]; *Tianyuraptor* -> -raptor

Tianzhen (Chin. place-name) Tiānzhèn 天鎮: a county of Shanxi Province (山西省), China.
e.g.: *Tianzhenosaurus*

tichosteus {(Gr.) τεῖχος/teichos, *n.*: a wall + ὀστέον/osteon, *n.*: a bone} walled bone.
e.g.: (*Tichosteus* Cope, 1877) *(nomen dubium)*

Tienshan (Chin. place-name) 天山 [Pinyin: Tiānshān] 'Heavenly Mountain' in Xinjiang Uyghur Autonomous Region, China.
e.g.: *Tienshanosaurus*

Tim (person's name) Timothy Rich and Timothy Flannery.
e.g.: *Timimus* -> -mimus

Timur Leng (Persian) Timūr(-i) Lang 'Timur the Lame' : Tamerlane, the fourteenth-century Central Asian ruler.
e.g.: *Timurlengia*

-Titan (Gr. myth.) Τῑτάν/Tītan (*nom.*), Τῑτᾶνος/Tītānos (*gen.*), *m.*: the Greek mythology giants.
e.g.: *Anatotitan*; *Angolatitan*; *Atacamatitan*; *Baurutitan*; *Blasilotitan*; *Bonatitan*; *Daxiatitan*; *Dongbeititan*; *Duriatitan*; *Elaltitan*; *Europatitan*; *Fukuititan*; *Giraffatitan*; *Gobititan*; *Gondwanatitan*; *Huanghetitan*; *etc.*

Titano-
e.g.: *Titanoceratops* -> -ceratops; *Titanosaurus* Lydekker, 1877; ("*Titanosaurus*" *montanus* Marsh, 1877) (preoccupied) >>> *Atlantosaurus*

-Titanis (Gr.) Τῑτᾱνίς/Tītānis, *fem.* of Τῑτάν.
e.g.: *Amargatitanis*; *Tambatitanis*

tochi- (Mong.) тохЬ/toki: ostrich.
e.g.: *Tochisaurus*

tom(o)-, -tom(o)- (Gr.) τομός/tomos, *m.*: a cut. τέμνω/temnō: to cut.
e.g.: (*Tomodon* Leidy, 1865) (preoccupied) >>> *Diplotomodon*

Tongan (Chin. place-name) Tōng'ān 通安: Huili, Sichuan Province (四川省), China.
e.g.: *Tonganosaurus*

Tongtian (Chin.) Tōngtiān 通天 'the road to heaven': referring to Tongtianyan (通天岩) of Ganzhou, Jiangxi, China.
e.g.: *Tongtianlong* -> -long

-tonitrus (Lat.) tonitrus (*nom.*), tonitrūs (*gen.*), *m.*: thunder.
e.g.: *Antetonitrus*

torilion
e.g.: (*Torilion* Carpenter & Ishida, 2010) >>> *Barilium*

Tornier (person's name) Gustav Tornier (1858–1938): a German zoologist and herpetologist.
e.g.: *Tornieria*

toro- (Gr.) τορός/toros: piercing, sharp, perforated. τορέω/toreō: pierce, perforate.
e.g.: *Torosaurus*

torv(i)-, torvo- (Lat.) torvus (*m.*), torva (*f.*), torvum (*n.*), *adj.*: staring, wild, grim, fierce, savage.
e.g.: *Torvosaurus*

tototl (Nahuatl) bird
e.g.: *Tototlmimus* -> -mimus

trach(o)- (Gr.) τραχύς/trachus: rugged, rough, savage.
e.g.: *Trachodon* -> -odon

-trachelo- (Gr.) τράχηλος/trachēlos, *m.*: the neck, throat.
e.g.: *Brachytrachelopan*

tralka (Mapuche) thunder.
e.g.: *Tralkasaurus*

Tratayen (place-name) a city in Neuquén, Argentina.
e.g.: *Tratayenia*

Trauku (Araukanian* legend) the Araukanian mountain spirit.
*a group of South American people.
e.g.: *Traukutitan* -> -titan

tri- (Gr.)τρῐ-/tri-, prefix, from τρίς/tris or τρίᾰ/tria (neut. of τρεῖς). τρεῖς/treis = (Lat.) trēs, tria: three.
e.g.: *Triceratops* -> -ceratops

trierarch- (Lat.) triērarchus (*nom.*), triērarchī (*gen.*) *m.*: a captain of a trireme, trierarch.
e.g.: *Trierarchuncus* -> -uncus

tri-gono- (Gr.)τρίγωνος/trigōnos: three-cornered, triangular. As *subst.*, *n.*: a triangle. | (place-name) Triângulo Mineiro: Trigónos (triângulo) in allusion to the region known as "Triângulo Mineiro" from Minas Gerais State.
e.g.: *Trigonosaurus*

trimucro {(Lat.) tri- = (Gr.) τρί-/tri-: three, three times + (Latin) mūcrō, *m.*: a sharp, point, edge, sword} three-pointed.
e.g.: (*Trimucrodon* Thulborn, 1975) *(nomen dubium)* -> -odon

Trini- (person's name) Trinidad Diaz, geologist.
e.g.: *Trinisaura* -> -saura

Triunfo (place-name) Triunfo Basin, in Paraíba State, Brazil.
e.g.: *Triunfosaurus*

tro- (Gr.) wounding [τρώγω/trōgō: to gnaw, nibble.]
e.g.: *Troodon* -> -odon

Tsaagan (Mong.) цагаан/tsagaan: white.
e.g.: *Tsaagan*

Tsagan-Teg (place-name) the Tsagan-Teg locality, Dzun-Bayan, in the southeastern Gobi Desert, Mongolia.
e.g.: *Tsagantegia*

Tsintao (place-name) Tsingtao 青島 [Pinyin: Qīngdǎo]: a city in eastern Shandong Province (山東省), on the east coast of China.
e.g.: *Tsintaosaurus*

Tugulu (Chin. stratal name) Tǔgǔlǔ 吐谷魯 | (stratal name) Tuguru Group (Early Cretaceous), Xinjiang, China.
e.g.: *Tugulusaurus*

Tuojiang (Chin. place-name) Tuójiāng 沱江. the Tuo River is one of the major tributaries of the upper Yangtze Jiang (揚子江).
e.g.: *Tuojiangosaurus*

Tupan (Tupi, myth.) thunder god.
e.g.: *Tupandactylus* -> dactylus [Pterosauria]

Turan (place-name) an old Persian name for Turkestan.
e.g.: *Turanoceratops* -> -ceratops

Turia (Lat. place-name) a river in eastern Spain. Called also Tūrium.
cf. Turia or Turium is famed for the proelium Turiense between Pompey and Sertorius (Plot. Pomp.18, Sert.19: Cie.p.Balb.2)
e.g.: *Turiasaurus*

tylo- (Gr.) τύλος/tylos, *m.* = τύλη/tylē, *f.*: a knot, callus, knob.
e.g.: *Tylocephale* -> -cephale; (*Tylosteus* Leidy, 1872) *(nomen rejectum)* >>> *Pachycephalosaurus (nomen conservandum)*

tyranno- (Gr.) τύραννος/tyrannos, *m.*: an absolute sovereign. As *adj.* kingly, royal.
e.g.: *Tyrannosaurus*; *Tyrannotitan* -> -titan

-tyrannus (Lat.) tyrannus (*nom.*), tyrannī (*gen.*), *m.*, = (Gr.) τύραννος/tyrannos, *m.*: a monarch, sovereign, tyrant.
e.g.: *Eotyrannus*; *Nanotyrannus*; *Siamotyrannus*; *Sinotyrannus*; *Suskityrannus*; *Yutyrannus*; *Zhuchengtyrannus*

-tyrannis
e.g.: *Aviatyrannis*

-tyrant (Eng.) ← (Gr.) τύραννος/tyrannos
e.g.: *Juratyrant*

U

Uberaba (place-name) a municipality in the west of Minas Gerais, Brazil.
e.g.: *Uberabatitan* -> -titan

ubirajara (Tupi) lord of the spear.
e.g.: *Ubirajara*

Udan (Mong. place-name) Udan-Sayr: the type locality in Ömnögovi Province, Mongolia.
e.g.: *Udanoceratops* -> -ceratops

ugro (Scandinavian) ugro: ugly.
e.g.: *Ugrosaurus*

Ugrunaaluk [oo-GREW-nah-luk] (Inupiaq / Inupiat*) ancient grazer.
*the language spoken by the Inuit natives of Alaska.
e.g.: *Ugrunaaluk*

Uinta (place-name) Uinta County, Utah. | Uinta Mountains: a chain of mountains in northeastern Utah, USA.
e.g.: (*Uintasaurus* Holland, 1919) >>> *Camarasaurus*

ultra- (Lat.) ultrā: beyond, farther, over.
e.g.: ("*Ultrasaurus*" Jensen, 1985) (preoccupied) >>> *Ultrasauros*; *Ultrasaurus* Kim, 1983

Umoona (Aborigine, place-name) the area in Coober Pedy, Australia.
e.g.: *Umoonasaurus* [Plesiosauria]

unay [u-na-hee] (Tupi*) unay 'black water': Aqua Negra.
*The Tupi people were one of indigenous people in Brazil.
e.g.: *Unaysaurus*

-uncus (Lat.) uncus (*nom.*), uncī (*gen.*) *m.*: a hook.
e.g.: *Trierarchuncus*

unenlag (Mapuche) {uñen: half + lag: bird} half bird.
e.g.: *Unenlagia*

UNESCO (acronym) United Nations Educational, Scientific, and Cultural Organization.
e.g.: *Unescoceratops* -> -ceratops

-unguis (Lat.) unguis (*nom.*), unguis (*gen.*), *m.*: claw.
e.g.: *Longusunguis* [Avialae]

Unquillo (place-name): Unquillo located in La Candelaria, Salta Province, Argentina.
e.g.: *Unquillosaurus*

-ura (Gr.) οὐρά/oura [ūrā], *f.*: the tail.
e.g.: *Archaeornithura* [Avialae]

-ur(o)-
e.g.: *Abdarainurus*; *Anurognathus* [Pterosauria]; *Chuandongocoelurus*; *Coelurus*; *Dacentrurus*; (*Kentrurosaurus*); *Macrurosaurus*; *Sinocoelurus*; *Talarurus*; (*Teinurosaurus*); (*Thecocoelurus*)

URBAC (acronym) Uzbek, Russian, British, American and Canadian scientists.
e.g.: *Urbacodon* -> -odon

Utah (place-name) a state in the Western United States.
e.g.: *Utahceratops* -> -ceratops; *Utahraptor* -> -raptor

Utatsu (Jpn. place-name) Utatsu-chō 歌津町*: a town located in Motoyoshi District, Miyagi Prefecture (宮城県), Japan.
*From 2005, Minamisanriku-chō 南三陸町.
e.g.: *Utatsusaurus* [Ichthyosauria]

Ute [yewt] (demonym) the Native American people who inhabit northeastern Utah.
e.g.: *Uteodon* -> -odon

V

vaga- (Lat.) vagus (*m.*), vaga (*f.*), vagum (*n.*), *adj.*: strolling, wandering. *cf.* vagātus, *part.* of [vagor: wander]
e.g.: *Vagaceratops* -> -ceratops

vahiny [va-heenh] (Malagasy) vahiny: foreigner, traveler.
e.g.: *Vahiny*

Valdo- (Lat. stratal name) Valdus 'Wealden': Wealden Group (Early Cretaceous), southern United Kingdom.
e.g.: *Valdoraptor* -> -raptor; *Valdosaurus*

Vallibona (place-name) a town located in Castellón Province, Spain.
e.g.: *Vallibonavenatrix* -> -venatrix

Vari- (Fr.) le Var: the Var, a river located in the southeast of France.
e.g.: *Variraptor* -> -raptor

-vates (Lat.) vātēs (*nom.*), vātis (*gen.*), *m./f.*: a seer, foreteller.

e.g.: *Rativates*

Vayu (Sanskrit) Vayu: God of Wind.
e.g.: *Vayuraptor* -> -raptor

Vect-, Vecti- (Lat. palce-name) Vectis = Wiht: the Isle of Wight, UK.
e.g.: *Vectaerovenator* -> -aero-, venator; *Vectidraco* -> -draco [Pterosauria]; (*Vectisaurus* Hulke, 1879) *(nomen dubium)*

vela- (Sp.) vela: a sail.
e.g.: *Velafrons* -> -frons

veloci- (Lat.) vēlōx (*nom.*), velōcis (*gen.*), *adj.*: fast, quick, rapid.
e.g.: *Velocipes* -> -pes; *Velociraptor* -> -raptor; *Velocisaurus*

-venator (Lat.) vēnātor (*nom.*), vēnātōris (*gen.*), *m.*: a hunter.
e.g.: *Afrovenator*; *Albertavenator*; *Arcovenator*; *Australovenator*; *Concavenator*; *Dracovenator*; *Duriavenator*; *Fosterovenator*; *Fukuivenator*; *Gobivenator*; *Ichthyovenator*; *Juravenator*; *Kayentavenator*; *Liaoningvenator*; *Linhevenator*; *Luchianovenator*; *Microvenator*; *Neovenator*; *Pandravenator*; *Philovenator*; *Phuwiangovenator*; *Powellvenator*; *Saltriovenator*; *Sinovenator*; *Skorpiovenator*; (*Stygivenator*); *Taurovenator*; *Teihivenator*; *Viavenator*; *Wiehenvenator*; *Xinjiangovenator*

-venatrix (Lat.) vēnātrīx, *f.* (feminine form of 'venator'): huntress.
e.g.: *Latenivenatrix*; *Vallibonavenatrix*

veneno- (Lat.) venēnum (*nom.*), venēnī (*gen.*), *n.*: a strong potion, poison. | (stratal name) the Poison Strip Member of the Cedar Mountain Formation (Early Cretaceous), Utah, USA.
e.g.: *Venenosaurus*

vesc- (Lat.) vēscus (*m.*), vesca (*f.*), vescum (*n.*), *adj.*: small, slender, thin.
e.g.: *Vescornis* -> -ornis [Avialae]

vesper (Lat.) vesper (*nom.*) vesperī (*gen.*), *m.*: evening, west.
e.g.: *Vespersaurus*

veterupristi- {(Lat.) vetus (*m./f./n.,nom.*), veteris (*gen.*): old, aged + pristis = (Gr.) πρίστις/ pristis: sawfish, a sea-monster}
e.g.: *Veterupristisaurus*

via (Lat.) via (*nom.*), viae (*gen.*), *f.*: a way, road, street, method.
e.g.: *Viavenator* -> -venator

-via
e.g.: *Arkharavia*

Vitakri (stratal name) Vitakri Member, Pab Formation (Late Cretaceous), western Pakistan.
e.g.: *Vitakridrinda* -> -drinda; *Vitakrisaurus*

-volans (Lat.) volāns: flyng.
e.g.: *Cryptovolans*

Volga (place-name) Volga River: a river flowing through central Russia.
e.g.: *Volgatitan* -> -titan

Volkheimer (person's name) Wolfgang Volkheimer: an Argentine paleontologist.
e.g.: *Volkheimeria*

-volucris (Lat.) volucris (*nom.*), volucris (*gen.*), *f.*: a bird, flying creature.
e.g.: *Yungavolucris* [Avialae]

-vora (Lat.) vora (*nom., sing., f.* of vorus) or (*nom.,pl., n.* of vorus) one that eat. vorō: to devour.
e.g.: *Eogranivora* [Avialae]

-vorax* (Lat.) vorāx (*nom.*), vorācis (*gen.*) *adj.*: devouring. *{vor- [vorō: devour] + -ax: 'inclined to'}.
e.g.: *Gnathovorax*

vorona (Malagasy) bird.
e.g.: *Vorona* [Avialae]

vouivre (Old French) vouivre = (Lat.) vīpera: viper.
e.g.: *Vouivria*

Vulcan- (Lat. myth.) Vulcānus (*nom.*), Vulcānī (*gen.*), *m.*: Vulcan (god of fire).
e.g.: *Vulcanodon* -> -odon

W

Wadhurst (stratal name) Wadhurst Clay Formation (Early Cretaceous), Weald Basin,

UK. | a town in East Sussex, England.
e.g.: (*Wadhurstia* Carpenter & Ishida, 2010) >>> *Hypselospinus*

Wakino (Jpn.) Wakino 脇野 | (stratal name) the Wakino Subgroup of the Kanmon Group (Early Cretaceous), northern Kyushu, Japan.
e.g.: *Wakinosaurus*

Walgett (place-name) a town in northern New South Wales, Australia.
e.g.: *Walgettosuchus* -> -suchus

Walker (person's name) Alick Walker: a British palaeontologist.
e.g.: (*Walkeria* Chatterjee, 1987) (preoccupied) >>> *Alwalkeria*; ("Walkersaurus" Welles & Powell *vide* Pickering, 1995) >>> *Duriavenator*

Wamwera (demonym) the Wamwera: among the Mwera people, the most populous tribe in the Lindi-region of Tanzania, the ones who live on the coast are called 'Wamwera' by other Mwera people. 'Mwera' means 'inland dweller' in Mwera language.
e.g.: *Wamweracaudia* -> caudia

Wangoni (demonym) the Wangoni: the residents living in Tendaguru, eastern Africa.
e.g.: ("Wangonisaurus" Maier, 2013) >>> *Giraffatitan*

Wannan (Chin. place-name) Wǎnnán 皖南 'southern Anhui': 皖 is the abbreviation for Anhui (安徽), China.
e.g.: *Wannanosaurus*

Wee Warra (place-name) Wee Warra: the fossil locality in New South Wales, Australia.
e.g.: *Weewarrasaurus*

Wellnhofer (person's name) Peter Wellnhofer: Chief Curator Emeritus, Bayerische Staatsammlung für Paläontologie und historische Geologie, Munich.
e.g.: *Wellnhoferia* [Avialae]

Wendi- (person's name) Wendy Sloboda: a Canadian fossil hunter.
e.g.: *Wendiceratops* -> -ceratops

Wiehen (Ger. place-name) Wiehengebirge 'Wiehen Hills': a hill range in North Rhine-Westphalia, Germany.
e.g.: *Wiehenvenator* -> -venator

Willinakaqe (Mapuche) {willi : south + iná: mimic + kaqe: duck} the duck-mimic of the South.
e.g.: *Willinakaqe*

Winton (place-name) a town in central western Queensland, Australia. | (stratal name) Winton Formation (Early Cretaceous), Queensland, Australia.
e.g.: *Wintonotitan* -> -titan

wu (Chin.) wǔ 舞: dance.
e.g.: *Wulong* -> -long

Wuerho (Chin. place-name) = Urho 烏爾禾 [Pinyin: Wūěrhé]: a district within the Xinjian Uyghur Autonomous Region, China.
e.g.: *Wuerhosaurus*

Wulaga (Chin. place-name) Wulaga, Heilongjiang, China.
e.g.: *Wulagasaurus*

Wulate (Chin. place-name) = Urad 烏拉特後旗 [Pinyin: Wūlātè Hòu Qí] (Urad Rear Banner): a banner of the Inner Mongolia Autonomous Region, China.
e.g.: *Wulatelong* -> -long

Wyley (person's name) J. F. Wyley.
e.g.: (*Wyleyia* Harrison & Walker, 1973) *(nomen dubium)* [Avialae]

Wyoming (place-name) a landlocked state in USA.
e.g.: ("Wyomingraptor" Bakker, 1997) -> -raptor

X

xeno- (Gr.) ξένος/xenos, *m.*: a guest-friend. As *adj.*, alien, strange, unusual.
e.g.: *Xenoceratops* -> -ceratops; *Xenoposeidon* -> -Poseidon; *Xenotarsosaurus*

xiang (Chin.) xiáng 翔: to fly.
e.g.: *Xiangornis* -> -ornis [Avialae]

Xianshan (Chin. place-name) Xiànshān 峴山, in Henan Province (河南省), China.

e.g.: *Xianshanosaurus*

xiao (Chin.) xiǎo 暁: dawn.
e.g.: *Xiaosaurus*

Xiaoting (Chin. person's name) Zhèng Xiǎotíng 鄭暁廷: a Chinese paleontologist.
e.g.: *Xiaotingia* [Anchiornithidae]

Xingtian (Chin. myth.) Xíngtiān刑 天 'Opposing Heaven': a Chinese deity, who continued the fight with a shield and an axe after he was beheaded.
e.g.: *Xingtianosaurus*

Xingxiu (Chin.) Xīngxiù Bridge 星宿橋: a bridge which was constructed during the Ming Dynasty of China, Lufeng County, Yunnan Province. | xingxiu 星宿: 'constellation'.
e.g.: *Xingxiulong* -> -long

Xinjiang (Chin. place-name) Xīnjiāng 新疆: the Xinjiang Uygur Autonomous Region, China.
e.g.: *Xinjiangovenator* -> -venator; *Xinjiangtitan* -> -titan

Xiongguan (Chin. place-name) Xióngguān 雄關 'Grand Pass': a historic name for nearby city of Jiāyùguān (嘉峪關市), Gansu Province (甘粛省), China.
e.g.: *Xiongguanlong* -> -long

Xixia (Chin. place-name) Xīxiá 西峡: a county in Nanyang, Henan Province (河南省), China.
e.g.: *Xixianykus* -> -onykus; *Xixiasaurus*

Xixipo (Chin. place-name) Xìxìpō 細細坡: a village in Yunnan Province (雲南省), China.
e.g.: *Xixiposaurus*

Xiyu (Chin. place-name) Xīyù 西域 'the western regions': Central Asia including Xinjiang.
e.g.: *Xiyunykus* -> -onykus

Xuanhan (Chin. place-name) Xuānhàn 宣漢: a county in Sichuan Province (四川省), China.
e.g.: *Xuanhanosaurus*

Xuanhua (Chin. place-name) Xuānhuà 宣化: a district in Zhangjiakou prefecture-level city, Hebei Province (河北省), China.
e.g.: ("Xuanhuasaurus" Zhao, 1985) >>> *Xuanhuaceratops* -> -ceratops

xunmeng (Chin.) xùnměng 迅猛: swift.
e.g.: *Xunmenglong*

Xuwu (Chin. person's name) Xùwǔ 敘五: the courtesy name of a Chinese geologist Wang Yue-lun (王曰倫).
e.g.: *Xuwulong* -> -long

Y

Yale (institution name) Yale Peabody Museum, New Haven, Connecticut, USA.
e.g.: (*Yaleosaurus* von Huene, 1932) >>> *Anchisaurus*

Yama (Sanscrit, Buddhism) a Tibetan Buddhist deity. *cf.* (Jpn.) Emma 閻魔. (Gr.) Hades.
e.g.: *Yamaceratops* -> -ceratops

Yamana (place-name) a locality in the Casanga Valley, Equador.
e.g.: *Yamanasaurus*

Yan (Chin.) Yān 燕: the ancient Chinese Yan Dynasty.
e.g.: *Yanornis* -> -ornis [Avialae]

Yandang (Chin. place-name) Yàndàng Mountain 雁蕩山: a mountain / a mountain range located in Zhejiang Province (浙江省), China.
e.g.: (*Yandangornis* Cai & Zhao, 1999) [Avialae?]

Yandu (Chin. place-name) Yándū 塩都 'salt capital': the ancient name for Zigong (自貢), Sichuan Province (四川省), China.
e.g.: *Yandusaurus*

Yang (person's name) Yang: late Chinese palaeontologist Yang Zhongjian 楊鍾健 (Chung Chien Young).
e.g.: *Yangavis* -> -avis [Avialae]

Yangchuan (Chin. place-name) 永川 [Pinyin: Yǒngchuān]: a county in Sichuan Province (四川省), China.
e.g.: *Yangchuanosaurus*

Yaverland (Eng. palce-name) a village on the Isle of Wight, UK. | Yaverland Battery: a battery on the Isle of Wight.
e.g.: *Yaverlandia*

yehuecauh [Ye-OO-ek-au] (Nahuatl) ancient.
e.g.: *Yehuecauhceratops* -> -ceratops

Yezo (place-name) Yezo = Ezo (蝦夷) = Hokkaido (北海道): the lands to the north of the Japanese island of Honshu.
e.g.: ("Yezosaurus" Obata & Muramoto, 1977) >>> (*Yezosaurus* Caldwell *et al.*, 2008) >>> (*Taniwhasaurus* Hector, 1874) [Mosasauridae]

yi [ee] (Chin.) yì 翼: wing.
e.g.: *Yi* [Scansoriopterygidae]

Yimen (Chin. place-name) Yìmén 易門: a county located in Yuxi (玉溪), Yunnan Province (雲南省), China.
e.g.: *Yimenosaurus*

yin (Chin.) yǐn 隠: hidden.
e.g.: *Yinlong* -> -long

Yingshan (Chin. place-name) Yíngshān 營山: a county of Sichuan Province (四川省), China.
e.g.: ("Yingshanosaurus" Zhou, 1985 or 1986) >>> *Yingshanosaurus*

Yixian (Chin. place-name) Yìxiàn 義縣: a county in west-central Liaoning Province. | (Chin. stratal name) Yìxiàn Formation (Early Cretaceous), Jinzgou (錦州), Liaoning Province (遼寧省), China, where *Yixianosaurus* was found. The formation is followed stratigraphically by the slightly younger Jiufotang Formation, where *Yixianornis* was found.
e.g.: *Yixianornis* -> -ornis [Avialae]; *Yixianosaurus* [Anchiornithidae]

Yizhou (Chin. place-name) Yìzhōu 益州: a jùn (郡) in Chuxiong Yi Autonomous Prefecture, Yunnan Province, China. *cf.* Yìzhōu 益州: a province of ancient China.
e.g.: *Yizhousaurus*

Yongjing (Chin. place-name) Yǒngjìng 永靖: a county in Linxia Hui Autonomous Prefecture, Gansu Province (甘肅省), China.
e.g.: *Yongjinglong* -> -long

Yu (Chin. place-name) Yù 豫: abbr. of Henan Province (河南省), China.
e.g.: *Yulong* -> -long

yu (Chin.) yǔ 羽: feather.
e.g.: *Yutyrannus* -> -tyrannus

-yu-
e.g.: *Changyuraptor*

Yuanmou (Chin. place-name) Yuánmóu 元謀: a county in Chuxiong Yi Autonomous Prefecture, Yunnan Province (雲南省), China.
e.g.: *Yuanmousaurus*

Yue (Chin. place-name) Yuè 越: a state in ancient China, in the modern provinces of Zhèjiāng (浙江), Shànghǎi (上海) and Jiāngsū (江蘇).
e.g.: *Yueosaurus*

Yumen (Chin. place-name) Yùmén 玉門: a city in Gansu Province (甘粛省), China.
e.g.: *Yumenornis* -> -ornis [Avialae]

Yunga- (place-name) the Yungas region, northwestern Argentina.
e.g.: *Yungavolucris* -> -volucris [Avialae]

Yungang (Chin. place-name) Yúngǎng 雲崗: a town or city of Datong, Shanxi Province (山西省), China. | Yungang Grottoes (雲崗石窟): ancient Chinese Buddhist temple grottes.
e.g.: *Yunganglong* -> -long

Yunmeng (Chin, place-name) Yúnmèng 雲夢.
e.g.: *Yunmenglong* -> -long

Yunnan (Chin. place-name) Yúnnán 雲南: a province in southwestern China.
e.g.: *Yunnanosaurus*

Yunyang (Chin. place-name) Yúnyáng 雲陽: a county in Chongqing, China.
e.g.: *Yunyangosaurus*

yurgovuch (Ute) coyote.
e.g.: *Yurgovuchia*

Z

Zalmoxes (legend) Ζάλμοξις/Zalmoxis: the Dacian deity Zalmoxis.
e.g.: *Zalmoxes*

Zanabazar (Tibetan Buddhism) Занабазар/Zanabazar.
e.g.: *Zanabazar*

Zapala (place-name) a city of Neuquén, Argentina.
e.g.: *Zapalasaurus*

zapsalis {(Gr.) ζά/za: insep. prefix: very + ψᾰλίς/psalis, *f.*: a pair of scissors}
e.g.: *Zapsalis*

zaraa (Mong.) зараа/zaraa: hedgehog.
e.g.: *Zaraapelta* -> -pelta

Zby [zee-bee] (person's name) Georges Zbyszewski: a Portuguese paleontologist and geologist of Russian origin.
e.g.: *Zby*

Zephyros (Gr. myth.) Ζέφῠρος/Zephyros, *m.*: God of the west wind, Zephyr.
e.g.: *Zephyrosaurus*

Zhang Heng (Chin. person's name) Zhāng Héng 張衡 (78–139): a Chinese polymathic scientist and statesman who lived during the Eastern Han Dynasty of China.
e.g.: *Zhanghenglong* -> -long

Zhejiang (Chin. place-name) Zhèjiāng 浙江: an eastern coastal province of China.
e.g.: *Zhejiangosaurus*

Zhenyuan (Chin. person's name) Zhènyuán Sūn 振元孫: Mr. Zhenyuan Sun, who secured the specimen for study.
e.g.: *Zhenyuanlong* -> -long

zhong (Chin.) zhōng 中: intermediate.
e.g.: *Zhongornis* -> -ornis [Avialae]

Zhongjian (Chin. person's name) Yáng Zhōngjiàn 楊鍾健 (C. C. Young), the IVPP's* founder.
*Institute of Vertebrate Paleontology and Paleoanthropology.
e.g.: *Zhongjianornis* -> -ornis [Avialae]; *Zhongjianosaurus*

Zhongyuan (Chin. place-name) Zhōngyuán 中原: the area on the lower reaches of the Yellow River (黄河) [Pinyin: Huáng Hé], China.
e.g.: *Zhongyuansaurus*

Zhou (person's name) Zhōu Zhōnghè 周忠和: a Chinese paleontologist.
e.g.: *Zhouornis* -> -ornis [Avialae]

Zhucheng (Chin. place-name) Zhūchéng 諸城: a county-level city in Shandong Province (山東省), China.
e.g.: *Zhuchengceratops* -> -ceratops; *Zhuchengosaurus*; *Zhuchengtitan* -> -titan; *Zhuchengtyrannus* -> *-tyrannus*

Zia (demonym) the Zia: an indigenous tribe living in New Mexico, USA. The people are a branch of the Pueblo community.
e.g.: *Ziapelta* -> -pelta

Zigong (Chin. place-name) Zìgòng 自貢: a prefecture-level city in Sichuan Province (四川省), China.
e.g.: *Zigongosaurus*

Zizhong (Chin. place-name) Zīzhōng 資中: a county of Sichuan Province (四川省), China.
e.g.: *Zizhongosaurus*

-zostron (Gr.) ζῶστρον/zostron, *n.*: a belt, girdle.
e.g.: (*Eozostrodon*) [Mammaliaformes]

-zui- (Chin.) zuǐ 嘴: beak.
e.g.: *Changzuiornis*

Zuni (demonym) the Zuni: a Native American tribe who live near the Zuni River.
e.g.: *Zuniceratops* -> -ceratops

Zuo (Chin. person's name) Zuǒ Zōngtáng 左宗棠: a Chinese statesman and military leader of the late Qing dynasty (清朝末期).
e.g.: *Zuolong* -> -long

Zuoyun (Chin. place-name) Zuǒyún 左雲: a county of Shanxi, China.
e.g.: *Zuoyunlong* -> -long

Zupay (Quechua*) Zupay = Supay: the Inca Death God, devil (not demon).
*a language spoken by the Quechua peoples in Andes and highlands of South America.
e.g.: *Zupaysaurus*

Zuul (movie) the Gatekeeper of Gozer, the dog appeared in *Gostbusters*.
e.g.: *Zuul*

Explanation

— etymology of type species —

A

Aardonyx celestae

> *Aardonyx celestae* Yates *et al*., 2010 [Hettangian; Ellliot Formation, Free State, South Africa]

Generic name: *Aardonyx* ← {(Afrikaans) aard-: Earth + (Gr.) ὄνυξ/onyx: claw}; referring to "the thick hematite encrustation of many of the bones, particularly the ungual phalanges, in the type quarry" (*1).

Specific name: *celestae* ← {(person's name) Celeste + -ae}; in honor of "Celeste Yates who prepared many of the bones" (*1).

Etymology: Celeste's **Earth claw**

Taxonomy: Saurischia: Sauropodomorpha: Anchisauria

Notes: According to Dr. Bonnan, Yates wanted to use Afrikaans language. The origin of the name is seemed to be an aardvark, which means an "earth pig". (*2)

[References: (*1) Yates AM, Bonnan MF, Neveling J, Chinsamy A, Blackbeard MG (2010). A new transitional sauropodomorph dinosaur from the Early Jurassic of South Africa and the evolution of sauropod feeding and quadrupedalism. *Proceedings of the Royal Society B: Biological Sciences* 277: 787–794.; (*2) *Arcusaurus pereirabdalorum*: the little sauropodomorph that could. The Evolving Paleontologist. Digital Home of Dr. Matthew F. Bonnan, Ph.D. 2009, 2012.〈https://matthewbonnan.wordpress.com/〉]

Abdarainurus barsboldi

> *Abdarainurus barsboldi* Averianov & Lopatin, 2020 [Late Cretaceous; Alagteeg Formation, Abdrant Nuru, Mongolia]

Generic name: *Abdarainurus* ← {(Rus. place-name) Abdarain Nuru: Abdrant Nuru locality + (Gr.) ουρά/ūra: tail} (*1)

Specific name: *barsboldi* ← {(person's name) Barsbold + -ī}; "in honour of Mongolian palaeontologist R. Barsbold" (*1).

Etymology: Barsbold's **tail from Abdarain (Abdrant) Nuru**

Taxonomy: Saurischia: Sauropoda: Titanosauria

[References: (*1) Averianov AO, Lopatin AV (2020). An unusual new sauropod dinosaur from the Late Cretaceous of Mongolia. *Journal of Systematic Palaeontology* 18: 1009–1032.]

Abelisaurus comahuensis

> *Abelisaurus comahuensis* Bonaparte & Novas, 1985 [Campanian; Anacleto Formation*, Río Negro, Argentina]
> *It was originally described as the Allen Formation which overlies the Anacleto Formation.

Generic name: *Abelisaurus* ← {(person's name) Abel + (Gr.) σαῦρος/sauros: lizard}; "in honor of Prof. Roberto Abel, Director of the Museo de Cipolletti and author of the discovery" (*1).

Specific name: *comahuensis* ← {(place-name) the Comahue region* + -ensis}

*Comahue: the northern part of Argentine Patagonia, including the provinces Neuquén and Río Negro.

Etymology: **Abel's lizard** from Comahue

Taxonomy: Theropoda: Ceratosauria: Abelisauridae

[References: (*1) Bonaparte JF, Novas FE (1985). *Abelisaurus comahuensis,* n. g., n. sp., Carnosauria of the Late Cretaceous of Patagonia. *Ameghiniana* 21: 259–265.]

Abrictosaurus consors

> *Lycorhinus consors* Thulborn, 1974 ⇒ *Abrictosaurus consors* (Thulborn, 1974) Hopson, 1975 [Hettangian; upper Elliot Formation, Qacha's Nek, Lesotho]

Generic name: *Abrictosaurus* ← {(Gr.) ἄ-βρικτος/a-briktos: awake + σαῦρος/sauros: lizard}; referring to Hopson's disagreement with Thulborn's suggestion that "heterodontosaurids underwent periods of aestivation (or hibernation)...during the yearly dry season" (*1) (*2).

Specific name: *consors* ← {(Lat.) consors: partner, spouse}; in allusion to suspected feminine gender of the holotype.

Etymology: partner's **awakened lizard**

Taxonomy: Ornithischia: Ornithopoda: Heterodontosauridae

[References: (*1) Thulborn RA (1974). A new heterodontosaurid dinosaur (Reptilia: Ornithischia) from the Upper Triassic Red Beds of Lesotho. *Zoological Journal of the Linnean Society* 55: 151–175.; (*2) Hopson JA (1975). On the generic separation of the ornithischian dinosaurs *Lycorhinus* and *Heterodontosaurus* from the Stormberg Series (Upper Triassic) of South Africa. *South African Journal of Science* 71: 302–305.]

Abrosaurus dongpoi

> *Abrosaurus dongpoi* Ouyang, 1989 [Callovian; Xiashaximiao Formation, Sichuan, China]

Generic name: *Abrosaurus* ← {abr- [(Gr.) ἁβρός/habros]: delicate, graceful + σαῦρος/sauros: lizard}; referring to "the main character of the skull" (*1).

Specific name: *dongpoi* ← {(Chin. person's name) Dōngpō東坡 + -ī}; referring to Su Dongpo (蘇東坡) that is "the stylized name of the distinguished Northern Sung Dynasty literary scholar from Sichuan named Sushi (1037–1101)" (*1).

Etymology: Dongpo's **delicate lizard**

Taxonomy: Saurischia: Sauropoda: Macronaria

cf. Saurischia: Sauropoda: Camarasauridae: Cetiosaurinae (*1)

[References: (*1) Ouyang H (1989). A new sauropod dinosaur from Dashanpu, Zigong County, Sichuan Province (*Abrosaurus dongpoensis* gen. *et* sp. nov). *Zigong Dinosaur Museum Newsletter* 2: 10–14.]

Abydosaurus mcintoshi

> *Abydosaurus mcintoshi* Chure *et al.*, 2010 [Albian; Cedar Mountain Formation, Uintah, Utah, USA]

Generic name: *Abydosaurus* ← {(Gr. place-name) Ἄβυδος/Abydos + σαῦρος/sauros: lizard}: referring to "Abydos, the Greek name for the city along the Nile River (now El Araba el Madfuna) that was the burial place of the head and neck of Osiris, Egyptian god of life, death, and fertility – an allusion to the type specimen, which is a skull and neck found in a quarry overlooking the Green River" (*1).

Specific name: *mcintoshi* ← {(person's name) McIntosh + -ī}: in honor of "Jack McIntosh* for his contributions to Dinosaur National Monument and to the study of sauropod dinosaurs" (*1).

*Jack McIntosh = John S. McIntosh

Etymology: McIntosh's **Abydos lizard**

Taxonomy: Saurischia: Sauropoda: Titanosauriformes: Brachiosauridae

Notes: Four skulls of juveniles were found. *Abydosaurus* had a long neck and a long tail.

[References: (*1) Chure D, Britt B, Whitlock JA, Wilson JA (2010). First complete sauropod dinosaur skull from the Cretaceous of Americas and the evolution of sauropod dentition. *Naturwissenschaften* 97: 379–391.]

Acantholipan gonzalezi

> *Acantholipan gonzalezi* Rivera-Sylva *et al.*, 2018 [Campanian; Pen Formation, Coahuila, Mexico]

Generic name: *Acantholipan* ← {(Gr.) acantho- [ἄκανθα/akantha: spine] + (demonym, Spanish contraction) "Lipan: *Lépai-Ndé* (gray people), a tribe of the Apaches from northern Mexico" (*1)}

Specific name: *gonzalezi* ← {(person's name) Gonzalez + -ī}; "in honor of Arturo H. González González, for his outstanding support to Mexican paleontology" (*1).

Etymology: Gonzalez's **spine of Lipan (Lépai-Ndé)**

Taxonomy: Ornithischia: Thyreophora: Ankylosauria: Nodosauridae

[References: (*1) Rivera-Sylva HE, Frey E, Stinnesbeck W, Carbot-Chanona G, Sanchez-Uribe IE, Guzmán-Gutiérrez JR (2018). Paleodiversity of Late Cretaceous Ankylosauria from Mexico and their phylogenetic significance. *Swiss Journal of Palaeontology* 138: 83–93.]

Acanthopholis horridus

> *Acanthopholis horridus** Huxley, 1867 (*nomen dubium*) [Albian–Cenomanian; Chalk Group, UK]

*In *The Dinosauria 2nd edition*, "*horridus*" is used, not "*horrida*".

Generic name: *Acanthopholis* ← {(Gr.) ἄκανθα/akantha: a thorn, prickle, the backbone or spine of animals + (Mod. Gr.) φολίς/pholis: scale}; referring to its armour.

Specific name: *horridus* ← {(Lat.) horridus: frightening, rough}

In 1890, Sir Arthur Smith Woodward emended the species name to *Acanthopholis horrida* because *pholis* is feminine (*1).

Etymology: frightening **spiny scales**

Taxonomy: Ornithischia: Thyreophora: Ankylosauria: Nodosauridae

Other species:

> *Acanthopholis macrocercus* Seeley, 1869 (*partim*) (*nomen dubium*) [early Cenomanian; upper Greensand Formation, England, UK]

> *Acanthopholis stereocercus* Seeley, 1869 (*partim*) (*nomen dubium*) [Albian; Cambridge Greensand, UK]

> *Acanthopholis platypus* Seeley, 1869 (*partim*) (*nomen dubium*) ⇒ *Macrurosaurus platypus* (Seeley, 1869) von Huene, 1956 (*nomen dubium*) [Cenomanian; upper Greensand Formation, England, UK]

> *Acanthopholis eucercus* Seeley, 1879 (*nomen dubium*) [Albian; Cambridge Greensand, UK]

Notes: According to *The Dinosauria 2nd edition* (2004), *Acanthopholis horridus* and other species were considered as *nomina dubia*. *Syngonosaurus* was synomised with *Acanthopholis* in 1999, but *Syngonosaurus* and *Eucercosaurus* were reinterpreted as basal iguanodontians, according to Barrett & Bonsor (2020).

[References: (*1) Woodward AS, Sherborn CD (1890). *A Catalogue of British Fossil Vertebrates.* Dulao & Company, London. 396p.]

Achelousaurus horneri

> *Achelousaurus horneri* Sampson, 1995 ⇒ *Centrosaurus horneri* (Sampson, 1995) Paul, 2010[Campanian; Two Medicine Formation, Montana, USA]

Generic name: *Achelousaurus* ← {(Gr. myth.) Ἀχελῷος/Achelōos: a river god + σαῦρος/sauros: lizard}; in allusion "to the apparently transitional morphology of the animals as well as to the ontogenetic and phylogenetic loss of horns characteristic of the genus."(*1).

Specific name: *horneri* ← {(person's name) Horner + -ī}; in honor of American paleontologist John ("Jack") R. Horner. (*1)

Etymology: Horner's **Achelous lizard**

Taxonomy: Ornithischia: Marginocephalia: Ceratopsia: Ceratopsidae: Centrosaurinae

[References: (*1) Sampson SD (1995). Two new horned dinosaurs from the Upper Cretaceous Two Medicine Formation of Montana; with a phylogenetic analysis of the Centrosaurinae (Ornithischia: Ceratopsidae). *Journal of Vertebrate Paleontology* 15: 743–760.]

Acheroraptor temertyorum

> *Acheroraptor temertyorum* Evans *et al.*, 2013 [Maastrichtian; Hell Creek Formation, Montana, USA]

Generic name: *Acheroraptor* ← {(Gr.) Ἀχέρων/Acherōn: "the River of Pain in the underworld" (*1) + (Lat.) raptor: thief, robber, plunderer}; in reference to the provenance from the Hell Creek Formation. A suffix 'raptor' is used in dromaeosaurids.

Specific name: *temertyorum* ← {(person's name) Temerty + -ōrum}; in honor of "James and Louise Temerty, for their enthusiastic service and contributions to the Royal Ontario Museum"(*1).

Etymology: James and Louise Temerty's **Acheron (underworld) thief**

Taxonomy: Theropoda: Maniraptora: Dromaeosauridae: Velociraptorinae

[References: (*1) Evans DC, Larson DW, Currie PJ (2013). A new dromaeosaurid (Dinosauria: Theropoda) with Asian affinities from the latest Cretaceous of North America. *Naturwissenschaften* 100: 1041–1049.]

Achillesaurus manazzonei

> *Achillesaurus manazzonei* Martinelli & Vera, 2007 [Santonian; Bajo de la Carpa Formation, Río Negro, Argentina]

Generic name: *Achillesaurus* ← {(Lat.) Achillēs = (Gr.) Ἀχιλλεύς/Achilleus: Achilles + σαῦρος/sauros: lizard}; "in reference to Achilles' heel, the week point of Achilles in the book 'Iliad' written by Homer, because the holotype has diagnostic features in this portion of the skeleton"(*1).

Specific name: *manazzonei* ← {(person's name) Manazzone + -ī}; "in honor of Prof. Rafael Manazzone, an amateur paleontologist who provided valuable data about Patagonian fossil localities, and assisted to several paleontological field trips". (*1)

Etymology: Manazzone's **Achilles lizard**

Taxonomy: Theropoda: Coelurosauria: Alvarezsauridae

[References: (*1) Martinelli AG, Vera EI (2007). *Achillesaurus manazzonei,* a new alvarezsaurid theropod (Dinosauria) from the Late Cretaceous Bajo de la Carpa Formation, Río Negro Province, Argentina. *Zootaxa.* 1582: 1–17.]

Achillobator giganticus

> *Achillobator giganticus* Perle *et al.*, 1999 [Cenomanian–Santonian; Bayan Shireh Formation, Burkhant, Mongolia]

Generic name: *Achillobator* ← {(Lat.) Achilles = (Gr. myth.) Ἀχιλλεύς/Achilleus: Achilles, a famous ancient Greek warrior + (Mong.) Ъаатар/baatar: hero}; referring to the calcaneal tendon.

Specific name: *giganticus* ← {giganticus: giant, gigantic}; in reference to *Achillobator*'s size.

According to Perle *et al.* (1999), *A. giganticus* is "nearly three times as big as *Deinonychus antirrhopus*".

Etymology: giant, **Achilles hero**

Taxonomy: Theropoda: Maniraptora: Dromaeosauridae

[References: Perle A, Norell M, Clarck JM (1999). A new maniraptoran theropod *Achillobator giganticus* (Dromaeosauridae), from the Upper Cretaceous of Burkhant, Mongolia. *Contributions of the Mongolian-American Paleontological Project* 101: 1–105.]

Acristavus gagslarsoni

> *Acristavus gagslarsoni* Gates *et al.*, 2011 [Campanian; Two Medicine Formation, Montana, USA]

Generic name: *Acristavus* ← {(Gr.) ἀ-/a-: non- + (Lat.) crista: crest + avus: grandfather}; "in reference to the absence of an osteological nasal crest, its stratigraphic position relative to other hadrosaurid taxa in the Two Medicine Formation, and the primitive nature of the skull"(*1).

Specific name: *gagslarsoni* ← {(nickname) Gags Larson + -ī}; in honor of "land owner Russell

Ellsworth Larson, on whose property the specimen was discovered"[*1].

Etymology: Gags Larson's **non-crested grandfather**

Taxonomy: Ornthischia: Ornithopoda: Hadrosauridae: Saurolophinae
cf. Ornithopoda: Hadrosauridae: Hadrosaurinae: Brachylophosaurini[*1]

[References: (*1) Gates TA, Horner JR, Hanna RR, Nelson CR (2011). New unadorned hadrosaurine hadrosaurid (Dinosauria, Ornithopoda) from the Campanian of North America. *Journal of Vetebrate Paleontology* 31: 798–811.]

Acrocanthosaurus atokensis

> *Acrocanthosaurus atokensis* Stovall & Langston, 1950 [Aptian–Albian; Antlers Formation, Atoka, Oklahoma, USA]

Generic name: *Acrocanthosaurus* ← {(Gr.) ἄκρος/akros: highest + ἀκανθο-/akantho-[ἄκανθα/akantha: thorn, spine] + σαῦρος/sauros: lizard}; referring to its elongate neural spines[*1].

According to Currie & Carpenter (2000), neural spines are more than 2.5 times corresponding presacral, sacral and proximal caudal lengths of the centra[*2].

Specific name: *atokensis* ← {(place-name) Atoka County + ensis}

Locality: **high-spined lizard** from Atoka County

Taxonomy: Theropoda: Carnosauria: Allosauroidea: Carcharodontosauridae
cf. Theropoda: Carnosauria: Allosauroidea: Allosauridae[*1][*2]

Other species:

> *Acrocanthosaurus altispinax* Paul, 1988 ⇒ *Becklespinax altispinax* (Paul, 1988) Olshevsky, 1991 ⇒ *Altispinax altispinax* (Paul, 1988) Rauhut, 2003 [Wealden Group, England, UK]

Notes: According to Currie & Carpenter (2000), *Acrocanthosaurus* is more closely related to Allosauridae than to Carcharodontosauridae.[*2]

[References: (*1) Stovall JW, Langston W (1950). *Acrocanthosaurus atokensis*, a new genus and species of Lower Cretaceous Theropoda from Oklahoma. *American Midland Naturalist* 43: 696–728.; (*2) Currie PJ, Carpenter K (2000). A new specimen of *Acrocanthosaurus atokensis* (Theropoda, Dinosauria) from the Lower Cretaceous Antlers Formation (Lower Cretaceous, Aptian) of Oklahoma, USA. *Geodiversitas* 22: 207–246.]

Acrotholus audeti

> *Acrotholus audeti* Evans *et al.*, 2013 [Santonian; Milk River Formation, Alberta, Canada]

Generic name: *Acrotholus* ← {(Gr.) ἄκρος/akros: highest + θόλος/tholos: dome}; "in reference to its greatly thickened cranial vault"[*1].

Specific name: *audeti* ← {(person's name) Audet + -ī}; in honor of "Roy Audet, for allowing access to his ranch where the holotype skull was found"[*1].

Etymology: Audet's **highest dome**

Taxonomy: Ornithischia: Marginocephalia: Pachycephalosauridae

[References: (*1) Evans DC, Schott RK, Larson DW, Brown CM, Ryan NJ (2013). The oldest North American pachycephalosaurid and the hidden diversity of small-bodied ornithischian dinosaurs. *Nature Communications* 4. 1828.]

Adamantisaurus mezzalirai

> *Adamantisaurus mezzalirai* Santucci & Bertini, 2006 [Campanian–Maastrichtian; Adamantina Formation, São Paulo State, Brazil]

Generic name: *Adamantisaurus* ← {(place-name) Adamantina Formation + (Gr.) σαῦρος/sauros: lizard}; referring to "the Adamantina Formation in western São Paulo State from which this specimen was collected"[*1].

Specific name: *mezzalirai* ← {(person's name) Mezzalira + -ī}; "in honor of Dr. Sérgio Mezzalira, the researcher who collected and first mentioned the remains in the literature"[*1].

Etymology: Mezzalira's **Adamantina lizard**

Taxonomy: Saurischia: Sauropodomorpha: Sauropoda: Titanosauriformes: Titanosauria

[References: (*1) Santucci RA, Bertini RJ (2006). A new titanosaur from western São Paulo State, Upper Cretaceous Bauru Group, south-east Brazil. *Palaeontology* 49: 59–66.]

Adasaurus mongoliensis

> *Adasaurus mongoliensis* Barsbold, 1983 [Maastrichtian; Nemegt Formation, Ömnögovi, Mongolia]

Generic name: *Adasaurus* ← {(Mong. myth.) Ada: "evil"[*1] + (Gr.) σαῦρος/sauros: lizard}

Specific name: *mongoliensis* ← {(place-name) Mongolia + -ensis}

Etymology: **Ada (evil spirit)'s lizard** from Mongolia

Taxonomy: Theropoda: Coelurosauria: Dromaeosauridae

[References: (*1) Barsbold R (1983). Carnivorous dinosaurs from the Cretaceous of Mongolia. *Transactions of the Joint Soviet-Mongolian Paleontological Expedition* (in Russian) 19: 5–119.]

Adelolophus hutchisoni

> *Adelolophus hutchisoni* Gates *et al.*, 2014

[Campanian; Wahweap Formation, Utah, USA]
Generic name: *Adelolophus* ← {(Gr.) ἄδηλος/adēlos: unknown + λοφος/lophos: crest}; "in reference to unknown morphology of the skull and crest of this taxon"(*1).
Specific name: *hutchisoni* ← {(person's name) Hutchison + -ī}; in honor of Dr. Howard Hutchison, discoverer of the type specimen and longtime proponent of southern Utahan vertebrate paleontology.(*1)
Etymology: Hutchison's **unknown crest**
Taxonomy: Ornithischia: Ornithopoda: Hadrosauridae: Lambeosaurinae
[References: (*1) Gates TA, Jinnah Z, Levitt C, Getty MA (2014). New hadrosaurid (Dinosauria, Ornithopoda) specimens from the lower-middle Campanian Wahweap Formation of southern Utah. In: Eberth DA, Evans DC (eds), *Hadrosaurus: Proceedings of the International Hadrosaur Symposium*. Indiana University Press, Bloomington: 156–173.]

Adeopapposaurus mognai
> *Adeopapposaurus mognai* Martínez, 2009 [Early Jurassic; Cañón del Colorado Formation, San Juan, Argentina]
Generic name: *Adeopapposaurus* ← {(Lat.) adeo: far + pappo- [pappo: to eat] + (Gr.) σαῦρος/sauros: lizard}; "far eating lizard", in reference to "the long neck characteristic of this taxon" (*1).
Specific name: *mognai* ← {(place-name) Mogna + -ī}; referring to "the locality of Mogna in San Juan Province, Argentina". (*1)
Etymology: Mogna's **far eating lizard**
Taxonomy: Saurischia: Sauropodomorpha: Prosauropoda: Massospondylidae
[References: (*1) Martínez, RN (2009). *Adeopapposaurus mognai*, gen. *et* sp. nov. (Dinosauria: Sauropodomorpha), with comments and adaptations of basal Sauropodomorpha. *Journal of Vertebrate Paleontology* 29: 142–164.]

Adratiklit boulahfa
> *Adratiklit boulahfa* Maidment *et al.*, 2019 [Bathonian; El Mers II Formation, Fès-Meknes, Morocco]
Generic name: *Adratiklit* ← {(Berber) Adras: mountain + tiklit: lizard} (*1)
The remains were discovered in the Middle Atlas Mountains of Morocco.
Specific name: *boulahfa* ← {(place-name) Boulahfa} (*1)
Etymology: **mountain lizard** from Boulahfa
Taxonomy: Ornithischia: Stegosauria
[References: (*1) Maidment SCR, Raven TJ, Quarhache D, Barrett PM (2019). North Africa's first stegosaur: implications for Gondwanan thyreophoran dinosaur diversity. *Gondwana Research* 77: 82–97.]

Adynomosaurus arcanus
> *Adynomosaurus arcanus* Prieto-Márquez *et al.*, 2019 [Maastrichtian; Conques Formation, Catalonia, Spain]
Generic name: *Adynomosaurus* ← {(Gr.) adyn- [ἀ-δύνᾰμος/a-dynamos: weak] + ὦμος/ōmos: shoulder + σαῦρος/sauros: lizard}; referring to "a scapula with a relatively unexpanded blade"(*1).
Specific name: *arcanus* ← {(Lat.) arcānus: secret, occult}
Etymology: secret, **weak shoulder lizard**
Taxonomy: Ornithischia: Ornithopoda: Hadrosauridae: Lambeosaurinae
[References: (*1) Prieto-Márquez A, Fondevilla V, Sellés AG, Wagner JR, Galobart À (2019). *Adynomosaurus arcanus*, a new lambeosaurine dinosaur from the Late Cretaceous Ibero-Armorican Island of the European archipelago. *Cretaceous Research* 96: 19–37. Abstract.]

Aegyptosaurus baharijensis
> *Aegyptosaurus baharijensis* Stromer, 1932 [Cenomanian; Baharîje Formation, Marsah Matruh, Egypt]
Generic name: *Aegyptosaurus* ← {(Gr.) Αἴγυπτος/Aigyptos: Egypt + σαῦρος/sauros: lizard}; according to the Egyptian deserts where remains were discovered.
Specific name: *baharijensis* ← {(place-name) Bahariya Formation + -ensis} (*1)
Etymology: **Egypt lizard** from Bahariya
Taxonomy: Saurischia: Sauropodomorpha: Sauropoda: Titanosauria
Notes: The fossils have been found in several locations in the Sahara Desert. In 1944 during World War II, the fossils, which were stored in Munich, were destroyed by the Allied bombing raid.(*2)
[References: (*1) Stromer, E. (1932). Ergebnisse der Forschungsreisen Prof. E. Stromers in den Wüsten Ägyptens. II.Wibeltierreste der Baharije-Stufe (unterstes Cenoman). 11. Sauropoda. *Abhandlungen der Bayerischen Akademie der Wissenschaften Mathematisch-Naturwissenschaftiche Abteliung, Neue Folge* 10: 1–21.; (*2) Smith JB, Lamanna MC, Lacovara KJ, Dodson P, Smith JR, Poole JC, Giegengack R, Attia Y. (2001). A giant sauropod dinosaur from an Upper Cretaceous mangrove deposit in Egypt. *Science* 292: 1704–1706.]

Aeolosaurus rionegrinus
> *Aeolosaurus rionegrinus* Powell, 1987 [Campanian–Maastrichtian; Angostura

Colorada Formation, Río Negro, Argentina]

Generic name: *Aeolosaurus* ← {(Gr. myth.) Αἴόλος/Aiolos: Aeolus, lord of the winds + σαῦρος/sauros: lizard}; referring to windy Patagonia.

Specific name: *rionegrinus* ← {(place-name) Río Negro + -inus}; referring to the fossil locality, in the Río Negro Province.

Etymology: **Aeolus' lizard** from Río Negro Province

Taxonomy: Saurischia: Sauropodomorpha: Sauropoda: Titanosauria: Aeolosaurini

Other species:

> *Aeolosaurus colhuehuapensis* Casal *et al.*, 2007 [Campanian; Lago Colhué Huapi Formation, Chubut, Argentina]

> *Aeolosaurus maximus* Santucci & De Arruda-Campos, 2011 ⇒ *Arrudatitan maximus* (Santucci & De Arruda-Campos, 2011) Silva Junior, 2021 [Campanian; Adamantina Formation, São Paulo, Brazil]

[References: Powell JE (1987). The Late Cretaceous fauna of Los Alamitos, Patagonia, Argentina part VI.The titanosaurids. *Revista del Museo Argentino de Siencias Naturales* 3: 111–142.]

Aepisaurus elephantinus

> *Aepisaurus elephantinus* Gervais, 1852 [Albian; Mont Ventoux Formation, Provence-Alpes-Côte-d'Azur, France]

This generic name has two spellings, *Aepisaurus* and *Aepysaurus*, the latter is used in *The Dinosauria 2nd edition.*

Generic name: *Aepisaurus* ← {(Gr.) αἰπεινός/aipeinos: lofty, high + σαῦρος/sauros: lizard}

Aepisaurus was named for an isolated humerus which was 90 cm long.

Specific name: *elephantinus*

cf. (Gr.) ἐλεφάντῐνος/elephantinos: of ivory [ἐλέφας/elephas: the elephant, the elephant's tusk, ivory]

Etymology: elephantine, **lofty lizard**

Taxonomy: Saurischia: Sauropodomorpha: Sauropoda

Notes: A single humerus, which Gervais described the genus based on, seems to be now lost. According to *The Dinosauria 2nd edition*, *Aepysaurus elephantinus* is considered to be a *nomen dubium*.

Aepyornithomimus tugrikinensis

> *Aepyornithomimus tugrikinensis* Tsogtbaatar *et al.*, 2017 [Campanian; Djadokhta Formation, Ömnögovi, Mongolia]

Generic name: *Aepyornithomimus* ← {*Aepyornis** + (Lat.) mīmus: 'as' or 'like'}; referring to "the largest ratite bird *Aepyornis*, which has similar pes structure" (*1).

*a gigantic extinct flightless bird found in Madagascar, called elephant bird.

Specific name: *tugrikinensis* ← {(place-name) Tögrögiin Shiree locality + -ensis}; "referring to the locality where the specimen was found" (*1).

Etymology: ***Aepyornis* mimic** from Tögrögiin Shiree locality

Taxonomy: Theropoda: Ornithomimosauria: Ornithomimidae

Notes: Tögrögiin Shiree is the locality where a famous find known as "Fighting Dinosaurs" and an abundance of fossils were discovered.

[References: (*1) Tsogtbaatar C, Kobayashi Y, Khishigjav T, Currie PJ, Watabe M, Rinchen B (2017). First ornithomimid (Theropoda, Ornithomimosauria) from the Upper Cretaceous Djadokhta Formation of Tögrögiin Shiree, Mongolia. *Scientific Reports* 7: 5853.]

Aerosteon riocoloradense

> *Aerosteon riocoloradense* Sereno *et al.*, 2009 [Santonian; Anacleto Formation, Mendoza, Argentina]

Generic name: *Aerosteon* ← {(Gr.) ἀήρ/aēr: air + ὀστέον/osteon (neuter gender): bone}; "named for the extreme development of pneumatic spaces* in skeletal bone" (*1) *spaces = cavities

Specific name: *riocoloradense* ← {(place-name) Río Colorado + -ense (neutral form of -ensis)}; "named for the site of discovery of the holotype" (*1)

Etymology: **air bone** from Río Colorado

Taxonomy: Theropoda: Megaraptora: Megaraptoridae

cf. Theropoda: Tetanurae: Allosauroidea (*1)

[References: (*1) Sereno PC, Martinez RN, Wilson JA, Varricchio DJ, Alcober OA, Larsson HCE (2008). Kemp T (ed), Evidence for avian intrathoracic air sacs in a new predatory dinosaur from Argentina. *PLoS ONE* 3(9): e3303.]

Afromimus tenerensis

> *Afromimus tenerensis* Sereno, 2017 [Aptian–Albian; Elrhaz Formation, Ténéré Desert, Agadez, Niger]

Generic name: *Afromimus* ← {(Lat. place-name) Āfro-: Africa + (Gr.) μῖμος/mīmos: mimic}

Specific name: *tenerensis* ← {(Fr. place-name) Ténéré Desert + (Lat.) -ensis: from (*1)}

Etymology: **African mimic** from Ténéré Desert

Taxonomy: Theropoda: Coelurosauria: Ornithomimosauria

[References: (*1) Sereno PC (2017). Early Cretaceous ornithomimosaurs (Dinosauria: Coelurosauria) from Africa. *Ameghiniana* 54(5): 576–616.]

Afrovenator abakensis

> *Afrovenator abakensis* Sereno *et al.*, 1994 [Bathonian; Tiourarén Formation, Abaka, Agadez, Niger]

Generic name: *Afrovenator* ← {Āfro- [(Lat. place-name) Āfer: African] + (Lat.) vēnātor: hunter}

Specific name: *abakensis* ← {(Tuareg, place-name) Abaka + -ensis}

Etymology: **African hunter** from Abaka

Taxonomy: Theropoda: Tetanurae:
Megalosauroidea: Megalosauridae

[References: Sereno PC, Wilson JF, Larsson HCE, Dutheil DB, Sues H-D (1994). Early Cretaceous dinosaurs from Sahara. *Science* 266: 267–270.; Rauhut OWM, Lopez-Arbarello A (2009). Consideration on the age of the Tiourarén Formation (Iullemmeden Basin, Niger, Africa): Implications for Gondwanan Mesozoic terrestrial vertebrate faunas. *Palaeogeography, Palaeoclimatology, Palaeoecology* 271: 259–267.]

Agathaumas sylvestris

> *Agathaumas sylvestris* Cope, 1872 (*nomen dubium*) [Maastrichtian; Lance Formation, Wyoming, USA]

Generic name: *Agathaumas* ← {(Gr.) ἄγαν/agan: great, much + θαῦμα/thauma: wonder}

Specific name: *sylvestris* ← {(Lat.) silvestris: of forest}.

Etymology: **great wonder** of forest (or **marvelous** forest-**dweller**)

Taxonomy: Ornithischia: Ceratopsia: Ceratopsidae:
Chasmosaurinae: Triceratopsini

Other species:

> *Thespesius occidentalis* Leidy, 1856 (*nomen dubium*) ⇒ *Hadrosaurus occidentalis* (Leidy, 1856) Cope, 1874 (*nomen dubium*) ⇒ *Trachodon occidentalis* (Leidy, 1856) Kuhn, 1936 (*nomen dubium*) [Maastrichtian; Lance Formation, South Dakota, USA]

= *Agathaumas milo* Cope, 1874 ⇒ *Hadrosaurus milo* (Cope, 1874) Hay, 1901 [Maastrichtian; Laramie Formation, Colorado, USA]

> *Triceratops horridus* Marsh, 1889 [Maastrichtian; Lance Formation, Wyoming, USA]

= *Polyonax mortuarius* Cope, 1874 (*nomen dubium*) ⇒ *Agathaumas mortuarius* (Cope, 1874) Hay, 1901 ⇒ *Triceratops mortuarius* (Cope, 1874) Kuhn, 1936 [Maastrichtian; Laramie Formation, Colorado, USA]

= *Triceratops flabellatus* Marsh, 1889 ⇒ *Sterrholophus flabellatus* (Marsh, 1889) Marsh, 1891 ⇒ *Agathaumas flabellatus* (Marsh, 1889) Burkhardt, 1892 [Maastrichtian; Lance Formation, Wyoming, USA]

> *Monoclonius crassus* Cope, 1876 [Campanian; Judith River Formation, Montana, USA]

?= *Monoclonius sphenocerus* Cope, 1889 ⇒ *Agathaumas sphenocerus* (Cope, 1889) Ballou, 1897 (*nomen dubium*) [Campanian; Judith River Formation, Montana, USA]

= *Agathaumas monoclonius* Breihaupt, 1994 (*nomen dubium*)

> *Triceratops prorsus* Marsh, 1890 ⇒ *Agathaumas prorsus* (Marsh, 1890) Lydekker, 1893 [Maastrichtian; Lance Formation, Wyoming, USA]

Notes: According to *The Dinosauria 2nd edition*, *Agathaumas sylvestris* is considered a *nomen dubium*.

Agilisaurus louderbacki

> *Agilisaurus louderbacki* Peng, 1990 [Bathonian–Callovian; lower Shaximiao (= Xiashaximiao) Formation, Dashanpu, Zigong, Sichuan, China]

Generic name: *Agilisaurus* ← {(Lat.) agilis: agile, nimble + (Gr.) σαῦρος/sauros: lizard}; "for an agile bipedal animal, as indicated by the light structure of the skeleton and limb ratios" (*1).

Specific name: *louderbacki* ← {(person's name) Louderback + -ī}; "in honor of the late U. S. geologist Dr. George D. Louderback, the first scientist to discover dinosaur fossils in the Sichuan Basin in 1915" (*1).

Etymology: Louderback's **agile lizard**

Taxonomy: Ornithischia: Neornithischia
cf. Ornithischia: Ornithopoda: Fabrosauridae (*1)

Other species:

> *Yandusaurus multidens* He & Cai, 1983 ⇒ *Agilisaurus multidens* (He & Cai, 1983) Peng, 1992 ⇒ *Othnielia multidens* (He & Cai, 1983) Paul, 1996 ⇒ *Hexinlusaurus multidens* (He & Cai, 1983) Barrett *et al.*, 2005 [Oxfordian; Xiashaximiao Formation, Sichuan, China]

Notes: The skeleton which was discovered by Louderback in 1915 was named *A. louderbacki* by Peng in 1990 and described in 1992. (*1)

[References: Peng G (1990). A new small ornithopod (*Agilisaurus louderbacki* gen. *et* sp. nov.) from Zigong, China. *Newsletter of the Zigong Dinosaur Museum* 2: 19–27.; (*1) Peng G (1992). Jurassic ornithopod *Agilisaurus louderbacki* (Ornithopoda: Fabrosauridae) from Zigong, Sichuan, China. *Vertebrata PalAsiatica* 30(1): 39–51.]

Agnosphitys cromhallensis [Silesauridae]

> *Agnosphitys cromhallensis* Fraser *et al.*, 2002 [Norian–Rhaetian; Penarth Group, Avon, UK]

Generic name: *Agnosphitys* ← {(Gr.) ἀγνώς/agnōs: unknown + phitys: begetter}; "with reference to the position of the new form relative to the

Dinosauria".

Specific name: *cromhallensis* ← {(place-name) Cromhall + -ensis}; referring to the locality.

Etymology: **unknown begetter** from Cromhall

Taxonomy: Dinosauromorpha: Silesauridae (according to Baron *et al.*, 2017)
cf. Saurischa: Sauropodomorpha: Guaibasauridae (according to Ezcurra, 2010)
cf. Archosauromorpha: Archosauria: Dinosauromorpha [*1]

Notes: According to *The Dinosauria 2nd edition*, *Agnosphitys cromhallensis* was regarded as a *nomen dubium*. Baron *et al.* (2017) recovered it as a member of the clade Silesauridae.

[References: (*1) Fraser NC, Padian K, Walkden GM, Davis ALM (2002). Basal dinosauriform remains from Britain and the diagnosis of the Dinosauria. *Paleontology* 45(1): 79–95.]

Agrosaurus macgillivrayi

> *Agrosaurus macgillivrayi* Seeley, 1891 ⇒ (*nomen dubium*) or *Thecodontosaurus macgillivrayi* (Seeley, 1891) von Huene, 1906 [Rhaetian; Magnesian Conglomerate Formation, Durdham Down, Bristol, UK]

Generic name: *Agrosaurus* ← {(Gr.) ἀγρός/agros: field + σαῦρος/sauros: lizard}

Specific name: *macgillivrayi* ← {(person's name) MacGillivray + -ī}; in honor of Mr. MacGillivray who collected the specimens during the voyage of the 'Fly' [*1].

Etymology: MacGillivray's **field lizard**

Taxonomy: Saurischia: Sauropodomorpha

Notes: The remains of *Agrosaurus* were originally believed to be collected in Australia by the crew of HMS Fly [*1]. But, the materials of *Agrosaurus macgillivrayi* came from Durdham Down, not the northeastern coast of Australia [*2]. According to *The Dinosauria 2nd edition*, *Agrosaurus* is considered to be a *nomen dubium*.

[References: (*1) Seeley HG (1891). On *Agrosaurus macgillivrayi* (Seeley), a saurischian reptile from the N. E. coast of Australia. *Quarterly Journal of the Geological Society of London* 47: 164–165.; (*2) Galton P (2007). Notes on the remains of archosaurian reptiles, mostly basal sauropodomorph dinosaurs, from the 1834 fissure fill (Rhaetian, Upper Triassic) at Clifton in Bristol, southwest England. *Revue de Paléobiologie* 26(2): 505–591.]

Agujaceratops mariscalensis

> *Chasmosaurus mariscalensis* Lehman, 1989 ⇒ *Agujaceratops mariscalensis* (Lehman, 1989) Lucas *et al.*, 2006 [Campanian; Aguja Formation, Texas, USA]

Generic name: *Agujaceratops* ← {(place-name) Aguja Formation + ceratops [(Gr.) κέρας/keras: horn + ὤψ/ōps: face]}

Specific name: *mariscalensis* ← {(place-name) Mariscal + -ensis}; named for Mariscal Mountain, in the southern part of Big Bend National Park.

Etymology: **Aguja horned face (ceratopsian)** from Mariscal

Taxonomy: Ornithischia: Ceratopsia: Ceratopsidae: Chasmosaurinae

Other species:

> *Agujaceratops mavericus* Lehman *et al.*, 2016 [Campanian; Aguja Formation, Texas, USA]

[References: Lucas SG, Sullivan RM, Hunt P (2006). Re-evaluation of *Pentaceratops* and *Chasmosaurus* (Ornithischia: Ceratopsidae) in the Upper Cretaceous of the Western Interior. *New Mexico Museum of Natural History and Science Bulletin* 35: 367–370.]

Agustinia ligabuei

> *Agustinia ligabuei* Bonaparte, 1999 [Aptian; Lohan Cura Formation, Neuquén, Argentina]

Generic name: *Agustinia* ← {(person's name) Agustin + -ia}; in honor of "the young student Agustin Martinelli, a member of the paleontological team and discoverer of the specimem" [*1].

Specific name: *ligabuei* ← {(person's name) Ligabue + -ī}; in honor of "Dr. Giancarlo Ligabue, from Venezia, an active philanthropist, who supported the 1997 expedition to Patagonia" [*1].

Etymology: Ligabue's one **for Agustin**

Taxonomy: Saurischia: Sauropodomorpha: Sauropoda: Agustiniidae

[References: (*1) Bonaparte JF (1999). An armoured sauropod from the Aptian of Northern Patagonia, Argentina. In: Tomida Y, Rich TH, Vickers-Rich P (eds), *Proceedings of the Second Gondwanan Dinosaur Symposium Tokyo, National Science Museum Monographs* No. 15: 1–12.]

Ahshislepelta minor

> *Ahshislepelta minor* Burns & Sullivan, 2011 [late Campanian; Kirtland Formation, San Juan Basin, New Mexico, USA]

Generic name: *Ahshislepelta* ← {(place-name) Ah-shi-sle-pah Wash + (Lat.) pelta* = (Gr.) πέλτη/peltē: shield}; "derived from the locality of the holotype, Ah-shi-sle-pah Wash (formerly Meyers Creek**), San Juan Basin, New Mexico" [*1].
*The term *pelta* is commonly used for the generic name of ankylosaurs.
**acc. Sullivan (2006).

Specific name: *minor* ← {(Lat.) minor: smaller}; "in reference to its small adult size relative to other North American ankylosaurids of similar age" [*1].

Etymology: lesser **shield (ankylosaur) from Ah-**

shi-sle-pah Wash

Taxonomy: Ornithischia: Ankylosauria: Ankylosauridae: Ankylosaurinae (*1)

[References: (*1) Burns M, Sullivan RM (2011). A new ankylosaurid from the Upper Cretaceous Kirtland Formation, San Juan Basin, with comments of the diversity of ankylosaurids in New Mexico. *Fossil Record 3. New Mexico Museum of Natural History and Science, Bulletin* 53: 169–178.]

Ajancingenia yanshini

> "*Ingenia*" *yanshini* Barsbold, 1981 (preoccupied) ⇒ *Oviraptor yanshini* (Barsbold, 1981) Paul, 1988* ⇒ *Ajancingenia yanshini* (Barsbold, 1981) Easter, 2013 ⇒ *Heyuannia yanshini*? (Barsbold, 1981) Funston *et al.*, 2018 [Campanian; Barun Goyot Formation, Ömnögovi, Mongolia]

*According to Yun (2019), this classification has not been supported by any subsequent authors.

Generic name: *Ajancingenia* ← {(Mong.) аянч/ajanc: traveler + Ingenia [(place-name) Ingen Khoboor + -ia]}; as a Western allusion of sticking one's thumb out for hitchhiking, in reference to the first manual ungula of *Ajancingenia* which is twice as large as the second. (*1)

Specific name: *yanshini* ← {(person's name) Yanshin + -ī}; in honor of A. L. Yanshin.

Etymology: Yanshin's **traveler of Ingen Khoboor**

Taxonomy: Theropoda: Coelurosauria: Oviraptorosauria

[References: (*1) Easter J (2013). A new name for the oviraptorid dinosaur "*Ingenia*" *yanshini* (Barsbold, 1981, preoccupied by Gerlach, 1957). *Zootaxa* 3737(2):184–190.]

Ajkaceratops kozmai

> *Ajkaceratops kozmai* Ősi *et al.*, 2010 [Santonian; Csehbánya Formation, Ajka, Veszprém, Hungary]

Generic name: *Ajkaceratops* ← {(place-name) Ajka + ceratops [(Gr.) κέρας/keras: horn + ὤψ/ōps: face]} (*1)

Specific name: *kozmai* ← {(person's name) Kozma + -ī}; in honor of Károly Kozma (*1).

Etymology: Kozma's **Ajka horned face (ceratopsian)**

Taxonomy: Ornithischia: Ceratopsia: Coronosauria: Bagaceratopidae

cf. Ceratopsia: Neoceratopsia: Coronosauria (*1)

[References: (*1) Ősi A, Butler RJ, Weishampel DB (2010). A Late Cretaceous ceratopsian dinosaur from Europe with Asian affinities. *Nature* 465(7297): 466–468.]

Akainacephalus johnsoni

> *Akainacephalus johnsoni* Wiersma & Irmis, 2018 [upper Campanian; Kaiparowits Formation, Utah, USA]

Generic name: *Akainacephalus* ← {(Gr.) ἄκαινα/akaina: thorn, spine + κεφαλή/kephalē: head}; "referring to the thorn-like cranial caputegulae of the holotype".

Specific name: *johnsoni* ← {(person's name) Johnson + -i}; "in honor of Randy Johnson, volunteer preparator at the Natural History Museum of Utah, who skillfully prepared the skull and lower jaws of UMNH VP 20202".

Etymology: Johnson's **thorny head**

Taxonomy: Ornithischia: Ankylosauria: Ankylosauridae: Ankylosaurini

[References: Wiersma JP, Irmis RB (2018). A new southern Laramidian ankylosaurid, *Akainacephalus johnsoni* gen. *et* sp. nov., from the upper Campanian Kaiparowits Formation of southern Utah, USA. *PeerJ* 6: e5016.]

Alamosaurus sanjuanensis

> *Alamosaurus sanjuanensis* Gilmore, 1922 [Maastrichtian; Ojo Alamo Formation, San Juan Basin, New Mexico, USA]

Generic name: *Alamosaurus* ← {(place-name) Ojo Alamo | (stratal name) Ojo Alamo Formation + (Gr.) σαῦρος/sauros: lizard}

Specific name: *sanjuanensis* ← {(place-name) San Juan + -ensis}

The first remains were discovered in the Barrel Spring Arroyo, one mile south of Ojo Alamo, San Juan County, New Mexico.

Etymology: **Ojo Alamo lizard** from San Juan

Taxonomy: Saurischia: Sauropoda: Titanosauria: Saltasauridae: Opisthocoelicaudiinae

cf. Sauropoda (*1)

[References: (*1) Gilmore CW (1922). A new sauropod dinosaur from the Ojo Alamo Formation of New Mexico. *Smithsonian Miscellaneous Collections* 72(14): 1–9.; Lucas SG, Sullivan RM (2000). The sauropod dinosaur *Alamosaurus* from the Upper Cretaceous of the San Juan Basin, New Mexico. *Dinosaurs of New Mexico Museum of Natural History and Science Bulletin* 17: 147–156.]

Alanqa saharica [Pterosauria]

> *Alanqa saharica* Ibrahim *et al.*, 2010 [Cenomanian; Kem Kem Beds, Errachidia, Morocco]

Generic name: *Alanqa* ← {(Arabic) Al Anqa}; a phoenix-like mythological flying creature from ancient tmes, similar to the Persian simurgh. (*1)

Specific name: *saharica* ← {(Arabic) Sahara: Desert

+ (Gr.) -ica: belonging to[*1]}
Etymology: **Al Anqa** of Sahara
Taxonomy: Pterosauria: Azhdarchoidea: Azhdarchidae
Notes: A wingspan of 6 m. [*1].

[References: (*1) Ibrahim N, Unwin DM, Martill DM, Baidder L, Zouhri S (2010). A new pterosaur (Pterodactyloidea: Azhdarchidae) from the Upper Cretaceous of Morocco. *PLoS ONE* 5(5): e10865.]

Alaskacephale gangloffi

> *Alaskacephale gangloffi* Sullivan, 2006 [Campanian–Maastrichtian; Prince Creek Formation, Alaska, USA]

Generic name: *Alaskacephale* ← {(place-name) Alaska + (Gr.) κεφαλή/kephalē: head}

A common suffix *cephale* is used to denote members of the Pachycephalosauridae.

Specific name: *gangloffi* ← {(person's name) Gangloff + -ī}; in honor of "Roland Gangloff who, in part, reported on the occurrence of the holotype and who has contributed significantly to the understanding of dinosaurs of the North American Arctic region"[*1].
Etymology: Gangloff's **Alaska head (Pachycephalosaur)**
Taxonomy: Ornithischia: Pachycephalosauridae
Notes: According to Sullivan, *Alaskacephale* had two divergent rows of nods on the squamosal, covering toward the midline of the skull. [*1]

[References: (*1) Sullivan RM (2006). A taxonomic review of the Pachycephalosauridae (Dinosauria: Ornithischia). *New Mexico Museum of Natural History and Science Bulletin* 35: 347–365.; Gangloff RA, Fiorillo AR, Norton DW (2005). The first pachycephalosaurine (Dinosauria) from the Paleo-Arctic of Alaska and its paleogeographic implications. *Journal of Paleontology* 79(5): 997–1001.]

Albalophosaurus yamaguchiorum

> *Albalophosaurus yamaguchiorum* Ohashi & Barrett, 2009 [Valanginian; Kuwajima Formation, Tetori Group, Hakusan, Ishikawa, Japan]

Generic name: *Albalophosaurus* ← {(Lat.) albus: white + (Gr.) λόφος/lophos: crest* + σαῦρος/sauros: lizard}; referring to the snow-covered crest of Mount Hakusan (白山) meaning 'white mountain' in Japanese and to the prominent ridges present on the maxillary and denary teeth of the holotype specimen, according to the authors.

*The original name of the village, Shiramine (白峰) meaning 'white crest' was amended to the new name of the city, 白山"white mountain" in February, 2005.

Specific name: *yamaguchiorum* ← {(Jpn. person's name) Yamaguchi + -ōrum}; in honor of Ichio Yamaguchi (山口一男) and Mikiko Yamaguchi (山口ミキ子), who have discovered and prepared many fossils from the Kuwajima Formation.
Etymology: Mr. Ichio Yamaguchi and Ms. Mikiko Yamaguchi's **white crest lizard**
Taxonomy: Ornithischia: Cerapoda
Notes: Mr. Yoshinori Kobayashi (小林美徳) discovered the remains at Kuwajima at Hakusan City, in June 1998.

[References: Ohashi T, Brrett PM (2009). A new ornithischian dinosaur from the Lower Cretaceous Kuwajima Formation of Japan. *Journal of Vertebrate Paleontology* 29(3): 748–757. Abstract.]

Albertaceratops nesmoi

> *Albertaceratops nesmoi* Ryan, 2007 [Campanian; Oldman Formation, Alberta, Canada]

Generic name: *Albertaceratops* ← {(place-name) Alberta: name of province where the holotype skull was discovered + ceratops: horned face [(Gr.) κερατ-/kerat- [κέρας/keras: horn] + ὤψ/ōps: face]} [*1]
Specific name: *nesmoi* ← {(person's name) Nesmo + -ī}; in honor of "Cecil Nesmo, a rancher from southern Alberta, whose assistance and hospitality has facilitated the collection of many important paleontological specimens, including the holotype of *Albertaceratops* n. gen." [*1].
Etymology: Nesmo's **Alberta horned face (ceratopsian)**
Taxonomy: Ornithischia: Marginocephalia: Ceratopsia: Neoceratopsia: Ceratopsidae: Centrosaurinae
Notes: *Albertaceratops* had long brow horns.

[References: (*1) Ryan MJ (2007). A new basal centrosaurine ceratopsid from the Oldman Formation, southeastern Alberta. *Journal of Paleontology* 81(2): 376–396.]

Albertadromeus syntarsus

> *Albertadromeus syntarsus* Brown *et al.*, 2013 [Campanian; Oldman Formation, Alberta, Canada]

Generic name: *Albertadromeus* ← {(place-name) Alberta + (Gr.) δρομεύς/dromeus: runner}
Specific name: *syntarsus* ← {(Gr.) σύν/syn: together with, joined + ταρσός/tarsos: ankle}; referring to the autapomorphic condition of the distal fibula.
Etymology: **Alberta runner** with fused tarsus
Taxonomy: Ornithischia: Ornithopoda: Thescelosauridae / Parksosauridae

[References: Brown CM, Evans DC, Ryan MJ. Russell AP (2013). New data on the diversity and abundance of small-bodied ornithopods (Dinosauria, Ornithischia) from the Belly River Group (Campanian) of Alberta. *Journal of Vertebrare Paleontology* 33: 495–520.]

Albertavenator curriei

> *Albertavenator currei* Evans *et al.*, 2017 [Maastrichtian; Horseshoe Canyon Formation, Alberta, Canada]

Generic name: *Albertavenator* ← {(place-name) Alberta + (Lat.) vēnātor: hunter}

Specific name: *curriei* ← {(person's name) Currie + -ī}; named after renowned Canadian palaeontologist Dr. Philip J. Currie.

Etymology: Currie's **Alberta hunter**

Taxonomy: Theropoda: Troodontidae

[References: Evans DC (2017). A new species of troodontid theropod (Dinosauria: Maniraptora) from the Horseshoe Canyon Formation (Maastrichtian) of Alberta, Canada. *Canadian Journal of Earth Sciences* 54(8): 813–826.]

Albertonykus borealis

> *Albertonykus borealis* Longrich & Currie, 2009 [early Maastrichtian; Horseshoe Canyon Formation, Alberta, Canada]

Generic name: *Albertonykus* ← {(place-name) Alberta + (Gr.) ὄνυξ/onyx: claw}

Specific name: *borealis* ← {(Latin) boreālis: northern}

Etymology: northern **Alberta claw**

Taxonomy: Theropoda: Maniraptora: Alvarezsauridae

Notes: It is seemed to be possible that "*Albertonykus* preyed on wood-nesting termites" with its forelimbs. (*1)

[References: (*1) Longrich NR, Currie PJ (2009). *Albertonykus borealis*, a new alvarezsaur (Dinosauria: Theropoda) from the early Maastrichtian of Alberta, Canada: implications for the systematic and ecology of the Alvarezsauridae. *Cretaceous Research* 30(1): 230–252.]

Albertosaurus sarcophagus

> *Albertosaurus sarcophagus* Osborn, 1905 ⇒ *Deinodon sarcophagus* (Osborn, 1905) Matthew & Brown, 1922 [Campanian; Horseshoe Canyon Formation, Alberta, Canada]

= *Albertosaurus arctunguis* Parks, 1928 ⇒ *Deinodon arctunguis* (Parks, 1928) Kuhn, 1939 [Campanian; Horseshoe Canyon Formation, Canada]

Generic name: *Albertosaurus* ← {(place-name) Alberta + (Gr.) σαῦρος/sauros: lizard}

Specific name: *sarcophagus* ← {(Gr.) σαρκο-φάγος/sarko-phagos: carnivorous}

Etymology: carnivorous, **Alberta lizard**

Taxonomy: Theropoda: Coelurosauria: Tyrannosauridae: Albertosaurinae

Other species:

> *Gorgosaurus libratus* Lambe, 1914 ⇒ *Albertosaurus libratus* (Lambe, 1914) Russell, 1970 [late Campanian; Dinosaur Park Formation, Alberta, Canada]

= *Gorgosaurus sternbergi* Matthew & Brown, 1923 ⇒ *Deinodon sternbergi* (Matthew & Brown, 1923) Kuhn, 1965 ⇒*Albertosaurus sternbergi* (Matthew & Brown, 1923) Russell, 1970 [late Campanian; Dinosaur Park Formation, Alberta, Canada]

> *Gorgosaurus lancensis* Gilmore, 1946 ⇒ *Albertosaurus lancensis* (Gilmore, 1946) Russell, 1970 ⇒ *Nanotyrannus lancensis* (Gilmore, 1946) Bakker *et al.*, 1988 [Maastrichtian; Hell Creek Formation, Montana, USA]

> *Tyrannosaurus rex* Osborn, 1905 [Maastrichtian; Lance Formation, Montana, USA]

= *Albertosaurus megagracilis* Paul, 1988 [Maastrichtian; Hell Creek Formation, Montana, USA]

> *Tyrannosaurus bataar* Maleev, 1955 ⇒ *Tarbosaurus bataar* (Maleev, 1955) Rozhdestvensky, 1965 ⇒ *Jenghizkhan bataar* (Maleev, 1955) Olshevsky vide Olshevsky *et al.*, 1995 [late Campanian; Nemegt Formation, Ömnögovi, Mongolia]

= *Gorgosaurus novojilovi* Maleev, 1955 ⇒ *Deinodon novojilovi* (Malleev, 1955) Maleev, 1964 ⇒ *Tarbosaurus novojilovi* (Maleev, 1955) Olshevsky, 1978 ⇒ *Maleevosaurus novojilovi* (Maleev, 1955) Pickering, 1984 ⇒ *Albertosaurus novojilovi* (Maleev, 1955) Mader & Bradley, 1989 ⇒ *Tyrannosaurus novojilovi* (Maleev, 1955) Glut, 1997 [Campanian; Nemegt Formation, Ömnögovi, Mongolia]

> *Albertosaurus periculosus* Riabinin, 1930 (*nomen dubium*) [late Maastrichtian; Yuliangze Formation, Heilongjiang, China]

> *Alectrosaurus olseni* Gilmore, 1933 ⇒ *Albertosaurus olseni* (Gilmore, 1933) Paul, 1988 [Campanian, Iren Dabasu Formation, Inner Mongolia, China]

[References: Osborn HF (1905). *Tyrannosaurus* and other Cretaceous carnivorous dinosaurs. *Bulletin of the American Museum of Natural History* 21(14): 259–265.]

Albinykus baatar

> *Albinykus baatar* Nesbitt *et al*., 2011 [Santonian; Javkhlant Formation, Khugenetslavkant, Dornogovi, Mongolia]

Generic name: *Albinykus* ← {Albin + (Gr.) ὄνυξ/onyx: claw}; "Albin, referring to 'wandering lights' as used by Mongolian Shamans to describe light phenomena in the Gobi Desert" (*1).

Specific name: *baatar* ← {(Mong.) баатар/baatar: hero}.

Etymology: heroic, **wandering light claw** [英雄游光爪龍]

Taxonomy: Theropoda: Coelurosauria: Alvarezsauridae

Notes: The holotype was discovered articulated in a seated position. (*1)

[References: (*1) Nesbitt SJ, Clarke JA, Turner AH, Norelle MA (2011). A small alvarezsaurid from the eastern Gobi Desert offers insight into evolutionary patterns in the Alvarezsauroidea. *Journal of Vertebrate Paleontology* 31:144–153.]

Albisaurus albinus [Archosauria]

> *Iguanodon albinus* Fritsch, 1893 ⇒ *Albisaurus albinus* (Fritsch, 1893) Fritsch, 1905 (*nomen dubium*) [Late Cretaceous; Czech]

= *Albisaurus scutifer* Fritsch, 1905 [Coniacian–Santonian; Priesener Formation, Vychodocesky, Czech]

Generic name: *Albisaurus* Fritsch, 1895*, 1905 ← {(place-name) River Albis [(Lat.) albus: white] + (Gr.) σαῦρος/sauros: lizard}

The Bílé Labe (=White Elbe) flows through Czech Republic.

*This taxon was first mentioned by Fritsch (Praha, 1895).

Specific name: *albinus* ← {Albis (=River Albis) + -inus}

Etymology: Albis' **River Albis lizard**

Taxonomy: Archosauria

Notes: *Albisaurus* was once thought to be a genus of dinosaur but is now thought to be a non-dinosaurian and considered a *nomen dubium*.

Alcmonavis poeschli [Avialae]

> *Alcmonavis poeschli* Rauhut *et al.*, 2019 [Tithonian; Mörnsheim Formation, Bavaria, Germany]

Generic name: *Alcmonavis* ← {(old Celtic, place-name) Alcmona: Altmühl River + (Lat.) avis: bird}; referring to "the Altmühl River, which flows through the principal region in which the famous 'Solnhofen limestones' are exposed" (*1).

Specific name: *poeschli* ← {(person's name) Pöschl + -ī}; in honor of Roland Pöschl as discoverer. (*1)

Etymology: Pöschl's **Alcmona bird**

Taxonomy: Theropoda: Maniraptora: Avialae

Notes: In 2018, a right wing was reported as a thirteenth specimen of *Archaeopteryx*, but according to Rauhut *et al.* (2019), the right wing (Mühlheim specimen) indicates to be a more derived avialan than *Archaeopteryx*. It was named *Alcmonavis poeschli*. (*1) (*2)

[References: (*1) Rauhut OWM, Foth C, Tischlinger H (2018). The oldest *Archaeopteryx* (Theropoda; Avialae): a new specimen from the Kimmeridgian/Tithonian boundary of Schamhaupten, Bavaria. *PeerJ* 6: e4191.; (*2) Rauhut OWM, Tischlinger H, Foth C (2019). A non-archaeopterygid avialan theropod from the Late Jurassic of southern Germany. *eLife* 8: e43789.]

Alcovasaurus longispinus

> *Stegosaurus longispinus* Gilmore, 1914 ⇒ *Alcovasaurus longispinus* (Gilmore, 1914) Galton & Carpenter, 2016 [Kimmeridgian; Morrison Formation, Alcova Quarry, Natrona, Wyoming, USA]

Generic name: *Alcovasaurus* ← {(place-name) Alcova: type locality + (Gr.) σαῦρος/sauros: lizard}

Specific name: *longispinus* ← {(Lat.) longus: long + spina: spine}; referring to the long tail spines.

Etymology: **Alcova lizard** with long spines

Taxonomy: Ornithischia: Thyreophora: Eurypoda: Stegosauria: Stegosauridae

Notes: A new genus 'Natronasaurus longispinus' proposed by Ulansky in 2014 for *Stegosaurus longispinus* failed to meet the requirements of the ICZN. (*1)

[References: (*1) Galton PM, Carpenter K (2016). The plated dinosaur *Stegosaurus longispinus* Gilmore, 1914 (Dinosauria: Ornithischia; Upper Jurassic, western USA), type species of *Alcovasaurus* n. gen. *Neus Jahrbuch für Geologie und Paläontologie - Abhandlungen* 279: 185–208.]

Alectrosaurus olseni

> *Alectrosaurus olseni* Gilmore, 1933 ⇒ *Albertosaurus olseni* (Gilmore, 1933) Paul, 1988 [Campanian; Iren Dabasu Formation, Inner Mongolia, China]

Generic name: *Alectrosaurus* ← {(Gr.) ἄλεκτρος/alectros: unbedded, unwedded, unmated + σαῦρος/sauros: lizard}; referring to the specimens of one or two individuals, which were found about 100 feet distant. (*1)

Specific name: *olseni* ← {(person's name) Olsen + -ī}; in honor of George Olsen, who collected the type specimens. [*1]

Etymology: Olsen's **unmated lizard**

Taxonomy: Theropoda: Coelurosauria:
Tyranosauroidea

[References: (*1) Gilmore CW (1933). On the dinosaurian fauna of the Iren Dabasu Formation. *Bulletin of the American Museum of Natural History* 67: 23–78.]

Aletopelta coombsi

> *Aletopelta coombsi* Kirkland & Ford, 2001 [Campanian; Point Loma Formation, California, USA]

Generic name: *Aletopelta* ← {(Gr.) ἀλήτης/alētēs: wanderer, wandering + πέλτη/peltē: small shield}; pelta, 'shield', a common suffix which is commonly used for ankylosaurs in reference to their armour, and 'aletes' was chosen because the tectonic plate has been wandering northward, carryng the specimen with it.

Specific name: *coombsi* ← {(person's name) Coombs + -ī}; in honor of paleontologist Walter P. Coombs, Jr.

Locality: Coombs' **wandering shield**

Taxonomy: Ornithischia: Thyreophora:
Ankylosauria: Ankylosauridae

[References: Ford TL, Kirkland JI (2001). Carlsbad ankylosaur (Ornithischia: Ankylosauria): an ankylosaurid and not a nodosaurid. In: Carpenter K (ed), *The Armored Dinosaurs*. Indiana University of Press, Bloomington: 239–260.]

Algoasaurus bauri

> *Algoasaurus bauri* Broom, 1904 [Berriasian; upper Kirkwood Formation, Eastern Cape, South Africa]

Generic name: *Algoasaurus* ← {(place-name) Algoa Bay + (Gr.) σαῦρος/sauros: lizard}

Specific name: *bauri* ← {(person's name) Baur + -ī}; "after the late George Baur" (1859–1898), "whose early death removed from the ranks of investigators one who could ill be spared" [*1].

Etymology: Baur's **Algoa Bay lizard**

Taxonomy: Saurischia: Sauropodomorpha: Sauropoda

Notes: According to *The Dinosauria 2nd edition*, *Algoasaurus* is regarded a *nomen dubium*. According to Broom (1904), a number of bones were discovered in the rock near Uitenhage. But many of the bones were made into bricks because they were not recognized as dinosaur specimens.

[References: (*1) Broom R (1904). On the occurrence of an opisthocoelian dinosaur (*Algoasaurus bauri*) in the Cretaceous beds of South Africa. *Geological Magazine*, decade 5, 1(483): 445–447.]

Alioramus remotus

> *Alioramus remotus* Kurzanov, 1976 [Maastrichtian; Nogon Tsav Formation, Bayankhongor, Mongolia]

Generic name: *Alioramus* ← {(Lat.) alius: other + rāmus: branch} [*1]

Specific name: *remotus* ← {(Lat.) remōtus: remote} [*1]

Etymology: remote, **other branch**

Taxonomy: Theropoda: Coelurosauria:
Tyrannosauroidea: Tyrannosauridae:
Tyrannosaurinae

Other species:

> *Alioramus altai* Brusatte *et al.*, 2009 [Maastrichtian; Nemegt Formation, Ömnögovi, Mongolia]

> *Qianzhousaurus sinensis* Lü *et al.*, 2014 ⇒ *Alioramus sinensis* (Lü *et al.*, 2014) Carr *et al.*, 2017 [Maastrichtian; Nanxiong Formation, Jiangxi, China]

[References: (*1) Kurzanov SM (1976). A new carnosaur from the Late Cretaceous of Nogon-Tsav, Mongolia. *The Joint Soviet-Mongolian Paleontological Expedition Transactions* 3: 93–104.]

Aliwalia rex

> *Eucnemesaurus fortis* van Hoepe, 1920 [Norian; lower Elliot Formation, Free State, South Africa]

= *Aliwalia rex* Galton, 1985 [Norian; lower Elliot Formation, Eastern Cape*, South Africa]
*Eastern Cape was previously part of Cape Province.

Generic name: *Aliwalia* ← {(place-name) Aliwal + -ia}

Specific name: *rex* ← {(Lat.) rēx: king, tyrant, despot, leader}

Etymology: **Aliwal** king

Taxonomy: Saurischia: Sauropodomorpha:
Prosauropoda: Riojasauridae
cf. Saurischia *incertae sedis* (acc. *The Dinosauria 2nd edition*)

Notes: According to *The Dinosauria 2nd edition* (2004), *Aliwalia rex* was classified as "Dinosauria *incertae sedis*". Yates (2006) indicates that *Eucnemesaurus* is valid and *Aliwalia* is considered a junior synonym of *Eucnemesaurus*.

[References: Yates AM (2006). Solving a dinosaurian puzzle: the identity of *Aliwalia rex* Galton. *Historical Biology* 19: 93–123.]

Allosaurus fragilis

> *Allosaurus fragilis* Marsh, 1877 ⇒ *Labrosaurus fragilis* (Marsh, 1877) Marsh, 1895⇒ *Antrodemus fragilis* (Marsh, 1877)

Gilmore & Stewart, 1945 [Kimmeridgian–Tithonian; Morrison Formation, Colorado, USA]
= *Poicilopleuron/Poecilopleuron valens* Leidy, 1870 ⇒ *Antrodemus valens* (Leidy, 1870) Leidy, 1870 (*nomen dubium*) ⇒ *Megalosaurus valens* (Leidy, 1870) Nopcsa, 1901 ⇒ *Allosaurus valens* (Leidy, 1870) Gilmore, 1920 [Kimmeridgian–Tithonian; Morrison Formation, Colorado, USA]
= *Laelaps trihedrodon* Cope, 1877 ⇒ *Antrodemus trihedrodon* (Cope, 1877) Hay, 1902 ⇒ *Dryptosaurus trihedrodon* (Cope, 1877) Kuhn, 1939 [Kimmeridgian–Tithonian; Morrison Formation, Colorado, USA]
= *Creosaurus atrox* Marsh, 1878 ⇒ *Antrodemus atrox* (Marsh, 1878) Gilmore, 1920 ⇒ *Allosaurus atrox* (Marsh, 1878) Paul, 1987 [Kimmeridgian–Titohnian; Morrison Formation, Wyoming, USA]
= *Allosaurus lucaris* Marsh, 1878 ⇒ *Labrosaurus lucaris* (Marsh, 1878) Marsh, 1879 ⇒ *Antrodemus lucaris* (Marsh, 1878) Hay, 1902 [Kimmeridgian–Tithonian; Morrison Formation, Wyoming, USA]
= *Epanterias amplexus* Cope, 1878 ⇒ *Allosaurus amplexus* (Cope, 1878) Paul, 1988 [Tithonian; Morrison Formation, Colorado, USA]
= *Camptonotus amplus* Marsh, 1879 ⇒ *Camptosaurus amplus* (Marsh, 1879) Marsh, 1885 ⇒ *Allosaurus amplus* (Marsh, 1879) Galton *et al.*, 2015 [Kimmeridgian; Morrison Formation, Wyoming, USA]
= *Labrosaurus ferox* Marsh, 1884 [Kimmeridgian–Tithonian; Morrison Formation, Colorado, USA]
= *Allosaurus ferox* Marsh, 1896 ⇒ *Antrodemus ferox* (Marsh, 1896) Ostrom & McIntosh, 1966 [Kimmeridgian–Tithonian; Morrison Formation, Wyoming, USA]

Generic name: *Allosaurus* ← {(Gr.) ἄλλος/allos: strange + σαῦρος/sauros: lizard}; referring to the vertebrae which are "distinguished from any known dinosaurs" [*1].

Specific name: *fragilis* ← {(Lat.) fragilis: fragile}; referring to the weight of the centra greatly reduced by deep pneumatic foramens in the side. [*1]

Etymology: fragile, **strange lizard**

Taxonomy: Theropoda: Carnosauria: Allosauridae

Other species:

?> *Allosaurus europaeus* Mateus *et al.*, 2006 [late Kimmeridgian; Lourinhã Formation, Estremadura, Portugal]
> "*Allosaurus jimmadseni*" Chure, 2000 ⇒ *Allosaurus jimmadseni* Chure & Loewen, 2020 [Kimmeridgian; Morrison Formation, Utah, USA]
> *Allosaurus lucasi* Dalman, 2014 (*nomen dubium*) [Tithonian; Morrison Formation, Colorado, USA]
> ?*Allosaurus medius* Marsh, 1888 (*nomen dubium*) [early Aptian; Arundel Clay Formation, Maryland, USA]
> ?*Allosaurus tendagurensis* Janensch, 1925 (*nomen dubium*) ⇒ *Antrodemus tendagurensis* (Janensch, 1925) von Huene, 1932 [late Kimmeridian; Tendaguru Formation, Lindi, Tanzania]
> "*Saurophagus maximus*" Stovall, 1941 ⇒ *Saurophaganax maximus* Chule, 1995 ⇒ *Allosaurus maximus* (Chule, 1995) Smith, 1998 [Kimmeridgian; Morrison Formation, Oklahoma, USA]

Notes: According to *The Dinosauria 2nd edition*, *Antrodemus* Leidy, 1870 is considered a synonym of *Allosaurus* Marsh, 1877 (*nomen conservandum*).

[References: (*1) Marsh OC (1877). Notice of new dinosaurian reptiles from the Jurassic formation. *American Journal of Science and Arts* 14: 514–516.]

Almas ukhaa

> *Almas ukhaa* Pei *et al.*, 2017 [Campanian; Djadochta Formation, Ömnögovi, Mongolia]

Generic name: *Almas* ← {(Mong. myth.) Almas; "in reference to the wild man or snowman of Mongolian mythology (Rincen, 1964)" [*1].

Specific name: *ukhaa* ← {(place-name) Ukhaa}; referring to "the locality of Ukhaa Tolgod, discovered in 1933, where the specimen was collected" [*1].

Etymology: **Almas (wild man)** from Ukhaa

Taxonomy: Theropda: Coelurosauria: Maniraptora: Troodontidae

[References: (*1) Pei R, Norell MA, Barta DE, Bever GS, Pittman M, Xu X (2017). Osteology of a new Late Cretaceous troodontid specimen from Ukhaa Tolgod, Ömnögovi Aimag, Mongolia. *American Museum Novitates* 3889: 1–47.]

Alnashetri cerropoliciensis

> *Alnashetri cerropoliciensis* Makovicky *et al.*, 2012 [Cenomanian; Candeleros Formation, Río Negro, Argentina]

Generic name: *Alnashetri* ← {("Günün-a-kunna dialect of the Tehuelche language") Alnashetri: "slender thighs"}; "in reference to long and slender hind limbs of the holotype" [*1].

Specific name: *cerropoliciensis* ← {(place-name) Cerro Policía + -ensis}; "in honor of the nearby hamlet of Cerro Policía, whose residents have generously assisted fieldwork efforts at La Buitrera since the locality was discovered by the second author in 1999" (*1).

Etymology: **slender thighs** from Cerro Policía

Taxonomy: Theropoda: Coelurosauria: Alvarezsauroidea

[References: (*1) Makovicky PJ, Apesteguía S, Gianechini FA (2012). A new coelurosaurian theropod from the La Buitrera fossil locality of Río Negro, Argentina. *Fieldiana Life and Earth Sciences* 5: 90.]

Alocodon kuehnei

> *Alocodon kuehnei* Thulborn, 1973 [Oxfordian; Cabaços Formation, Beira Litoral, Portugal]

Generic name: *Alocodon* ← {(Gr.) ἄλοξ/alox: furrow + ὀδών/odōn: tooth}

Specific name: *kuehnei* ← {(person's name) Kühne + -ī}; in honor of German paleontologist Georg Kühne.

Etymology: Kühne's **furrowed tooth**

Taxonomy: Ornithischia: Neornithischia

Notes: According to *The Dinosauria 2nd edition* (2004), *Alocodon kuehnei* is considered to be a *nomen dubium*.

Altirhinus kurzanovi

> *Altirhinus kurzanovi* Norman, 1998 [Albian; Khuren Dukh Formation, Dornogovi, Mongolia]

Generic name: *Altirhinus* ← {(Lat.) alti- [altus: high] + (Gr.) ῥῖνός/rhinos [ῥίς/rhīs: nose or snout]}; "in recognition of the highly arched nasal bones of the skull which give the snout of this animal a distinctively elevated profile" (*1).

Specific name: *kurzanovi* ← {(person's name) Kurzanov + -ī}; in honor of Russian paleontologist Sergei M. Kurzanov.

Etymology: Kurzanov's **high snout**

Taxonomy: Ornithischia: Ornithopoda: Iguanodontia: Iguanodontoidea

cf. Ornithischia: Ornithopoda: Euornithopoda: Iguanodontidae (*1)

[References: (*1) Norman DB (1998). On Asian ornithopods (Dinosauria, Ornithischia). *Zoological Journal of the Linnean Society* 122: 291–348.]

Altispinax dunkeri

> *Megalosaurus dunkeri* von Huene, 1923 ⇒ *Altispinax dunkeri* (von Huene, 1923) von Huene, 1923, Kuhn, 1939 [Valanginian–Hauterivian; Wealden Group, East Sussex, England, UK]

(=) (objective synonym) *Acrocanthosaurus altispinax* Paul, 1988 ⇒ *Becklespinax altispinax* (Paul, 1988) Olshevsky, 1991 ⇒ *Altispinax altispinax* (Paul, 1988) Rauhut, 2003 (according to Maisch, 2016)

(=) (objective synonym) *Altispinax* "lydekkerhueneorum" Pickering, 1984 (*nomen nudum*) ⇒ *Becklespinax* "lydekkerhueneorum" (Pickering, 1984) Olshevsky, 1991 (*nomen nudum*)

cf. Megalosaurus dunkeri Dames, 1884 (*nomen dubium*) [Berriasian; Bückeberg Formation, Deisters (locality), Niedersachsen, Germany]

Generic name: *Altispinax* ← {(Lat.) altus: high + spina: spine}; referring to a series of three dorsal vertebrae with very high spines.

Specific name: *dunkeri* ← {(person's name) Dunker + -ī}; in honor of paleontologist Wilhelm Dunker.

Etymology: Dunker's **one with high spines**

Taxonomy: Theropoda

Other species:

> *Megalosaurus oweni* Lydekker, 1889 ⇒ *Altispinax oweni* (Lydekker, 1889) von Huene, 1923 ⇒ *Valdoraptor oweni* (Lydekker,1889) Olshevsky, 1991 (*nomen dubium*) [Valanginian–Hauterivian; Tunbridge Wells Sand Formation, UK]

> *Megalosaurus parkeri* von Huene, 1923 ⇒ *Altispinax parkeri* (von Huene, 1923) von Huene, 1932 ⇒ *Metriacanthosaurus parkeri* (von Huene, 1923) Walker, 1964 [early Oxfordian; Oxford Clay Formation, UK]

Notes: According to *The Dinosauria 2nd edition* (2004), *Megalosaurus dunkeri* Dames, 1884 (type of *Altispinax* von Huene, 1923) was considered a *nomen dubium*. But, in 2016, *Altispinax dunkeri* von Huene, 1923 was regarded as a valid taxon according to the ICZN, based on von Huene's original description (*1).

[References: (*1) Maisch MW (2016). The nomenclature status of the carnivorous dinosaur genus *Altispinax* v. Huene, 1923 (Saurischia, Theropoda) from the Lower Cretaceous of England. *Neues Jahrbuch für Geologie und Paläontologie – Abhandlungen* 280(2): 215–219.]

Alvarezsaurus calvoi

> *Alvarezsaurus calvoi* Bonaparte, 1991 [Santonian; Bajo de la Carpa Formation, Neuquén, Argentina]

Generic name: *Alvarezsaurus* ← {(person's name) Álvarez + (Gr.) σαῦρος/sauros: lizard}; in honor of Don Gregorio Álvarez, noted historian of Neuquén.

Specific name: *calvoi* ← {(person's name) Calvo + -ī}; in honor of Dr. Jorge Orlando Calvo.

Etymology: Calvo's **Álvarez's lizard**

Taxonomy: Theropoda: Coelurosauria: Alvarezsauridae

Alwalkeria maleriensis

> *Walkeria maleriensis* Chatterjee, 1987 (preoccupied) ⇒ *Alwalkeria maleriensis* (Chatterjee, 1987) Chatterjee & Creisler, 1994 [Carnian; lower Maleri Formation, Andhra Pradesh, India]

Generic name: *Alwalkeria* ← {(person's name) Al Walker + -ī}; "in honor of British palaeontologist Alick D. Walker for his valuable contribution to Mesozoic vertebrates" (*1).

Specific name: *maleriensis* ← {(place-name) Maleri Formation + -ensis}

Etymology: **for Alick Walker** from Maleri Formation

Taxonomy: Saurischia

Notes: Small size. As the original generic name had been preoccupied by a bryozoan, *Walkeria uva* (Linnaeus, 1758), *Alwalkeria* (new name) was proposed. (*1)

[References: (*1) Chatterjee S, Creisler BS (1994). *Alwalkeria* (Theropda) and *Morturneria* (Plesiosauria), new names for preoccupied *Walkeria* Chatterjee, 1987 and *Turneria* Chatterjee and Small, 1989. *Journal of Vertebrate Paleontology* 14: 142.]

Alxasaurus elesitaiensis

> *Alxasaurus elesitaiensis* Russell & Dong, 1993 [Aptian; Bayan Gobi Formation, Inner Mongolia, China]

Generic name: *Alxasaurus* ← {(place-name) Alxa*: (Alashan) Desert + (Gr.) σαῦρος/sauros: lizard}
*Alxa = Ālāshàn (阿拉善)

Specific name: *elesitaiensis* ← {(place-name) Elesitai + -ensis}; named after the village of Elesitai.

Etymology: **Alxa Desert lizard** from Elesitai

Taxonomy: Theropoda: Coelurosauria: Therizinosauroidea

[References: Russell DA, Dong Z-M (1993). The affinities of a new theropod from the Alxa Desert, Inner Mongolia, People's Republic of China. *Canadian Journal of Earth Sciences* 30: 2107–2127. Abstract.]

Amanzia greppini

> *Megalosaurus meriani* Greppin, 1870 (*partim*) ⇒ *Ornithopsis greppini* von Huene, 1922 ⇒ *Cetiosauriscus greppini* (von Huene, 1922) von Huene, 1927 ⇒ *Amanzia greppini* (von Huene, 1922) Schwarz *et al.*, 2020 [early Kimmeridgian; Reuchenette Formation, Moutier, Bern, Switzerland]

Generic name: *Amanzia* ← {(person's name) Amanz + -ia}; "in honor of the well-known Swiss geologist Amanz Gressly (1814–1865) who introduced the term "facies" into geology and discovered the first dinosaur fossil from Switzerland in 1856".

Specific name: *greppini* ← {(person's name) Greppin + -ī}; in honor of the Swiss geologist Jean-Baptiste Greppin.

Etymology: **After Amanz** and Greppin

Taxonomy: Saurischia: Sauropodomorpha: Sauropoda: Eusauropoda

Amargasaurus cazaui

> *Amargasaurus cazaui* Salgado & Bonaparte, 1991 [Barremian–Aptian; La Amarga Formation, Neuquén, Argentina]

Generic name: *Amargasaurus* ← {(place-name) La Amarga + (Gr.) σαῦρος/sauros: lizard}; "in reference to La Amarga, in Neuquén Province, that produces the holotype" (*1).

Specific name: *cazaui* ← {(person's name) Cazau + -ī}; "in recognition of Dr. Luis B. Cazau, of Yacimientos Petroliferos Fiscales Exploration who, in 1983, interested the authors in the paleontological exploration of the La Amarga Formation, offering data on discoveries" (*1).

Etymology: Cazau's **lizard of La Amarga** (*1)

Taxonomy: Saurischia: Sauropodomorpha: Sauropoda: Dicraeosauridae

[References: (*1) Salgado L, Bonaparte JF (1991). Un nuevo sauropodo Dicraeosauridae, *Amargasaurus casaui* gen. *et* sp. nov., de la Formacion La Amarga, Neocomiano de la Provincia Neuquén, Argentina. *Ameghiniana* 28(3–4): 333–346.]

Amargatitanis macni

> *Amargatitanis macni* Apesteguía, 2007 [Barremian; La Amarga Formation, Neuquén, Argentina]

Generic name: *Amargatitanis* ← {(place-name) "Amarga: the fossil locality" + "(Gr.) titanis*: titan, giant"}; "because of the affinities of the specimen with the titanosaurs" (*1).
**cf.* Τιτᾱνίς/Tītānis: Titaness.

Specific name: *macni* ← {(acronym) MACN + -ī}; "in honor of the Museo Argentino de Ciencias Naturales (MACN) for the sustained contribution and human resources formed in Vertebrate Paleontology during the 19th and 20th centuries" (*1).

Etymology: MACN's **La Amarga Titaness**

Taxonomy: Saurischia: Sauropodomorpha: Sauropoda: Dicraeosauridae (*2)
cf. Titanosauria (*1)

[References: (*1) Apesteguía S (2007). The sauropod diversity of the La Amarga Formation (Barremian), Neuquén (Argentina). *Gondwana Research* 12(4): 533–546.; (*2) Gallina PA (2016). Reappraisal of the Early Cretaceous sauropod dinosaur *Amargatitanis macni* (Apesteguía, 2007), from northwestern Patagonia, Argentina. *Cretaceous Research* 64: 79–87. Abstract.]

Amazonsaurus maranhensis

> *Amazonsaurus maranhensis* Carvalho *et al.*, 2003 [Aptian–Albian; Itapecuru Formation, Maranhão, Brazil]

Generic name: *Amazonsaurus* ← {(place-name) Amazon: the Brazilian Legal Amazon region + (Gr.) σαῦρος/sauros: lizard, reptile} [*1]

Specific name: *maranhensis* ← {(place-name) Maranhão: the Brazilian state + -ensis} [*1]

Etymology: **Amazon lizard** from Maranhão

Taxonomy: Saurischia: Sauropodomorpha: Sauropoda: Diplodocoidea: Rebbachisauridae

[References: (*1) Carvalho IS, Avilla LS, Salgado L (2003). *Amazonsaurus maranhensis* gen. *et* sp. nov. (Sauropoda, Diplodocoidea) from the Lower Cretaceous (Aptian–Albian) of Brazil. *Cretaceous Research* 24: 697–713.]

Ambiortus dementjevi [Avialae]

> *Ambiortus dementjevi* Kurochkin, 1982 [Barremian; Ondorukhaa Formation, Bayankhongor, Mongolia]

Generic name: *Ambiortus* ← {(Lat.) ambiguus: uncertain + ortus: origin, beginning}

Specific name: *dementjevi* ← {(person's name) Dementjev + -ī}

Kurochkin named this oldest ornithurine after his peer Georgi Petrovich Dementjev (1898–1969).

Etymology: Dementjev's **uncertain origin**

Taxonomy: Theropoda: Avialae: Ornithuromorpha: Ornithurae

Ambopteryx longibrachium [Scansoriopterygidae]

> *Ambopteryx longibrachium* Wang *et al.*, 2019 [Oxfordian; Haifanggou Formation, Yanliao Biota, Liaoning, China]

Generic name: *Ambopteryx* ← {(Lat.) ambo: both + (Gr.) πτέρυξ/pteryx: wing}; "referring to the pterosaur-like wing that is present in this non-avian dinosaur" [*1].

Specific name: *longibrachium* ← {(Lat.) longus: long + brachium: upper arm} [*1]

Etymology: **both wings** with long upper arms

Taxonomy: Theopoda: Maniraptora: Scansoriopterygidae

[References: (*1) Wang M, O'Connor JK, Xu X, Zhou Z (2019). A new Jurassic scansoriopterygid and the loss of membranous wings in theropod dinosaurs. *Nature* 569: 256–259.]

Ampelosaurus atacis

> *Ampelosaurus atacis* Le Loeuff, 1995 [Maastrichtian; Marnes Rouges Inférieures Formation, Aude, France]

Generic name: *Ampelosaurus* ← {(Gr.) ἄμπελος/ampelos: vine (the vineyard of Elien, De la Nature des Animaux, 11, 32) + σαῦρος/sauros: reptile}; referring to the layer situated in the southern border of the Blanquette de Limoux vineyard. [*1]

cf. (Gr.) ἀμπελών/ampelōn: vineyard | ἄμπελος/ampelos: vine

Specific name: *atacis* ← {(Lat.) genitive form of Atax: from the Atax (=Aude River)}

Etymology: **vineyard reptile** from Atax (Aude River) [reptile des vignobles de l'Aude [*2]]

Taxonomy: Saurischia: Sauropodomorpha: Sauropoda: Titanosauria: Nemegtosauridae
cf. Sauropoda: Titanosauridae [*1]

Notes: *Ampelosaurus* had a small head.

[References: (*1) Le Loeuff J (2001). *Ampelosaurus atacis* (nov. gen., nov. sp.), a new titanosaurid (Dinosauria, Sauropoda) from the Late Cretaceous of the Upper Aude Valley (France). *Comptes Rendus de l'Académie des Sciences Paris, ser.IIa* 321: 693–699.; (*2) Original citation: Le Loeuff J. (1995). *Ampelosaurus atacis* (nov. gen., nov. sp.), un nouveau Titanosauridae (Dinosauria, Sauropoda) du Crétacé supérieur de la Haute Vallée de l'Aude (France). *Comptes Rendus de l'Académie des Sciences Paris, ser. IIa* 321: 693–699.]

Amphicoelias altus

> *Amphicoelias altus* Cope, 1877 [Kimmeridgian; Brushy Basin Member, Morrison Formation, Colorado, USA]

Generic name: *Amphicoelias* ← {(Gr.) ἀμφί/amphi: on both side + κοῖλος/koilos: hollow, hollowed}; referring to the large pneumatic cavity on each side of the vertebral axis. [*1]

Specific name: *altus* ← {(Lat.) high, elevated, deep, profound}.

Etymology: deep, **bi-con cave**

Taxonomy: Saurischia: Sauropodomorpha: Sauropoda: Diplodocoidea

Other species:

> *Amphicoelias brontodiplodocus* Galiano & Albersdörfer 2010 (*nomen nudum*) [Oxfordian; Morrison Formation, Wyoming, USA]

> *Amphicoelias fragillimus* Cope, 1878 ⇒ *Maraapunisaurus fragillimus* (Cope, 1878) Carpenter, 2018 [Tithonian; Morrison

Formation, Colorado, USA]
> *Camarasaurus supremus* Cope, 1877 [Kimmeridgian; Brushy Basin Member, Morrison Formation, Colorado, USA]
= *Amphicoelias latus* Cope, 1878 [Kimmeridgian; Salt Wash Member, Morrison Formation, Colorado, USA]

[References: (*1) Cope ED (1878). On the Vertebrata of the Dakota Epoch of Colorado. *Proceedings of the American Philosophical Society* 17: 233–247.; Woodruff DC, Foster JR (2015). The fragile legacy of *Amphicoelias fragilimus* (Dinosauria: Sauropoda: Morrison Formation – Latest Jurassic) *Peer J* 3: e838v1.]

Amtocephale gobiensis

> *Amtocephale gobiensis* Watabe *et al.*, 2011 [Cenomanian–late Santonian; Baynshire Formation, Ömnögovi, Mongolia]

Generic name: *Amtocephale* ← {(place-name) Amtgai/Amtgay + (Gr.) κεφαλή/kephalē: head}
Specific name: *gobiensis* ← {(place-name) Gobi Desert + -ensis}
Etymology: **Amtgai head** from Gobi Desert
Taxonomy: Ornithischia: Marginocephalia: Pachycephalosauridae

[References: Watabe M, Tsogtbaatar K, Sullivan RM (2011). A new pachycephalosaurid from the Baynshire Formation (Cenomanian–late Santonian), Gobi Desert, Mongolia. *Fossil Record 3. New Mexico Museum of Natural History and Science, Bulletin* 53: 489–497.]

Amtosaurus magnus

> *Amtosaurus magnus* Kurzanov & Tumanova 1978 (*nomen dubium*) [Cenomanian; Bayanshiree Svita / Baynshire Formation, Ömnögovi, Mongolia]

Generic name: *Amtosaurus* ← {(place-name) Amtgay + (Gr.) σαῦρος/sauros: lizard}
Specific name: *magnus* ← {(Lat.) māgnus: great}.
Etymology: great **Amtgay lizard**
Taxonomy: Ornithischia
Other species:

> *Amtosaurus archibaldi* Averianov, 2002 ⇒ *Bissektipelta archibaldi* (Averianov, 2002) Parish & Barrett, 2004 [Turonian; Bissekty Formation, Navoi, Uzbekistan]

Notes: According to *The Dinosauria 2nd edition* (2004), *Amtosaurus magnus* and *A. archibaldi* were considered to be "provisional Ankylosauria *incertae sedis*", but now *A. magnus* is considered as a *nomen dubium* (*1) and *A. archibaldi* was given a new name, *Bissektipelta archibaldi*.

[References: Kurzanov SM, Tumanova TA (1978). On the structure on the endocranium in some ankylosaurs from Mongolia. *Paleontological Journal* 1978: 90–96.; (*1) Parish JC, Barrett PM (2004). A reappraisal of the ornithischian dinosaur *Amtosaurus magnus* Kurzanov and Tumanova 1978, with comments on the status of *A. archibaldi* Averianov, 2002. *Canadian Journal of Earth Sciences* 41(3): 299–306.]

Amurosaurus riabinini

> *Amurosaurus riabinini* Bolotsky & Kurzanov, 1991 [Maastrichtian; Udurchukan Formation, Amur Oblast, Russia]

Generic name: *Amurosaurus* ← {(place-name) Amur River + (Gr.) σαῦρος/sauros: lizard} (*1)
The remains were discovered in far eastern area of Russia near Chinese border.
Specific name: *riabinini* ← {(person's name) Riabinin + -ī}; in honor of Russian paleontologist Anatoly Riabinin. He led the paleontological expeditions to the Amur region in the 1910s.
Etymology: Riabinin's **Amur lizard**
Taxonomy: Ornithischia: Ornithopoda: Hadrosauridae: Lambeosaurinae

Amygdalodon patagonicus

> *Amygdalodon patagonicus* Cabrera, 1947 [Toarchian–Bajocian; Cerro Carnerero Formation, Chubut, Argentina]

Generic name: *Amygdalodon* ← {Gr.) ἀμύγδἄλον/amygdalon: almond + (Gr.) ὀδών/odōn [= ὀδούς/odous [odūs]]: tooth}
Specific name: *patagonicus* ← {patagonicus; of Patagonia}
Etymology: **almond-shaped tooth** from Patagonia
Taxonomy: Saurischia: Sauropodomorpha: Sauropoda

[References: Cabrera A (1947). A new sauropod from the Jurassic of Patagonia. *Notas del Museo de la Plata. Paleontologia* 12(95): 1–17.]

Anabisetia saldiviai

> *Anabisetia saldiviai* Coria & Calvo, 2002 [Cenomanian–Turonian; Cerro Lisandro Formation, Neuquén, Argentina]

Generic name: *Anabisetia* ← {(person's name) Ana Biset + -ia}; "in honor of the late Ana María Biset, archaeologist of Dirección General de Cultura de Neuquén, for her important contribution to the provincial fossil legislation" (*1).
Specific name: *saldiviai* ← {(person's name) Saldivia + -ī}; in honor of "Mr. Roberto Saldivia, who found the first remains and kindly helped in field work" (*1).
Etymology: **for Ana Biset** and Saldivia
Taxonomy: Ornithischia: Ornithopoda: Iguanodontia

[References: (*1) Coria RA, Calvo JO (2002).

A new iguanodontian ornithopod from Neuquen Basin, Patagonia, Argentina. *Journal of Vertebrate Paleontology* 22(3): 503–509.]

Analong chuanjieensis

> *Analong chuanjieensis* Ren *et al*., 2020 [Middle Jurassic; Chuanjie Formation, Yunnan, China]

Generic name: *Analong* ← {(Chin. place-name) Ana 阿納 [Pinyin: Ēnà] + lóng龍: dragon}
Specific name: *chuanjieensis* ← {(place-name) Chuānjiē川街 + -ensis}
Etymology: **Ana dragon** from Chuanjie
Taxonomy: Saurischia: Sauropodomorpha: Sauropoda: Mamenchisauridae

Anasazisaurus horneri

> *Anasazisaurus horneri* Hunt & Lucas, 1993 ⇒ *Kritosaurus horneri* (Hunt & Lucas, 1993) Prieto-Márquez, 2016 [Campanian; Kirtland Formation, New Mexico, USA]

Generic name: *Anasazisaurus* ← {(demonym) Anasazi + (Gr.) σαῦρος/sauros: lizard}; referring to Anasazi, the prehistoric people who lived in Chaco Canyon near the type locality.
Specific name: *horneri* ← {(person's name) Horner + -ī}; in honor of paleontologist Jack R. Horner.
Etymology: Horner's **Anasazi lizard**
Taxonomy: Ornithischia: Ornithopoda: Iguanodontia: Hadrosauridae: Saurolophinae

[References: Hunt AP, Lucas SG (1993). Cretaceous vertebrates of New Mexico. *Dinosaurs of New Mexico. New Mexico Museum of Natural History and Science Bulletin* 2: 77–91.]

Anatotitan copei

> *Claosaurus annectens* Marsh, 1892 ⇒ *Trachodon annectens* (Marsh, 1892) Hatcher, 1902 ⇒ *Thespesius annectens* (Marsh, 1892) Gilmore, 1924 ⇒*Anatosaurus annectens* (Marsh, 1892) Lull & Wright, 1942 ⇒ *Edmontosaurus annectens* (Marsh, 1892) Brett-Surman, 1990 [Maastrichtian; Lance Formation, Wyoming, USA]

= *Anatosaurus copei* Lull & Wright, 1942 ⇒ *Anatotitan copei* (Lull & Wright, 1942) Brett-Surman *vide* Chapman & Brett-Surman, 1990 [Maastrichtian; Hell Creek Formation, South Dakota, USA]

Generic name: *Anatotitan* ← {(Lat.) anat- [anas: duck] + (Gr. myth.) Τῑτάν/Tītan: Giant}; referring to its duck-like bill.
Specific name: *copei* ← {(person's name) Cope + -ī}; in honor of Edward Drinker Cope.
He discovered a total of 56 new dinosaur species. As a herpetologist and paleontologist, he described over 1,000 species of fossil vertebrates including fishes in total.
Etymology: Cope's **duck titan**
Taxonomy: Ornithischia: Ornithopoda: Hadrosauridae
Notes: According to *The Dinosauria 2nd edition* (2004), *Anatosaurus* and *Anatotitan* are considered to be synonyms of *Edmontosaurus*.

Anchiceratops ornatus

> *Anchiceratops ornatus* Brown, 1914 [Campanian–Maastrichtian; Horseshoe Canyon Formation, Alberta, Canada]

= *Anchiceratops longirostris* Sternberg, 1929 [Maastrichtian; Horseshoe Canyon Formation, Alberta, Canada]

Generic name: *Anchiceratops* ← {(Gr.) ἄγχῐ/anchi: near, nigh, close-by + ceratops [κέρας/keras: horn + ὤψ/ōps: eye, face, countenance]}
Barnum Brown named this dinosour *Anchiceratops* because it represented a transitional form related to both *Monoclonius* and *Triceratops*.
Specific name: *ornatus* ← {(Lat.) ornātus: ornate, decorated}; referring to the ornate margin of its frill.
Etymology: ornated, **near horned face (ceratopsian)**
Taxonomy: Ornithischia: Marginocephalia: Ceratopsia: Ceratopsidae: Chasmosaurinae
Notes: *Anchiceratops* had two long brow horns and a short horn on the nose.

[References: Brown B (1914). *Anchiceratops*, a new genus of horned dinosaurs from the Edmonton Cretaceous of Alberta. With a discussion of the origin of the ceratopsian crest and the brain casts of *Anchiceratops* and *Trachodon*. *Bulletin of the American Museum of Natural History* 33: 539–548.]

Anchiornis huxleyi [Anchiornithidae]

> *Anchiornis huxleyi* Xu *et al*., 2009 [Oxfordian; Tiaojishan Formation; Jianchang, Liaoning, China]

Generic name: *Anchiornis* ← {(Gr.) ἄγχῐ/anchi: nearby + ὄρνις/ornis: bird}; "referring to the animal's being very closely related to birds" (*1).
Specific name: *huxleyi* ← {(person's name) Huxley + -ī}; in honor of"T. H. Huxley, who pioneered research into avian origins" (*1).
Etymology: Huxley's **near-bird**
Taxonomy: Theropoda: Avialae?: Anchiornithidae
cf. Theropoda: Coelurosauria: Troodontidae (acc. Zheng *et al*. (2014))(*2)
cf. Theropoda: Maniraptora: Avialae(*1)
Notes: *Anchiornis* is more closely related to avialans than to deinonychosaurian or troodontid

according to Rui *et al.* (2017).

[References: (*1) Xu X, Zhao Q, Norell M, Sullivan C, Hone D, Erickson G, Wang X, Hans F, Guo Y (2009). A new feathered maniraptoran dinosaur fossils that fills a morphological gap in avian origin. *Chinese Science Bulltin* 54 (3)430–435.; (*2) Zheng X, O'Connor J, Wang X, Wang M, Zhang X, Zhou Z (2014). On the absence of sternal elements in *Anchiornis* (Paraves) and *Sapeornis* (Aves) and the complex early evolution of the avian sternum. *Proceedings of the National Academy of Sciences* 111(38): 13900–13905.]

Anchisaurus polyzelus

> *Megadactylus polyzelus* Hitchcock, 1865 (preoccupied) ⇒*Amphisaurus polyzelus* (Hitchcock, 1865) Marsh, 1882 (preoccupied) ⇒ *Anchisaurus polyzelus* (Hitchcock, 1865) Marsh, 1885 ⇒ *Thecodontosaurus polyzelus* (Hitchcock, 1865) von Huene, 1914 [Hettangian–Sinemurian; Portland Formation, Massachusetts, USA]

= *Anchisaurus major* Marsh, 1889 ⇒ *Ammosaurus major* (Marsh, 1889) Marsh, 1891 [Hettangian; Portland Formation, Connecticut, USA]

= *Anchisaurus colurus* Marsh, 1891 ⇒ *Yaleosaurus colurus* (Marsh, 1891) von Huene, 1932 [Hettangian–Sinemurian; Portland Formation, Connecticut, USA]

= *Anchisaurus solus* Marsh, 1892 ⇒ *Ammosaurus solus* (Marsh, 1892) Kuhn, 1939 [Hettangian–Sinemurian; Portland Formation, Connecticut, USA]

Generic name: *Anchisaurus* ← {(Gr.) ἄγχῐ/anchi: near, nigh + σαῦρος/sauros: lizard}

Specific name: *polyzelus* ← {polyzelus; much coveted, much sought for}

Etymology: much coveted, **near lizard**

Taxonomy: Saurischia: Sauropodomorpha: Anchisauria: Anchisauridae

Notes: According to *The Dinosauria 2nd edition*, *Anchisaurus major* and *Anchisaurus solus* were considered as synonyms of *Ammosaurus major.* Later *A. major* and *A. solus* were admitted as other species of *Anchisaurus polyzelus*, but now considered synonyms of *Anchisaurus polyzelus.* (*1)

The bones first discovered in 1818 were assumed to be those of a human (*2).

[References: (*1) ICZN (2015). Opinion 2361 (Case 3561): *Anchisaurus Marsh*, 1885 (Dinosauria, Sauropodomorpha): Usage conserved by designation of a neotype for its type species *Megadactylus polyzelus* Hitchcock, 1865. *Bulletin of Zoological Nomenclature* 72(2): 176–177.; (*2) Nathan S (1820). Fossil bones found in red sandstones. *American Journal of Science* 2: 146–47.]

Andesaurus delgadoi

> *Andesaurus delgadoi* Calvo & Bonaparte, 1991 [Albian–Cenomanian; Candeleros Formation, Neuquén, Argentina]

Generic name: *Andesaurus* ← {(place-name) Andes Mountains + (Gr.) σαῦρος/sauros: lizard}; "in reference to the Andes Mountains, for their proximity to the place of discovery" (*1).

Specific name: *delgadoi* ← {(person's name) Delgado + -ī}; "in honor of Sr. Alejandro Delgado, discoverer of the material described here" (*1).

Etymology: Delgado's **Andes lizard**

Taxonomy: Saurischia: Sauropodomorpha: Sauropoda: Titanosauridae: Andesaurinae

Notes: *Andesaurus*, a basal titanosaur, had a long neck and a long tail.

[References: Calvo JO, Bonaparte JF (1991). *Andesaurus delgadoi* n. g. n. sp. (Saurischia–Sauropoda) dinosaurio Titanosauridae de la Formación Río Limay Formation (Albiano–Cenomaniano), Neuquén, Argentina. *Ameghiniana* 28(3–4): 303–310.; (*1) Calvo JO, Bonaparte JF (1991). *Andesaurus delgadoi* gen. *et* sp. nov. (Saurischia-Sauropoda), Titanosaurid dinosaur from the Río Limay Formation (Albian–Cenomanian), Neuquén, Argentina. (translated by Wilson JA, University of Chicago, Winter 1999); Mannion PD, Calvo JO (2011). Anatomy of the basal titanosaur (Dinosauria, Sauropoda) *Andesaurus delgadoi* from the mid-Cretaceous (Albian–early Cenomanian) Río Limay Formation, Neuquén Province, Argentina: implications for titanosaur systematics. *Zoological Journal of the Linnean Society* 163(1): 155–181.]

Angaturama limai

> ?*Irritator challengeri* Martill *et al.*, 1996* [Albian; Santana Formation, Ceará, Brazil]

= *Angaturama limai* Kellner & Campos, 1996 [Albian; Santana Formation, Ceará, Brazil]

**Irritator challengeri* was published one month prior to *Angaturama limai.*

Generic name: *Angaturama* ← {(Tupi) Angaturama: "noble, brave"}; named after Angaturama, a protective spirit in the aboriginal Tupi Indian culture of Brazil.

Specific name: *limai* ← {(person's name) Lima + -ī}; alluding to paleontologist Murilo R. de Lima.

Etymology: Lima's **brave (Angaturama)**

Taxonomy: Theropoda: Spinosauridae: Spinosaurinae

Notes: *Angaturama* had a well developed premaxillary sagittal crest. According to *The Dinosauria 2nd edition*, *Angaturama* was regarded as a valid genus, but is now considered a synonym of *Irritator*.

[References: Kellner AWA, Campos DA (1996). First Early Cretaceous dinosaur from Brazil with comments on Spinosauridae. *Neues Jahrbuch für Geologie und Paläontologie. Abhandlungen* 199(2): 151–166. Abstract.]

Angolatitan adamastor

> *Angolatitan adamastor* Mateus *et al.*, 2011 [Coniacian*; Itombe Formation, Bengo, Angola] *Originally regarded as Turonian.

Generic name: *Angolatitan* ← {(place-name) Angola + (Gr.) Τιτάν/Tītan: giant}

Specific name: *adamastor* ← {Lat.) Adamāstor: a giant}; in reference to "a mythological sea giant from the South Atlantic feared by the Portuguese sailors" (*1).

Etymology: Adamastor-like **Angolan giant**

Taxonomy: Saurischia: Sauropodomorpha: Sauropoda: Titanosauriformes: Somphospondyli

[References: (*1) Mateus O, Jacobs LL, Schulp AS, Polcyn MJ, Tavares TS, Neto AB, Morais ML, Antunes MT (2011). *Angolatitan adamastor*, a new sauropod dinosaur and the first record from Angola. *Anais da Academia Brasileira de Ciências* 83(1): 221–233.]

Angulomastacator daviesi

> *Angulomastacator daviesi* Wagner & Lehman, 2009 [Campanian; Aguja Formation, Texas, USA]

Generic name: *Angulomastacator* ← {(Lat.) angulus: corner, angle + (Gr.) μάσταξ/mastax: jaw, mouth + (Lat.) -tor: signifying agency}; in reference both to the unusual morphology of the maxilla and the Big Bend of the Rio Grande. Preferred translation 'bend chewer' (*1).

Specific name: *daviesi* ← {(person's name) Davies + -ī}; in honor of Kyle L. Davies.

Etymology: Davies' **bend chewer** (*1)

Taxonomy: Ornithischia: Sauropoda: Hadrosauridae: Lambeosaurinae

[References: (*1) Wagner JR, Lehman TM (2009). An enigmatic new lambeosaurine hadrosaur from the Upper Shale Member of the Aguja Formation, Texas. *Journal of Vertebrate Paleontology* 29(2): 605–611.]

Anhuilong diboensis

> *Anhuilong diboensis* Ren *et al.*, 2018 [Middle Jurassic; Hongqin Formation, Anhui, China]

Generic name: *Anhuilong* ← {(Chin. place-name) Ānhuī Province安徽省 + lóng 龍: dragon}

Specific name: *diboensis* ← {(Chin. place-name) Dìbó地博+ -ensis}

Etymology: **Anhui dragon** from Dibo

Taxonomy: Saurischia: Eusauropoda: Mamenchisauridae

Notes: The Hongqin Formation contains also *Huangshanlong anhuiensis*.

[References: Ren X-X, Huang J-D, You H-L (2018). The second mamenchisaurid dinosaur from the Middle Jurassic of eastern China. *Historical Biology, An International Journal of Paleobiology* 32: 602–610.]

Aniksosaurus darwini

> *Aniksosaurus darwini* Martínez & Novas, 2006 [Cenomanian–Turonian; Bajo Barreal Formation, Chubut, Argentina]

Generic name: *Aniksosaurus* ← {(Mod. Gr.) άνοιξη: spring + (Gr.) σαῦρος/sauros: lizard}; referring to September 21st (*i.e.*, the beginning of the spring in the Southern Hemisphere) , the day when the theropod was found (*1).

Specific name: *darwini* ← {(person's name) Darwin + -ī}; in honor of the great naturalist Charles Darwin, who visited Patagonia in 1832–1833, inspiring him to reach clearer interpretations of the evolution of life that changed human thought forever (*1).

Etymology: Darwin's **spring lizard**

Taxonomy: Theropoda: Tetanurae: Coelurosauria

Notes: *Aniksosaurus* was a small tetanurine.

[References: (*1) Martinez RD, Novas FE (2006). *Aniksosaurus darwini* gen. *et* sp. nov., a new coelurosaurian theropod from the early Late Cretaceous of Central Patagonia, Argentina. *Revista del Museo Argentino de Ciencias Naturales, n.s.* 8(2): 243–259.]

Animantarx ramaljonesi

> *Animantarx ramaljonesi* Carpenter *et al.*, 1999 [early Cenomanian; Cedar Mountain Formation, Utah, USA]

Generic name: *Animantarx* ← {(Lat.) animant- [animans: living] + arx: fortress or citadel} (*1)

This name is based on Richard Swann Lull's observation regarding ankylosaurs (1914).

Specific name: *ramaljonesi* ← {(person's name) Ramal Jones + -ī}; in honor of Ramal Jones who discovered the specimen using a modified scintillometer in an area with no bones exposed (see Jones & Burge, 1995) (*1).

Etymology: Ramal Jones' **living fortress**

Taxonomy: Ornithischia: Ankylosauria: Nodosauridae: Nodosaurinae

[References: (*1) Carpenter K, Kirkland JI, Burge D, Bird J (1999). Ankylosaurs (Dinosauria: Ornithischia) of the Cedar Mountain Formation, Utah, and their stratigraphic distribution. : In (Gilette DD ed), *Vertebrate Paleontology in Utah*. Utah Geological Survey, Salt Lake City: 243–251.]

Ankylosaurus magniventris

> *Ankylosaurus magniventris* Brown, 1908 [Maastrichtian; Hell Creek Formation, Montana, USA]

Generic name: *Ankylosaurus* ← {ankylo-: fused, coössified [(Gr.) [med.] ἀγκυλωσις/ankylōsis | ankylo- [(Gr.) ἀγκυλος/ankylos: crooked, curved] + σαῦρος/sauros: lizard}

Specific name: *magniventris* ← {(Lat.) māgnus: great + ventris [venter: belly]}

Etymology: **fused (and curved) lizard** with great belly

Taxonomy: Ornithischia: Thyreophora: Ankylosauria: Ankylosauridae: Ankylosaurinae: Ankylosaurini (acc. Arbour & Malllon, 2017)

[References: Brown B, Kaisen PC (1908). The Ankylosauridae, a new family of armored dinosaurs from the Upper Cretaceous. *Bulletin of the American Museum of Natural History* 24: 187–201.]

Anodontosaurus lambei

> *Anodontosaurus lambei* Sternberg, 1929 [Campanian–Maastrichtian; Horseshoe Canyon Formation, Alberta, Canada]

Generic name: *Anodontosaurus* ← {(Gr.) ἀν-/an-: non- + ὀδών/odōn: tooth + σαῦρος/sauros: lizard}

Specific name: *lambei* ← {(person's name) Lambe + -ī}; in honor of Lawrence Morris Lambe.

Etymology: Lambe's **toothless lizard**

Taxonomy: Ornithischia: Thyreophora: Ankylosauria: Ankylosauridae: Ankylosaurinae

Other species:

> *Anodontosaurus inceptus* Penkalski, 2018 [late Campanian; Dinosaur Park Formation, Alberta, Canada]

Notes: Sternberg thought *Anodontosaurus* did not have teeth, because the teeth had been removed by the compressed damage to the specimen, according to Vickaryous & Russell (2003). *Anodontosaurus* is actually seemed to have had teeth.

According to *The Dinosauria 2nd edition* (2004), *Anodontosaurus lambei* was considered to be a synonym of *Euoplocephalus tutus* (Lambe, 1902), but now *A. lambei* is valid(*1).

[References: (*1) Arbour VM, Currie PJ (2013). *Euoplocephalus tutus* and the diversity of ankylosaurid dinosaurs in the Late Cretaceous of Alberta, Canada, and Montana, USA. *PLoS ONE* 8(5): e62421.]

Anomalipes zhaoi

> *Anomalipes zhaoi* Yu *et al.*, 2018 [Campanian; Wangshi Group, Kugou, Zhucheng, Shandong, China]

Generic name: *Anomalipes* ← {(Late Latin) anomalus: peculiar, abnormal + pēs: a foot}; "referring to the unusual shape of the foot"(*1).

Specific name: *zhaoi* ← {(person's name) Zhao + -ī}; "in honour of Xijin Zhao, a Chinese paleontologist who has made great contributions to research on Zhucheng dinosaur fossils" (*1).

Etymology: Zhao's **unusual foot**

Taxonomy: Theropoda: Oviraptorosauria: Caenagnathidae

[References: (*1) Yu Y, Wang K, Chen S, Sullivan C, Wang S (2018). A new caenagnathid dinosaur from the Upper Cretaceous Wangshi Group of Shandong, China, with comments on size variation among oviraptorosaurs. *Scientific Reports* 8 (5030).]

Anoplosaurus curtonotus

> *Anoplosaurus curtonotus* Seeley, 1879 [late Albian–early Cenomanian; Cambridge Greensand, Cambridgeshire, UK]

Generic name: *Anoplosaurus* ← {(Gr.) ἄνοπλος/anoplos: without the ὅπλον/hoplon or large shield, unarmored, unarmed + σαῦρος/sauros: lizard}; referring to the fact that no armour plates had been found.

Specific name: *curtonotus* ← {(Lat.) curtus: shortened, mutilated, broken, short + (Gr.) νῶτον/nōton: the back (*1)}

Etymology: **unarmed lizard** with a short back

Taxonomy: Ornithischia: Thyreophora: Ankylosauria: Nodosauridae

Other species:

> *Anoplosaurus major* Seeley, 1879 (*partim*) (*nomen dubium*) ⇒ *Acanthopholis major* (Seeley, 1879) Nopsca, 1902 (*nomen dubium*) [late Albian; Cambridge Greensand, Cambridgeshire, England, UK]

> *Syngonosaurus macrocercus* Seeley, 1879 (*nomen dubium*) ⇒ *Anoplosaurus macrocercus* (Seeley, 1879) Kuhn, 1964 (*nomen dubium*) [early Cenomanian; upper Greensand Formation, England, UK]

> *Eucercosaurus tanyspondylus* Seeley, 1879 (*nomen dubium*) ⇒ *Anoplosaurus tanyspondylus* (Seeley, 1879) Steel, 1969 (syntype) [late Albian; Upper Greensand, Cambridge County, England, UK]

Notes: According to Barrett & Bonsor (2020), *Syngonosaurus* and *Eucerocosaurus* were reinterpreted as basal iguanodontians.

[References: Seeley HG (1879). On the dinosaurian of the Cambridge Greensand. *Quarterly Journal of the Geological Society* 35: 591–636.]

Anserimimus planinychus

> *Anserimimus planinychus* Barsbold, 1988

[Maastrichtian; Nemegt Formation; Ömnögovi, Mongolia]
Generic name: *Anserimimus* ← {(Lat.) ānser: goose + (Gr.) μῖμος/mīmos: mimic}
Specific name: *planinychus* ← {(Lat.) planus: flat + (Gr.) ὄνυξ/onyx: claw}
Etymology: flat-clawed **goose mimic**
Taxonomy: Theropoda: Coelurosauria: Ornithomimosauria: Ornithomimidae

Antarctopelta oliveroi

> *Antarctopelta oliveroi* Salgado & Gasparini, 2006 [Campanian; Snow Hill Island Formation, Santa Marta Cove, James Ross Island (Antarctica)]

Generic name: *Antarctopelta* ← {(Gr.) ἀντ-ἄρκτο-/ant-arkto-: Antarctica + (Lat.) pelta = (Gr.) πέλτη/peltē: shield} [*1]
Specific name: *oliveroi* ← {(person's name) Olivero + -ī}; "in honor of Eduardo Olivero, an outstanding Argentine geologist and paleontologist specializing in Antarctica, who discovered the holotype" [*1].
Etymology: Olivero's **shield of Antarctica**
Taxonomy: Ornithischia: Thyreophora: Ankylosauria: Nodosauridae
cf. Ornithischia: Ankylosauria [*1]
Notes: The teeth of *Antarctopelta oliveroi* are leaf-like, similar to many nodosaurids [*1].

[References: (*1) Salgado L, Gasparini Z (2006). Reappraisal of an ankylosaurian dinosaur from the Upper Cretaceous of James Ross Island (Antarctica). *Geodiversitas* 28(1): 119–135.]

Antarctosaurus wichmannianus

> *Antarctosaurus wichmannianus* von Huene, 1929 [Campanian–Maastrichtian; Anacleto Formation, Río Negro, Argentina]

Generic name: *Antarctosaurus* ← {(Gr.) Antarcto-[ἀντ-/ant-: against, opposite + ἄρκτος/arktos: north]: not northern + σαῦρος/sauros: lizard}; referring to the southern-most continent where the remains were found.
Specific name: *wichmannianus* ← {(person's name) Wichmann + -ianus}; in honor of geologist Ricardo Wichmann.
Etymology: Wichmann's **southern lizard**
Taxonomy: Saurischia: Sauropodomorpha: Sauropoda: Titanosauria: Lithostrotia: Antarctosauridae
Other species:

?> *Antarctosaurus giganteus* von Huene, 1929 (*nomen dubium*) [Coniacian–Santonian; Plottier Formation, Neuquén Basin, Argentina]

> *Antarctosaurus septentrionalis* von Huene & Matley, 1933 (*nomen dubium*) ⇒ *Jainosaurus septentrionalis* (von Huene & Matley, 1933) Hunt *et al.*, 1995 [Maastrichtian; Lameta Formation, Madhya Pradesh, India] [*1]

> *Antarctosaurus jaxartensis* Riabinin, 1939 (*nomen dubium*) [Coniacian–Santonian; Dabrazhin Formation, Kazakhstan]

> *Antarctosaurus brasiliensis* Arid & Vizotto, 1971 (*nomen dubium*) [Campanian–Maastrichtian; Adamantina Formation, Bauru Basin, São Paulo, Brazil]

[References: (*1) Wilson JA, D'Emic M, Curry Rogers CA, Mohabey DM, Sen S (2009). Reassessment of the sauropod dinosaur *Jainosaurus* (= "*Antarctosaurus*") *septentrionalis* from the Upper Cretaceous of India. *Contributions from the Museum of Paleontology, University of Michigan* 32(2): 17–40.; Hunt AP, Lockley M, Lucas SG, Meyer CA (1994). The global sauropod fossil record. *Gaia* 10: 261–279.]

Antetonitrus ingenipes

> *Antetonitrus ingenipes* Yates & Kitching, 2003 [Hettangian; upper Elliot Formation, Free State, South Africa]

Generic name: *Antetonitrus* ← {(Lat.) ante-: before + tonitrus: thunder}; referring to the early occurrence of this sauropod relative to *Brontosaurus* (Greek, thunder lizard) one of the most familiar sauropod names [*1].
cf. (Eng.) thunder, (Lat.) tonitrus, (Gr.) βροντη/brontē, (Jpn.) 雷kaminari.
Specific name: *ingenipes* ← {(Lat.) ingens: massive + pēs: paw, foot}; referring to its robust hands and feet [*1].
Etymology: **one before the thunder (*Brontosaurus*)** with massive feet
Taxonomy: Saurischia: Sauropodomorpha: Sauropoda
Notes: *Antetonitrus* is considered an earliest sauropod dinosaur [*1].

[References: (*1) Yates AM, Kitching JW (2003). The earliest known sauropod dinosaur and the first steps towards sauropod locomotion. *Proceedings of the Royal Society of London B: Biological Sciences* 270:1753–1758.]

Antrodemus valens

> *Poicilopleuron valens* Leidy, 1870 (*nomen dubium*) ⇒ *Antrodemus valens* (Leidy, 1870) Leidy, 1870 (*nomen dubium*) ⇒ *Megalosaurus valens* (Leidy, 1870) Nopcsa, 1901 (*nomen dubium*) ⇒ *Allosaurus valens* (Leidy, 1870) Gilmore, 1920 (*nomen dubium*) [Kimmeridgian–Tithonian; Morrison

Formation, Colorado, USA]
Generic name: *Antrodemus* ← {(Gr.) ἄντρον/antron: cave + δέμας/demas: body}
Specific name: *valens* ← {(Lat.) valēns: strong, stout}
Etymology: stout, **cavity-bodied**
Taxonomy: Saurischia: Theropoda: Allosauridae
Other species:
> *Allosaurus fragilis* Marsh, 1877 ⇒ *Labrosaurus fragilis* (Marsh, 1877) Nopsca, 1901 ⇒ *Antrodemus fragilis* (Marsh, 1877) Lapparent & Zbydzewski, 1957 [Kimmeridgian–Tithonian; Morrison Formation, Colorado, USA]
> *Laelaps trihedrodon* Cope, 1877 ⇒ *Dryptosaurus trihedrodon* (Cope, 1877) Hay, 1902 ⇒ *Antrodemus trihedrodon* (Cope, 1877) Kuhn, 1939 [Kimmeridgian; Morrison Formation, Colorado, USA]
> *Allosaurus ferox* Marsh, 1896 ⇒ *Antrodemus ferox* (Marsh, 1896) Ostrom & McIntosh 1966 [Kimmeridgian; Morrison Formation, Wyoming, USA]
> *Allosaurus lucaris* Marsh, 1878 ⇒ *Antrodemus lucaris* (Marsh, 1878) Hay, 1902 [Kimmeridgian; Morrison Formation, Wyoming, USA]
> *Creosaurus atrox* Marsh, 1878 ⇒ *Antrodemus atrox* (Marsh, 1878) Gilmore, 1920 ⇒*Allosaurus atrox* (Marsh, 1878) Paul, 1987 [Kimmeridgian; Morrison Formation, Wyoming, USA]
> *Allosaurus medius* Marsh, 1888 ⇒ *Antrodemus medius* (Marsh, 1888) Hay, 1901 [early Aptian; Arundel Clay Formation, Maryland, USA]
> *Allosaurus tendagurensis* Janensch, 1925 ⇒ *Antrodemus tendagurensis* (Janensch, 1925) von Huene, 1932 [late Kimmeridgian; Tendaguru Formation, Lindi, Tanzania]
> *Megalosaurus meriani* Greppin, 1870 (*nomen dubium*) ⇒ *Labrosaurus meriani* (Greppin, 1870) Janensch, 1920 (*nomen dubium*) ⇒ *Antrodemus meriani* (Greppin, 1870) Steel, 1970 ⇒ *Allosaurus meriani* (Greppin, 1870) Olshevsky, 1978 (*nomen dubium*) ⇒ ?*Ceratosaurus meriani* (Greppin, 1870) *emend.* Madsen & Welles, 2000 (*nomen dubium*) [early Kimmeridgian; Reuchenette Formation, Bern, Switzerland]
> *Allosaurus sibiricus* Riabinin, 1914 ⇒ *Antrodemus sibiricus* (Riabinin, 1914) Steel, 1970 ⇒ ?*Chilantaisaurus sibiricus* (Riabinin, 1914) Molnar *et al.*, 1990 (*nomen dubium*) [Berriasian–Hauterivian; Turgin Formation, Udinsk, Siberia, Russia]
Notes: According to *The Dinosauria 2nd edition*, *Antrodemus* is regarded as a synonym of *Allosaurus*.

Anurognathus ammoni [Pterosauria]
> *Anurognathus ammoni* Döderlein, 1923 [Tithonian; Solnhofen Limestone, Bayern, Germany]
Generic name: *Anurognathus* ← {(Gr.) ἄν-/an-: without + οὐρά/oura [ūrā]: tail + γνάθος/gnathos: jaw}; in reference to its unusually small tail.
Specific name: *ammoni* ← {(person's name) Ammon + -ī}; in honor of Ludwig von Ammon.
Etymology: Ammon's **tailless jaw**
Taxonomy: Pterosauria: Rhamphorhynchoidea: Anurognathidae
Notes: *Anurognathus* had a short head with pin-like teeth. Its tail was comparatively short.

[References: Döderlein L (1923). *Anurognathus ammoni,* ein neuer Flugsaurier. Sitzungsberichte der *Mathematisch-Physikalischen Klasse der Bayerischen Akademie der Wissenschaften zu, München* 1923: 306–307.]

Anzu wyliei
> *Anzu wyliei* Lamanna *et al.*, 2014 [Maastrichtian; Hell Creek Formation, South Dakota, USA]
Generic name: *Anzu* ← {(myth.) Anzu: "a feathered demon in ancient Mesopotamian (Sumerian and Akkadian) mythology"}; "alluding to the distinctive appearance of this large, presumably feathered dinosaur" (*1).
Specific name: *wyliei* ← {(person's name) Wylie + -ī}; "in honor of Mr. Wylie J. Tuttle, grandson of Mr. and Mrs. Lee B. Foster, in recognition of Mr. and Mrs. Foster's generous support of the scientific research and collections activities at Carnegie Museum of Natural History" (*1).
Etymology: Wylie's **Anzu**
Taxonomy: Theropoda: Oviraotorosauria: Caenagnathidae: Caenagnathinae
Notes: *Anzu* was recovered from the Hell Creek Formation, and therefore it is called "Chicken from hell".

[References: (*1) Lamanna MC, Sues H-D, Schachner ER, Lyson TR (2014). A new large-bodied oviraptorosaurian theropod dinosaur from the latest Cretaceous of western North America. *PLoS ONE* 9(3): e92022.]

Aoniraptor libertatem
> *Aoniraptor libertatem* Motta *et al.*, 2016 [Cenomanian–Turonian; Huincul Formation, Río Negro, Argentina]
Generic name: *Aoniraptor* ← {(Tehuelche) aoni: south + (Lat.) raptor: thief}; referring to its discovery in South America (*1).
Specific name: *libertatem* ← {(Lat.) lībertās: independence}; "due to the 200 years anniversary

of 9th July 1810, which led to the declaration of independence of Argentina from the Spanish government, thus consitituting one of the most important chapters in the history of this South American country" (*1).

Etymology: independent, **southern thief**

Taxonomy: Theropoda: Tyrannosauroidea: Megaraptora

[References: (*1) Motta MJ, Aranciaga Rolando AM, Rozadilla S, Agnolin F, Chimento N, Egli FB, Novas FE (2016). New theropod fauna from the Upper Cretaceous (Huincul Formation) of northwestern Patagonia, Argentina. *New Mexico Museum of Natural History and Science Bulletin* 71: 231–253.]

Aorun zhaoi

> *Aorun zhaoi* Choiniere *et al*., 2013 [Callovian; Shishugou Formation, Xinjiang, China]

Generic name: *Aorun* ← {(Chin.) Āo Rùn敖閏: 西海龍王 "the Dragon King of the West Sea in the epic *Journey to the west*"} (*1).

Specific name: *zhaoi* ← {(person's name) Zhao + -ī}; in honor of Professor Zhào Xǐjìn, who led several important vertebrate palaeontological expeditions to the Junggar Basin and introduced Xu X, Forster C. A and Choiniere J. N. (researchers) to the field area (*1).

Etymology: Zhao's **Aorun**

Taxonomy: Theropoda: Coelurosauria

[References: (*1) Choiniere JN, Clark JM, Forster CA, Norell MA, Eberth DA, Erickson GM, Chu H, Xu X (2013) A juvenile specimen of a new coelurosaur (Dinosauria: Theropoda) from the Middle-Late Jurassic Shishugou Formation of Xinjiang, People's Republic of China. *Journal of Systematic Palaeontology* 12(2): 177–215.]

Apatoraptor pennatus

> *Apatoraptor pennatus* Funston & Currie, 2016 [Campanian; Horseshoe Canyon Formation, Alberta, Canada]

Generic name: *Apatoraptor* ← {Apato- (Gr. myth.) Ἀπάτη/Apate: "an evil spirit released from Pandora's Box, used as a personification of deceit" + (Lat.) raptor: thief} (*1)

"The name is appropriate because the holotype deceived the collectors, who believed that it was an ornithomimid" (*1).

Specific name: *pennatus* ← {(Lat.) pennātus: feathered, winged}; referring to the feathered or 'winged' appearance of the arms (*1).

Etymology: feathered, **deceived thief**

Taxonomy: Theropoda: Coelurosauria: Maniraptora: Oviraptorosauria: Caenagnathidae: Elmisaurinae

[References: (*1) Funston GF, Currie PJ (2016). A new caenagnathid (Dinosauria: Oviraptorosauria) from the Horseshoe Canyon Formation of Alberta, Canada, and a reevaluation of the relationships of Caenagnathidae. *Journal of Vertebrate Paleontology* 36(4): e1160910.]

Apatornis celer [Avialae]

> *Ichthyornis celer* Marsh, 1873 ⇒ *Apatornis celer* (Marsh, 1873) Marsh, 1873 [Santonian–Campanian; Niobrara Formation, Kansas, USA]

Generic name: *Apatornis* ← {(Gr.) ἀπάτη/apatē: trick, fraud, deceit + ὄρνις/ornis: bird}

Specific name: *celer* ← {(Lat.) celer: swift, fleet}.

Etymology: swift, **deceptive bird**

Taxonomy: Theropoda: Avialae: Ornithurae

Other species:

> *Cimolopteryx retusa* Marsh, 1892 ⇒ *Apatornis retusus* (Marsh, 1892) Brodkorb, 1963 ⇒ *Palintropus retusus* (Marsh, 1892) Brodkorb, 1970 [Maastrichtian; Lance Formation, Wyoming, USA]

Apatosaurus ajax

> *Apatosaurus ajax* Marsh, 1877 ⇒ *Atlantsaurus ajax* (Marsh, 1877) Steel, 1970 ⇒ *Brontosaurus ajax* (Marsh, 1877) Bakker, 1986 [Kimmeridgian–Tithonian; Morrison Formation, Colorado, USA]

?= *Titanosaurus montanus* Marsh, 1877 (preoccupied) ⇒ *Atlantosaurus montanus* (Marsh, 1877) Marsh, 1877 (*nomen dubium*) [Kimmeridgian–Tithonian; Morrison Formation, Colorado, USA]

Generic name: *Apatosaurus* ← {(Gr.) ἀπάτη/apatē: deceit + σαῦρος/sauros: lizard}

Marsh and researchers thought the chevron bones, which differed from those of most known dinosaurs, were similar to those of mosasaurs (Squamata).

Specific name: *ajax* ← {(Lat.) Ajax = (Gr. myth.) Αἴας/Aias: a Greek hero and legendary king of Salamis}

Etymology: Ajax's **deceptive lizard**

Taxonomy: Saurischia: Sauropodomorpha: Sauropoda: Diplodocoidea: Diplodocidae: Apatosaurinae

Other species:

> *Apatosaurus louisae* Holland, 1915 ⇒ *Atlantosaurus louisae* (Holland, 1915) Steel, 1970 [Tithonian; Morrison Formation, Utah, USA]

= *Apatosaurus laticollis* Marsh, 1879 ⇒ *Atlantosaurus laticollis* (Marsh, 1879) Steel, 1970 [Kimmeridgian-Tithonian; Morrison Formation, Colorado, USA]

> "*Apatosaurus*" *minimus* Mook, 1917 ⇒ *Atlantosaurus minimus* (Mook, 1917) Steel, 1970 [Kimmeridgian; Morrison Formation, Wyoming, USA]
> *Apatosaurus grandis* Marsh, 1877 ⇒ *Camarasaurus grandis* (Marsh, 1877) Gilmore, 1925 [Kimmeridgian; Morrison Formation, Wyoming, USA]
> *Brontosaurus excelsus* Marsh, 1879 ⇒ *Apatosaurus excelsus* (Marsh, 1879) Riggs, 1903 ⇒ *Camarasaurus excelsus* (Marsh, 1879) Osborn *vide* von Huene, 1929 ⇒*Atlantosaurus excelsus* (Marsh, 1879) Steel, 1970 [Kimmeridgian; Morrison Formation, Wyoming, USA]
> *Elosaurus parvus* Peterson & Gilmore, 1902 ⇒ *Apatosaurus parvus* (Peterson & Gilmore, 1902) Upchurch *et al.*, 2004 ⇒ *Brontosaurus parvus* (Peterson & Gilmore, 1902) Tschopp *et al.*, 2015 [Kimmeridgian; Morrison Formation, Wyoming, USA]
> *Apatosaurus alenquerensis* Lapparent & Zbyszewski, 1957 ⇒*Atlantosaurus alenquerensis* (Lapparent & Zbyszewski, 1957) Steel, 1970 ⇒ *Camarasaurus alenquerensis* (Lapparent & Zbyszewski, 1957) McIntosh, 1990 ⇒ *Lourinhasaurus alenquerensis* (Lapparent & Zbyszewski, 1957) Dantas *et al.*, 1998 [late Kimmeridgian; Lourinhã Formation, Lisboa, Portugal]
> *Apatosaurus yahnahpin* Filla & Redman, 1994 ⇒ *Eobrontosaurus yahnahpin* (Filla & Redman, 1994) Bakker, 1998 ⇒ *Brontosaurus yahnahpin* (Filla & Redman, 1994) Tschopp *et al.*, 2015 [Kimmeridgian; Morrison Formation, Wyoming, USA]

Notes: Tschopp *et al.* (2015) concluded that *Brontosaurus* is a valid genus of sauropod distinct from *Apatosaurus*. (*1).

[References: Marsh OC (1877). Notice of new dinosaurian reptiles from the Jurassic formation. *American Journal of Sciences* 3(14): 514–516.; (*1) Tschopp E, Mateus O, Benson RBJ (2015). A specimen-level phylogenetic analysis and taxonomic revision of Diplodocidae (Dinosauria, Sauropoda). *PeerJ* 3: e857.]

Appalachiosaurus montgomeriensis

> *Appalachiosaurus montgomeriensis* Carr *et al.*, 2005 [Campanian; Demopolis Chalk Formation, Alabama, USA]

Generic name: *Appalachiosaurus* ← {(place-name) Appalachio-: Appalachia* | Appalachian Mountains + (Gr.) σαῦρος/sauros: lizard}; in reference to "the occurrence of the specimen in eastern North America during the Late Cretaceous" (*1).

*Appalachia was an island mass separated from Laramidia in the Mesozoic Era.

Specific name: *montgomeriensis* ← {(place-name) Montgomery County + (Lat.) -ensis: from} (*1)

Etymology: **Appalachian lizard** from Montgomery County

Taxonomy: Theropoda: Tyrannosauroidea

[References: (*1) Carr TD, Williamson TE, Schwimmer DR (2005). A new genus and species of tyrannosauroid from the Late Cretaceous Demopolis Formation of Alabama. *Jornal of Vertebrate Paleontology* 25(1): 119–143.]

Aquilarhinus palimentus

> *Aquilarhinus palimentus* Prieto-Márquez *et al.*, 2019 [Campanian; Aguja Formation, Texas, USA]

Generic name: *Aquilarhinus* ← {(Lat.) aquila: eagle + (Gr.) ῥῖνός/rhinos [ῥίς/rhis: nose]}; referring to "the morphology of the rostrum" (*1).

Specific name: *palimentus* ← {(Lat.) pāla: spade, shovel + mentus [mentum: chin]}; referring to "the assumed resembrance of the predentary to a spade or shovel given to dorsomedial projection of the symphyseal process of the dentary" (*1).

Etymology: shovel-billed **eagle nose**

Taxonomy: Ornithischia: Ornithopoda: Hadrosauridae

[References: (*1) Prieto-Márquez A, Wagner JR, Lehman T (2019). An unusual 'shovel-billed' dinosaur with trophic specializations from the early Campanian of Trans-Pecos Texas, and the ancestral hadrosaurian crest. *Journal of Systematic Palaeontology* 18: 461–498.]

Aquilops americanus

> *Aquilops americanus* Farke *et al.*, 2014 [Aptian–Albian; Cloverly Formation, Montana, USA]

Generic name: *Aquilops* ← {(Lat.) aquila: eagle + (Gr.) ὤψ/ōps: face}; "referring to the hooked beak on the skull of the animal" (*1).

Specific name: *americanus* ← {americanus: American}; "reflecting the species' status as the earliest unequivocal neoceratopsian in North America" (*1).

Etymology: American **eagle face**

Taxonomy: Ornithischia: Ceratopsia: Neoceratopsia

[References: (*1) Farke AA, Maxwell WD, Cifelli RL, Wedel MJ (2014). A ceratopsian dinosaur from the Lower Cretaceous of western North America, and the biogeography of Neoceratopsia. *PLoS ONE* 9(12): e112055.]

Aragosaurus ischiaticus

> *Aragosaurus ischiaticus* Sanz *et al.*, 1987

[Berriasian; Villar del Arzobispo Formation, Teruel, Aragon, Spain]

Generic name: *Aragosaurus* ← {(Spanish, place-name) Aragón: an autonomous community in Spain + (Gr.) σαῦρος/sauros: lizard}; "dedicado a Aragón" (*1) [dedicated to Aragón].

Specific name: *ischiaticus* ← {(Spanish) isquiático [ischion: ischium + -atico: -aticus]} ; "hace referencia al especial desarrollo dorso-ventral de la apófisis púbica del isquion" (*1) [referring to the special dorso-ventral development of the pubic process of the ischium].

Etymology: **Aragón lizard** with large ischium [**lagarto de Aragón** con gran isquion]

Taxonomy: Saurischia: Sauropodomorpha: Sauropoda
cf. Saurischia: Sauropodomorpha: Eusauropoda (acc. *The Dinosauria 2nd edition*)

[References: (*1) Sanz JL, Buscalioni AD, Casanovas ML, Santafé JV (1987). Dinosaurios del Cretácico Inferior de Galve (Teruel, España). *Estudios Geológicos, volume extraordinario Galve-Tremp*: 45–64.]

Aralosaurus tuberiferus

> *Aralosaurus tuberiferus* Rozhdestvensky, 1968 [Coniacian–Santonian; Bostobe Formation, Central Kazakhstan]

Generic name: *Aralosaurus* ← {(place-name) Aral Sea + (Gr.) σαῦρος/sauros: lizard}

Specific name: *tuberiferus* ← {(Lat.) tuber: tuber, lump + ferō: bear, carry}; "bearing a tuber" on its snout. (*1)

Etymology: **Aral lizard** bearing a tuber

Taxonomy: Ornithischia: Ornithopoda:
Hadrosauridae: Lambeosaurinae: Aralosaurini

Notes: A medium-sized hadrosaur with a helmet-like crest had 1000 small teeth in 30 rows.

[References: (*1) Rozhdestvensky AK (1968). Gadrozavry Kazakhstana [Hadrosaurs of Kazakhstan]. In: *Verkhnepaleozoiskie i Mezozoiskie Zemnovodnye i Presmykaiuschchiesia* [*Upper Paleozoic and Mesozoic Amphibians and Reptiles*] Akademia Nauk SSSR, Moscow: 97–141.]

Aratasaurus museunacionali

> *Aratasaurus museunacionali* Sayão *et al.*, 2020 [Aptian; Romualdo Formation, Ceará, Brazil]

Generic name: *Aratasaurus* ← {(Tupi) arata [ara: born + atá: fire] + (Gr.) σαῦρος/sauros: lizard}

Specific name: *museunacionali* ← {(institution name) Museu Nacional + -ī}; in honor of "the Museu Nacional / Universidade Federal do Rio de Janeiro, which is the oldest scientific institution of Brazil and was recently devasted by a fire".

Etymology: Museu Nacional's **lizard born of fire**

Taxonomy: Theropoda: Tetanurae: Coelurosauria

Archaeoceratops oshimai

> *Archaeoceratops oshimai* Dong & Azuma, 1997 [Early Cretaceous; Xinminbao Group, Gansu, China]

Generic name: *Archaeoceratops* ← {(Gr.) ἀρχαῖο-/archaio- [ἀρχαῖος/archaios: ancient] + ceratops [κερατ-/kerat- (←κέρας/keras: horn) + ὤψ/ōps: face]}

Specific name: *oshimai* ← {(Jpn. person's name) Oshima + -ī}; in honor of Mr. Oshima, the director of Chunichi-Shinbun, Japanese newspaper publisher, who supported the expedition.

Etymology: Oshima's **ancient horned face (ceratopsian)**

Taxonomy: Ornithischa: Cerapoda:
Marginocephalia: Ceratopsia: Neoceratopsia: Archaeoceratopsidae

Other species:

> *Archaeoceratops yujingziensis* You *et al.*, 2010 [Aptian; Xiagou Formation, Gansu, China]

Notes: Estimated to be 1 m in total length. *Archaeoceratops* had no horns, with a small bony frill.

[References: You H-L, Tanoue K, Dodson P (2010). A new species of *Archaeoceratops* (Dinosauria: Neoceratopsia) from the Early Cretaceous of the Mazongshan area, northwestern China. In: Ryan MJ, Chinnery-Allgeier B, Eberth DA (eds), *New Perspectives on Horned Dinosaurs: The Royal Tyrrell Museum Ceratopsian Symposium.* Indiana University Press, Broomington: 59–67.]

Archaeodontosaurus descouensi

> *Archaeodontosaurus descouensi* Buffetaut, 2005 [Bajocian–Bathonian; Isalo III Formation, Madagascar]

Generic name: *Archaeodontosaurus* ← {(Gr.) ἀρχαῖος/archaios: ancient + ὀδών/odōn: tooth + σαῦρος/sauros: lizard}

Specific name: *descouensi* ← {(person's name) Descouens + -ī}; in honor of Dr. Didier Descouens.

Etymology: Descouens' **ancient-toothed lizard**

Taxonomy: Saurischia: Sauropodomorpha:
Sauropoda

[References: Buffetaut E (2005). A new sauropod dinosaur with prosauropod-like teeth from the Middle Jurassic of Madagascar. *Bulletin de la Société Géglogique de France* 176(5): 467–473. Abstract.]

Archaeopteryx lithographica [Avialae]

> *Archaeopteryx lithographica* von Meyer, 1861 (*nomen conservandum*) [Tithonian; Solnhofen

Limestone, Bavaria (Bayern), Germany]
= *Griphornis longicaudatus* Owen *vide* Woodword, 1862 (*nomen rejectum*) ⇒ *Griphosaurus longicaudatus* (Owen *vide* Woodword, 1862) Owen *vide* Brodkorb, 1963
= *Griphosaurus problematicus* Wagner *vide* Woodword, 1862 (*nomen rejectum*)
= *Archaeopteryx macrura* Owen, 1863 (London specimen) (*nomen rejectum*) [Tithonian; Solnhofen Formation, Bayern, Germany]
= *Archaepoteryx siemensii* Dames, 1897 ⇒ *Archaeornis siemensii* (Dames, 1897) Petronievics *vide* Petronievics & Woodward, 1917 (Berlin specimen) [Tithonian; Solnhofen Formation, Bayern, Germany]
= *Archaeopteryx owenii* Petronievics, 1917 (*nomen rejectum*) [Tithonian; Solnhofen, Bayern, Germany]
= *Archaeopteryx oweni* Petronievics, 1921 (*nomen oblitum*) (London specimen)
= *Archaeopteryx recurva* Howgate, 1984 ⇒ *Jurapteryx recurva* (Howgate, 1984) Howgate, 1985 [Tithonian; Solnhofen Formation, Bayern, Germany]
= *Archaeopteryx bavarica* Wellnhofer, 1993 [Tithonian; Altmühltal Formation, Solnhofen, Bayern, Germany]
= ?*Wellnhoferia grandis* Elzanowski, 2001 [Tithonian; upper member, Solnhofen Formation, Bayern, Germany]

Generic name: *Archaeopteryx* ← {(Gr.) ἀρχαῖος/archaios: ancient + πτέρυξ/pteryx: feather, wing}
Specific name: *lithographica* ← {(Gr.) λιθογραφικός/litho-graphikos}; lithographic.
Etymology: lithographic, **ancient feather**
Taxonomy: Theropoda: Avialae: Archaeopterygidae
Other species:
> *Archaeopteryx albersdoerferi* Kundrát, 2018 [Tithonian; Mörnsheim Formation, Germany]
> *Pterodactylus crassipes* von Meyer, 1857 ⇒ *Rhamphorhynchus crassipes* (von Meyer, 1857) von Meyer, 1857 ⇒ *Scaphognathus crassipes* (von Meyer, 1857) Wagner, 1861 ⇒ *Archaeopteryx crassipes* (von Meyer, 1857) Ostrom, 1970 ⇒ *Ostromia crassipes* (von Meyer, 1857) Foth & Rauhut, 2017 (Haarlem specimen) [Tithonian; Painten Formation, Bayern, Germany]

Archaeorhynchus spathula [Avialae]

> *Archaeorhynchus spathula* Zhou & Zhang, 2006 [Barremian–Aptian; Yixian Formation, Liaoning, China]

Generic name: *Archaeorhynchus* ← {(Gr.) archae- [ἀρχαῖος/archaios: ancient] + ῥύγχος/ rhynchos: beak}
Specific name: *spathula* ← {spathula [(Gr.) σπάθη/spathē]}: indicating the spathulate dentary[*1].
Etymology: spatulate, **ancient beak**
Taxonomy: Theropoda: Avialae: Ornithothoraces: Euornithes

[References: (*1) Zhou Z, Zhang F (2006). A beaked basal ornithurine bird (Aves, Ornithurae) from the Lower Cretaceous of China. *Zoologica Scripta* 35: 363–373.]

Archaeornithoides deinosauriscus

> *Archaeornithoides deinosauriscus* Elzanowski & Wellnhofer, 1992 [Campanian; Djadokhta Formation, Ömnögovi, Mongolia]

Generic name: *Archaeornithoides* ← {*Archaeornis** [(Gr.) ἀρχαῖος/archaios: ancient + ὄρνις/ornis: bird] + -oides: resembling, like}[*1]
*a junior synonym of *Archaeopteryx*.
Specific name: *deinosauriscus* ← {(Gr.) δεινός/deinos + σαῦρος/sauros + (dim.) -iscus}; of little dinosaur.[*1]
Etymology: ***Archaeornis*-like**, of little dinosaur
Taxonomy: Theropoda: Coelurosauria: Troodontidae *cf.* Theropoda: Archaeornithoididae[*1]

[References: (*1) Elzanowski A, Wellnhofer P (1993). A new link between theropods and birds from the Cretaceous of Mongolia. *Nature* 359: 821–823.]

Archaeornithomimus asiaticus

> *Ornithomimus asiaticus* Gilmore, 1933 ⇒ *Archaeornithomimus asiaticus* (Gilmore, 1933) Russell, 1972 [Campanian; Iren Dabasu Formation, Inner Mongolia, China]

Generic name: *Archaeornithomimus* ← {(Gr.) ἀρχαῖος/archaios: ancient + *Ornithomimus* [ὄρνις/ornis: bird + μιμος/mīmos: mimic]}
Specific name: *asiaticus* ← {(Lat.) asiāticus: of Asia}.
Etymology: **ancient *Ornithomimus*** (bird mimic) of Asia
Taxonomy: Theropoda: Coelurosauria: Ornithomimosauria: Ornithomimidae
Other species:
?> *Archaeornithomimus bissektensis* Nesov, 1995 [Turonian; Bissekty Formation, Navoi, Uzbekistan]

[References: Gilmore CW (1933). On the dinosaurian fauna of the Iren Dabasu Formation. *Bulletin of the American Museum of Natural History* 67: 23–78.]

Archaeornithura meemannae [Avialae]

> *Archaeornithura meemannae* Wang *et al.*, 2015 [Hauterivian; Huajiying Formation,

Fengning, Hebei, China]

Generic name: *Archaeornithura* ← {(Gr.) ἀρχαῖος/archaios: ancient + ornithura [ὄρνις/ornis + οὐρά/ūra]: ornithuromorph} (*1)

Specific name: *meemannae* ← {(person's name) Meemann + -ae}; "in honour of Dr. Meemann Chang for her continuous support of the study of the Jehol Biota" (*1).

Etymology: Dr. Meemann's **ancient ornithuromorph**

Taxonomy: Theropoda: Avialae: Ornithothoraces: Ornithuromorpha: Hongshanornithidae

[References: (*1) Wang M, Zheng X, O'Connor JK, Lloyd GT, Wang X, Wang Y, Zhang X, Zhou Z (2015). The oldest record of Ornithuromorpha from the Early Cretaceous of China. *Nature Communications* 6: 6987.]

Arcovenator escotae

> *Arcovenator escotae* Tortosa *et al.*, 2013 [late Campanian; Lower Argiles Rutilantes Formation, south-eastern France]

Generic name: *Arcovenator* ← {(place-name) river Arc + (Lat.) vēnātor: hunter}; referring to the basin of the Arc river (*1).

Specific name: *escotae* ← {(company) Escota + -ae}; in honor of the motorway company ESCOTA, which has funded the excavations at the locality since 2006 (*1).

Etymology: Escota's **Arc hunter**

Taxonomy: Theropoda: Ceratosauria: Abelisauroidea: Abelisauridae

[References: (*1) Tortosa T, Buffetaut E, Vialle N, Dutour Y, Turini E, Cheylan G (2013). A new abelisaurid dinosaur from the Late Cretaceous of southern France: Palaeobiogeographical implications. *Annales de Paléontologie* 100(1): 63–86.]

Arcusaurus pereirabdalorum

> *Arcusaurus pereirabdalorum* Yates *et al.*, 2011 [Hettangian–Sinemurian; Elliot Formation, Free State, South Africa]

Generic name: *Arcusaurus* ← {(Lat.) arcus: rainbow + (Gr.) σαῦρος/sauros: lizard}; referring to the Rainbow Nation* (*1).

*Rainbow Nation is a term coined by Archbishop Desmond Tutu to describe post-apartheid South Africa.

Specific name: *pereirabdalorum* ← {(person's name) Pereira + Abdala + -ōrum}; in honor of Lucille Pereira & Fernando Abdala, who discovered the fossils (*1)

Etymology: Pereira and Abdala's **rainbow lizard**

Taxonomy: Saurischia: Sauropodomorpha

[References: Yates AM, Bonnan MF, Neveling J (2011). A new basal sauropodomorpha dinosaur from the Early Jurassic of South Africa. *Journal of Vertebrate Paleontology* 31(3): 610–625. Abstract.; (*1) *Arcusaurus pereirabdalorum*: the little sauropodomorph that could. The Evolving Paleontologist Digital Home of Dr. Matthew F. Bonnan, Ph. D.〈https://matthewbonnan.wordpress.com/〉]

Arenysaurus ardevoli

> *Arenysaurus ardevoli* Pereda-Suberbiola *et al.*, 2009 [Maastrichtian; Tremp Formation, Huesca, Aragón, Spain]

Generic name: *Arenysaurus* ← {(place-name) Arén (= Areny de Noguera in Catalonian language): the village of Huesca province + (Gr.) σαῦρος/sauros: lizard} (*1)

Specific name: *ardevoli* ← {(person's name) Ardèvol + -ī}; "in honour of Lluís Ardèvol (Geoplay, Tremp, Lleida), who discovered the Blasi site" (*1).

Etymology: Ardèvol's **Areny (Arén) lizard**

Taxonomy: Ornithischia: Ornithopoda: Hadrosauridae: Lambeosaurinae

[References: (*1) Pereda-Suberbiola X, Canudo JI, Cruzado-Caballero P, Barco JL, López-Martínez N, Oms O, Ruiz-Omeñaca JI (2009). The last hadrosaurid dinosaurs of Europe: A new lambeosaurine from the uppermost Cretaceous of Aren (Huesca, Spain). *Comptes Rendus Palevol* 8(6): 559–572.]

Argentinosaurus huinculensis

> *Argentinosaurus huinculensis* Bonaparte & Coria, 1993 [Albian–Cenomanian; Huincul Formation, Plaza Huincul, Neuquén, Argentina]

Generic name: *Argentinosaurus* ← {(place-name) Argentina + σαῦρος/sauros: lizard} (*1)

Specific name: *huinculensis* ← {(place-name) Plaza Huincul : the town where the holotype was found + -ensis} (*1)

The Huincul Formation where the holotype was found is near the town of Plaza Huincul.

Etymology: **Argentine lizard** from Plaza Huincul

Taxonomy: Saurischia: Sauropodomorpha: Sauropoda: Titanosauria

[References: (*1) Bonaparte JF, Coria R (1993). A new and huge titanosaur sauropod from the Limay Formation (Albian–Cenomanian) of Neuquén Province, Argentina. *Ameghiniana* 30(3): 271–282.]

Argyrosaurus superbus

> *Argyrosaurus superbus* Lydekker, 1893 [Campanian; Lago Volhué Huapi Formation, Chubut, Argentina]

Generic name: *Argyrosaurus* ← {(Gr.) ἄργυρος/argyros: silver + σαῦρος/sauros: lizard}; because the type species was discovered in Argentina, which means 'silver land'.

Specific name: *superbus* ← {(Lat.) superbus: proud}.
Etymology: proud **silver lizard**
Taxonomy: Saurischia: Sauropodomorpha:
Sauropoda: Titanosauria
[References: Mannion PD, Otero A (2012). A reappraisal of the Late Cretaceous Argentinean sauropod dinosaur *Argyrosaurus superbus*, with a description of a new titanosaur genus. *Journal of Vertebrate Paleontology* 32(3): 614–638.]

Aristosuchus pusillus
> *Poekilopleuron pusillus* Owen, 1876 ⇒ *Aristosuchus pusillus* (Owen, 1876) Seeley, 1887 [Barremian; Wessex Formation, Isle of Wight, UK]
= *Poekilopleuron minor* Owen *vide* Cope, 1878
Generic name: *Aristosuchus* ← {(Gr.) ἄριστος/aristos: bravest, best, superior, noblest + Σοῦχος/Souchos: the name of the Egyptian crocodile-headed god Sobek}
Specific name: *pusillus* ← {(Lat.) pusillus: vert little, very small}.
Etymology: very small, **superior crocodile-headed god**
Taxonomy: Theropoda: Coelurosauria:
Compsognathidae
[References: Seeley HG (1887). *On Aristosuchus pusillus* (Owen), being further notes on the fossils described by Sir R. Owen as *Poikilopleuron pusillus*, Owen. *Quarterly Journal of the Geological Society* 43: 221–228.]

Arkansaurus fridayi
> *Arkansaurus fridayi* Hunt, 2003 (*nomen nudum*) ⇒ *Arkansaurus fridayi* Hunt & Quinn, 2018 (officially described) [Aptian–Albian; Cedar Mountain Formation, Arkansas, USA]
Generic name: *Arkansaurus* ← {(place-name) Arkansas + (Gr.) σαῦρος/sauros: lizard}
Specific name: *fridayi* ← {(person's name) Friday + -ī}; in honor of Joe B. Friday, who discovered the fossil remains.
Etymology: Friday's **Arkansas lizard**
Taxonomy: Theropoda: Coelurosauria:
Ornithomimosauria
[References: Hunt RK, Quinn JH (2018). A new ornithomimosaur from the Lower Cretaceous Trinity Group of Arkansas. *Journal of Vertebrate Paleontology* 38(1): e1421209.]

Arkharavia heterocoelica
> *Arkharavia heterocoelica* Alifanov & Bolotsky, 2010 [Maastrichtian; Udurchukan Formation, Amur, Russia]
Generic name: *Arkharavia* ← {(Russian, place-name) Архара (Arkhara/Arhara) + (Lat.) via: road}
Specific name: *heterocoelica* ← {hetero- [(Gr.) ἕτερος/heteros = (Lat.) alter] + κοῖλος/koilos: hollow, hollowed}; referring to the saddle-shaped centrum and high neutral spine of the anterior caudal vertebrae. [*1]
Etymology: hetero-hollowed **Arkhara road**
Taxonomy: Ornithischia: Ornithopoda:
Hadrosauridae *cf.* Saurischia: Sauropoda [*1]
[References: (*1) Alifanov VR, Bolotsky YL (2010). *Arkharavia heterocoelica* gen. *et* sp. nov., a new sauropod dinosaur from the Upper Cretaceous of the Far East of Russia. *Paleontological Journal* 44(1): 84–91.]

Arrhinoceratops brachyops
> *Arrhinoceratops brachyops* Parks, 1925 [Campanian–Maastrichtian; Horseshoe Canyon Formation, Alberta, Canada]
Generic name: *Arrhinoceratops* ← {(Gr.) ἀ-/a-: not, without + ῥῖνο-/rhīno- [ῥις/rhis: nose] + κερατ-/kerat- [κέρας/keras: horn] + ὤψ/ōps: face}
Parks thought it was lacking a nose horn [*1], however, Tyson pointed that a nasal horn core was present [*2].
Specific name: *brachyops* ← {(Gr.) βραχύς/brachys: short + ὤψ/ōps: face}
Etymology: short-faced, **ceratopsian having no nose horn**
Taxonomy: Ornithischia: Marginocephalia:
Ceratopsia: Ceratopsidae: Chasmosaurinae
[References: (*1) Parks WA (1925). *Arrhinoceratops brachyops*, a new genus and species of Ceratopsia from the Edmonton Formation of Alberta. *University of Toronto Studies, Geology Series* 19: 1–15.; (*2) Tyson H (1981). The structure and relationships of the horned dinosaur *Arrhinoceratops Parks* (Ornithischia: Ceratopsidae). *Canadian Journal of Earth Scieces* 18(8): 1241–1247.]

Arstanosaurus akkurganensis
> *Arstanosaurus akkurganensis* Shilin & Suslov, 1982 [Santonian–Campanian; Bostobe Formation, Akkurgan-Boltyk, Kyzylorda/Qyzylorda, Kazakhstan]
Generic name: *Arstanosaurus* ← {(place-name) Arstan + (Gr.) σαῦρος/sauros: lizard}; referring to the Arstan-ancient region of Kazakhstan nearby where the fossil was found.
Specific name: *akkurganensis* ← {(place-name) Akkurgan + -ensis}
Etymology: **Arstan lizard** from Akkurgan
Taxonomy: Ornithischia: Hadrosauroidea:

Hadrosauridae: Lambeosaurinae

Notes: According to *The Dinosauria 2nd edition*, *Arstanosaurus* is considered as a *nomen dubium*.

Asfaltovenator vialidadi

> *Asfaltovenator vialidadi* Rauhut & Pol, 2019 [late Toarchian–Bajocian; Cañadón Asfalto Formation, Chubut, Argentina]

Generic name: *Asfaltovenator* ← {(place-name) Cañadón Asfalto Formation + (Lat.) vēnātor: hunter}

Specific name: *vialidadi* ← {Vialidad + -ī}; in honour of "the Administración de Vialidad Provincial of Chubut and the Dirección Nacional de Vialidad, for their aid to paleontological expeditions of the Museo Paleontológicico Egidio Feruglio" (*1).

Etymology: Vialidad's **hunter from Cañadón Asfalto Formation**

Taxonomy: Saurischia: Theropoda: Tetanurae: Allosauroidea

[References: (*1) Rauhut OWM, Pol D (2019). Probable basal allosaurid from the early Middle Jurassic Cañadón Asfalto Formation of Argentina highlights phylogenetic uncertainty in tetanuran theropod dinosaurs. *Scientific Reports* 9: 18826.]

Asiaceratops salsopaludalis

> *Asiaceratops salsopaludalis* Nessov *et al*., 1989 [Cenomanian; Khodzhakul Formation, Navoi, Uzbekistan]

Generic name: *Asiaceratops* ← {(place-name) Asia + ceratops [(Gr.) κέρας/keras: horn + ὤψ/ōps: face]}

Specific name: *salsopaludalis* ← {(Lat.) salsus: salt + palūs: swamp, marsh}

Etymology: **Asian horned face (ceratopsian)** of the salt marsh

Taxonomy: Ornithischia: Marginocephalia: Ceratopsia: Leptoceratopsidae

Other species:

> *Microceratops sulcidens* Bolin, 1953 ⇒ *Asiaceratops sulcidens* (Bolin, 1953) Nessov *et al*., 1989 [Barremian–Aptian; Xinminbao Group, Gansu, China]

Notes: According to *The Dinosauria 2nd edition*, *Asiaceratops salsopaludalis* and also *Microceratops sulcidens* were regarded as *nomina dubia*.

Asiatosaurus mongoliensis

> *Asiatosaurus mongoliensis* Osborn, 1924 [Early Cretaceous; Öösh Formation, Guangxi, China]

Generic name: *Asiatosaurus* ← {(Gr.) Ἀσια: Asia + (Gr.) σαύρα/saura: lizard}

Specific name: *mongoliensis* ← {(place-name) Mongolia + -ensis}

Etymology: **Asian lizard** from Mongolia

Taxonomy: Saurischia: Sauropodomorpha: Sauropoda

Other species:

> *Asiatosaurus kwangshiensis* Hou *et al*., 1975 [Aptian; Xinlong Formation, Guangxi, China]

Notes: According to *The Dinosauria 2nd edition* (2004). *A. mongoliensis* and *A. kwangshiensis* are considered as *nomina dubia*.

[References: Osborn HF (1924). Sauropoda and Theropoda from the Lower Cretaceous of Mongolia. *American Museum Novitates* 128: 1–7.]

Asilisaurus kongwe [Silesauridae]

> *Asilisaurus kongwe* Nesbitt *et al*., 2010 [Anisian; Manda Formation, Ruhuhu Basin, Tanzania]

Generic name: *Asilisaurus* ← {(Swahili) asili: ancestor or foundation + (Gr.) σαῦρος/sauros: lizard} (*1)

Specific name: *kongwe* ← {(Swahili) kongwe: ancient} (*1)

Etymology: ancient, **ancestor lizard**

Taxonomy: Archosauria: Ornithodira: Dinosauriformes: Silesauridae

[References: (*1) Nesbitt SJ, Sidor CA, Irmis RB, Angielczyk KD, Smith RMA, Tsuji L A (2010). Ecologically distinct dinosaurian sister group shows early diversification of Ornithodira. *Nature* 464(7285): 95–98.]

Astrodon johnstoni

> *Astrodon johnstoni* Leidy, 1865 [Aptian; Arundel Clay Formation, Prince George's County, Maryland, USA]

= *Pleurocoelus nanus* Marsh, 1888 (juvenile) [Aptian; Arundel Clay Formation, Maryland, USA]

= *Pleurocoelus altus* Marsh, 1888 1921 (adult) [Aptian–Albian; Arundel Clay Formation, Maryland, USA]

Generic name: *Astrodon* Johnston, 1859* ← {(Gr.) αστρ-/astr- [ἀστηρ/aster: star] + ὀδών/odōn (←ὀδούς/odous [odūs]: tooth)}; referring to a cross section of its tooth.

*Christopher Johnston, a dentist, received two teeth from Philip Tyson, a chemist. Johnston cut one tooth in half and discovered a star pattern. He named it *Astrodon* without a specific epithet in 1859. So, Leidy named it *Astrodon johnstoni* in 1865 and presented it in *American Journal of Dental Science* (*1).

Specific name: *johnstoni* ← {(person's name)

Johnston + -ī}; in honor of the dentist Christopher Johnston.

Etymology: Johnston's **star tooth**

Taxonomy: Saurischia: Sauropodoorpha: Sauropoda: Titanosauriformes [*2]

Other species:

> *Cetiosaurus conybearei* Merville, 1849 ⇒ *Pelorosaurus conybearei* (Merville, 1849) Mantell, 1850 [late Valanginian; Tunbridge Wells Sand Formation, England, UK]

= *Pleurocoelus valdensis* Lydekker, 1889 (*nomen dubium*) ⇒ *Astrodon valdensis* (Lydekker, 1889) Swinton, 1936 [Isle of Wight, UK]

> *Omosaurus armatus* Owen, 1875 ⇒ *Dacentrurus armatus* (Owen, 1875) Lucas, 1902 [late Kimmeridgian; Kimmeridge Clay Formation, Wiltshire, UK]

= *Astrodon pussilus* Lapparent & Zbyszewski, 1957 [late Kimmeridgian; Lisboa, Portugal]

> *Apatosaurus grandis* Marsh, 1877 ⇒ *Morosaurus grandis* (Marsh, 1877) Williston, 1898 ⇒ *Camarasaurus grandis* (Marsh, 1877) Gilmore, 1925 [Kimmeridgian; Morrison Formation, Wyoming, USA]

= *Pleurocoelus montanus* Marsh, 1896 ⇒ *Astrodon montanus* (Marsh, 1896) Kuhn, 1936 [Kimmeridgian; Morrison Formation, Wyoming, USA]

Notes: According to *The Dinosauria 2nd edition*, *Astrodon johnstoni* was regarded as a *nomen dubium*. D'Emic (2013) considered *Astrodon johnstoni*, *Pleurocoelus altus* and *Pleurocoelus nanus* to be *nomina dubia*.

[References: (*1) *Astrodon johnstoni*, Maryland State Dinosaur. ⟨https://msa.maryland.gov/msa/mdmanual/01glance/html/symbols/dino.html⟩ [2018/02/08]; (*2) Carpenter K, Tidwell V (2005). Reassessment of the Early Cretaceous sauropod *Astrodon johnsoni* Leidy, 1856 (Titanosauriformes). In: Carpenter K and Tidwell V (eds), *Thunder Lizards: The Sauropodomorph Dinosaurs*. Indiana University Press. Broomington: pp. 38–77.]

Astrophocaudia slaughteri

> *Astrophocaudia slaughteri* D'Emic, 2012 [Albian; Trinity Group, Texas, USA]

Generic name: *Astrophocaudia* ← {(Gr.) ἀ-/a-: non- + (Gr.) στροφο-/stropho-: twisting or turning + (Lat.) cauda: tail}; referring to "the tightly articulating hyposphene-hypantrum system in the anterior and middle caudal vertebrae, which also resembles a star (astron; Greek) in posterior view, and also referring to *Astrodon*, the first Early Cretaceous North American sauropod" [*1].

Specific name: *slaughteri* ← {(person's name) Slaughter + -ī}; in honor of "Dr. Robert H. Slaughter, who excavated the specimen in the 1960s" [*1].

Etymology: Slaughter's **non-twisting tail**

Taxonomy: Saurischia: Sauropodomorpha: Sauropoda: Neosauropoda: Titanosauriformes [*1]

[References: (*1) D'Emic MD (2012). Revision of the sauropod dinosaurs of the Lower Cretaceous Trinity Group, southern USA, with the description of a new genus. *Journal of Systematic Palaeontology* 11(6): 1–20.]

Asylosaurus yalensis

> *Asylosaurus yalensis* Galton, 2007 [Rhaetian; Magnesian Conglomerate Formation, Clifton, Bristol, UK]

Generic name: *Asylosaurus* ← {(Gr.) ἄσῦλος/asylos: unharmed, safe from violence or asylon = refuge, sanctuary + σαῦρος/sauros: lizard} [*1]

Specific name: *yalensis* ← {(university) Yale + -ensis}; "of Yale College (now University), where O. C. Marsh stored the specimen, so it was unharmed in air raids on BCM* in November, 1940" [*1].

*Bristol City Museum and Art Gallery, Bristol, UK.

Etymology: **unharmed lizard** of Yale University

Taxonomy: Saurischia: Sauropodomorpha

[References: (*1) Galton P (2007). Notes on the remains of archosaurian reptiles, mostly basal sauropodomorph dinosaurs, from the 1834 fissure fill (Rhaetian, Upper Triassic) at Clifton in Bristol, southwest England. *Revue de Paléobiologie* 26(2): 505–591.]

Atacamatitan chilensis

> *Atacamatitan chilensis* Kellner *et al.*, 2011 [Turonian–Coniacian; Tolar Formation, Atacama Desert (Antofagasta Region), Chile]

Generic name: *Atacamatitan* ← {(place-name) Atacama Desert + (Gr. myth.) Τῑτάν/Tītan}; "Atacama, from the desert where the specimen was found, and Titan, which relates to a group of Greek divinities" [*1].

Specific name: *chilensis* ← {(place-name) Chile + -ensis}; "in allusion to Chile, the country where the specimen was found" [*1].

Etymology: **Atacama Desert titan** from Chile

Taxonomy: Saurischia: Sauropodomorpha: Sauropoda: Titanosauria: Titanosauriformes: Titanosauridae [*1]

[References: (*1) Kellner AWA, Rubilar -Rogers D, Vargas A, Suárez M (2011). A new titanosaur sauropod from the Atacama Desert, Chile. *Anais da Academia Brasileira de Ciências* 83(1): 211–219.]

Atlantosaurus montanus

> *Titanosaurus montanus* Marsh, 1877 (preoccupied) ⇒ *Atlantosaurus montanus* (Marsh, 1877) Marsh, 1877 (?*nomen dubium*) [Kimmeridgian–Tithonian; Morrison Formation, Colorado, USA]

Generic name: *Atlantosaurus* ← {(Gr. myth.) Ἄτλας/Atlas: a mythical Titan + σαῦρος/sauros: lizard}
Specific name: *montanus* ← {(Lat.) mōntānus: of mountains}.
Etymology: **Atlas lizard** of mountains
Taxonomy: Saurischia: Sauropodomorpha: Diplodocidae
Other species:

> *Atlantosaurus immanis* Marsh, 1878 [Kinmeridgian; Morrison Formation, Colorado, USA]

?= *Apatosaurus ajax* Marsh, 1877 ⇒ *Atlantosaurus ajax* (Marsh, 1877) Steel, 1970 ⇒ *Brontosaurus ajax* (Marsh, 1877) Bakker, 1986 [Kimmeridgian; Morrison Formation, Colorado, USA]

> *Apatosaurus alenquerensis* Lapparent & Zbyszewski, 1957 ⇒*Atlantosaurus alenquerensis* (Lapparent & Zbyszewski, 1957) Steel, 1970 ⇒ *Camarasaurus alenquerensis* (Lapparent & Zbyszewski, 1957) McIntosh, 1990 ⇒ *Lourinhasaurus alenquerensis* (Lapparent & Zbyszewski, 1957) Dantas *et al.*, 1998 [late Kimmeridgian; Lourinhã Formation, Lisboa, Portugal] (acc. *The Dinosauria 2nd edition*)

Notes: According to *the Dinosauria 2nd edition*, *Titanosaurus montanus* Marsh, 1877 (type of *Atlantosaurus*) is considered to be a *nomen dubium*. Some researchers think *Atlantosaurus montanus* is a synonym of *Apatosaurus ajax*.

[References: (*1) Marsh OC (1877). Notice of new dinosaurian reptiles from the Jurassic formation. *American Journal of Science* 14(84): 514–516.]

Atlasaurus imelakei

> *Atlasaurus imelakei* Monbaron *et al.*, 1999 [Bathonian–Callovian; Tiouguit Formation, Azilal, Morocco]

Generic name: *Atlasaurus* ← {(place-name) Atlas Mountains + (Gr.) σαῦρος/sauros: lizard}
Specific name: *imelakei* ← {(Arabic) Imelake: the name of a giant + -ī}
Etymology: Imelake's (giant) **Atlas lizard**
Taxonomy: Saurischia: Sauropodomorpha: Sauropoda

[References: Monbaron M, Russell DA, Taquet P (1999). *Atlasaurus imelakei* n.g., n. sp., a brachiosaurid-like sauropod from the Middle Jurassic of Morocco. *Comptes Rendus de l'Académie des Sciences à Paris, Séries IIAa. Sciences de la Terre et des Planètes* 329(7): 519–526. Abstract.]

Atlascopcosaurus loadsi

> *Atlascopcosaurus loadsi* Rich *et al.*, 1989 [Aptian–Albian; Eumeralla Formation, Victoria, Australia]

Generic name: *Atlascopcosaurus* ← {(institution name) Atlas Copco + (Gr.) σαῦρος/sauros: lizard}; in honor of the Atlas Copco (Compagnie Pneumatique Commerciale), for their support of the excavations at Dinosaur Cove in Victoria.
Specific name: *loadsi* ← {(person's name) Loads + -ī}; in honor of William Loads, the Atlas Copco regional manager.
Etymology: Loads' **Atlas Copco lizard**
Taxonomy: Ornithischia: Ornithopoda

[References: Rich T, Vickers-Rich P (1989). Polar dinosaurs and biotas of the Early Cretaceous of southeastern Australia. *National Geographic Research* 5(1): 15–53.]

Atrociraptor marshalli

> *Atrociraptor marshalli* Currie & Varricchio, 2004 [late Campanian or early Maastrichtian; Horseshoe Canyon Formation, Alberta, Canada]

Generic name: *Atrociraptor* ← {(Lat.) atrōx: savage + raptor: robber}
Specific name: *marshalli* ← {(person's name) Marshall + -ī}; in honor of "Wayne Marshall of East Coulee, Alberta, who discovered the type specimen" (*1).
Etymology: Marshall's **savage robber**
Taxonomy: Theropoda: Coelurosauria: Dromaeosauridae

[References: (*1) Currie PJ, Varricchio DJ (2004). A new dromaeosaurid from the Horseshoe Canyon Formatin (Upper Cretaceous) of Alberta, Canada. In: Currie PJ, Koppelhus EB, Shugar MA, Wright JL (eds), *Feathered Dragons*. Indiana University Press, Bloomington and Indianapolis. pp.: 112–132.]

Atsinganosaurus velauciensis

> *Atsinganosaurus velauciensis* Garcia *et al.*, 2010 [Campanian; Argiles et Grès à Reptiles Formation, Provence-Alpes-Côte d'Azur, France]

Generic name: *Atsinganosaurus* ← {(Byzantine Greek) ατσινγανος [αθίγγανος/athinganos: gypsy] + σαῦρος/sauros: lizard}; "referring to the existence of Late Cretaceous migrations between western and eastern Europe revealed by these remains" (*1).

Specific name: *velauciensis* ← {(place-name) Velaux-La Bastide Neuve + -ensis}; "from the Latin Velaucio, the name of the city–Velaux–where the material was collected"(*1).
Etymology: **gypsy lizard** from Velaux-La Bastide Neuve
Taxonomy: Saurischia: Sauropodomorpha:
Sauropoda: Titanosauria

[References: (*1) Garcia G, Amico S, Fournier F, Thouand E, Valentin X (2010). A new titanosaur genus (Dinosauria, Sauropoda) from the Late Cretaceous of southern France and its paleobiogeographic implications. *Bulletin de la Société Géologique de France* 181(3): 269–277.]

Aublysodon mirandus

> *Aublysodon mirandus* Leidy, 1868 (*nomen dubium*) ⇒ *Ornithomimus mirandus* (Leidy, 1868) Hay, 1930 [Campanian; Judith River Formation, Montana, USA]

= *Aublysodon amplus* Marsh, 1892 ⇒ *Deinodon amplus* (Marsh, 1892) Kuhn, 1939 ⇒*Tyrannosaurus amplus* (Marsh, 1892) Hay, 1930 ⇒ *Manospondylus amplus* (Marsh, 1892) Olshevsky, 1978 ⇒ *Stygivenator amplus* (Marsh, 1892) Olshevsky *et al.*, 1995 [Maastrichtian; Lance Formation, Wyoming, USA]

= *Aublysodon cristatus* Marsh, 1892 ⇒*Deinodon cristatus* (Marsh, 1892) ⇒ *Stygivenator cristatus* (Marsh, 1892) Olshevsky *et al.*, 1995 [Maastrichtian; Lance Formation, Wyoming, USA]

Generic name: *Aublysodon* ← {(Gr.) αὖ/au: again, backwards + blyso-[βλύζω/blyzō: flow, spout forth] + ὀδών/odōn: tooth}
Specific name: *mirandus* ← {(Lat.) mīrandus: wonderful, strange, singular}.
Etymology: wonderful ***Aublysodon***
Taxonomy: Theropoda: Coelurosauria:
Tyrannosauroidea: Tyrannosauridae
Other species:

> *Deinodon horridus* Leidy, 1856 (*nomen dubium*) ⇒ *Megalosaurus horridus* (Leidy, 1856) Leidy, 1857 (*nomen dubium*) [Campanian; Judith River Formation, Montana, USA]

= *Aublysodon lateralis* Cope, 1876 (*nomen dubium*) ⇒ *Deinodon lateralis* (Cope, 1876) Hay, 1902 [Fort Union Beds, Montana, USA]

> *Tyrannosaurus bataar* Maleev, 1955 ⇒ *Tarbosaurus bataar* (Maleev, 1955) Rozhdestvensky, 1965 ⇒ *Jenghizkhan bataar* (Maleev, 1955) Olshevsky *vide* Olshevsky *et al.*, 1977 [Campanian; Nemegt Formation, Ömnögovi, Mongolia]

= *Gorgosaurus lancinator* Maleev, 1955 ⇒ *Deinodon lancinator* (Maleev, 1955) Kuhn, 1965 ⇒ *Aublysodon lancinator* (Maleev, 1955) Charig, 1967 [Maastrichtian; Nemegt Formation, Ömnögovi, Mongolia]

= *Gorgosaurus novojilovi* Maleev, 1955 ⇒ *Deinodon novojilovi* (Maleev, 1955) Kuhn, 1964 ⇒ *Aublysodon novojilovi* (Maleev, 1955) Charig, 1967 ⇒ *Maleevosaurus novojilovi* (Maleev, 1955) Carpenter, 1992 [Campanian; Nemegt Formation, Ömnögovi, Mongolia]

= *Shanshanosaurus huoyanshanensis* Dong, 1977 ⇒ *Aublysodon huoyanshanensis* (Dong, 1977) Paul, 1988 [Campanian; Subashi Formation, Xinjiang, China]

> *Tyrannosaurus rex* Osborn, 1905 [Maastrichtian; Lance Formation, Montana, USA]

= *Aublysodon molnari* Paul, 1988 ⇒ *Stygivenator molnari* (Paul, 1988) Olshevsky, 1995 [Maastrichtian; Hell Creek Formation, Montana, USA]

Notes: According to *The Dinosauria 2nd edition* (2004), *Aublysodon mirandus* and the other species were considered as *nomina dubia*.

[References: Cope ED (1876). Descriptions of some vertebrate remains from the Fort Union Beds of Montana. *Baleontological Bulletin* 22: 1–14.]

Aucasaurus garridoi

> *Aucasaurus garridoi* Coria *et al.*, 2002 [Late Cretaceous; Anacleto Formation, Neuquén, Argentina]

Generic name: *Aucasaurus* ← {(Mapuche, place-name) Auca Mahuevo, a Lagerstätte in Anacleto Formation: the fossil locality + (Gr.) σαῦρος/sauros: lizard}
Specific name: *garridoi* ← {(person's name) Garrido + -ī}; "in homage to Mr. Alberto Garrido, who discovered the holotype"(*1).
Etymology: Garrido's **lizard from Auca Mahuevo**
Taxonomy: Theropoda: Abelisauridae:
Carnotaurinae: Carnotaurini

[References: (*1) Coria RA, Chiappe LM, Dingus L (2002). A new close relative of *Carnotaurus sastrei*, Bonaparte 1985, from the Late Cretaceous of Patagonia. *Journal of Vertebrate Paleontology* 22(2): 460–465.]

Augustynolophus morrisi

> *Saurolophus morrisi* Prieto-Márquez & Wagner 2013 ⇒ *Augustynolophus morrisi* (Prieto-Márquez & Wagner, 2013) Prieto-Márquez *et al.*, 2014 [Maastrichtian; Moreno Formation, California, USA]

Generic name: *Augustynolophus* ← {(person's name) Augustyn + lophus [Gr. λόφος/lophos: crest]}; "in recognition of Mrs. Gretchen Augustyn and her family, who have provided instrumental support to the scientific and educational programs of the Dinosaur Institute of the Natural History Museum of Los Angeles County. The suffix '-lophus' refers to the phylogenetic affinities of this taxon with members of the Saurolophine tribe"(*1).
Specific name: *morrisi* ← {(person's name) Morris + -ī}; "named for paleontologist William J. Morris (1923–2000), in recognition of his substantial contributions to our understanding of the functional morphology and evolutionary history of the hadrosaurid dinosaurs of the Pacific coast and Western Interior of North America"(*2).
Etymology: Morris' **Augustyn crest**
Taxonomy: Ornithischia: Ornithopoda: Hadrosauridae: Saurolophinae
Notes: This genus was originally described as a species of *Saurolophus*. However, it was determined to be a separate genus because its cranial structure was different from other species of *Saurolophus*.

[References: (*1) Prieto-Márquez A, Wagner JR, Bell Phil R, Chiappe LM (2014). The late-surviving 'duck-billed' dinosaur *Augustynolophus* from the upper Maastrichtian of western North America and crest evolution in Saurolophini. *Geological Magazine* 152(2): 225–241.]; (*2) Prieto-Márquez A, Wagner JR (2013). A new species of saurolophine hadrosaurid dinosaur from the Late Cretaceous of the Pacific coast of North America. *Acta Palaeontologica Polonica* 58(2): 255–268.]

Auroraceratops rugosus

> *Auroraceratops rugosus* You *et al.*, 2005 [Aptian; Xinminpu Group, Gansu, China]

Generic name: *Auroraceratops* ← {(Lat.) aurōra: dawn + (Gr.) κέρας/keras: horn + ὤψ/ōps: face}; referring to "its status as an early neoceratopsian, but also honoring Dawn Dodson (aurora = dawn), wife of 37 years to Peter Dodson, one of the authors who described the specimen and gracious hostess to several generations of paleontologists". (*1)
Specific name: *rugosus* ← {(Lat.) rūgōsus: rugose; referring to "the rugose nature of the skull and jaws"(*1).
Etymology: rugose, **Dawn's dawn ceratopsian** (neoceratopsian)
Taxonomy: Ornithischia: Marginocephalia: Ceratopsia: Neoceratopsia

[References: (*1) You H, Li D, Ji, Q, Lamanna M, Dodson P (2005). On a new genus of basal neoceratopsian dinosaur from the Early Cretaceous of Gansu Province, China. *Acta Geologica Sinica* 79(5): 593–597.]

Aurornis xui [Anchiornithidae]

> *Aurornis xui* Godefroit *et al.*, 2013 [Oxfordian; Tiaojishan Formation, Jianchang, Liaoning, China]

Generic name: *Aurornis* ← {(Lat.) aurōra: daybreak, dawn + (Gr.) ὄρνις/ornis: bird}
Specific name: *xui* ← {(person's name) Xu + -ī}; "in honor of Xú Xīng, for his exceptional and contribution to our understanding of the evolution and biology of feathered dinosaurs"(*1).
Etymology: Xu's **daybreak** (primitive) **bird**
Taxonomy: Theropoda: Avialae?: Anchiornithidae*
cf. Theropoda: Maniraptora: Paraves: Avialae (*1)
*Anchiornithidae Xu *et al.*, 2016 *sensu* Foth & Rauhut, 2017
Notes: According to Pei *et al.* (2017), *Aurornis xui* may be a junior synonym of *Anchiornis huxleyi* Xu *et al.*, 2009.

[References: (*1) Godefroit P, Andrea C, Hu D.-Y, Escuillié F, Wu W, Dyke G (2013). A Jurassic avialan dinosaur from China resolves the early phylogenetic history of birds. *Nature* 498(7454): 359–362.]

Aussiedraco molnari [Pterosauria]

> *Aussiedraco molnari* Kellner *et al.*, 2011 [Albian; Toolebuc Formation, Queensland, Australia]

Generic name: *Aussiedraco* ← {(place-name) Aussie: a shortened form of Australian + (Lat.) draco: dragon} (*1)
Specific name: *molnari* ← {(person's name) Molnar + -ī}; in honor of "Ralph E. Molnar, who made many important contributions to our knowledge of Australian vertebrate fossils"(*1).
Etymology: Molnar's **Australian dragon**
Taxonomy: Pterosauria: Pterodactyloidea: Pteranodontoidea: Targaryendraconidae

[References: (*1) Kellner AWA, Rodrigues T, Costa FR (2011). Short note on a pteranodontoid pterosaur (Pterodactyloidea) from western Queensland, Australia. *Anais da Academia Brasileira de Ciências* 83(1): 301–308.]

Australodocus bohetii

> *Australodocus bohetii* Remes, 2007 [Tithonian; Tendaguru Formation, Tendaguru, Lindi, Tanzania]

Generic name: *Australodocus* ← {(Lat.) austrālis: southern + (Gr.) δοκός/dokos: beam}; "southern, with reference to the Gondwanan provenance, and beam, alluding to the close relationship of this genus to the North American *Diplodocus*" (*1).
Specific name: *bohetii* ← {(person's name) Boheti

+ -ī}; in honor of "Boheti bin Amrani, the native African crew supervisor and chief preparatory of the German Tendaguru Expedition, whose excellent work was essential for the success of the European researchers (Janensh, 1914)."(*1)

Etymology: Boheti's **southern beam**

Taxonomy: Saurischia: Sauropodomorpha: Sauropoda: Titanosauriformes (*2)
cf. Saurischia: Sauropodomorpha: Sauropoda: Diplodocoidea: Flagellicaudata: Diplodocidae: Diplodocinae (*1)

[References: (*1) Remes K (2007). A second Gondwanan diplodocid dinosaur from the Upper Jurassic Tendaguru Beds of Tanzania, East Africa. *Palaeontology* 50(3): 653–667.; (*2) Whitlock JA (2011). Re-evaluation of *Australodocus bohetii*, a putative diplodocoid sauropod from the Tendaguru Formation of Tanzania, with comment on Late Jurassic sauropod faunal diversity and palaeoecology. *Palaeogeography, Palaeoclimatology, Palaeoecology* 309(3–4): 333–341.]

Australovenator wintonensis

> *Australovenator wintonensis* Hocknull *et al.*, 2009 [late Albian–Turonian; Winton Formation, Winton, Queensland, Australia]

Generic name: *Australovenator* ← {(Lat.) australis: southern + vēnātor: hunter}; in reference to the locality being in the Southern Hemisphere, Australia, and its carnivorous diet.

Specific name: *wintonensis* ← {(place-name) Winton: the township + -ensis}

Etymology: **southern hunter** from Winton

Taxonomy: Theropoda: Tetanurae: Megaraptora
cf. Theropoda: Tetanurae: Allosauroidea (*1)

[References: (*1) Hocknull SA, White MA, Tischler TR, Cook AG, Calleja ND, Sloan T, Elliot DA (2009). New mid-Cretaceous (latest Albian) dinosaurs from Winton, Queensland, Australia. *PLoS ONE* 7(6): e39364]

Austriadactylus cristatus [Pterosauria]

> *Austriadactylus cristatus* Dalla Vecchia *et al.*, 2002 [middle Norian (= late Alaunian); Seefelder Schichten, Tyrol, NW Austria] (*1)

Generic name: *Austriadactylus* ← {(place-name) Austria + (Gr.) δάκτυλος/daktylos: finger} (*1)

Specific name: *cristatus* ← {(Lat.) cristātus: crested}.

Etymology: crested **Austrian finger**

Taxonomy: Pterosauria: Rhamphorhynchoidea

Notes: Estimated wingspan is about 120 cm.

[References: (*1) Dalla Vecchia FM, Rupert W, Hopf H, Reitner J (2002). A crested rhamphorhynchid pterosaur from the Late Triassic of Austria. *Journal of Vertebrate Paleontology* 22(1): 196–199.]

Austrocheirus isasii

> *Austrocheirus isasii* Ezcurra *et al.*, 2010 [Cenomanian (←early Maastrichtian); middle Mata Amarilla Formation (previously considered to be the Pari Aike Formation), Santa Cruz, Argentina]

Generic name: *Austrocheirus* ← {austro- [(Lat.) auster: the south] + cheirus [(Gr.) χειρός/cheiros [χείρ/cheir: hand, manus]]}; "in allusion to the non-atrophied manus of this taxon which constitutes the first example of a medium-sized Cretaceous abelisauroid with this condition" (*1).

Specific name: *isasii* ← {(person's name) Isasi + -ī}; "in honor of Mr. Marcelo Isasi, for his outstanding work in the last 25 years as a technician in palaeontology and discoverer of the holotype specimen of *Austrocheirus*" (*1).

Etymology: Isasi's **southern hand**

Taxonomy: Theropoda: Ceratosauria: Abelisauroidea

[References: (*1) Ezcurra MD, Agnolin FL, Novas FE (2010). An abelisauroid dinosaur with a non-atrophied manus from the Late Cretaceous Pari Aike Formation of southern Patagonia. *Zootaxa* 2450:1–25.]

Austroposeidon magnificus

> *Austroposeidon magnificus* Bandeira *et al.*, 2016 [Campanian–Maastrichtian; Presidente Prudente Formation, São Paulo State, Brazil]

Generic name: *Austroposeidon* ← {Austro-: Southern [(Lat.) auster: the south wind, the south]: in allusion to South America + (Gr. myth.) Ποσειδῶν/Poseidōn: "the Greek God responsible for earthquakes"(*1)}

Specific name: *magnificus* ← {(Lat.) maginificus: great, elevated, noble}; "in allusion to the large size of the specimen" (*1).

Etymology: great **Poseidon from south (South America)**

Taxonomy: Saurischia: Sauropodomorpha: Sauropoda: Titanosauria

[References: (*1) Bandeira KLN, Medeiros Simbras F, Machado EB, de Almeida Campos D, Oliveira GR, Kellner AWA (2016). A new giant Titanosauria (Dinosauria: Sauropoda) from the Late Cretaceous Bauru Group, Brazil. *PLoS ONE* 11(10): e0163373.]

Austroraptor cabazai

> *Austroraptor cabazai* Novas, 2008 [Campanian–Maastrichtian; Allen Formation, Río Negro, Argentina]

Generic name: *Austroraptor* ← {(Lat.) austro- [australis] + (Lat.) raptor: thief}; "in reference to southern South America" (*1).

Specific name: *cabazai* ← {(person's name) Cabaza + -ī}; "in honour to the late Héctor Cabaza,

founder of the Museo Municipal de Lamarque" (*1).
Etymology: Cabaza's **southern thief**
Taxonomy: Theropoda: Dromaeosauridae: Unenlagiinae

[References: (*1) Novas FE (2009). A bizarre Cretaceous theropod dinosaur from Patagonia and the evolution of Gondwanan dromaeosaurids. *Proceedings of the Royal Society B* 276: 1101–1107.]

Austrosaurus mckillopi

> *Austrosaurus mckillopi* Longman, 1933 [Albian–Turonian; Winton Formation, Queensland, Australia]

Generic name: *Austrosaurus* ← {Austro-* [(Lat.) auster: the south] + (Gr.) σαῦρος/sauros: lizard}
*The prefix "Austro-" is used to denote Australia.
Specific name: *mckillopi* ← {(person's name) Mckillop + -ī}; in honor of H. J. McKillop, manager of Clutha Station.
Etymology: Mckillop's **southern (Australian) lizard**
Taxonomy: Saurischia: Sauropodomorpha: Sauropoda: Titanosauriformes: Somphospndyli
cf. Saurischia: Sauropodomorpha: Sauropoda: Titanosauria (acc. *The Dinosauria 2nd edition*)

[References: Longman HA (1933). A new dinosaur from the Queensland Cretaceous. *Memoirs of the Queensland Museum* X (III): 131–144.]

Avaceratops lammersi

> *Avaceratops lammersi* Dodson, 1986 [Campanian; Judith River Formation, Montana, USA]

Generic name: *Avaceratops* ← {(person's name) Ava + (Gr.) κέρας/keras: horn + ὤψ/ōps: face}; in honor of Ava Cole, the wife of Eddie Cole who "discovered a major bonebed in Judith River sediments" (*1).
Specific name: *lammersi* ← {(person's name) the Lammers + -ī}; in honor of the Lammers family, owners of the land. (*1)

In 1990 George Olshevsky emended the name to *A. lammersorum*, however, Dodson objected because the genitive singular might also refer to a single family name.
Etymology: Lammers and **Ava's horned face (ceratopsian)**
Taxonomy: Ornithischia: Marginocephalia: Ceratopsia: Ceratopsidae: Centrosaurinae: Nasutoceratopsini

[References: (*1) Dodson P (1986). *Avaceratops lammersi*: a new ceratopsid from the Judith River Formation of Montana. *Proceedings of the Academy of Natural Sciences of Philadelphia* 138(2): 305–317.]

Aviatyrannis jurassica

> *Aviatyrannis jurassica* Rauhut, 2003 [Kimmeridgian; Alcobaça Formation, Guimarota locality, Leiria, Portugal]

Generic name: *Aviatyrannis* ← {(Lat.) avia: grandmother + tyrannis [tyrannus: tyrant]}
Specific name: *jurassica* ← {(Lat.) jurassica (feminine form of jurassicus)}; referring to the Jurassic age of the taxon.
Etymology: **tyrant's grandmother** from the Jurassic (*1)
Taxonomy: Theropoda: Coelurosauria: Tyrannosauroidea

[References: (*1) Rauhut OWM (2003). A tyrannosauroid dinosaur from the Upper Jurassic of Portugal. *Paleontology* 46(5): 903–910.]

Avimaia schweitzerae [Avialae]

> *Avimaia schweitzerae* Bailleul *et al.*, 2019 [Aptian; Xiagou Formation, Changma, Yumen, Gansu, China]

Generic name: *Avimaia* ← {(Lat.) avis: bird + (Gr.) μαῖα/maia: mother}; "referring to the fact the specimen is a female preserved with an egg in the body cavity" (*1).
Specific name: *schweitzerae* ← {(person's name) Schweitzer + -ae}; "in honor of Mary Higby Schweitzer for her ground-breaking works on MB and for her role in establishing the field of molecular paleontology" (*1).
Etymology: Schweitzer's **bird mother**
Taxonomy: Dinosauria: Theropoda: Enantiornithes
cf. Aves: Pygostylia: Ornithothoraces: Enantiornithes (*1)

[References: (*1) Bailleul AM, O'Connor J, Zhang S, Li Z, Wang Q, Lamanna MC, Zhu X, Zhou Z (2019). An Early Cretaceous enantiornithine (Aves) preserving an unlaid egg and probable medullary bone. *Nature Communications* 10 (1275).]

Avimimus portentosus

> *Avimimus portentosus* Kurzanov, 1981 [Campanian; Barun Goyot Formation, Ömnögovi, Mongolia]

Generic name: *Avimimus* ← {(Lat.) avis: bird + (Gr.) μῖμος/mīmos: mimic}
Specific name: *portentosus* ← {(Lat.) portentōsus: unusual}
Etymology: unusual, **bird mimic**
Taxonomy: Theropoda: Maniraptoriformes: Oviraptorosauria: Caenagnathoidea: Avimimidae
Other species:

> *Avimimus nemegtensis* Funston *et al.*, 2017

[Maastrichtian or Campanian?; Nemegt Formation, Nemegt Basin, Mongolia]

[References: Kurzanov SM (1981). An unusual theropod from the Upper Cretaceous of Mongolia. Iskopayemyye pozvonochnyye Mongolii (Fossil Vertebrates of Mongolia). In: "*Trudy Sovmestnay Sovetsko-Mongolskay Paleontologyeskay Ekspeditsiy* (*Joint Soviet-Mongolian Paleontological Expedition*). *vol. 15*. Nauka, Moscow. pp.: 39–49.]

Avisaurus archibaldi [Avialae]

> *Avisaurus archibaldi* Brett-Surman & Paul, 1985 [Maastrichtian; Hell Creek Formation, Montana, USA]

Generic name: *Avisaurus* ← {(Lat.) avis: bird + (Gr.) σαῦρος/sauros: lizard}

Specific name: *archibaldi* ← {(person's name) Archibald + -ī}; in honor of J. David Archibald, its discoverer.

Etymology: Archibald's **bird lizard**

Taxonomy: Theropoda: Avialae: Enantiornithes: Avisauridae

Other species:

> *Avisaurus gloriae* Varricchio & Chiappe, 1995 ⇒ *Gettyia gloriae* (Varricchio & Chiappe, 1995) Atterholt *et al.*, 2018 [Campanian; Two Medicine Formation, Montana, USA]

[References: Brett-Surman MK, Paul GS (1985). A new family of bird-like dinosaurs linking Laurasia and Gondwanaland. *Jouenal Vertebrate Paleontology* 5(2): 133–138.]

Azhdarcho lancicollis [Pterosauria]

> *Azhdarcho lancicollis* Nessov, 1984 [Turonian; Bissekty Formation, Navoi, Uzbekistan]

= *Azhdarcho imparidens* Nessov, 1981 (*nomen nudum*) [Turonian; Bissekty Formation, Navoi, Uzbekistan]

Generic name: *Azhdarcho* ← {(Uzbek) "azhdarkho: name of mythical dragon"}

Specific name: *lancicollis* ← {(Lat.) lancea: lance, spear + collum: neck}

Etymology: **Azhdarcho (dragon)** with a lance-like neck

Taxonomy: Pterosauria: Azhdarchidae

[References: Nesov LA (1984). *Azhdarcho lancicollis* original description. Upper Cretaceous pterosaurs and birds from central Asia. *Paleontologicheskii Zhurnal* 1: 47–57.]

Azendohsaurus laaroussii [Archosauromorpha]

> *Azendohsaurus laaroussii* Dutuit, 1972 [middle Carnian; Timezgadiouine Formation, Argana Group, Marrakech, Morocco]

Generic name: *Azendohsaurus* ← {(place-name) Azendoh: the village in the Atlas Mountains region + (Gr.) σαῦρος/sauros: lizard}

Specific name: *laaroussii* ← {(person's name) Laaroussi + -ī}

Etymology: Laaroussi's **Azendoh lizard**

Taxonomy: Archosauromorpha: Azendohsauridae (*2)

cf. Saurischia: Sauropodomorpha: Prosauropoda (acc. *The Dinosauria 2nd edition*)

Other species:

> *Azendohsaurus madagaskarensis* Flynn *et al.*, 2010 [Ladinian; Isalo II Formation, Toliara, Madagascar]

Notes: Estimated to be 2–3 m in total length. *Azendohsaurus laaroussii* was regarded as an ornithischian dinosaur by Dutuit in 1972 (*1). According to *The Dinosauria 2nd edition* (2004), it was considered as a member of Prosauropoda, but new research suggests that *Azendohsaurus laaroussii* and *A. madagaskarensis* are not dinosaurs. (*2)

[References: (*1) Dutuit J-M (1972). Decouverte d'un Dinosaure ornithischian dans le Trias supérieur de l'Atlas occidental marocain. *Comptes Rendus de l'Académie des Sciences à Paris, Série D* 275: 2841–2844.; (*2) Flynn JJ, Nesbitt SJ, Parrish JM, Ranivoharimanana L, Wyss AR (2010). A new species of *Azendohsaurus* (Diapsida: Archosauromorpha) from the Triassic Isalo Group of southwestern Madagascar: cranium and mandible. *Palaeontology* 53(3): 669–688.]

B

Baalsaurus mansillai

> *Baalsaurus mansillai* Calvo & Gonzalez Riga, 2018 [Turonian–Coniacian; Portezuelo Formation, Neuquén Group, Neuquén, Argentina]

Generic name: *Baalsaurus* ← {(place-name) Baal: the name of dinosaur site | (Phoenician / Canaanite myth.) Baal: the fertility god + (Gr.) σαῦρος/sauros: reptile}

Specific name: *mansillai* ← {(person's name) Mansilla + -ī}; "in honor of Mr. Juan Eduardo Mansilla, a technician at the Geology and Paleontology Museum of the National University of Comahue, Parque Natural Geo-Paleontológico Proyecto Dino, Barreales Lake, who discovered the material" (*1).

Etymology: Mansilla's **Baal reptile**

Taxonomy: Saurischia: Sauropoda: Titanosauriformes: Titanosauria

[References: (*1) Calvo JO, Gonzalez Riga B (2018). *Baalsaurus mansillai* gen. *et* sp. nov. a new titanosaurian sauropod (Late Cretaceous) from Neuquén, Patagonia, Argentina. *Anais da Academia*

Brasileira de Ciências. 91(2): e20180661:-11.]

Bactrosaurus johnsoni

> *Bactrosaurus johnsoni* Gilmore, 1933 [Campanian; Iren Dabasu Formation, Inner Mongolia, China]

Genus name: *Bactrosaurus* ← {(Gr.) βάκτρον/ baktron: club + σαῦρος/sauros: lizard}; referring to "posterior dorsal vertebra with tall spines, club-shaped" (*1).

Specific name: *johnsoni* ← {(person's name) Johnson + -ī}; in honor of Albert F. Johnson.

The Johnson Quarry was named in honor of A. F. Johnson by Barnum Brown.

Etymology: Johnson's **club lizard**

Taxonomy: Ornithischia: Ornithopoda: Hadrosauroidea: Hadrosauromorpha

cf. Ornithischia: Ornithopoda: Hadrosauridae (acc. *The Dinosauria 2nd edition*)

Other species:

> *Bactrosaurus prynadai* Riabinin, 1939 (*nomen dubium*) ⇒ *Tanius prynadai* (Riabinin, 1939) Young 1958 (*nomen dubium*) [Coniacian–Santonian; Dabrazhin Formation, Tajikistan]

> *Cionodon kysylkumensis* Riabinin, 1931 (*nomen dubium*) ⇒ ?*Bactrosaurus kysylkumensis* (Riabinin, 1931) Nessov, 1995 [Late Cretaceous; Kysyl-kum desert, Central Asia]

[References: (*1) Gilmore CW (1933). On the dinosaurian fauna of the Iren Dabasu Formation. *American Museum of Natural History, Bulletin* 67 (2): 23–78.]

Bagaceratops rozhdestvenskyi

> *Bagaceratops rozhdestvenskyi* Maryańska & Osmólska, 1975 [Campanian; Barun Goyot Formation, Nemegt Basin, Gobi Desert, Mongolia]

= *Lamaceratops tereschenkoi*? Alifanov, 2003 [Campanian; Barun Goyot Formation, Mongolia]

= *Magnirostris dodsoni*? You & Dong, 2003 [Campanian; Bayan Mandahu, Inner Mongolia, China]

= *Platyceratops tatarinovi*? Alifanov, 2003 [Campanian; Barun Goyot Formation, Mongolia]

= *Gobiceratops minutus*? Alifanov, 2008 [Campanian; Barun Goyot Formation, Mongolia]

Generic name: *Bagaceratops* ← {(Mong.) бага/ baga: small + ceratops [(Gr.) κερατ-/ kerat- (κέρας/keras: horn)] + ὤψ/ōps: face}; referring to its "probably smaller body size than other protoceratopsids" (*1).

Specific name: *rozhdestvenskyi* ← {(person's name) Rozhdestvensky + -ī}; "in honor of Dr. A. K. Rozhdestvensky in recognition of his work on dinosaurs" (*1).

Etymology: Dr. Rozhdestvensky's **small horned face (ceratopsian)**

Taxonomy: Ornithischia: Marginocephalia: Ceratopsia: Neoceratopsia: Coronosauria: Protoceratopsidae; Bagaceratopidae*

*Bagaceratopidae was named by Alifanov (2003) (*2), but, it was considered inactive by Sereno (2005).

Notes: *Bagaceratops* has a small horn-like prominence on the snout. According to *The Dinosauria 2nd edition* (2004), *Breviceratops kozlowskii* was considered a synonym of *Bagaceratops rozhdestvenskyi,* but according to Crepiński (2019), *Breviceratops kozlowskii* is considered a distinct taxon.

[References: (*1) Maryañska T, Osmólska H (1975). Protoceratopsidae (Dinosauria) of Asia. *Palaeontologia Polonica* 33: 133–181.; (*2) Alifanov VR (2003). Two new dinosaur of the infraorder Neoceratopsia (Ornithischia) from the Upper Cretaceous of the Nemegt Depression, Mongolian People's Republic. *Paleontological Journal* 37(5): 524–535.]

Bagaraatan ostromi

> *Bagaraatan ostromi* Osmólska, 1996 [early Maastrichtian; Nemegt Formation, Ömnögovi, Mongolia]

Generic name: *Bagaraatan* ← {(Mong.) бага/baga: small + араатан/araatan: predator} (*1).

Specific name: *ostromi* ← {(person's name) Ostrom + -ī}; in honor of Dr. John H. Ostrom (*1).

Etymology: Ostrom's **small predator**

Taxonomy: Theropoda: Coelurosauria: Tyrannosauroidea (acc. Loewen *et al.* 2013)

cf. Theropoda: Tetanurae (*1)

[References: (*1) Osmólska H (1996). An unusual theropod dinosaur from the Late Cretaceous Nemegt Formation of Mongolia. *Acta Palaeontologica Polonica* 41 (1): 1–38.]

Bagualosaurus agudoensis

> *Bagualosaurus agudoensis* Pretto *et al.*, 2018 [Carnian; Santa Maria Formation, Agudo, Rio Grande do Sul, Brazil]

Generic name: *Bagualosaurus* ← {(a term employed regionally in southern Brazil) bagual: referring to an animal or person of strong build or valour + saurus: lizard} (*1)

Specific name: *agudoensis* ← {(place-name) Agudo: the town where the holotype was collected +

-ensis} [*1]
Etymology: **strongly built lizard** from Agudo
Taxonomy: Saurischia: Sauropodomorpha

[References: (*1) Pretto FA, Langer MC, Schultz CL (2018). A new dinosaur (Saurischia: Sauropodomorpha) from the Late Triassic of Brazil provides insights on the evolution of sauropodomorph body plan. *Zoological Journal of the Linnean Society* 20: 1–29.]

Bahariasaurus ingens

> *Bahariasaurus ingens* Stromer, 1934 [early Cenomanian; Baharîje Formation, Marsa Matruh, Egypt]

Generic name: *Bahariasaurus* ← {(place-name) Bahariya Formation + (Gr.) σαῦρος/ sauros: lizard}
Specific name: *ingens* ← {(Lat.) ingēns: not natural, huge, massive}.
Etymology: huge, **Bahariya lizard**
Taxonomy: Theropoda: Avetheropoda: Megaraptora
Notes: The type specimen was destroyed during World War II.

[References: Stromer E (1934). Ergebnisse der Forschungsreisen Prof. E. Stromers in den Wüsten Ägyptens. II. Wirbeltier-Reste der Baharije-Stufe (unterstes Cenoman). 13. Dinosauria. *Abhandlungen der Bayerischen Akademie der Wissenschaften, Mathematisch-Naturwissenschaftliche Abteilung n. f.*, 22: 1–79.]

Bainoceratops efremovi

> *Bainoceratops efremovi* Tereschenko & Alifanov, 2003 [Campanian; Djadokhta Formation, Ömnögivi, Mongolia]

Generic name: *Bainoceratops* ← {(Mong. place-name) Баянзаг/Bain-Dzak + *Ceratops* [(Gr.) κερατ-/cerat- (←κέρας/keras: horn) + ὤψ/ōps: face]}; "after the Bain-Dzak locality and the generic name *Ceratops*" [*1].
Specific name: *efremovi* ← {(person's name) Efremov + -ī}; "in honor of the paleontologist I. A. Efremov" [*1].
Etymology: Efremov's **Bain-Dzak horned face (ceratopsian)**
Taxonomy: Ornithischia: Marginocephalia: Ceratopsia

[References: (*1) Tereschenko VS, Alifanov VR (2003). *Bainoceratops efremovi*, a new protoceratopid dinosaur (Protoceratopidae, Neoceratopsia) from the Bain-Dzak locality (south Mongolia). *Paleontological Journal* 37(3): 293–302.]

Bajadasaurus pronuspinax

> *Bajadasaurus pronuspinax* Gallina *et al.*, 2019 [late Berriasian–Valanginian; Bajada Colorada Formation, Neuquén Basin, Neuquén, Argentina]

Generic name: *Bajadasaurus* ← {(place-name) Bajada: the locality Bajada Colorada | (Sp.) bajada: downhill + (Gr.) σαῦρος/sauros: lizard} [*1]
Specific name: *pronuspinax* ← {(Lat.) pronus: bent over + "spinax: spine" [*1]}; "referring to the anteriorly pointed, curved, neural spines of the cervical vertebrae" [*1].
Etymology: **Bajada lizard** with curved spines
Taxonomy: Saurischia: Sauropoda: Diplodocoidea: Dicraeosauridae

[References: (*1) Gallina PA, Apesteguía S, Canale JI, Haluza A (2019). A new long-spined dinosaur from Patagonia sheds light on sauropod defense system. *Scientific Reports* 9(1): 2700.]

Balaur bondoc [Avialae?]

> *Balaur bondoc* Csiki *et al.*, 2010 [Maastrichtian; Sebes Formation, Alba, Romania]

Generic name: *Balaur** ← {(Romanian folklore) Balaur; a dragon}.
*"Balaur is an archaic Romanian term that designates a mythical ophidian, dragon-like creature, and this word is often used as synonym for dragon in contemporary Romanian" (partly cited from Etymology in Supporting Information Appendix [Csiki, 2010]).
Specific name: *bondoc** ← {(Romanian) bondoc: stocky}
*"a Romanian folk word used to designate a clumsy, chubby creature (human or animal" (partly cited from Appendix [Csiki, 2010]).
Etymology: stocky **dragon (winged reptile)**
Taxonomy: Theropoda: Paraves: Eumaniraptora: Avialae [*2]
cf. Theropoda: Coelurosauria: Maniraptora: Dromaeosauridae [*1]
Notes: *Balaur bondoc* is reinterpreted as "a basal avialan rather than as a dromaeosaurid" [*2].

[References: (*1) Csiki Z, Vremir M, Brusatte SL, Norell MA (2010). An aberrant island-dwelling theropod dinosaur from the Late Cretaceous of Romania. *Proceedings of the National Academy of Sciences of the United States of America* 107(35): 15357–15361.; (*2) Cau A, Brougham T, Naish D (2015). The phylogenetic affinities of the bizarre Late Cretaceous Romanian theropod *Balaur bondoc* (Dinosauria, Maniraptora): dromaeosaurid or flightless bird? *PeerJ* 3: e1032.]

Balochisaurus malkani

> *Balochisaurus malkani* Malkani, 2006

[Maastrichtian; Pab Formation, Balochistan, Pakistan]

Generic name: *Balochisaurus* ← {(demonym) Balochi; Baloch tribes + (Gr.) σαῦρος/sauros; reptile}; "honoring to the Baloch tribes of Pakistan, as they host the Kachi Bohri locality from Central Sulaiman Range" (*1).

Specific name: *malkani* ← {(person's name) Malkani}; for M. Sadiq Malkani.

Etymology: Malkani's **Balochi lizard**

Taxonomy: Saurischa: Sauropodomorpha: Titanosauria: Balochisauridae

Notes: *Balochisaurus* was thought to be valid.

[References: (*1) Malkani MS (2006). Biodiversity of saurischian dinosaurs from the latest Cretaceous park of Pakistan. *Journal of Applied and Emerging Sciences* 1(3): 108–140.]

Bambiraptor feinbergi

> *Bambiraptor feinbergi* Burnham *et al.*, 2000 ⇒ *Bambiraptor feinbergorum* (Burnham *et al.*, 2000) Norrell & Makovicky, 2004 [Campanian; Two Medecine Formation, Montana, USA]

Generic name: *Bambiraptor* ← {(Disney movie character) Bambi | (Italy) bambino: baby, child + (Lat.) raptor: robber}; in reference to the young age specimen, which was discovered by fourteen-year-old Wesley Linster in 1933. Bambi was a nickname for the holotype, originally coined by the Linster Family.

Specific name: *feinbergi* ← {(person's name) Feinberg + -ī}; in honor of Michael and Ann Feinberg who acquired the specimen from a fossil dealer and made it available to science.

Though Norell & Makovicky in *The Dinosauria 2nd edition* (2004) preferred to use *Bambiraptor feinbergorum* instead of *B. feinbergi*, the grammatical emendation such as "feinbergi" to "feinbergorum" are no longer admitted by ICZN.*

*ICZN = International Commission Zoological Nomenclature / International Code of Zoological Nomenclature.

Etymology: Feinberg's **Bambi robber (baby raptor)**

Taxonomy: Theropoda: Maniraptora: Dromaeosauridae

Notes: *Bambiraptor* had the largest brain-size in relation to the body-size of any known dinosaur. The type specimen appears to be a juvenile. 95 percent of *Bambiraptor*'s bones were recovered (*1).

[References: Burnham DA, Derstler KL, Currie PJ, Bakker RT, Zhou Z, Ostrom JH (2000). Remarkable new birdlike dinosaur (Theropoda; Maniraptora) from the Upper Cretaceous of Montana. *University of Kansas Paleontological Contributions* 13: 1–14.; (*1) Turner AH, Makovicky PJ, Norell MA (2012). A review of dromaeosaurid systematics and paravian phylogeny. *Bulletin of the American Museum of Natural History* 371: 1–206.]

Banji long

> *Banji long* Xu & Han, 2010 [Maastrichtian; Nanxiong Formation, Ganzhou, Jiangxi, China]

Generic name: *Banji* ← {(Chin.) bān: speckle, but sometimes referring to stripes + jǐ: crest}; "referring to the animal's bearing a crest with distinctive striations over the snout" (*1).

Specific name: *long* ← {(Chin.) lóng: dragon}.

Etymology: **striped-crest** of dragon (**stripe-crested** dragon)

Taxonomy: Theropoda: Oviraptorosauria: Oviraptoridae

[References: (*1) Xu X, Han F-L (2010). A new oviraptorid dinosaur (Theropoda: Oviraptorosauria) from the Upper Cretaceous of China. *Vertebrata PalAsiatica* 48(1): 11–18.]

Bannykus wulatensis

> *Bannykus wulatensis* Xu *et al.*, 2018 [Aptian; Bayin-Gobi Formation, Wulatehouqi, Inner Mongolia, China]

Generic name: *Bannykus* ← {(Chin.) ban: half + (Gr.) ὄνυξ/onyx: claw}; "referring to the transitional features seen in this animal" (*1).

Specific name: *wulatensis* ← {(place-name) Wulate + -ensis}; "derived from Wulatehouqi (Wulate Rear Banner), the country-level administrative division in which the type locality is situated" (*1).

Etymology: **half claw** from Wulate

Taxonomy: Theropoda: Alvarezsauria

Notes: The holotype was probably a sub-adult (*1).

[References: (*1) Xu X, Choiniere J, Tan Q, Benson RBJ, Clark J, Sullivan C, Zhao Q, Han F, Ma Q, He Y, Wang S, Xing H, Tan L (2018). Two early Cretaceous fossils document transitional stages in alvarezsaurian dinosaur evolution. *Current Biology* 28: 1–8.]

Baotianmansaurus henanensis

> *Baotianmansaurus henanensis* Zhang *et al.*, 2009 [Turonian; Gaogou Formation, Henan, China]

Generic name: *Baotianmansaurus* ← {(Chin. place-name) Bǎotiānmàn寶天曼 + (Gr.) σαῦρος/sauros: lizard}

Specific name: *henanensis* ← {(Chin. place-name) Hénán河南 + -ensis}

Etymology: **Baotianman lizard** from Henan Province

Taxonomy: Saurischia: Sauropodomorpha:

Sauropoda: Titanosauriformes

[References: Zhang X, Lü J, Xu L, Li J, Hu W, Jia S, Ji Q, Zhang C (2009). A new sauropod dinosaur from the Late Cretaceous Gaogou Formation of Nanyang, Henan Province. *Acta Geologica Sinica* 83(2): 212–221.]

Baptornis advenus [Avialae]

> *Baptornis advenus* Marsh, 1877 [Coniacian–Campanian; Niobrara Formation, Kansas, USA]

= *Parascaniornis stensioei* Lambrecht, 1933 [Campanian; Belemnellocamax mammillatus ammonoid zone, Sweden]

Generic name: *Baptornis* ← {(Gr.) bapto- [βάπτω/baptō: dive, dip in water] + ὄρνις/ornis: bird}

Specific name: *advenus* ← {(Lat.) advenus: foreign, migrant, unskilled}

Etymology: migratory, **diving bird**

Taxonomy: Theropoda: Avialae: Hesperornithes (= Hesperornithiformes)

Other species:

> *Baptornis varneri* Martin & Cordes-Person, 2007 ⇒ *Brodavis varneri* (Martin & Cordes-Person, 2007) Martin *et al.*, 2012 [Campanian; Pierre Shale Formation, South Dakota, USA]

Notes: *Baptornis* is a genus of flightless aquatic birds.

Barapasaurus tagorei

> *Barapasaurus tagorei* Jain *et al.*, 1975 [Sinemurian–Pliensbachian; Kota Formation, Andhra Pradesh, India]

Generic name: *Barapasaurus* ← {(Indian language, Bengali) bara; big + pā; leg + (Gr.) σαῦρος/sauros; lizard}; referring to the femur over 1.7 m long.

Specific name: *tagorei* ← {(person's name) Tagore + -ī}; in honor of the great Bengali poet Rabindranath Tagore (1861–1941). The first year of fieldwork was carried out in 1961, in the centenary year of Tagore's bith. He is the winner of the Nobel Prize for Literature in 1913.

Etymology: Tagore's **big-legged lizard**

Taxonomy: Saurischia: Sauropodomorpha: Sauropoda: Gravisauria

cf. Saurischia: Sauropodamorpha: Sauropoda: Cetiosauridae (acc. *The Dinosauria 2nd edition*)

Notes: *Kotasaurus* also comes from the same Kota Formation, Andhra Pradesh States, India.

[References: Jain SL, Kutty TS, Roy-Chowdhurry T, Chatterjee S, (1975). The sauropod dinosaur from the Lower Jurassic Kota Formation of India. *Proceeding of the Royal Society A* 188: 221–228.]

Barilium dawsoni

> *Iguanodon dawsoni* Lydekker, 1888 ⇒ *Barilium dawsoni* (Lydekker, 1888) Norman, 2010 ⇒ *Torilion dawsoni* (Lydekker, 1888) Carpenter & Ishida, 2010 [Valanginian; Wadhurst Clay Formation, East Sussex, UK]

= *Kukufeldia tilgatensis* McDonald *et al.*, 2010 [Valanginian; Grinstead Clay Member, Tunbridge Wells Sand Formation, West Sussex, UK]

?= *Sellacoxa pauli* Carpenter & Ishida, 2010 (*nomen dubium*) [Valanginian; Wadhurst Clay Formation, East Sussex, UK]

Generic name: *Barilium* ← {(Gr.) βαρύς/barys: heavy + (Lat.) ilium: a bone of the hip, [Eng. anat.]

Specific name: *dawsoni* ← {(person's name) Dawson + -ī}; in honor of the discoverer Charles Dawson.

Etymology: Dawson's **heavy ilium**

Taxonomy: Ornithischia: Ornithopoda: Iguanodontia: Iguanodontidae

[References: Lydekker R (1888). Note on a new Wealden iguanodont and other dinosaurs. *Quarterly Journal of the Geological Society of London* 44: 46–61.; Norman DB (2010). A taxonomy of iguanodontians from the lower Wealden Group (Cretaceous: Valanginian) of southern England. *Zootaxa* 2489: 47–66.]

Barosaurus lentus

> *Barosaurus lentus* Marsh, 1890 [Kimmeridgian–Tithonian; Morrison Formation, South Dakota, USA]

= *Barosaurus affinis* Marsh, 1899 [Kimmeridgian–Tithonian; Morrison Formation, South Dakota, USA] (acc. *The Dinosauria 2nd edition*)

Generic name: *Barosaurus* ← {(Gr.) βάρο-/baro- [βᾰρύς/barys; heavy] + σαῦρος/sauros: lizard}

Specific name: *lentus* ← {(Lat.) lentus; slow}

Etymology: slow, **heavy lizard**

Taxonomy: Saurischia: Sauropodomorpha: Sauropoda: Diplodocidae: Diplodocinae

Other species:

> *Gigantosaurus africanus* Fraas, 1908 ⇒ *Tornieria africana* (Fraas, 1908) Sternfeld, 1911 ⇒ *Barosaurus africanus* (Fraas, 1908) Janensch, 1922 [Tithonian; Tendaguru Formation, Tanzania]

> *Barosaurus africanus* "gracilis" Janensch, 1961 (*nomen nudum*) ⇒ *Barosaurus* "gracilis" (Janensch, 1961) Russell *et al.*, 1980 (*nomen nudum*) ⇒ *Tornieria* "gracilis" (Janensch, 1961) Olshevsky, 1991 (*nomen nudum*) [Tendaguru, Tanzania]

Notes: *Barosaurus* is a giant, long-tailed, long-necked dinosaur. According to Remes (2006),

Gigantosaurus africanus is recombined as *Tornieria africana*.

[References: Marsh OC (1890). Description of new dinosaurian reptiles. *American Journal of Science* 3(39): 81–86.]

Barrosasaurus casamiquelai

> *Barrosasaurus casamiquelai* Salgado & Coria, 2009 [Campanian; Anacleto Formation, Neuquén, Argentina]

Generic name: *Barrosasaurus* ← {(place-name) the Sierra Barrosa locality + (Gr.) σαῦρος/sauros: lizard}

Specific name: *casamiquelai* ← {(person's name) Casamiquela + -ī}; in honor of the Argentine paleontologist Rodolfo Magín Casamiquela.

Etymology: Casamiquela's **Sierra Barrosa lizard**

Taxonomy: Saurischia: Sauropodomorpha: Sauropoda

[References: Salgado L, Coria RA (2009). *Barrosasaurus casamiquelai* gen. *et* sp. nov., a new titanosuar (Dinosauria, Sauropoda) from the Anacleto Formation (Late Cretaceous: early Campanian) of Sierra Barrosa (Neuquén, Argentina). *Zootaxa* 2222: 1–16.]

Barsboldia sicinskii

> *Barsboldia sicinskii* Maryańska & Osmólska, 1981 [Maastrichtian; Nemegt Formation, Ömnögovi, Mongolia]

Generic name: *Barsboldia* ← {(person's name) Barsbold + -ia}; "in honor of Dr. Rinchen Barsbold, the eminent Mongolian paleontologist" (*1).

Specific name: *sicinskii* ← {(person's name) Siciński + -ī}; "in honour of Mr. Wojciech Siciński, technical assistant in the Institute of Paleobiology, Polish Academy of Sciences, Warsaw" (*1).

Etymology: **One in honor of Barsbold** and Siciński

Taxonomy: Ornithischia: Ornithopoda: Iguanodontia: Hadrosauridae: Saurolophinae (*2)
cf. Ornithischia: Ornithopoda: Iguanodontia: Hadrosauridae: Lambeosaurinae (*1)

[References: (*1) Maryańska T, Osmólska H (1981). First lambeosaurine dinosaur from the Nemegt Formation, Upper Cretaceous, Mongolia. *Acta Palaeontologica Polonica* 26: 243–255.; (*2) Prieto-Márquez A (2011). A reappraisal of *Barsboldia sicinskii* (Dinosauria: Hadrosauridae) from the Late Cretaceous of Mongolia. *Journal of Paleontology* 85(3): 468–477.]

Baryonyx walkeri

> *Baryonyx walkeri* Charig & Milner, 1986 [Barremian; Weald Clay Formation, Surrey, UK]

Generic name: *Baryonyx* ← {(Gr.) βᾰρύς/barys: heavy + ὄνυξ/onyx: claw}

Specific name: *walkeri* ← {(person's name) Walker + -ī}; in honor of the discoverer, amateur fossil hunter William J. Walker.

Etymology: Walker's **heavy claw**

Taxonomy: Theropoda: Tetanurae: Megalosauroidea: Spinosauridae: Baryonychinae

Other species:

> *Suchomimus tenerensis* Sereno *et al.*, 1998 ⇒ *Baryonyx tenerensis* (Sereno *et al.*, 1998) Sues *et al.*, 2002 [Aptian; Elrhaz Formation, Agadez, Niger]

?= *Cristatusaurus lapparenti* Taquet & Russell, 1998 (*nomen dubium*) [Aptian–Albian; Elrhaz Formation, Agadez, Niger]

Notes: Acid-etched scales of *Scheenstia* (*1) was found in the stomach region of a fossil *Baryonyx walkeri* specimen.

[References: Charig AJ & Milner AC (1986). *Baryonyx*, a remarkable new theropod dinosaur. *Nature* 324(6095): 359–361.; (*1) López-Arbarello A (2012). Phylogenetic interrelationships of ginglymodian fishes (Actinopterygii: Neopterygii). *PLoS ONE* 7(7): e39370.]

Batrachognathus volans [Pterosauria]

> *Batrachognathus volans* Ryabinin, 1948 [Oxfordian–Kimmeridgian; Karabastau Formation, Turkistan (prev. South Kazakhstan), Kazakhstan]

Generic name: *Batrachognathus* ← {(Gr.) βάτραχος/batrachos: frog + γνάθος/gnathos: jaw}; in reference to the short wide head.

Specific name: *volans* ← {(Lat.) volāns: flying}.

Etymology: flying **frog jaw**

Taxonomy: Archosauria: Ornithodira: Pterosauria: Anurognathidae

Batyrosaurus rozhdestvenskyi

> *Batyrosaurus rozhdestvenskyi* Godefroit *et al.*, 2012 [Santonian–Campanian; Bostobe Formation, Qyzylorda, Kazakhstan]

Generic name: *Batyrosaurus* ← {(Kazakh) batyrs: heroic knights (*1) + (Gr.) σαῦρος/sauros: lizard}

Specific name: *rozhdestvenskyi* ← {(person's name) Rozhdestvensky + -ī}; in honor of the Russian paleontologist Anatoly Konstantinovich Rozhdestvensky for his pioneering works on Middle Asian Iguanodontia. (*1)(*2)

Etymology: Rozhdestvensky's **heroic lizard**

Taxonomy: Ornithischia: Ornithopoda: Iguanodontia: Iguanodontoidea: Hadrosauroidea

[References: (*1) Godefroit P, Escuillié F, Bolotsky YL, Lauters P (2012). A new basal hadrosauroid dinosaur from Upper Cretaceous of Kazakhstan. In: Godefroit

P (ed), *Bernissart Dinosaurs and Early Cretaceous Terrestrial Ecosystems.* Indiana University Press, Bloomington: 335–358.; (*2) Rozhdestvenskiy AK (1974). *The Joint Soviet-Mongolian Paleontological Expedition, Transaction* 1: 107–131.]

Baurutitan britoi

> *Baurutitan britoi* Kellner *et al.*, 2005 [Maastrictian; Marília Formation, Bauru Group, Minas Gerais, Brazil]

Generic name: *Baurutitan* ← {(stratal name) Bauru Group + titan [(Gr. myth.) Τῑτάν/Titan]: the giants} (*1)

Specific name: *britoi* ← {(person's name) Brito + -ī}; "in honor of Ignacio Aureliano Machado Brito (1938–2001), an important Brazilian paleontologist", who supervised several students including two authers of the paper (D. A. Campos and A. W. A. Kellner) (*1).

Etymology: Brito's **Bauru giant**

Taxonomy: Sauischia: Sauropodomorpha: Sauropoda: Titanosauria: Titanosauridae (*1)

[References: (*1) Kellner AWA, Campos DA, Trotta MNF (2005). Description of a titanosaurid caudal series from the Bauru Group, Late Cretaceous of Brazil. *Arquivos do Museu Nacional, Rio de Janeiro* 63: 529–564.]

Bauxitornis mindszentyae [Avialae]

> *Bauxitornis mindszentyae* Dyke & Ősi, 2010 [Santonian; Csehbánya Formation, Veszprém, Hungary]

Generic name: *Bauxitornis* ← {(place-name) a bauxite mine [(mineral.) bauxite] + (Gr.) ὄρνις/ornis: bird}; referring to the locality where the type specimen was discovered.

Specific name: *mindszentyae* ← {(person's name) Mindszenty + -ae}; in honor of Andrea Mindszenty, Ősi's advisor.

Etymology: Mindszenty's **bird of a bauxite mine**

Taxonomy: Theropoda: Aialae: Enantiornithes: Avisauridae

Bayannurosaurus perfectus

> *Bayannurosaurus perfectus* Xu *et al.*, 2018 [Aptian; Bayingebi Formation, Bayannur, Inner Mongolia, China]

Generic name: *Bayannurosaurus* ← {(place-name) Bayannur + (Gr.) σαῦρος/sauros: lizard} (*1)

Specific name: *perfectus* ← {(Lat.) perfectus: perfect}; "in reference to the perfect preservation of the skeleton designated as the holotypic specimen" (*1).

Etymology: perfect **Bayannur lizard**

Taxonomy: Ornithischia: Ornithopoda: Iguanodontia: Ankylopollexia

[References: (*1) Xu X, Tan Q, Gao Y, Bao Z, Yin Z, Guo B, Wang J, Tan L, Zhang Y, Xing H (2018). A large-sized basal ankylopollexian from East Asia, shedding light on early biogeographic history of Iguanodontia. *Science Bulletin* 63(9): 556–563.]

Becklespinax altispinax

> *Megalosaurus dunkeri* von Huene, 1923* ⇒ *Altispinax dunkeri* von Huene, 1923 [Valanginian; Wealden Group, East Sussex, England, UK]

= (objective synonym) *Acrocanthosaurus altispinax* Paul, 1988 ⇒ *Becklespinax altispinax* (Paul, 1988) Olshevsky, 1991 ⇒ *Altispinax altispinax* (Paul, 1988) Rauhut, 2003

= (objective synonym) *Altispinax* "lydekkerhueneorum" Pickering, 1984 (*nomen dubium*) ⇒ *Becklespinax* "*lydekkerhueneorum*" (Pickering, 1984) Olshevsky, 1991 (*nomen dubium*)

**cf. Megalosaurus dunkeri* Dames, 1884 [Berriasian; Bückeberg Formation, Niedersachsen, Germany]

Generic name: *Becklespinax* ← {(person's name) Beckles + spinax [(Lat.) spīna: a thorn, spine]; in honor of fossil collector Samuel H. Beckles (1814–1890).

Specific name: *altispinax* ← {(Lat.) altus: high + spina: the backbone, spine}; in reference to the high-spined vertebrae.

Etymology: high-spined, **Beckles' spine**

Taxonomy: Theropoda: Avetheropoda: Carnosauria: Tetanurae

Notes: According to *The Dinosauria 2nd edition*, *Becklespinax* was considered to be valid, but now it is thought to be a junior objective synonym of *Altispinax*. (*1)

[References: (*1) Maisch MW (2016). The nomenclatural status of the carnivorous dinosaur genus *Altispinax* v. Huene, 1923 (Saurischia, Theropoda) from the Lower Cretaceous of England. *Neues Jahrbuch für Geologie und Paläontologie – Abhandlungen* 280(2): 215–219(5).]

Beg tse

> *Beg tse* Yu *et al.*, 2020 [Albian–Cenomanian; Ulaanoosh Formation, Ömnögovi, Mongolia]

Generic name + Specific name: *Beg tse* ← {(Mongolian culture) Beg-tse*: Himalayan deity}; referring to "a pre-Buddhist god of war. Beg-tse is commonly portrayed as heavily armored with large rugosities on its body, which refers to the rugose structures on the jugal and surangular". *(Mong.) begder: 'coat of mail'.

Etymology: **Beg**-tse (a god of war of the Himalayan deity)
Taxonomy: Ornithischia: Neoceratopsia

[References: Yu C, Prieto-Marquez A, Chinzorig T, Badamkhatan Z, Norell M (2020). A neoceratopsian dinosaur from the Early Cretaceous of Mongolia and the early evolution oc ceratopsia. *Communications Biology* 3(1): 1–8.]

Beibeilong sinensis
> *Beibeilong sinensis* Pu *et al.*, 2017 [Cenomanian–Turonian; Gaogou Formation, Henan, China]

Generic name: *Beibeilong* ← {beibei + (Chin.) lóng 龍: dragon}; based on a nest with an embryo (nicknamed "Baby Louie") and eggs. Baby Louie = 路易貝貝 [Pinyin: lù路yì易bèi貝bèi貝] (*1).

According to Pu *et al.* (2017), the generic name is derived from Chinese Pinyin 'beibei' for baby and 'long' for dragon (*2). *cf.* 貝: precious treasure.
Specific name: *sinensis* ← {(place-name) Sin + -ensis}; referring to its discovery in China.
Etymology: **baby dragon** from China [中華貝貝龍]
Taxonomy: Theropoda: Oviraptorosauria: Caenagnathidae

[References: (*1) 科學家發現全新恐龍物種命名「中華貝貝龍」每日頭條2017-05-10〈https://kknews.cc/science/294jxqz.html〉; (*2) Pu H, Zelenitsky DK, Lü J, Currie PJ, Carpenter K, Xu L, Koppelhus EB, Jia S, Xiao L, Chuang H, Li T, Kundrát M, Shen C (2017). Perinate and eggs of a giant caenagnathid dinosaur from the Late Cretaceous of central China. *Nature Communications* 8: 14952.]

Beipiaognathus jii
> *Beipiaognathus jii* Hu *et al.*, 2016 (*nomen dubium*) [Barremian–Aptian: Yixian Formation, Liaoning, China]

Generic name: *Beipiaognathus* ← {(Chin. place-name) Běipiào北票 + (Gr.) γνάθος/gnathos: jaw}
Specific name: *jii* ← {(person's name) Jì季 + -ī}
Etymology: Ji's **Beipiao jaw**
Taxonomy: Theropoda: Coelurosauria: Compsognathidae

[References: Hu Y, Wang X, Huang J (2016). A new species of compsognathid from the Early Cretaceous Yixian Formation of western Liaoning, China. *Journal of Geology* 40: 191–196.]
;

Beipiaosaurus inexpectus
> *Beipiaosaurus inexpectus* Xu *et al.*, 1999 [Aptian; Jianshangou Beds, Yixian Formation, Liaoning, China]

Generic name: *Beipiaosaurus* ← {(Chin. place-name) Běipiào北票 + (Gr.) σαῦρος/sauros: lizard}
Specific name: *inexpectus* (*1) ← {(Lat.) inexpetatus, inexspectatus: unlooked for, unexpected}; referring to the unexpected features.
Etymology: **Beipiao lizard** with unexpected features
Taxonomy: Theropoda: Coelurosauria: Therizinosauroidea

[References: Xu X, Tang Z-L, Wang X-L (1999). A therizinosauroid dinosaur with integumentary structures from China. *Nature* 399(6734): 350–354.; (*1) Xu X, Zheng X, You H (2008). A new feather type in a nonavian theropod and the early evolution of feathers. *Proceedings of the National Academy of Sciences* 106(3): 832–834.]

Beishanlong grandis
> *Beishanlong grandis* Makovicky *et al.*, 2010 [Aptian; Xiagou Formation, Xinminbao Group, Gansu, China]

Generic name: *Beishanlong* ← {(Chin. place-name) Běi Shān北山 'Northern Mountain' + lóng龍: dragon} (*1)
Specific name: *grandis* ← {(Lat.) grandis: large}; in reference to the large size of this taxon.
Etymology: large **Beishan dragon**
Taxonomy: Theropoda: Tetanurae: Coelurosauria: Ornithomimosauria: Deinocheiridae

[References: (*1) Makovicky PJ, Li D, Gao K-Q, Lewin M, Erickson GM, Norell MA (2009). A giant ornithomimosaur from the Early Cretaceous of China. *Proceedings of the Royal Society B: Biological Sciences* 277(1679): 191–198.]

Bellubrunnus rothgaengeri [Pterosauria]
> *Bellubrunnus rothgaengeri* Hone *et al.*, 2012 [Kimmeridgian; Brunn, Upper Palatinate, Germany]

Generic name: *Bellubrunnus* ← {(Lat.) bellus: beautiful + (place-name) Brunn} (*1)
Specific name: *rothgaengeri* ← {(person's name) Rothgaenger + -ī}; in honor of Monika Rothgaenger for finding the holotype (*1).
Etymology: Rothgaenger's **beautiful one of Brunn**
Taxonomy: Archosauria: Ornithodira: Pterosauria: Rhamphorhynchidae (*1)
Notes: The wing-span of the holotype is around 30 cm and it is considered to be a juvenile pterosaur.

[References: (*1) Hone DW, Tischlinger H, Frey E, Röper M (2012). A new non-pterodactyloid pterosaur from the Late Jurassic of southern Germany. *PLoS ONE*. 7(7): e39312.]

Bellusaurus sui
> *Bellusaurus sui* Dong, 1990 [Middle Jurassic; Shishugou Formation, northeastern Junggar

Basin, Xinjiang, China]

Generic name: *Bellusaurus* ← {(Lat.) bellus: fine, beautiful + (Gr.) σαῦρος/sauros: lizard}; referring to its gracile body structure.

Specific name: *sui* ← {(Chin. person's name) Sū 蘇 + -ī}; "in honor of Chinese preparatory Su, Youling.

Etymology: Su's **fine lizard** [蘇氏巧龍]

Taxonomy: Saurischia: Sauropodomorpha: Sauropoda: Camarasauridae

Notes: Seventeen, probably juvenile, individuals were found in a single quarry (*1).

[References: (*1) Dong Z (1990). Sauropoda from the Kelameili Region of the Junggar Basin, Xinjiang Autonomous Region. *Vertebrata PalAsiatica* 28(1): 43–58.]

Berberosaurus liassicus

> *Berberosaurus liassicus* Allain *et al.*, 2007 [Pliensbachian–Toarcian; Toundoute Continental Series, Ouarzazate, Morocco]

Generic name: *Berberosaurus* ← {(demonym) the Berbers + (Gr.) σαῦρος/sauros: lizard}; "from the Berbers who live mainly in Morocco".

Specific name: *liassicus* ← {[geol.] Lias: the Lias epoch, Early Jurassic + -icus}; "referring to the statigraphic epoch of the specimen".

Etymology: Liassic **Berber lizard**

Taxonomy: Theropoda: Ceratosauria

[References: Allain R, Tykoski R, Aquesbi N, Jalil NE, Monbaron M, Russell D, Taquet P (2007). An abelisauroid (Dinosauria: Theropoda) from the Early Jurassic of the High Atlas Mountains, Morocco, and the radiation of ceratosaurs. *Journal of Vertebrate Paleontology* 27(3): 610–624.]

Betasuchus bredai

> *Megalosaurus bredai* Seeley, 1883 ⇒ Ornithomimidorum genus b *bredai* (Seeley, 1883) von Huene, 1932 (*nomen oblitum*) ⇒ *Betasuchus bredai* (Seeley, 1883) von Huene, 1932 (*nomen dubium*) [Maastrichtian; Maastricht Formation, Limburg, Netherlands]

Generic name: *Betasuchus* ← {(Gr.) β = (Lat.) B: genus b of the ornithomimids + suchus: crocodile}

Von Huene reallocated the second of two *Megalosaurus* species as "b" and the first one, *M. lonzeensis*, as "Ornithomimidorum genus a".

Specific name: *bredai* ← {(person's name) Breda + -ī}; in honor of Jacob Gijsbertus Samuël van Breda, the late Dutch biologist and geologist.

Etymology: Breda's **genus B crocodile**

Taxonomy: Theropoda: Abelisauroidea

Notes: According to *The Dinosauria 2nd edition*, *Megalosaurus bredai* Seeley, 1883 is considered a *nomen dubium*.

[References: Seeley HG (1883). On the dinosaurs from the Maastricht Beds. *Journal of the Geological Society* 39: 246–253.]

Bicentenaria argentina

> *Bicentenaria argentina* Novas *et al.*, 2012 [Cenomanian; Candeleros Formation, Río Negro, Argentina]

Generic name: *Bicentenaria* ← {bi-: two + centenary: hundred}; "from bicentenary, due to the 200 years of the Revolution of May (25th May, 1810), which lead to the first Argentinean autonomous government, thus constituting one of the most important chapters in the history of this South American country" (*1).

Specific name: *argentina* ← {(place-name) Argentina}; "in honor of Argentina, the country where the specimens were recovered" (*1).

Etymology: Argentine **bicentenary one**

Taxonomy: Theropoda: Coelurosauria

[References: (*1) Novas FE, Ezcurra MD, Agnolin FL, Pol D, Ortíz R (2012). New Patagonian Cretaceous theropod sheds light about the early radiation of Coelurosauria. *Revista del Museo Argentino de Ciencias Naturales, Nueva Serie* 14 (1): 57–81.]

Bienosaurus lufengensis

> *Bienosaurus lufengensis* Dong, 2001 [Sinemurian?; lower Lufeng Formation, Yunnan, China]

Generic name: *Bienosaurus* ← {(Chin. person's name) Bien卞 [Pinyin: Biàn] + (Gr.) σαῦρος/sauros: lizard}; in honor of Dr. Bian Meinian who collected the holotype.

Specific name: *lufengensis* ← {(Chin. place-name) Lùfēng禄豊 + -ensis}

Etymology: **Dr. Bian's lizard** from Lufeng [禄豊卞氏龍]

Taxonomy: Ornithischia: Thyreophora: Ankylosauria: Scelidosauridae

[References: Dong Z (2001). Primitive armored dinosaur from the Lufeng Basin, China. In: Tanke D, Carpenter K (eds), *Mesozoic Vertebrate Life*. Indiana University Press, Bloomington: 237–243.]

Bissektipelta archibaldi

> *Amtosaurus archibaldi* Averianov, 2002 ⇒ *Bissektipelta archibaldi* (Averianov, 2002) Parish & Barrett, 2004 [Turonian; Bissekty Formation, Navoi, Uzbekistan]

Generic name: *Bissektipelta* ← {(place name) Bissekty Formation + (Lat.) pelta = (Gr.) πέλτη/peltē: shield}

Specific name: *archibaldi* ← {(person's name)

Archibald + -ī}; in honor of Prof. J. David Archibald (San Diego State University), the leader of the URBAC* project.

*Uzbekistan-Russia-Britain-America-Canada

Geological age: Archibald's **Bissekty shield**

Taxonomy: Ornithischia: Thyreophora: Ankylosauria: Ankylosauridae

Notes: *Amtosaurus archibaldi* became a valid ankylosaurid taxon, *Bissektipelta* (*1).

[References: (*1) Parish JC, Barrett PM (2004). A reappraisal of the ornithischian dinosaur *Amtosaurus magnus* Kurzanov and Tumanova 1978, with comments on the status of *A. Archibaldi* Averianov 2002. *Canadian Journal of Earth Sciences* 41(3): 299–306.]

Bistahieversor sealeyi

> *Bistahieversor sealeyi* Carr & Williamson, 2010 [Campanian; Hunter Wash Member, Kirtland Formation, New Mexico, USA]

Generic name: *Bistahieversor* ← {(Navajo, place-name) "Bistahí: place of adobe formations, in reference to the Bisti Wilderness Area" + (Lat.) "ēversor: destroyer, in reference to the presumed predatory habits of the animal"} (*1).

Specific name: *sealeyi* ← {(person's name) Sealey + -ī}; in honor of "Paul Sealey, Research Associate at the New Mexico Museum of Natural History, in recognition of his discovery of the holotype specimen" (*1).

Etymology: Sealey's **Bistahí destroyer**

Taxonomy: Theropoda: Coelurosauria: Tyrannosauridae: Tyrannosaurinae

[References: (*1) Carr TD, Williamson TE (2010). *Bistahieversor sealeyi,* gen. *et* sp. nov., a new tyrannosauroid from New Mexico and the origin of deep snouts in Tyrannosauroidea. *Journal of Vertebrate Paleontology* 30 (1): 1–16.]

Blasisaurus canudoi

> *Blasisaurus canudoi* Cruzado-Caballero *et al.*, 2010 [Maastrichtian; Arén Formation, Arén, Huesca, Aragón, Spain]

Generic name: *Blasisaurus* ← {(place-name) Blasi-1 site + (Gr.) σαῦρος/sauros: lizard}; "from the Blasi sites in Arén, a village in the province of Huesca (Aragon, Spain)" (*1).

Specific name: *canudoi* ← {(person's name) Canudo + -ī}; "in honor of paleontologist José Ignacio Canudo, one of the leaders of the Grupo *Aragosaurus* IUCA (Zaragoza)*, who excavated the Blasi sites and other dinosaur localities of Aragon, for his contribution to the study of Iberian dinosaurs" (*1).

*Instituto Universitario de Investigación en Ciencias Ambientales de Aragón Universidad Zaragoza

Etymology: Canudo's **lizard from Blasi-1 site**

Taxonomy: Ornithischia: Ornithopoda: Hadrosauridae: Lambeosaurinae

[References: (*1) Cruzado-Caballero P, Pereda-Suberbiola X, Ruiz-Omeñanca JI (2010). *Blasisaurus canudoi* gen. *et* sp. nov., a new lambeosaurine dinosaur (Hadrosauridae) from the latest Cretaceous of Arén (Huesca, Spain). *Canadian Journal of Earth Sciences* 47(12): 1507–1517.]

Blikanasaurus cromptoni

> *Blikanasaurus cromptoni* Galton & Heerden, 1985 [Norian–Rhaetian; lower Elliot Formation, Eastern Cape, South Africa]

Generic name: *Blikanasaurus* ← {(place-name) Blikana + (Gr.) σαῦρος/sauros: lizard}; derived from Blikana Mountain, close to the discovery site.

Specific name: *cromptoni* ← {(person's name) Crompton + -ī}; in honor of Alfred Walter Crompton.

Etymology: Crompton's **Blikana lizard**

Specific name: Saurischia: Sauropodomorpha: Sauropoda: Blikanasauridae

[References: Galton PM, van Heerden J (1985). Partial hindlimb of *Blikanasaurus cromptoni* n. gen. and n. sp., representing a new family of prosauropod dinosaurs from the Upper Triassic of South Africa. *Geobios* 18(4): 509–516. ; Yates AM (2008). A second specimen of *Blikanasaurus* (Dinosauria: Sauropoda) and the biostratigraphy of the lower Elliot Formation. *Palaeontogia Africana* 43: 39–43.]

Bohaiornis guoi [Avialae]

> *Bohaiornis guoi* Hu *et al.*, 2011 [Aptian; Jiufotang Formation, Liaoning, China]

Generic name: *Bohaiornis* ← {(Chin. place-name) Bóhǎi渤海 + (Gr.) ὄρνις/ornis: bird}; "derived from the name of the Bohai Sea, an inland sea close to the type locality" (*1).

Specific name: *guoi* ← {(Chin. person's name) Guō郭 + -ī}; "in honor of Mr. Guo Chen, who collected the holotype specimen" (*1).

Etymology: Guo's **bird from Bohai Sea** [郭氏渤海鳥]

Taxonomy: Theropoda: Avialae: Enantiornithes: Bohaiornithidae *cf.* Aves: Enantiornithes (*1)

[References: (*1) Hu D, Li L, Hou L, Xu X (2011). A new enantiornithine bird from the Lower Cretaceous of western Liaoning, China. *Journal of Vertebrate Paleontology* 31(1): 154–161.]

Bolong yixianensis

> *Bolong yixianensis* Wu *et al.*, 2010 [Aptian; Yixian Formation, Liaoning, China]

Generic name: *Bolong* ← {(Chin. person's name) Bó 薄 + lóng龍: dragon}; "in honor of Bó Hăichén and Bó Xué, who discovered and excavated the holotype".
Specific name: *yixianensis* ← {(Chin. place-name) Yìxiàn 義縣 + -ensis}; referring "both to the Yixian Formation, where the holotype was discovered, and to the city of Yixian, where the holotype is housed and displayed".
Etymology: **Bo's dragon** from Yixian Formation [義縣薄氏龍]
Taxonomy: Ornithischia: Ornithopoda: Iguanodontia: Iguanodontoidea: Hadrosauroidea
[References: Wu W, Godefroit P, Hu D-Y (2010). *Bolong yixianensis* gen. *et* sp. nov.: a new iguanodontoid dinosaur from the Yixian Formation of Western Liaoning, China. *Geology and Resources* 19(2): 127–133.]

Boluochia zhengi [Avialae]
> *Boluochia zhengi* Zhou, 1995 [Aptian; Jiufotang Formation, Boluochi, Chaoyang, Liaoning, China]
Generic name: *Boluochia* ← {(Chin. place-name) Bōluóchì波羅赤 + -ia}
Specific name: *zhengi* ← {(Chin. person's name) Zhèng鄭 + -ī}; in honor of Professor Zheng Zuoxin.
Etymology: Zheng's **one from Boluochi**
Taxonomy: Theropoda: Avialae: Enantiornithes: Longipterygidae

Bonapartenykus ultimus
> *Bonapartenykus ultimus* Agnolin *et al.*, 2012 [Campanian–Maastrichtian; Allen Formation, Río Negro, Argentina]
Generic name: *Bonapartenykus* ← {(person's name) Bonaparte + (Gr.) ὄνυξ/onyx: claw}; in honor of paleontologist José Fernando Bonaparte.
Specific name: *ultimus* ← {(Lat.) ultimus: latest}
Etymology: latest **Bonaparte's claw**
Taxonomy: Theropoda: Coelurosauria: Alvarezsauridae
Notes: Skeletal remains of a bird-like dinosaur, *Bonapartenykus,* was found near two broken eggshells.
[References: Agnolin FL, Powell JE, Novas FE, Kundrát M (2012). New alvarezsaurid (Dinosauria, Theropoda) from uppermost Cretaceous of north-western Patagonia with associated eggs. *Cretaceous Research* 35: 33–56.]

Bonapartesaurus rionegrensis
> *Bonapartesaurus rionegrensis* Cruzado-Caballero & Powell, 2017 [Campanian–Maastrichtian; Allen Formation, Río Negro, Argentina]
Generic name: *Bonapartesaurus* ← {(person's name) José Bonaparte + (Gr.) σαῦρος/sauros: lizard}; "in honor of the paleontologist José Fernando Bonaparte, for his contribution to paleontology in Argentina" (*1).
Specific name: *rionegrensis* ← {(place-name) Río Negro Province + -ensis}
Etymology: **Bonaparte lizard** from Río Negro Province
Taxonomy: Ornithischia: Ornithopoda: Hadrosauridae: Saurolophinae: Saurolophini
cf. Ornithischia: Ornithopoda: Hadrosauridae: Hadrosaurinae (*1)
[References: (*1) Cruzado-Caballero P, Powell JE (2017). *Bonapartesaurus rionegrensis*, a new hadrosaurine dinosaur from South America: implications for phylogenetic and biogeographic relations with North America. *Journal of Vertebrate Paleontology* 37: 1–16.]

Bonatitan reigi
> *Bonatitan reigi* Martinelli & Forasiepi, 2004 [Campanian–Maastrichtian; Allen Formation, Bajo de Santa Rosa, Río Negro, Argentina]
Generic name: *Bonatitan* ← {(person's name) Bonaparte + (Gr.) Τῑτάν/Titan: giant}; "in honor of Dr. José Fernando Bonaparte, due to his immense contribution to the knowledge of Mesozoic vertebrates of South America" (*1).
Specific name: *reigi* ← {(person's name) Reig + -ī}; "in honor to Dr. Osvaldo Reig for his contribution to South American paleontology" (*1).
Etymology: Reig's **Bonaprte giant**
Taxonomy: Saurischia: Sauropodomorpha: Sauropoda: Titanosauria: Saltasauridae
cf. Titanosauria: Titanosauridae: Saltasaurinae (*1)
[References: (*1) Martinelli A, Forasiepi AM (2004). Late Cretaceous vertebrates from Bajo de Santa Rosa (Allen Formation), Río Negro province, Argentina, with the description of a new sauropod dinosaur (Titanosauridae). *Revista del Museo Argentino de Ciencias Naturales, Nueva Serie* 6(2): 257–305.]

Bonitasaura salgadoi
> *Bonitasaura salgadoi* Apesteguía, 2004 [Santonian; Bajo de la Carpa Formation, Río Negro, Argentina]
Generic name: *Bonitasaura* ← {(place-name) "La Bonita" hill: the name of the quarry + (Gr.) σαύρα/saura: a female reptile} (*1)
Specific name: *salgadoi* ← {(person's name) Salgado + -ī}; "in honor of Leonardo Salgado, the Argentinian paleontologist who gave new perspectives to sauropod research" (*1).
Etymology: Salgado's **La Bonita lizard**

Taxonomy: Saurischia: Sauropodomorpha: Sauropoda: Sauropoda: Titanosauria

[References: (*1) Apesteguía S (2004). *Bonitasaura salgadoi* gen. *et* sp. nov.: a beaked sauropod from the Late Cretaceous of Patagonia. *Naturwissenschaften* 91(10): 493–497.]

Borealopelta markmitchelli

> *Borealopelta markmitchelli* Brown *et al*., 2017 [Aptian; Clearwater Formation, Alberta, Canada]

Generic name: *Borealopelta* ← {(Lat.) boreālis: northern + (Lat.) pelta = (Gr.) πέλτη/peltē: shield}; "in reference to the northern locality and the preserved epidermal scales and dermal osteoderms" (*1).

Specific name: *markmitchelli* ← {(person's name) Mark Mitchell + -ī}; "in honor of Mark Mitchell for his more than 7,000 hours of patient and skilled preparation of the holotype" (*1).

Etymology: Mark Mitchell's **northern shield**

Taxonomy: Ornithischia: Ankylosauria: Nodosauridae

[References: (*1) Brown CM, Henderson DM, Vinther J, Fletcher I, Sistiaga A, Herrera J, Summons RE (2017). An exceptionally preserved three-dimensional armored dinosaur reveals insights into coloration and Cretaceous predator-prey dynamics. *Current Biology* 27(16): 2514–2521.]

Borealosaurus wimani

> *Borealosaurus wimani* You *et al*., 2004 [Cenomanian–Turonian; Sunjiawan Formation, Liaoning, China]

Generic name: *Borealosaurus* ← {(Lat.) boreālis: northern [(Gr.) βορέας/Boreās: the north wind] + σαῦρος/sauros: lizard}; "referring to the location of the fossil site in northern China, and the Northern Hemisphere" (*1).

Specific name: *wimani* ← {(person's name) Wiman + -ī}; "in honor of Swedish paleontologist Carl Wiman, who named the first Chinese dinosaur (*Euhelopus*) in 1929" (*1).

Etymology: Wiman's **northern lizard**

Taxonomy: Saurischia: Sauropoda: Titanosauriformes: Titanosauria

[References: (*1) You H, Ji Q, Lamanna MC, Li J, Li Y (2004). A titanosaurian sauropod dinosaur with opisthocoelous caudal vertebrae from the Early Late Cretaceous of Liaoning Province, China. *Acta Geologica Sinica* 78(4): 907–911.]

Boreonykus certekorum

> *Boreonykus certekorum* Bell & Currie, 2015 [Campanian; Wapiti Formation, Alberta, Canada]

Generic name: *Boreonykus* ← {(Lat.) boreās [= (Gr.) βορέας: the north wind] + onychos [(Gr.) ὄνυξ/onyx: claw]}; "referring to the modern day boreal forest where the type specimen was found" (*1).

Specific name: *certekorum* ← {(institution name) Certek + -ōrum}; in honor of "Certek Heating Solutions and the Barendregt family (Wembley, Alberta) for their continued support of paleontology in the Peace Region" (*1).

Etymology: Certek's **boreal claw**

Taxonomy: Theropoda: Dromaeosauridae: Velociraptorinae

[References: (*1) Bell PR, Currie PJ (2016). A high-latitude dromaeosaurid, *Boreonykus certekorum*, gen. *et* sp. nov. (Theropoda), from the upper Campanian Wapiti Formation, west-central Alberta. *Journal of Vertebrate Paleontology* 36(1): e1034359.]

Borogovia gracilicrus

> *Borogovia gracilicrus* Osmólska, 1987 [Maastrichtian; Nemegt Formation, Ömnögovi, Mongolia]

Generic name: *Borogovia* ← {(novel) Borogove + -ia}; "borogove – the name of a fantastic creature from "*Alice in Wonderland*" by Lewis Carroll" (*1).

Specific name: *gracilicrus* ← {(Lat.) gracilis: slender + crus: shin}; "because of very long and slender shin" (*1).

Etymology: **Borogove** with slender shins

Taxonomy: Theropoda: Coelurosauria: Troodontidae

[References: (*1) Osmólska H (1987). *Borogovia gracilicrus* gen. *et* sp. n. a new troodontid dinosaur from the Late Cretaceous of Mongolia. *Acta Palaeontologica Polonica* 32: 133–150.]

Bothriospondylus suffossus

> *Bothriospondylus suffossus* Owen, 1875 (*nomen dubium*) ⇒ *Astrodon suffossus* (Owen, 1875) Hatcher, 1903 ⇒ *Ornithopsis suffossa* (Owen, 1875) von Huene, 1922 [Kimmeridgian; Kimmeridge Clay Formation, Wiltshire, UK]

Generic name: *Bothriospondylus* ← {bothrion, *dim.* [(Gr.) βόθρος/bothros: furrow, excavation] + σπόνδυλος/spondylos: vertebra}; in reference to "the lateral excavations of the centrum, undermining, as it were, the base of the neural arch" (*1).

Specific name: *suffossus* ← {(Lat.) suffossus: undermined (←suffodiō: undermine)}

Etymology: undermined, **excavated vertebra**

Taxonomy: Saurischia: Sauropodomorpha: Sauropoda: Neosauropoda

Other species:

> *Bothriospondylus robustus* Owen, 1875 (*nomen dubium*) ⇒ *Marmarospondylus robustus* (Owen, 1875) Owen, 1875 [late

Bathonian; Forest Marble Formation, UK]

> *Ornithopsis hulkei* Seeley, 1870 (Lectotype) ⇒ *Bothriospondylus magnus* (Seeley, 1870) Owen, 1875 ⇒ *Chondrosteosaurus magnus* (Seeley, 1870) Owen,1876 ⇒ *Pelorosaurus hulkei* (Seeley, 1870) Lydekker, 1893 ⇒ *Eucamerotus hulkei* (Seeley, 1870) Sebaschan, 2005 [Barremian; Wessex Formation, Isle of Wight, UK]

> *Bothriospondylus elongatus* Owen, 1875 (*nomen dubium*) ⇒ *Ornithopsis hulkei* Lydekker, 1889 (*partim*) [Valanginian; Tunbridge Wells Sand Formation, England, UK]

> *Bothriospondylus madagascariensis* Lydekker, 1895 (*nomen dubium*) ⇒ *Lapparentosaurus madagascariensis* (Lydekker, 1895) Bonaparte, 1986 [Bathonian; Isalo III Formation, Mahajanga, Madagascar]

Notes: According to Mannion (2010), *Bothriospondylus* was concluded to be a *nomen dubium*.

[References: (*1) Owen R (1875). Monographs on the British fossil Reptilia of the Mesozoic formations. Part II. (Genera *Bothriospondylus*, *Cetiosaurus*, *Omosaurus*) *Paleontographical Society Monographs* 29: 15–26.]

Brachiosaurus altithorax

> *Brachiosaurus altithorax* Riggs, 1903 ⇒ *Giraffatitan altithorax* Paul, 1988* [Kimmeridgian; Morrison Formation, Colorado, USA] *according to The *Dinosauria 2nd edition, Giraffatitan* is considered a junior synonym of *Brachiosaurus*.

Generic name: *Brachiosaurus* ← {(Gr.) βρᾰχίων/brachiōn: arm + σαῦρος/sauros: lizard}; referring to the length of the arms which was unusual for a sauropod.

Specific name: *altithorax* ← {(Lat.) altus: high, deep + thorax: thorax, breastplate}

Etymology: **arm lizard** having high thorax

Taxonomy: Saurischia: Sauropodomorpha: Sauropoda: Brachiosauridae

Other species:

> *Brachiosaurus brancai* Janensch, 1914 ⇒ *Giraffatitan brancai* (Janensch, 1914) Paul, 1988 [Kimmeridgian; Tendaguru Formation, Lindi, Tanzania]

= *Brachiosaurus fraasi* Janesch, 1914 [Kimmeridgian; Tendaguru Formation, Lindi, Tanzania]

> *Brachiosaurus atalaiensis* Lapparent & Zbyszewski, 1957 ⇒ *Lusotitan atalaiensis* (Lapparent & Zbyszewski, 1957) Antunes & Mateus, 2003 [Kimmeridgian; Lourinhã, Portugal]

> *Brachiosaurus nougaredi* Lapparent, 1960 (*nomen dubium*) [Late Jurassic; Taouratine Group, Illizi, Algeria]

[References: Taylor MP (2009). A re-evaluation of *Brachiosaurus altithorax* Riggs, 1903 (Dinosauria: Sauropoda) and its generic separation from *Giraffatitan brancai* (Janensch 1914). *Journal of Vertebrate Paleontology* 31(3): 787–806.]

Brachyceratops montanensis

> *Brachyceratops montanensis* Gilmore, 1914 [Campanian; Two Medicine Formation, Montana, USA]

Generic name: *Brachyceratops* ← {(Gr.) βραχύς/brachys: short + κερατ-/kerat- [κέρας/ keras: horn] + ὤψ/ōps: face}; in reference to the short snout.

Specific name: *montanensis* ← {(place-name) Montana + -ensis}

Etymology: **short horned-face** from Montana

Taxonomy: Ornithischia: Marginocephalia: Ceratopsia: Ceratopsidae: Centrosaurinae

Notes: The incomplete and jumbled remains of five juvenile individuals have been thought to be *Brachyceratops*, however, one specimen is reclassified as *Rubeosaurus*. Now *Brachyceratops* is considered a *nomen dubium*, according to McDonald (2011).

[References: Gilmore CW (1914). A new ceratopsian dinosaur from the Upper Cretaceous of Montana, with note on *Hypacrosaurus*. *Smithsonian Miscellaneous Collections* 63(3): 1–10.]

Brachylophosaurus canadensis

> *Brachylophosaurus canadensis* Sternberg, 1953 [Campanian; Oldman Formation, Alberta, Canada]

= *Brachylophosaurus goodwini* Horner, 1988 [Campanian; Judith River Formation, Montana, USA]

Generic name: *Brachylophosaurus* ← {(Gr.) βραχύς/brachys: short + λόφος/lophos: crest + σαῦρος/sauros: lizard}; referring to the shape of the crest on its skull.

Specific name: *canadensis* ← {(place-name) Canada + -ensis}

Etymology: **short-crested lizard** from Canada

Taxonomy: Ornithischia: Ornithopoda: Hadrosauridae: Saurophinae

Brachypodosaurus gravis

> *Brachypodosaurus gravis* Chakravarti, 1934 [Maastrichtian; Lameta Formation, Jubbulpore, India]

Generic name: *Brachypodosaurus* ← {(Gr.) βραχύς/brachys: short + ποδος/podos [πούς/pūs: foot] + σαῦρος/sauros: lizard}
Specific name: *gravis* ← {(Lat.) gravis: heavy}.
Etymology: heavy, **short-footed lizard**
Taxonomy: ?Ornithischia: ?Thyreophora: ?Ankylosauria (*1)
Notes: According to *The Dinosauria 2nd edition*, this genus is considered to be a *nomen dubium* as so few remains have been found.
[References: (*1) Maryańska T (1977). Ankylosauridae (Dinosauria) from Mongolia. *Palaeontologia Polonica* 37: 85–151.]

Brachytrachelopan mesai
> *Brachytrachelopan mesai* Rauhut *et al*., 2005 [Oxfordian–Tithonian; Cañadón Calcáreo Formation, Chubut, Argentina]
Generic name: *Brachytrachelopan* ← {(Gr.) βραχύς/brachys: short + τράχηλος/trachēlos: neck + (Gr.myth.) Πάν/Pan: the god the shepherds}; in reference to the fact that the specimen was found by Daniel Mesa, a local shepherd, while looking for stray sheep. (*1)
Specific name: *mesai* ← {(person's name) Mesa + -ī}; in honor of Daniel Mesa and his family, who found the specimen (*1).
Etymology: Mesa's **short-necked Pan**
Taxonomy: Saurischia: Sauropodomorpha: Sauropoda: Dicraeosauridae
[References: (*1) Rauhut OWM, Remes K, Fechner R, Cladera G, Puerta P (2005). Discovery of a short-necked sauropod dinosaur from the Late Jurassic period of Patagonia. *Nature* 435: 670–672.]

Bradycneme draculae
> *Bradycneme draculae* Harrison & Walker, 1975 [Maastrichtian; Sânpetru Formation, Transylvania, Romania]
Generic name: *Bradycneme* ← {(Gr.) βρᾰδύς*/ "bradys (= heavy or massive)" (*1) (*sic*) + κνήμη/knēmē: leg} The gender is feminine.
*βρᾰδυς: slow. Βρᾰδυ-σκελής, ές, (σκέλος) slow of leg.
Specific name: *draculae* ← {(Romanian) dracul: evil one + -ae} (*1)
cf. (Lat.) dracō = (Gr.) δράκων, a dragon, or serpent of huge size.
Etymology: evil's **heavy / slow leg**
Taxonomy: Theropoda: Coelurosauria: Alvarezsauridae
Notes: The original describers believed the animal a new famiy of owls (*1). According to *The Dinosauria 2nd edition*, *Bradycneme dracurae* is considered a *nomen dubium*.
[References: (*1) Harrison CJO, Walker CA (1975). The Bradycnemidae, a new family of owls from the Upper Cretaceous of Romania. *Palaeontology* 18(3): 563–570.]

Brasilotitan nemophagus
> *Brasilotitan nemophagus* Machado *et al*., 2013 [Maastrichtian; Adamantina Formation, São Paulo, Brazil]
Generic name: *Brasilotitan* ← {(Port. place-name) Brasil: Brazil + (Gr. myth.) Τιτάν/Titan: "a group of Greek divinities" (*1)}
Specific name: *nemophagus* ← {(Gr.) νέμος/nemos: pasture, wood + φαγός/phagos [φᾰγέιν/phagein: to eat, devour]}; "in allusion to the herbivorous nature of this species" (*1).
Etymology: plant-eating, **Brazil Titan**
Taxonomy: Saurischia: Sauropodomorpha: Sauropoda: Titanosauria
[References: (*1) Machado EB, Avilla LS, Nava WR, Campos DA, Kellner AWA (2013). A new titanosaur sauropod from the Late Cretaceous of Brazil. *Zootaxa* 3701(3): 301–321.]

Bravoceratops polyphemus
> *Bravoceratops polyphemus* Wick & Lehman, 2013 [Maastrichtian; Javelina Formation, Texas, USA]
Generic name: *Bravoceratops* ← {(Sp. / Port.) bravo: brave, wild [(place-name) Río Bravo del Norte (= Rio Grande)] + *Ceratops* [(Gr.) κερατ-/kerat- [κέρας/keras: horn] + ὤψ/ōps: face]}
Specific name: *polyphemus* ← {(Gr. myth.) Πολύφημος/Polyphēmos: the giant son of Poseidon and Thoosa}
Etymology: Polyphemus' **wild-horned face (from Río Bravo del Norte)**
Taxonomy: Ornithischia: Ceratopsia: Ceratopsidae: Chasmosaurinae
[References: Wick SL, Lehman TM (2013). A new ceratopsian dinosaur from the Javelina Formation (Maastrichtian) of West Texas and implications for chasmosaurine phylogeny. *Naturwissenschaften* 100(7): 667–682.]

Breviceratops kozlowskii
>*Protoceratops kozlowskii* Mariańska & Osmólska, 1975 ⇒ *Breviceratops kozlowskii* (Mariańska & Osmólska, 1975) Kurzanov, 1990 [Campanian; Barun Goyot Formation, Ömnögovi, Mongolia]
Generic name: *Breviceratops* ← {(Lat.) brevis: short + (Gr.) κερατ-/kerat- [κέρας/keras: horn] + ὤψ/ōps: face}
Specific name: *kozlowskii* ← {(person's name)

Kozlowski + -ī}; in honor of Polish paleontologist Roman Kozłowski.
Etymology: Kozłowski's **short horned- face**
Taxonomy: Ornithischia: Cerapoda: Ceratopsia: Bagaceratopidae
Notes: Paul Sereno (2000) explained that the juvenile *Breviceratops* would grow into a mature *Bagaceratops*.

Brodavis americanus [Avialae]
> *Brodavis americanus* Martin *et al.*, 2012 [Maastrichtian; Frenchman Formation, Saskatchewan, Canada]
Generic name: *Brodavis* ← {(person's name) Brodkorb + (Lat.) avis: bird}; in memory of the late Professor Pierce Brodkorb, an American ornithologist and paleontologist.
Specific name: *americanus* ← {americanus: of America}
Etymology: **Brodkorb's bird** from America
Taxonomy: Theropoda: Avialae: Ornithurae: Hesperornithes
Other species:
> *Brodavis baileyi* Martin *et al.*, 2012 [Maastrichtian; Hell Creek Formation, South Dakota, USA]
> *Brodavis mongoliensis* Martin *et al.*, 2012 [Maastrichtian; Nemegt Formation, Ömnögovi, Mongolia]
> *Baptornis varneri* Martin & Cordes-Person, 2007 ⇒*Brodavis varneri* (Martin & Cordes-Person, 2007) Martin *et al.*, 2012 [Campanian; Pierre Shale Formation, South Dakota, USA]
[References: Martin LD (2012). A new evolutionary lineage of diving birds from the Late Cretaceous of North America and Asia. *Palaeoworld* 21(1): 59–63.]

Brohisaurus kirthari
> *Brohisaurus kirthari* Malkani, 2003 [Kimmeridgian; Sembar Formation, Balochistan, Pakistan]
Generic name: *Brohisaurus* ← {(demonym) Brohi: the Brohi tribe + (Gr.) σαῦρος/sauros: reptile} (*1)
Specific name: *kirthari* ← {(place-name) Kirthar + -ī}; referring to the Kirthar Range which is the host of Jurassic dinosaur. (*1)
Etymology: **Brohi's reptile** from Kirthar Range
Taxonomy: Saurischia: Sauropodomorpha: Sauropoda: Titanosauridae
[References: (*1) Malkani MS (2003). First Jurassic dinosaur fossils found from Kirthar Range, Khuzdar District, Balochistan, Pakistan. *Geological Bulletin of the University of Peshawar* 36: 73–83.]

Brontomerus mcintoshi
> *Brontomerus mcintoshi* Taylor *et al.*, 2011 (? *nomen dubium*) [Aptian–Albian; Cedar Mountain Formation, Utah, USA]
Generic name: *Brontomerus* ← {(Gr.) βροντή/brontē: thunder + μηρός/mēros: thigh}; "'thunder-thighs', in reference to the substantial femoral musculature implied by the morphology of the ilium" (*1).
Specific name: *mcintoshi* ← {(person's name) McIntosh + -ī}; "in honor of veteran sauropod worker John S. McIntosh, for his seminal paleontological work, done mostly unfunded and on his own time, which has been an inspiration to all of us who follow" (*1).
Etymology: McIntosh's thunder thighs
Taxonomy: Saurischia: Sauropodomorpha: Sauropoda: Neosauropoda: Camarasauromorpha
[References: (*1) Taylor MP, Wedel MJ, Cifelli RL (2011). A new sauropod dinosaur from the Lower Cretaceous Cedar Mountain Formation, Utah, USA. *Acta Palaeontologica* 56(1): 75–98.]

Brontosaurus excelsus
> *Brontosaurus excelsus* Marsh, 1879 ⇒ *Apatosaurus excelsus* (Marsh, 1879) Riggs, 1903 ⇒ *Camarasaurus excelsus* (Marsh, 1879) Osborn *vide* von Huene, 1929 ⇒ *Atlantosaurus excelsus* (Marsh, 1879) Steel, 1970 [Kimmeridgian; Morrison Formation, Wyoming, USA]
= *Brontosaurus amplus* Marsh, 1881 ⇒ *Apatosaurus amplus* (Marsh, 1881) Riggs, 1903 ⇒ *Atlantosaurus amplus* (Marsh, 1881) Steel, 1970 [Kimmeridgian; Morrison Formation, Wyoming, USA]
Generic name: *Brontosaurus* ← {(Gr.) βροντή/brontē: thunder + σαῦρος/sauros: lizard}; for its great size.
Specific name: *excelsus* ← {(Lat.) excelsus: noble, high}.
Etymology: noble **thunder lizard**
Taxonomy: Saurischia: Sauropodomorpha: Sauropoda: Diplodocidae: Apatosaurinae
Other species:
> *Elosaurus parvus* Peterson & Gilmore, 1902 ⇒ *Apatosaurus parvus* (Peterson & Gilmore, 1902) Upchurch *et al.*, 2004 ⇒ *Brontosaurus parvus* (Peterson & Gilmore, 1902) Tschopp *et al.*, 2015 [Kimmeridgian; Morrison Formation, Wyoming, USA]
> *Apatosaurus yahnahpin* Filla & Redman, 1994 ⇒ *Eobrontosaurus yahnahpin* (Filla & Redman, 1994) Bakker, 1998 ⇒ *Brontosaurus*

yahnahpin (Filla & Redman, 1994) Tschopp *et al.*, 2015 [Kimmeridgian; Morrison Formation, Wyoming, USA]

Notes: *Brontosaurus* had long been considered to be a synonym of *Apatosaurus* (Riggs, 1903) (*1). However, in recent research *Brontosaurus* was separated from *Apatosaurus. Brontosaurus* contains three species: *B. excelsus*, *B. yahnahpin,* and *B. parvus* (*2).

[References: (*1) Riggs ES (1903). Structure and relationships of opisthocoelian dinosaurs. *Publications of the Field Columbian Museum Geographical Series*. 2 (4): 165–196.; (*2) Tschopp E, Mateus O, Benson RBJ (2015). A specimen-level phylogenetic analysis and taxonomic revision of Diplodocidae (Dinosauria, Sauropoda). *PeerJ* 3: e857.]

Bruhathkayosaurus matleyi

> *Bruhathkayosaurus matleyi* Yadagiri & Ayyasami, 1989 [Maastrichtian; Kallamedu Formation, Tamil Nadu, India]

Generic name: *Bruhathkayosaurus* ← {(Sanskrit) bruhathkaya: huge body + (Gr.) σαῦρος/sauros: lizard}

Specific name: *matleyi* ← {(person's name) Matley + -ī}; in honor of British paleontologist and geologist Charles Alfred Matley (1866–1947).

Etymology: Matley's **huge-bodied lizard**

Taxonomy: Saurischia: Sauropodomorpha: Sauropoda

Notes: The authors originally classified *Bruhathkayosaurus* as a theropod (Yadagiri & Ayyasami, 1989). It was later recognized as a sauropod (Mortimer, 2001). The validity of this genus seems to be questionable. According to *The Dinosauria 2nd edition*, *Bruhathkayosaurus matleyi* is considered as a *nomen dubium*.

[References: Yadagiri P, Ayyasami K (1989). A carnosaurian dinosaur from the Kallamedu Formation (Maastrichtian horizon), Tamilnadu. In: Sastry MVA, Satry VV, Ramanujam CGK, Kapoor HM, Jagannatha Rao BR, Satsangi PP, Mathur UB (eds), *Simposium on the Three Decades of Development in Palaeontology and Stratigraphy in India. Volume 1. Precambrian to Mesozoic, Geological Society of India Special Publication* 11(1): 523–528.]

Bugenasaura infernalis

> *Thescelosaurus garbanii* Morris, 1976 [late Maastrichtian; Hell Creek Formation, Montana, USA]

= *Bugenasaura infernalis* Galton, 1995 [Maastrichtian; Hell Creek Formation, South Dakota, USA]

Generic name: *Bugenasaura* ← {(Lat. prefix) bu- [(Gr.) βου-/bū-: huge] + (Lat.) gena: cheek + (Gr.) σαῦρα/saura (feminine form): lizard}; "in allusion to the uniquely massive ridges on the maxilla and dentary for the attachment of cheeks" (*1).

Specific name: *infernalis* ← {(Lat.) infernālis: infernal, "belonging to the lower regions"}; "an allusion to the occurrence in the Hell Creek Formation" (*1).

Etymology: **huge-cheeked lizard** from Hell Creek Formation

Taxonomy: Ornithischia: Parksosauridae: Thescelosaurinae (*2)

cf. Ornithischia: Ornithopoda (acc. *The Dinosauria 2nd edition*)

cf. Ornithischia: Ornithopoda: Hypsilophodontidae: Thescelosaurinae (*1)

Notes: According to *The Dinosauria 2nd edition*, *Bugenasaura infernalis* was valid, but now *Bugenasaura* Galton, 1995 is considered to be a synonym of *Thescelosaurus* Gilmore, 1913.

[References: (*1) Galton PM (1999). Cranial anatomy of the hypsilophodontid dinosaur *Bugenasaura infernalis* (Ornithischia: Ornithopoda) from the Upper Cretaceous of North America. *Revue Paléobiologie, Genève* 18(2): 517–534.; (*2) Boyd CA (2015). The systematic relationships and biogeographic history of ornithischian dinosaurs. *PeerJ* 3: e1523.; Boyed CA, Brown CM, Scheetz RD, Clarke JA (2009). Taxonomic revision of the basal neornithischian taxa *Thescelosaurus* and *Bugenasaura. Journal of Vertebrate Paleontology* 29(3): 758–770.]

Buitreraptor gonzalezorum

> *Buitreraptor gonzalezorum* Makovicky *et al.*, 2005 [Cenomanian–Turonian; Candeleros Formation, Río Negro, Argentina]

Generic name: *Buitreraptor* ← {(place-name) La Buitrera: 'vulture roost', the type locality | (Sp.) buitre: vulture + (Lat.) raptor: robber} (*1)

Specific name: *gonzalezorum* ← {(person's name) González + -ōrum}; in honor of the brothers Fabián and Jorge González, who discovered the holotype, for their dedicated participation in 'La Buitrera' fieldwork over the years (*1).

Etymology: González brothers' **robber from La Buitrera**

Taxonomy: Theropoda: Maniraptora: Dromaeosauridae: Unenlagiinae (*1) (*2)

[References: (*1) Makovicky PJ, Agnolin FL, Apesteguía S (2005). The earliest dromaeosaurid theropod from South America. *Nature* 437: 1007–1011.; (*2) Gianechini FA, Apesteguía S (2011). Unenlagiinae revisited: dromaeosaurid theropods from South America. *Anais da Academia Brasileira de Ciências* 83(1): 163–195.]

Burianosaurus augustai

> *Burianosaurus augustai* Madzia *et al.*, 2017 [late Cenomanian; Peruc-Korycany Formation, Kutná Hora, Czech Republic]

Generic name: *Burianosaurus* ← {(person's name) Burian + (Gr.) σαῦρος/sauros: reptile, lizard}; in honor of the Czech palaeoartist Zdeněk Burian.

Specific name: *augustai* ← {(person's name) Josef Augusta + -ī}; in honor of the Czech palaeontologist Josef Augusta.

Etymology: Augusta and **Burian's lizard**

Taxonomy: Ornithischia: Neornithischia: Ornithopoda

[References: Madzia D, Boyd CA, Mazuch M (2017). A basal ornithopod dinosaur from the Cenomanian of the Czech Republic. *Journal of Systematic Palaeontology* 16(11): 967–979.]

Buriolestes schultzi

> *Buriolestes schultzi* Cabreira *et al.*, 2016 [Carnian; Santa Maria Formation, São João do Polêsine, Brazil]

Generic name: *Buriolestes* ← {(person's name) the Buriol family: the type-locality owners + (Gr.) ληστής/lēstēs: robber} (*1)

Specific name: *schultzi* ← {(person's name) Schultz + -ī}; in honor of the paleontologist Cesar Schultz. (*1)

Etymology: Schultz's **Buriol robber**

Taxonomy: Saurischia: Sauropodomorpha

[References: (*1) Cabreira SF, Kellner AWA, Dias-da-Silva S, da Silva LR, Bronzati M, de Almeida Marsola JC, Müller RT, de Souza Bittencourt J, Batista BJ, Raugust T, Carrilho R, Brodt A, Langer MC (2016). A unique Late Triassic dinosauromorph assemblage reveals dinosaur ancestral anatomy and diet. *Current Biology* 26(22): 3090–3095.]

Byronosaurus jaffei

> *Byronosaurus jaffei* Norell *et al.*, 2000 [late Campanian; Djadochta Formation, Gobi Desert, Mongolia]

Generic name: *Byronosaurus* ← {(person's name) Byron + (Gr.) σαῦρος/sauros: lizard}; "in honor of Byron Jaffe" (*1).

Specific name: *jaffei* ← {(person's name) Jaffe + -ī}; "in recognition of Byron Jaffe's family's support for the Mongolian Academy of Sciences-American Museum of Natural History Paleontological Expeditions" (*1).

Etymology: **Byron** Jaffe's **lizard**

Taxonomy: Theropoda: Deinonychosauria: Troodontidae

[References: (*1) Norell MA, Makovicky PJ, Clark JM (2000). A new troodontid theropod from Ukhaa Tolgod, Mongolia. *Journal of Vertebrate Paleontology* 20(1): 7–11.]

C

Caenagnathasia martinsoni

> *Caenagnathasia martinsoni* Currie *et al.*, 1993 [Turonian–Coniacian; Bissekty Formation, Bukhoro, Uzbekistan]

Generic name: *Caenagnathasia* ← {(Gr.) καινός/kainos: recent + γνάθος/gnathos: jaw + Ἀσία/Asia: Asia}

Specific name: *martinsoni* ← {(person's name) Martinson + -ī}; in honor of the paleontologist Gerbert Genrikhovich Martinson.

Etymology: Martinson's **caenagnathid (new jaw) from Asia**

Taxonomy: Theropoda: Oviraptorosauria: Caenagnathidae

Notes: *Caenagnathasia* seems to be the smallest known oviraptorosaur.

[References: Currie PJ, Godfrey SJ, Nessov L (1994). New caenagnathid (Dinosauria, Theropoda) specimens from the Upper Cretaceous of North America and Asia. *Canadian Journal of Earth Sciences* 30(10–11): 2255–2272.]

Caenagnathus collinsi

> *Caenagnathus collinsi* Sternberg, 1940 [Campanian; Dinosaur Park Formation, Alberta, Canada]

Generic name: *Caenagnathus* ← {(Gr.) καινός/kainos: recent + γνάθος/gnathos: jaw}

Specific name: *collinsi* ← {(person's name) Collins + -ī}; "in honor of the late Dr. William Henry Collins (1878–1937), for many years director of the Geological Survey of Canada" (*1).

Etymology: Collins' **recent jaws**

Taxonomy: Theropoda: Coerulosauria: Caenagnathidae: Caenagnathinae

Other species:

> *Chirostenotes pergracilis* Gilmore, 1924 [Campanian; Dinosaur Park Formation, Alberta, Canada]

= *Caenagnathus sternbergi*? Cracraft, 1971 ⇒ *Chirostenotes sternbergi* (Cracraft, 1971) Eberth *et al.*, 2001 [Campanian; Dinosaur Park Formation, Alberta, Canada]

= *Macrophalangia canadensis* Sternberg, 1932 [Campanian; Dinosaur Park Formation, Alberta, Canada]

Notes: According to *The Dinosauria 2nd edition*, *Caenagnathus* was considered a synonym of *Chirostenotes*. New material was found to belong to *Caenagnathus collinsi,* according to Funston *et al.* (2015).

[References: (*1) Sternberg R. M. (1940). A toothless bird from the Cretaceous of Alberta. *Journal of Paleontology* 14(1):81–85.; Funston GF, Persons WS, Bradley GJ, Currie PJ (2015). New material of the large-bodied caenagnathid *Caenagnathus collinsi* from the Dinosaur Park Formatin of Alberta, Canada. *Cretaceous Research* 54: 179–187.]

Caihong juji [Anchiornithidae]

> *Caihong juji* Hu *et al.*, 2018 [Oxfordian; Tiaojishan Formation, Hebei, China]

Generic name: *Caihong* ← {(Chin.) cǎihóng彩虹: rainbow}; "in reference to the beautiful preservation of the holotype specimen of the animal and the array of insights it offers into paravian evolution" (*1).

Specific name: *juji* ← {(Chin.) jù: big + jí: crest}; referring to the animal's prominent lacrimal crests (*1).

Etymology: **color of rainbow** with the large crest

Taxonomy: Theropoda: Maniraptora: Paraves: Avialae? : Anchiornithidae

cf. Theropoda: Maniraptora: Paraves (*1)

[References: (*1) Hu D, Clarke JA, Eliason CM, Qiu R, Li Q, Shawkey MD, Zhao C, D'Alba L, Jiang J, Xu X (2018). A bony-crested Jurassic dinosaur with evidence of iridescent plumage highlights complexity in early paravian evolution. *Nature Communications* 9: 217.]

Calamosaurus foxi

> *Calamospondylus foxi* Lydekker, 1889 (preoccupied) ⇒ *Calamosaurus foxi* (Lydekker, 1889) Lydekker, 1891 [Barremian; Wessex Formation, Isle of Wight, UK]

Generic name: *Calamosaurus* ← {(Gr.) κάλᾰμος/kalamos: reed, quill + σαῦρος/sauros: lizard}

Specific name: *foxi* ← {(person's name) Fox + -ī}; in honor of Reverend William Fox.

Etymology: Fox's **reed (quill vertebrae) lizard**

Taxonomy: Theropoda: Coelurosauria: Compsognathidae

Calamospondylus oweni

> *Calamospondylus oweni* Fox, 1866 (*nomen dubium*) [Aptian; Vectis Formation, Isle of Wight, UK]

Generic name: *Calamospondylus* ← {(Gr.) κάλᾰμος/kalamos: quill (= quil), reed + σπόνδυλος/spondylos: vertebrae}

Specific name: *oweni* ← {(person's name) Owen + -ī}; in honor of the paleontologist Richard Owen.

Etymology: Owen's **quill vertebrae**

Taxonomy: Theropoda: Oviraptorosauria

Other species:

> *Calamospondylus foxi* Lydekker, 1889 (*nomen dubium*) ⇒ *Calamosaurus foxi* (Lydekker, 1889) Lydekker, 1891 [Barremian; Wessex Formation, Isle of Wight, UK]

Notes: *According to The Dinosauria 2nd edition, Calamospondylus oweni* and *C. foxi* are considered as *nomina dubia*.

Callovosaurus leedsi

> *Camptosaurus leedsi* Lydekker, 1889 ⇒ *Callovosaurus leedsi* (Lydekker, 1889) Galton, 1980 [Callovian; Oxford Clay Formation, Fletton, Peterborough, Cambridgeshire, UK]

Generic name: *Callovosaurus* ← {Callovian: Middle Jurassic-age + (Gr.) σαῦρος/sauros: lizard}

The bone-bearing layer is from the lower Oxford Clay, which is middle Callovian in age.

Specific name: *leedsi* ← {(person's name) Leeds + -ī}; in honor of Alfred Nicholson Leeds (1847–1917), an English amateur paleontologist.

Etymology: Leeds' **Callovian lizard**

Taxonomy: Ornithischia: Ornithopoda: Iguanodontia: Dryosauridae

Notes: The femur was found in a brick pit near Fletton. *Callovosaurus* is proposed to be a valid genus (*1).

[References: (*1) Ruis-Omeñaca JI, Ignacio J, Pereda Suberbiola X, Galton PM (2007). *Callovosaurus leedsi*, the earliest dryosaurid dinosaur (Ornithischa: Euornithopoda) from the Middle Jurassic of England. In: Carpenter K (ed), *Horns and Beaks: Ceratopsian and Ornithopod Dinosaurs.* Indiana University Press, Bloomington and Indianapolis: 3–16.]

Camarasaurus supremus

> *Camarasaurus supremus* Cope, 1877 [Kimmeridgian; Brushy Basin Member, Morrison Formation, Colorado, USA]

= ?*Amphicoelias latus* Cope, 1877 [Kimmeridgian; Salt Wash Member, Morrison Formation, Colorado, USA]

= *Caulodon diversidens* Cope, 1877 (*nomen dubium*) [Kimmeridgian–early Tithonian; Morrison Formation, Colorado, USA]

= *Caulodon leptoganus* Cope, 1878 (*nomen dubium*) [Kimmeridgian–early Tithonian; Morrison Formation, Colorado, USA]

= *Camarasaurus leptodirus* Cope, 1879 (*nomen dubium*) [Tithonian; Brushy Basin Member, Morrison Formation, Colorado, USA]

Generic name: *Camarasaurus* ← {(Gr.) κᾰμάρα/kamara: chamber + σαῦρος/sauros: lizard}; in reference to the hollow chambers in its vertebrae.

Specific name: *supremus* ← {(Lat.) suprēmus: largest}.

Etymology: largest **chambered lizard**

Taxonomy: Saurischia: Sauropodomorpha: Sauropoda: Camarasauromorpha: Camarasauridae

Other species:

> *Apatosaurus grandis* Marsh, 1877 ⇒ *Morosaurus grandis* (Marsh, 1877) Williston, 1898 ⇒ *Camarasaurus grandis* (Marsh, 1877) Gilmore, 1925 [Kimmeridgian; Brushy Basin Member, Morrison Formation, Wyoming, USA]

= *Morosaurus impar* Marsh, 1878 ⇒ *Camarasaurus impar* (Marsh, 1878) Steel, 1970 [Kimmeridgian; Brushy Basin Member, Morrison Formation, Wyoming, USA]

= *Morosaurus robustus* Marsh, 1878 ⇒ *Camarasaurus robustus* (Marsh, 1878) White, 1958 [Kimmeridgian; Brushy Basin Member, Morrison Formation, Wyoming, USA]

= *Pleurocoelus montanus* Marsh, 1896 ⇒ *Astrodon montanus* (Marsh, 1896) Kuhn, 1939 [Kimmeridgian; Wyoming, USA]

> *Morosaurus lentus* Marsh, 1889 ⇒ *Camarasaurus lentus* (Marsh, 1889) Gilmore, 1925 [Kimmeridgian; Morrison Formation, Wyoming, USA]

= *Camarasaurus annae* Ellinger, 1950 [early Tithonian; Morrison Formation, Utah, USA]

= *Uintasaurus douglassi* Holland, 1919 [early Tithonian; Morrison Formation, Utah, USA]

> *Cathethosaurus lewsi* Jensen, 1988 ⇒ *Camarasaurus lewsi* (Jensen, 1988) McIntosh *et al.*, 1995 [Kimmeridgian; Morrison Formation, Colorado, USA]

> *Morosaurus agilis* Marsh, 1889 ⇒ *Camarasaurus agilis* (Marsh, 1889) Kuhn, 1939 ⇒ *Smitanosaurus agilis* (Marsh, 1889) Whitlock & Wilson, 2020 [Kimmeridgian–Tithonian; Morrison Formation, Colorado, USA]

> *Apatosaurus alenquerensis* de Lapparent & Zbyszewski, 1957 ⇒ *Camarasaurus alenquerensis* (de Lapparent & Zbyszewski, 1957) McIntosh, 1990 ⇒ *Lourinhasaurus alenquerensis* (de Lapparent & Zbyszewski, 1957) Dantas *et al.*, 1998 [Kimmeridgian; Lourinhã Formation, Lisboa, Portugal]

> *Brontosaurus excelsus* Marsh, 1879 ⇒ *Apatosaurus excelsus* (Marsh, 1879) Riggs, 1903 ⇒ *Camarasaurus excelsus* (Marsh, 1879) Osborn *vide* von Huene, 1929 ⇒ *Atlantosaurus excelsus* (Marsh, 1879) Steel, 1970 [Kimmeridgian; Morrison Formation, Wyoming, USA]

[References: Cope ED (1877). On a gigantic saurian from the Dakota epoch of Colorado. *Palaeontological bulletin* 25: 5–10.]

Camarillasaurus cirugedae

> *Camarillasaurus cirugedae* Sánchez-Hernández & Benton, 2014 [Barremian; Camarillas Formation, Teruel, Aragón, Spain]

Generic name: *Camarillasaurus* ← {(place-name) Camarillas: the name of the locality | (stratal name) Camarillas Formation + (Gr.) σαῦρος/sauros: reptile} (*1)

Specific name: *cirugedae* ← {(person's name) Cirugeda + -e}; in honor of discoverer Pedro Cirugeda Buj. (*1)

Etymology: Cirugeda's **Camarillas reptile**

Taxonomy: Theropoda: Ceratosauria

[References: (*1) Sánchez-Hernández B, Benton MJ (2014). Filling the ceratosaur gap: A new ceratosaurian theropod from the Early Cretaceous of Spain. *Acta Palaeontologica Polonica* 59(3): 581–600.]

Camelotia borealis

> *Camelotia borealis* Galton, 1985 [Rhaetian; Westbury Formation, Somerset, UK]

= (referred material) *Avalonia sanfordi* Seeley, 1898 (*partim*) (preoccupied) ⇒ *Avalonianus sanfordi* (Seeley, 1898) Kuhn, 1961 (*partim*) [Rhaetian; Westbury Formation, Somerset, UK]

Generic name: *Camelotia* ← {(legend) Camelot + -ia}; with reference to Camelot, the seat of King Arthur's legendary court in Avalon that was probably in Somerset, England (*1).

Specific name: *borealis* ← {(Lat.) boreālis: northern}; with reference to it being the only record of this family from the northern hemisphere (*1).

Etymology: **Camelot's one** from the Northern Hemisphere

Taxonomy: Saurischia: Sauropodomorpha: Anchisauria: Melanorosauridae

[References: (*1) Galton PM (1985). Notes on the Melanorosauridae, a family of large prosauropod dinosaurs (Saurischia: Sauropodomorpha). *Géobios* 18(5): 671–676.]

Camposaurus arizonensis

> *Camposaurus arizonensis* Hunt *et al.*, 1998 [Norian; Bluewater Creek Formation, Arizona, USA]

Generic name: *Camposaurus* ← {(person's name) Camp + (Gr.) σαῦρος/sauros: lizard}; in honor of the paleontologist and zoologist Charles Lewis Camp (1893–1975).

Specific name: *arizonensis* ← {(place-name) Arizona + -ensis}
Etymology: **Camp's lizard** from Arizona
Taxonomy: Theropoda: Coelophysidae

[References: Hunt AP, Lucas SG, Heckert AB, Sullivan RM, Lockley MG (1998). Late Triassic dinosaurs from the western United States. *Géobios* 31(4): 511–531.]

Camptodontornis yangi [Avialae]

> *Camptodontus yangi* Li *et al.*, 2010 (preoccupied) ⇒ *Camptodontornis yangi* (Li *et al.*, 2010) Demirjian, 2019 [Aptian; Jiufotang Formation, Liaoning, China]

Generic name: *Camptodontornis* ← {(Gr) campt- [κάμπτω/camptō: bend] + odont- [ὀδών/odōn: tooth] +ὄρνις/ornis: bird}
Specific name: *yangi* ← {(person's name) Yang + -ī}
Etymology: Yang's **bent-toothed bird**
Taxonomy: Dinosauria: Theropoda: Enantiornithes: Longipterygidae
Notes: Wang *et al.* (2015) consider *Camptodontornis* to be a probable synonym of *Longipteryx*.

Camptosaurus dispar

> *Camptonotus dispar* Marsh, 1879 (preoccupied) [*1] ⇒ *Camptosaurus dispar* (Marsh, 1879) Marsh, 1885 [Kimmeridgian–Tithonian; Salt Wash Member, Morrison Formation, Wyoming, USA]

= *Brachyrophus altarkansanus* Cope, 1878 (*nomen dubium*) [Kimmeridgian–early Tithonian; Morrison Formation, Colorado, USA]

= *Symphyrophus musculosus* Cope, 1878 (*nomen dubium*) [Kimmeridgian; Salt Wash Member, Morrison Formation, Colorado, USA]

= *Camptosaurus medius* Marsh, 1894 [Kimmeridgian; Salt Wash Member, Morrison Formation, Wyoming, USA]

= *Camptosaurus nanus* Marsh, 1894 [Kimmeridgian; Salt Wash Member, Morrison Formation, Wyoming, USA]

= *Camptosaurus browni* Gilmore, 1909 [Kimmeridgian; Salt Wash Member, Morrison Formation, Wyoming, USA]

Generic name: *Camptosaurus* ← {(Gr.) καμπτός/kamptos: flexible [κάμπτω/kamptō: to bend] + σαῦρος/sauros: lizard}

cf. Camptonotus; in reference to the "character of the sacral vertebrae". According to Marsh (1879), "the sacral vertebrae are not coösified [*sic*]. That this is not merely a character of immaturity is shown by some of the other vertebrae in the type specimen, which have their neural arches so completely united to the centra that the suture is nearly or quite obliterated." [*2].

Specific name: dispar ← {(Lat.) dispār; unlike, different}.
Etymology: different, **flexible lizard**
Taxonomy: Ornithischia: Ornithopoda: Iganodontia: Ankylopollexia: Camptosauroidea: Camptosauridae
Other species:

> *Allosaurus fragilis* Marsh, 1877 [Kimmeridgian–Tithonian; Morrison Formation, Wyoming, USA]

= *Camptonotus amplus* Marsh, 1879 ⇒ *Camptosaurus amplus* (Marsh, 1879) Marsh,1885 ⇒ *Allosaurus amplus* (Marsh, 1879) Paul, 1988 [Kimmeridgian; Morrison Formation, Wyoming, USA]

> *Iguanodon prestwichii* Hulke, 1880 ⇒ *Cumnoria prestwichii* (Hulke, 1880) Seeley, 1888 ⇒ *Camptosaurus prestwichi* (Hulke, 1880) Lydekker, 1889 ⇒ *Camptosaurus prestwichii* (Hulke, 1880) Gilmore, 1909 [Kimmeridgian; Kimmeridge Clay Formation, UK]

> *Iguanodon hoggi* Owen, 1874 ⇒ *Camptosaurus hoggi* (Owen, 1874) Norman & Barrett, 2002 ⇒ *Owenodon hoggi* (Owen, 1874) Galton, 2009 [Berriasian; Lulworth Formation, England, UK]

> *Camptosaurus aphanoecetes* Carpenter & Wilson, 2008 ⇒ *Uteodon aphanoecetes* (Carpenter & Wilson, 2008) McDonald, 2011 [Tithonian; Morrison Formation, Utah, USA]

> *Camptosaurus depressus* Gilmore, 1909 ⇒ *Planicoxa depressa* (Gilmore, 1909) Di Croce & Carpenter, 2001 ⇒ *Osmakasaurus depressus* (Gilmore, 1909) McDonald, 2011 [Barremian; Lakota Formation, South Dakota, USA]

> *Camptosaurus inkeyi* Nopcsa, 1900 (*nomen dubium*) [Late Cretaceous; Transylvania, Romania]

> *Camptosaurus leedsi* Lydekker, 1889 ⇒ *Callovosaurus leedsi* (Lydekker, 1889) Galton, 1980 [Callovian; Oxford Clay Formation, Cambridgeshire, UK]

> *Camptosaurus valdensis* Lydekker, 1889 (*nomen dubium*) [late Barremian; Isle of Wight, UK]

Notes: According to *The Dinosauria 2nd edition*, *Camptosaurus inkeyi* is considered a synonym of *Zalmoxes robustus*.

[References: (*1) Marsh OC (1885). Names of extinct reptiles. *American Journal of Science* 29: 169.; (*2) Marsh OC (1879). Notice of new Jurassic reptiles. *American Journal of Science and Arts* 18: 501–505.]

Campylodoniscus ameghinoi

> *Campylodon ameghinoi* von Huene, 1929 (preoccupied) ⇒ *Campylodoniscus ameghinoi* (von Huene, 1929) Kuhn, 1961 [Cenomanian–Turonian; Bajo Barreal Formation, Golfo San Jorge Basin, Argentina]

Generic name: *Campylodoniscus* ← {(Gr.) καμπύλος/kampylos: bent + ὀδών/odōn [ὀδούς/odūs: tooth] + -iscus}

Campylodon was pre-occupied by a fish and the genus was renamed into *Campylodoniscus* in 1961.

Specific name: *ameghinoi* ← {(person's name) Ameghino + -ī}; in honor of Argentine naturalist Florentino Ameghino (1853–1911).

Etymology: Ameghino's **bent tooth**

Taxonomy: Saurischia: Sauropodomorpha:
Sauropoda: Titanosauria?

Notes: According to *The Dinosauria 2nd edition*, *Campylodoniscus ameghinoi* is considered to be *a nomen dubium*.

Canardia garonnensis

> *Canardia garonnensis* Prieto-Márquez *et al.*, 2013 [Maastrichtian; Marnes d'Auzas Formation, Haute-Garonne, France]

Generic name: *Canardia* ← {(Fr.) canard: duck + -ia}; "alluding to the hadrosaurian nature of this animal (hadrosauroids are also informally known as "duck-billed" dinosaurs)" (*1).

Specific name: *garonnensis* ← {(place-name) Garonne + -ensis}; "in reference to Haute-Garonne, the department in southern France where this lambeosaurine has been found" (*1).

Etymology: **duck-billed dinosaur** from Haute-Garonne

Taxonomy: Ornithischia: Ornithopoda:
Hadrosauridae: Lambeosaurinae: Aralosaurini*
*Aralosaurini Prieto-Márquez *et al.*, 2013 is a tribe, which contains *Aralosaurus* and *Canardia*.

[References: (*1) Prieto-Màrquez A (2013). Diversity, relationships, and biogeography of the Lambeosaurine dinosaurs from the European Archipelago, with description of the new alarosaurin *Canardia garonnensis*. *PLoS ONE* 8(7): e69835.]

Carcharodontosaurus saharicus

> *Megalosaurus saharicus* Depéret & Savornin, 1925 ⇒ *Megalosaurus* (*Dryptosaurus*) *saharicus* (Depéret & Savornin, 1925) Depéret & Savornin, 1928 (described based on teeth) [Cenomanian; Kem Kem Beds, Algeria]

= *Carcharodontosaurus saharicus* (Depéret & Savornin, 1925) Stromer, 1931 (described based on skull and skeleton excavated in 1914) [late Aptian–Cenomanian; Bahariya Formation, Egypt]

= *Megalosaurus africanus* von Huene, 1956 [Albian; Continental Intercalaire Formation, Adrar, Algeria]

Generic name: *Carcharodontosaurus* ← {*Carcharodon*: the shark genus [(Gr.) κάρχἄρος/karcharos: jugged, sharp + ὀδοντο-/odonto- [ὀδούς/odūs: tooth]] + σαῦρος/sauros: lizard}; according to the gross similarity with teeth of *Carcharodon*, one of the shark genus. (*1)

Specific name: *saharicus* ← {(place-name) the Sahara Desert + -icus}

Etymology: **shark-toothed lizard** from Sahara Desert

Taxonomy: Theropoda: Carnosauria:
Allosauroidea: Carcharodontosauridae:
Carcharodontosaurinae

Other species:

> *Carcharodontosaurus iguidensis* Brusatte & Sereno, 2007 [Cenomanian; Echkar Formation, Agadez, Niger]

Notes: The fossils were destroyed by the Allied air strike in World War II in 1944, but, the other fossils Sereno found in 1995 was designated as neotype in 2007.

[References: (*1) Stromer E (1931). Results of the research expedition of Prof. E. Stromer in the Egyptian Desert. II Vertebrate remains from Bahariya Beds (lowermost Cenomanian). 10. A skeletal remain of *Carcharodontosaurus* nov. gen. *Abhandlungen der Bayerischen Akademie der Wissenschaften Mathematisch-Naturwissenschaftliche Abteilung. Neue Foge* 9: 1–23.]

Cardiodon rugulosus

> *Cardiodon rugulosus* Owen, 1840–1845 ⇒ *Cetiosaurus rugulosus* (Owen, 1840–1845) Steel, 1970 (*nomen dubium*) [Bathonian; Forest Marble Formation, Wiltshire, UK] (*1)

Generic name: *Cardiodon* ← {(Gr.) καρδία/kardia: heart + ὀδών/odōn: tooth}; in reference to its heart-shaped profile.

Specific name: *rugulosus* ← {(Lat.) rugulōsus: wrinkled}.

The species name *rugulosus* was added by Owen (1844) (*1).

Etymology: wrinkled **heart-shaped tooth**

Taxonomy: Saurischia: Sauropodomorpha:
Sauropoda: Turiasauria?

Notes: The tooth, which was found near Bradford-on-Avon, is now lost.

[References: (*1) Upchurch P, Martin J (2003). The

anatomy and taxonomy of *Cetiosaurus* (Saurischia, Sauropoda) from the Middle Jurassic of England. *Journal of Vertebrate Paleontology* 23(1): 208–231.]

Carnotaurus sastrei

> *Carnotaurus sastrei* Bonaparte, 1985 [Maastrichtian; La Colonia Formation, Chubut, Argentina]

Generic name: *Carnotaurus* ← {(Lat.) carn- [carō: flesh] + taurus: bull}; in reference to the bull-like horns.

Specific name: *sastrei* ← {(person's name) Sastre + -ī}; in honor of Angel Sastre, owner of the ranch.

Etymology: Sastre's **carnivorous bull**

Taxonomy: Theropoda: Ceratosauria: Abelisauridae: Carnotaurinae

Notes: The skeleton is preserved with extensive skin impressions. (*1)

[References: Bonaparte JF (1985). A horned Cretaceous carnosaur from Patagonia. *National Geograpgic Research* 1(1): 149–151.; Bonaparte JF, Novas FE, Coria RA (1990). *Carnotaurus sastrei* Bonaparte, the horned, lightly built carnosaur from the middle Cretaceous of Patagonia. *Natural Museum of Los Angeles County Contributions in Science* 416: 1–41.; (*1) Czerkas ST, Czerkas SJ (1997). The integument and life restoration of *Carnotaurus*. In: Wolberg DI, Stump E, Rosenberg GD (eds), *Dinofest International*. Academy of Natural Sciences, Philadelphia: 155–158.]

Caseosaurus crosbyensis

> *Caseosaurus crosbyensis* Hunt *et al.*, 1998 [Carnian; Tecovas Formation, Texas, USA]

Generic name: *Caseosaurus* ← {(person's name) Case + (Gr.) σαῦρος/sauros: lizard}; in honor of the scientist Ermine Cowles Case who discovered it.

Specific name: *crosbyensis* ← {(place-name) Crosby + -ensis}; in reference to Crosby County.

Etymology: **Case's lizard** from Crosby County

Taxonomy: Saurischia: Herrerasauridae

Notes: In *The Dinosauria 2nd edition*, *Caseosaurus* is regarded as a synonym of *Chindesaurus*. According to Baron & Williams (2018), *Caseosaurus crosbyensis* was considered to be valid.

[References: Hunt AP, Lucas SG, Heckert AB, Sullivan RM, Lockley MG (1998). Late Triassic dinosaurs from the western United States. *Geobios* 31(4): 511–531.]

Cathartesaura anaerobica

> *Cathartesaura anaerobica* Gallina & Apesteguía, 2005 [Cenomanian; Huincul Formation, "La Buitrera" locality, Río Negro, Argentina]

Generic name: *Cathartesaura* ← {"*Cathartes*: the extant vulture genera abundant in the quarry area* + (Gr.) σαύρα/saura, *fem.*: lizard, reptile"}; "implying the combination of the generic and specifical names of those vultures (*Cathartes aura* Linnaeus, 1758)" (*1).

*The name of the locality is "La Buitrera" (= the Vulture Colony).

Specific name: *anaerobica*: *fem.* "for ANAEROBICOS S. A., an Argentine company of adhesives that provided fieldwork and lab support in the extraction and protection of the fossile materials" (*1).

Etymology: ANAEROBICOS S.A.'s **vulture lizard** (from La Buitrera)

Taxonomy: Saurischia: Sauropodomorpha: Sauropoda: Rebbachisauridae

[References: (*1) Gallina PA, Apesteguía S (2005). *Cathartesaura anaerobica* gen. *et* sp. *nov.*, a new rebbachisaurid (Dinosauria, Sauropoda) from the Huincul Formation (Upper Cretaceous), Río Negro, Argentina. *Revista Museo Argentino Ciencias Naturales, n. s.* 7(2): 153–166.]

Cathayornis yandica [Avialae]

> *Cathayornis yandica* Zhou *et al.*, 1992 [Aptian; Jiufotang Formation, Chaoyang, Liaoning, China]

= *Cuspirostrisornis houi* Hou, 1997 [Aptian; Jiufotang Formation, Liaoning, China]

= *Largirostrornis sexdentornis* Hou, 1997 [Aptian, Jiufotang Formation, Liaoning, China]

Generic name: *Cathayornis* ← {(place-name) Cathay: the old poetic name of China + (Gr.) ὄρνις/ornis: bird}

Specific name: *yandica* ← {(Chin. place-name) Yāndū燕都: the old name of Chaoyang + -ica [-icus]}

Etymology: **China bird** of Chaoyang

Taxonomy: Theropoda; Avialae; Enantiornithes; Euenantiornithes

Other species:

> *Cathayornis aberransis*? Hou *et al.*, 2002 [Barremian; Jiufotang Formation, Liaoning, China]

> *Cathayornis chabuensis*? Li *et al.*, 2008 (*nomen dubium*) [Barremian–Aptian; Jingchuan Formation, Chabu Sumu, Inner Mongolia] (*1)

> *Cathayornis caudatus* Hou, 1997 ⇒ *Hournis caudatus* (Hou, 1997) Wang & Liu, 2015 [Aptian; 2nd Member, Jiufotang Formation, Liaoning, China]

[References: (*1) Wang M, Liu D (2015). Taxonomical reappraisal of Cathayornithidae (Aves: Enantiornithes). *Journal of Systematic Palaeontology* 14(1): 29–47.]

Cathetosaurus lewisi

> *Cathetosaurus lewisi* Jensen, 1988 ⇒

Camarasaurus lewisi (Jensen, 1988) McIntosh *et al.*, 1995 [Kimmeridgian, Morrison Formation, Colorado, USA]

Generic name: *Cathetosaurus* ← {(Gr.) κάθετος/ cathetos: perpendicular + σαῦρος/ sauros: lizard}; "referring to an ability to stand erect on its rear legs" (*1).

Specific name: *lewisi* ← {(person's name) Lewis + -ī}; "honoring Mr. Arnold D. Lewis, stout companion of the trail, who patiently trained me (Jim Jensen) in laboratory and field work". (*1)

Etymology: Lewis' **perpendicular lizard**

Taxonomy: Saurischia: Sauropodomorpha: Sauropoda: Camarasauridae

Notes: According to *The Dinosauria 2nd edition*, *Cathetosaurus lewisi* was considered as a synonym of *Camarasaurus lewisi*. But, according to the recent research based on the recognition of the second specimen from Wyoming, *Cathetosaurus* is reported as a valid genus, according to Mateus & Tschopp (2013) (*2).

[References: (*1) Jensen JA (1988). A forth new sauropod dinosaur from the Upper Jurassic of the Colorado Plateau and sauropod bipedalism. *Great Basin Naturalist* 48(2): 121–145.; (*2) Mateus O, Tschopp E (2013). *Cathetosaurus* as a valid sauropod genus and comparisons with *Camarasaurus*. *Journal of Vertebrate Paleontology, Program and Abstracts* 2013: 173.]

Caudipteryx zoui

> *Caudipteryx zoui* Ji *et al.*, 1998 [Aptian; Yixian Formation, Liaoning, China]

cf. The authors who described this taxa use "Jiulongsong Member of Chaomidianzi Formation", but this corresponds to lower part of Yixian.

Generic name: *Caudipteryx* ← {(Lat.) cauda: tail + (Gr.) πτέρυξ/pteryx: feather} (*1).

Specific name: *zoui* ← {(person's name) Zōu鄒 + -ī}; in honor of "Zou Jiahua, vice-premier of China and an avid supporter of the scientific work in Liaoning" (*1).

Etymology: Zou's **tail feather**

Taxonomy: Theropoda: Maniraptoriformes: Oviraptorosauria: Caudipteridae (=Caudipterygidae)

Other species:

> *Caudipteryx dongi* Zhou & Wang, 2000 [late Barremian; Yixian Formation, Liaoning, China]

Notes: Gastroliths have been found in two specimens of *Caudipteryx*.

[References: (*1) Ji Q, Currie PJ, Norell MA, Ji S (1998). Two feathered dinosaurs from northestern China. *Nature.* 393: 753–761.; Dyke GJ, Norell MA (2005). *Caudipteryx* as a non-avialan theropod rather than a flightless bird. *Acta Palaeontologica Polonica* 50(1): 101–116.]

Cedarosaurus weiskopfae

> *Cedarosaurus weiskopfae* Tidwell *et al.*, 1999 [Barremian; Cedar Mountain Formation, Utah, USA.

Generic name: *Cedarosaurus* ← {Cedar + (Gr.) σαῦρος/sauros: lizard}; "named for the Cedar Mountain Formation from which the type specimen was collected" (*1).

Specific name: *weiskopfae* ← {(person's name) Weiskopf + -ae}; named "for the late Carol Weiskopf for her hard work in the field and lab" (*1).

Etymology: Weiskopf's **lizard from Cedar Mountain Formation**

Taxonomy: Saurischia: Sauropodomorpha: Sauropoda: Brachiosauridae

[References: (*1) Tidwell V, Carpenter K, Brooks W (1999). New sauropod from the Lower Cretaceous of Utah, USA. *Oryctos* 2: 21–37.]

Cedarpelta bilbeyhallorum

> *Cedarpelta bilbeyhallorum* Carpenter *et al.*, 2001 [Barremian; Cedar Mountain Formation, Utah, USA]

Generic name: *Cedarpelta* ← {(Eng.) cedar + (Gr.) πέλτη/peltē: shield}; in reference to the Cedar Mountain Formation and its armour.

Specific name: *bilbeyhallorum* ← {(person's name) Bilbey & Hall + -ōrum}; in honor of "Sue Ann Bilbey and Evan Hall, who discovered the locality" (*1).

Etymology: Bilbey and Hall's **Cedar Mountain shield**

Taxonomy: Ornithischia: Ankylosauria: Ankylosauridae

[References: (*1) Carpenter K, Kirkland JI, Birge D, Bird J (2001). Disarticulated skull of a new primitive ankylosaurid from the Lower Cretaceous of Utah. In: Carpenter K (ed). *The Armored Dinosaurs*. Indiana University Press, Bloomington: 211–238.]

Cedrorestes crichtoni

> *Cedrorestes crichtoni* Gilpin *et al.*, 2007 [Barremian; Cedar Mountain Formation, Utah, USA]

Generic name: *Cedrorestes* ← {(Lat.) cedr- [cedrus: cedar] + orestes: "mountain dweller" ← [(Gr. Myth.) Ὀρέστης/Orestēs [όρος/oros: mountain + ἵστημι/histēmi: to stand]]}"'Cedar Mountain dweller'; referring to the Cedar Mountain Formation in eastern Utah where this specimen was recovered" (*1).

Specific name: *crichtoni* ← {(person's name) Crichton + -ī}; "in honor of Michael Crichton for promoting the public's interest in dinosaurs through his *Jurassic Park* novels" [*1].
Etymology: Crichton's **Cedar Mountain dweller**
Taxonomy: Ornithischia: Ornithopoda:
Iguanodontia: Ankylopollexia: Styracosterna
cf. Ornithischia: Ornithopoda: Hadrosauridae [*1]
[References: (*1) Gilpin D, DiCroce T, Carpenter K (2007). A possible new basal hadrosaur from the Lower Cretaceous Cedar Mountain Formation of eastern Utah. In: Carpenter K (ed). *Hornes and Beaks: Ceratopsian and Ornithopod Dinosaurs.* Indiana University Press, Bloomington: 79–89.]

Centrosaurus apertus

> *Centrosaurus apertus* Lambe, 1904 ⇒ *Monoclonius* (*Centrosaurus*) *apertus* (Lambe, 1904) Lambe & Lull, 1933 ⇒ *Monoclonius apertus* (Lambe, 1904) Kuhn, 1936 ⇒ *Eucentrosaurus apertus* (Lambe, 1904) Chure & McIntosh, 1989 [Campanian; Dinosaur Park Formation, Alberta, Canada]

= *Monoclonius dawsoni* Lambe, 1902 ⇒ *Brachyceratops dawsoni* (Lambe, 1902) Lambe, 1915 ⇒ *Centrosaurus dawsoni* (Lambe, 1902) Sternberg, 1940 [Campanian; Dinosaur Park Formation, Alberta, Canada]

= *Monoclonius flexus* Brown, 1914 ⇒ *Monoclonius* (*Centrosaurus*) *flexus* (Brown, 1914) Brown & Lull, 1933 ⇒ *Eucentrosaurus flexus* (Brown, 1914) Chure & McIntosh, 1989 [Campanian; Dinosaur Park Formation, Alberta, Canada]

= *Monoclonius cutleri* Brown, 1917 ⇒ *Centrosaurus cutleri* (Brown, 1917) Russell, 1930 ⇒ *Monoclonius* (*Centrosaurus*) *cutleri* (Brown, 1917) Brown & Lull, 1933 [Campanian; Oldman Formation, Alberta, Canada]

= *Monoclonius nasicornus* Brown, 1917 ⇒ *Centrosaurus nasicornus* (Brown, 1917) Lull, 1933 ⇒ *Eucentrosaurus nasicornus* (Brown, 1917) Chure & McIntosh, 1989 [Campanian; Dinosaur Park Formation, Alberta, Canada] (according to Frederickson & Tumarkin-Deratzian, 2014)

= *Centrosaurus longirostris* Sternberg, 1940 ⇒ *Monoclonius longirostris* (Sternberg, 1940) Kuhn, 1964 ⇒ *Eucentrosaurus longirostris* (Sterberg, 1940) Chure & McIntosh, 1989 [Campanian; Dinosaur Park Formation, Alberta, Canada]

Generic name: *Centrosaurus* ← {(Gr.) κέντρον/kentron: point + σαῦρος/sauros: lizard}; referring to the series of small hornlets placed along the margin of their frills.
Specific name: *apertus* ← {(Lat.) apertus: open}
Etymology: open, **pointed lizard**
Taxonomy: Ornithischia: Marginocephalia:
Ceratopsia: Ceratopsidae: Centrosaurinae
(*cf.* Stegosaurian *Kentrosaurus* is also derived from the same Greek word.)
Other species:

> *Monoclonius recurvicornis* Cope, 1889 (*nomen dubium*) ⇒ *Ceratops recurvicornis* (Cope, 1889) Hatcher *vide* Stanton & Hatcher, 1905 (*nomen dubium*) ⇒ *Centrosaurus recurvicornis* (Cope, 1889) Chure & McIntosh, 1989 (*nomen dubium*) [Campanian; Judith River Formation, Montana, USA]

> *Styracosaurus albertensis* Lambe, 1913 ⇒ *Centrosaurus albertensis* (Lambe, 1913) Paul, 2010 [Campanian; Dinosaur Park Formation, Alberta, Canada]

> *Styracosaurus ovatus* Gilmore, 1930 ⇒ *Centrosaurus ovatus* (Gilmore, 1930) Paul, 2010 ⇒ *Rubeosaurus ovatus* (Gilmore, 1930) Horner & McDonald, 2010 [Campanian; Two Medicine Formation, Montana, USA]

> *Pachyrhinosaurus canadensis* Sternberg, 1950 ⇒ *Centrosaurus canadensis* (Sternberg, 1950) Paul, 2010 [Maastrichtian; lower Horseshoe Canyon Formation, Alberta, Canada]

> *Einiosaurus procurvicornis* Sampson, 1995 ⇒ *Centrosaurus procurvicornis* (Sampson, 1995) Paul, 2010 [Campanian; Two Medicine Formation, Montana, USA]

> *Achelousaurus horneri* Sampson, 1995 ⇒ *Centrosaurus horneri* (Sampson, 1995) Paul, 2010 [Campanian; Two Medicine Formation, Montana, USA]

> *Centrosaurus brinkmani* Ryan & Russell, 2005 ⇒ *Coronosaurus brinkmani* (Ryan & Russell, 2005) Ryan *et al.*, 2012 [middle Campanian; Oldman Formation, Alberta, Canada]

> *Pachyrhinosaurus lakustai* Currie *et al.*, 2008 ⇒ *Centrosaurus lakustai* (Currie *et al.*, 2008) Paul, 2010 [Campanian; Wapiti Formation, Alberta, Canada]

Notes: *Centrosaurus* bore single large horns over their noses.

Cerasinops hodgskissi

> *Cerasinops hodgskissi* Chinnery & Horner, 2007 [Campanian; Two Medicine Formation, Montana, USA]

Generic name: *Cerasinops* ← {(Lat.) cerasinus:

of cherry + (Gr.) ὤψ/ōps: face}; "referring to the red beds of the type locality and the dark red tinge of colour on the specimen. This name also incorporates the nickname Cera, by which the specimen has been referred for over two decades" (*1).
Specific name: *hodgskissi* ← {(person's name) Hodgskiss + -ī}; in honor of Wilson Hodgskiss, the land owner (*1).
Etymology: Hodgskiss' **cherry-red face**
Taxonomy: Ornithischia: Ceratopsia: Leptoceratopsidae

[References: (*1) Chinnery BJ, Horner JR (2007). A new neoceratopsian dinosaur linking North American and Asian taxa. *Journal of Vertebrate Paleontology* 27(3): 625–641.]

Ceratonykus oculatus

> *Ceratonykus oculatus* Alifanov & Barsbold, 2009 [Maastrichtian; Barun Goyot Formation, Mongolia]

Generic name: *Ceratonykus* ← {(Gr.) κεράτῐνος/keratinos: horned, of horn [κέρας/keras: horn] + ὄνυξ/onyx: claw}
Specific name: *oculatus* ← {(Lat.) oculātus: with eyes}; "sharp-sighted, big-eyed" (*1); referring to the large orbits of the animal.
Etymology: sharp-sighted **horned claw**
Taxonomy: Theropoda: Alvarezsauridae

[References: (*1) Alifanov VR, Barsbold R (2009). *Ceratonykus oculatus* gen. *et* sp. nov., a new dinosaur (? Theropoda, Alvarezsauria) from the Late Cretaceous of Mongolia. *Paleontological Journal* 43(1): 94–106.]

Ceratops montanus

> *Ceratops montanus* Marsh, 1888 (*nomen dubium*) ⇒ *Proceratops montanus* (Marsh, 1888) Lull, 1906 (*nomen dubium*) [Campanian; Judith River Formation, Montana, USA]

Generic name: *Ceratops* ← {(Gr.) κερατ-/kerat- [κέρας/keras: horn] + ὤψ/ōps: face}
Specific name: *montanus* ← {(Lat.) mōntānus: of mountains}; referring to Montana.
Etymology: **horned face** from Montana
Taxonomy: Ornithischia: Ceratopsia: Ceratopsidae: Ceratopsinae
Other species:

> *Ceratops horridus* Marsh, 1889 ⇒ *Triceratops horridus* (Marsh, 1889) Marsh, 1889 [Maastrichtian; Lance Formation, Wyoming, USA]

> *Bison alticornis* Marsh, 1887 (*nomen dubium*) ⇒ *Ceratops alticornis* (Marsh, 1887) Marsh, 1889 (*nomen dubium*) ⇒ *Polyonax alticornis* (Marsh, 1887) Cope, 1889 ⇒ ?*Triceratops alticornis* (Marsh, 1887) Hatcher, 1907 *vide* Hatcher *et al.*, 1907 (*nomen dubium*) [Maastrichtian; Denver Formation, Colorado, USA]

> *Hadrosaurus paucidens* Marsh, 1889 (*nomen dubium*) ⇒ *Ceratops paucidens* (Marsh, 1889) Hay, 1901 (*nomen dubium*) [Campanian; Judith River Formation, Montana, USA]

> *Monoclonius recurvicornis* Cope, 1889 (*nomen dubium*) ⇒ *Ceratops recurvicornis* (Cope, 1889) Hatcher *vide* Stanton & Hatcher, 1905 (*nomen dubium*) ⇒ *Centrosaurus recurvicornis* (Cope, 1889) Sternbergi, 1940 (nomen dubium) ⇒ *Eucentrosaurus recurvicornis* (Cope, 1889) Chure & McIntosh, 1989 (*nomen dubium*) [Campanian; Judith River Formation, Montana, USA]

> *Monoclonius belli* Lambe, 1902 ⇒ *Ceratops belli* (Lambe, 1902) Hatcher *vide* Stanton & Hatcher, 1905 ⇒ *Chasmosaurus belli* (Lambe, 1902) Lambe, 1914 [Campanian; middle-upper Dinosaur Park Formation, Alberta, Canada]

?= *Monoclonius canadensis* Lambe, 1902 ⇒ *Ceratops canadensis* (Lull, 1902) Hatcher *et al.*, 1907 ⇒ *Eoceratops canadensis* (Lambe, 1902) Lambe, 1915 ⇒ *Chasmosaurus canadensis* (Lambe, 1902) Lehman, 1989 [Campanian; Belly River Group, Alberta, Canada]

Notes: According to *The Dinosauria 2nd edition*, *Ceratops montanus* is considered a *nomen dubium*.

[References: Marsh OC (1888). A new family of horned Dinosauria, from the Cretaceous. *The American Journal of Science Series 3* 36: 477–478.]

Ceratosaurus nasicornis

> *Ceratosaurus nasicornis* Marsh, 1884 ⇒ *Megalosaurus nasicornis* (Marsh, 1884) Cope, 1892 [Kimmeridgian–Tithonian; Morrison Formation, Colorado, USA]

?= *Ceratosaurus dentisulcatus* Madsen & Welles, 2000 [Kimmeridgian; Morrison Formation, Utah, USA]

?= *Ceratosaurus magnicornis* Madsen & Welles, 2000 [Kimmeridgian–Tithonian; Morrison Formation, Colorado, USA]

Generic name: *Ceratosaurus* ← {(Gr.) κερατο-/kerato- [κέρας/keras: horn] + σαῦρος/sauros: lizard}
Specific name: *nasicornis* ← {(Lat.) nasus: snout + cornu: horn}; referring to elevated, trenchant horn-core, situated on the nasals. (*1)
Etymology: **horn lizard** with snout-horn
Taxonomy: Theropoda: Ceratosauria: Ceratosauridae

Other species:

> ?*Megalosaurus meriani* Greppin, 1870 (*nomen dubium*) ⇒ *Labrosaurus meriani* (Greppin, 1870) von Huene, 1926 (*nomen dubium*) ⇒ *Antrodemus meriani* (Greppin, 1870) Steel, 1970 (*nomen dubium*) ⇒ *Allosaurus meriani* (Greppin, 1870) Olshevsky, 1991 (*nomen dubium*) ⇒ *Ceratosaurus meriani* (Greppin, 1870) emend Madsen & Welles, 2000 (*nomen dubium*) [Kimmeridgian; Reuchenette Formation, Moutier, Bern, Switzerland]

> ?*Labrosaurus sulcatus* Marsh, 1896 ⇒ *Antrodemus sulcatus* (Marsh, 1896) ⇒ *Ceratosaurus sulcatus* (Marsh, 1896) emend Madsen & Welles, 2000, Bakker & Bir, 2004 (*nomen dubium*) [Kimmeridgian; Morrison Formation, Wyoming, USA]

> ?*Ceratosaurus roechlingi* Janensch, 1925 (*nomen dubium*) [Tithonian; Tendaguru Formation, Mtwara, Tanzania]

> ?*Labrosaurus stechowi* Janensch, 1920 (*nomen dubium*) ⇒ *Antrodemus stechowi* (Janensch, 1920) Chabli, 1986 (*nomen dubium*) ⇒ *Ceratosaurus stechowi* (Janensch, 1920) Rauhut, 2011 (*nomen dubium*) [Tendagulu Formation, Tanzania]

> ?*Megalosaurus ingens* Janensch, 1920 (*nomen dubium*) ⇒ *Ceratosaurus ingens* (Janensch, 1920) Paul, 1988 [Tithonian; Tendaguru Formation, Mandawa Basin, Tanzania]

[References: (*1) Marsh OC (1884). Principal characters of American Jurassic dinosaurs. Part VIII: The order Theropoda. *American Journal of Science* 27(160): 320–340.]

Cerebavis cenomanicus [Avialae]

> *Cerebavis cenomanicus* Kurochkin *et al.*, 2006 [Cenomanian; Melovatka Formation, Volgograd Region, Russia]

Generic name: *Cerebavis* ← {(Lat.) cerebrum: brain + avis (feminine gender): bird} (*1)

Cerebavis is described based on a brain mold. The cerebrum is relatively large.

Specific name: *cenomanicus*: "from the Cenomanian Stage" (*1).

Etymology: **brain bird** from the Cenomanian Stage

Taxonomy: Ornithurae (*2) *cf.* Enantiornithes (*1)

Notes: *Cerebavis* had well-developed senses of smell, eyesight, and hearing. (*1)

[References: (*1) Kurochkin EN, Saveliev SV, Postnov AA, Pervushov EM, Popov EV (2006). On the brain of a primitive bird from the Upper Cretaceous of European Russia. *Paleontological Journal* 40(6): 655–667.; (*2) Walsh SA, Milner AC, Bourdon E (2016). A reappraisal of *Cerebavis cenomanica* (Aves, Ornithurae), from Melovatka, Russia. *Journal of Anatomy* 229(2): 215–227.]

Cetiosauriscus stewarti

> *Cetiosauriscus stewarti* Charig, 1980 [Callovian; Oxford Clay Formation, Cambridgeshire, UK]

Generic name: *Cetiosauriscus* von Huene, 1927 ← {(Gr.) κήτειος/kēteios: of sea monsters, the whale-like + σαῦρος/sauros: lizard + -iscus: resembling, like}

Specific name: *stewarti* ← {(person's name) Stewart + -ī}; in honor of Sir Ronald Stewart, the chairman of the London Brick Company that owned the clay pit where the fossils were found.

Etymology: Stewart's ***Cetiosaurus*** **- like**

Taxonomy: Saurischia: Sauropodomorpha: Sauropoda: Diplodocoidea

Other species:

> *Cetiosaurus longus* Owen, 1842 (*nomen dubium*) ⇒ *Cetiosauriscus longus* (Owen, 1842) McIntosh, 1990 [Kimmeridgian; Portland Stone Formation, Dorset, England, UK]

> *Cetiosaurus glymptonensis* Phillips, 1871 ⇒ *Cetiosauriscus glymptonensis* (Phillips, 1871) McIntosh, 1990 [late Bathonian; Forest Marble Formation, England, UK]

> *Ornithopsis leedsii* Hulke, 1887 ⇒ *Pelorosaurus? leedsi* (Hulke, 1887) Lydekker, 1895 ⇒ *Cetiosaurus leedsi* (Hulke, 1887) Woodward, 1905* ⇒ *Cetiosauriscus leedsi* (Hulke, 1887) von Huene, 1927* [Callovian-Oxfordian; Oxford Clay, Peterborough, UK] *referred specimen

> *Megalosaurus meriani* Greppin, 1870 (*partim*) ⇒ *Ornithopsis greppini* von Huene, 1922 ⇒ *Cetiosauriscus greppini* (von Huene, 1922) von Huene, 1932 ⇒ *Amanzia greppini* (von Huene, 1922) Schwarz *et al.*, 2020 [Kimmeridgian; Reuchenette Formation, Moutier, Switzerland]

[References: von Huene F (1927). Sichtung der Grundlagen der jetzigen kenntnis der sauropoden [Short review of the present knowledge of the Sauropoda]. *Eclogae Geologica Helveticae* 20: 444–470.]

Cetiosaurus oxoniensis

> *Cetiosaurus oxoniensis* Phillips, 1871 ⇒ *Ornithopsis oxoniensis* (Phillips, 1871) Seeley, 1889 [Bathonian; Forest Marble Formation, Oxfordshire, UK]

= *Cetiosaurus giganteus* Owen, 1842 (*nomen nudum*) ⇒ *Cetiosaurus giganteus* Owen *vide* Huxley 1870 [Bathonian; Great Oolite Group, England, UK] (acc. *The Dinosauria 2nd edition*)

Generic name: *Cetiosaurus* Owen, 1841 ← {(Gr.)

κήτειος/kēteios [κῆτος/kētos: sea monster] + σαῦρος/sauros: lizard}; derived from a relationship of the animal to which it belonged with the Cetacea*.

*Cetacea is derived from (Lat.) cētos (= κῆτος/kētos: a sea-monster).

Specific name: *oxoniensis* ← {(Lat. place-name) Oxon (=Oxford) + -ensis}

Etymology: **sea-monster lizard** from Oxford

Taxonomy: Saurischia: Sauropodomorpha: Sauropoda: Cetiosauridae

Other species:

> *Cetiosaurus longus* Owen, 1842 (*nomen dubium*) ⇒ *Cetiosauriscus longus* (Owen, 1842) McIntosh, 1990 [Kimmeridgian; Portland Stone Formation, England, UK]

> *Cetiosaurus medius* Owen, 1842 (*nomen dubium*) [Bajocian; Inferior Oolite Group, England, UK]

> *Cetiosaurus brachyurus* Owen, 1842 (*nomen dubium*) [Valanginian; Wadhurst Clay Formation, England, UK]

> *Cetiosaurus brevis* Owen, 1842 (*partim*) (*nomen dubium*) ⇒ *Cetiosaurus conybearei* Melville, 1849 (Syntype of *Pelorosaurus conybearei*) (*nomen dubium*) ⇒ *Morosaurus brevior* (Owen, 1842) Sauvage, 1898 ⇒ *Morosaurus brevis* (Owen, 1842) Lydekker, 1889 ⇒ *Pelorosaurus brevis* (Owen, 1842) von Huene, 1927 (*cf. Pelorosaurus* Mantell, 1850) [Valanginian; Tunbridge Wells Sand Formation, Wealden Group, England, UK]

> *Cetiosaurus conybearei* Melville, 1849 ⇒ *Pelorosaurus conybearei* (Melville, 1849) Mantell, 1850 [Valanginian; Tunbridge Wells Sand Formation, England, UK]

= *Cetiosaurus conybearei* Melville, 1849 (*nomen dubium*) (syntype) [Early Cretaceous; upper Wealden Formation, West Sussex, UK]

> *Cardion rugulosus* Owen, 1840–1845 (*nomen dubium*) ⇒ *Cetiosaurus rugulosus* (Owen, 1840–1845) Steel, 1970 [late Bathonian; Forest Marble Formation, Wiltshire, England, UK]

> *Cetiosaurus humerocristatus* Hulke, 1874 ⇒ *Pelorosaurus humerocristatus* (Hulke, 1874) Sauvage, 1887 ⇒ *Ornithopsis humerocristatus* (Hulke, 1874) Lydekker, 1888 ⇒ *Duriatitan humerocristatus* (Hulke, 1874) Barrett *et al.*, 2010 [late Kimmeridgian; Kimmeridge Clay Formation, UK]

See also *Duriatitan humerocristatus*.

> *Cetiosaurus rigauxi* Sauvage, 1874* [middle Titonian; Pas-de-Calais, France] * Pliosaurid.

> *Ornithopsis leedsii* Hulke, 1887 ⇒ *Pelorosaurus? leedsi* (Hulke, 1887) Lydekker, 1895 ⇒ *Cetiosaurus leedsi* (Hulke, 1887) Woodward, 1905* ⇒ *Cetiosauriscus leedsi* (Hulke, 1887) von Huene, 1927* (*nomen dubium*) [Callovian-Oxfordian; Oxford Clay, Peterborough, UK] *referred specimen

> *Cetiosaurus leedsi* von Huene, 1927, non Hulke, 1887 ⇒ *Cetiosauriscus stewarti* Charig, 1980 [Callovian; Oxford Clay, England, UK]

n.b. von Huene, 1927 named *Cetiosauriscus leedsi* for *Cetiosaurus leedsi*, but not the type of *C. leedsi*, but for the referred specimen published by Woodward, 1905. The ICZN ruled in favor of CHARIG in 1995, therefore the correct species is *stewarti* and not *leedsi*.

> *Ornithopsis greppini* von Huene, 1922 ⇒ *Cetiosauriscus greppini* (von Huene, 1922) von Huene, 1929 ⇒ *Amanzia greppini* (von Huene, 1922) Schwarz *et al.*, 2020 [early Kimmeridgian; Reuchenette Formation, Bern, Switzerland]

> *Cetiosaurus mogrebiensis* de Lapparent, 1955 (*nomen dubium*) [middle Bathonian; El Mers Formation, Boulemane, Morocco]

Notes: When Owen named the Dinosauria in 1842, he did not include *Cetiosaurus* into Dinosauria. In 2014, *Cetiosaurus oxoniensis* Phillips, 1871 was placed as the type species.

[References: (*1) Owen R (1841). A description of a portion of the skeleton of the *Cetiosaurus*, a gigantic extinct saurian reptile occurring in the oolitic formations of different portions of England. *Proceedings of the Geological Society of London* 3: 457–462.]

Changchengornis hengdaoziensis [Avialae]

> *Changchengornis hengdaoziensis* Ji *et al.*, 1999 [Barremian–Aptian; Hengdaozi Member, Chaomidianzi Formation*, Liaoning, China]

*"a newly established unit formerly regarded as the lower part of the Yixian Formation"(*1).

Generic name: *Changchengornis* ← {(Chin. place-name) Chángchéng長城: the Great Wall of China + (Gr.) ὄρνις/ornis: bird}

Specific name: *hengdaoziensis* ← {(Chin. place-name) Héngdàozǐ 横道子 + -ensis}; "referring to the stratigraphic horizon where it was collected" (*1).

Etymology: **Changcheng bird** from Hengdaozi [横道子長城鳥]

Taxonomy: Theropoda: Avialae: Confuciusornithidae

[References: (*1) Ji Q, Chiappe LM, Ji S-A (1999). A new late Mesozoic confuciusornithid bird from China. *Journal of Vertebrate Paleontology* 19(1): 1–7.]

Changchunsaurus parvus

> *Changchunsaurus parvus* Zan *et al.*, 2005 [Aptian–Cenomanian; Quantou Formation, Jilin, China]

Generic name: *Changchunsaurus* ← {(Chin. place-name) Chángchūn長春: the capital of Jilin Province + (Gr.) σαῦρος/sauros: lizard}

Specific name: *parvus* ← {(Lat.) parvus: small}.

Etymology: small **Changchun lizard**

Taxonomy: Ornithischia: Ornithopoda: Jeholosauridae

[References: Zan S-Q, Chen J, Jin L-Y, Li T (2005). A primitive ornithopod from the Early Cretaceous Quantou Formation of central Jilin, China. *Vertebrata PalAsiatica* 43(3): 182–193.]

Changmaornis houi [Avialae]

> *Changmaornis houi* Wang *et al.*, 2013 [Aptian; Xiagou Formation, Yumen, Gansu, China]

Generic name: *Changmaornis* ← {(Chin. place-name) Chāngmǎ昌馬: the name of the town + (Gr.) ὄρνις/ornis: bird} [*1]

Specific name: *houi* ← {(Chin. person's name) Hou + -ī}; "in honor of Professor Lian-Hai Hou, who named *Gansus yumenensis*" [*1].

Etymology: Hou's **Changma bird**

Taxonomy: Theropoda: Avialae: Pygostylia: Ornithothoraces: Ornithuromorpha

Notes: *Changmaornis houi*, *Yumenornis huangi* and *Jiuquanornis niui* have been discovered from Changma Basin in Gansu Province. [*1]

[References: (*1) Wang Y-M, O'Connor JK, Li D-Q, You H-L (2013). Previously unrecognized ornithuromorph bird diversity in the Early Cretaceous Changma Basin, Gansu Province, northwestern China. *PLoS ONE* 8(10): e77693.]

Changmiania liaoningensis

> *Changmiania liaoningensis* Yang *et al.*, 2020 [Berremian; Yixian Formation, Liaoning, China]

Generic name: *Changmiania* ← { (Chin.) chángmián 長眠: eternal sleep + -ia}

Specific name: *liaoningensis* ← {(place-name) Liáoníng遼寧+ -ensis}

Etymology: **eternal sleep** from Liaoning

Taxonomy: Ornithischia: Neornithischia: Ornithopoda

Notes: Two nearly complete articulated skeletons were discovered. It is hypothesized that they were buried suddenly, because of the absence of weathering and scavenging trace.

[References: Yang Y, We W, Dieudonné P, Godefroit P (2020). A new basal ornithopod dinosaur from the Lower Cretaceous of China. *PeerJ.* 8: e9832.]

Changyuraptor yangi

> *Changyuraptor yangi* Han *et al.*, 2014 [Barremian–Aptian; Yixian Formation, Liaoning, China]

Generic name: *Changyuraptor* ← {(Chin.) cháng 長: long + yǔ羽: feather + (Lat.) raptor: robber, plunderer}; "referring to the predatory habits inferred for the holotype" [*1]. It had a long and feathered tail.

Specific name: *yangi* ← {(Chin. person's name) Yáng楊 + -ī}; in honor of "Prof. Yang Yandong, Chairman of the Bohai University (Liaoning Province, China), who provided the financial support to acquire the specimen" [*1].

Etymology: Yang's **long-feathered thief**

Taxonomy: Theropoda: Dromaeosauridae: Microraptoria

Notes: Having feathers all over the body, its forelimbs and hindlimbs give the appearance of having two pairs of wings.

[References: (*1) Han G, Chiappe LM, Ji S-A, Habib M, Turner AH, Chinsamy A, Liu X, Han L (2014). A new raptorial dinosaur with exceptionally long fethering provides insights into dromaeosaurid flight performance. *Nature Communications* 5: 4382.]

Changzuiornis ahgmi [Avialae]

> *Changzuiornis ahgmi* Huang *et al.*, 2016 [Aptian; Jiufotang Formation, Lingyuan, Liaoning, China]

Generic name: *Changzuiornis* ← {(Chin.) changzui; referring to the long beak + (Gr.) ὄρνις/ornis: bird} [*1]

Specific name: *ahgmi* ← {(acronym) AHGM (= Anhui Geological Museum) + -ī}; "referring to the place where the holotype specimen is housed" [*1].

Etymology: Anhui Geological Museum's **long-beaked bird**

Taxonomy: Theropoda: Avialae: Ornithurae

cf. Aves: Ornithurae [*1]

[References: (*1) Huang J, Wang X, Hu Y, Liu J, Peteya JA, Clarke JA (2016). A new ornithurine from the Early Cretaceous of China sheds light on the evolution of early ecological and cranial diversity in birds. *PeerJ* 4: e1765.]

Chaoyangia beishanensis [Avialae]

> *Chaoyangia beishanensis* Hou & Zhang, 1993 [Aptian; Jiufotang Formation, Chaoyang, Liaoning, China]

Generic name: *Chaoyangia* ← {(Chin. place-name) Cháoyáng朝陽: Chaoyang County + -ia}

Specific name: *beishanensis* ← {(Chin. place-name) Běishān北山: Beishan Quarry + -ensis}

Etymology: One from **Chaoyang,** Beishan [北山朝

陽鳥]

Taxonomy: Theropoda: Avialae: Euornithes

Notes: *Chaoyangia* was regarded as a member of the Ornithurae, according to *The Dinosauria 2nd edition*.

Chaoyangsaurus youngi

> *Chaoyangsaurus youngi* Zhao *et al*., 1999 [Tithonian; Tuchengzi Formation, Liaoning, China]

Generic name: *Chaoyangsaurus* ← {(Chin. place-name) Cháoyáng朝陽 + (Gr.) σαῦρος/sauros: lizard}

Specific name: *youngi* ← {(person's name) Young 楊 [Pinyin: Yáng] + -ī}; in honor of the Chinese paleontologist C. C. Young楊鍾健.

Etymology: Young's **Chaoyang lizard**

Taxonomy: Ornithischia: Ceratopsia: Chaoyangsauridae

[References: Zhao X, Cheng Z, Xu X (1999). The earliest ceratopsian from the Tuchengzi Formation of Liaoning, China. *Journal of Vertebrate Paleontology* 19(4): 681–691.]

Charonosaurus jiayinensis

> *Charonosaurus jiayinensis* Godefroit *et al*., 2000 [Maastrichtian; Yuliangze Formation, Heilongjiang, China]

Generic name: *Charonosaurus* ← {(Gr. myth.) Χάρων/Charōn: Charon + σαῦρος/sauros: lizard}; derived from Charon, boatswain of Styx River in Greek and Roman mythology. *Charonosaurus* was found along the banks of the Amur River.(*1)

Specific name: *jiayinensis* ← {(Chin. place-name) Jiāyìn嘉陰 + -ensis}

Etymology: **Charon's lizard** from Jiayin

Taxonomy: Ornithischia: Ornithopoda: Hadrosauridae: Lambeosaurinae: Parasaurolophini

[References: (*1) Godefroit P, Zan S, Jin L (2000). *Charonosaurus jiayinensis* n.g., n. sp., a lambeosaurine dinosaur from the late Maastrichtian of northeastern China. *Earth and Planetary Sciences* 330: 875–882.]

Chasmosaurus belli

> *Monoclonius belli* Lambe, 1902 ⇒ *Ceratops belli* (Lambe, 1902) Hatcher *vide* Stanton & Hatcher, 1905 ⇒ *Chasmosaurus belli* (Lambe, 1902) Lambe, 1914 ⇒ *Protosaurus belli* (Lambe, 1902) Lambe, 1914 [Campanian; Dinosaur Park Formation, Alberta, Canada]

= *Monoclonius canadensis* Lambe, 1902 ⇒ *Ceratops canadensis* (Lambe, 1902) Hatcher *vide* Stanton & Hatcher, 1905 ⇒ *Eoceratops canadensis* (Lambe, 1902) Lambe, 1915 ⇒ *Chasmosaurus canadensis* (Lambe, 1902) Lehman, 1990 [middle-Campanian; Belly River Group, Alberta, Canada]

= *Chasmosaurus kaiseni* Brown, 1933 [Campanian; Oldman Formation, Alberta, Canada]

= *Chasmosaurus brevirostris* Lull, 1933 [Campanian; Dinosaur Park Formation, Alberta, Canada]

Generic name: *Chasmosaurus* ← {(Gr.) χάσμα/chasma: chasm, wide opening + σαῦρος/sauros: lizard}; referring to the the large parietal fenestrae in the skull frill.

Specific name: *belli* ← {(person's name) Bell + -ī}

Etymology: Bell's **chasm lizard**

Taxonomy: Ornithischia: Marginocephalia: Ceratopsia: Ceratopsidae: Chasmosaurinae

Other species:

> *Chasmosaurus russelli* Sternberg, 1940 [Campanian; Dinosaur Park Formation, Alberta, Canada]

?= *Mojoceratops perifania* Longrich, 2010 [Campanian; Dinosaur Park Formation, Alberta, Canada]

> *Chasmosaurus mariscalensis* Lehman, 1989 ⇒ *Agujaceratops mariscalensis* (Lehman, 1989) Lucas *et al*., 2006 [Judithian; Aguja Formation, Texas, USA]

> *Chasmosaurus irvinensis* Holmes *et al*. 2001 ⇒ *Vagaceratops irvinensis* (Holmes *et al*., 2001) Sampson *et al*., 2010 [late Campanian; Dinosaur Park Formation, Alberta, Canada]

Notes: *Chasmosaurus* had one horn on the nose and two on the brow.

[References: Godfrey SJ, Holmes R (1995). Cranial morphology and systematics of *Chasmosaurus* (Dinosauria: Ceratopsidae) from the Upper Cretaceous of western Canada. *Journal of Vertebrate Paleontology* 15(4): 726–742.]

Chebsaurus algeriensis

> *Chebsaurus algeriensis* Mahammed *et al*., 2005 [Callovian; Aïssa Formation, Occidental Saharan Atlas, Algeria]

Generic name: *Chebsaurus* ← {(Arab.) cheb: teenager + (Gr.) σαῦρος/sauros: lizard}; referring to the fact that this animal died when it was juvenile(*1).

Specific name: *algeriensis* ← {(place-name) Algeria + -ensis} (*1)

Etymology: **young lizard** from Algeria

Taxonomy: Saurischia: Sauropodomorpha: Sauropoda: Cetiosauridae?

[References: (*1) Mahammed F, Läng É, Mami L,

Mekahli L, Benhamou M, Bouterfa B, Kacemi A, Chérief S-A, Chaouati H (2005). The 'Giant of Ksour', a Middle Jurassic sauropod dinosaur from Algeria. *Comptes-Rendus Palevol* 4(8): 707–714.]

Chenanisaurus barbaricus

> *Chenasaurus barbaricus* Longrich *et al*., 2017 [Maastrichtian; Ouled Abdoun Basin, Morocco]

Generic name: *Chenanisaurus* ← {(place-name) Chennane + (Gr.) σαῦρος/sauros: lizard}; named after Sidi Chennane mines in the Ouled Abdoun Basin, Morocco.

Specific name: *barbaricus* ← {(Gr.) βαρβᾰρικός/barbaricos: barbaric}; referring to Barbary.

Etymology: **Chennane lizard** from Barbary

Taxonomy: Saurischia: Theropoda: Abelisauridae

[References: Longrich NR, Pereda Suberbiola X, Jalil N-E, Khaldoune F, Jourani E (2017). An abelisaurid from the latest Cretaceous (Maastrichtian) of Morocco, North Africa. *Cretaceous Research* 76: 40–52.]

Chialingosaurus kuani

> *Chialingosaurus kuani* Young, 1959 [Bathonian–Oxfordian; upper Shaximiao Formation, Sichuan, China]

Generic name: *Chialingosaurus* ← {(Chin. place-name) Chialing嘉陵 [Pinyin: Jiālíng] + (Gr.) σαῦρος/sauros: lizard}; referring to Chialing River.

Specific name: *kuani* ← {(Chin. person's name) Kuān關 [Pinyin: Guān] + -ī}; in honor of geologist Yaowu Guan of the Sichuan Regional Petroleum Exploration Office from Taipingzhai.

Etymology: Kuan's **lizard from Chialing** [關氏嘉陵龍]

Taxonomy: Ornithischia: Thyreophora: Stegosauria: Stegosauridae

Other species:

> *Chialingosaurus* "guangyuanensis" Li *et al*., 1999 (*nomen nudum*) [Oxfordian; upper Shaximiao Formation, China]

Notes: According to Maidment *et al*. (2006), *Chialingosaurus kuani* is regarded as a *nomen dubium* because the holotype specimen is "a juvenile, bearing no diagnostic characters".

[References: Young CC (1959). On a new Stegosauria from Szechuan, China. *Vertebrata PalAsiatica* 3(1): 1–8.]

Chilantaisaurus tashuikouensis

> *Chilantaisaurus tashuikouensis* Hu, 1964 [Turonian; Dashuigou Formation, Inner Mongolia, China]

Generic name: *Chilantaisaurus* ← {(Chin. place-name) Chilantai吉蘭泰 [Pinyin: Jílántài] + (Gr.) σαῦρος/sauros: lizard}

Specific name: *tashuikouensis* ← {(Chin. place-name) Tashuikou 大水溝 [Pinyin: Dàshuǐgōu] + -ensis}

Etymology: **Jilantai lizard** from Dashuikou

Taxonomy: Theropoda: Tetanurae

Other species:

> *Chilantaisaurus maortuensis* Hu, 1964 ⇒ "Alashansaurus" *maortuensis* (Hu, 1964) Chure, 2000 ⇒ *Shaochilong maortuensis* (Hu, 1964) Brusatte *et al*., 2009 [Turonian; Ulansuhai Formation, Inner Mongolia, China]

> *Allosaurus sibiricus* Riabinin, 1914 ⇒ *Antrodemus sibiricus* (Riabinin, 1914) Steel, 1970 ⇒ ?*Chilantaisaurus sibiricus* (Riabinin, 1914) Molnar *et al*., 1990 (*nomen dubium*) [Berriasian–Hauterivian; Turgin Formation, Zabaykal'ye, Russia]

> *Chilantaisaurus zheziangensis* Dong, 1979 (*nomen dubium*) [Santonian; Tangshang Formation, Zhejiang, China]

Notes: According to Zanno (2010) and Qian *et al*. (2012), *Chilantaisaurus zheziangensis* is a therizinosaur.

[References: Hu S-Y [胡寿永] (1964). Carnosaurian remains from Alashan, Inner Mongolia [内蒙古阿拉善旗肉食類化石]. *Vertebrata PalAsiatica* 8(1):42–63.]

Chilesaurus diegosuarezi

> *Chilesaurus diegosuarezi* Novas *et al*., 2015 [Tithonian; Toqui Formation, Aysén, Chile]

Generic name: *Chilesaurus* ← {(place-name) Chile + (Gr.) σαῦρος/sauros: lizard} (*1)

Specific name: *diegosuarezi* ← {(person's name) Diego Suárez + -ī}; "in honor of Diego Suárez, who at the age of 7, discovered the bone remains in the Toqui Formation" (*1).

Etymology: Diego Suárez's **Chile lizard**

Taxonomy: Theropoda: Tetanurae

[References: (*1) Novas FE, Salgado L, Suárez M, Agnolín FL, Ezcurra MD, Chimento NR, de la Cruz R, Isasi MP, Vargas AO, Rubilar-Rogers D (2015). An enigmatic plant-eating theropod from the Late Jurassic period of Chile. *Nature* 522(7556): 331–334.]

Chindesaurus bryansmalli

> *Chindesaurus bryansmalli* Long & Murry, 1995 [Carnian–Norian; Chinle Formation, Arizona, USA]

Generic name: *Chindesaurus* ← {(place-name) Chinde Point | (Navajo) chiindii: ghost or evil spirit + σαῦρος/sauros: lizard}; referring to the Chinde Point, Petrified Forest National Park,

Arizona[*1].
Specific name: *bryansmalli* ← {(person's name) Bryan Small + -ī}; in honor of Bryan Small, a discoverer.
Etymology: Bryan Small's **Chinde lizard**
Taxonomy: Saurischia: Herrerasauridae

[References: (*1) Long RA, Murry PA (1995). Late Triassic (Carnian and Norian) tetrapods from the Southwestern United States. *New Mexico Museum Natural History Science Bulletin* 4: 1–254.]

Chingkankousaurus fragilis

> *Chingkankousaurus fragilis* Young, 1958 [Campanian; Jingangkou Formation, Shandong, China]

Generic name: *Chingkankousaurus* ← {(Chin. place-name) Chingkankou金剛口 [Pinyin: Jīngāngkǒu] + σαῦρος/sauros: lizard}
Specific name: *fragilis* ← {(Lat.) fragilis: fragile}.
Etymology: fragile **Jingangkou lizard**
Taxonomy: Theropoda: Coelurosauria: Tyrannosauroidea
Notes: *Chingkankousaurus* was considered to be a synonym of ?*Tarbosaurus bataar*, according to *The Dinosauria 2nd edition* (2004), and recently this genus is considered to be a *nomen dubium*, according to Brusatte *et al.* (2013).

[References: Young CC (1958). The dinosaurian remains of Laiyang, Shantung. *Palaeontologia Sinica, New Series C,* Whole Number 42(16): 1–138.]

Chinshakiangosaurus chunghoensis

> *Chinshakiangosaurus chunghoensis* Ye *vide* Dong, 1992* [Hettangian; Fengjiahe Formation, Yunnan, China]
> *Ye mentioned this specimen in 1975, but it was a *nomen nudum* until Dong published an official description in 1992.

= *Chinshakiangosaurus zhonghoensis* Ye, 1975 (*nomen nudum*) (acc. Zhao, 1985)

Generic name: *Chinshakiangosaurus* ← {(Chin. place-name) Chinshakiang金沙江 [Pinyin: Jīnshājiāng] + σαῦρος/sauros: lizard}
Specific name: *chunghoensis* ← {(Chin. place-name) Chungho中和 [Pinyin: Zhōnghè] + -ensis}
Etymology: **Jinshajiang lizard** from Zhonghe [中和金沙江龍]
Taxonomy: Saurischia: Sauropodomorpha: Sauropoda
Notes: According to *The Dinosauria 2nd edition* (2004), *Chinshakiangosaurus chunghoensis* was classified as a *nomen dubium*. Upchurch *et al.* (2007) declared *Chinshakiangosaurus* as a valid taxon.

Chirostenotes pergracilis

> *Chirostenotes pergracilis* Gilmore, 1924 [middle Campanian–early Maastrichtian; Dinosaur Park Formation, Alberta, Canada]

= *Macrophalangia canadensis* Sternberg, 1932 [Campanian; Dinosaur Park Formation, Alberta, Canada]
= *Caenagnathus sternbergi* Cracraft, 1971 ⇒ *Chirostenotes sternbergi* (Cracraft, 1971) Snively *et al.*, 2001 [Campanian; Dinosaur Park Formation, Alberta, Canada]

Generic name: *Chirostenotes* ← {(Gr.) χείρ/cheir: hand + στενότης/stenotēs: narrowness [στενός/stenos: narrow]}
Specific name: *pergracilis* ← {(Lat.) per: through, very + gracilis: gracile, slender}
Etymology: very slender, **narrow handed one**
Taxonomy: Theropoda: Coelurosauria: Oviraptorosauria: Caenagnathidae: Elmisaurinae
Other species:

> *Ornithomimus elegans*, Parks, 1933 ⇒ *Macrophalangia elegans* (Parks, 1933) Koster *et al.*, 1987 ⇒ *Elmisaurus elegans* (Parks, 1933) Currie, 1989 ⇒ *Chirostenotes elegans* (Parks, 1933) Sues, 1998 ⇒ *Leptorhynchos elegans* (Parks, 1933) Longrich *et al.*, 2013 ⇒ *Citipes elegans* (Parks, 1933) Funston, 2020 [Campanian; Dinosaur Park Formation, Alberta, Canada]

Notes: According to *The Dinosauria 2nd edition*, *Caenagnathus collinsi* R. M. Sternberg, 1940 was considered to be a synonym of *Chirostenotes pergracilis*, but it turnd out to be valid because more complete fossils were found in 2014 and 2015.

[References: Gilmore CW (1924). A new coelurid dinosaur from the Belly River Cretaceous of Alberta. *Canada Department of Mines Geologocal Series* 38(43): 1–12.]

Choconsaurus baileywillisi

> *Choconsaurus baileywillisi* Simón *et al.*, 2017 [Cenomanian; Huincul Formation, Neuquén, Argentina]

Generic name: *Choconsaurus* ← {(place-name) Villa El Chocón + (Gr.) σαῦρος/sauros: lizard}; "in reference to the locality of Villa El Chocón (Neuquén Province, Argentina), where the type species comes from".
Specific name: *baileywillisi* ← {(person's name) Bailey Willis + -ī}; in honor of "Bailey Willis (1857–1949), a North American geologist who explored the Argentinean Patagonia during the

first decades of the 20th century. As a consultant of the National Government of Argentina, Willis carried out multiple productive projects for the Patagonian region, including the utilization of the Limay River for generation of hydraulic energy".
Etymology: Bailey Willis' **Chocón lizard**
Taxonomy: Saurischia: Sauropodomorpha:
Sauropoda: Titanosauria
[References: Simón E, Salgado L, Calvo JO (2018). A new titanosaur sauropod from the Upper Cretaceous of Patagonia, Neuquén Province, Argentina. *Ameghiniana* 55(1): 1–29.]

Chondrosteosaurus gigas
> *Chondrosteosaurus gigas* Owen, 1876 (*nomen dubium*) [Berriasian–Barremian; Wessex Formation, Isle of Wight, UK]

Generic name: *Chondrosteosaurus* ← {(Mod. Gr.) χόνδρος/chondros: cartilage + (Gr.) ὀστέον/osteon: bone + σαῦρος/sauros: lizard}; referring to having pneumatic air sacs.
Specific name: *gigas* ← {(Gr. myth.) Γίγας/Gigas [in pl. Γίγαντες/Gigantes: the Giants]}
cf. γίγας/gigas: mighty. Γῐγάντειος/Giganteios: gigantic.
Etymology: gigantic **cartilage bone lizard**
Taxonomy: Saurischia: Sauropodomorpha:
Sauropoda
Other species:
> *Ornithopsis hulkei* Seeley, 1870 (Lectotype) ⇒ *Bothriospondylus magnus* (Seeley, 1870) Owen, 1875 ⇒ *Chondrosteosaurus magnus* (Seeley, 1870) Owen,1876 ⇒ *Pelorosaurus hulkei* (Seeley, 1870) Lydekker, 1893 ⇒*Eucamerotus hulkei* (Seeley, 1870) Sebaschan, 2005 [Barremian; Wessex Formation, Isle of Wight, UK]

Notes: According to *The Dinosauria 2nd edition*, *Chondrosteosaurus gigas* is considered to be a *nomen dubium*.

Chongmingia zhengi [Avialae]
> *Chongmingia zhengi* Wang *et al.*, 2016 [Aptian; Jiufotang Formation, Liaoning, China]

Generic name: *Chongmingia* ← {(Chin. myth.) Chóngmíng重明* + -ia}; "referring to a Chinese mythological bird"(*1), Chongmingniao 重明鳥.
*Chongming (double-pupil) Bird重明鳥, which has two pupils in each of its eyes, is shaped like a chicken and sings like a phoenix.
Specific name: *zhengi* ← {(person's name) Zhèng鄭 + -ī}; "in honor of Mr. Xiaoting Zheng (鄭曉廷) for his generous contribution in the establishment of the Shandong Tianyu Museum of Nature"(*1).
Etymology: Zheng's ***Chongmingniao* (mythological double-pupil bird)**
Taxonomy: Theropoda: Avialae: Euavialae:
Avebrevicauda: Pygostylia: Jinguofortisidae
cf. Aves(*1)
[References: (*1) Wang M, Wang X, Wang Y, Zhou Z (2016). A new basal bird from China with implications for morphological diversity in early birds. *Scientific Reports* 6: 19700.]

Choyrodon barsboldi
> *Choyrodon barsboldi* Gates *et al.*, 2018 [Albian; Khuren Dukh Formation, Dorngovi, Mongolia]

Generic name: *Choyrodon* ← {(place-name) Choyr: a city near Khuren Dukh locality + (Gr.) ὀδών/odon: tooth, a common ending for ornithopod dinosaur taxa} (*1)
Specific name: *barsboldi* ← {(person's name) Barsbold + -ī}; in honor of "Dr. Rinchen Barsbold, a leading dinosaur paleontologist of Mongolia and leader of the paleontology expedition that discovered the first remains of this species"(*1).
Etymology: Barsbold's **Choyr tooth**
Taxonomy: Ornithischia: Ornithopoda: Iguanodontia
[References: (*1) Gates TA, Tsogtbaatar K, Zanno LE, Chinzorig T, Watabe M (2018). A new iguanodontian (Dinosauria, Ornithopoda) from the Early Cretaceous of Mongolia. *PeerJ* 6: e5300.]

Chromogisaurus novasi
> *Chromogisaurus novasi* Ezcurra, 2010 [Carnian; Ischigualasto Formation, San Juan, Argentina]

Generic name: *Chromogisaurus* ← {(Gr.) χρῶμα/chrōma: colour, paint + gi- [γῆ/gē: ground, earth] + σαῦρος/sauros: reptile}; "in allusion to the Valle Pintado (Painted Valley) locality of the Ischigualasto Formation in which the new taxon was found" (*1).
Specific name: *novasi* ← {(person's name) Novas + -ī}; "in honor of Dr. Fernando Novas, for his outstanding research on early dinosaur evolution"(*1).
Etymology: Novas' **lizard from Valle Pintado**
Taxonomy: Saurischia: Sauropodomorpha:
Guaibasauridae: Saturnaliinae
[References: (*1) Ezcurra MD (2010). A new early dinosaur (Saurischia: Sauropodomorpha) from the Late Triassic of Argentina: a reassessment of dinosaur origin and phylogeny. *Journal of Systematic Paleontology* 8: 371–425.]

Chuandongocoelurus primitivus
> *Chuandongocoelurus primitivus* He, 1984

[Bathonian–Callovian; lower Shaximiao Formation, Sichuan, China]

Generic name: *Chuandongocoelurus* ← {(Chin. place-name) Chuāndōng川東: 'eastern Sichuan' + *Coelurus* [(Gr.) κοῖλος/koilos: hollow + οὐρά/ūrā: tail]}

Specific name: *primitivus* ← {(Lat.) prīmitīvus: primitive}

Etymology: primitive, ***Coelurus* (hollow tail) from Chuandong** [原始川東虚骨龍]

Taxonomy: Theropoda

Notes: The animal was immature at the time of death. According to *The Dinosauria 2nd edition*, *Chuandongocoelurus primitivus* is considered a *nomen dubium*.

Chuanjiesaurus anaensis

> *Chuanjiesaurus anaensis* Fang *et al*., 2000 [Bathonian; Chuanjie Formation, Lufeng, Yunnan, China]

Generic name: *Chuanjiesaurus* ← {(Chin. place-name) Chuānjiē川街: the township where the specimen was excavated + (Gr.) σαῦρος/sauros: lizard} (*1). The fossils were excavated from the Chuanjie Formation.

Specific name: *anaensis* ← {(Chin. place-name) A'na 阿納 [Pinyin: Ēnà]: the village at the locality + -ensis} (*1)

Etymology: **Chuanjie lizard** from Ana village

Taxonomy: Saurischia: Sauropodomorpha:
Sauropoda: Eusauropoda: Mamenchisauridae (*2)
cf. Saurischa: Cetiosauridae (*1)

[References: (*1) Fang X, Long Q, Lu L, Zhang Z, Pan S, Wang Y, Li X, Cheng Z (2000). Lower, Middle and Upper Jurassic subdivision in the Lufeng region, Yunnan Province. In; *Proceedings of the Third National Stratigraphical Congress of China*. Geological Publishing House, Beijing : 208–214.; (*2) Sekiya T (2011). Re-examination of *Chuanjiesaurus anaensis* (Dinosauria: Sauropoda) from the Middle Jurassic Chuanjie Formation, Lufeng County, Yunnan Province, southwest China. *Memoir of the Fukui Prefectural Dinosaur Museum* 10: 1–54.]

Chuanqilong chaoyangensis

> *Chuanqilong chaoyangensis* Han *et al*., 2014 [Aptian; Jiufotang Formation, Lingyuan, Liaoning, China]

Generic name: *Chuanqilong* ← {(Chin.) chuánqí 伝奇: legendary + lóng龍: dragon}; "referring to western Liaoning providing a spectacular assemblage of Mesozoic terrestrial fossils" (*1).

Specific name: *chaoyangensis* ← {(Chin. place-name) Cháoyáng朝陽 + -ensis}; "derived from the broader geographical area including the type locality" (*1).

Etymology: **legendary dragon** from Chaoyang

Taxonomy: Ornithischia: Thyreophora:
Ankylosauria: Ankylosauridae

[References: (*1) Han F, Zheng W, Hu D, Xu X, Barrett PM (2014). A new basal ankylosaurid (Dinosauria: Ornithischia) from the Lower Cretaceous Jiufotang Formation of Liaoning Province, China. *PLoS ONE* 9(8): e104551.]

Chubutisaurus insignis

> *Chubutisaurus insignis* Corro, 1974 [Albian; Cerro Barcino Formation, Chubut, Argentina]

Generic name: *Chubutisaurus* ← {(place-name) Chubut Province + (Gr.) σαῦρος/sauros: lizard}

Specific name: *insignis* ← {(Lat.) īnsīgnis: notable, distinguished}; "in reference to the size of the limbs and of the vertebrae" (*1).

Etymology: notable **Chubut lizard**

Taxonomy: Saurischia: Sauropodpmorpha:
Sauropoda: Titanosauria
cf. Saurischia: Sauropodomorpha: Sauropoda: Chubutisauridae (*1)

[References: (*1) del Corro G (1974). Un Nuevo saurópodo del Cretácico *Chubutisaurus insignis* gen. *et* sp. nov. (Saurischia–Chubutisauridae nov.) del Cretácico Superior (Chubutiano), Chubut, Argentina. *Actas I Congreso Argentino de Paleontología y Bioestratigrafía* 2: 229–240.]

Chungkingosaurus jiangbeiensis

> *Chungkingosaurus jiangbeiensis* Dong *et al*., 1983 [Oxfordian; Shangshaximiao Formation, Sichuan, China]

= *Chungkingosaurus* "giganticus" Ulansky, 2014 *vide* Galton & Carpenter, 2016 (*nomen nudum*) [Oxfordian; Shangshaximiao Formation, Sichuan, China]

= *Chungkingosaurus* "magnus" Ulansky, 2014 *vide* Galton & Carpenter, 2016 (*nomen nudum*) [Oxfordian; Shangshaximiao Formation, Sichuan, China]

Generic name: *Chungkingosaurus* ← {(Chin. place-name) Chungking重慶 [Pinyin: Chòngqìng] + (Gr.) σαῦρος/sauros: lizard}

Specific name: *jiangbeiensis* ← {(Chin. place-name) Jiāngběi江北 + -ensis}; referring to the Jiangbei district of Chungking municipality (*1).

Etymology: **Chungking (= Chongqing) lizard** from Jiangbei

Taxonomy: Ornithischia: Thyreophora: Stegosauria:
Huayangosauridae (acc. Maidment *et al*. 2008)
cf. Ornithischia: Thyreophora: Stegosauridae (acc. *The Dinosauria 2nd edition*)

Notes: According to Maidment & Wei (2006),

Chungkingosaurus is considered to be a basal member of the Stegosauria.

[References: (*1) Dong Z, Zhou S, Zhang Y (1983). Dinosaurs from the Jurassic of Sichuan. *Palaeontologica Sinica, New Series C* 162(23): 1–136.]

Chupkaornis keraorum [Avialae]

> *Chupkaornis keraorum* Tanaka *et al*., 2017 [Coniacian–Santonian; Kashima Formation, Mikasa, Hokkaido, Japan]

Generic name: *Chupkaornis* ← {(Jpn. Ainu*) cupka: eastern** + (Gr.) ὄρνις/ornis: bird}

*Ainu is a language spoken by the Ainu ethnic group of Hokkaido.

**This specimen was discovered in Hokkaido, Japan, Far East.

Specific name: *keraorum* ← {(Jpn. person's name) Kera + -ōrum}; "in honor of Masatoshi and Yasuji Kera, who discovered the specimen and contributed greatly to the Mikasa City Museum" (*1).

Etymology: Kera's **eastern bird**

Taxonomy: Theropoda: Avialae: Ornithuromorpha: Hesperornithiformes

[References: (*1) Tanaka T, Kobayashi Y, Kurihara K, Fiorillo AR, Kano M (2017). The oldest Asian hesperornithiform from the Upper Cretaceous of Japan, and the phylogenetic reassessment of Hesperornithiformes. *Journal of Systematic Palaeontology* 16(8): 689–709.]

Chuxiongosaurus lufengensis

> *Jingshanosaurus xinwaensis* Zhang & Yang, 1995 [Hettangian; Lufeng Formation, Yunnan, China]

= *Chuxiongosaurus lufengensis* Lü *et al*., 2010 [Hettangian; Lufeng Formation, Yunnan, China]

Generic name: *Chuxiongosaurus* ← {(Chin. place-name) Chŭxióng楚雄 + (Gr.) σαῦρος/sauros: lizard}

Specific name: *lufengensis* ← {(Chin. place-name) Lùfēng禄豐 + -ensis}; referring to Lufeng County, the classic dinosaur locality in Yunnan Province.

Etymology: **Chuxiong lizard** from Lufeng [禄豐楚雄龍]

Taxonomy: Saurischia: Sauropodomorpha: Massopoda

Notes: According to Zhang *et al*. (2019), *Chuxiongosaurus lufengensis* is considered to be a junior synonym of *Jingshanosaurus xinwaensis*.

[References: Lü J, Kobayashi Y, Li T, Zhong S (2010). A new basal sauropod dinosaur from Lufeng Basin, Yunnan Province, Southwestern China. *Acta Geologica Sinica* (English Edition) 84(6): 1336–1342.]

Cionodon arctatus

> *Cionodon arctatus* Cope, 1874 (*nomen dubium*) [Maastrichtian; Laramie Formation, Colorado, USA]

Generic name: *Cionodon* ← {(Gr.) κίων/kiōn: column + ὀδών/odōn: tooth}

Specific name: *arctatus* ← {(Lat.) arctātus [arctō: tighten, compress]}

Etymology: compressed, **column tooth**

Taxonomy: Ornithischia: Ornithopoda: Hadrosauridae

Other species:

> *Cionodon stenopsis* Cope, 1875 (*nomen dubium*) [Maastrichtian; Frenchman Formation, Alberta, Canada]

> *Cionodon kysylkumensis* Riabinin, 1931 (*nomen dubium*)⇒?*Bactrosaurus kysylkumensis* (Riabinin, 1931) Nessov, 1995 [Cenomanian; Kysyl-kum desert, central Asia]

Notes: According to *The Dinosauria 2nd edition*, *Cionodon arctatus*, *C. stenopsis* and *C. kysylkumensis* were regarded as *nomina dubia*.

Citipati osmolskae

> *Citipati osmolskae* Clark *et al*., 2001 [Campanian; Djadokhta Formation, Gobi Desert, Mongolia]

Generic name: *Citipati* ← {(Sanskrit) citi: funeral pyre + pati: lord}; the lord of cemeteries in Tantric Buddhist tradition, typically depicted as a human skeleton (*1).

Specific name: *osmolskae* ← {(person's name) Osmólska + -ae}; in honor of Polish paleontologist Halszka Osmólska for her work on oviraptorids and other Mongolian theropod dinosaurs (*1).

Etymology: Osmólska's **lord of funeral pyre**

Taxonomy: Theropoda: Coelurosauria: Oviraptorosauria: Oviraptoridae

Notes: *Citipati* is considered to be the link of non-avian dinosaurs and birds. The nickname of the specimen announced in 1995 was 'Big Mamman', whose brooding posture resembles those of birds.

[References: (*1) Clark JM, Norell MA, Barsbold R (2001). Two new oviraptorids (Theropoda; Oviraptorosauria), upper Cretaceous Djadokhta Formation, Ukhaa Tolgod, Mongolia. *Journal of Vertebrate Paleontology* 21(2): 209–213.]

Citipes elegans

> *Ornithomimus elegans* Parks, 1933 ⇒ *Elmisaurus elegans* (Parks, 1933) Currie, 1989 ⇒ *Chirostenotes elegans* (Parks, 1933) Sues,

1997 ⇒ *Leptorhynchos elegans* (Parks, 1933) Longrich *et al.*, 2013 ⇒ *Citipes elegans* (Parks, 1933) Funston, 2020 [Campanian; Dinosaur Park Formation, Alberta, Canada]

Generic name: *Citipes* ← {(Lat.) citus: swift + pēs: foot}

Specific name: *elegans* ← {(Lat.) ēlegāns: elegant}

Etymology: elegant, **swift foot**

Taxonomy: Saurischia: Theropoda: Caenagnathidae: Elmisaurinae

Claorhynchus trihedrus

> *Claorhynchus trihedrus* Cope, 1892 (*nomen dubium*) [Maastrichtian; Laramie Formation, Colorado, USA]

Generic name: *Claorhynchus* ← {(Gr.) κλάω-/klaō (= claō): broken + ῥύγχος /rhynchos: beak}

Specific name: *trihedrus* ← {(Gr.) τρί-/ tri- [τρεῖς/ treis]*: three + hedrus} trihedral.

*τρῐ-, Prefix, from τρίς or τρίᾰ [neut. of τρεῖς]

Etymology: trihedral, **broken beak**

Taxonomy: Ornithischa: Genasauria: Neornithischia: Cerapoda

Notes: According to *The Dinosauria 2nd edition*, *Claorhynchus* is considered a *nomen dubium*.

[References: Cope ED (1892). Fourth note on the Dinosauria of the Laramie. *The American Naturalist* 26: 756–758.]

Claosaurus agilis

> *Hadrosaurus agilis* Marsh, 1872 ⇒ *Claosaurus agilis* (Marsh, 1872) Marsh, 1890 ⇒ *Trachodon agilis* (Marsh, 1872) Kuhn, 1936 [Coniacian–Campanian; Niobrara Chalk Formation, Kansas, USA]

Generic name: *Claosaurus* ← {(Gr.) κλάω-/klaō-*: broken + σαῦρος/sauros: lizard}; referring to the odd position when discovered. *k=c

Marsh (1890) changed the genus name to *Claosaurus* (broken lizard) when major differences between it and *Hadrosaurus* became apparent. The remains found in Kansas indicated a 12 to 15 feet long animal (*1).

Specific name: *agilis* ← {(Lat.) agilis: nimble}.

Etymology: nimble, **broken lizard**

Taxonomy: Ornithischia: Ornithopoda: Hadrosauroidea: Hadrosauromorpha

Other species:

> *Claosaurus affinis* Wieland, 1903 [Campanian; Pierre Shale Formation, South Dakota, USA]

> *Claosaurus annectens* Marsh, 1892 ⇒ *Trachodon annectens* (Marsh, 1892) Hatcher, 1902 ⇒ *Thespesius annectens* (Marsh, 1892) Gilmore, 1924 ⇒*Anatosaurus annectens* (Marsh, 1892) Lull & Wright, 1942 ⇒ *Edmontosaurus annectens* (Marsh, 1892) Brett-Surman, 1990 [Maastrichtian; Lance Formation, Wyoming, USA]

Notes: *Claosaurus* is a genus of primitive hadrosaurian.

[References: (*1) *Claosaurus agilis* Marsh ⟨http://oceansofkansas.com/Marsh1890.html⟩; Marsh OC (1872). Notice of a new species of *Hadrosaurus*. *American Journal of Science, Series 3* 16: 301.]

Coahuilaceratops magnacuerna

> *Coahuilaceratops magnacuerna* Loewen *et al.*, 2010 [Campanian; Cerro del Pueblo Formation, Coahuila, Mexico]

Generic name: *Coahuilaceratops* ← {(place-name) Coahuila + *Ceratops* [(Gr.) κέρας/keras: horn + ὤψ/ōps: face]}; referring to the Mexican state of Coahuila.

Specific name: *magnacuerna* ← {(Lat.) magnus: great + (Sp.) cuerna: horn}; "in reference to the very large supraorbital horncores of this taxon" (*1).

Etymology: **Coahuila horned face** with large horns

Taxonomy: Ornithischia: Ceratopsia: Ceratopsidae: Chasmosaurinae

[References: (*1) Loewen MA, Sampson SD, Lund EK, Farke AA, Aguillón-Martinez MC, de Leon CA, Rodríguez-de la Rosa RA, Getty MA, Eberth DA (2010). Horned dinosaurs (Ornithischia: Ceratopsidae) from the Upper Cretaceous (Campanian) Cerro del Pueblo Formation, Coahuila, Mexico. In: Ryan MJ, Chinnery-Allgeier BJ, Eberth DA (eds), *New Perspectives on Horned Dinosaurs: The Royal Tyrell Museum Ceratopsian Syposium*. Indiana University Press, Bloomington: 99–116.]

Coelophysis bauri

> *Coelurus bauri* Cope, 1887 ⇒ *Tanystrophaeus bauri* (Cope, 1887) Cope, 1887 ⇒ *Coelophysis bauri* (Cope, 1887) Cope, 1889 (*nomen conservandum*) [Rhaetian; Chinle Formation, New Mexico, USA]

= *Coelurus longicollis* Cope, 1887 ⇒ *Tanystropheus longicollis* (Cope, 1887) Cope, 1887 ⇒ *Coelophysis longicollis* (Cope, 1887) Cope, 1889 ⇒ *Longosaurus longicollis* (Cope, 1887) Welles, 1984 [Norian; Chinle Formation, New Mexico, USA]

= *Tanystropheus willistoni* Cope, 1887 ⇒ *Coelophysis willistoni* (Cope, 1887) Cope, 1889 [Norian; Chinle Formation, New Mexico, USA]

= *Rioarribasaurus colberti* Hunt & Lucas, 1991 (*nomen rejectum*) ⇒ *Syntarsus colberti* (Hunt & Lucas, 1991) Paul, 1993 (preoccupied)

[Rhaetian; Chinle Formation, New Mexico, USA]

Generic name: *Coelophysis* ← {(Gr.) κοῖλος/koilos: hollow + φύσις/physis: form}; referring to its hollow vertebrae.

Specific name: *bauri* ← {(person's name) Baur + -ī}; in honor of a vertebrate paleontologist George Baur (1859–1898).

Etymology: Baur's **hollow form**

Taxonomy: Theropoda: Neotheropoda: Coelophysoidea: Coelophysidae

cf. Theropoda: Ceratosauria: Coelophysoidea (acc. *The Dinosauria 2nd edition*)

Other species:

> *Syntarsus rhodesiensis* Raath, 1969 (preoccupied) ⇒ *Coelophysis rhodesiensis* (Raath, 1969) Paul, 1988 ⇒ *Megapnosaurus rhodesiensis* (Raath, 1969) Ivie *et al.*, 2001 [Hettangian–Sinemurian; Forest Sandstone Formation, Zimbabwe]

> *Syntarsus kayentakatae* Rowe, 1989 ⇒ *Megapnosaurus kayentakatae* (Rowe, 1989) Ivie *et al.*, 2001 ⇒ *Coelophysis kayentakatae* (Rowe, 1989) Bristowe & Raath, 2004 [Sinemurian–Pliensbachian; Kayenta Formation, Arizona, USA]

> *Tanystrosuchus posthumus* von Huene, 1908 ⇒ *Coelophysis posthumus* (von Huene, 1908) Kuhn, 1965 [Norian; Baden–Wurttemberg, Germany]

> *Podokesaurus holyokensis* Talbot, 1911 ⇒ *Coelophysis holyokensis*? (Talbot, 1911) Colbert, 1964 [Hettangian–Sinemurian; Portland Formation, Massachusetts, USA]

Notes: According to Gregory (1988), *Coelophysis* may have hunted in a group.

[References: Colbert EH (1964). The Triassic dinosaur genera *Podokesaurus* and *Coelophysis*. *American Museum Novitates* 2168: 1–12.]

Coelurus fragilis

> *Coelurus fragilis* Marsh,1879 [Kimmeridgian; Lake Como Member, Morrison Formation, Wyoming, USA]

= *Coelurus agilis* Marsh, 1884 ⇒ *Elaphrosaurus agilis* (Marsh, 1884) Russell *et al.*, 1980 [Kimmeridgian; Lake Como Member, Morrison Formation, Wyoming, USA]

Generic name: *Coelurus* ← {(Gr.) κοῖλος/koilos: hollow + οὐρά/ūrā: tail + -us}

According to Marsh (1879), "the most characteristic specimens" are "vertebrae, which in the dorsal and lumbar region have their centra so much excavated that the wall are reduced to a thin shell"(*1).

Specific name: *fragilis* ← {(Lat.) fragilis: fragile}

Etymology: fragile **hollow tail**

Taxonomy: Theropoda: Coelurosauria: Coeluridae

Other species:

> *Coelurus gracilis* Marsh, 1888 (*nomen dubium*) ⇒ *Dromaeosaurus gracilis* (Marsh, 1888) Matthew & Brown, 1922 (*nomen dubium*) [Aptian; Arundel Clay Formation, Maryland, USA]

> *Coelurus bauri* Cope, 1887 ⇒ *Coelophysis bauri* (Cope, 1887) Cope, 1889 [Rhaetian; Chinle Formation, New Mexico, USA]

= *Coelurus longicollis* Cope, 1887 ⇒ *Coelophysis longicollis* (Cope, 1887) Cope, 1889 ⇒ *Longosaurus longicollis* (Cope, 1887) Welles, 1984 [Norian; Chinle Formation, New Mexico, USA]

> *Thecospondylus daviesi* Seeley, 1888 (*nomen dubium*) ⇒ *Coelurus daviesi* (Seeley, 1888) Nopcsa, 1901 (*nomen dubium*) ⇒ *Thecocoelurus daviesi* (Seeley, 1888) von Huene, 1923 (*nomen dubium*) [Barremian; Wessex Formation, Isle of Wight, UK]

> *Ornitholestes hermanni* Osborn, 1903 ⇒ *Coelurus hermanni* (Osborn, 1903) Hay, 1930 [Kimmeridgian; Wyoming, USA]

Notes: *Coelurus* is a small bipedal carnivore.

[References: (*1) Marsh OC (1879). Notice of new Jurassic reptiles. *American Journal of Science, Series 3* 18: 501–505.]

Colepiocephale lambei

> *Stegoceras lambei* Sternberg, 1945 ⇒ *Colepiocephale lambei* (Sternberg, 1945) Sullivan, 2003 [Campanian; Foremost Formation, Alberta, Canada]

Generic name: *Colepiocephale* ← {(Lat.) colepium: "knuckle (of beef or pork) used in reference to the curled 'knuckle' –like appearance (in lateral view) of the posterior part of the frontoparietal dome" (*1) + (Gr.) κεφᾰλή/cephale: head}

Specific name: *lambei* ← {(person's name) Lambe + -ī}

Etymology: Lambe's **knuckle (fist) head**

Taxonomy: Ornithischia: Pachycephalosauria: Pachycephalosauridae: Pachycephalosaurinae

Notes: According to *The Dinosauria 2nd edition*, *Colepiocephale lambei* was a synonym of *Stegoceras validum* Lambe, 1918 and was renamed in 2003.

[References: (*1) Sullivan RM (2003). Revision of the dinosaur Stegoceras lambe (Ornithischia, Pachycephalosauridae). *Journal of Vertebrate*

Paleontology 23(1): 181–207.]

Coloradisaurus brevis

> *Coloradia brevis* Bonaparte, 1978 (preoccupied) ⇒ *Coloradisaurus brevis* (Bonaparte, 1978) Lambert, 1983 [Norian–Rhaetian; Los Colorados Formation, La Rioja, Argentina]

Generic name: *Coloradisaurus* ← {(place-name) Los Colorados Formation + (Gr.) σαῦρος/sauros: lizard} (*1)

Specific name: *brevis* ← {(Lat.) brevis: short}; referring to its shorter snout. (*1)

Etymology: short, **Los Colorados lizard**

Taxonomy: Saurischia: Sauropodomorpha: Massospondylidae

[References: (*1) Bonaparte JF (1978). *Coloradia brevis* n. g. *et* n. sp. (Saurischia–Prosauropoda), dinosaurio Plateosauridae de la Formacion Los Colorados, Triasico Superior de la Rioja, Argentina. *Ameghiniana* 15(3–4): 327–332.]

Comahuesaurus windhauseni

> *Comahuesaurus windhauseni* Carballido *et al.*, 2012 [Aptian–Albian; Lohan Cura Formation, Neuquén Basin, Argentina]

Generic name: *Comahuesaurus* ← {(Mapuche, place-name) Comahue: the region in North Patagonia + (Gr.) σαῦρος/sauros: lizard}

Comahue means "place of abundance or perhaps 'where the water hurt'" (*1).

Specific name: *windhauseni* ← {(person's name) Windhausen + -ī}; in honor of "Anselmo Windhausen for his contribution to the geological knowledge of the Neuquén Basin" (*1).

Etymology: Windhausen's **Comahue lizard**

Taxonomy: Saurischia: Sauropoda: Neosauropoda: Diplodocoidea: Rebbachisauridae

[References: (*1) Carballido JL, Salgado L, Pol D, Canudo JI, Garrido A (2012). A new basal rebbachisaurid (Sauropoda, Diplodocoidea) from the Early Cretaceous of the Neuquén Basin, evolution and biogeography of the group. *Historical Biology: An International Journal of Paleobiology* 24(6): 631–654.]

Compsognathus longipes

> *Compsognathus longipes* Wagner, 1859 [Tithonian; Solnhofen Formation, Bayern (Bavaria), Germany]

= *Compsognathus corallestris* Bidar *et al.*, 1972 [Tithonian; mucronatum ammonoid zone, Provence-Alpes-Côte d'Azur, France]

Generic name: *Compsognathus* ← {(Gr.) κομψός/kompsos: elegant + γνάθος/gnathos: jaw}

Specific name: *longipes* ← {(Lat.) longus: long + pēs: foot, leg}.

Etymology: long-legged, **elegant jaw**

Taxonomy: Theropoda: Coelurosauria: Compsognathidae: Compsognathinae

Notes: *Compsognathus* was a small, bipedal animal with long hind legs and a longer tail.

Compsosuchus solus

> *Compsosuchus solus* von Huene & Matley, 1933 [Maastrichtian; Lameta Formation, Madhya Pradesh, India]

Generic name: *Compsosuchus* ← {(Gr.) κομψός/kompsos: delicate, elegant + Σοῦχος/ Souchos: crocodile}

Specific name: *solus* ← {(Lat.) sōlus: alone, only}.

Etymology: alone, **delicate crocodile**

Taxonomy: Theropoda: Ceratosauria: Noasauridae

Concavenator corcovatus

> *Concavenator corcovatus* Ortega *et al.*, 2010 [Barremian; La Huérguina Formation, Cuenca, Spain]

Generic name: *Concavenator* ← {(Lat. place-name) Conca: the Spanish province of Cuenca + (Lat.) vēnātor: hunter} (*1)

Specific name: *corcovatus* ← {corcovatus*: hump-backed}; "referring to the hump-like structure formed by the elongation of two presacral vertebrae" (*1). * (Sp.) corcova: hump.

Etymology: hump-backed, **Cuenca hunter**

Taxonomy: Theropoda: Carnosauria: Carcharodontosauridae

Notes: *Concavenator* had structures which look like quill knobs on its ulna (*1).

[References: (*1) Ortega F, Escaso F, Sanz JL (2010). A bizarre, humped Carcharodontosauria (Theropoda) from the Lower Cretaceous of Spain. *Nature* 467(7312): 203–206.]

Conchoraptor gracilis

> *Conchoraptor gracilis* Barsbold, 1986 [Maastrichtian; Nemegt Formation, Ömnögovi, Mongolia]

Generic name: *Conchoraptor* ← {(Gr.) κόγχη/ konchē: conch, mussel + (Lat.) raptor: plunderer}

Specific name: *gracilis* ← {(Lat.): gracilis: gracile, slender}.

Etymology: gracile, **conch plunderer**

Taxonomy: Theropoda: Coelurosauria: Oviraptoridae

[References: Barsbold, R. (1986). Raubdinosaurier Oviraptoren. In: Vorobyeva EL (ed). *Herpetologische Untersuchungen in der Mongolischen Volksrepublik.* Akademia Nauk SSSR, Moscow: 210–223.]

Concornis lacustris [Avialae]

> *Concornis lacustris* Sanz & Buscalioni, 1992 [Barremian; La Huérguina Formation, Cuenca, Spain]

Generic name: *Concornis* ← {(Lat. place-name) Conca: the name of the Cuenca region + (Gr.) όρνις/ornis: bird} (*1)

Specific name: *lacustris* ← {(Lat.) lacus; lake, pond}; referring to its possible lacustrine habitat (*1).

Etymology: lacustrine, **bird from Cuenca province**

Taxonomy: Theropoda: Avialae: Enantiornithes: Avisauridae

[References: (*1) Sanz JL, Buscalioni AD (1992). A new bird from the Early Cretaceous of Las Hoyas, Spain, and the early radiation of birds. *Palaeontology* 35(4): 829–845.]

Condorraptor currumili

> *Condorraptor currumili* Rauhut, 2005 [Toarcian–Bajocian; Cañadón Asfalto Formation, Chubut, Argentina]

Generic name: *Condorraptor* ← {(place-name) Condor: the village of Cerro Cóndor + (Lat.) raptor: robber, snatcher} (*1)

Specific name: *currumili* ← {(person's name) Currumil + -ī}; in honor of Hipólito Currumil, the land-owner and discoverer of the locality (*1).

Etymology: Currumil's **robber from Cerro Cóndor village**

Taxonomy: Theropoda: Tetanurae: Megalosauroidea: Piatnitzkysauridae

cf. Theropoda: Tetanurae (*1)

Notes: Piatnitzkysauridae Carrano *et al.*, 2012 consists of *Condorraptor*, *Marshosaurus* and *Piatnitzkysaurus*.

[References: (*1) Rauhut OWM (2005). Osteology and relationships of a new theropod dinosaur from the Middle Jurassic of Patagonia. *Paleontology* 48(1): 87–110.]

Confuciusornis sanctus [Avialae]

> *Confuciusornis sanctus* Hou *et al.*, 1995 [Barremian–Aptian; Yixian Formation, Liaoning, China]

= *Confuciusornis chuonzhous* Hou, 1997* [Barremian–Aptian; Yixian Formation; Liaoning, China] *(based on a paratype of *C. sanctus*)

= *Confuciusornis suniae* Hou, 1997 [Barremian–Aptian; Yixian Formation, Liaoning, China]

= *Jinzhouornis yixianensis* Hou *et al.*, 2002 [Barremian; Yixian Formation, Liaoning, China] (according to Chiappe *et al.*, 2018)

= *Jinzhouornis zhangjiyingia* Hou *et al.*, 2002 [Barremian; Yixian Formation, Liaoning, China] (according to Chiappe *et al.*, 2018)

Generic name: *Confuciusornis* ← {(Lat. Person's name) Cōnfūcius 孔夫子 [Pinyin: Kǒng fúzǐ]: Confucius, Chinese moral philosopher + (Gr.) ὄρνις/ornis: bird}

Specific name: *sanctus* ← {(Lat.): sānctus: holy}.

cf. 聖賢 [Pinyin: shèngxiàn]

Etymology: holy **Confucius' bird** [聖賢孔子鳥]

Taxonomy: Theropoda: Avialae: Confuciusornithidae

Other species:

> *Confuciusornis dui* Hou *et al.* 1999 [Barremian–Aptian; Yixian Formation, Liaoning, China]

> *Confuciusornis feducciai* Zhang *et al.*, 2009 [Barremian–Aptian; Yixian Formation, Liaoning, China]

> *Confuciusornis jianchangensis* Li *et al.*, 2010 [Aptian; Jiufotang Formation, Liaoning, China]

Notes: According to the study by Chinsamy *et al.* (2013), a short-tailed specimen was confirmed that it was female (*1).

[References: Hou L, Zhou Z, Gu Y, Zhang H (1995). *Confuciusornis sanctus*, a new Late Jurassic sauriurine bird from China. *Chinese Science Bulletin* 40(18): 1545–1551.; (*1) Chinsamy A, Chiappe LM, Marugán-Lobón JS, Chunling G, Fengjiao Z (2013). Gender identification of the Mesozoic bird *Confuciusornis sanctus*. *Nature Communications* 22(4): 1381.]

Convolosaurus marri

> *Convolosaurus marri* Andrzejewski *et al.*, 2019 [Aptian; Twin Mountains Formation, Texas, USA]

Generic name: *Convolosaurus* ← {convolo-: "flocking" (*1) [(Lat.) convolō: to fly together] + (Gr.) σαῦρος/sauros: lizard}; "referring to clusters of juvenile specimens" (*1).

Specific name: *marri* ← {(person's name) Marr + -ī}; "in honor of Dr. Ray H. Marr who produced the Society of Vertebrate Paleontology video "We are SVP" and "About the SVP Logo" posted on the SVP website (vertpaleo.org), and who is a strong proponent of students at Southern Methodist University (SMU)" (*1).

Etymology: Marr's **flocking lizard**

Taxonomy: Ornithischia: Neornithischia: Cerapoda: Ornithopoda

[References: (*1) Andrzejewski KA, Winkler DA, Jacobs LL (2019). A new basal ornithopod (Dinosauria: Ornithischia) from the Early Cretaceous of Texas. *PLoS ONE* 14(3): e0207935.]

Coronosaurus brinkmani

> *Centrosaurus brinkmani* Ryan & Russell, 2005 ⇒ *Coronosaurus brinkmani* (Ryan & Russell,

2005) Ryan *et al.*, 2012 [Campanian; Oldman Formation, Alberta, Canada]

Generic name: *Coronosaurus* ← {(Lat.) corōna: crown + (Gr.) σαῦρος/sauros: lizard}; "in reference to the multiple occurrences of extra epiparietals that cover the posterior margin of the parietal, giving it a crown-like appearance" (*2).

Specific name: *brinkmani* ← {(person's name) Brinkman + -ī}; "in honor of Donald Brinkman for his research illuminating the palaeoecology of the Late Cretaceous environments of Alberta" (*1).

Etymology: Brinkman's **crown lizard**

Taxonomy: Ornithischia: Coronosauria: Ceratopsia: Ceratopsidae: Centrosaurinae

Notes: A group Coronosauria Sereno, 1986 has not been named after *Coronosaurus* Ryan & Russell, 2005.

[References: (*1) Ryan MJ, Russell AP (2005). A new centrosaurine ceratopsid from the Oldman Formation of Alberta and its implications for centrosaurine taxonomy and systematics. *Canadian Journal of Earth Sciences* 42(7): 1369–1387. ; (*2) Ryan MJ, Evans DC, Shepherd KM (2012). A new ceratopsid from the Foremost Formation (middle Campanian) of Alberta. *Canadian Journal of Earth Sciences* 49(10): 1251–1262.]

Corythoraptor jacobsi

> *Corythoraptor jacobsi* Lü *et al.*, 2017 [Campanian–Maastrichtian; Nanxiong Formation, Ganzhou area, Jiangxi, China]

Generic name: *Corythoraptor* ← {(Gr.) κόρυς/korys*: helmet + (Lat.) raptor: robber}; "referring to a raptor bearing a 'cassowary-like crest' on its head" (*1). *k=c

Specific name: *jacobsi* ← {(person's name) Jacobs + -ī}; "in honor of Professor Louis L. Jacobs, who has contributed to dinosaur research and has given excellent mentoring to three authors (Lü J, Lee Y and Kobayashi Y) when they were Ph.D. students at Southern Methodist University, Dallas, Texas, USA" (*1).

Etymology: Jacob's **helmet robber**

Taxonomy: Theropoda: Oviraptorosauria: Oviraptoridae

[References: (*1) Lü J, Li G, Kundrát M, Lee Y-N, Sun Z, Kobayashi Y, Shen C, Teng F, Liu H (2017). High diversity of the Ganzhou oviraptorid fauna increased by a new "cassowary-like" crested species. *Scientific Reports* 7: 6393.]

Corythosaurus casuarius

> *Corythosaurus casuarius* Brown, 1914 [Campanian; Dinosaur Park Formation, Alberta, Canada]

= *Corythosaurus excavatus* Gilmore, 1923 [Campanian; Belly River Group, Alberta, Canada]

= *Corythosaurus bicristatus* Parks, 1935 [Campanian; Dinosaur Park Formation, Alberta, Canada]

=*Corythosaurus brevicristatus* Parks, 1935 [Campanian; Dinosaur Park Formation, Alberta, Canada]

= *Tetragonosaurus erectofrons* Parks, 1931 ⇒ *Procheneosaurus erectofrons* (Parks, 1931) [Campanian; Dinosaur Park, Formation, Alberta, Canada]

= *Tetragonosaurus cranibrevis* Sternberg, 1935 ⇒ *Procheneosaurus cranibrevis* (Sternberg, 1935) [Campanian; Dinosaur Park Formation, Alberta, Canada]

Generic name: *Corythosaurus* ← {(Gr.) κόρῠθο-/korytho- [κόρῠς/korys: helmet] + σαῦρος/sauros: lizard}; referring to "the extraordinary crest which rises above the brain-case like a Corinthian helmet and crest of a cassowary resembled" (*1).

Specific name: *casuarius* ← {casuarius: cassowary}; referring to the cassowary, a bird with a similar skull crest(*1).

Etymology: **lizard with Corinthian helmet** similar to cassowary's crest

Taxonomy: Ornithischia: Ornithopoda: Hadrosauridae: Lambeosaurinae: Lambeosaurini

Other species:

> *Stephanosaurus intermedius* Parks, 1923 ⇒ *Corythosaurus intermedius* (Parks, 1923) Parks, 1935 [Campanian; Dinosaur Park Formation, Alberta, Canada]

> *Lambeosaurus lambei* Parks, 1923 [Campanian; Dinosaur Park Formation, Alberta, Canada]

= *Corythosaurus frontalis* Parks, 1995 [late Campanian; Dinosaur Park Formation, Alberta, Canada]

[References: (*1) Brown B (1914). *Corythosaurus casuarius*, a new crested dinosaur from the Belly River Cretaceous, with provisional classification of the family Trachodontidae. *American Museum of Natural History Bulletin*. 33(35): 559–565.]

Craspedodon lonzeensis

> *Craspedodon lonzeensis* Dollo, 1883 (*nomen dubium*) [Santonian; Glauconie argileuse, Namur, Belgium]

Generic name: *Craspedodon* ← {(Gr.) κράσπεδον/kraspedon: edge + ὀδών/odōn (←ὀδούς/odūs: tooth)}; referring to the edge of its teeth.

Specific name: *lonzeensis* ← {(place-name) Lonzée

+ -ensis}

Etymology: **edged-tooth** from Lonzée

Taxonomy: Ornithischia: Marginocephalia: Ceratopsia: Neoceratopsia [*1]

cf. Ornithopoda: Iguanodontoidea (acc. *The Dinosauria 2nd edition*)

Craspedodon was long thought to be an iguanodontian.

Notes: *Craspedodon* is considered a *nomen dubium* since it is based on only a few teeth.

[References: Dollo ML (1883). Note sur les restes de dinosauriens rencontrés dans le Crétacé supérieur de la Belgique. *Bulletin du Musée Royale d'Histoire Naturelle de Belgique* 2: 205–221.; (*1) Godefroit P, Lambert O (2007). A re-appraisal of *Craspedodon lonzeensis* Dllo, 1883 from the Upper Cretaceous of Belgium: the first record of a neoceratopsian dinosaur in Europe? *Bulletin de l'Institut Royal des Sciences Naturelles de Belgique, Sciences de la Terre* 77: 83–93.]

Craterosaurus pottonensis

> *Craterosaurus pottonensis* Seeley, 1874 [Valanginian–Barremian; Woburn Sands Formation, UK]

Generic name: *Craterosaurus* ← {(Gr.) κρᾱτήρ/krater: bowl + σαῦρος/sauros: lizard}

Seeley mistook the fossil for the base of a cranium. It was identified as the front part of a neural arch, according to Nopcsa (1912).

Specific name: *pottonensis* ← {(place-name) Potton Sands or Potton bonebed + -ensis}

Etymology: **bowl lizard** from the Potton Sands

Taxonomy: Ornithischia: Thyreophora: Stegosauria

[References: Seeley HG (1874). On the base of a large lacertian cranium from Potton Sands, presumably dinosaurian. *Quarterly Jounal of the Geological Society of London.* 30: 690–692.]

Cratoavis cearensis [Avialae]

> *Cratoavis cearensis* Carvalho *et al*., 2015 [Aptian; Crato Formation, Araripe Basin, Ceará, Brazil]

Generic name: *Cratoavis* ← {(stratal name) Crato + (Lat.) avis: a bird}; referring to "the Crato Member lithostratigraphic unit", and "the zoological group Aves" [*1].

Specific name: *cearensis* ← {(place-name) Ceará State + ensis} [*1]

Etymology: Crato's **Aves (birds) from Ceará**

Taxonomy: Theropoda: Avialae: Ornithothoraces: Enantiornithes: Euenantiornithes

[References: (*1) Carvalho IS, Novas FE, Agnolín FL, Isasi MP, Freitas FI, Andrade JA (2015). A new genus and species of enantiornithine bird from the Early Cretaceous of Brazil. *Brazilian Journal of Geology* 45(2): 161–171.]

Crichtonpelta benxiensis

> *Crichtonsaurus benxiensis* Lü *et al*., 2007 (*nomen dubium*) ⇒ *Crichtonpelta benxiensis* (Lü *et al*., 2007) Arbour & Currie, 2015 [Cenomanian–Turonian; Sunjiawan Formation, Liaoning, China]

Generic name: *Crichtonpelta* ← {(person's name) Crichton + (Lat.) pelta = (Gr.) πέλτη/pelte: a small shield}; after Michael Crichton, author of *Jurassic Park* (1990).

Specific name: *benxiensis* ← {(Chin. place-name) Běnxī + -ensis}; referring to the Benxi Geological Museum.

Etymology: **Crichton's pelta**, from Benxi

Taxonomy: Ornithischia: Ankylosauria: Ankylosauridae

Notes: *Crichtonsaurus benxiensis* includes a well-preserved skull that can be differentiated from other ankylosaur species. The new combination *Crichtonpelta benxiensis* is proposed to receive the diagnostic material of '*Crichtonsaurus*' *benxiensis* [*1].

[References: (*1) Arbour VM, Currie PJ (2016). Systematics, phylogeny and palaeobiogeography of the ankylosaurid dinosaurs. *Journal of Systematic Palaeontology* 14(5): 385–444.]

Crichtonsaurus bohlini

> *Crichtonsaurus bohlini* Dong, 2002 [Cenomanian–Turonian, Sunjiawan Formation, Liaoning, China]

Generic name: *Crichtonsaurus* ← {(person's name) Michael Crichton + (Gr.) σαῦρος/sauros: lizard}; "dedicated to Michael Crichton, the auther of the book *Jurassic Park* (1990)" [*1].

Specific name: *bohlini* ← {(person's name) Bohlin + -ī}; "dedicated to Birger Bohlin, a well-known Swedish collector of vertebrate fossil including several ankylosaurs along the Silk Road, Northwest China" [*1].

Etymology: Bohlin's **Crichton's lizard**

Taxonomy: Ornithischia: Ankylosauria: Ankylosauridae: Ankylosaurinae

Other species:

> *Crichtonsaurus benxiensis* Lü *et al*., 2007 (*nomen dubium*) ⇒ *Crichtonpelta benxiensis* (Lü *et al*., 2007) Arbour & Currie, 2015 [Cenomanian; Sunjiawan Formation, Liaoning, China]

Notes: The type species of *Crichtonsaurus bohlini* lacks diagnostic characters and it is considered a *nomen dubium* [*2].

[References: (*1) Dong Z-M (2002). A new armored dinosaur (Ankylosauria) from Beipiao Basin, Liaoning Province, northeast China. *Vetebrata PalAsiatica* 40(4): 276–285.; (*2) Arbour VM, Currie PJ (2016). Systematics, phylogeny and palaeobiogeography of the ankylosaurid dinosaurs. *Journal of Systematic Palaeontology* 14(5): 385–444.]

Cristatusaurus lapparenti

> *Cristatusaurus lapparenti* Taquet & Russell, 1998 [Aptian; Elrhaz Formation, Agadez, Niger]

?= *Suchomimus tenerensis* Sereno *et al.*, 1998 ⇒ *Baryonyx tenerensis* (Sereno *et al.*, 1998) Sues *et al.*, 2002 [Aptian; Elrhaz Formation, Agadez, Niger]

Generic name: *Cristatusaurus* ← {(Lat.) crista: crest + (Gr.) σαῦρος/sauros: lizard}; referring to its sagittal crest on the snout.

Specific name: *lapparenti* ← {(person's name) Lapparent + -ī}; in honor of the late French paleontologist Albert-Félix de Lapparent.

Etymology: Lapparent's **crested lizard**

Taxonomy: Theropoda: Megalosauroidea: Spinosauridae

Notes: According to *The Dinosauria 2nd edition* (p.98), "Should *Cristatusaurus lapparenti* prove to be from the same species as *Suchomimus tenerensis*, the former species would have priority (on the basis of less than one month)".

[References: Taquet P, Russell DA (1998). New data on spinosaurid dinosaurs from the Early Cretaceous of the Sahara. *Comptes Rendus de l'Académie des Sciences, Série IIA. Sciences de la Terre et des Planètes* 327: 347–353.]

Crittendenceratops krzyzanowskii

> *Crittendenceratops krzyzanowskii* Dalman *et al.*, 2018 [Campanian; Fort Crittenden Formation, Arizona, USA]

Generic name: *Crittendenceratops* ← {(place-name) Crittenden + ceratops: 'horned-face'}; "referring to the Fort Crittenden Formation, the stratum from which the specimen came" (*1).

Specific name: *krzyzanowskii* ← {(person's name) Krzyzanowski + -ī}; "in honor of the late Stan Krzyzanowski, who discovered and collected the specimen" (*1).

Etymology: Krzyzanowski's **ceratopsian from Fort Crittenden Formation**

Taxonomy: Ornithischia: Ceratopsidae: Centrosaurinae: Nasutoceratopsini

[References: (*1) Dalman SG, Hodnett J-PM, Lichtig AJ, Lucas SG (2018). A new ceratopsid dinosaur (Centrosaurinae: Nasutoceratopsoni) from the Fort Crittenden Formation, Upper Cretaceous (Campanian) of Arizona. *New Mexico Museum of Natural History and Science Bulletin* 79: 141–164.]

Cruxicheiros newmanorum

> *Cruxicheiros newmanorum* Benson & Radley, 2010 [Bathonian; Norton Limestone Formation, Warwichshire, UK]

Generic name: *Cruxicheiros* ← {(Lat.) crux: cross + (Gr.) cheiros [χείρ/cheir: hand]}; "intended as 'cross hand', a version of the locality name" (*1).

Specific name: *newmanorum* ← {(person's name) Newman + -ōrum}; "after the Newman family, owners of Cross Hands Quarry, Warwickshire, United Kingdom" (*1).

Etymology: Newman family's **cross hand**

Taxonomy: Saurischia: Theropoda: Tetanurae

[References: (*1) Benson RBJ, Radley JD (2010). A new large-bodied theropod dinosaur from the Middle Jurassic of Warwickshire, United Kingdom. *Acta Paleontologica Polonica* 55(1): 35–42.]

Cryolophosaurus ellioti

> *Cryolophosaurus ellioti* Hammer & Hickerson, 1994 [Sinemurian–Pliensbachian; Hanson Formation, Transantarctic Mountains, Antarctica]

Generic name: *Cryolophosaurus* ← {(Gr.) κρύος/kryos: icy cold, frost + λόφος/lophos: crest + σαῦρος/sauros: lizard}

Specific name: *ellioti* ← {(person's name) Elliot + -ī}; in honor of the Ohio State University geologist David Elliot.

Etymology: Elliot's **frozen-crested lizard**

Taxonomy: Theropoda: Neotheropoda: Averostra: Tetanurae

cf. Theropoda: Avetheropoda: Carnosauria (acc. *The Dinosauria 2nd edition*)

Notes: *Cryolophosaurus* is characterized by a unique cranial crest. Its nickname is Elvisaurus.

[References: Hammer WR, Hickerson WJ (1994). A crested theropod dinosaur from Antarctica. *Science* 264(5160): 828–830.]

Cryptosaurus eumerus

> *Cryptosaurus eumerus* Seeley, 1869*, 1875 (*nomen dubium*) ⇒ *Cryptodraco eumerus* (Seeley, 1875) Lydekker, 1889 [late Oxfordian; Ampthill Clay Formation (Oxford Clay), England, UK]

*This name had been a *nomen nudum* until 1875.

Generic name: *Cryptosaurus* Seeley, 1869 ← {(Gr.) κρύπτός/kryptos: hidden + σαῦρος/sauros: lizard}

Specific name: *eumerus* ← {(Gr.) εὖ-/eu-: good + μηρός/mēros: thigh}

Etymology: **hidden lizard** with well-formed thigh
Taxonomy: Ornithischia: Thyreophora: Ankylosauria
Notes: *Cryptodraco* was unnecessary replacement name, because "*Cryptosaurus*" had not been preoccupied by another animal (*Cystosaurus*).

Cumnoria prestwichii

> *Iguanodon prestwichii* Hulke, 1880 ⇒ *Cumnoria prestwichii* (Hulke, 1880) Seeley, 1888 ⇒ *Camptosaurus prestwichii* (Hulke, 1880) Lydekker, 1889 [Kimmeridgian; Kimmeridge Clay Formation, Oxfordshire, UK]

Generic name: *Cumnoria* ← {(place-name) Cumnor Hurst + -ia}
Specific name: *prestwichii* ← {(person's name) Prestwich + -iī}; in honor of English geologist Joseph Prestwich.
Etymology: Prestwich's **Cumnor's one**
Taxonomy: Ornithischia: Ornithopoda: Styracosterna

Cuspirostrisornis houi [Avialae]

> *Cuspirostrisornis houi* Hou, 1997 [Aptian; Jiufotang Formation, Liaoning, China]

Generic name: *Cuspirostrisornis* ← {(Lat.) "Cuspirostris: sharp-beaked" (*1) [cuspis: point, pointed end + rōstrum: beak] + (Gr.) ὄρνις/ornis: bird}
Specific name: *houi* ← {(Chin. person's name) Hóu侯 + -ī}; "in honor of the collector of the specimen, Mr. Jinfeng Hou, who is also the premier illustrator at the Institute of Vertebrate Paleontology and Paleoanthropology" (*1), Beijng.
Etymology: Hou's **sharp-beaked bird** [侯氏尖嘴鳥]
Taxonomy: Theropoda: Avialae: Ornithothoraces: Enantiornithes: Avisauridae

[References: (*1) Hou L (1997). *Mesozoic Birds of China.* Phoenix Valley Provincial Aviary of Taiwan, Taipei [In Chinese]. Published by the Phoenix Valley Provincial Aviary, Taiwan.]

D

Daanosaurus zhangi

> *Daanosaurus zhangi* Ye *et al.*, 2005 [Oxfordian–Tithonian; Dashanpu Formation, Sichuan, China]

Generic name: *Daanosaurus* ← {(Chin. place-name) Dàān大安 + (Gr.) σαῦρος/sauros: lizard}
Specific name: *zhangi* ← {(Chin. person's name) Zhāng張 + -ī}; in honor of Chinese paleontologist Fucheng Zhang.
Etymology: Zhang's **Daan lizard** [張氏大安龍]
Taxonomy: Saurischia: Sauropodomorpha: Brachiosauridae

[References: Ye Y, Gao Y, Jiang S (2005). A new genus of sauropod from Zigong, Sichuan [四川自貢蜥足類—新属]. *Vertebrata PalAsiatica* 43(3):180–181.]

Dacentrurus armatus

> *Omosaurus armatus* Owen, 1875 (preoccupied) ⇒ *Stegosaurus armatus* (Owen, 1875) Lydekker, 1890 ⇒ *Dacentrurus armatus* (Owen, 1875) Lucas, 1902 ⇒ *Dacentrurosaurus armatus* (Owen, 1875) Hennig, 1925 [Kimmeridgian; Kimmeridge Clay Formation, Wiltshire, UK]

= *Omosaurus lennieri* Nopcsa, 1911 ⇒ *Dacentrurus lennieri* (Nopcsa, 1911) Hennig, 1915 [Kimmeridgian; Argiles d'Octeville Formation, Haute-Normandie, France]

= *Astrodon pusillus* Lapparent & Zbyszewski, 1957 [Kimmeridgian; Lisboa, Portugal]

Generic name: *Dacentrurus* ← {(Gr.) δᾰ-/da-: *intensive Prefix*, very + κέντρον/kentron: sharp point + οὐρά/oura [ūrā]: tail}; referring to the tail spike.
Specific name: *armatus* ← {(Lat.) armātus: equipped with armour}
Etymology: **tail with very sharp points**, in armour
Taxonomy: Ornitischia: Thyreophora: Stegosauria: Stegosauridae
Other species:

> *Omosaurus hastiger* Owen, 1877 ⇒ *Stegosaurus hastiger* (Owen, 1877) Lydekker, 1890 ⇒ *Dacentrurus hastiger* (Owen, 1877) Hennig, 1915 (*nomen dubium*) [Kimmeridgian; Kimmeridge Clay Formation, England, UK]

> *Omosaurus phillipsi* Seeley, 1893 ⇒ *Dacentrurus phillipsi* (Seeley, 1893) Hennig, 1915 (*nomen dubium*) [Oxfordian; Calcareous Grit Formation, England, UK]

> *Omosaurus durobrivensis* Hulke, 1887 (*nomen occupatum*) ⇒ *Stegosaurus durobrivensis* (Hulke, 1887) Hulke, 1887 ⇒ *Dacentrurus durobrivensis* (Hulke, 1887) Hennig, 1915 ⇒ *Lexovisaurus durobrivensis* (Hulke, 1887) Hoffstetter, 1957 (*nomen dubium*) [middle Callovian; Oxford Clay Formation, Cambridgeshire, England, UK]

Notes: Since *Omosaurus* had been preoccupied by a phytosaur, it was renamed in 1902.

According to *The Dinosauria 2nd edition*, *Omosaurus hastiger* Owen, 1877 and *Astrodon pusillus* Lapparent & Zbyszewski, 1957 were regarded as synonyms of *Dacentrurus armatus*. Cobos *et al.* (2010) proposed that *Miragaia* is a junior synonym of *Dacentrurus*, but Coste & Mateus (2019) affirmed the validity of *Miragaia longicollum*.

[References: Owen R (1875). *Monographs on the Fossil Reptilia of the Mesozoic Formations Part II (Genera Bothriospondylus, Cetiosaurus, Omosaurus). The Palaeontographical Society*, London: 15–93.]

Daemonosaurus chauliodus

> *Daemonosaurus chauliodus* Sues *et al*., 2011 [Rhaetian; Chinle Formation, New Mexico, USA]

Generic name: *Daemonosaurus* ← {(Gr.) δάιμων/daimōn: evil spirit + σαῦρος/sauros: reptile}; "in allusion to legends about evil spirits at Ghost Ranch, New Mexico" [*1].

Specific name: *chauliodus* ← {(Gr.) χαυλιόδους/chauliodūs: prominent toothed} "with prominent teeth" [*1].

Etymology: prominent-toothed **evil reptile**

Taxonomy: Saurischia: Theropoda

[References: (*1) Sues H-D, Nesbitt SJ, Berman DS, Henrici CH (2011). A late-surviving basal theropod dinosaur from the latest Triassic of North America. *Proceedings of the Royal Society B: Bilogical Sciences* 278 (1723): 3459–3464.]

Dahalokely tokana

> *Dahalokely tokana* Farke & Sertich, 2013 [Turonian; Ambolafotsy Formation, Madagascar]

Generic name: *Dahalokely* ← {(Malagasy) dahalo: bandit + kely: small}; "referring to the small size of the animal relative to many abelisauroids" [*1].

Specific name: *tokana* ← {(Malagasy) tokana: lonely}; "referring to the organism's isolation on the landmass of Indo-Madagascar" [*1].

Etymology: lonely **small bandit**

Taxonomy: Theropoda: Abelisauroidea

[References: (*1) Farke AA, Sertich JJW (2013). An abelisauroid theropod dinosaur from the Turonian of Madagascar. *PLoS ONE* 8 (4): e62047.]

Dakosaurus maximus [Loricata]

> *Geosaurus maximus* Plieninger, 1846 ⇒ *Dakosaurus maximus* (Plieninger, 1846) von Quenstedt, 1856 [Kimmeridgian; Kimmeridge Clay Formation, UK]

= *Dakosaurus gracilis* Quenstedt, 1885 ⇒ *Megalosaurus gracilis* (Quenstedt, 1885) Douville, 1885 [uncertain]

= *Dakosaurus lissocephalus* Seeley, 1869 [Kimmeridgian; Kimmeridge Clay Formation, UK]

= *Megalosaurus schnaitheimii* Bunzel, 1871 [Kimmeridgian; Mergelstätten Formation, Germany]

Generic name: *Dakosaurus* ← {(Gr.) δάκος/dakos: an animal of which the bite is dangerous, a noxious beast + σαῦρος/sauros: lizard}

Specific name: *maximus* ← {(Lat.) māximus (*sup.* of [magnus: large, great])}

Etymology: greatest **biter lizard**

Taxonomy: Reptilia: Thalattosuchia: Metriorhynchidae

Other species:

> *Dakosaurus andiniensis* Vignaud & Gasparini, 1996 [Tithonian; Vaca Muerta Formation, Mendoza, Argentina]

> *Dacosaurus* /*Dakosaurus lapparenti* Debelmas & Strannoloubsky, 1957 ⇒ *Geosaurus lapparenti* (Debelmas & Strannoloubsky, 1957) Young & de Andrade, 2009 [Valanginian; France]

> *Dakosaurus carpenteri* Wilkinson *et al*., 2008 ⇒ *Geosaurus carpenteri* Young & de Andrade, 2009 [Kimmeridgian; Kimmeridge Clay Formation, Westbury, UK]

> *Aggiosaurus nicaeensis* Ambayrac, 1913 ⇒ *Megalosaurus nicaeensis* (Ambayrac, 1913) Romer, 1956 ⇒ *Dakosaurus nicaeensis* (Ambayrac, 1913) Young & de Andrade, 2009 [Oxfordian; Cap d'Aggio-La Turbie, France]

> *Liodon paradoxus* Wagner, 1853 ⇒ *Dakosaurus paradoxus* (Wagner, 1853) Fraas, 1902 [Tithonian; Solnhofen Formation, Bayern, Germany]

Dakotadon lakotaensis

> *Iguanodon lakotaensis* Weishampel & Bjork, 1989 ⇒ *Dakotadon lakotaensis* (Weishampel & Bjork, 1989) Paul, 2008 [Barremian; Lakota Formation, South Dakota, USA]

Generic name: *Dakotadon* ← {(place-name) Dakota: "the State of the holotype's origin" + (Gr.) ὀδών/odōn: tooth} [*1]

Specific name: *lakotaensis* ← {(stratal name) Lakota Formation + -ensis}

Etymology: **Dakota tooth** from Lakota Formation

Taxonomy: Ornithischia: Ornithopoda: Iguanodontia: Ankylopollexia: Styracosterna

[References: Paul GS (2008). A revised taxonomy of the iguanodont dinosaur genera and species. *Cretaceous Research* 29(2): 192–216.]

Dakotaraptor steini

> *Dakotaraptor steini* DePalma *et al.,* 2015 [Maastrichtian; Hell Creek Formation, Harding, South Dakota, USA]

Generic name: *Dakotaraptor* ← {(place-name / demonym) Dakota + (Lat.) raptor: plunderer};

"referring to the geographic location of the discovery as well as the Dakota First Nations Tribe" (*1).
Specific name: *steini* ← {(person's name) Stein + -ī}; "in honor of paleontologist Walter W. Stein" (*1).
Etymology: Stein's **Dakota plunderer**
Taxonomy: Theropoda: Maniraptora: Dromaeosauridae
[References: (*1) DePalma RA, Burnham DA, Martin LD, Larson PL, Bakker RT (2015). The first giant raptor (Theropoda: Dromaeosauridae) from the Hell Creek Formation. *Paleontological Contributions* 2015(14): 1–16.]

Daliansaurus liaoningensis
> *Daliansaurus liaoningensis* Shen *et al.*, 2017 [Barremian, Yixian Formation, Liaoning, China]
Generic name: *Daliansaurus* ← {(Chin. place-name) Dàlián大連 + (Gr.) σαῦρος/sauros: lizard}; referring to the city of Dalian in the south of Liaoning Province (*1).
Specific name: *liaoningensis* ← {(Chin. place-name) Liáoníng遼寧 +-ensis}
Etymology: **Dalian lizard** from Liaoning
Taxonomy: Theropoda: Maniraptora; Troodontidae; Sinovenatorinae (*1)
Notes: Troodontidae is a group of small, feathered non-avian theropod dinosaurs (*1).
[References: (*1) Shen C, Lü J, Liu S, Kundrát M, Brusatte SL, Gao H (2017). A new troodontid dinosaur from the Lower Cretaceous Yixian Formation of Liaoning Province, China. *Acta Geologica Sinica* (English Edition) 91(3): 763–780.]

Dandakosaurus indicus
> *Dandakosaurus indicus* Yadagiri, 1982 [Toarcian; Kota Formation, Andhra Pradesh, India]
Generic name: *Dandakosaurus* ← {(place-name) Dandaka-aranya [(Sanscrit) aranya: forest] + (Gr.) σαῦρος/sauros: lizard}
Specific name: *indicus* ← {(Lat.) Indicus: of India}.
Etymology: **Dandak forest lizard** from India
Taxonomy: Theropoda: Averostra
Notes: According to *The Dinosauria 2nd edition*, this genus as a *nomen dubium* because little is known about it.

Danubiosaurus anceps
> *Struthiosaurus austriacus* Bunzel, 1871 (*1) [early Campanian; Grünbach Formation, Gosau Group, Niederosterreich, Austria]
= *Danubiosaurus anceps* Bunzel, 1871 (*1) [early Campanian; Grünbach Formation, Gosau Group, Niederosterreich, Austria]
Generic name: *Danubiosaurus* ← {(Lat. place-name) Danubio- [Dānubius: the Danube River] + (Gr.) σαῦρος/sauros: lizard}
Specific name: *anceps* ← {(Lat.) anceps: two-headed, doubtful}
Etymology: uncertan **lizard from the Danube River**
Taxonomy: Ornithischia: Thyreophora: Ankylosauria: Nodosauridae
Notes: According to *The Dinosauria 2nd edition*, *Danubiosaurus anceps* is considered a synonym of *Struthiosaurus*, which has priority.
[References: (*1) Bunzel E (1871). Die Reptilfauna der Gosauformation in der Neuen Welt bei Wiener-Neustadt. *Abhandlungen der Kaiserlich-Königlichen Geologischen Reichsanstalt* 5: 1–18.]

Darwinopterus modularis [Pterosauria]
> *Darwinopterus modularis* Lü *et al.*, 2010 [Bathonian–Oxfordian; Tiaojishan Formation, Liaoning, China]
Generic name: *Darwinopterus* ← {(person's name) Darwin + (Gr.) πτέρυξ/pteryx = πτερόν/pteron: wing}; in honor of biologist Charles Darwin for "the anniversaries of his birth (200 years) and the publication of *On the origin of species* (150 years)" (*1).
Specific name: *modularis* ← {(Lat.) modulāris: meaning "composed of interchangeable units" (*1)}.
Etymology: modular **Darwin's wing**
Taxonomy: Archosauria: Ornithodira: Pterosauria
Other species:
> *Darwinopterus linglongtaensis* Wang *et al.*, 2010 [Bathonian–Kimmeridgian; Tiaojishan Formation, Liaoning, China]
> *Darwinopterus robustodens* Lü *et al.*, 2011 [Bathonian–Kimmeridgian; Tiaojishan Formation, Liaoning, China]
Notes: Over 30 fossil specimens have been collected from the Tiaojishan Formation. *Darwinopterus* was described as a transitional fossil between two pterosaurs, rhamphorhyncoid and pterodactyloid (*2).
[References: (*1) Lü J, Unwin DM, Jin X, Liu Y, Ji Q (2010). Evidence for modular evolution in a long-tailed pterosaur with a pterodactyloid skull. *Proceeding of the Royal Society B. Biological Sciences* 277(1680): 383–389.; (*2) Hecht J (2011). Did pterosaurs fly out of their eggs? New Scientist online edition, 20 Jan 2011. Accessed online 21 Jan 2011.〈https://www.newscientist.com/article/dn20011-did-pterosaurs-fly-out-of-their-eggs.html〉]

Darwinsaurus evolutionis
> *Iguanodon fittoni* Lydekker, 1889 ⇒

Hypselospinus fittoni (Lydekker, 1889) Norman, 2010 ⇒ *Wadhurstia fittoni* (Lydekker, 1889) Carpenter & Ishida, 2010 [Valanginian; Wadhurst Clay Formation, East Sussex, UK]
= *Darwinsaurus evolutionis* Paul, 2012 [Valanginian; Wadhurst Clay Formation, East Sussex, UK]

Generic name: *Darwinsaurus* ← {(person's name) Darwin + (Gr.) σαῦρος/sauros: lizard}; "for Charles Darwin" (*1).

Specific name: *evolutionis* ← {(Lat.) ēvolūtiōnis *gen.* [ēvolūtiō: an unrolling]}; "in recognition of Darwin's theory, with the diverse iguanodonts standing as an example of complex evolution via rapid speciation" (*1).

Etymology: evolutional **Darwin's lizard**

Taxonomy: Ornithischia: Ornithopoda: Styracosterna

Notes: According to Norman (2013), *Darwinsaurus evolutionis* is seen as a junior synonym of *Hypselospinus fittoni*.

[References: (*1) Paul G (2011). Notes on the rising diversity of iguanodont taxa, and iguanodonts named after Darwin, Huxley and evolutionary science. *Actas de V Jornadas Internacionales sobre Paleontologia de Dinosaurios y su Entorno, Salas de los Infantes, Burgos.* Colectivo de Arqueologico-Paleontologico de Salas de los Infantes (Burgos): 121–131.]

Dashanpusaurus dongi

> *Dashanpusaurus dongi* Peng *et al.*, 2005 [Bathonian–Oxfordian; Dashanpu Formation, Sichuan, China]

Generic name: *Dashanpusaurus* ← {(Chin. place-name) Dàshānpū大山鋪 + (Gr.) σαῦρος/sauros: lizard}

Specific name: *dongi* ← {(Chin. person's name) Dong 董 + -ī}; in honor of the Chinese paleontologist Dŏng Zhīmíng董枝明.

Etymology: Dong's **Dashanpu lizard**

Taxonomy: Saurischia: Sauropodomorpha: Sauropoda: Camarasauridae

Daspletosaurus torosus

> *Daspletosaurus torosus* Russell, 1970 ⇒ *Tyrannosaurus* (*Daspletosaurus*) *torosus* Paul, 1988* [Campanian; Oldman Formation, Alberta, Canada]
> **Tyrannosaurus torosus* Paul, 1988 has not been generally accepted.

Generic name: *Daspletosaurus* ← {(Gr.) δασπλῆτο-/dasplēto-: frightful [δασπλής/dasplēs] + σαῦρος/sauros: lizard}

Specific name: *torosus* ← {(Lat.) torōsus: muscular, fleshy}; "with reference to the large body" (*1).

Etymology: brawny, **frightful lizard**

Taxonomy: Theropoda: Coelurosauria: Tyrannosauridae: Tyrannosaurinae: Daspletosaurini

Other species:
?> *Daspletosaurus horneri* Carr *et al.*, 2017 [Campanian; Two Medicine Formation, Montana, USA]

Notes: An apex predator. The type specimen of *Daspletosaurus torosus* was discovered in Alberta by Charles M. Sternberg in 1921, and described by Dale Russell in 1970. Originally it was thought to be a specimen of *Gorgosaurus*.

[References: (*1) Russell DA (1970). Tyrannosaurs from the Late Cretaceous of western Canada. *National Museum of Natural Sciences, Publications in Paleontology* 1: 1–34.]

Datanglong guangxiensis

> *Datanlong guangxiensis* Mo *et al.*, 2014 [Aptian–Albian; Xinlong Formation, Datang, Guanxi, China]

Generic name: *Datanglong* ← {(Chin. place-name) Dàtáng 大塘: Datang Town and Datang Basin + lóng龍: dragon}

Specific name: *guangxiensis* ← {(Chin. place-name) Guăngxī広西 + -ensis}

Etymology: **Datang dragon** from Guangxi

Taxonomy: Theropoda: Allosauroidea: Allosauria: Carcharodontosauria

[References: Mo J, Zhou F, Li G, Zhen H, Cao C (2014). A new Carcharodontosauria (Theropoda) from the Early Cretaceous of Guanxi, Southern China. *Acta Geologica Sinica* (English Edition) 88(4): 1051–1059.]

Datonglong tianzhenensis

> *Datonglong tianzhenensis* Xu *et al.*, 2016 [Cenomanian–Campanian; Huiquanpu Formation, Tianzhen, Datong, Shanxi, China]

Generic name: *Datonglong* ← {(Chin. place-name) Dàtóng 大同: Datong City + lóng 龍: dragon}

Specific name: *tianzhenensis* ← {(Chin. place-name) Tiānzhèn天鎮: Tianzhen County + -ensis}

Etymology: **Datong dragon** from Tianzhen

Taxonomy: Ornithischia: Ornithopoda: Iguanodontia: Styracosterna: Hadrosauroidea

[References: Xu S-C, You H-L, Wang J-W, Wang S-Z, Yu J, Jia L (2016). A new hadrosauroid dinosaur from the Late Cretaceous of Tianzhen, Shanxi Province, China. *Vertebrata Pal Asiatica* 54(1): 67–78.]

Datousaurus bashanensis

> *Datousaurus bashanensis* Dong & Tang, 1984 [Bathonian–Oxfordian; lower Shaximiao Formation, Sichuan, China]

Generic name: *Datousaurus* ← {(Chin.) dàtóu大頭: big head | (Malay) 酋首: chieftain + (Gr.) σαῦρος/sauros: lizard}

Specific name: *bashanensis* ← {(Chin. place-name) Bāshān巴山* + -ensis}

*BāshānShŭshuĭ巴山蜀水is the idiom meaning mountains and rivers of Sichuan. Bashan is the beautiful name for Sichuan. 四川俗有巴山蜀水之美称 (*1).

Etymology: **big head (or chieftain) lizard** from Sichuan [巴山酋龍]

Taxonomy: Saurischia: Sauropodomorpha: Sauropoda: ?Mamenchisauridae

[References: (*1) Dong Z, Tang Z (1984). Note on a new Mid-Jurassic sauropod (*Datousaurus bashanensis* gen. *et* sp. nov.) from Sichuan Basin, China. *Vertebrata PalAsiatica* 22(1): 69–75.]

Daxiatitan binglingi

> *Daxiatitan binglingi* You *et al.*, 2008 [Aptian; Hekou Group, Gansu, China]

Generic name: *Daxiatitan* ← {(Chin. place-name) Dàxià大夏: Daxia River + (Gr. myth.) Τῑτάν/Tītan: giant}; Daxia: "name of a branch of the Yellow River running along the Linxia area of Gansu Province". Titan: "Greek mythological giants, symbolic of its great size" (*1).

Specific name: *binglingi* ← {(Chin.) Bĭnglíng炳靈: "one hundred thousand Buddhas" + -ī}; "referring to Bingling Temple炳靈寺, a famous attraction near the Liujiaxia Dam along the Yellow River, about 80 km southwest of Lanzhou, the capital city of Gansu Province" (*1).

Etymology: Bingling's **Daxia River giant** [炳靈大夏巨龍]

Taxonomy: Saurischia: Sauropoda: Titanosauriformes (*1)

[References: (*1) You HL, Li D-Q, Zhou L-Q, Ji Q (2008). *Daxiatitan binglingi*: a giant sauropod dinosaur from the Early Cretaceous of China. *Gansu Geology* 17(4): 1–10.]

Deinocheirus mirificus

> *Deinocheirus mirificus* Osmólska & Roniewicz, 1970 [Maastrichtian; Nemegt Formation, Nemegt Basin, Gobi Desert, Mongolia]

Generic name: *Deinocheirus* ← {(Gr.)δεινός/deinos: horrible + χειρός/cheiros [χείρ/cheir: hand]}; referring to "the large fore limbs and strong claws" (*1).

Specific name: *mirificus* ← {(Lat.) mīrificus: unusual, peculiar}; "because of the unusual structure of the fore limbs" (*1).

Etymology: unusual, **horrible hand**

Taxonomy: Theropoda: Coelurosauria: Ornithomimosauria: Deinocheiridae

cf. Theropoda: Carnosauria: Megalosauroidea: Deinocheiridae (*1)

Notes: An unusual, incomplete skeleton of a gigantic carnosaurian dinosaur was recovered in 1965 (*1) and the specimen was named in 1970. Most of the body skeletons of this dinosaur has been revealed by Lee *et al.* (2014). (*2)

[References: (*1) Osmólska H, Roniewicz E (1970). Deinocheiridae, a new family of theropod dinosaurs. *Palaeontologica Polonica* (21): 5–19.; (*2) Lee YN, Barsbold R, Currie PJ, Kobayashi Y, Lee HJ, Godefroit P, Escuillié FO, Chinzorig T (2014). Resolving the long-standing enigmas of a giant ornithomimosaur *Deinocheirus mirificus*. *Nature* 515(7526): 257–260.]

Deinodon horridus

> *Deinodon horridus* Leidy, 1856 (*nomen dubium*) ⇒ *Megalosaurus horridus* (Leidy, 1856) Leidy, 1857 (*nomen dubium*) [Campanian; Judith River Formation, Montana, USA]

= *Aublysodon lateralis* Cope, 1876 ⇒ *Deinodon lateralis* (Cope, 1876) Hay, 1902 [Fort Union Beds, Montana, USA]

= *Ornithomimus grandis* Marsh, 1890 ⇒ *Deinodon grandis* (Marsh, 1890) Osborn, 1916 ⇒ *Aublysodon grandis* (Marsh, 1890) von Huene, 1932 [Campanian; Eagle Sandstone Formation, Montana, USA]

= *Dryptosaurus kenabekides* Hay, 1899 ⇒ *Deinodon kenabekides* (Hay, 1899) Olshevsky, 1995 [Campanian; Judith River Formation, Montana, USA]

Generic name: *Deinodon* ← {(Gr.) δεινός/deinos: terrible + ὀδών/odōn: tooth}

Specific name: *horridus* ← {(Lat.) horridus: rough, terribe, frightful}

Etymology: frightful, **terrible tooth**

Taxonomy: Theropoda: Coelurosauria: Tyrannosauridae: Deinodontinae

Other species:

> *Laelaps explanatus* Cope, 1876 ⇒ *Dryptosaurus explanatus* (Cope, 1876) Marsh, 1877 ⇒ *Deinodon explanatus* (Cope, 1876) Lambe, 1902 ⇒ *Dromaeosaurus explanatus* (Cope, 1876) (*nomen dubium*) Hay, 1930 [Ford Union Beds, Montana, USA]

> *Laelaps cristatus* Cope, 1877 ⇒ *Dryptosaurus cristatus* (Cope, 1877) Hay, 1902 ⇒*Deinodon cristatus* (Cope, 1877) Osborn, 1902 ⇒ *Dromaeosaurus cristatus* (Cope, 1877) Matthew & Brown, 1922 ⇒ *Troodon cristatus* (Cope, 1877) Olshevsky, 1955 [Montana, USA]

> *Laelaps laevifrons* Cope, 1877 ⇒ *Deinodon*

laevifrons (Cope, 1877) Osborn, 1902 ⇒ *Dromaeosaurus laevifrons* (Cope, 1877) Matthew & Brown, 1922 (reclassified) [Campanian; Dinosaur Park Formation, Alberta, Canada]
> *Aublysodon amplus* Marsh, 1892 (*nomen dubium*) ⇒ *Deinodon amplus* (Marsh, 1892) Kuhn, 1939 ⇒ *Aublysodon amplus* (reclassified)
> *Aublysodon cristatus* Marsh, 1892 (*nomen dubium*) ⇒ *Deinodon cristatus* (Marsh, 1892) Hay, 1902 ⇒ *Aublysodon cristatus* (reclassified)
> *Gorgosaurus libratus* Lambe, 1914 ⇒ *Deinodon libratus* (Lambe, 1914) Matthew & Brown, 1922 ⇒ *Albertosaurus libratus* (Lambe, 1914) Russell, 1970 [Campanian, Dinosaur Park Formation, Alberta, Canada]
= *Laelaps incrassatus* Cope, 1876 ⇒ *Dryptosaurus incrassatus* (Cope, 1876) Hay, 1902 ⇒ *Deinodon incrassatus* (Cope, 1876) Osborn, 1902 (*nomen dubium*) ⇒ "*Albertosaurus*" *incrassatus* (Cope, 1876) von Huene, 1932 [Campanian; Judith River Formation, Montana, USA]
= *Laelaps falculus* Cope, 1876 ⇒ *Deinodon falculus* (Cope, 1876) Osborn, 1902 ⇒ *Dromaeosaurus falculus* (Cope, 1876) Olshevsky, 1978-1979 (*nomen dubium*) ⇒ *Dryptosaurus falculus* (Cope, 1876) [Campanian; Judith River Formation, Montana, USA]
= *Laelaps hazenianus* Cope, 1877 ⇒ *Deinodon hazenianus* (Cope, 1877) Osborn, 1902 (*nomen dubium*) ⇒ *Dryptosaurus hazenianus* (Cope, 1877) Hay, 1902 (*nomen dubium*) [Campanian; Judith River Formation, Montana, USA]
= *Gorgosaurus sternbergi* Matthew & Brown, 1923 ⇒ *Deinodon sternbergi* (Matthew & Brown, 1923) Kuhn, 1965 ⇒ *Albertosaurus sternbergi* (Matthew & Brown, 1923) Russell, 1970 [Campanian; Dinosaur Park Formation, Alberta, Canada]
> *Albertosaurus sarcophagus* Osborn, 1905 ⇒ *Deinodon sarcophagus* (Osborn, 1905) Matthew & Brown, 1922 [Campanian; Horseshoe Canyon Formation, Alberta, Canada]
= *Albertosaurus arctunguis* Parks, 1929 ⇒ *Deinodon arctunguis* (Parks, 1928) Kuhn, 1939 [Campanian; Horseshoe Canyon Formation, Alberta, Canada]
> *Tyrannosaurus bataar* Maleev, 1955 ⇒ *Tarbosaurus bataar* (Maleev, 1955) Rozhdestvensky, 1965 ⇒ *Jenghizkhan bataar* (Maleev, 1955) Olshevsky *vide* Olshevsky *et al.*, 1995 [late Campanian; Nemegt Formation, Ömnögovi, Mongolia]
= *Albertosaurus periculosus* Riabinin, 1930 ⇒ *Deinodon periculosus* (Riabinin, 1930) Kuhn, 1965 ⇒ *Alectrosaurus periculosus* (Riabinin, 1930) Olshevsky, 1991 ⇒ *Jenghizkhan periculosus* (Riabinin, 1930) Olshevsky, 1995 ⇒ *Tarbosaurus*? *periculosus* (Riabinin, 1930) Olshevsky, 1995 [late Maastrichtian; Yuliangze Formation, Heilongjiang, China]
= *Gorgosaurus lancinator* Maleev, 1955 ⇒ *Deinodon lancinator* (Maleev, 1955) Kuhn, 1965 ⇒ *Aublysodon lancinator* (Maleev, 1955) Charig, 1967 [Maastrichtian; Nemegt Formation, Ömnögovi, Mongolia]
= *Gorgosaurus novojilovi* Maleev, 1955 ⇒ *Deinodon novojilovi* (Maleev, 1955) Maleev, 1964 or Kuhn, 1965 ⇒ *Maleevosaurus novojilovi* (Maleev, 1955) Pickering, 1984 ⇒ *Maleevosaurus novojilovi* (Maleev, 1955) Carpenter, 1992 [Maastrichtian; Nemegt Formation, Ömnögovi, Mongolia]
> *Gorgosaurus lancensis* Gilmore, 1946 ⇒ *Deinodon lancensis* (Gilmore, 1946) Kuhn, 1965 ⇒ *Aublysodon lancensis* (Gilmore, 1946) ⇒ *Albertosaurus lancensis* (Gilmore, 1946) Russell, 1970 ⇒*Nanotyrannus lancensis* (Gilmore, 1946) Bakker *et al.*, 1988 (reclassified) [Maastrichtian; Hell Creek Formation, Montana, USA]

Notes: According to *the Dinosauria 2nd edition*, *Deinodon horridus* was considered a probable synonym of *Gorgosaurus libratus*.

[References: (*1) Sahni A (1972). The vertebrate fauna of the Judith River Formation, Montana. *Bulletin of the American Museum of Natural History* 147(6): 352–412.; Matthew WD, Brown B (1922). The family Deinodontidae, with notice of a new genus from the Cretaceous of Alberta. *Bulletin of the American Museum of Natural History* 46(6): 367–385.]

Deinonychus antirrhopus

> *Deinonychus antirrhopus* Ostrom, 1969 ⇒ *Velociraptor antirrhopus* (Ostrom, 1969) Paul, 1988 [late Aptian–middle Albian; Cloverly Formation, Montana, USA]

Generic name: *Deinonychus* ← {(Gr.) δεινός/deinos: terrible + ὄνυχος/onychos [ὄνυξ/onyx: claw or talon]} [*1]; named for the sickle claw on its foot.

Specific name: *antirrhopus* ← {(Gr.) ἀντίρροπος/antirrhopos: counterbalancing}; in reference to the unusual adaptation of the caudal vertebrae [*1].

Etymology: counterbalancing, **terrible claw**

Taxonomy: Theropoda: Coelurosauria: Dromaeosauridae

Notes: Paleontologist Ostrom's study of *Deinonychus* ignited the debate on whether dinosaurs were warm-blooded or cold-blooded.

[References: (*1) Ostrom JH (1969). A new theropod dinosaur from the Lower Cretaceous of Montana.

Postilla 128: 1–17.; Ostrom JH (1969). Osteology of *Deinonychus antirrhopus*, an unusual theropod from the Lower Cretaceous of Montana. *Peabody Museum of Natural History Bulletin* 30: 1–165.]

Delapparentia turolensis

> *Iguanodon bernissartensis* Boulenger, 1881 [Barremian; Sainte-Barbe Clays Formation, Bernissart, Hainaut, Belgium]

= *Delapparentia turolensis* Ruiz-Omeñaca, 2011 [Barremian; Camarillas Formation, Teruel, Aragon, Spain]

Generic name: *Delapparentia* ← {(person's name) de Lapparent + -ia}; in honor of French paleontologist Albert-Félix de Lapparent (1905–1975).

Specific name: *turolensis* ← {(Lat. place-name) Turia [= Teruel] + -ensis}

Etymology: **de Lapparent's one** from Turia (Teruel)

Taxonomy: Ornithischia: Ornithopoda: Hadrosauriformes: Iguanodontidae

[References: Ruiz-Omeñaca JI (2011). *Delapparentia turolensis* nov. gen *et* sp., un nuevo dinosaurio iguanodontoideo (Ornithischia: Ornithopoda) en el Cretácico Inferior de Galve. *Estudios Geológicos* 67(1): 83–110.]

Deltadromeus agilis

> *Deltadromeus agilis* Sereno *et al.*, 1996 [Cenomanian; Aoufous Formation, Er Rachida, Morocco]

Generic name: *Deltadromeus* ← {(Gr.) δέλτα/delta: delta + δρομεύς/dromeus: runner}; referring to "the nonmarine deltaic facies, Kem Kem beds, where the fossils were found" [*1].

Specific name: *agilis* ← {(Lat.) agilis: quick}

Etymology: swift, **deltaic runner**

Taxonomy: Theropoda: Coelurosauria

[References: (*1) Sereno PC, Dutheil DB, Iarochene M, Larsson HCE, Lyon GH, Magwene PM, Sidor CA, Varricchio DJ, Wilson JA (1996). Predatory dinosaurs from the Sahara and Late Cretaceous faunal differentiation. *Science* 272(5264): 986–991.]

Demandasaurus darwini

> *Demandasaurus darwini* Fernández-Baldor *et al.*, 2011 [Barremian–Aptian; Castrillo dela Reina Formation, Burgos, Spain]

Generic name: *Demandasaurus* ← {(place-name) "the Sierra de la Demanda: the mountain chain" + (Gr.) σαῦρος/sauros: "lizard, reptile"} [*1]

Specific name: *darwini* ← {(person's name) Darwin + -ī}; "in honor of the naturalist Charles R. Darwin (1809–1882)" [*1].

Etymology: Darwin's **Demanda lizard**

Taxonomy: Saurischia: Sauropodomorpha: Rebbachisauridae

[References: (*1) Fernández-Baldor FT, Canudo JI, Huerta P, Montero D, Suberbiola XP, Salgado L (2011). *Demandasaurus darwini*, a new rebbachisaurid sauropod from the Early Cretaceous of the Iberian Peninsula. *Acta Palaeontologica Polonica* 56(3): 535–552.]

Denversaurus schlessmani

> *Denversaurus schlessmani* Bakker, 1988 [Maastrichtian; Lance Formation, South Dakota, USA]

Generic name: *Denversaurus* ← {(place-name) Denver + (Gr.) σαῦρος/sauros: lizard}; referring to the Denver Museum of National History.

Specific name: *schlessmani* ← {(person's name) Schlessman + -ī}; in honor of Lee E. Schlessman, a benefactor of the museum and the founder of the Schlessman Family Foundation.

Etymology: Schlessman's **Denver lizard**

Taxonomy: Ornithischia: Thyreophora: Ankylosauria: Nodosauridae

Notes: According to *The Dinosauria 2nd eition* (2004), *Denversaurus schlessmani* was considered as a synonym of *Edmontonia longiceps* Sternberg, 1928, but now it seems that *Denversaurus* is thought to be a distinct nodosaurid genus. According to Burns (2015), *Denversaurus* is a valid taxon.

[References: Bakker RT (1988). Review of the Late Cretaceous nodosauroid Dinosauria: *Denversaurus schlessmani*, a new armor-plated dinosaur from the Latest Cretaceous of South Dakota, the last survivor of the nodosaurians, with comments on Stegosaur-Nodosaur relationships. *Hunteria* 1(3): 1–23.]

Diabloceratops eatoni

> *Diabloceratops eatoni* Kirkland *et al.*, 2010 [Campanian; Wahweap Formation, Utah, USA]

Generic name: *Diabloceratops* ← {(Sp.) diablo: devil + (Latinized Greek) ceratops: horned-face [(Gr.) κέρας/keras: horn + ὤψ/ōps: face]}; "in reference to the pair of long sweeping spines on the back of the frill" [*1].

Specific name: *eatoni* ← {(person's name) Eaton + -ī}; "a patronym in honor of Jeffrey G. Eaton, a paleontologist at Weber State University in Ogden, Utah, in recognition for his extensive work on the Cretaceous vertebrate faunas of southern Utah, and his role in the establishment of Grand Staircase-Escalante National Monument" [*1].

Etymology: Eaton's **devil horned-face (ceratopsian)**

Taxonomy: Ornithischia: Ceratopsia: Ceratopsidae:

Centrosaurinae

[References: (*1) Kirkland JI, DeBlieux DD (2010). New basal centrosaurine ceratopsian skulls from the Wahweap Formation (middle Campanian), Grand Staircase-Escalante National Monument, southern Utah. In: Ryan MJ, Chinnery-Allgeier BJ, Eberth DA (eds), *New Perspectives on Horned Dinosaurs. The Royal Tyrrell Museum Ceratopsian Symposium*. Indiana University Press, Bloomington: 117–140.]

Diamantinasaurus matildae

> *Diamantinasaurus matildae* Hocknull *et al.*, 2009 [Cenomanian; Winton Formation, Winton, Queensland, Australia]

Generic name: *Diamantinasaurus* ← {(place-name) Diamantina + (Gr.) σαῦρος/sauros: lizard}; "in reference to the Diamantina River which runs near the type locality" (*1).

Specific name: *matildae* ← {(person's name) Matilda + -ae}; "in reference to the song "Waltzing Matilda", one of Australia's National songs, written by Banjo Patterson in Winton ("Matilda Country") in 1895" (*1).

Etymology: Matilda's **Diamantina lizard**

Taxonomy: Saurischia: Sauropodomorpha: Titanosauria: Antarctosauridae

Notes: Its nickname is 'Matilda'.

[References: (*1) Hocknull SA, White MA, Tischler TR, Cook AG, Calleja ND, Sloan T, Elliot DA (2009). Sereno P (ed). *New mid-Cretaceous (latest Albian) dinosaurs from Winton, Queensland, Australia. PLoS ONE* 4(7): e6190.; *Diamantinasaurus matildae* – Australian Museum. 〈https://australian.museum/learn/dinosaurs/fact-sheets/diamantinasaurus-matildae/〉]

Diclonius pentagonus

> *Diclonius pentagonus* Cope, 1876 (*nomen dubium*) ⇒ *Trachodon pentagonus* (Cope, 1876) [Late Cretaceous; Judith River Formation, Montana, USA]

Generic name: *Diclonius* ← {(Gr.) δι-/di- [δίς/dis: twice, doubly] + κλών/klon: sprout, twig}; referring to the number of sprouting teeth in comparison to *Monoclonius*, which used only one set of teeth at a time.

Specific name: *pentagonus* ← {(Gr.) πεντα-/penta-: five- + γωνία/gonia:angle}; based on a single tooth.

Etymology: five-angled, **double sprout**

Taxonomy: Ornithischia: Ornithopoda: Hadrosauridae

Other species:

> *Diclonius calamarius* Cope, 1876 (*nomen dubium*) [Campanian; Judith River Formation, Montana, USA]

> *Diclonius perangulatus* Cope, 1876 (*nomen dubium*) ⇒ *Thespesius perangulatus* (Cope, 1876) Steel, 1969 (*nomen dubium*) ⇒ *Trachodon perangulatus* (Cope, 1876) [Campanian; Judith River Formation, Montana, USA]

> *Trachodon mirabilis* Leidy, 1856 (*nomen dubium*) ⇒ *Hadrosaurus mirabilis* (Leidy, 1856) Leidy, 1868 ⇒ *Diclonius mirabilis* (Leidy, 1856) Cope, 1883 [middle Campanian; Judith River Formation, Montana, USA]

Dicraeosaurus hansemanni

> *Dicraeosaurus hansemanni* Janensch, 1914 [late Kimmeridgian; Tendaguru Formation, Lindi, Tanzania]

Generic name: *Dicraeosaurus* ← {(Gr.) δικραιος/dikraios*: bifurcated + σαῦρος/sauros: lizard}; in reference to the Y-shaped spines of the cervical (neck) vertebrae.

*"Der neue Name *Dicraeosaurus* trägt der tiefgehenden Zweispaltigkeit der Dornfortsätze Rechnung (δικραιος = Zweispaltig)" (*1).

Specific name: *hansemanni* ← {(person's name) Hansemann + -ī}; in honor of Dr. David von Hansemann for his expedition support.

Etymology: Hansemann's **bifurcated lizard**

Taxonomy: Saurischia: Sauropodomorpha: Sauropoda: Dicraeosauridae

Other species:

> *Dicraeosaurus sattleri* Janensch, 1914 [Tithonian; Tendaguru Formation, Lindi, Tanzania]

[References: (*1) Janensch W (1914). Übersicht über die Wirbeltierfauna der Tendaguru-Schichten, nebst einer kurzen Charakterisierung der neu aufgeführten Arten von sauropoden [Overview of the vertebrate fauna of the Tendaguru Beds]. *Sonderabdruck aus em Archiv für Biontologie. Bd. III*. 1: 81–110.]

Dilong paradoxus

> *Dilong paradoxus* Xu *et al.*, 2004 [Barremian–Aptian; Yixian Formation, Liaoning, China]

Generic name: *Dilong* ← {(Chin.) dì帝: emperor + lóng龍: dragon} (*1)

Specific name: *paradoxus* ← {(Gr.) παράδοξος/paradoxos: contrary to expectation}; "surprising characters of this animal" (*1).

Etymology: incredible, **emperor dragon**

Taxonomy: Theropoda: Coelurosauria: Tyrannosauroidea

Notes: *Dilong* is the first fossil evidence of feathers for tyrannosauroids and it has relatively long arms with three-fingered hands (*1).

[References: (*1) Xu X, Norell MA, Kuang X, Wang X, Zhao Q, Jia C (2004). Basal tyrannosauroids from China and evidence for protofeathers in tyrannosauroids. *Nature* 431: 680–684.]

Dilophosaurus wetherilli

> *Megalosaurus wetherilli* Welles, 1954 ⇒ *Dilophosaurus wetherilli* (Welles, 1954) Welles, 1970 [Sinemurian–Pliensbachian; Kayenta Formation, Arizona, USA]

Generic name: *Dilophosaurus* ← {(Gr.) δι-/ di-: two, double + λόφος/lophos: crest + σαῦρος/sauros: lizard}; referring to a pair of crests on its skull.

Specific name: *wetherilli* ← {(person's name) Wetherill + -ī}; in honor of John Wetherill.

Etymology: Wetherill's **double-crested lizard**

Taxonomy: Theropoda: Neotheropoda: Dilophosauridae

cf. Theropoda: Ceratosauria: Coelophysoidea (acc. *The Dinosauria 2nd edition*)

Other species:

> *Dilophosaurus sinensis* Hu, 1993 ⇒ *Sinosaurus sinensis* (Hu, 1993) Wang *et al.*, 2017 [Hettangian; Lufeng Formation, Yunnan, China]

[References: Welles SP (1954). New Jurassic dinosaur from the Kayenta Formation of Arizona. *Geological Society of America Bulletin* 65(6): 591–598.; Welles SP (1970). *Dilophosaurus* (Reptilia: Saurischia), a new name for a dinosaur. *Journal of Paleontology* 44(5): 989.; *Dilophosaurus*! A Narrated Exhibition – "guided tour" narrated by Samuel P. Welles. 〈https://ucmp.berkeley.edu/dilophosaur/intro.html〉]

Diluvicursor pickeringi

> *Diluvicursor pickeringi* Herne *et al.*, 2018 [Albian; Eumeralla Formation, Victoria, Australia]

Generic name: *Diluvicursor* ← {(Lat.) diluvi- [diluvium: deluge, flood] + cursor: runner}; "in reference to the deep high-energy palaeo-river within which the type material was deposited and the palaeo-floodplain upon which the river extended" (*1).

Specific name: *pickeringi* ← {(person's name) Pickering + -ī}; "to acknowledge the significant contribution of David A. Pickering to Australian palaeontology and in memory of his passing during the production of the work" (*1).

Etymology: Pickering's **flood runner**

Taxonomy: Ornithischia: Cerapoda: Ornithopoda

[References: (*1) Herne MC, Tait AM, Weisbecker V, Hall M, Nair JP, Cleeland M, Salisbury SW (2018). A new small-bodied ornithopod (Dinosauria, Ornithischia) from a deep, high-energy Early Cretaceous river of the Australian-Antarctic rift system. *PeerJ* 5: e4113.]

Dimorphodon macronyx [Pterosauria]

> *Pterodactylus macronyx* Buckland, 1829 ⇒ *Dimorphodon macronyx* (Buckland, 1829) Owen, 1859 [Hettangian–Sinemurian; Blue Lias, South Wales, UK]

Generic name: *Dimorphodon* ← {(Gr.) δι-/di-: double + μορφή/morphē: shape + ὀδών/odōn (←ὀδούς/odūs: tooth)}

Specific name: *macronyx* ← {(Gr.) μακρός/makros: large + ὄνυξ/onyx: claw}

Etymology: **two-shaped tooth**, with large claw

Taxonomy: Archosauria: Ornithodira: Pterosauria: Dimorphodontidae

Other species:

> *Dimorphodon weintraubi* Clark *et al.*, 1998 [Pliensbachian; La Boca Formation, Tamaulipas, Mexico]

Dineobellator notohesperus

> *Dineobellator notohesperus* Jasinski *et al.*, 2020 [Maastrichtian; Naashoibito Member, Ojo Alamo Formation, New Mexico, USA]

Generic name: *Dineobellator* ← {(Navajo) Diné: the people of the Navajo Nation + (Lat.) bellātor: warrior} (*1)

Specific name: *notohesperus* ← {(Gr.) νότο-/noto-: southern, south + ἕσπερος/hesperos: western | Hesperus: the personification of the evening star}; "in reference to the American Southwest" (*1).

Etymology: southwestern **Diné warrior (= Navajo warrior** from Southwest)

Taxonomy: Theropoda: Coelurosauria: Dromaeosauridae: Velociraptorinae

[References: (*1) Jasinski SE, Sullivan RM, Dodson P (2020). New dromaeosaurid dinosaur (Theropoda, Dromaeosauridae) from New Mexico and biodiversity of dromaeosaurids at the end of the Cretaceous. *Scientific Reports* 10(1): 5105.]

Dingavis longimaxilla [Avialae]

> *Dingavis longimaxilla* O'Connor *et al.*, 2015 [Aptian; Jiufotang Formation, Jehol Group, Liaoning, China]

Generic name: *Dingavis* ← {(Chin. person's name) Dīng丁 + (Lat.) avis: bird}; "in honour of the late distinguished Chinese geologist Wenjiang Ding (丁文江), often considered the 'father of Chinese geology', who brought Amadeus William Grabau to Beijing University in 1920. Professor Ding was the first person to teach palaeontology in China and he also served as the chief editor of *Palaeontologia Sinica*, one of the earliest Chinese journals to receive international recognition" (*1).

Specific name: *longimaxilla* ← {(Lat.) longus: long

+ maxilla}; "referring to the elongate maxilla that distinguishes this taxon from all other Jehol ornithuromorphs" (*1).

Etymology: **Ding's bird** with elongate maxillae (long lostrum)

Taxonomy: Theropoda: Ornithuromorpha

cf. Aves: Pygostylia: Ornithothoraces: Ornithuromorpha (*1)

[References: (*1) O'Connor JK, Wang M, Hu H (2016). A new ornithuromorph (Aves) with an elongate rostrum from the Jehol Biota, and the early evolution of rostralization in birds. *Journal of Systematic Palaeontology* 14(11): 939–948.]

Dinheirosaurus lourinhanensis

> *Dinheirosaurus lourinhanensis* Bonaparte & Mateus, 1999 ⇒ ?*Supersaurus lourinhanensis* (Bonaparte & Mateus, 1999) Tshopp *et al.*, 2015 [Tithonian; Lourinhã Formation, Laurinhã, Portugal]

Generic name: *Dinheirosaurus* ← {(Port.) Porto Dinheiro + (Gr.) σαῦρος/sauros: lizard}

Specific name: *lourinhanensis* ← {(place-name) Laurinhã: the name of the Municipality + -ensis} (*1)

Etymology: **Porto Dinheiro lizard** from Laurinhã

Taxonomy: Saurischia: Sauropodomorpha: Sauropoda: Diplodocidae

Notes: According to Tschopp *et al.* (2015), *Dinheirosaurus lourinhanensis* may be a species of *Supersaurus*.

[References: (*1) Bonaparte JF, Mateus O (1999). A new diplodocid, *Dinheirosaurus lourinhanensis* gen. *et* sp. nov, from the Late Jurassic Beds of Portugal. *Revista del Museo Argentino de Ciencias Naturales* 5(2): 13–29.]

Dinocephalosaurus orientalis [Archosauromorpha]

> *Dinocephalosaurus orientalis* Li, 2003 [Anisian; Guanling Formation, Guizhou, China]

Generic name: *Dinocephalosaurus* ← {(Gr.) δεινός/deinos: terrible + κεφᾰλή/kephalē: head + σαῦρος/sauros: lizard}

Specific name: *orientalis* ← {(Lat.) orientālis: eastern}.

Etymology: eastern, **terrible-headed reptile**

Taxonomy: Archosauromorpha: Protorosauria: Prolacertiformes: Tanystropheidae

Notes: *Dinocephalosaurus* was an extremely long necked, aquatic protorosaur.

[References: Li C (2003). First record of protorosaurid reptile (Order Protorosauria) from the Middle Triassic of China. *Acta Geologica Sinica* 77(4):419–423.]

Dinodocus mackesoni

> *Dinodocus mackesoni* Owen, 1884 (*nomen dubium*) ⇒ *Pelorosaurus mackesoni* (Owen, 1884) Steel, 1970, McIntosh, 1990* (*nomen dubium*) [Aptian–Albian; Lower Greensand Group, Kent, UK]

*accrding to *The Dinosauria 2nd edition.*

Generic name: *Dinodocus* ← {(Gr.) δεινός/deinos: terrible + δοκός/dokos: beam}

Specific name: *mackesoni* ← {(person's name) Mr. H. B. Mackeson + -ī}

Etymology: Mackeson's **terrible beam**

Taxonomy: Saurischia: Sauropodomorpha: Sauropoda

Notes: According to *The Dinosauria 2nd edition*, *Dinodocus* is considered a *nomen dubium*.

Dinosauria ~~~~~~~~~~~~~~~

〈(Gr.) δεινός/deinos: fearfully-great + σαῦρος/sauros: lizard (reptile)〉

Dinosauria; the word which was coined by Richard Owen in 1842. He contained three fossil reptiles *Megalosaurus*, *Iguanodon*, and *Hilaeosaurus* in Dinosauria.

~~~~~~~~~~~~~~~~~~~~~~

Diodorus scytobrachion [Silesauridae]

> *Diodorus scytobrachion* Kammerer *et al.*, 2012 [?Carnian–Norian; Timezgadiouine Formation, Imziln, Morocco]

Generic name: *Diodorus* ← {(person's name) Diodorus}; "named after Diodorus, legendary king of Berber people and son of Sufax, the founder of Tangier. Also named in honour of Diodorus Siculus, a 1st century Greek historian, who wrote about North Africa".

Specific name: *scytobrachion* ← {(Gr.) scyto-[σκῦτος/scytos: a skin, a leather thong] + βρᾰχίων/ brachion: arm} leathery arm; referring "both to a possible integument for this taxon and the classical mythographer Dionysius Scytobrachion, who chronicled the mythical history North Africa".

Etymology: leather armed ***Diodorus***

Taxonomy: Dinosauriformes: Silesauridae

Diplodocus longus

> *Diplodocus longus* Marsh, 1878 (*nomen dubium**) [Kimmeridgian–Tithonian; Morrison Formation, Colorado, USA]

*according to Tschopp *et al.* (2015)

Generic name: *Diplodocus* ← {(Gr.) διπλόος/diploos: double + δοκός/dokos: beam}; in reference to "the caudal vertebrae, which are elongated, deeply excavated below, and have double chevrons, with both anterior and posterior rami" (*1).
~~~~~~~~~~~~~~~~~~~~~~

Specific name: *longus* ← {(Lat.) longus: long, extended}.
Etymology: long, **double beams**
Taxonomy: Saurischia: Sauropodomorpha: Sauropoda: Diplodocoidea: Diplodochidae: Diplodocinae
Other species:
> *Diplodocus lacustris* Marsh, 1884 (*nomen dubium*) [Kimmeridgian–Tithonia; Morrison Formation, Colorado, USA]
> *Diplodocus carnegii* Hatcher, 1901 [Kimmeridgian–Tithonian; Morrison Formation, Wyoming, USA]
> *Seismosaurus halli* Gillette, 1991 ⇒ *Seismosaurus hallorum* (Gillette, 1991) emend. Olshevsky *vide* Gillette, 1994 ⇒ *Diplodocus hallorum* (Gillette, 1991) Lucas *et al.*, 2006 (*2) [Kimmeridgian–Tithonian; Morrison Formation, New Mexico, USA]
> *Diplodocus hayi* Holland, 1924 ⇒ *Galeamopus hayi* (Holland, 1924) Tschopp *et al.*, 2015 [Kimmeridgian–Tithonian; Morrison Formation, Wyoming, USA]

Notes: *Diplodocus* is a sauropod with a long neck and an extremely long tail. According to *The Dinosauria 2nd edition*, *Diplodocus longus* Marsh, 1878 was regarded as the type species. Tschpp & Mateus (2016) proposed to replace the type species with *D. carnegii* Hatcher, 1901, but The International Commission on Zoological Nomen Clature (2018) declined the proposal.

[References: (*1) Marsh OC (1878). Principal characters of American Jurassic dinosaurs. Part I.*American Journal of Science* 3: 411–416.; (*2) Lucas SG, Spielmann JA, Rinehart LF, Heckert AB, Herne MC, Hunt AP, Foster JR, Sullivan RM (2006). Taxonomic status of *Seismosaurus hallorum*, a Late Jurassic sauropod dinosaur from New Mexico. In: Foster JR & Lucas SG (eds), *Paleontology and Geology of the Upper Jurassic Morrison Formation. New Mexico Museum of the Natural History and Science Bulletin* 36: 149–161.]

Diplotomodon horrificus

> *Tomodon horrificus* Leidy, 1865 (preoccupied) ⇒ *Diplotomodon horrificus* (Leidy, 1865) Leidy, 1868 (*nomen dubium*) [Maastrichtian; Navesink Formation, New Jerzey, USA]

Generic name: *Diplotomodon* ← {(Gr.) διπλόος/diploos: double + τομός/tomos: cutting, sharp + ὀδών/odōn: tooth}
Specific name: *horrificus* ← {(Lat.) horrificus: dreadful}.
Etymology: dreadful **double-sharped tooth**
Taxonomy: Theropoda: Tyrannosauroidea
Notes: This species was considered to be a marine reptile and named *Tomodon,* but this name had been used for that of a snake. The genus *Tomodon* was renamed in 1868. According to *The Dinosauria 2nd edition*, *Diplotomodon* is considered as a *nomen dubium*.

Dolichosuchus cristatus

> *Dolichosuchus cristatus* von Huene, 1932 [Norian; Löwenstein Formation, Baden-Württenburg, Germany]

Generic name: *Dolichosuchus* ← {(Gr.) δολιχός/dolichos: long + Σοῦχος/Sūchos: Sobec, (crocodile)}
Specific name: *cristatus* ← {(Lat.) cristātus: crested}
Etymology: crested, **long crochodile**
Taxonomy: Theropoda: Neotheropoda: Coelophysoidea
Notes: According to *The Dinosauria 2nd edition*, *Dolichosuchus cristatus* is considered to be a *nomen dubium*. *Dolichosuchus* was originally thought to be a member of Hallopodidae.

Dollodon bampingi

> *Iguanodon atherfieldensis* Hooley, 1925 (*1) ⇒ *Mantellisaurus atherfieldensis* (Hooley, 1925) Paul, 2007 [late Hauterivian; Wealden Group, UK]
= *Dollodon bampingi* Paul, 2008 (*nomen dubium*) [Barremian; Saint-Barbe Clays Formation, Bernissart, Belgium]

Generic name: *Dollodon* ← {(person's name) Louis Dollo + (Gr.) ὀδών/odōn: tooth}; in honor of "Louis Dollo, who first described this and other Bernissart iguanodonts" (*2).
Specific name: *bampingi* ← {(person's name) Bamping + -ī}; "in honor of Mr. D. Bamping for his support of this research" (*2).
Etymology: Bamping's **Dollo's tooth**
Taxonomy: Ornithischia: Ornithopoda: Hadrosauriformes
Other species:
> *Iguanodon bernissartensis* Boulenger, 1881 [Barremian; Sainte-Barbe Clays Formation, Bergium]
?= *Iguanodon seelyi* Hulke, 1882 ⇒ *Dollodon seelyi* (Hulke, 1882) Carpenter & Ishida, 2010 [late Hauterivian; Wealden Group, England, UK]

Notes: The validity of *Dollodon* has been disputed. *D. bampingi* is considered as a junior synonym of *Mantellisaurus atherfieldensis*, according to McDonald (2012). (*2)

[References: (*1) McDonald AT (2012). The status of *Dollodon* and other basal iguanodonts (Dinosauria:

Ornithischia) from the Lower Cretaceous of Europe. *Cretaceous Research* 33(1): 1–6.; (*2) Paul GS (2008). A revised taxonomy of the iguanodont dinosaur genera and species. *Cretaceous Research* 29(2): 192–216.]

Dongbeititan dongi

> *Dongbeititan dongi* Wang *et al.*, 2007 [Barremian–Aptian; Yixian Formation, Liaoning, China]

Generic name: *Dongbeititan* ← {(Chin. place-name) Dōngběi東北: "northern region of China, which includes Liaoning, Jilin, Heilongjiang provinces" + (Gr. myth.) Τῑτάν/Tītan: "giants, symbolic of great size"} (*1)

Specific name: *dongi* ← {(Chin. person's name) Dǒng Zhīmíng董枝明 + -ī}; "in honor of Prof. Dong Zhiming who has contributed greatly to research and education on Chinese dinosaurs" (*1).

Etymology: Dong's **Dongbei giant**

Taxonomy: Saurischia: Sauropodomorpha: Sauropoda: Titanosauriformes: Somphospondyli (*1)

[References: (*1) Wang X, You H, Meng Q, Gao C, Cheng X, Liu J (2007). *Dongbeititan dongi*, the first sauropod dinosaur from the Lower Cretaceous Jehol Group of western Liaoning Province, China. *Acta Geologica Sinica* 81(6): 911–916.]

Dongyangopelta yangyanensis

> *Dongyangopelta yangyanensis* Chen *et al.*, 2013 [Albian–Cenomanian; Chaochuan Formation, Zhejiang, China]

Generic name: *Dongyangopelta* ← {(Chin. place-name) Dōngyáng東陽; Dongyang City + (Lat.) pelta = (Gr.) πέλτη/peltē: shield}

Specific name: *yangyanensis* ← {(place-name) Yángyán楊岩 + -ensis}

Etymology: **Dongyang shield** from Yangyan village [楊岩東陽盾龍]

Taxonomy: Ornithischia: Ankylosauria: Nodosauridae

[References: Chen R, Zheng W, Azuma Y, Shibata M, Lou T, Jin Q, Jin X (2013). A new nodosaurid ankylosaur from the Chaochuan Formation of Dongyang, Zhejiang Province, China. *Acta Geologica Sinica* 87(3): 658–671.]

Dongyangosaurus sinensis

> *Dongyangosaurus sinensis* Lü *et al.*, 2008 [Cenomanian; Fangyan Formation, Dongyang, Zhejiang, China]

Generic name: *Dongyangosaurus* ← {(Chin. place-name) Dōngyáng東陽: Dongyang County + (Gr.) σαῦρος/sauros: lizard}

Specific name: *sinensis* ← {(place-name) Sin-: China + -ensis}

Etymology: **Dongyang lizard** from China

Taxonomy: Saurischia: Sauropodomorpha: Sauropoda: Titanosauria: Saltasauridae

[References: Lü J, Azuma Y, Chen R, Zheng W, Jin X (2008). A new titanosauriform sauropod from the early Late Cretaceous of Dongyang, Zhejiang Province. *Acta Geologica Sinica* 82 (2): 225–235.]

Draconyx loureiroi

> *Draconyx loureiroi* Mateus & Antunes, 2001 [Tithonian; Lourinhã Formation, Lourinhã, Portugal]

Generic name: *Draconyx* ← {(Lat.) draco: dragon + ὄνυξ/onyx: claw}; in recognition of the claw material (*1).

Specific name: *loureiroi* ← {(person's name) Loureiro + -ī}; after João de Loureiro (1717–1791), Portuguese jesuit, pioneer in Palaeontology in Portugal, also an excellent botanist, astronomer and medical doctor, well-known for his "Flora Cochinchinensis" (he spent a large part of his life in Southeast Asia) (*1).

Etymology: Loureiro's **dragon claw**

Taxonomy: Ornithischia: Ornithopoda: Camptosauridae

[References: (*1) Mateus O, Antunes MT (2001). *Draconyx loureiroi,* a new camptosauridae (Dinosauria, Ornithopoda) from the Late Jurassic of Lourinhã, Portugal. *Annales de Paléontologie* 87(1): 61–73.]

Dracopelta zbyszewskii

> *Dracopelta zbyszewskii* Galton, 1980 [Tithonian; Lourinhã Formation, Lourinhã, Portugal]

Generic name: *Dracopelta* ← {(Lat.) draco: dragon + (Gr.) πέλτη/peltē: small shield}

Specific name: *zbyszewskii* ← {(person's name) Zbyszewski + -ī}; in honor of paleontologist Georges Zbyszewski.

Etymology: Zbyszewski's **dragon shield**

Taxonomy: Ornithischia: Thyreophora: Ankylosauria

[References: Galton PM (1980). Partial skeleton of *Dracopelta zbyszewskii*, n.gen. and n. sp., an ankylosaurian dinosaur from the Upper Jurassic of Portugal. *Geobios* 13(3): 451–457.]

Dracoraptor hanigani

> *Dracoraptor hanigani* Martill *et al.*, 2016 [Hettangian; Blue Lias Formation, Wales, UK]

Generic name: *Dracoraptor* ← {(Lat.) draco: the dragon of Wales* + raptor: robber} (*1)

*the red dragon appearing on the flag of Wales.

Specific name: *hanigani* ← {(person's name) Hanigan + -ī}; "in honor of Nick & Rob Hanigan, who discovered the skeleton and generously donated it to

Amgueddfa Cymru-National Museum of Wales" (*1).
Etymology: Hanigan's **dragon robber**
Taxonomy: Theropoda: Neotheropoda

[References: (*1) Martill DM, Vidovic SU, Howells C, Nudds JR (2016). The oldest Jurassic dinosaur: a basal neotheropod from the Hettangian of Great Britain. *PLoS ONE* 11(1): e0145713.]

Dracorex hogwartsia

> *Troodon wyomingensis* Gilmore, 1931 ⇒ *Pachycephalosaurus wyomingensis* (Gilmore, 1931) Brown & Schlaikjer, 1943 (*nomen conservandum*) [Maastrichtian; Lance Formation, Wyoming, USA]

= *Dracorex hogwartsia* Bakker *et al.*, 2006 [Maastrichtian; Hell Creek Formation, South Dakota, USA]

Generic name: *Dracorex* ← {(Lat.) draco: dragon + (Lat.) rex: king} (*1)
Specific name: *hogwartsia* ← {(novel) Hogwarts + -ia}; "after the fictional 'Hogwarts Academy,' invention of auther J. K. Rowling, the species named in honor of her contribution to children's education and the joy of exploration" (*1).
Etymology: **dragon-king** of Hogwarts Academy
Taxonomy: Ornithischia: Marginocephalia: Pachycephalosauria: Pachycephalosauridae: Pachycephalosaurinae: Pachycephalosaurini
Notes: *Dracorex* has two half-rings of pyramidal spikes on the snout. Horner & Goodwin (2009) suspect that *Dracorex* is a juvenile *Pachycephalosaurus* (*2).

[References: (*1) Bakker RT, Sullivan RM, Porter V, Larson PL, Saulsbury SJ (2006). *Dracorex hogwartsia*, n. gen., n. sp., a piked, flat-headed pachycephalosaurid dinosaur from the Upper Cretaceous Hell Creek Formation of South Dakota. In: Lucas SG, Sullivan RM (eds), *Late Cretaceous Vertebrates from the Western Interior. New Mexico Museum of National History and Science Bulletin* 35: 331–345.]

Dracovenator regenti

> *Dracovenator regenti* Yetes, 2006 [Hettangian–Sinemurian; upper Elliot Formation, Eastern Cape, South Africa]

Generic name: *Dracovenator* ← {(Lat.) draco: dragon + vēnātor: hunter}; "referring to both its probable habits of preying on prosauropod dinosaurs and its location in the foothills of the Drakensberg (Dutch: Dragon's Mountain) Range" (*1).
Specific name: *regenti* ← {(person's name) Regent + -ī}; "in honor of the late Regent 'Lucas' Huma, Professor Kitching's long-term field assistant and friend" (*1).
Etymology: Regent's **dragon hunter from Drakensberg**
Taxonomy: Theropoda: Neotheropoda: Dilophosauridae

[References: (*1) Yates AM (2005). A new theropod dinosaur from the Early Jurassic of South Africa and its implications for the early evolution of theropods. *Palaeontologia Africana* 41: 105–122.]

Dravidosaurus blanfordi

> *Dravidosaurus blanfordi* Yadagiri & Ayyasami, 1979 [Coniacian; Trichinopoly Group, Tamil Nadu, India]

Generic name: *Dravidosaurus* ← {(place-name) Dravida Nadu + (Gr.) σαῦρος/sauros: lizard}
Specific name: *blanfordi* ← {(person's name) Blanford + -ī}; in honor of William Thomas Blanford CIE, who was an English geologist and naturalist.
Etymology: Blanford's **Dravida Nadu lizard**
Taxonomy: Ornithischia: Stegosauria (*1) (*3)
cf. Diapsida: Sauropterygia: Plesiosauria (*2)
Notes: *Dravidosaurus* was first described as stegosaurian in 1979 (*1). However, it was claimed to be a plesiosaur (Chatterjee, 1996) (*2). Galton & Ayyasami (2017) reaffirmed the stegosaurian classification of *Dravidosaurus* (*3).

[References: (*1) Yadagiri P, Ayyasami K (1979). A new stegosaurian dinosaur from Upper Cretaceous sediments of south India. *Journal of the Geological Society of India* 20(11): 521–530.; (*2) Chatterjee S, Rudra DK (1996). KT events in India: impact, rifting, volcanism and dinosaur extinction. In: Novas FA, Molnar RE (eds), *Proceedings of the Gondwanan Dinosaur Symposium, Brisbane, Memoirs of the Queensland Museum* 39(3): 489–532.; (*3) Galton PM, Ayyasami K (2017). Purported latest bone of a plated dinosaur (Ornithischia: Stegosauria), a "dermal plate" from the Maastrichtian (Upper Cretaceous) of southern India". *Neues Jahrbuch für Geologie und Paläontologie – Abhandlungen* 285(1): 91–96.]

Dreadnoughtus schrani

> *Dreadnoughtus schrani* Lacovara *et al.*, 2014 [Campanian–Maastrichtian; Cerro Fortaleza Formation, Santa Cruz, Argentina]

Generic name: *Dreadnoughtus* ← {(O.E.) Dreadnought: fearing nothing + -us}; "alluding to the gigantic body size of the taxon (which presumably rendered healthy adult individuals nearly impervious to attack) and the predominant battleships of the early 20th century (two of which, ARA [Armada dela República Argentina] *Rivadavia* and ARA *Moreno*, were part of the Argentinean navy)" (*1).
Specific name: *schrani* ← {(person's name) Schran + -ī}; "in honor of the American entrepreneur

Adam Schran for his support of this research"[*1].
Etymology: Schran's **Dreadnought**
Taxonomy: Saurischia: Titanosauriformes: Titanosauria

[References: (*1) Lacovara KJ, Ibiricu LM, Lamanna MC, *et al.* (2014). A gigantic, exceptionally complete titanosaurian sauropod dinosaur from southern Patagonia, Argentina. *Scientific Reports* 4: 6196.]

Drinker nisti

> *Nanosaurus agilis* Marsh, 1877 [Kimmeridgian–Tithonian; Morrison Formation, Colorado, USA]
= *Drinker nisti* Bakker *et al.*, 1990 [Kimmeridgian–Tithonian; Morrison Formation, Wyoming, USA]

Generic name: *Drinker* ← {(person's name) Drinker (masculine)}; "in honor of paleontologist Edward Drinker Cope" [*1].
Specific name: *nisti* ← {(acronym) NIST (= "National Institute of Standards and Technology" [*1]) + -ī}
Etymology: NIST's **Drinker**
Taxonomy: Ornithischia: Neornithischia: Nanosauridae
Notes: According to *The Dinosauria 2nd edition*, *Drinker nisti* was valid and *Nanosaurus agilis* was a *nomen dubium*, but now, according to Galton & Carpenter (2018), *Drinker nisti Bakker et al.* (1990), and *Othnielosaurus consors* (Galton, 2007), are both junior subjective synonyms of *Nanosaurus agilis* Marsh, 1877.

[References: (*1) Bakker RT, Galton PM, Siegwarth J, Filla J (1990). A new latest Jurassic vertebrate fauna, from the highest levels of the Morrison Formation at Como Bluff, Wyoming. Part IV. The dinosaurs: A new *Othnielia*-like hypsilophodontoid. *Hunteria* 2(6): 8–14.]

Dromaeosauroides bornholmensis

> *Dromaeosauroides bornholmensis* Christiansen & Bonde, 2003 [Berriasian–Valanginian; Jydegaard Formation, island of Bornholm, Denmark]

Generic name: *Dromaeosauroides* ← {*Dromaeosaurus* + -oides}
Specific name: *bornholmensis* ← {(place-name) Bornholm in the Baltic Sea + -ensis}
Etymology: ***Dromaeosaurus*-like** from Bornholm
Taxonomy: Theropoda: Dromaeosauridae: Dromaeosaurinae

[References: Bonde N, Christiansen P (2003). New dinosaurs from Denmark. *Comptes Rendus Palevol* 2: 13–26.]

Dromaeosaurus albertensis

> *Dromaeosaurus albertensis* Matthew & Brown, 1922 [late Campanian; Dinosaur Park Formation, Alberta, Canada]

Generic name: *Dromaeosaurus* ← {(Gr.) δρομαῖος/dromaios: swift-running, fleet [δρομεύς/dromeus: runner] + σαῦρος/sauros: lizard}
Specific name: *albertensis* ← {(place-name) Alberta + -ensis}
Etymology: **swift-running lizard** from Alberta
Taxonomy: Theropoda: Maniraptora: Dromaeosauridae: Dromaeosaurinae
Other species:

> *Laelaps explanatus* Cope, 1876 ⇒ *Dryptosaurus explanatus* (Cope, 1876) Hay, 1902 ⇒ *Deinodon explanatus* (Cope, 1876) Lambe, 1902 ⇒ *Dromaeosaurus explanatus* (Cope, 1876) Kuhn, 1939 (*nomen dubium*) [Montana, USA]
> *Laelaps laevifrons* Cope, 1876 ⇒ *Deinodon laevifrons* (Cope, 1876) Osborn, 1902 ⇒*Dryptosaurus laevifrons* (Cope, 1876) Hay, 1902 ⇒ *Dromaeosaurus laevifrons* (Cope, 1876) Matthew & Brown, 1922 (*nomen dubium*) [Montana, USA]
> *Laelaps cristatus* Cope, 1876 ⇒ *Dryptosaurus cristatus* (Cope, 1876) Hay, 1902 ⇒*Deinodon cristatus* (Cope, 1876) Osborn, 1902 ⇒ *Dromaeosaurus cristatus* (Cope, 1876) Matthew & Brown, 1922 (*nomen dubium*) ⇒ *Troodon cristatus* (Cope, 1876) Olshevsky, 1995 [Montana, USA]
> *Coelurus gracilis* Marsh, 1888 ⇒ *Dromaeosaurus gracilis* (Marsh, 1888) Matthew & Brown, 1922 (*nomen dubium*) [Aptian; Arundel Clay Formation, Maryland, USA]

[References: Matthew WD, Brown B (1922). The family Deinodontidae, with notice of a new genus from the Cretaceous of Alberta. *Bulletin of American Museum of National History* 46: 367–385.]

Dromiceiomimus brevitertius

> *Ornithomimus edmontonicus* Sternberg, 1933 [Campanian; Horseshoe Canyon Formation, Alberta, Canada]
= *Struthiomimus brevitertius* Parks, 1926 ⇒ *Dromiceiomimus brevitertius* (Parks, 1926) Russell, 1972 [Maastrichtian; Horseshoe Canyon Formation, Alberta, Canada]

Generic name: *Dromiceiomimus* ← {(Generic name) Dromiceius / Dromaius [(Gr.) δρομαῖος/dromaios: swift] + μῖμος/mimos: mimic}

In 1816, French ornithologist L. J. P. Vieillot used two generic names, *Dromaius* and *Dromiceius*, to describe the emu in a paper. *Dromaius* is used as the genus name for the emu, and in 1972, Dale Russell used *Dromiceius* for the name of *Dromiceiomimus*.

Specific name: *brevitertius* ← {(Lat.) brevitertius: short third-fingered}.

Etymology: **emu mimic** having a short third finger
Taxonomy: Theropoda: Coelurosauria: Ornithomimosauria: Ornithomimidae
Other species:

> *Struthiomimus samueli* Parks, 1928 ⇒ *Ornithomimus samueli* (Parks, 1928) Kuhn, 1965 ⇒ *Dromiceiomimus samueli* (Parks, 1928) Russell, 1972 [Campanian; Dinosaur Park Formation, Canada]

Notes: *Dromiceiomimus* Russel, 1972 is considered a synonym of *Ornithomimus* Marsh, 1890, in *The Dinosauria 2nd edition* (2004).

[References: Russell D (1972). Ostrich dinosaurs from the Late Cretaceous of western Canada. *Canadian Journal of Earth Sciences* 9: 375–402.]

Drusilasaura deseadensis

> *Drusilasaura deseadensis* Navarrete *et al.*, 2011 [Cenomanian–Turonian; Bajo Barreal Formation, Santa Cruz, Argentina]

Generic name: *Drusilasaura* ← {(person's name) Drusila + (Gr.) σαῦρα/saura, *f.*: lizard}; "in honor of Drusila Ortiz de Zárate, a young female member of the family who owns the María Aika Ranch where the fossil was found" (*1).
Specific name: *deseadensis* ← {(place-name) Deseado + -ensis}; "referring to the valley of the Deseado River where the sauropod was found" (*1).
Etymology: **Drusila's lizard** from Deseado River
Taxonomy: Saurischia: Sauropodomorpha: Sauropoda: Titanosauria: Titanosauridae (*1)

[References: (*1) Navarrete C, Casal G, Martínez R (2011). *Drusilasaura deseadensis* gen. *et* sp. nov., a new titanosaur (Dinosauria-Sauropoda), of the Bajo Barreal Formation, Upper Cretaceous of north of Santa Cruz, Argentina. *Revista Brasileira de Paleontologia* 14(1): 1–14.]

Dryosaurus altus

> *Laosaurus altus* Marsh, 1878 ⇒ *Dryosaurus altus* (Marsh, 1878) Marsh, 1894 [Tithonian; upper Brushy Basin Member, Morrison Formation, Wyoming, USA]

Generic name: *Dryosaurus* ← {(Gr.) δρῦς/drys: tree, oak + σαῦρος/sauros: lizard}
Specific name: *altus* ← {(Lat.) altus: tall}; Marsh named the remains of euornithopods *Laosaurus altus*, referring to it being larger than *Laosaurus celer* in 1878 (*1).
Etymology: tall, **tree lizard**
Taxonomy: Ornithischia: Ornithopoda: Iguanodontia: Dryosauroidea: Dryosauridae
Other species:

> *Dryosaurus elderae* Carpenter & Galton, 2018 [Late Jurassic; Morrison Formation, Utah, USA]

> *Dryosaurus canaliculatus* Galton, 1975 ⇒ *Valdosaurus canaliculatus* (Galton, 1975) Galton, 1977 [late Barremian; Wessex Formation, England, UK]

> *Dysaltosaurus lettowvorbecki* Virchow, 1919 ⇒ *Dryosaurus lettowvorbecki* (Virchow, 1919) Galton, 1977 [late Kimmeridgian; Tendaguru Formation, Lindi, Tanzania]

[References: (*1) Marsh OC (1878). Principal characters of American Jurassic dinosaurs. Part 1. *American Journal of Science and Arts* 16: 411–416.]

Dryptosaurus aquilunguis

> *Laelaps aquilunguis* Cope, 1866 (preoccupied)* ⇒ *Dryptosaurus aquilunguis* (Cope, 1866) Marsh, 1877 ⇒ *Megalosaurus aquilunguis* (Cope, 1866) Osborn, 1898 [Maastrichtian; New Egypt Formation, New Jersey, USA]
> *The genus name of *Laelaps* was already preoccupied by a mite and later renamed.

Generic name: *Dryptosaurus* ← {(Gr.) drypto- [δρύπτω/dryptō: tear] + σαῦρος/sauros: lizard}
Specific name: *aquilunguis* ← {(Lat.) aquila: eagle + unguis: claw}
Etymology: eagle-clawed **tearing lizard**
Taxonomy: Theropoda: Coelurosauria: Tyrannosauroidea: Dryptosauridae
Other species:

> *Laelaps macropus* Cope, 1868 (*nomen dubium*) ⇒ *Dryptosaurus macropus* (Cope, 1868) Hay, 1902 ⇒ *Teihivenator macropus* (Cope, 1868) Yun, 2017 (*nomen dubium*) [early Maastrichtian; Navesink Formation, New Jersey, USA]

> *Laelaps explanatus* Cope, 1876 (*nomen dubium*) ⇒ *Dryptosaurus explanatus* (Cope, 1876) Hay, 1902 ⇒ *Dromaeosaurus explanatus* (Cope, 1876) Hay, 1930 [Campanian; Dinosaur Park Formation, Alberta, Canada]

> *Laelaps laevifrons* Cope, 1876 (*nomen dubium*) ⇒ *Dryptosaurus laevifrons* (Cope, 1876) Hay, 1902 ⇒ *Dromaeosaurus laevifrons* (Cope, 1876)

> *Laelaps incrassatus* Cope, 1876 ⇒ *Dryptosaurus incrassatus* (Cope, 1876) Hay, 1902

> *Laelaps trihedrodon* Cope, 1877 ⇒ *Megalosaurus trihedrodon* (Cope, 1877) Nopcsa, 1901 ⇒ *Antrodemus trihedrodon* (Cope, 1877) Kuhn, 1939 ⇒ *Dryptosaurus trihedrodon* (Cope, 1877) Kuhn, 1939 (*nomen dubium*) [Kimmeridgian–Tithonian; Morrison Formation, Colorado, USA]

> *Megalosaurus crenatissimus* Depéret, 1896 ⇒ *Dryptosaurus crenatissimus* (Depéret, 1896) Depéret & Savornin, 1928 ⇒ *Majungasaurus*

crenatissimus (Depéret, 1896) Lavocat, 1955 [Maastrichtian; Maevarano Formation, Mevarana, Madagascar]

Notes: *Dryptosaurus* has been placed, in the theropod families Coeluridae, Deinodontidae, Megalosauridae, Tyrannosauridae, and Dryptosauridae. Carpenter *et al.* consider that it should be placed in its own family, Dryptosauridae (*1).

[References: (*1) Carpenter K, Russell D, Baired D, Denton R (1997). Redescription of the holotype of *Dryptosaurus aquilunguis* (Dinosauria: Theropoda) from the Upper Cretaceous of New Jersey. *The Journal of Vertebrate Paleontology* 17(3): 561–573.]

Dsungaripterus weii [Pterosauria]

> *Dsungaripterus weii* Young,1964 [Barremian–Aptian; Lianmuqin Formation, Dsungari (= Junggar) Basin, Sinkiang (Xinjiang), China]

Generic name: *Dsungaripterus* ← {(place-name) Dzungarian* Basin + (Gr.) πτέρον/ pteron: feather, wing}; referring to the Junggar Basin where the first fossil was found.

*(Mong. dialect) зуун гар/ Zuun Gar (=left hand)

Specific name: *weii* ← {(Chin. person's name) Wèi 魏 + -ī}; in honor of paleontologist C. M. Wei.

Etymology: Wei's **wing from Junggar Basin**

Taxonomy: Archosauria: Ornithodira: Pterosauria: Dsungaripteridae

[References: Young C-C (楊鍾健) (1964). On a new pterosaurian from Sinkiang, China. *Vertebrata PalAsiatica*. 8(3): 221–255.]

Dubreuillosaurus valesdunensis

> *Poekilopleuron? valesdunensis* Allain, 2002 ⇒ *Dubreuillosaurus valesdunensis* (Allain, 2002) Allain, 2005 [Bathonian; Calcaires de Caen Formation, Basse-Normandie, France]

Generic name: *Dubreuillosaurus* ← {(person's name) André Dubreuil: the mayor of Conteville + (Gr.) σαῦρος/sauros: lizard}; "in honor of the Dubreuil family, who discovered the specimen in 1994" (*1).

Specific name: *valesdunensis* ← {(place-name) Val-ès-Dunes + (Lat.) -ensis: from}; "from Val-ès-Dunes, the Norman name of the battlefield near Conteville where the rebellious barons of the Bessin and the Cotentin were defeated by William the Conqueror in 1047 and the holotype was found" (*2).

Etymology: **Dubreuil's lizard** from Val-ès Dunes

Taxonomy: Theropoda: Megalosauridae

[References: (*1) Allain R (2005). The postcranial anatomy of the megalosaur *Dubreuillosaurus valesdunensis* (Dinosauria Theropoda) from the Middle Jurassic of Normandy, France. *Journal of Vertebrate Paleontology*. 25(4): 850–858.]; (*2) Allain R (2002). Discovery of megalosaur (Dinosauria, Theropoda) in the middle Bathonian of Normandy (France) and its implications for the phylogeny of basal Tetanurae. *Journal of Vertebrate Paleontology* 22(3): 548– 563.]

Duriatitan humerocristatus

> *Cetiosaurus humerocristatus* Hulke, 1874 ⇒ *Pelorosaurus humerocristatus* (Hulke, 1874) Sauvage, 1887 ⇒ *Ornithopsis humerocristatus* (Hulke, 1874) Lydekker, 1888 ⇒ *Duriatitan humerocristatus* (Hulke, 1874) Barrett *et al.*, 2010 [Kimmeridgian; Kimmeridge Clay, Dorset, UK]

Generic name: *Duriatitan* ← {(Lat. place-name) Duria = Dorset + (Gr.) Τῑτάν/Tītan}

Specific name: *humerocristatus* ← {(Lat.) humerus: humerus, the upper arm bone + cristātus [crista: crest]}

Etymology: **Dorset Titan**, crested on the upper arm bone

Taxonomy: Saurischia: Sauropodomorpha: Titanosauriformes

[References: Barrett PM, Benson RBJ, Upchurch P (2010). Dinosaurs of Dorset: Part II, the sauropod dinosaurs (Saurischia, Sauropoda) with additional coments on the theropods. *Proceedings of the Dorset Natural History and Archaeological Society* 131: 113–126.; Hulke JW (1874). Note on a very large saurian limb-bone adapted for progression upon land, from the Kimmeridge Clay of Weymouth, Dorset. *Quartely Journal of the Geological Society of London* 30: 16–17.]

Duriavenator hesperis

> *Megalosaurus hesperis* Waldman, 1974 ⇒ *Walkersaurus hesperis* (Waldman, 1974) Welles & Powell, 1995 *vide* Welles *et al.*, 1994 (*nomen nudum*) ⇒ *Duriavenator hesperis* (Waldman, 1974) Benson, 2008 [Bajocian; upper Inferior Oolite Formation, Sherborne, Dorset, UK]

Generic name: *Duriavenator* ← {(place-name) Duria = Dorset + (Lat.) vēnātor: hunter}

Specific name: *hesperis* ← {(Gr.) ἕσπερος/hesperos: the West, western} (*1).

Etymology: western, **Dorset hunter**

Taxonomy: Theropoda: Megalosauridae: Megalosaurinae

[References: (*1) Waldman M (1974). Megalosaurids from the Bajocian (Middle Jurassic) of Dorset. *Palaeontology* 17(2): 325–339.; Benson RBJ (2008). A redescription of *Megalosaurus hesperis* (Dinosauria, Theropoda) from the Inferior Oolite (Bajocian, Middle

Jurassic) of Dorset, United Kingdom. *Zootaxa* 1931: 57–67.]

Dynamosaurus imperiosus

> *Tyrannosaurus rex* Osborn, 1905 [Maastrichtian; Lance Formation, Montana, USA] (actual holotype)

= *Dynamosaurus imperiosus* Osborn, 1905 ⇒ *Tyrannosaurus imperiosus* (Osborn, 1905) Osborn, 1906 [Maastrichtian; Lance Formation, Wyoming, USA] (first discovery in 1900, by Barnum Brown)

Generic name: *Dynamosaurus* ← {(Gr.) δύνᾰμις/dynamis: power + σαῦρος/sauros: lizard}

Specific name: *imperiosus* ← {(Lat.) imperiōsus: mighty, powerful, tyrannical}

Etymology: tyrannical, **powerful lizard**

Taxonomy: Theropoda: Tyrannosauridae: Tyrannosaurinae

[References: Osborn HF (1905). *Tyrannosaurus* and other Cretaceous carnivorous dinosaurs. *Bulletin of the American Museum of Natural History* 21(14): 259–265.]

Dynamoterror dynastes

> *Dynamoterror dynastes* McDonald *et al.*, 2018 [Campanian; Allison Member, Menefee Formation, San Juan, New Mexico, USA]

Generic name: *Dynamoterror* ← {(Gr.) δύνᾰμις/dynamis: power + (Lat.) terror: terror}

Specific name: *dynastes* ← {(Lat.) dynastēs: ruler}; refering to "the binomen '*Dynamosaurus imperiosus*' (Osborn, 1905), which is a junior synonym of *Tyrannosaurus rex* (Osborn, 1905, 1906), but a particular childhood favorite of the lead author" (*1).

Etymology: **powerful terror,** ruler (*1)

Taxonomy: Theropoda: Coelurosauria: Tyrannosauroidea: Tyrannosauridae: Tyrannosaurinae

[References: (*1) McDonald AT, Wolfe DG, Dooley Jr AC (2018). A new tyrannosaurid (Dinosauria: Theropoda) from the Upper Cretaceous Menefee Formation of New Mexico. *PeerJ* 6: e5749.]

Dyoplosaurus acutosquameus

> *Dyoplosaurus acutosquameus* Parks, 1924 [Campanian; Dinosaur Park Formation, Alberta, Canada]

Generic name: *Dyoplosaurus* ← {(Gr.) δύ-/dy-: 2, two + ὅπλον/hoplon: weapon + σαῦρος/sauros: lizard}

Specific name: *acutosquameus* ← {(Lat.) acutus: sharp, pointed + squameus: scaly}

Etymology: sharp-pointed scaly, **double-armed lizard**

Taxonomy: Ornithischia: Thyreophora: Ankylosauria: Ankylosauridae: Ankylosaurinae

Other species:

> *Dyoplosaurus giganteus* Maleev, 1956 ⇒*Tarchia gigantea* (Maleev, 1956) Tumanova, 1977 (*nomen dubium*) [Campanian; Nemegt Formation, Ömnögovi, Mongolia]

Notes: *Dyoplosaurus acutosquameus* was proposed that it was a junior synonym of *Euoplocephalus tutus* (Coombs, 1971). However, *Dyoplosaurus* is considered to be a valid taxon (Arbour, 2009)(*1).

[References: (*1) Arbour VM, Burns ME, Sissons RL (2009). A redescription of the ankylosaurid dinosaur *Dyoplosaurus acutosquameus* Parks, 1924 (Ornithischia: Ankylosauria) and a revision of the genus. *Journal of Vertebrate Paleontology* 29(4): 1117–1135.]

Dysalotosaurus lettowvorbecki

> *Dysalotosaurus lettowvorbecki* Virchow, 1919 ⇒ *Dryosaurus lettowvorbecki* (Virchow, 1919) Galton, 1977 [Kimmeridgian; Tendaguru Formation, Lindi, Tanzania]

Generic name: *Dysalotosaurus* ← {(Gr.) δυσάλωτος/dysalōtos: hard to catch or take, difficult to be captured + σαῦρος/sauros: lizard}

Specific name: *lettowvorbecki* ← {(person's name) Lettow-Vorbeck + -ī}; in honor of the Imperial German Army Officer, Paul von Lettow-Vorbeck.

Etymology: Lettow-Vorbeck's **uncatchable lizard**

Taxonomy: Ornithischia: Ornithopoda: Dryosauridae

Notes: According to *The Dinosauria 2nd edition*, *Dysalotosaurus* Virchow, 1919 was considered as a synonym of *Dryosaurus* Marsh, 1894, but recent studies reject this synonymy (Hübner & Rauhut, 2010).

Dyslocosaurus polyonychius

> *Dyslocosaurus polyonychius* McIntosh *et al.*, 1992 [Maastrichtian; Lance Formation, Wyoming, USA]

Generic name: *Dyslocosaurus* ← {(Gr.) Prefix, δύς-/dys-: bad, poor + (Lat.) locus: place + (Gr.) σαῦρος/sauros: lizard}; "in reference to the inadequate provenance information" (*1).

Specific name: *polyonychius* ← {(Gr.) πολύς/polys: many + ὄνυξ/onyx: claw}; "in reference to the claws of digits IV and possibly V" (*1).

Etymology: **bad place** lizard with many claws

Taxonomy: Saurischia: Sauropodomorpha: Sauropoda: Dicraeosauridae (*2)
cf. Sauropoda: Diplodocidae (*1)

[References: (*1) McIntosh J, Coombs WP, Russell DA (1992). A new diplodocid sauropod (Dinosaura) from Wyoming, U.S.A. *Journal of Vertebrate Paleontology*

12(2): 158–167.; (*2) Tschopp E, Mateus OV, Benson RBJ (2015). A specimen-level phylogenetic analysis and taxonomic revision of Diplodocidae (Dinosaurua, Sauropoda). *PeerJ* 3: e857.]

Dystrophaeus viaemalae

> *Dystrophaeus viaemalae* Cope, 1877 [Kimmeridgian; Morrison Formation, Utah, USA]

Generic name: *Dystrophaeus* ← {(Gr.) δῦσ-/dys-: bad, ill + στροφεύς/stropheus: joint, vertebra, socket}

Specific name: *viaemalae* ← {(Lat.) viaemalae: of the bad road}.

This fossil is the first one found in the Triassic beds of the Rocky Mountain region, and was derived from an inhospitable region [*1].

Etymology: **coarse joint**, of the bad road

Taxonomy: Saurischia: Sauropodomorpha: Sauropoda

Notes: A modern analysis (Gillette, 1996) concluded the taxon to be a member of Diplodocidae. Many researchers consider the taxon to be a *nomen dubium.*

[References: (*1) Cope ED (1877). On a dinosaurian from the Trias of Utah. *Proceeings of the American Philosophical Society* 16: 579–584.]

Dystylosaurus edwini

> *Supersaurus vivianae* Jensen, 1985 [Kimmeridgian–Tithonian; Morrison Formation, Colorado, USA] [*1]

= *Dystylosaurus edwini* Jensen, 1985 [*2] [Kimmeridgian–Tithonian; Morrison Formation, Colorado, USA]

= *Ultrasaurus macintoshi* Jensen, 1985 ⇒ *Ultrasauros macintoshi* (Jensen, 1985) Olshevsky, 1991 [Kimmeridgian–Tithonian; Morrison Formation, Colorado, USA]

Generic name: *Dystylosaurus* ← {(Gr.) δύ-/dy-: 2, two + στύλος/stylos: beam + σαῦρος/sauros: lizard} [*2]

Specific name: *edwini* ← {(person's name) Edwin + -ī}; in honor of the late Daniel Edwin (Eddie) Jones, who, with his wife, Vivian, brought more new dinosaur taxa to science than any other two amateurs while providing 20 years of logistic support for fieldwork on the Uncompahgre "Plateau." [*2]

Etymology: Edwin's **double-beamed lizard**

Taxonomy: Saurischia: Sauropodomorpha: Sauropoda: Diplodocidae: Diplodocinae

cf. Saurischia: Sauropodomorpha: Sauropoda: Brachiosauridae (acc. *The Dinosauria 2nd edition*)

Notes: *Dystylosaurus edwini*, Brachiosauridae, was considered to be valid, according to *The Dinosauria 2nd edition* (2004). But, this genus is now considered to be a synonym of *Supersaurus*, Diplodocidae [*1].

[References: (*1) Curtice B, Stadtman K (2001). The demise of *Dystylosaurus edwini* and a revision of *Supersaurus vivianae*. In: McCord RD, Boaz D (eds), *Western Association of Vertebrate Paleontologists and Southwest Paleontological Symposium-Proceedings 2001. Mesa Southwest Museum Bulletin* (8): 33–40.; (*2) Jensen JA (1985). Three new sauropod dinosaurs from the Upper Jurassic of Colorado. *Great Basin Naturalist* 45: 697–709.]

E

Echinodon becklesii

> *Echinodon becklesii* Owen, 1861 [Berriasian; Lulworth Formation, Purbeck Beds, Durdleston Bay, Dorset, UK]

Generic name: *Echinodon* ← {(Gr.) ἐχῖνος/echīnos: hedgehog + ὀδών/odōn [= ὀδούς/odūs: tooth]} "prickly tooth". [*1]

"The teeth of this genus are distinguished by the marginal serrations of the apical half of the crown, which increase in size from the apex to the base of that angular part of the tooth, the two basal points resembling spines, and terminating respectively, or formig the confluence of, the two thickened ridges bounding the fore and hind borders of the basal half of the crown" [*1].

Specific name: *becklesii* ← {(Lat. person's name) genitive form of Becklesius}; in honor of the discoverer Samuel Husbands Beckles (1814–1890).

Etymology: Beckles' **prickly tooth**

Taxonomy: Ornithischia: Ornithopoda: Heterodontosauridae

Notes: Owen originally thought *Echinodon becklesii* as a kind of lizard, Lacertilia [*1].

[References: (*1) Owen R (1861). *The Fossil Reptilia of the Liassic formations. British Fossil Reptiles. Supplement No. II. Mesozoic Lizards*: 126–128.]

Edmontonia longiceps

> *Edmontonia longiceps* Sternberg, 1928 [Campanian–Maastrichtian; Horseshoe Canyon Formation, Edmonton Group, Alberta, Canada]

= *Denversaurus schlessmani* Bakker, 1988* ⇒ *Edmontonia schlessmani* (Bakker, 1988) Hunt & Lucas, 1992 [Maastrichtian; Lance Formation, South Dakota, USA] *(acc. *The Dinosauria 2nd edition*)

Generic name: *Edmontonia* ← {(place-name) Edmonton + -ia}; after the Edmonton Formation, Alberta.

Specific name: *longiceps* ← {(Lat.) longus: long +

-ceps [caput: head]}

Etymology: long-headed **one from Edmonton Formation**

Taxonomy: Ornithischia: Thyreophora: Ankylosauria: Nodosauridae

Other species:

> *Palaeoscincus rugosidens* Gilmore, 1930 ⇒ *Edmontonia rugosidens* (Gilmore, 1930) Russell, 1940 ⇒ *Panoplosaurus* (*Edmontonia*) *rugosidens* (Gilmore, 1930) Coombs, 1971 ⇒ *Palaeoscincus* (*Chassternbergia*) *rugosidens* (Gilmore, 1930) Bakker, 1988 ⇒ *Chassternbergia rugosidens* (Gilmore, 1930) Olshevsky, 1991* [Campanian; Two Medicine Formation, Montana, USA]

*Some researchers think *Chassternbergia* to be a synonym of *Edmontonia*, but some think *Chassternbergia* is a valid genus with a type species *Chassternbergia rugosidens.*

> *Glyptodontopelta mimus* Ford, 2000 [Maastrichtian; Naashoibito Member, Ojo Alamo Formation, San Juan, New Mexico, USA]

= *Edmontonia australis* Ford, 2000 (*nomen dubium*) [Maastrichtian; Kirtland Formation, New Mexico, USA]

Notes: *Edmontonia* had many sharp spikes along its sides. The four largest spikes jutted out from the shoulders on each side. According to Burns (2015), *Denversaurus schlessmani* is a valid species.

[References: Sternberg CM (1928). A new armored dinosaur from the Edmonton Formation of Alberta. *Transactions of the Royal Society of Canada, Series 3* 22: 93–106.]

Edmontosaurus regalis

> *Edmontosaurus regalis* Lambe, 1917 [Maastrichtian; Horseshoe Canyon Formation (former lower Edmonton Formation), Edmonton Group, Alberta, Canada]

= *Trachodon atavus* Cope, 1871 (according to *The Dinosauria 2nd edition*) [Maastrichtian; Navesink Formation, New Jersy, USA] See also *Hadrosaurus foulkii.*

= *Agathaumas milo* Cope, 1874 (according to *The Dinosauria 2nd edition*) [Maastrichtian; Laramie Formation, Colorado, USA]

cf. Agathaumas milo Cope, 1874 (*nomen dubium*) (according to Cope, 1874) [Maastrichtian; Denver Formation, Colorado, USA] See also *Thespesius occidentalis*.

Generic name: *Edmontosaurus* ← {(place-name) Edmonton + (Gr.) σαῦρος/sauros: lizard}

According to Lambe (1917), the skeleton was discovered from the Edmonton Formation (upper Cretaceous), Red Deer River, Alberta, Canada [*1]. He did not mention the etymology [*1].

Specific name: *regalis* ← {(Lat.) rēgālis: regal}

Etymology: regal **Edmonton lizard**

Taxonomy: Ornithischia: Ornithopoda: Hadrosauridae: Saurolophinae: Edmontosaurini

Other species:

> *Claosaurus annectens* Marsh, 1892 ⇒ *Trachodon annectens* (Marsh, 1892) Hatcher, 1902 ⇒ *Thespesius annectens* (Marsh, 1892) Gilmore, 1924 ⇒ *Anatosaurus annectens* (Marsh, 1892) Lull & Wright, 1942 ⇒ *Edmontosaurus annectens* (Marsh, 1892) Brett-Surman *vide* Chapman & Brett-Surman, 1990 [Maastrichtian; Lance Formation, Wyoming, USA]

= *Trachodon longiceps* Marsh, 1890 ⇒ *Hadrosaurus longiceps* (Marsh, 1890) Nopcsa, 1900 ⇒ *Anatotitan longiceps* (Marsh, 1890) Olshevsky, 1991 [Maastrichtian, Lance Formation, Wyoming, USA]

= *Thespesius edmontoni / edmontonensis* Gilmore, 1924 ⇒ *Anatosaurus edmontoni / edmontonensis* (Gilmore, 1924) Lull & Wright, 1942 [Campanian; Horseshoe Canyon Formation, Alberta, Canada]

= *Trachodon marginatus* Lambe, 1913 (in part) [Campanian; Dinosaur Park Formation, Alberta, Canada]

= *Claosaurus affinis* Wieland, 1903 (*nomen dubium*) [Campanian; Pierre Shale Formation, South Dakota, USA]

= *Thespesius saskatchewanensis* Sternberg, 1926 ⇒ *Anatosaurus sascatchewanensis* (Sternberg, 1926) Lull & Wright, 1942 ⇒ *Edmontosaurus saskatchewanensis* (Sternberg, 1926) Brett-Surman *vide* Chapman & Brett-Surman, 1990 [late Maastrichtian; Frenchman Formation, Sascatchewan, Canada]

= *Anatosaurus copei* Lull & Wright, 1942 ⇒ *Anatotitan copei* (Lull & Wright, 1942) Brett-Surman *vide* Chapman & Brett-Surman, 1990 ⇒ *Edmontosaurus copei* (Lull & Wright, 1942) [Maastrichtian; Hell Creek Formation, South Dakota, USA]

= *Diclonius mirabilis* Cope, 1876 (*nomen dubium*, in part) [Maastrichtian; Lance Formation, South Dakota, USA]

cf. Trachodon mirabilis Leidy, 1856 [Campanian; Judith River Formation, Montana, USA]

> *Hadrosaurus minor* Cope, 1869 (*nomen dubium*) ⇒ *Edmontosaurus minor* (Cope, 1874) Baird & Horner, 1977 (*nomen dubium*)

[Maastrichtian; Navesink Formation (= Hornerstown Formation), New Jersey, USA]

Notes: According to *The Dinosauria 2nd edition*, *Trachodon atavus* Cope, 1871 (*atavus cavatus*) and *Agathaumas milo* were considered as synonyms of *Edmontosaurus regalis*, and *Thespesius edmontonensis* was regarded as a synonym of *Edmontosaurus annectens*. Campione and Evans (2011) found *Thespesius edmontoni* to be a synonym of *Edmontosaurus regalis*.

[References: (*1) Lambe LM (1917). A new genus and species of crestless hadrosaur from the Edmonton Formation of Alberta. *The Ottawa Naturalist* 31(7): 65–73.; Ostrom JH (1964). A reconsideration of the paleoecology of hadrosaurian dinosaurs. *American Journal of Science* 262(8): 975–997.]

Efraasia minor

> *Teratosaurus minor* von Huene, 1907–1908 (*nomen dubium*) ⇒ *Efraasia minor* (von Huene, 1907–1908) Yates, 2003 [middle Norian; Löwenstein Formation, Baden-Württemberg, Germany]

= *Thecodontosaurus diagnosticus* Fraas, 1913 (*nomen nudum*) ⇒ *Palaeosaurus diagnosticus* von Huene, 1932 ⇒ *Efraasia diagnostica* (von Huene, 1932) Galton, 1973 [middle Norian; Löwenstein Formation, Baden-Württemberg, Germany]

= *Sellosaurus fraasi* von Huene, 1907–1908 (*nomen dubium*) [middle Norian; Löwenstein Formation, Baden-Württemberg, Germany]

= *Plateosaurus gracilis* (*partim*) von Huene, 1932 ⇒ *Sellosaurus gracilis* (*partim*) (von Huene, 1932) Galton, 1984 [Norian; Löwenstein Formation, Baden-Württemberg, Germany]

Generic name: *Efraasia* Galton, 1973 ← {(person's name) E. Fraas + -ia}; "in honor of Professor E. Fraas who found the material" (*1).

Specific name: *minor* ← {(Lat.) minor: smaller}; referring to the fact that *E. minor* was smaller than the *Teratosaurus suevicus* Meyer, 1861*.

*Loricata, Rauisuchidae.

Etymology: smaller **E. Fraas' one**

Taxonomy: Saurischia: Sauropodomorpha

Notes: Three species: *Teratosaurus minor* von Huene, 1908; *Sellosaurus fraasi* von Huene, 1908; *Palaeosaurus diagnosticus* von Huene, 1932 (*nomen nudum*) were combined into *Efraasia minor* von Huene, 1908 (*2).

[References: (*1) Galton PM (1973). On the anatomy and relationships of *Efraasia diagnostica* (Huene) n. gen., a prosauropod dinosaur (Reptilia: Saurischia) from the Upper Triassic of Germany. *Paläontologische Zeitschrift* 47(3): 229–255.; (*2) Yates AM (2003). The species taxonomy of the sauropodomorph dinosaurs from the Löwenstein Formation (Norian, Late Triassic) of Germany. *Palaeontology* 46(2): 317–337.]

Einiosaurus procurvicornis

> *Einiosaurus procurvicornis* Sampson, 1995 ⇒ *Centrosaurus procurvicornis* (Sampson, 1995) Paul, 2010 [Campanian; Two Medicine Formation, Montana, USA]

Generic name: *Einiosaurus* ← {(Blackfoot) eini [pron.: eye-knee]: buffalo + (Gr.) σαῦρος/sauros: lizard}; named since the fossils were recovered from the lands on the Black-feet Reservation (*1).

Specific name: *procurvicornis* ← {(Lat.) prōcurvus: forward-curving + -cornis [cornū: horn]}

Etymology: **buffalo lizard** with forward-curving horn

Taxonomy: Ornithischia: Marginocephalia: Ceratopsia: Ceratopsidae: Centrosaurinae

Notes: *Einiosaurus* was a herbivorous dinosaur with a complex dental battery.

[References: (*1) Sampson SD (1995). Two new horned dinosaurs from the Upper Cretaceous Two Medicine Formation of Montana; with a phylogenetic analysis of the Centrosaurinae (Ornithischia: Ceratopsidae). *Journal of Vertebrate Paleontology* 15(4): 743–760.]

Ekrixinatosaurus novasi

> *Ekrixinatosaurus novasi* Calvo *et al.*, 2004 [Cenomanian–Turonian; Candeleros Formation, Neuquén, Argentina]

Generic name: *Ekrixinatosaurus* ← {(Gr.) ekrixi-: explosion + (Lat.) nato-: born [nāscor: to be born] + (Gr.) σαῦρος/sauros: reptile or lizard}; "referred to the fact that the fossil was discovered after its rocky tomb was dynamited, and the Greek saurus, meaning 'reptile' or 'lizard' " (*1).

Specific name: *novasi* ← {(person's name) Novas + -ī}; "in honor of Dr. Fernando Novas for his important contributions to the study of abelisaurid theropods" (*1).

Etymology: Novas' **expolsion-born lizard**

Taxonomy: Saurischia: Theropoda: Abelisauridae

[References: (*1) Calvo JO, Rubilar-Rogers D, Moreno K (2004). A new Abelisauridae (Dinosauria: Theropoda) from northwest Patagonia. *Ameghiniana* 41(4): 555–563.]

Elaltitan lilloi

> *Elaltitan lilloi* Mannion & Otero, 2012 [Cenomanian–Turonian; Bajo Barreal Formation, Chubut, Argentina]

Generic name: *Elaltitan* ← {(myth.) Elal [pron.: "ee-lal" (*1)] + (Gr.myth.) Τῑτάν/Tītan : giant}; referring to "the creator god of the Tehuelche people of Chubut Province" (*1).

Specific name: *lilloi* ← {(person's name) Lillo + -ī}; "in honor of Miguel Lillo, for his contribution and legacy to natural sciences in Tucumán"[*1].
Etymology: Lillo's **Elal titanosaur**
Taxonomy: Saurischia: Sauropodomorpha: Titanosauria
cf. Sauropoda: Neosauropoda: Titanosauria: Lithostrotia [*1]

[References: (*1) Mannion PD, Otero A (2012). A reappraisal of the Late Cretaceous Argentinean sauropod dinosaur *Argyrosaurus superbus*, with a description of a new titanosaur genus. *Journal of Vertebrate Paleontology* 32(3): 614–638.]

Elaphrosaurus bambergi

> *Elaphrosaurus bambergi* Janensch, 1920 [Kimmeridgian; Tendaguru Formation, Mtwara, Tanzania]

Generic name: *Elaphrosaurus* ← {(Gr.) ἐλαφρός/elaphros: light, light-weight, nimble + σαῦρος/sauros: lizard}
Specific name: *bambergi* ← {(person's name) Bamberg + -ī}; in honor of Paul Bamberg who was an expedition patron.
Etymology: Bamberg's **fleet lizard**
Taxonomy: Theropoda: Noasauridae: Elaphrosaurinae
Other species:

> *Elaphrosaurus iguidiensis* Lapparent, 1960 (*nomen dubium*) [early Berriasian; Irhazer Shales Formation, Agadez, Niger]
> *Elaphrosaurus gautieri* Lapparent, 1960 ⇒ *Spinostropheus gautieri* (Lapparent, 1960) Sereno *et al.*, 2004 [Berriasian–Barremian; Tiourarén Formation, Irhazer Group, Agadez, Niger]
> *Coelurus fragilis* Marsh, 1879 [Kimmeridgian; Lake Como Member, Morrison Formation, Wyoming, USA]
= *Coelurus agilis* Marsh, 1884 ⇒ *Elaphlosaurus agilis* (Marsh, 1884) Russell *et al.*, 1980 [Kimmeridgian; Lake Como Member, Morrison Formation, Wyoming, USA]
> *Tanycolagreus topwilsoni* Carpenter *et al.*, 2005 [Kimmeridgian; Morrison Formation, Wyoming, USA]
= *Elaphrosaurus philtippettensis* / *philtippettorum* Pickering, 1995 (*nomen nudum*) [Kimmeridgian; Morrison Formation, Colorado, USA]

Elasmosaurus platyurus [Plesiosauria]

> *Elasmosaurus platyurus* Cope, 1868 [Campanian; Niobrara Formation, Kansas, USA]

Generic name: *Elasmosaurus* ← {(Mod. Gr.) ἔλασμα/elasma: metal plate + (Gr.) σαῦρος/sauros: lizard}; from the caudal laminae, and the great plate bones of the sternal and pelvic regions[*1].
Specific name: *platyurus* ← {(Gr.) πλατύς/platys: wide, broad + οὐρά/ūrā: tail}
Etymology: **plate lizard** with a broad tail
Taxonomy: Sauropterygia: Plesiosauria: Elasmosauridae
Notes: Cope erroneously placed the head of the skeleton of *Elasmosaurus* at the end of the tail*. Marsh made light of the error, which became part of their "Bone Wars" rivalry.
*File: Cope Elasmosaurus.jpg

[References: (*1) Cope E D (1868). Remarks on a new large enaliosaurian, *Elasmosaurus platyurus*. *Proceedings of the Academy of Natural Sciences of Philadelphia* 20: 92–93.]

Elmisaurus rarus

> *Elmisaurus rarus* Osmólska, 1981 ⇒ *Chirostenotes rarus* (Osmólska, 1981) Paul, 1988 [Maastrichtian; Nemegt Formation, Ömnögovi, Mongolia]

Generic name: *Elmisaurus* ← {(Mong.) elmyi: pes + (Gr.) σαῦρος/sauros: lizard} [*1]
Specific name: *rarus* ← {(Lat.) rārus: rare} [*1]
Etymology: rare **foot lizard**
Taxonomy: Theropoda: Oviraptorosauria: Caenagnathoidea: Caenagnathidae / Elmisauridae: Elmisaurinae [*2]
cf. Theropoda: ?Coelurosauria: Elmisauridae [*1]
Other species:

> *Ornithomimus elegans* Parks, 1933 ⇒ *Macrophalangia elegans* (Parks, 1933) Koster *et al.*, 1987 ⇒ *Elmisaurus elegans* (Parks, 1933) Currie, 1989 ⇒ *Chirostenotes elegans* (Parks, 1933) Sues, 1997 ⇒ *Leptorhynchos elegans* (Parks, 1933) Longrich *et al.*, 2013 ⇒ *Citipes elegans* (Parks, 1933) Funston, 2020 [Campanian; Dinosaur Park Formation, Alberta, Canada]

[References: (*1) Osmólska H (1981). Coossified tarsometatarsi in theropod dinosaurs and their bearing on the problem of bird origins. *Palaeontologia Polonica* 42: 79–95.; (*2) Currie PJ, Funston GF, Osmólska H (2016). New specimens of the crested theropod dinosaur *Elmisaurus rarus* from Mongolia. *Acta Palaeontologica Polonica* 61(1): 143–157.]

Elopteryx nopcsai

> *Elopteryx nopcsai* Andrews, 1913 [Maastrichtian; Sânpetru Formation, Transylvania, Romania]

Generic name: *Elopteryx* ← {(Gr.) ἕλος/helos: marsh + πτέρυξ/pteryx: wing}
Specific name: *nopcsai* ← {(person's name) Nopcsa

+ -ī}; in honor of Hungarian paleontologist Franz Nopcsa von Felsö - Szilvás.

Etymology: Nopcsa's **marsh wing**

Taxonomy: Theropoda: Maniraptoriformes: Maniraptora

cf. Theropoda: Tetanurae: Avetheropoda: Coelurosauria: Maniraptoriformes: Maniraptora: ?Alvarezsauridae (*2)

Notes: Charles William Andrews originally considered fragments of limb-bones to be those of a bird (*1). Many paleontologists consider *Elopteryx* to be a *nomen dubium*.

[References: (*1) Andrews CW (1913). On some bird remains from the Upper Cretaceous of Transylvania. *Geological Magazine* 5: 193–196.; (*2) Kessler E, Grigorescu D, Csiki Z (2005). *Elopteryx* revised—A new bird-like specimen from the Maastrichtian of the Haţeg basin (Romania). *Acta Palaeontologica Romaniae* 5: 249–258.]

Elrhazosaurus nigeriensis

> *Valdosaurus nigeriensis* Galton & Taquet, 1982 ⇒ *Elrhazosaurus nigeriensis* (Galton & Taquet, 1982) Galton, 2009 [Aptian–Albian; Elrhaz Formation, Agadez, Niger]

Generic name: *Elrhazosaurus* ← {(place-name) Elrhaz Formation + (Gr.) σαῦρος/sauros: lizard}

Specific name: *nigeriensis* ← {(place-name) Niger + -ī + -ensis}

Etymology: **Elrhaz Formation lizard** from Niger

Taxonomy: Ornithischia: Ornithopoda: Iguanodontia: Dryosauridae

Elsornis keni [Avialae]

> *Elsornis keni* Chiappe *et al.*, 2007 [Campanian; Djadokhta Formation, Ömnögovi, Mongolia]

Generic name: *Elsornis* ← {(Mong.) элс/els: sand + (Gr.) ὄρνις/ornis: bird}

Specific name: *keni* ← {(person's name) Ken + -ī}; in honor of Mr. Ken Hayashibara.

Etymology: Ken's **sand bird**

Taxonomy: Theropoda: Avialae: Enatiornithes

[References: Chiappe LM, Suzuki S, Dyke GJ, Watabe M, Tsogtbaatar K, Barsbold R (2007). A new enantiornithine bird from the Late Cretaceous of the Gobi desert. *Journal of Systematic Palaeontology* 5(2): 193–208. Abstract.]

Emausaurus ernsti

> *Emausaurus ernsti* Haubold, 1990 [Toarcian; unnamed unit, Grimmen, Mecklenburg-Vorpommern, Germany]

Generic name: *Emausaurus* ← {(acronym) EMAU (= Ernst Moritz Arndt University of Greifswald) + (Gr.) σαῦρος/sauros: lizard}

Specific name: *ernsti* ← {(person's name) Ernst + -ī}; in honor of geologist Werner Ernst, who found the fossils.

Etymology: Ernst's **Ernst Moritz Arndt University of Greifswald lizard**

Taxonomy: Ornithischia: Thyreophora

Notes: The specimen, which is estimated 2 m long, is considered to be a juvenile.

[References: Haubold H (1990). Ein neuer dinosaurier (Ornithischia, Thyreophora) aus dem unteren Jura des nördlichen Mitteleuropa. *Revue de Paleobiologie* 9(1): 149–177.]

Embasaurus minax

> *Embasaurus minax* Riabinin,1931 [Berriasian; Mount Koi-Kara, Kazakhstan]

Generic name: *Embasaurus* ← {(place-name) Emba River + (Gr.) σαῦρος/sauros: lizard}

Specific name: *minax* ← {(Lat.) minā̆x: projecting, menacing}.

Etymology: menacing, **Emba River lizard**

Taxonomy: Theropoda

Notes: According to *The Dinosauria 2nd edition*, *Embasaurus minax* is considered to be a *nomen dubium*.

Enaliornis barretti [Avialae]

> *Pelagornis barretti* Seeley, 1866 (preoccupied) ⇒ *Enaliornis barretti* Seeley, 1876 [Cenomanian; Cambridge Greensand, Cambridge, UK]

Generic name: *Enaliornis* ← {(Gr.) ἐνάλιος/enalios: belonging to the sea, of the sea + ὄρνις/ornis: bird}

Specific name: *barretti* ← {(person's name) Barrett + -ī}; in honor of "Lucas Barrett of the Woodwardian Museum, the discoverer of the first bones of *Enaliornis*" (*1).

Etymology: Barrett's **sea bird**

Taxonomy: Theropoda: Ornithurae: Hesperornithiformes: Hesperornithes

Other species:

> *Enaliornis sedgwicki* Seeley, 1876 [Albian; Cambridge Greensand, Cambridge, UK]

> *Enaliornis seeleyi* Galton & Martin, 2002 [Albian; Cambridge Greensand, Cambridge, UK]

Notes: The original name, *Pelagornis* named by Seeley in 1866, was preoccupied by a Miocene bird. *Enaliornis barretti* is the largest of the three species. (*1)

[References: (*1) Galton P, Martin LD (2002). Postcranial anatomy and systematics of *Enaliornis* Seeley, 1876, a foot-propelled diving bird (Aves: Ornithurae: Hesperornithiformes) from the Early Cretaceous of England. *Revue de Paleobiologie* 21(2): 489–538.]

Enantiophoenix electrophyla [Avialae]

> *Enantiophoenix electrophyla* Cau & Arduini, 2008 [Cenomanian; Beirut, Lebanon]

Generic name: *Enantiophoenix* ← {(Gr.) ἐναντίος/ enantios: opposite + (myth.) phoenix}; "referring to the enatiornithine status of MSNM* V3882 (Dalla Vecchia & Chiappe, 2002) and to the mythological bird "Phoenix", which alludes to the ancient name of Lebanon" (*1).

* Museo di Storia Naturale di Milano

Specific name: *electrophyla* ← {(Gr.) ἤλεκτρον/ electron: electron, amber + φυλα/phyla} "that likes amber" (*1); "referring to the presence of amber corpuscles scattered between its bones (Dalla Vecchia & Chiappe, 2002)" (*1).

Etymology: **opposite phoenix** with amber corpuscles

Taxonomy: Theropoda: Avialae: Enantiornithes: Avisauridae

cf. Theropoda: Aves: Ornithothoraces: Enantiornithes: Euenantiornithes: Avisauridae (*1)

[References: (*1) Cau A, Arduini P (2008). *Enantiophoenix electrophyla* gen. *et* sp. nov. (Aves, Enantiornithes) from the Upper Cretaceous (Cenomanian) of Lebanon and its phylogenetic relationships. *Atti della Societa Italiana di Scienze Naturali e del Museo ivico di Storia Naturale in Milano* 149(2): 293–324.]

Enantiornis leali [Avialae]

> *Enantiornis leali* Walker, 1981 [Maastrichtian; Lecho Formation, Jujuy and Salta, Argentina]

Generic name: *Enantiornis* ← {(Gr.) ἐναντίος/ enantios: opposite + ὄρνις/ornis: bird}; referring to the way the joint articulates is reversed.

Walker did not describe the etymology clearly but he explained that *Enantiornis* was different from modern birds in the nature of the articulation between the scapula (shoulder blade) and the coracoid (*1).

Specific name: *leali*

Etymology: Leal's **opposite bird**

Taxonomy: Theropoda: Maniraptora: Avialae: Enantiornithes

Other species:

> *Enantiornis martini* Nessov & Panteleyev, 1993 ⇒ *Incolornis martini* (Nessov & Panteleyev, 1993) Panteleyev, 1998 [Late Cretaceous; Uzbekistan]

> *Enantiornis walkeri* Nessov & Panteleyev, 1993 ⇒ *Explorornis walkeri* (Nessov & Panteleyev, 1993) Panteleyev, 1998 [Late Cretaceous; Uzbekistan]

[References: (*1) Walker CA (1981). New subclass of birds from the Cretaceous of South America. *Nature* 292: 51–53. Abstract.]

Enigmosaurus mongoliensis

> *Enigmosaurus mongoliensis* Barsbold & Perle., 1983 [Cenomanian–Santonian; Bayan Shireh Formation, Dornogovi, Mongolia]

Generic name: *Enigmosaurus* ← {(Gr.) enigmo-: mysterious (*1) [(Gr.) αἴνιγμα/ ainigma: riddle] + σαῦρος/sauros: lizard}; after the puzzling and unusual shape of its pelvis.

Specific name: *mongoliensis* ← {(place-name) Mongolia + -ensis}

Etymology: **mysterious lizard** from Mongolia

Taxonomy: Theropoda: Coelurosauria: Therizinosauroidea

[References: (*1) Barsbold R (1983). Хищные динозавры Мела Монголии [Carnivorous dinosaurs from the Cretaceous of Mongolia]. *Transactions: The Joint Soviet-Mongolian Paleontological Expedition.* 19: 5–119.]

Eoabelisaurus mefi

> *Eoabelisaurus mefi* Pol & Rauhut, 2012 [Aalenian–Bajocian; Cañadón Asfalto Formation, Chubut, Argentina]

Generic name: *Eoabelisaurus* ← {(Gr.) ἠώς/ēōs: dawn + *Abelisaurus* [(person's name) Abel + (Gr.) σαῦρος/sauros: lizard}; "for the early occurrence of the new taxon, and *Abelisaurus*, type genus of the Abelisauridae" (*1).

Specific name: *mefi* ← {(acronym) MEF + -ī}; referring to "the popular abbreviation of the Museo of Paleontológico Egidio Feruglio (MEF)" (*1).

Etymology: MEF's **dawn Abel's lizard (*Abelisaurus*)**

Taxonomy: Theropoda: Ceratosauria: Abelisauridae

[References: (*1) Pol D, Rauhut OWM (2012). A Middle Jurassic abelisaurid from Patagonia and the early diversification of theropod dinosaurs. *Proceedings of the Royal Society B. Biological Sciences* 279 (1741): 3170–3175.]

Eoalulavis hoyasi [Avialae]

> *Eoalulavis hoyasi* Sanz *et al.*, 1996 [Barremian; La Huérguina Formation, Las Hoyas, Cuenca, Spain]

Generic name: *Eoalulavis* ← {(Gr.) Ἠώς/Ēōs: dawn + (Lat.) alula*: bastard wing + (Lat.) avis: bird}; "because it provides the oldest evidence of an alula" (*1).

*alula: winglet, diminutive of (Lat.) ala: wing.

Specific name: *hoyasi* ← {(Spanish) Hoyas + -ī}; referring to "Las Hoyas fossil site" (*1).

Etymology: **dawn bastard wing bird** from Las Hoyas

Taxonomy: Theropoda: Avialae: Enantiornithes

cf. Aves: Ornithothoraces: Enantiornithes (*1)

Notes: Wingspan of about 17 cm.

[References: (*1) Sanz JL, Chiappe LM, Pérez-Moreno BP, Buscalioni AD, Moratalla JJ, Ortega F, Poyato-Ariza FJ (1996). An early Cretaceous bird from Spain and its implications for the evolution of avian flight. *Nature* 382 (6590): 442–445.]

Eobrontosaurus yahnahpin

> *Apatosaurus yahnahpin* Filla & Redman, 1994 ⇒ *Eobrontosaurus yahnahpin* (Filla & Redman, 1994) Bakker, 1998 ⇒*Brontosaurus yahnahpin* (Filla & Redman, 1994) Tschopp *et al.*, 2015 [Kimmeridgian–Tithonian; Morrison Formation, Wyoming, USA]

Generic name: *Eobrontosaurus* ← {(Gr.) ἠώς/ēōs: dawn + *Brontosaurus* [βροντή/brontē: thunder + σαῦρος/sauros: lizard]}

Specific name: *yahnahpin* ← {(Lakota) mah-koo yah-nah-pin: 'breast necklace'}; referring to the pairs of sternal ribs.

Etymology: **dawn *Brontosaurus* (thunder lizard)** with breast necklace

Taxonomy: Saurischia: Sauropodomorpha: Diplodocidae: Apatosaurinae

Notes: Bakker (1998) coined the new generic name *Eobrontosaurus* (*1), but according to Tschopp *et al.* (2015), *Eobrontosaurus yahnahpin* is considered as a species of *Brontosaurus* (*2).

[References: (*1) Bakker RT (1998). Dinosaur mid-life crisis: the Jurassic-Cretaceous transition in Wyoming and Colorado. In: Lucas SG, Kirkland JI, Estep JW (eds), *Lower and Middle Cretaceous Terrestrial Ecosystems*, New Mexico Museum of NaturalHistory and Science Bulletin: 67–77.; (*2) Tschopp E, Mateus O, Benson RBJ (2015). A specimen-level phylogenetic analysis and taxonomic revision of Diplodocidae (Dinosauria, Sauropoda). *PeerJ* 3: e857.]

Eocarcharia dinops

> *Eocarcharia dinops* Sereno & Brusatte, 2008 [Aptian–Albian; Elrhaz Formation, Ténéré, Niger]

Generic name: *Eocarcharia* ← {(Gr.) ἠώς/ēōs: dawn + (Mod. Gr.) καρχαρίας/karcharias : shark}; "in reference to its basal position in the "shark-toothed" theropod clade Carcharodontosauridae" (*1).

Specific name: *dinops* ← {(Gr.) δεινός/deinos: fierce + ὤψ/ōps: eye}; "in reference to the massive ornamented brow above the orbit" (*1).

Etymology: fierce-eyed, **dawn shark's one (member of carcharodontosaurs)**

Taxonomy: Theropoda: Carnosauria: Allosauroidea: Carcharodontosauridae

[References: (*1) Sereno PC, Brusatte SL (2008). Basal abelisaurid and carcharodontosaurid theropods from the Lower Cretaceous Elrhaz Formation of Niger. *Acta Palaeontologica Polonica* 53(1): 15–46.]

Eocathayornis walkeri [Avialae]

> *Eocathayornis walkeri* Zhou, 2002 [Aptian; Jiufotang Formation, Liaoning, China]

Generic name: *Eocathayornis* ← {(Gr.) eo- [ἠώς/ēōs: dawn] + Cathay: old poetic name for China +ὄρνις/ornis: bird}

Specific name: *walkeri* ← {(person's name) Walker + -ī}; "dedicated to C. A. Walker who first published and recognized the significance of the Enantiornithes" (*1).

Locality: Walker's **dawn Cathay (China) bird**

Taxonomy: Theropoda: Avialae: Enantiornithes

cf. Aves: Enantiornithes: Cathayornithiformes: Cathayornithidae (*1)

[References: (*1) Zhou Z (2002). A new and primitive enantiornithine bird from thr Early Cretaceous of China. *Journal of Vertebrate Paleontology* 22(1): 49–57.]

Eoconfuciusornis zhengi [Avialae]

> *Eoconfuciusornis zhengi* Zang *et al.*, 2008 [Hauterivian; Dabeigou Formation, Hebei, China]

Generic name: *Eoconfuciusornis* ← {(Gr.) ἠώς/ēōs: dawn + *Confuciusornis* [(Lat. person's name) Confucius + ὄρνις/ornis: bird]}; "indicating that some features of this new bird are more primitive than other confuciusornithid birds" (*1).

Specific name: *zhengi* ← {(Chin. person's name) Zhèng鄭 + -ī}; "dedicated to the distinguished Chinese ornithologist Zhèng Guāngměi" (*1).

Locality: Zheng's **dawn *Confuciusornis***

Taxonomy: Theropoda: Avialae: Confuciusornithidae

[References: (*1) Zhang F-C, Zhou Z-H, Benton M J (2008). A primitive confuciusornithid bird from China and its implications for early avian flight. *Science in China Series D: Earth Sciences* 51(5): 625–639.]

Eocursor parvus

> *Eocursor parvus* Butler *et al.*, 2007 [Hettangian; upper Elliot Formation*, Free State, South Africa]

*According to Butler *et al.* (2007), originally, *Eocursor* was thought to be from the lower Elliot Formation, of Late Triassic age, but McPhee *et al.* (2017) reinterpreted it.

Generic name: *Eocursor* ← {(Gr.) ἠώς/ēōs: dawn + (Lat.) cursor: runner}; "in reference to the early occurrence of this ornithischian, its apparent locomotory abilities" (*1).

Specific name: *parvus* ← {(Lat.) parvus: little,

small}
Etymology: small, **dawn runner**
Taxonomy: Ornithischia

[References: (*1) Butler RJ, Smith RMH, Norman DB (2007). A primitive ornithischian dinosaur from the Late Triassic of South Africa, and the early evolution and diversification of Ornithischia. *Proceedings of the Royal Society B: Biological Sciences* 274 (1621): 2041–2046.]

Eodromaeus murphi

> *Eodromaeus murphi* Martinez *et al.*, 2011 [Carnian; Ischigualasto Formation, San Juan, Argentina]

Generic name: *Eodromaeus* ← {(Gr.) ἠώς/ēōs: dawn + δρομεύς/dromeus: runner}; "in allusion to its early age, and slender axial and appendicular proportions"(*1).
Specific name: *murphi* ← {(person's name) Murphy + -ī}; in honor of Jim Murphy, "the Earthwatch volunteer, who discovered the holotype specimen" (*1).
Etymology: Murphy's **dawn runner**
Taxonomy: Theropoda

[References: (*1) Martinez RN, Sereno PC, Alcober OA, Colombi CE, Renne PR, Montañez IP, Currie BS (2011). A basal dinosaur from the dawn of the dinosaur era in southwestern Pangaea. *Science* 331(6014): 206–210.]

Eoenantiornis buhleri [Avialae]

> *Eoenantiornis buhleri* Hou *et al.*, 1999 [Barremian–Aptian; Yixian Formation, Liaoning, China]

Generic name: *Eoenantiornis* ← {(Gr.) ἠως/ēōs: dawn + *Enantiornis* [(Gr.) ἐναντίος/enatios: opposite + ὄρνις/ornis: bird]} (*1)
Specific name: *buhleri* ← {(person's name) Buhler + -ī}; in honor of the late Paul Bühler, a distinguished German functional morphologist and paleornithologist (*1).
Etymology: Bühler's **dawn *Enantiornis*** [歩氏始反鳥]
Taxonomy: Theropoda: Avialae: Enantiornithes
Notes: Moderate sized toothed bird.

[References: (*1) Hou L, Martin LD, Zhou Z, Feduccia A (1999). *Archaeopteryx* to opposite birds–missing link from the Mesozoic of China. *Vertebrata PalAsiatica* 37(2): 88–95.]

Eogranivora edentulata [Avialae]

> *Eogranivora edentulata* Zheng *et al.*, 2018 [Barremian–Aptian; Yixian Formation, Lingyuan, Liaoning, China]

Generic name: *Eogranivora* ← {(Gr.) ἠως/eos: early + (Lat.) granivora [granum: a grain, seed + vorō: to devour] seed-eater}; "referring to the crop contents preserved in the holotype" (*1).
Specific name: *edentulata* ← {(Lat.) e-dentulata: edentulous, without teeth}; "referring to the toothless rostrum that characterizes this taxon"(*1).
Etymology: toothless, **early seed-eater**
Taxonomy: Theropoda: Avialae: Ornithothoraces: Ornithuromorpha

[References: (*1) Zheng X, O'Connor JK, Wang X, Wang Y, Zhou Z (2018). Reinterpretation of a previously described Jehol bird clarifies early trophic evolution in the Ornithuromorpha. *Proceedings of the Royal Society B* 285: 20172494.]

Eolambia caroljonesa

> *Eolambia caroljonesa* Kirkland, 1998 [Cenomanian; Cedar Mountain Formation, Utah, USA]

Generic name: *Eolambia* ← {(Gr.) ἠώς/ēōs: dawn + (person's name) Lambe + -ia}; referring to "the early occurrence of this lambeosaurine" and being "a contraction for 'dawn lambeosaurine' "(*1).
Specific name: *caroljonesa* ← {(person's name) Carol Jones + -a}; in honor of "Carol Jones of Salt Lake City, Utah, who discovered the site"(*1).
Etymology: Carol Jones' **dawn lambeosaurine**
Taxonomy: Ornithischia: Ornithopoda: Iguanodontia: Hadrosauroidea
cf. Ornithopoda: Hadrosauridae: ? Lambeosaurinae(*1)

[References: (*1) Kirkland JI (1998). A new hadrosaurid from the upper Cedar Mountain Formation (Albian–Cenominian: Cretaceous) of eastern Utah—the oldest known hadrosaurid (lambeosaurine?) In: Lucas SG, Kirkland JI, Estep JW (eds), *Lower and Middle Cretaceous Terrestrial Ecosystems. New Mexico Museum of Natural History and Science Bulletin.* 14: 283–295.]

Eomamenchisaurus yuanmouensis

> *Eomamenchisaurus yuanmouensis* Lü *et al.*, 2008 [Bajocian; Zhanghe Formation, Yunnan, China]

Generic name: *Eomamenchisaurus* ← {(Gr.) eo- [ἠώς/ēōs: dawn] + *Mamenchisaurus* [(Chin. place-name) Mamenchi馬門溪 + (Gr.) σαῦρος/sauros: lizard]}
Specific name: *yuanmouensis* ← {(Chin. place-name) Yuánmóu元謀 + -ensis}
Etymology: **dawn *Mamenchisaurus*** from Yuanmou
Taxonomy: Saurischia: Sauropodomorpha: Eusauropoda: Mamenchisauridae

[References: Lü J, Li T, Zhong S, Ji Q, Li S (2010). A new mamenchisaurid dinosaur from the Middle

Jurassic of Yuanmou, Yunnan Province, China. *Acta Geologica Sinica* 82(1): 17–26.]

Eoplophysis vetustus

> *Omosaurus vetustus* von Huene, 1910 (preoccupied) ⇒ *Dacentrurus vetustus* (von Huene, 1910) Hennig, 1915 ⇒ *Lexovisaurus vetustus* (von Huene, 1910) Galton *et al.*, 1980 (*nomen dubium*) ⇒ *Eoplophysis vetustus* (von Huene, 1910) Ulansky, 2014 (*nomen dubium*) [Bathonian; Cornbrash Formation, Oxfordshire, UK]

Generic name: *Eoplophysis* ← {(Gr.) ἠως/ēōs: dawn + ὅπλον/hoplon: a heavy shield + φύσις/physis: form}

Specific name: *vetustus* ← {(Lat.) vetustus: ancient}

Etymology: ancient **dawn armed shield**

Taxonomy: Ornithischia: Stegosauria: Stegosauridae

Eoraptor lunensis

> *Eoraptor lunensis* Sereno *et al.*, 1993 [Carnian; Ischigualasto Formation, San Juan, Argentina]

Generic name: *Eoraptor* ← {(Gr.) ἠώς/ēōs: dawn + (Lat.) raptor: plunderer}; "in reference to its primitive structure and early temporal occurrence" and "in reference to its carnivorous habits and grasping hand" (*1).

Specific name: *lunensis* ← {(Lat.) luna: moon + -ensis}; "in reference to the type locality" (*1). The place of discovery is called the 'Valle de la Luna'.

Etymology: **dawn plunderer** from the Valley of the Moon

Taxonomy: Saurischia: Eusaurischia

cf. Saurischia: Sauropodomorpha (*2)

cf. Theropoda (*1) (*3)

Notes: In 2017, Baron *et al.* found *Eoraptor* to be earliest diverging member of Theropoda (*3).

[References: (*1) Sereno PC, Forster CA, Rogers RR, Monetta AM (1993). Primitive dinosaur skeleton from Argentina and the early evolution of Dinosauria. *Nature* 361: 64–66.; (*2) Sereno PC, Martínez RN, Alcober OA (2013). Osteology of *Eoraptor lunensis* (Dinosauria, Sauropodomorpha). *Journal of Vertebrate Paleontology* 32(1)12: 83–179.; (*3) Baron MG, Norman DB, Barrett PM (2017). A new hypothesis of dinosaur relationships and early dinosaur evolution. *Nature* 543: 501–506.]

Eosinopteryx brevipenna [Anchiornithidae]

> *Eosinopteryx brevipenna* Godefroit *et al.*, 2013 [Oxfordian; Tiaojishan Formation, Liaoning, China]

Generic name: *Eosinopteryx* ← {(Gr.) eo- [ἠως/eos: daybreak, dawn] + (place-name) Sino-: Chinese + (Gr.) πτέρυξ/pteryx: wing, feather (*1)}

Specific name: *brevipenna* ← {(Lat.) brevi-: short + penna: feather} (*1)

Etymology: **dawn Chinese wing** with short feather

Taxonomy: Theropoda: Avialae (?): Anchiornithidae*

*Anchiornithidae Xu *et al.*, 2016 *sensu* Foth & Rauhut, 2017

cf. Theropoda: Maniraptora: Troodontidae (*1)

[References: (*1) Godefroit P, Demuynck H, Dyke G, Hu D, Escuillié F, Claeys P (2013). Reduced plumage and flight ability of a new Jurassic paravian theropod from China. *Nature Communications* 4: 1394.]

Eotrachodon orientalis

> *Eotrachodon orientalis* Prieto-Márquez *et al.*, 2016 [Santonian; Mooreville Chalk Formation, Montgomery, Alabama, USA]

Generic name: *Eotrachodon* ← {(Gr.) ἠώς/ēōs: dawn + *Trachodon* [(Gr.) τραχύς/trachys: rough + ὀδών/odōn: tooth}; paying "homage to the first hadrosaurid genus described (Leidy, 1856), long considered a *nomen dubium* (Lambe, 1918)" (*1).

Specific name: *orientalis* ← {(Lat.) orientālis: eastern}; "in reference to the southeastern occurrence in North America of this hadrosaurid" (*1).

Etymology: **dawn *Trachodon*** from the east

Taxonomy: Ornithischia: Ornithopoda: Iguanodontia: Hadrosauridae

[References: (*1) Prieto-Márquez A, Erickson GM, Ebersole JA (2016). A primitive hadrosaurid from southeastern North America and the origin and early evolution of 'duck-billed' dinosaurs. *Journal of Vertebrate Paleontology* 36(2): e1054495.]

Eotriceratops xerinsularis

> *Eotriceratops xerinsularis* Wu *et al.*, 2007 [Maastrichtian; Horseshoe Canyon Formation, Alberta, Canada]

Generic name: *Eotriceratops* ← {(Gr.) ἠώς/ēōs: dawn + *Triceratops* [(Gr.) τρι-/tri-: three + κέρας/keras: horn + ὤψ/ōps: face]}; implying that "it is an early member of the '*Triceratops*' group" (*1).

Specific name: *xerinsularis* ← {(Gr.) ξηρός/xēros: dry + (Lat.) insularis: of island}; "referring to the Dry Island Buffalo Jump Provincial Park where the specimen was collected" (*1).

Etymology: **dawn member of *Triceratops* group** of the Dry Island

Taxonomy: Ornithischia: Ceratopsia: Ceratopsidae: Chasmosaurinae

[References: (*1) Wu X-C, Brinkman DB, Eberth DA, Braman DR (2007). A new ceratopsid dinosaur (ornithischian) from the uppermost Horseshoe Canyon Formation (upper Maastrichtian), Alberta, Canada. *Canadian Journal of Earth Sciences* 44(9): 1243–1265.]

Eotyrannus lengi

> *Eotyrannus lengi* Hutt *et al.*, 2001 [Berriasian–Barremian; Wessex Formation, Isle of Wight, UK]

Generic name: *Eotyrannus* ← {(Gr.) ἠώς/ēōs: dawn, early + (Lat.) tyrannus: tyrant}; "in allusion to tyrannosauroids as 'tyrant dinosaurs' " (*1).

Specific name: *lengi* ← {(person's name) Leng + -ī}; "after Mr. Gavin Leng, the discoverer" (*1).

Etymology: Leng's **early tyrant**

Taxonomy: Theropoda: Thetanurae: Avetheropoda: Coelurosauria: Tyrannosauroidea

Notes: The specimen is estimated 4 meters long and considered a juvenile.

[References: (*1) Hutt S, Naish D, Martill DM, Barker MJ, Newbery P (2001). A preliminary account of a new tyrannosauroid theropod from the Wessex Formation (Early Cretaceous) of southern England. *Cretaceous Research* 22(2): 227–242.]

Eousdryosaurus nanohallucis

> *Eousdryosaurus nanohallucis* Escaso *et al.*, 2014 [Kimmeridgian; Alcobaça Formation, Lourinhã, Portugal]

Generic name: *Eousdryosaurus* ← {(Lat.) Ēōus: eastern + *Dryosaurus* [(Gr.) δρῦς/drys: a tree, the oak + σαῦρος/sauros: lizard]}

Specific name: *nanohallucis* ← {(Lat.) nānus = (Gr.) νᾶνος/nānos: a dwarf + [biol.] hallux [(Lat.) (h) allus]: big toe}

Etymology: **eastern *Dryosaurus*** with a small hallux

Taxonomy: Ornithischia: Ornithopoda: Dryosauridae

Epachthosaurus sciuttoi

> *Epachthosaurus sciuttoi* Powell, 1990 [Campanian–Maastrichtian; Bajo Barreal Formation, Chubut, Argentina]

Generic name: *Epachthosaurus* ← {(Gr.) ἐπαχθής/epachthēs: heavy + σαῦρος/sauros: lizard}

Specific name: *sciuttoi* ← {(person's name) Sciutto + -ī}; in honor of geologist Juan Carlos Sciutto.

Etymology: Sciutto's **heavy lizard**

Taxonomy: Saurischia: Sauropodomorpha: Sauropoda: Titanosauria

Epichirostenotes curriei

> *Epichirostenotes curriei* Sullivan *et al.*, 2011 [Campanian; Horseshoe Canyon Formation, Alberta, Canada]

Generic name: *Epichirostenotes* ← {(Gr.) ἐπί/epi: upon, after + *Chirostenotes* [χείρ/cheir: hand + στενότης/stenotēs: narrowness]}

Specific name: *curriei* ← {(person's name) Currie + -ī}; "in honor of Philip J. Currie for his scholarly work concerning Late Cretaceous theropod dinosaurs" (*1).

Etymology: Currie's, **after-*Chirostenotes***

Taxonomy: Theropoda: Oviraptorosauria: Caenagnathidae

[References: (*1) Sullivan RM, Jasinski SE, Mark PA, van Tomme MPA (2011). A new caenagnathid *Ojoraptorosaurus boerei*, n. gen., n. sp. (Dinosauria Oviraptorosauria), from the Upper Cretaceous Ojo Alamo Formation (Naashoibito Member), San Juan Basin, New Mexico. *Fosil Record 3. New Mexico Museum of Natural History and Science, Bulletin* 53: 418–428.]

Epidendrosaurus ningchengensis [Scansoriopterygidae]

> *Scansoriopteryx heilmanni* Czerkas & Yuan, 2002(*1) [uncertain; uncertain, Liaoning, China]

= *Epidendrosaurus ningchengensis* Zhang *et al.*, 2002 [Callovian–Kimmeridgian; Daohugou Beds, Tiaojishan Formation, Ningcheng, Inner Mogolia, China]

Generic name: *Epidendrosaurus* ← {(Gr.) ἐπί/epi: upon + δένδρον/dendron: tree + σαῦρος/sauros: lizard}; "derived from the obvious arboreal adaptation of this animal" (*2).

Specific name: *ningchengensis* ← {(Chin. place-name) Níngchéng寧城 + -ensis}

Etymology: **tree-dweling lizard** from Ningcheng

Taxonomy: Theropoda: Coelurosauria: Maniraptora: Scansoriopterygidae

cf. Theropoda: Coelurosauria: Avialae *incertae sedis* (acc. *The Dinosauria 2nd edition*)

cf. Theropoda: Coelurosauria: Maniraptora (*2)

Notes: According to *The Dinosauria 2nd edition*, *Scansoriopteryx* was regarded as a junior synonym of *Epidendrosaurus*, but Article 21 of ICZN seems to give priority to *Scansoriopteryx*.

[References: (*1) Padian K (2004). Basal Avialae. In: *The Dinosauria 2nd edition*. University of California Press, Berkeley: 210–231.; (*2) Zhang F, Zhou Z, Xu X, Wang X (2002). A juvenile coelurosaurian theropod from China indicates arboreal habits. *Naturwissenschaften* 89: 394–398.]

Epidexipteryx hui [Scansoriopterygidae]

> *Epidexipteryx hui* Zhang *et al.*, 2008 [Callovian; Daohugou Beds, Inner Mongolia, China]

Generic name: *Epidexipteryx* ← {(Gr.) ἐπίδειξις*/epideixis: display + πτέρυξ/pteryx: wing, feather}
* = ἐπίδεξις

Specific name: *hui* ← {(Chin. person's name) Hú胡 + -ī}; "in honour of the late young palaeontologist Yaoming Hu, who contributed significantly to the study of Mesozoic mammals from China" (*1).

In China *Epidexipteryx* is called Hushiyaolong.
Etymology: Hu's **display feather** [胡氏耀龍]
Taxonomy: Theropoda: Coelurosauria: Maniraptora: Avialae: Scansoriopterygidae [*1]

[References: (*1) Zhang F, Zhou Z, Xu X, Wang X, Sullivan C (2008). A bizarre Jurassic maniraptoran from China with elongate ribbon-like feathers. *Nature* 455: 1105–1108.]

Equijubus normani

> *Equijubus normani* You *et al.*, 2003 [Aptian; Zhonggou Formation, Gansu, China]

Generic name: *Equijubus* ← {(Lat.) equus: horse + juba: mane}; "Horse Mane is what 'Ma Zong' means in Chinese, and Ma Zong Mountain is where the fossil was discovered" [*1].
Specific name: *normani* ← {(person's name) Norman + -ī}; "in honor of Dr. David B. Norman for his work on ornithopod dinosaurs" [*1].
Etymology: Norman's **one from Mazong (horse mane) Mountain**
Taxonomy: Ornithischia: Ornithopoda: Iguanodontia: Hadrosauroidea

[References: (*1) You H-l, Luo Z-X, Shubin NH, Witmer LM, Tang Z-l, Tang F (2003). The earliest-known duck-billed dinosaur from deposits of late Early Cretaceous age in northwest China and hadrosaurid evolution. *Cretaceous Research* 24: 347–355.]

Erectopus superbus

> *Megalosaurus superbus* Sauvage, 1882 ⇒ *Erectopus superbus* (Sauvage, 1882) von Huene, 1923 [early Albian; La Penthiève Beds, France]

Generic name: *Erectopus* ← {(Lat.) ērēctus: upright, erect + (Gr.) πούς/pous [pūs]: foot}
Specific name: *superbus* ← {(Lat.) superbus: proud}
Etymology: proud, **upright foot**
Taxonomy: Theropoda: Tetanurae: Allosauroidea
Other species:

> *Erectopus sauvagei* von Huene, 1932 [Albian; France] (a species or synonym of *E. superbus*)

> *Megalosaurus insignis* Eudes-Deslongchamps & Lennier *vide* Lennier, 1870 ⇒ *Erectopus insignis* (Eudes-Deslongchamps & Lennier *vide* Lennier, 1870) Stromer, 1931 (*nomen dubium*) [Kimmeridgian; Marnes à Deltoideum delta Formation, Normandy, France]

Erketu ellisoni

> *Erketu ellisoni* Ksepka & Norell, 2006 [Cenomanian–Santonian*; Bayan Shireh Formation, Dornogovi, Mongolia]

*late Early Cretaceous. The exact age of this locality is uncertain. (according to Ksepka & Norell, 2006)

Generic name: *Erketu* (Mong. shamanistic tradition) Erketü Tengri; "the Mighty Tengri, a creator-god who called Yesügei, the father of Chingis Khan, into being" [*1].
Specific name: *ellisoni* ← {(person's name) Ellison + -ī}; "in honor of Mick Ellison, for his contributions to ongoing American Museum of Natural History AMNH* dinosaur research" [*1]
*American Museum of Natural History
Etymology: Ellison's **Erketü Tengri**
Taxonomy: Saurischia: Sauropodomorpha: Sauropoda: Titanosauriformes: Somphospondyli
Notes: Erketu had an exteremely long neck.

[References: (*1) Ksepka DT, Norell MA (2006). *Erketu ellisoni*, a long-necked sauropod from Bor Guvé (Dornogov Aimag, Mongolia). *American Museum Novitates* 3508: 1–16.]

Erliansaurus bellamanus

> *Erliansaurus bellamanus* Xu *et al.*, 2002 [Santonian; Iren Dabasu Formation, Sunitezuoqi, Inner Mongolia, China]

Generic name: *Erliansaurus* ← {(place-name) Erlian = Erenhot (二連浩特) + (Gr.) σαῦρος/sauros: lizard}
Specific name: *bellamanus* ← {(Lat.) bellus: beautiful + manus: hand}; for the well preserved manus of the holotype [*1].
Etymology: **Erlian lizard** with a beautiful forelimb
Taxonomy: Theropoda: Therizinosauroidea: Therizinosauridae

[References: (*1) Xu X, Zhang Z-H, Sereno PC, Zhao X-J, Kuang X-W, Han J, Tan L (2002). A new therizinosauroid (Dinosauria, Theropoda) from the Upper Cretaceous Iren Dabasu Formation of Nei Mongol. *Vertebrata PalAsiatica* 40: 228–240.]

Erlikosaurus andrewsi

> *Erlikosaurus andrewsi* Perle, 1981 [Cenomanian–Santonian; Bayan Shireh Formation, Ömnögovi, Mongolia] (acc. *The Dinosauria 2nd edition*)

E. andrewsi was described by Perle in 1981 [*2], however it had been also described by Barsbold & Perle in 1980 [*1].
Generic name: *Erlikosaurus* ← {(Mong.) Erlik: "the lamaist deity, king of the dead" [*1] | (Mong. folk myth.) "an evil spirit" [*2] + (Gr.) σαῦρος/sauros: lizard}
Specific name: *andrewsi* ← {(person's name) Andrews + -ī}; "in honor of Dr. Roy Chapman Andrews, leader of the American Asiatic Expeditions in 1922–1930" [*1] [*2].
Etymology: Andrews' **Erlik lizard**
Taxonomy: Theropoda: Coelurosauria:

Therizinosauridae*
*= Segnosauridae [*2] / Enigmosauridae / Nanshiungosauridae

[References: (*1) Barsbold R, Perle A (1980). Segnosauria, a new infraorder of carnivorous dinosaurs. *Acta Palaeontologica Polonica* 25(2): 187–195.; (*2) Perle A (1981). New Segnosauridae from the Upper Cretaceous of Mongolia. *Trudy – Sovmestnaya Sovetsko – Mongol'skaya Paleontologicheskaya Ekspeditsiya* 15: 50–59.]

Eshanosaurus deguchiianus

> *Eshanosaurus deguchiianus* Xu *et al.*, 2001 [Sinemurian; lower Lufeng Formation, Yunnan, China]

Generic name: *Eshanosaurus* ← {(Chin. place-name) Éshān峨山: Eshan Yi Autonomous County + (Gr.) σαῦρος/sauros: lizard}

Specific name: *deguchiianus* ← {(person's name) Deguchi + -ianus}; in honor of Hikaru Deguchi, who gave encouragement and support to the first author in studying dinosaurs [*1].

Etymology: Deguchi's **Eshan County lizard**

Taxonomy: Theropoda: Therizinosauria

Notes: The classification of *Eshanosaurus* is seemed to be controvertial. [*2]

[References: (*1) Xu X, Clark JM (2001). A new therizinosaur from the Lower Jurassic Lower Lufeng Formation of Yunnan, China. *Journal of Vertebrate Paleontology* 21(3): 477–483.; (*2) Barrett PM (2009). The affinities of the enigmatic dinosaur *Eshanosaurus deguchiianus* from the Early Jurassic of Yunnan Province, People's Republic of China. *Palaeontology* 52(4): 681–688.]

Eucamerotus foxii

> *Ornithopsis* Hulke, 1872* (no species name) ⇒ *Eucamerotus foxii* (Hulke, 1872) Blows, 1995 [Barremian; Wessex Formation, Isle of Wight, UK]
*cf. *Ornithopsis hulkei* Seeley, 1870 is considered as a lectotype of *Eucamerotus foxii*.

Generic name: *Eucamerotus* ← {(Gr.) εὐ-/ eu-: well- + kamerotus [καμάρα/kamara: chamber]}; in reference to the hollows of the vertebrae. [*1]

Specific name: *foxii* ← {(person's name) Fox + -ī}; in honor of paleontologist William Fox who collected most of the paratypes [*1].

Etymology: Fox's **well-chambered**

Taxonomy: Saurischia: Sauropoda: Neosauropoda: Macronaria: Titanosauriformes
cf. Saurischia: Sauropodomorpha: Sauropoda: Brachiosauridae [*1]

Notes: According to *The Dinosauria 2nd edition* (2004), *Eucamerotus* was thought to be a *nomen dubium*. However, Upchurch *et al.* (2011) considered *Eucamerotus* as a valid genus because of a more recent review.

[References: (*1) Blows WT (1995). The Early Cretaceous brachiosaurid dinosaurs *Ornithopsis* and *Eucamerotus* from the Isle of Wight, England. *Palaeontology* 38 (1): 187–197.]

Eucnemesaurus fortis

> *Eucnemesaurus fortis* van Hoepen, 1920 [Norian; lower Elliot Formation, Free State, South Africa]
= *Aliwalia rex* Galton, 1985 [Norian; lower Elliot Formation, Eastern Cape, South Africa]

Generic name: *Eucnemesaurus* ← {(Gr.) εὐ-/eu-: good + κνήμη/knēmē: tibia + σαῦρος/sauros: lizard}

Specific name: *fortis* ← {(Lat.) fortis: strong}

Etymology: strong, **good shanked lizard**

Taxonomy: Saurischia: Sauropodomorpha: Riojasauridae

Other species:

> *Eucnemesaurus entaxonis* McPhee *et al.*, 2015 [Norian; lower Elliot Formation, Eastern Cape, South Africa]

Notes: According *to The Dinosauria 2nd edition*, *Eucnemesaurus fortis* was regarded as a *nomen dubium*. Yates (2007) indicates that *Eucnemesaurus* is the senior synonym of *Aliwalia rex* [*1].

[References: van Hoepen ECN (1920). Contributions to the knowledge of the reptiles of the Karoo Formation. 6. Further dinosaurian material in the Transvaal Museum. *Annals of the Transvaal Museum* 7(2): 93–141.; (*1) Yates AM (2007). Solving a dinosaurian puzzle: the identity of *Aliwalia rex* Galton. *Historical Biology* 19: 92–123. Abstract.]

Eudimorphodon ranzii [Pterosauria]

> *Eudimorphodon ranzii* Zambelli, 1973 [Norian; Zorzino Limestone, Cene, Italy]

Generic name: *Eudimorphodon* ← {(Gr.) εὐ-/eu-: true + δι-: two + μορφή/morphē: form, shape + ὀδών/odōn (←ὀδούς/odous [odūs]: tooth)}

Specific name: *ranzii* ← {(person's name) Ranzi + -ī}; in honor of Italian zoologist Silvio Ranzi.

Etymology: Ranzi's **true dimorphic tooth**

Taxonomy: Pterosauria: Eudimorphodontidae

Other species:

> *Eudimorphodon rosenfeldi* Dalla Vecchia, 1995 ⇒ *Carniadactylus rosenfeldi* (Dalla Vecchia, 1995) Dalla Vecchia, 2009 [Norian; Friuli, Italy]
> *Eudimorphodon cromptonellus* Jenkins *et al.*, 2001 ⇒ *Arcticodactylus cromptonellus* (Jenkins *et al.*, 2001) Kellner, 2015 [Norian;

Fleming Fjord Formation, East Greenland, Greenland]

Notes: *Eudimorphodon* had multi-cusped teeth. Its wingspan was about 75–90 cm and is known as one of the oldest pterosaurs. In 2015 *Eudimorphodon cromptonellus* was given a new generic name *Arcticodactylus* by Kellner.

Euhelopus zdanskyi

> *Helopus zdanskyi* Wiman, 1929 (preoccupied) ⇒ *Euhelopus zdanskyi* (Wiman, 1929) Romer, 1956 [Berriasian–Valanginian; Mengyin Formation, Shandong, China]

Helopus turned out to be pre-occupied by a bird, *Helopus* (currently *Hydroprogne*). Therefore, the name had to be changed.

Generic name: *Euhelopus* ← {(Gr.) εὐ-/eu-: good, true + ἕλος/helos: marsh, swamp + πούς/pous [pūs]: foot}

According to Wiman, the series where the dinosaurs were found in eastern Shantung includes lake mollusks. (*1)

Specific name: *zdanskyi* ← {(person's name) Zdansky + -ī}; in honor of paleontologist Otto Zdansky who excavated most of its remains.

Father R. Mertens discovered and excavated parts of the exemplar. (*1)

Etymology: Zdansky's **true marsh foot**

Taxonomy: Saurischia: Sauropoda: Titanosauriformes: Somphospondyli: Euhelopodidae

cf. Saurischia: Sauropodomorpha: Sauropoda: Eusauropoda* (according to *The Dinosauria 2nd edition*)

*Eusauropoda was coined by Paul Upchurch in 1995 (*2).

[References: (*1) Wiman C (1929). Die Kreide-Dinosaurier aus Shantung [The Cretaceous dinosaur from Shantung]. *Palaeontologia Sinica Series C* 6(1): 1–67.; (*2) Upchurch P (1995). The evolutionary history of sauropod dinosaurs. *Philosophical Transactions of the Royal Society of London B*. 349: 365–390.]

Euoplocephalus tutus

> *Stereocephalus tutus* Lambe, 1902 (preoccupied)* ⇒ *Euoplocephalus tutus* (Lambe, 1902), Lambe, 1910 [Campanian; Dinosaur Park Formation, Alberta, Canada]

**Stereocephalus* was preoccupied by an insect (*1).

Generic name: *Euoplocephalus* ← {(Gr.) εὔ-οπλος*/ eu-oplos: well-armed + κεφαλή/ kephalē: head} (*1)

*cf. ὅπλον/ hoplon: weapon

Specific name: *tutus* ← {(Lat.) tūtus: safe, secure}; protected from danger or harm. (*1)

Etymology: guarded, **well-armed head**

Taxonomy: Ornithischia: Thyreophora: Ankylosauria: Ankylosauridae: Ankylosaurinae

Notes: *Euoplocephalus* was largely covered by bony armor plates and had a heavy club-like tail end. Similar to *Ankylosaurus magniventris*, *Euoplocephalus* had small teeth (Matthew *et al.*, 2002). According to *the Dinosauria 2nd edition*, *Dyoplosaurus acutosquameus* Parks, 1924, *Scolosaurus cutleri* Nopcsa, 1929, *Anodontosaurus lambei* Sternberg, 1929 were considered to be as synonyms of *Euoplocephalus tutus*. However, *D. acutosquameus* and *S. cutleri* were classified into new own genus. (*2)

[References: (*1) Lambe LM (1910). Note on the parietal crest of *Centrosaurus apertus* and a proposed new generic name for *Stereocephalus tutus. The Ottawa Naturalist* 24: 149–51.; (*2) Penkalski P, Blows WT (2013). *Scolosaurus cutleri* (Ornithischia: Ankylosauria) from the Upper Cretaceous Dinosaur Park Formation of Alberta, Canada. *Canadian Journal of Earth Sciences* 50(2): 171–182.]

Euparkeria capensis [Archosauromorpha]

> *Euparkeria capensis* Broom, 1913 [Anisian; Burgersdorp Formation, Eastern Cape, South Africa]

= *Browniella africana* Broom, 1913 [Anisian; Burgersdorp Formation, Eastern Cape, South Africa]

Generic name: *Euparkeria* ← {(Gr.) εὐ-/eu-: good + (person's name) Parker + -ia}; in honor of W. K. Parker.

Specific name: *capensis* ← {(place-name) Cape + -ensis}

Etymology: **good Parker's one** from Eastern Cape (former Cape Province)

Taxonomy: Archosauromorpha: Euparkeriidae

Eurhinosaurus longirostris [Ichthyosauria]

> *Ichthyosaurus longirostris* Mantell, 1851 ⇒ *Eurhinosaurus longirostris* (Mantell, 1851) Abel, 1909 [Toarcian; Staffelegg Formation, Switzerland]

Generic name: *Eurhinosaurus* ← {(Gr.) εὔ-ρῖνος/ eu-rhīnos: good-nosed + σαῦρος/sauros: lizard}; referring to its long upper jaw.

Specific name: *longirostris* ← {(Lat.) longus: long + rostrum: snout}

Etymology: **good-nosed lizard** with long snout

Taxonomy: Ichthyosauria: Leptonectidae

[References: Maisch MW, Matzke AT (2000). The Ichthyosauria. *Stuttgarter Beiträge zur Naturkunde Serie B (Geologie und Paläntologie)* 298: 1–159.; Reisdorf A, Maisch MW, Wetzel A (2011). First record of the leptonectid ichthyosaur *Eurhinosaurus longirostris* from the Early Jurassic of Switzerland and its stratigraphic framework. *Swiss Journal of geosciences* 104(2): 212–224.]

Euronychodon portucalensis

> *Euronychodon portucalensis* Antunes & Sigogneau-Russell, 1991 [Campanian–Maastrichtian; Lourinhã Formation, Portugal]

Generic name: *Euronychodon* ← {(place-name) Euro-: European [(Gr.) Εὐρωπη/ Eurōpē] + (Gr.) ὄνυξ/onyx: claw + ὀδών/odōn: tooth}; a contraction of Europe & *Paronychodon*.

Specific name: *portucalensis* ← {(place-name) Portucale (= Portugal) + -ensis}

Etymology: **European *Paronychodon* (near claw tooth)** from Portugal

Taxonomy: Theropoda: Troodontidae

Other species:

> *Euronychodon asiaticus* Nesov, 1995 [middle Tronian; Bissekty Formation, Navoi, Uzbekistan]

Europasaurus holgeri

> *Europasaurus holgeri* Mateus *et al.*, in Sander *et al.*, 2006 [Kimmeridgian; Langenberg Formation, Niedersachsen, Germany]

Generic name: *Europasaurus* ← {(place-name) Europa: Europe [(Gr.) Εὐρώπη/Eurōpē] + (Gr.) σαῦρος/sauros: reptile}

Specific name: *holgeri* ← {(person's name) Holger + -ī}; in honor of discoverer Holger Lüdtke.

Etymology: Holger's **reptile from Europe**

Taxonomy: Saurischia: Sauropodomorpha: Sauropoda: Brachiosauridae

Notes: *Europasaurus* is a small sauropod.

[References: Sander PM, Mateus O, Laven T, Knötschke N (2006). Bone histology indicates insular dwarfism in a new Late Jurassic sauropod dinosaur. *Nature* 441 (7094): 739–741.]

Europatitan eastwoodi

> *Europatitan eastwoodi* Fernández-Baldor *et al.*, 2017 [Barremian–Aptian; Castrillo de la Formation, Salas de los Infantes, Burgos, Spain]

Generic name: *Europatitan* ← {(place-name) Europa: Europe + (Gr. myth.) Τιτᾶνες/ Titānes: the Titans}; "in reference to Europe, the continent where it was found, and the titans, ancient Greek deities known for their gigantic size, endowed with great power" (*1).

Specific name: *eastwoodi* ← {(person's name) Clint Eastwood + -ī}; "dedicated to US actor Clint Eastwood, the protagonist of the film "The Good, the Bad and the Ugly", which was partially filmed near Salas de los Infantes" (*1).

Etymology: Eastwood's **European Giant (titanosaurs)**

Taxonomy: Saurischia: Sauropoda: Neosauropoda: Titanosauriformes: Somphospondyli

[References: (*1) Fernández-Baldor FT, Canudo JI, Huerta P, Moreno-Azanza M, Montero D (2017). *Europatitan eastwoodi*, a new sauropod from the Lower Cretaceous of Iberia in the initial radiation of somphospondylans in Laurasia. *PeerJ* 5: e3409.]

Europejara olcadesorum [Pterosauria]

> *Europejara olcadesorum* Vullo *et al.*, 2012 [Barremian; Calizas de la Huérgina Formation, Cuenca, Spain]

Generic name: *Europejara* ← {(place-name) Euro-: Europe + *Tapejara* [(Tupi) "the old being"]}

Specific name: *olcadesorum* ← {(demonym) the Olcades + -ōrum}; "named after the Olcades, Celtiberians who were the first inhabitants of the Cuenca region" (*1).

Etymology: Olcades' **tapejarids from Europe**

Taxonomy: Pterosauria: Pterodactyloidea: Tapejaridae: Tapejarinae

[References: (*1) Vullo R, Marugán-Lobón J, Kellner AWA, Buscalioni AD, Gomez B, de la Fuente M, Moratalla JJ (2012). A new crested pterosaur from the Early Cretaceous of Spain: The first European tapejarid (Pterodactyloidea: Azhdarchoidea) *PLoS ONE* 7(7): e38900.]

Europelta carbonensis

> *Europelta carbonensis* Kirkland *et al.*, 2013 [Albian; Escucha Formation, Teruel, Aragón, Spain]

Generic name: *Europelta* ← {(place-name) Euro-: a contraction for Europe + pelta (= (Gr.) πέλτη/ peltē: shield, a common root for ankylosaurian genera)} (*1)

Specific name: *carbonensis* ← {(Lat.) carbō: coal + -ensis}; "in honor of access to the fossil locality in the Santa María coal mine provided by Sociedad Anónima Minera Catalano-Aragonesa (SAMCA Group), which has been extracting coal in Ariño (Teruel) since 1919" (*1).

Etymology: **Europe's shield** from Santa María coal mine

Taxonomy: Ornithischia: Ankylosauria: Nodosauridae

cf. Ornithischia: Ankylosauria: Nodosauridae: Struthiosaurinae (*1)

Notes: *Struthiosaurinae* contains *Europelta* but not *Cedarpelta*, *Peloroplites*, *Sauropelta* or *Edmontonia* (*1).

[References: (*1) Kirkland JI, Alcalá L, Loewen MA, Espílez E, Mampel L, Wiersma JP (2013). The basal nodosaurid ankylosaur *Europelta carbonensis* n. gen., n. sp. from the Lower Cretaceous (lower Albian) Escucha Formation of northeastern Spain. *PLoS ONE* 8(12): e80405.]

Euskelosaurus browni

> *Euskelosaurus browni* Huxley, 1866 [Norian; lower Elliot Formation, Aliwal North, Eastern Cape, South Africa]

Generic name: *Euskelosaurus* ← {(Gr.) εὐ-/eu-: good + σκέλος/skelos: leg, limb + σαῦρος/sauros: lizard}

Specific name: *browni* ← {(person's name) Brown + -ī}; in honor of Alfred Brown, a discoverer.

Etymology: Brown's **good-legged lizard**

Taxonomy: Saurischia: Sauropodomorpha: Prosauropoda: Plateosauridae

Other species:

> *Euskelosaurus africanus* Haughton, 1924 (*nomen dubium*) [Norian; lower Elliot Formation, Eastern Cape, South Africa]

Notes: Thin bone of *Euskelosaurus* is twisted (van Heerden, 1979). (*1)

[References: (*1) van Heerden J (1979). The morphology and taxonomy of *Euskelosaurus* (Reptilia: Saurischia; Late Triassic) from South Africa. *Navorsinge van die Nasionale Museum* 4(2): 21–84.]

Eustreptospondylus oxoniensis

> *Eustreptospondylus oxoniensis* Walker, 1964 ⇒ *Magnosaurus oxoniensis* (Walker, 1964) Rauhut, 2003 [late Callovian; Oxford Clay Formation, Oxfordshire, UK]

Generic name: *Eustreptospondylus* ← {(Gr.) εὐ-/eu-: true, good + στρεπτός/streptos: turned, reversed + σπόνδῦλος/spondylos [= σφόνδῦλος: vertebra]}

Specific name: *oxoniensis* ← {(place-name) Oxon (=Oxford) + -ensis}

Etymology: **true *Streptospondylus* (well curved vertebrae)** from Oxford

Taxonomy: Theropoda: Megalosauridae

Other species:

> *Eustreptospodylus divesensis* Walker, 1964 ⇒ *Piveteausaurus divesensis* (Walker, 1964) Taquet & Welles, 1977 [Callovian; Marnes de Dives, Normandy, France]

Excalibosaurus costini [Ichthyosauria]

> *Excalibosaurus costini* McGowan, 1986 [Sinemurian; Somerset, UK]

Generic name: *Excalibosaurus* ← {(legend) Excalibur: legendary sword + (Gr.) σαῦρος/sauros: lizard}; referring to its elongated snout.

Specific name: *costini* ← {(person's name) Costin + -ī}; in honor of David Costin.

Etymology: Costin's **Excalibur lizard**

Taxonomy: Ichthyosauria: Leptonectidae

Notes: *Excalibosaurus* looks like a swordfish. According to Maisch & Matzke (2000), *Excalibosaurus* was thought to be a junior synonym of *Eurhinosaurus.*

[References: McGowan C (2003). A new Specimen on *Excalibosaurus* from the English Lower Jurassic. *Journal of Vertebrate Paleontology* 23(4): 950–956.]

F

Fabrosaurus australis

> *Fabrosaurus australis* Ginsburg, 1964 [Hettangian–Sinemurian; upper Elliot Formation, Mafeteng, Lesotho]

Generic name: *Fabrosaurus* ← {(person's name) Fabre + (Gr.) σαῦρος/sauros: lizard}; in honor of the geologist Jean Fabre.

Specific name: *australis* ← {(Lat.) austrālis: southern}; named for the location, Lesotho, Southern Africa.

Etymology: southern **Fabre's lizard**

Taxonomy: Ornithischia

Notes: *Fabrosaurus* is based upon a single undiagnostic dentary, and regarded as a *nomen dubium* (*1).

[References: (*1) Butler RJ (2005). The 'fabrosaurid' ornithischian dinosaurs of the Upper Elliot Formation (Lower Jurassic) of South Africa and Lesotho. *Zoological Journal of Linnean Society* 145(2): 175–218.]

Falcarius utahensis

> *Falcarius utahensis* Kirkland *et al.*, 2005 [Barremian; Cedar Mountain Formation, Utah, USA]

Generic name: *Falcarius* ← {(Lat.) falcārius*: a sickle-maker}; named for its sharp, curved claws (*1).

*falcārius: gladiateur armé de faux. (Dictionnaire Latin-Français, Français-Latin. Larousse, 2008)

Specific name: *utahensis* ← {(place-name) Utah + -ensis}

Etymology: **sickle-cutter** from Utah / **gradiator** armed with sickles from Utah

Taxonomy: Theropoda: Coelurosauria: Therizinosauria

Notes: Authors described *Falcarius* was the missing link between predatory dinosaurs and plant-eating therizinosaurs. (*1)

[References: (*1) Kirkland JI, Zanno LE, Sampson SD, Clark JM, DeBlieux DD (2005). A primitive therizinosauroid dinosaur from the Early Cretaceous of Utah. *Nature* 435: 84–87.]

Ferganasaurus verzilini

> *Ferganasaurus verzilini* Alifanov & Averianov, 2003 [Callovian; Balabansai Formation,

Fergana Valley, Jalal-Abad, Kirgyzstan]
Generic name: *Ferganasaurus* ← {(place-name) Fergana Valley + (Gr.) σαῦρος/sauros: lizard} (*1)
Specific name: *verzilini* ← {(place-name) Verzilin + -ī}; "in honor of Prof. Nikita N. Verzilin, who found the holotype in 1966" (*1).
Etymology: Verzilin's **lizard from Fergana Valley**
Taxonomy: Saurischia: Sauropodomorpha: Sauropoda: Neosauropoda (*1)

[References: (*1) Alifanov VR, Averianov AO (2003). *Ferganasaurus verzilini*, gen. *et* sp. nov., a new neosauropod (Dinosauria, Saurischia, Sauropoda) from the Middle Jurassic of Fergana Valley, Kirghizia. *Journal of Vertebrate Paleontology* 23(2): 358–372.]

Ferganocephale adenticulatum

> *Ferganocephale adenticulatum* Averianov *et al.*, 2005 [Callovian; Balabansai Formation, Fergana Valley, Jalal-Abad, Kirgyzstan]

Generic name: *Ferganocephale* ← {(place-name) Fergana Valley + (Gr.) κεφαλή/kephalē: head}; "head, the common suffix for pachycephalosaurid genera; gender neutral" (*1).
Specific name: *adenticulatum* ← {(Gr.) a: denoting absence + denticulum: denticle}; meaning absence or easy loss of vestigial marginal crown denticles by wear, (*1) that is, "without tooth serrations".
Etymology: **Fergana Valley's head (pachycephalosaur)**, without tooth serrations
Taxonomy: Ornithischia: Neornithischia
cf. Ornithischia; Marginocephalia; Pachycephalosauria; Pachycephalosauridae (*1)
Notes: *Ferganocephale* was originally classified in the group Pachycephalosauridae (*1). Sullivan RM (2006) considers the taxon a *nomen dubium*.

[References: (*1) Averianov AO, Martin T, Bakirov AA (2005). Pterosaur and dinosaur remains from the Middle Jurassic Balabansai Svita in the northern Fergana Depression, Kyrgyzstan (Central Asia). *Palaeontology* 48(1): 135–155.]

Ferrisaurus sustutensis

> *Ferrisaurus sustutensis* Arbour & Evans, 2019 [Maastrichtian; Tango Creek Formation, British Columbia, Canada]

Generic name: *Ferrisaurus* ← {(Lat.) ferrum: iron + (Gr.) σαῦρος/sauros: lizard}; "in reference to the specimen's discovery along a railway line" (*1).
Specific name: *sustutensis* ← {(place-name) Sustut + -ensis};"in reference to its provenance near Sustut River and within the Sustut Basin" (*1).
Etymology: **Iron lizard** from Sustut Basin
Taxonomy: Ornithischia: Neornithischia: Neoceratopsia: Coronosauria: Leptoceratopsidae

[References: (*1) Arbour VM, Evans DC (2019). A new leptoceratopsid dinosaur from Maastrichtian-aged deposits of the Sustut Basin, northern British Columbia, Canada. *PeerJ* 7: e7926.]

Flexomornis howei [Aves]

> *Flexomornis howei* Tykoshi & Fiorillo, 2010 [Cenomanian; Woodbine Formation, Texas, USA]

Generic name: *Flexomornis* ← {(Lat.) flex-: bend, curve, or turn (*1) + (Gr.) om- [ὦμος/ōmos: shoulder] + ὄρνις/ornis: bird}
cf. flexus, *adj.*: bent, winding.
Specific name: *howei* ← {(person's name) Howe + -ī}; "in recognition of Kris Howe, the local fossil enthusiast who brought the site and its vertebrate fossils to authors' attention" (*1).
Etymology: Howe's **flexed shoulder bird**
Taxonomy: Aves: Ornithurae: Ornithothoraces: Enantiornithes (*1)

[References: (*1) Tykoshi RS, Fiorillo AR (2010). An enantiornithine bird from the Lower middle Cenomanian of Texas. *Journal of Vertebrate Paleontology* 30(1): 288–292.]

Foraminacephale brevis

> *Stegoceras breve* Lambe, 1918 ⇒ *Prenocephale brevis* (Lambe, 1918) Sullivan, 2000 ⇒ *Foraminacephale brevis* (Lambe, 1918) Schott & Evans, 2016 [Campanian; Dinosaur Park Formation, Alberta, Canada]

Generic name: *Foraminacephale* ← {(Lat.) foramina [forāmen: hole] + (Gr.) κεφᾰλή/cephalē: head}; referring to the numerous foramina (many pits) that cover the dorsal surface of the skull. (*1)
Specific name: *brevis* ← {(Lat.) brevis: small, short}
Etymology: small, **foramina head**
Taxonomy: Ornithischia: Pachycephalosauria: Pachycephalosauridae

[References: (*1) Schott RK (2011) Systematics and ontogeny of *Foraminacephale brevis* gen. nov. (Ornithischia: Pachycephalosauria). In: *Ontogeny, Diversity, and Systematics of Pachycephalosaur Dinosaurs from the Belly River Group of Alberta, Master's Thesis*, University of Toronto. Toronto.]

Fosterovenator churei

> *Fosterovenator churei* Dalman, 2014 [Kimmeridgian–Tithonian; Morrison Formation, Wyoming, USA]

Generic name: *Fosterovenator* ← {(person's name) Foster + (Lat.) vēnātor: hunter}; "in honor of John R. Foster in recognition of his contributions to the study of the vertebrate fauna of the

Morrison Formation" (*1).
Specific name: *churei* ← {(person's name) Chure + -ī}; "in honor of Daniel J. Chure in recognition of his contributions to the study of the vertebrate fauna of the Morrison Formation" (*1).
Etymology: Chure and **Foster's hunter**
Taxonomy: Theropoda: Ceratosauridae

[References: (*1) Dalman SG (2014). New data on a small theropod dinosaurs from the Upper Jurassic Morrison Formation of Como Bluff, Wyoming, USA. *Volumina Jurassica* 12(2): 181–196.]

Fostoria dhimbangunmal

> *Fostoria dhimbangunmal* Bell *et al.*, 2019 [Cenomanian; Griman Creek Formation, New South Wales, Australia]

Generic name: *Fostoria* ← {(person's name) Foster = (Old English) Fostor + -ia}; "in honor of Robert Foster, the miner who discovered the bone bed" (*1).
Specific name: *dhimbangunmal* [pron: dim-baan goon-mal] ← {(the language of the Yuwaalaraay / Yuwaalayaay / Gamilaraay peoples) dhimba: sheep + ngunmal: yard}; "after the Sheepyard opal field where the bone bed is located" (*1).
Etymology: **For Foster** from the Sheepyard
Taxonomy: Ornithischia: Ornithopoda: Iguanodontia

[References: Bell PR, Brougham T, Herne MC, Frauenfelder T, Smith ET (2019). *Fostoria dhimbangunmal*, gen. *et* sp. nov., a new iguanodontian (Dinosauria, Ornithopoda) from the mid-Cretaceous of Lightning Ridge, New South Wales, Australia. *Journal of Vertebrate Paleontology*: e1564757.]

Frenguellisaurus ischigualastensis

> *Herrerasaurus ischigualastensis* Reig, 1963 [Carnian; Ischigualasto Formation, San Juan, Argentina]

= *Ischisaurus cattoi* Reig, 1963 [Carnian; Ischigualasto Formation, San Juan, Argentina]

= *Frenguellisaurus ischigualastensis* Novas, 1986 [Norian; Ischigualasto Formation, San Juan, Argentina]

Generic name: *Frenguellisaurus* ← {(person's name) Frenguelli + σαῦρος/sauros: lizard}; "in honor of Dr. Joaquin Frenguelli, who realized an important paleontological and geological work in the Triassic Ischigualasto-Villa Unión Valley" (*1).
Specific name: *ischigualastensis* ← {(place-name) Ischigualasto + -ensis}; referring to "Ischigualasto, locality where the described material was discovered" (*1).
Etymology: **Frenguelli's lizard** from Ischigualasto
Taxonomy: Herrerasauridae
Notes: According to *The Dinosauria 2nd edition*, *Frenguellisaurus* is considered a synonym of *Herrerasaurus*.

[References: (*1) Novas FE (1986). Un probable terópodo (Saurischia) de la Formación Ischigualasto (Triásico Superior), San Juan, Argentina [A probable theropod (Saurischia) from the Ischigualasto Formation (Upper Triassic), San Juan, Argentina]. *IV Congreso Argentino de Paleontologia y Bioestratigrafia*: 1–6. Translated by Carrano M.]

Fruitadens haagarorum

> *Fruitadens haagarorum* Butler *et al.*, 2010 [Tithonian; Morrison Formation, Colorado, USA]

Generic name: *Fruitadens* ← {(place-name) Fruita, Colorado, USA + (Lat.) dēns: tooth}
Specific name: *haagarorum* ← {(person's name) Haaga + - ōrum}; "in honor of Paul Haaga Jr., Heather Haaga, Blythe Haaga, Paul Haaga III, and Catalina Haaga for their support of the Natural History Museum of Los Angeles County (LACM, Los Angeles, USA)" (*1).
Etymology: Haaga's **Fruita tooth**
Taxonomy: Ornithischia: Ornithopoda: Heterodontosauridae

[References: (*1) Butler RJ, Galton PM, Porro LB, Chiappe LM, Henderson DM, Erickson GM (2009). Lower limits of ornithischian dinosaur body size inferred from a new Upper Jurassic heterodontosaurid from North America. *Proceedings of the Royal Society B: Biological Sciences* 277(1680): 375–381.]

Fukuiraptor kitadaniensis

> *Fukuiraptor kitadaniensis* Azuma & Currie, 2000 [Barremian; Kitadani Formation, Katsuyama, Fukui, Japan]

Generic name: *Fukuiraptor* ← {(Jpn. place-name) Fukui Prefecture (福井県), Japan + (Lat.) raptor: robber, plunderer}
Specific name: *kitadaniensis* ← {(Jpn. place-name) Kitadani 北谷 + -ensis}; in reference to the Kitadani Formation (*1).
Etymology: **Fukui thief** from Kitadani Formation
Taxonomy: Theropoda: Carnosauria
Notes: The immature specimen of *Fukuiraptor*, which is about 4.2 m long with strongly curved, sharp claws indicates a basal allosauroid (*2).

[References: (*1) Suzuki S, Shibata M, Azuma Y, Yukawa H, Sekiya T, Masaoka Y (2015). Sedimentary environment of dinosaur fossil bearing successions of the Lower Cretaceous Kitadani Formation, Tetori Group, Katsuyama City, Fukui, Japan. *Memoir of the Fukui Prefectual Dinosaur Museum*. 14: 1–9.; (*2) Azuma Y, Currie P (2000). A new carnosaur (Dinosauria: Theropoda) from the Lower Cretaceous of Japan. *Canadian Journal of Earth Sciences* 37(12): 1735–1753.]

Fukuisaurus tetoriensis

> *Fukuisaurus tetoriensis* Kobayashi & Azuma, 2003 [Barremian; Kitadani Formation, Katsuyama, Fukui, Japan]

Generic name: *Fukuisaurus* ← {(Jpn. place-name) Fukui Prefecture (福井県), Japan + (Gr.) σαῦρος/sauros: lizard}

Specific name: *tetoriensis* ← {(Jpn. place-name) Tetori Group (手取層群) + -ensis}

Etymology: **Fukui lizard** from Tetori Group

Taxonomy: Ornithischia: Ornithopoda: Hadrosauriformes: Hadrosauroidea (acc. Ramirez-Velasco *et al.* 2012)

cf. Ornithischia: Ornithopoda: Iguanodontia: Iguanodontoidea (acc. *The Dinosauria 2nd edition*)

Notes: Its nickname is Fukuiryū (福井竜) in Japanese.

[References: Kobayashi Y, Azuma Y (2003). A new iguanodontian (Dinosauria: Ornithopoda), from the Lower Cretaceous Kitadani Formation of Fukui Prefecture, Japan. *Journal of Vertebrate Paleontology* 23(1): 166–175.]

Fukuititan nipponensis

> *Fukuititan nipponensis* Azuma & Shibata, 2010 [Barremian; Kitadani Formation, Katsuyama, Fukui, Japan]

Generic name: *Fukuititan* ← {(Jpn. place-name) Fukui 福井, Japan + (Gr.) Τῑτάν/Tītan: giant}

Specific name: *nipponensis* ← {(Jpn. place-name) Nippon 日本: Japan + -ensis}

Etymology: **Fukui giant (titanosaur)** from Japan

Taxonomy: Saurischia: Sauropodomorpha: Sauropoda: Titanosauriformes

[References: Azuma Y, Shibata M (2010). *Fukuititan nipponensis*, a new titanosauriform sauropod from the Eary Cretaceous Tetori Group of Fukui Prefecture, Japan. *Acta Geologica Sinica* 84(3): 454–462.]

Fukuivenator paradoxus

> *Fukuivenator paradoxus* Azuma *et al.*, 2016 [Barremian; Kitadani Formation, Katsuyama, Fukui, Japan]

Generic name: *Fukuivenator* ← {(Jpn. place-name) Fukui 福井: a prefecture in the central Japan + (Lat.) vēnātor: hunter}

Specific name: *paradoxus* ← {(Gr.) παράδοξος/paradoxos: contrary to opinion, incredible, paradoxical}; "referring to the surprising combination of characters in this theropod dinosaur" (*1).

Etymology: paradoxical, **Fukui hunter**

Taxonomy: Theropoda: Maniraptora

[References: (*1) Azuma Y, Xu X, Shibata M, Kawabe S, Miyata K, Imai T (2016). A bizarre theropod from the Early Cretaceous of Japan highlighting mosaic evolution among coelurosaurians. *Scientific Reports* 6: 20478.]

Fulgurotherium australe

> *Fulgurotherium australe* von Huene, 1932 (*partim*) [Albian*; Griman Creek Formation, New South Wales, Australia]
> *(Raza *et al.*, 2009). *cf.* Cenomanian (Bell *et al.*, 2019)

Generic name: *Fulgurotherium* ← {(Lat.) fulgur: lightning + (Gr.) θηρίον/thērion: beast}; referring to the Lightning Ridge in New South Wales.

Specific name: *australe* ← {(Lat.) austrāle, *n.* (austrālis, *m.* / *f.*): southern}

Etymology: southern **lightning beast**

Taxonomy: Ornithischia: Neornithischia

cf. Ornithischia: Ornithopoda: Euornithopoda (according to *The Dinosauria 2nd edition*)

Notes: Von Huene (1932) thought *Fulgurotherium australe* was a small theropod dinosaur. Molner and Galton (1986) described it as a primitive ornithopod. Most researchers consider it a *nomen dubium*. The original specimen is the opalised distal end of a femur (*1).

[References: (*1) *Fulgurotherium australe* -Australian Museum 〈https://australian.museum/learn/dinosaurs/fact-sheets/fulgurotherium-australe/〉]

Fumicollis hoffmani [Avialae]

> *Fumicollis hoffmani* Bell & Chiappe, 2015 [Coniacian–Santonian; Smoky Hill Chalk Member, Niobrara Formation, Longan, Kansas, USA]

Generic name: *Fumicollis* ← {(Lat.) fumi- [fūmus: smoke] + collis: hill}; "in reference to the Smoky Hill Member of the Niobrara Chalk" (*1).

Specific name: *hoffmani* ← {(person's name) Hoffman + -ī}; "in recognition of Karen and Jim Hoffman, whose generous support has greatly enhanced the programs of the Natural History Museum of Los Angeles County, including research at the Dinosaur Institute" (*1).

Etymology: Hoffman's **smoky hill**

Taxonomy: Theropoda: Avialae: Ornituromorpha: Ornuthurae: Hesperornithiformes

cf. Aves: Ornithuromorpha: Ornithurae: Hesperornithiformes (*1)

[References: (*1) Bell A, Chiappe LM (2015). Identification of a new hesperornithiform from the Cretaceous Niobrara Chalk and implications for ecologic diversity among early diving birds. *PLoS ONE* 10(11): e0141690.]

Fusuisaurus zhaoi

> *Fusuisaurus zhaoi* Mo *et al.*, 2006 [Aptian; Napai Formation, Fusui, Guangxi, China]

Generic name: *Fusuisaurus* ← {(Chin. place-name) Fúsuí扶綏 + (Gr.) σαῦρος/sauros: lizard}

Specific name: *zhaoi* ← {(Chin. person's name) Zhào趙 + -ī}; in honor of Chinese paleontologist Zhao Xijin.

Etymology: Zhao's **Fusui lizard**

Taxonomy: Saurischia: Sauropodomorpha: Sauropoda: Titanosauriformes

[References: Mo J, Wang W, Huang Z, Huang X, Xu X (2006). A basal titanosauriform from the Early Cretaceous of Guangxi, China. *Acta Geologica Sinica* 80(4): 486–489.]

Futabasaurus suzukii [Plesiosauria]

> *Futabasaurus suzukii* Sato *et al.*, 2006 [Santonian; Tamayama Formation, Iwaki, Fukushima, Japan]

Generic name: *Futabasaurus* ← {(Jpn. place-name) Futaba Group + (Gr.) σαῦρος/sauros: lizard}

Specific name: *suzukii* ← {(Jpn. person's name) Suzuki鈴木 + -ī}; in honor of discoverer Tadashi Suzuki who was a high school student at that time.

Etymology: Suzuki's **lizard fromFutaba Group**

Taxonomy: Plesiosauria: Elasmosauridae

Notes: *Futabasaurus* is the first elasmosaurid found in Japan. [*1] It is commonly known as "Futabasuzukiryū (双葉鈴木竜 Futaba-Suzuki-dragon)". This skeleton is exhibited in the National Museum of Nature and Science, Tokyo.

[References: (*1) Sato T, Hasegawa Y, Manabe M (2006). A new elasmosaurid plesiosaur from the Upper Cretaceous of Fukushima, Japan. *Palaeontology* 49(3): 467–484.]

Futalognkosaurus dukei

> *Futalognkosaurus dukei* Calvo *et al.*, 2007 [Coniacian; Portezuelo Formation, Neuquén, Argentina]

Generic name: *Futalognkosaurus* ← {(Mapuche) futa: giant + lognko: chief + (Gr.) σαῦρος/sauros: lizard}

Specific name: *dukei* ← {(institution name) Duke + -ī}; "in honor of the Duke Energy Argentina Company that sponsored the excavation (2002–2003)"[*1].

Etymology: Duke's **giant chief lizard**

Taxonomy: Saurischia: Sauropodomorpha: Sauropoda: Titanosauria: Lognkosauria [*1]

Notes: An estimated length is 32– 34 m [*1].

[References: (*1) Calvo JO, Porfiri JD, González-Riga BJ, Kellner AWA (2007). A new Cretaceous terrestrial ecosystem from Gondwana with the description of a new sauropod dinosaur. *Anais da Academia Brasileira de Ciências* 79(3): 529–541.]

G

Galeamopus hayi

> *Diplodocus hayi* Holland, 1924 ⇒ *Galeamopus hayi* (Holland, 1924) Tschopp *et al.*, 2015 [Kimmeridgian–Tithonian; Morrison Formation, Wyoming, USA]

Generic name: *Galeamopus* ← {(Lat.) galeam [galea: helmet] + opus: need, necessity}; literally translating to "the German name Wilhelm (meaning 'want helmet, protection') and its English translation William", and in honor of two Williams: "William H. Utterback found the genoholotype specimen HMNS175 in 1902 and Willian J. Holland described its braincase in 1906", and "alluding to "the fact that the fragile braincase is the only described part of the holotype skeleton to date"[*1], that is, the specimen is in need of a helmet that protects its braincase.

Specific name: *hayi* ← {(person's name) Hay + -ī}; in honor of Oliver Perry Hay.

Etymology: Hay's **helmet in need (Williams)**

Taxonomy: Saurischia: Sauropodomorpha: Diplodocidae: Diplodocinae

Other species:

> *Galeamopus pabsti* Tschopp & Mateus, 2017 [Kimmeridgian; Morrison Formation, Wyoming, USA]

[References: (*1) Tschopp E, Mateus O, Benson RBJ (2015). A specimen-level phylogenetic analysis and taxonomic revision of diplodocidae (dinosaurian, Sauropoda) *PeerJ* 3: e857.]

Galleonosaurus dorisae

> *Galleonosaurus dorisae* Herne *et al.*, 2019 [upper Barremian; Wonthaggi Formation, Gippsland Basin, Victoria, Australia]

Generic name: *Galleonosaurus* ← {(Eng.) galleon: a type of large sailing ship + (Gr.) σαῦρος/sauros: lizard}; "in reference to the appearance of the maxilla to the upturned hull of a galleon" [*1].

Specific name: *dorisae* ← {(person's name) Doris + -ae}; "in recognition of Doris Seegets-Villiers for her geological, palynological, and taphonomic work on the Flat Rocks fossil vertebrate locality"[*1].

Etymology: Doris' **galleon lizard**

Taxonomy: Ornithischia: Neornithischia: Ornithopoda

[References: (*1) Herne MC, Nair JP, Evans AR, Tait AM (2019). New small-bodied ornithopods (Dinosauria, Neornithischia) from the Early Cretaceous Wonthaggi Formation (Strzelecki Group) of the

Australian-Antarctic rift system, with revision of *Qantassaurus intrepidus* Rich and Vickers-Rich, 1999. *Journal of Paleontology* 93(3): 543–584.]

Gallimimus bullatus

> *Gallimimus bullatus* Osmólska *et al.*, 1972 ⇒ *Ornithomimus bullatus* (Osmólska *et al.*, 1972) Paul, 1988 [Maastrichtian; Nemegt Formation, Ömnögovi, Mongolia]

Generic name: *Gallimimus* ← {(Lat.) gallus: chicken + (Gr.) μῖμος/mīmos: mimic}; referring to "a chicken, because of strikingly similar structure of anterior portion of neck to that in representatives of the Galliformes*" (*1).

*Galliformes is an order of ground-feeding birds including turkey, grouse, chicken and so on.

Specific name: *bullatus* ← {(Lat.) bullātus: capsuled, wearing a bulla*}; referring to "the unusual capsule occurring in the skull base" (*1).

*bulla: "the capsule of gold worn on the neck by young boys of noble Roman families" (*1) and that means bubble.

According to the authors, a strange bulbous structure which is hollowed is presumed representing Rathke's pouch.

Etymology: capsuled, **chicken mimic**

Taxonomy: Theropoda: Coelurosauria: Ornithomimosauria: Ornithomimidae

Notes: *Gallimimus* is seemed to be the largest member of the family Ornithomimidae.

[References: (*1) Osmólska H, Roniewicz E, Barsbold R (1972). A new dinosaur, *Gallimimus bullatus* n. gen., n. sp. (Ornithomimidae) from the Upper Cretaceous of Mongolia. *Palaeontologica Polonica*.27: 103–143.]

Galveosaurus herreroi

> *Galveosaurus herreroi* Sánchez-Hernández, 2005 (*1) [Kimmeridgian–Tithonian; Villar del Arzobispo Formation, Garve, Teruel, Aragon, Spain]

Galvesaurus herreroi

> *Galvesaurus herreroi* Barco *et al.*, 2005 (*2) [Kimmeridgian–Tithonian; Villar del Arzobispo Formation, Galve, Teruel, Aragon, Spain]

Generic name: *Galveosaurus* (*1) / *Galvesaurus* (*2) ← {(place-name) Galve + (Gr.) σαῦρος/sauros: lizard}

Specific name: *herreroi* ← {(person's name) Herrero + -ī}; dedicated to José María Herrero, who discovered the first remains.

Etymology: Herrero's **Galve lizard**

Taxonomy: Saurischia: Sauropodomorpha: Sauropoda

Notes: Two groups of scientists studied the same specimen and published almost simultaneously. Due to Sanchez-Hernández, the name *Galveosaurus* has priority (*3).

[References: (*1) Sánchez-Hernández B (2005). *Galveosaurus herreroi*, a new sauropod dinosaur from Villar del Arzobispo Formation (Tithonian–Berriasian) of Spain. *Zootaxa* 1034: 1–20.; (*2) Barco JL, Canundo JI, Cuenca-Bescós G, Ruiz-Omeñaca JI (2005). Un nuevo dinosaurio saurópodo, *Galvesaurus herreroi* gen. nov., sp. nov., del tránsito Jurásico-Cretácico en Galve (Teruel, NE de España). *Naturaleza Aragonesa* 15: 4–17.; (*3) Sánchez-Hernández B (2006). The new sauropod from Spain: *Galveosaurus* or *Galvesaurus*? *Zootaxa* 1201: 63–68.]

Gannansaurus sinensis

> *Gannansaurus sinensis* Lü *et al.*, 2013 [Maastrichtian; Nanxiong Formation, Gannan, Ganzhou, Jiangxi, China]

Generic name: *Gannansaurus* ← {(Chin. place-name) Gànnán: the district area belonging to Ganzhou of Jiangxi Province + (Gr.) σαῦρος/sauros: lizard} (*1)

Specific name: *sinensis* ← {(Gr. place-name) Sin-: China + -ensis}

Etymology: **Gannan lizard** from China

Taxonomy: Saurischia: Sauropodomorpha: Sauropoda: Neosauropoda: Titanosauriformes: Somphospondyli

[References: (*1) Lü J, Yi L, Zhong H, Wei X (2013). A new somphospondylan sauropod (Dinosauria, Titanosauriformes) from the Late Cretaceous of Ganzhou, Jiangxi Province of Southern China. *Acta Geologica Sinica* 87 (3): 678–685.]

Gansus yumenensis [Avialae]

> *Gansus yumenensis* Hou & Liu, 1984 [Aptian; Xiagou Formation, Gansu, China]

Generic name: *Gansus* ← {(Chin. place-name) Gānsù甘肅 + -us}

Specific name: *yumenensis* ← {(Chin. place-name) Yùmén玉門 + -ensis}

Etymology: **Gansu** from Yumen

Taxonomy: Theropoda: Avialae: Ornithuromorpha

Other species:

> *Gansus zheni* Liu *et al.*, 2014 [Aptian; Jiufotang Formation, Liaoning, China]

Notes: *Gansus* was about the size of a pigeon.

[References: Hou L, Liu Z (1984). A new fossil bird from the Lower Cretaceous of Gansu and early evolution of birds. *Scientia Sinica, series B* 27: 1296–1301.]

Ganzhousaurus nankangensis

> *Ganzhousaurus nankangensis* Wang *et al.*, 2013 [Maastrichtian; Nanxiong Formation,

Nankang, Ganzhou, Jiangxi, China]
Generic name: *Ganzhousaurus* ← {(Chin. place-name) Gànzhōu贛州 + (Gr.) σαῦρος/sauros: lizard}
Specific name: *nankangensis* ← {(Chin. place-name) Nánkāng南康 + -ensis}
Etymology: **Ganzhou lizard** from Nankang County
Taxonomy: Theropoda: Oviraptorosauria: Oviraptoridae

[References: Wang S, Sun C, Sullivan C, Sun C, Xu X (2013). A new oviraptorid (Dinosauria: Theropoda) from the Upper Cretaceous of southern China. *Zootaxa* 3640(2): 242–257.]

Gargantuavis philoinos [Avialae]
> *Gargantuavis philoinos* Buffetaut & Le Loeuff, 1998 [Campanian–Maastrichtian; Marnes Rouges Inférieures Formation, Aude, France]
Generic name: *Gargantuavis* ← {(Fr. folklore) Gargantua: the giant of French folklore made by François Rabelais + (Lat.) avis: bird} [*1]
Specific name: *philoinos* ← {(Gr.) φίλοινος/philoinos: "one who likes wine" [*1]}; "because the sites which have yielded remains of this bird are in the midst of vineyards" [*1].
Etymology: **Gargantua's bird**, fond of wine
Taxonomy: Theropoda: Avialae: Ornithothoraces: Euornithes

[References: (*1) Buffetaut E, Le Loeuff J (1998). A new giant ground bird from the Upper Cretaceous of southern France. *Journal of the Geological Society, London* (155): 1–4.]

Gargoyleosaurus parkpinorum
> *Gargoyleosaurus parkpinorum* Carpenter *et al.*, 1998 [Kimmeridgian–Tithonian; Morrison Formation, Wyoming, USA]
Generic name: *Gargoyleosaurus* ← {[archit.] gargoyle* | (legend) Gargoyle + (Gr.) σαῦρος/sauros: lizard}; "in reference to the gargoyle-like appearance of the skull in profile" [*1].
*an ugly figure of a person or an animal that made of stone and through which water is carried away from the roof of a building, especially a church. (Oxford Advanced Learner's Dictionary, 2000.)
Specific name: *parkpinorum** ← {(person's name) J. Parker & T. Pinegar: discoverers + -ōrum}
*The species epithet was amended in 2001 from *parkpini* to *parkpinorum*, in accordance with ICZN (International Commission on Zoological Nomenclature) art. 31. 1. 2A.
Etymology: Parker and Pinegar's **Gargoyle lizard**
Taxonomy: Ornithischia: Ankylosauria: Nodosauridae: Polacanthinae
cf. Ornithischia: Thyreophora: Ankylosauria: Ankylosauridae (according to *The Dinosauria 2nd edition*)
Notes: *Gargoyleosaurus* shows a mixture of characters seen in Ankylosauridae and Nodosauridae [*1].

[References: (*1) Carpenter K, Miles C, Cloward K, (1998). Skull of a Jurassic ankylosaur (Dinosauria). *Nature* 393: 782–783.]

Garudimimus brevipes
> *Garudimimus brevipes* Barsbold, 1981 [Cenomanian; Bayan Shireh Formation, Ömnögovi, Mongolia]
Generic name: *Garudimimus* ← {(myth.) Garudi- [Garuda: a legendary bird or bird-like creature] + (Gr.) μῖμος/mīmos: mimic} [*1]
Specific name: *brevipes* ← {(Lat.) brevis: short + pēs: foot} [*1]
Etymology: **Garuda mimic** with short feet
Taxonomy: Theropoda: Coelurosauria: Ornithomimosauria: Garudimimidae (= Deinocheiridae)

[References: (*1) Barsbold R (1981). Bezzubyye khischchnyye dinozavry Mongolii. *Sovmestnaia Sovetsko-Mongol'skaia Paleontologicheskaia Ekspeditsiia* Trudy 15: 28–39.; Kobayashi Y, Barsbold R (2005). Reexamination of a primitive ornithomimosaur, *Garudimimus brevipes* Barsbold, 1981 (Dinosauria: Theropoda), from the Late Cretaceous of Mongolia. *Canadian Journal of Earth Sciences* 42: 1501–1521.]

Gasosaurus constructus
> *Gasosaurus constructus* Dong & Tang, 1985 [Bathonian–Oxfordian; lower Shaximiao Formation, Sichuan, China]
Generic name: *Gasosaurus* ← {(Eng.) gas + (Gr.) σαῦρος/sauros: lizard}; alluding to the gas-minig company.
Specific name: *constructus* ← {(Lat.) constrūctus [cōnstruō: pile up, build]}; referring to the fact that fossils were recovered during the construction.
Etymology: constructed, **gas lizard** [建設気龍]
Taxonomy: Theropoda: Tetanurae: Avetheropoda
cf. Theropoda: Carnosauria: Megalosauridae [*1]

[References: (*1) Dong Z, Tang Z (1985). A new mid-Jurassic theropod (*Gasosaurus constructus* gen. *et* sp. nov.) from Dashanpu, Zigong, Sichuan Province, China. *Vertebrata PalAsiatica* 23(1): 77–83.]

Gasparinisaura cincosaltensis
> *Gasparinisaura cincosaltensis* Coria & Salgado, 1996 [Campanian; Anacleto Formation*, Río Negro, Patagonia, Argentina]
*cf. Coniacian–Santonian; Anacleto Member, Río Colorado Formation [*1]

Generic name: *Gasparinisaura* ← {(person's name) Gasparini + (Gr.) σαῦρα/saura, *f.*: lizard}; "in honor of Dr. Zulma Brandoni de Gasparini, for her contribution to the study of Mesozoic reptiles from Patagonia" (*1).

Specific name: *cincosaltensis* ← {(place-name) Cinco Saltos City + -ensis}

Etymology: **Gasparini's lizard** from Cinco Saltos

Taxonomy: Ornithischia: Neornithischia: Elasmaria
  cf. Ornithischia: Ornithopoda (= Euornithopoda): Euornithopoda (= Ornithopoda): Iguanodontia: Euiguanodontia (*1)

[References: (*1) Coria RA, Salgado L (1996). A basal iguanodontian (Ornithischia: Ornithopoda) from the Late Cretaceous of South America. *Journal of Vertebrate Paleontology* 16: 445–457.]

Gastonia burgei

> *Gastonia burgei* Kirkland, 1998 [Barremian–Cenomanian; Yellow Cat Member, Cedar Mountain Formation, Utah, USA]

Generic name: *Gastonia* ← {(person's name) Gaston + -ia}; in honor of "Robert Gaston, who discovered the type locality and has contributed greatly to the research" (*1).

Specific name: *burgei* ← {(person's name) Burge + -ī}; in honor of "Donald L. Burge, director of the College of Eastern Utah Prehistoric Museum, in recognition of his ongoing contributions to dinosaur paleontology in eastern Utah" (*1).

Etymology: Burge and **Gaston's one**

Taxonomy: Ornithischia: Thyreophora: Ankylosauria: Nodosauridae: Polacanthinae
  cf. Ornithischia: Thyreophora: Ankylosauria: Ankylosauridae (acc. *The Dinosauria 2nd edition*)
  cf. Ornithischia: Thyreophora: Ankylosaurindae: Polacanthinae (*1)

Other species:

> *Gastonia lorriemcwhinneyae** Kinneer *et al.*, 2016 [Barremian–Aptian; Ruby Ranch Member, Cedar Mountain Formation, Utah, USA] *named for Lorrie McWhinney

Notes: *Gastonia* is a medium-sized ankylosaur with a sacral shield and large shoulder spikes. This is often considered a nodosaurid, however, recovered as a non-polacanthine basal member of the Ankylosauridae (Arbour, 2014).

[References: (*1) Kirkland JI (1998). A polacanthine ankylosaur (Ornithischia: Dinosauria) from the Early Cretaceous (Barremian) of eastern Utah. In: Lucas SG, Kirkland JI, Estep JW (eds), *Lower and Middle Cretaceous Terrestrial Ecosystems, New Mexico Museum of Natural History and Science Bulletin* 14: 271–281.]

Geminiraptor suarezarum

> *Geminiraptor suarezarum* Senter *et al.*, 2010 [Berriasian–Hauterivian (formerly considered Barremian?); Yellow Cat Member, Cedar Mountain Formation, Utah, USA]

Generic name: *Geminiraptor* ← {(Lat.) gemini-[geminae: twins]: the Suarez sisters + (Lat.) raptor: "one who seizes or takes by force, a common part of deinonychosaurian genus names" (*1)}

Specific name: *suarezarum* ← {(person's name) Suarez + -ārum}; in honor of Drs. Celina and Marina Suarez, the twin geologists who discovered the Suarez site* (*1).
  *Suarez site: the lower Yellow Cat Member of the Cedar Mountain Formation, in western Grand County, Utah.

Etymology: Suarez's **twins seizer**

Taxonomy: Theropoda: Toroodontidae

[References: (*1) Senter P, Kirkland JI, Bird J, Bartlett JA (2010). A new troodontid theropod dinosaur from the Lower Cretaceous of Utah. *PLoS ONE* 5(12): e14329.]

Genusaurus sisteronis

> *Genusaurus sisteronis* Accarie *et al.*, 1995 [Albian; Marnes Bleues Formation, Alpes de Haute Province, France]

Generic name: *Genusaurus* ← {(Lat.) genū: knee + (Gr.) σαῦρος/sauros: lizard}; referring to the cnemial crest in front of the proximal end of the tibia.

Specific name: *sisteronis* ← {(place-name) Sisteron: the town near which the specimen found + -ōnis}

Etymology: **knee lizard** from Sisteron

Taxonomy: Theropoda: Abelisauridae: Majungasaurinae

[References: Accarie H, Beaudoin B, Dejax J, Friès G, Michard JC, Taquet P (1995). Découverte d'un dinosaure théropode nouveau (*Genusaurus sisteronis* n. g., n. sp.) dans l'Albien marin de Sisteron (Alpes de Haute-Province, France) et extension au Crétacé inférieur de la lignée cératosaurienne [Discovery of a new theropod dinosaur *Genusaurus sisteronis* n. g., n. sp. in the marine Albian of Sisteron (Alpes de Haute-Province, France) and the extension into the lower Cretaceous of the ceratosaur lineage]. *Compte Rendu hebdomadaire des scéances de l'Académie des Sciences á Paris* 320(2): 327–334. Abstract.]

Genyodectes serus

> *Genyodectes serus* Woodward, 1901 [Aptian; Cerro Barcino Formation, Chubut, Argentina]

Generic name: *Genyodectes* ← {(Gr.) γένῦς/genys: jaw + δήκτης/dēktēs [δάκνω/daknō: bite]}

Specific name: *serus* ← {(Lat.) sērus: late}

Etymology: late **jaw biter**

Taxonomy: Theropoda: Ceratosauria [*1]

Notes: The holotype had been variously referred to as a megalosaurid, a tyrannosaurid, Theropoda *incertae sedis*, and so on. According to *The Dinosauria 2nd edition*, *Genyodectes* was regarded as a *nomen dubium*.

[References: (*1) Rauhut OWM (2004). Provenance and anatomy of *Genyodectes serus*, a large-toothed ceratosaur (Dinosauria: Theropoda) from Patagonia. *Journal of Vertebrate Paleontology* 24(4): 894–902.]

Geranosaurus atavus

> *Geranosaurus atavus* Broom, 1911 [Sinemurian–Pliensbachian; Clarens Formation, Eastern Cape, South Africa]

Generic name: *Geranosaurus* ← {(Gr.) γέρανος/geranos: crane + σαῦρος/sauros: lizard}; in reference to "the slender birdlike hind-limb bones" [*1].

Specific name: *atavus* ← {(Lat.) atavus: ancestor}

Etymology: ancestral **crane lizard**

Taxonomy: Ornithischia

Notes: According to *The Dinosauria 2nd editoin*, *Geranosarus* is considered to be a *nomen dubium*.

[References: (*1) Broom R (1911). On the dinosaurs of the Stormberg, South Africa. *Annals of the South African Museum* 7(4): 291–308.]

Gettyia gloriae [Avialae]

> *Avisaurus gloriae* Varricchio & Chiappe, 1995 ⇒ *Gettyia gloriae* (Varricchio & Chiappe, 1995) Atterholt *et al.*, 2018 [Campanian; Two Medicine Formation, Montana, USA]

Generic name: *Gettyia* ← {(person's name) Getty + -ia}; "in honor of Mike Getty, a great friend, technician, and field paleontologist, who is dearly missed" [*1].

Specific name: *gloriae* ← {(person's name) Gloria + -ae}; "in recognition of Gloria Siebrecht, who has contributed countless volunteer hours to the Museum of the Rockies both in the field and in the preparation of specimens and whose keen eye discovered this specimen" [*2].

Etymology: **for Getty,** of Gloria

Taxonomy: Aves: Ornithothoraces: Enantiornithes: Avisauridae [*1]

[References: (*1) Atterholt J, Hutchison JH, O'Connor JK (2018). The most complete enantiornithine from North America and a phylogenetic analysis of the Avisauridae. *PeerJ* 6: e5910.; (*2) Varricchio DJ, Chiappe LM (2010). A new enantiornithine bird from the Upper Cretaceous Two Medicine Formation of Montana. *Journal of Vertebrate Paleontology* 15(1): 201–204.]

Gideonmantellia amosanjuanae

> *Gideonmantellia amosanjuanae* Ruiz-Omeñaca *et al.*, 2012 [Barremian; Camarillas Formation, Galve, Teruel, Aragon, Spain]

Generic name: *Gideonmantellia* ← {(person's name) Gideon Mantell + -ia}; "in honor of Gideon Algernon Mantell, the first author to describe and to figure "hypsilophodontid" remains, in 1849, as those of a very young *Iguanodon* (subsequently regarded as the paratype of *Hypsilophodon foxii* Huxley, 1869; Galton, 1974)" [*1].

Specific name: *amosanjuanae* ← {(person's name) Amo Sanjuan + -ae}; "dedicated to Olga María Amo Sanjuán, a fellow of the Department of Palaeontology at the University of Zaragoza who was doing her thesis on eggshell fragments of vertebrates from the Lower Cretaceous of Galve when she died prematurely in October 2002" [*1].

Etymology: Amo Sanjuán's **one for Gideon Mantell**

Taxonomy: Ornithischia: Neornithischia: Cerapoda: Ornithopoda

[References: (*1) Ruiz-Omeñaca JI, Canudo JI, Cuenca-Bescós G, Cruzado-Caballero P, Gasca JM, Moreno-Azanza M (2012). A new basal ornithopod dinosaur from the Barremian of Galve, Spain. *Comptes Rendus Palevol* 11: 453–444.]

Giganotosaurus carolinii

> *Giganotosaurus carolinii* Coria & Salgado, 1995 [Cenomanian; Candeleros Formation, Neuquén, Argentina]

Generic name: *Giganotosaurus* ← {(Gr.) γίγας/gigas: giant [Γίγας/Gigas] + νότος/notos: the south + σαῦρος/sauros: lizard}

Specific name: *carolinii* ← {(person's name) Carolini + -ī}; in honor of the discoverer, Ruben Carolini.

Etymology: Carolini's **giant southern lizard**

Taxonomy: Theropoda: Carnosauria: Allosauroidea: Carcharodontosauridae: Giganotosaurini

Notes: *Giganotosaurus carolinii* from the Southern Hemisphere is probably the world's biggest predatory dinosaur, having a body 12.5 m in length (Coria and Salgado, 1995) [*1].

[References: (*1) Coria RA, Salgado L (1995). A new giant carnivorous dinosaur from the Cretaceous of Patagonia. *Nature* 377(6546): 224–226.]

Gigantoraptor erlianensis

> *Gigantoraptor erlianensis* Xu *et al.*, 2007 [Santonian; Iren Dabasu Formation, Inner Mongolia, China]

Generic name: *Gigantoraptor* ← {(Gr.) Γιγαντο/Giganto- [Γίγας/Gigas: giant] + (Lat.) raptor:

seizer}; "referring to the animal being a gigantic raptor dinosaur" [*1].

Specific name: *erlianensis* ← {(place-name) Erlian + -ensis}; "from the Erlian Basin" [*1].

Etymology: **gigantic thief (raptor)** from Erlian Basin

Taxonomy: Theropoda: Caenagnathidae

cf. Theropoda: Oviraptorosauria [*1]

Notes: The *Gigantoraptor* holotype is estimated to be 8 m in total length and 3.5 m high. The holotype is considered probably to have died at the age of eleven [*1]. The foot has large strongly curved toe claws. Xu *et al.* (2007) assigned *Gigantoraptor* to the *Oviraptoridae* [*1]. Longrich *et al.* (2010) found *Gigantoraptor* to be a caenagnathid [*2].

[References: (*1) Xu X, Tan Q, Wang J, Zhao X, Tan L (2007). A gigantic bird-like dinosaur from the Late Cretaceous of China. *Nature* 447: 844–847. (*2) Longrich NR, Currie PJ, Don Z-M (2010). A new oviraptorid (Dinosauria: Theropoda) from the Upper Cretaceous of Bayan Mandahu, Inner Mongolia. *Palaeontology* 53(5): 945-960.]

Gigantosaurus megalonyx

> *Gigantosaurus megalonyx* Seeley,1869 [Kimmeridgian–Tithonian, Kimmeridge Clay Formation, Ely, Cambridgeshire, UK]

Generic name: *Gigantosaurus* ← {(Gr.) Γιγαντο-/Giganto- [Γίγας/Gigas: giant] + σαῦρος/sauros: lizard}

Specific name: *megalonyx* ← {(Gr.) μεγάλος/megalos: great + ὄνυξ/onyx: claw}

Etymology: **giant lizard** with a great claw

Taxonomy: Saurischia: Sauropodomorpha: Sauropoda

Other species:

> *Gigantosaurus africanus* Fraas, 1908 ⇒*Tornieria africana* (Fraas,1908) Sternfeld, 1911 ⇒ *Barosaurus africanus* (Fraas, 1908) Janensch, 1922 [Tithonian; Tendaguru Formation, Lindi, Tanzania]

> *Gigantosaurus dixeyi* Haughton, 1928 (preoccupied) ⇒ *Tornieria dixeyi* (Haughton, 1928) Sternfeld, 1911 ⇒ *Malawisaurus dixeyi* (Haughton, 1928) Jacobs *et al.*, 1993 [Aptian; Dinosaur Beds Formation, Northern, Malawi]

> *Gigantosaurus robustus* Fraas, 1908 ⇒ *Tornieria robusta* (Fraas, 1908) Sternfeld, 1911 ⇒*Barosaurus robustus* (Fraas, 1908) Haughton, 1928 ⇒ *Janenschia robusta* (Fraas,1908) Wild, 1991 [Tithonian; Tendaguru Formation, Lindi, Tanzania]

Notes: Seeley (1869) described that these bones were found at different times in different localities [*1]. *Gigantosaurus megalonyx* was synonymised to *Ornithopsis humerocristatus* and to *Pelorosaurus* by Lydekker (1888) and von Huene (1909), respectively. According to *The Dinosauria 2nd edition*, *G. megalonyx* is considered a *nomen dubium*.

[References: (*1) Seeley HG (1869). *Gigantosaurus megalonyx*, a terrestrial reptile, from the Kimmeridge Clay. In: *Index to the Fossil Remains of Aves, Ornithosauria, and Reptilia from the Secondary System of Strata, arranged in the Woodwardian Museum of the University of Cambridge.* Deighton, Bell and Co., Cambridge: 94–95.]

Gigantspinosaurus sichuanensis

> *Gigantspinosaurus sichuanensis* Ouyang, 1992 [Oxfordian; upper Shaximiao Formation, Sichuan, China]

Generic name: *Gigantspinosaurus* ← {(Lat.) Gigās = (Gr.) γίγας: a giant | (Lat.) Gigantēus = (Gr.) γιγάντειος: of the giants, enormous + (Lat.) spina: spine + (Gr.) σαῦρος/sauros: lizard}; in reference to the gigantic shoulder spines.

Specific name: *sichuanensis* ← {(Chin. place-name) Sìchuān四川 + -ensis}

Etymology: **giant-spined lizard** from Sichuan

Taxonomy: Ornithischia: Thyreophora: Stegosauria

Notes: Maidment *et al.* (2006) concluded that *G. sichuanensis* are considered a valid taxon [*1].

[References: (*1) Maidment SCR, Wei G (2006). A review of the Late Jurassic stegosaurs (Dinosauria, Stegosauria) from the People's Republic of China. *Geological Magazine* 143(5): 621–634.]

Gilmoreosaurus mongoliensis

> *Mandschurosaurus mongoliensis* Gilmore, 1933 ⇒ *Gilmoreosaurus mongoliensis* (Gilmore, 1933) Brett-Surman, 1979 [Santonian; Iren Dabasu Formation, Inner Mongolia, China]

Generic name: *Gilmoreosaurus* ← {(person's name) Gilmore + (Gr.) σαῦρος/sauros: lizard}; "after Charles W. Gilmore, the original author" [*1].

Specific name: *mongoliensis* ← {(place-name) Mongolia + -ensis}

Etymology: **Gilmore's lizard** from Mongolia

Taxonomy: Ornithischia: Ornithopoda: Hadrosauroidea: Hadrosauromorpha

Other species:

?> *Gilmoreosaurus atavus* Nesov, 1995 [Cenomanian; Khodzhakul Formation, Navoi, Uzbekistan]

?> *Gilmoreosaurus arkhangelskyi* Nesov, 1995 [Turonian; Bissekty Formation, Navoi, Uzbekistan]

[References: (*1) Brett-Surman MK (1979). Phylogeny and palaeobiogeography of hadrosaurian dinosaurs. *Nature* 277: 560–562.]

Giraffatitan brancai

> *Brachiosaurus brancai* Janensch, 1914 ⇒ *Giraffatitan brancai* (Janensch, 1914) Paul, 1988 [Callovian–Hauterivian; Tendaguru Formation, Lindi, Tanzania]

= *Brachiosaurus fraasi* Janensch, 1914 [Kimmeridgian; Tendaguru Formation, Lindi, Tanzania]

Generic name: *Giraffatitan* ← {(New Latin) giraffa: giraffe + (Gr.) Τιτάν/Tītan}; in recognition of the taxon's giraffe-like form (*1).

Specific name: *brancai* ← {(person's name) Branca + -ī}; in honor of German paleontologist Wilhelm von Branca.

Etymology: Branca's **gigantic giraffe**

Taxonomy: Saurischia: Sauropodomorpha: Sauropoda: Brachiosauridae

Notes: According to The *Dinosauria 2nd edition, Giraffatitan* Paul, 1988 is considered a junior synonym of *Brachiosaurus* Riggs, 1903.

[References: (*1) Paul GS (1988). The brachiosaur giants of the Morrison and Tendaguru with a description of a new subgenus, *Giraffatitan*, and a comparison of the world's largest dinosaurs. *Hunteria* 2(3): 1–14.]

Glacialisaurus hammeri

> *Glacialisaurus hammeri* Smith & Pol, 2007 [Hettangian–Pliensbachian; Hanson Formation, Beardmore Glacier, Antarctica]

Generic name: *Glacialisaurus* ← {(Lat.) glacialis: icy, frozen + (Gr.) σαῦρος/sauros: lizard}; "in reference of the geographic location of the type species, which is from Beardmore Glacier region in the Central Transantarctic Mountains" (*1).

Specific name: *hammeri* ← {(person's name) Hammer + -ī}; "in honor of Dr. William R. Hammer (Augustana College, Rock Island, USA), for his contributions to vertebrate palaeontology and Antarctic research" (*1).

Etymology: Hammer's **lizard from glacier**

Taxonomy: Saurischia: Sauropodomorpha: Massospondylidae

[References: (*1) Smith ND, Pol D (2007). Anatomy of a basal sauropodomorph dinosaur from the Early Jurassic Hanson Formation of Antarctica. *Acta Palaeontologica Polonica* 52(4): 657–674.]

Glishades ericksoni

> *Glishades ericksoni* Prieto-Marquez, 2010 [Campanian; Two Medicine Formation, Montana, USA]

Generic name: *Glishades* ← {(Lat.) glis*: mud + (Gr. myth.) Ἅιδης/Haidēs | ἀ-ϊδής/aidēs: unseen} "concealed in mud"; in reference to its being found in sedimentary strata. Also, "Hades" was the dark lord of the underworld in Greek mythology, here metaphorically referring to the "world" beneath the surface where fossils occur (*1).

*glis (*nom.*), glitis (*gen.*), *f.*, terra tenax (acc. in Gaffiot, 1934)

*(mineralogy) a tenacious kind of earth. *cf.* glis (← glūtus: tenacious)

Specific name: *ericksoni* ← {(person's name) Erickson + -ī}; in honor of Dr. Gregory M. Erickson for his important contributions to the knowledge of archosaur paleobiology (*1).

Etymology: Erickson's **Hades concealed in mud**

Taxonomy: Saurischia: Ornithopoda: Hadrosauroidea

Notes: Campione *et al.* (2012) consider *Glishades ericksoni* to be a *nomen dubium*.

[References: (*1) Prieto-Márquez A (2010). *Glishades ericksoni*, a new hadrosauroid (Dinosauria: Ornithopoda) from the Late Cretaceous of North America. *Zootaxa* 2452: 1–17.]

Glyptodontopelta mimus

> *Glyptodontopelta mimus* Ford, 2000 [Campanian–Maastrichtian; Ojo Alamo Formation, New Mexico, USA]

= *Edmontonia australis* Ford, 2000 [Maastrichtian; Kirtland Formation, New Mexico, USA]

Generic name: *Glyptodontopelta* ← {*Glyptodon** [(Gr.) γλυπτός/glyptos: carved, sculptured + ὀδών/odōn: tooth] + (Gr.) πέλτη/peltē: shield}; "in reference to the shape of the pelvic shield that forms a shield similar to that of a glyptodont" (*1).

*a genus of large, armored mammals.

Specific name: *mimus* ← {(Gr.) μῖμος/mīmos: mimic}

Etymology: **Glyptodon shield** mimic

Taxonomy: Saurischia: Thyreophora: Ankylosauria: Nodosauridae

[References: (*1) Ford TL (2000). A review of ankylosaur osteoderms from New Mexico and a preliminary review of ankylosaur armour. In: Lucas SG, Hecket AB (eds), *Dinosaurs of New Mexico New Mexico Museum of Natural History and Science Bulletin* 17: 157–176.]

Gnathovorax cabreirai

> *Gnathovorax cabreirai* Pacheco *et al.*, 2019 [Carnian; Santa Maria Formation, Rio Grande do Sul, Brazil]

Generic name: *Gnathovorax* ← {(Gr.) γνάθος/gnathos: jaw + (Lat.) vorō: devour + -āx ("inclined to")} (*1)

Specific name: *cabreirai* ← {(person's name) Cabreira + -ī}; in honor of "Dr. Sérgio Furtado Cabreira, the palaeontologist that found the

specimen described here" [*1].
Etymology: Cabreira's ravenous jaws
Taxonomy: Saurischia: Herrerasauridae
Notes: It is seemed that *Gnathovorax* had powerful jaws.
[References: (*1) Pacheco C, Müler RT, Langer M, Pretto FA, Kerber L, Dias da Silva S (2019). *Gnathovorax cabreirai*: a new early dinosaur and the origin and initial radiation of predatory dinosaurs. *PeerJ* 7: e7963.]

Gobiceratops minutus

> *Bagaceratops rozhdestvenskyi* Mariańska & Osmólska, 1975 [Campanian–Maastrichtian; Barun Goyot Formation, Ömnögovi, Mongolia]
= *Gobiceratops minutus* Alifanov, 2008 [Campanian–Maastrichtian; Barun Goyot Formation, Khermin Tsav locality, Ömnögovi, Mongolia]
For the other synonyms, see also *Bagacertatops rozhdestvenskyi*.
Generic name: *Gobiceratops* ← {(place-name) Gobi Desert + (Generic name) *Ceratops* [(Gr.) κέρας/keras: horn + ὤψ/ōps: eye, face]} [*1]
Specific name: *minutus* ← {(Lat.) minūtus: small}
Etymology: small **Gobi horned-face (ceratopsian)**
Taxonomy: Ornithischia: Ceratopsia: Neoceratopsia: Coronosauria: Bagaceratopsidae / Bagaceratopidae
Notes: According to Czepiński (2020), *Gobiceratops minutus*, *Lamaceratops tereschenkoi*, *Platyceratops tatarinovi* and *Magnirostris dodsoni* are considered to be synonyms of *Bagaceratops rozhdestvenskyi*. [*2]
[References: (*1) Alifanov V (2008). The tiny horned dinosaur *Gobiceratops minutus* gen. *et* sp. nov. (Bagaceratopidae, Neoceratopsia) from the Upper Cretaceous of Mongolia. *Paleontological Journal* 42(6): 621–633.]; (*2) Czepiński Ł (2020). Ontogeny and variation of a protoceratopsid dinosaur *Bagaceratops rozhdestvenskyi* from the Late Cretaceous of the Gobi Desert. *Historical Biology: An International Journal of Paleobiology* 32(10): 1394–1421.]

Gobihadros mongoliensis

> *Gobihadros mongoliensis* Tsogtbaatar *et al.*, 2019 [Cenomanian; Bayan Shireh Formation, Ömnögovi, Mongolia]
Generic name: *Gobihadros* ← {(place-name) Gobi Desert + hadros: hadrosauroidea}
Specific name: *mongoliensis* ← {(place-name) Mongolia + -ensis}
Etymology: **hadrosauroid from the Gobi Desert** of Mongolia
Taxonomy: Ornithischia: Ornithopoda: Iguanodontia: Hadrosauroidea

Gobipteryx minuta [Avialae]

> *Gobipteryx minuta* Elżanowski, 1974 [Campanian; Barun Goyot Formation, Gobi Desert, Ömnögovi, Mogolia]
= *Nanantius valifanovi* Kurochkin, 1996 [Campanian; Barun Goyot Formation, Ömnögovi, Mongolia]
Generic name: *Gobipteryx* ← {(place-name) Gobi Desert + (Gr.) πτέρυξ/pteryx, *f.* = πτερόν/pteron, *n.*: wing}
Specific name: *minuta* ← {(Lat.) minūta (feminine form of 'minūtus'): small}
Etymology: small, **Gobi wing**
Taxonomy: Theropoda: Avialae: Enantiornithes: Gobipterygidae
cf. Palaeognathae: Gobipterygiformes: Gobipterygidae [*1]
[References: (*1) Elżanowski A (1974). Preliminary note on the palaeognathous bird from the Upper Cretaceous of Mongoia. *Palaeontologia Polonica* 30: 103–109.]

Gobiraptor minutus

> *Gobiraptor minutus* Lee *et al.*, 2019 [Maastrichtian; Nemegt Formation, Ömnögovi, Mongolia]
Generic name: *Gobiraptor* ← {(place-name) Gobi Desert + (Lat.) raptor: thief}
Specific name: *minutus* ← {(Lat.) minūtus: small}; referring to "the small size of the holotype specimen".
Etymology: small **thief from Gobi Desert**
Taxonomy: Saurischia: Theropoda: Oviraptorosauria: Oviratoridae

Gobisaurus domoculus

> *Gobisaurus domoculus* Vickaryous *et al.*, 2001 [*1] [Turonian*; Ulansuhai Formation, Inner Mongolia, China]
= *Zhongyuansaurus luoyangensis* Xu *et al.*, 2007 [Cenomanian; Mangchuan Formation, Henan, China]
*The Ulansuhai Formation where *Gobisaurus* found was presumed the Aptian stage but later studies indicate that it cannot be older than the Turonian stage of the Late Cretaceous [*2].
Generic name: *Gobisaurus* ← {(place-name) Gobi Desert + (Gr.) σαῦρος/sauros: lizard}
Specific name: *domoculus* ← {(Lat.) domo: subdue + oculus: the eye}
Etymology: overlooked, **Gobi lizard**
Taxonomy: Ornithischia: Ankylosauria: Ankylosauridae

Notes: *Gobisaurus domoculus* closely resembles *Shamosaurus scutatus* (*1). According to Arbour (2014), *Zhongyuansaurus luoyangensis* is probably a junior synonym of *Gobisaurus domoculus*.

[References: (*1) Vickaryous MK, Russell AP, Currie PJ, Zhao X-Z (2001). A new ankylosaurid (Dinosauria: Ankylosauria) from the Lower Cretaceous of China, with comments on ankylosaurian relationships. *Canadian Journal of Earth Sciences* 38(12): 1767–1780.; (*2) Kobayashi Y, Lü J-C (2003). A new ornithomimid dinosaurian with gregarious habits from the Late Cretaceous of China. *Acta Palaeontologica Polonica* 48: 235–259.]

Gobititan shenzhouensis

> *Gobititan shenzhouensis* You *et al.*, 2003 [Barremian; Xinminbao Group, Gansu, China]

Generic name: *Gobititan* ← {(place-name) Gobi: the geographical region in Mongolia and Inner Mongolia + (Gr.myth.) Τῑτάν/Tītan: giants, symbolic of great size} (*1)

Specific name: *shenzhouensis* ← {(Chin. place-name) Shénzhōu神州 (= an ancient name for China) + -ensis}

Etymology: **Gobi titanosaur** (Titans) from China

Taxonomy: Saurischia: Sauropodomorpha: Sauropoda: Titanosauriformes: Titanosauria

[References: (*1) You H, Tang F, Luo Z (2003). A new basal titanosaur (Dinosauria: Sauropoda) from the Early Cretaceous of China. *Acta Geologica Sinica* 77(4): 424–429.]

Gobivenator mongoliensis

> *Gobivenator mongoliensis* Tsuihiji *et al.*, 2014 [late Campanian; Djadokhta Formation, Ömnögovi, Mongolia]

Generic name: *Gobivenator* ← {(place-name) Gobi Desert + (Lat.) vēnātor: hunter}

Specific name: *mongoliensis* ← {(place-name) Mongolia + (Lat. suffix) -ensis: from}

Etymology: **Gobi hunter** from Mongolia

Taxonomy: Theropoda: Troodontidae

[References: Tsuihiji T, Barsbold R, Watabe M, Tsogtbaatar K, Chinzorig T, Fujiyama Y, Suzuki S (2014). An exquisitely preserved troodontid theropod with new information on the palatal structure from the Upper Cretaceous of Mongolia. *Naturwissenschaften* 101(2): 131–142.]

Gojirasaurus quayi

> *Gojirasaurus quayi* Carpenter, 1997 [Norian; Cooper Canyon Formation, New Mexico, USA]

Generic name: *Gojirasaurus* ← {((Jpn.) Gojira*: Godzilla + (Gr.) σαῦρος/sauros: lizard}

*"Gojira, a large fictional monster of the Japanese cinema, in reference to the large size of this Triassic theropod" (*1).

Specific name: *quayi* ← {(place-name) Quay + -ī}; "referring to Quay County, New Mexico" (*1).

Etymology: Quai County's **Gojira (Godzilla) lizard**

Taxonomy: Theropoda: Ceratosauria: Coelophysoidea

[References: (*1) Carpenter K (1997). A giant coelophysoid (Ceratosauria) theropod from the Upper Triassic of New Mexico, USA. *Neues Jahrbuch für Geologie und Paläontologie, Abhandlungen* 205(2): 189–208.]

Gondwanatitan faustoi

> *Gondwanatitan faustoi* Kellner & de Azevedo, 1999 [Maastrichtian; Adamantina Formation, São Paulo, Brazil]

Generic name: *Gondwanatitan* ← {(place-name) Gondwana + (Gr. myth) Τῑτάν/Tītan: giant}; named after Gondwana, the continental mass that once united all southern continents (and India) and Titan, which relates to a group of Greek divinities (*1).

Specific name: *faustoi* ← {(place-name) Fausto + -ī}; in honor of Dr. Fausto L. de Souza Cunha, former curator at the Museu Nacional / UFRJ*, who collected and recognized the importance of this specimen (*1).

*Universidade Federal do Rio de Janeiro

Etymology: Fausto's **giant from Gondwana**

Taxonomy: Saurischia: Sauropoda: Titanosauria: Aeolosauridae

[References: (*1) Kellner AWA, de Azevedo SAK (1999). A new sauropod dinosaur (Titanosauria) from the Late Cretaceous of Brazil. *National Science Museum Monographs*. 15: 111–142.]

Gongbusaurus shiyii

> *Gongbusaurus shiyii* Dong *et al.*, 1983 [Oxfordian; upper Shaximiao (= Shangshaximiao) Formation, Sichuan, China]

Generic name: *Gongbusaurus* ← {(Chin.) gōngbù工部: "the Ministry of Works in feudal China" + (Gr.) σαῦρος/sauros: lizard}; commemorating the famous Tang唐 Dynasty poet Dufu杜甫 who was an adjunct member in this ministry, and it was thereby commonly referred to as the "Du Ministry of works". (*1)

Specific name: *shiyii* ← {(Chin.) shíyí拾遺 + -ī}; referring to "an official post during the Tang Dynasty with the responsibility to counsel the emperor upon potential error, a post that was also held by the poet Minister Dufu. As these specimens were found among a mixed assemblage of material in Sichuan, or were seemingly lost and then recovered to rectify the error, the acknowledgement of Dufu through this double entendre commemorates the great poet of the Tang Dynasty, by the erection of *Gongbusaurus shii* gen. *et* sp. nov" (*1).

Etymology: Shiyi's **Gongbu lizard** [拾遺工部龍]

Taxonomy: Ornithischia: Neornithischia

cf. Ornithischia: Fabrosauridae (*1)

Other species:

?> *Gongbusaurus wucaiwanensis* Dong, 1989 [Callovian; Shishugou Formation, Xinjiang, China]

Notes: According to *The Dinosauria 2nd edition*, *Gongbusaurus shii* is considered a *nomen dubium*, and "*Gongbusaurus*" *wucaiwanensis* is an unnamed euornithopodan.

[References: (*1) Dong Z, Zhou S, Zhang Z (1983). Dinosaurs from the Jurassic of Sichuan. *Palaeontologica Sinica, New Series C* 162(23): 1–145.]

Gongpoquansaurus mazongshanensis

> *Probactrosaurus mazongshanensis* Lü, 1997 ⇒ *Gongpoquansaurus mazongshanensis* (Lü, 1997) You *et al.*, 2014 [Barremian–Albian; Zhonggou Formation, Gansu, China]

Generic name: *Gongpoquansaurus* ← {(Chin. place-name) Gōngpóquán公婆泉: the name of the basin + (Gr.) σαῦρος/sauros: lizard}

Specific name: *mazongshanensis* ← {(Chin. place-name) Mǎzōngshān + -ensis}

Etymology: **Gongpoquan lizard** from Mazongshan

Taxonomy: Ornithischia: Ornithopoda: Iguanodontia: Ankylopolexia: Styracosterna: Hadrosauriformes: Hadrosauroidea

[References: You H-I, Li D-Q, Dodson P (2014). *Gongpoquansaurus mazongshanensis* (Lü, 1997) *comb. nov.* (Ornithischia: Hadrosauroidea) from the Early Cretaceous of Gansu Province, Northwestern China. In: Eberth DA, Evans DC (eds), Hadrosaurs. Indiana University Press, Bloomington: 73–76.]

Gongxianosaurus shibeiensis

> *Gongxianosaurus shibeiensis* He *et al.*, 1998 [Toarcian; Ziliujing Formation, Shibeixiang, Gonxian, Yibin, Sichuan, China]

Generic name: *Gongxianosaurus* ← {(Chin. place-name) Gǒngxiàn珙縣 + (Gr.) σαῦρος/sauros}; "derived from the county and village from which the specimens were excavated" (*1).

Specific name: *shibeiensis* ← {(Chin. place-name) Shíbēi石碑 + -ensis}

Etymology: **Gongxian lizard** from Shibeixiang Village

Taxonomy: Saurischia: Sauropodomorpha: Sauropoda

Notes: Adult specimens are approximately 14 m in length (*1).

[References: (*1) He X, Wang C, Liu S, Zhou F, Liu T, Cai K, Dai B (1998). A new species of sauropod from the Early Jurassic of Gongxian Co., Sichuan. *Acta Geologica Sichuan* 18(1): 1–7.]

Gorgosaurus libratus

> *Gorgosaurus libratus* Lambe, 1914 ⇒ *Deinodon libratus* (Lambe, 1914) Matthew & Brown, 1922 (*nomen dubium*) ⇒ *Albertosaurus libratus* (Lambe, 1914) Russell, 1970 [Campanian; Dinosaur Park Formation, Alberta, Canada]

= *Gorgosaurus sternbergi* Matthew & Brown, 1923 ⇒ *Deinodon sternbergi* (Matthew & Brown, 1923) Kuhn, 1965 ⇒ *Albertosaurus sternbergi* (Matthew & Brown, 1923) Russell, 1970 [Campanian; Dinosaur Park Formation, Alberta, Canada]

= *Laelaps incrassatus* Cope, 1876 (*nomen dubium*) [Campanian; Judith River Formation, Montana, USA]

?= *Laelaps falculus* Cope, 1876 (*nomen dubium*) [Campanian; Judith River Formation, Montana, USA]

?= *Laelaps hazenianus* Cope, 1876 (*nomen dubium*) [Campanian; Judith River Formation, Montana, USA]

?= *Dryptosaurus kenabekides* Hay, 1899 (*nomen dubium*) [Campanian; Judith River Formation, Montana, USA]

According to *The Dinosauria 2nd edition*, *Gorgosaurus sternbergi*, *Laelaps incrassatus* Cope, 1876, ?*Laelaps falculus* Cope, 1876, ?*Laelaps hazenianus* Cope, 1876, ?*Dryptosaurus kenabekides* Hay, 1899 and ?*Deinodon horridus* Leidy, 1856 were considered synonyms of *Gorgosaurus libratus*.

Generic name: *Gorgosaurus* ← {(Gr.) γοργός/gorgos: fierce + σαῦρος/sauros: lizard}

Specific name: *libratus* ← {(Lat.) librātus: balanced}; "in reference to the animal's probable well-balanced and easy gait" (*1).

Etymology: balanced, **fierce lizard**

Taxonomy: Theropoda: Coelurosauria: Tyrannosauroidea: Tyrannosauridae: Albertosaurinae

Other species:

> *Gorgosaurus lancensis* Gilmore, 1946 ⇒ *Albertosaurus* (*Nanotyrannus*) *lancensis* (Gilmore, 1946) Paul, 1988 ⇒ *Nanotyrannus lancensis* (Gilmore, 1946) Bakker *et al.*, 1988 [Maastrichtian; Hell Creek Formation, Montana, USA]

> *Tyrannosaurus bataar* Maleev, 1955 ⇒ *Tarbosaurus bataar* (Maleev 1955) Maleev, 1955 ⇒ *Jenghizkhan bataar* (Maleev, 1955) Olshevsky *et al.*, 1995 [Campanian; Nemegt Formation, Ömnögovi, Mongolia]
= *Gorgosaurus novojilovi* Maleev, 1955 ⇒ *Deinodon novojilovi* (Maleev, 1955) Kuhn, 1965 ⇒ *Aublysodon novojilovi* (Maleev, 1955) Charig, 1967 ⇒ *Albertosaurus novojilovi* (Maleev, 1955) Mader & Bradley, 1989 ⇒ *Maleevosaurus novojilovi* (Maleev, 1955) Pickering, 1984 [Campanian; Nemegt Formation, Ömnögovi, Mongolia]
= *Gorgosaurus lancinator* Maleev, 1955 ⇒ *Deinodon lancinator* (Maleev, 1955) Kuhn, 1965 ⇒ *Aublysodon lancinator* (Maleev, 1955) Charig, 1967 [Maastrichtian; Nemegt Formation, Ömnögivi, Mongolia]

Notes: *Gorgosaurus* is a carnivorous dinosaur of large size, reaching a length of about twenty-nine feet (Lambe, 1914) (*1). *Gorgosaurus* and *Albertosaurus* are thought to be extremely similar.

[References: (*1) Lambe LM (1914). On a new genus and species of carnivorous dinosaur from the Belly River Formation of Alberta, with a description of the skull of *Stephanosaurus marginatus* from the same horizon. *Ottawa Naturalist* 28: 13–20.]

Goyocephale lattimorei

> *Goyocephale lattimorei* Perle *et al.*, 1982 [Campanian–Maastrichtian; unnamed unit, Övörkhangai, Mongolia]

Generic name: *Goyocephale* ← {(Mong.) гоё/goyo: decorated, elegant + (Gr.) κεφαλή/kephalē: head}

Specific name: *lattimorei* ← {(person's name) Lattimore + -ī}; "in honor of the eminent American mongolist Prof. Owen Lattimore" (*1).

He is the author of *The Desert Road to Turkestan* (1928).

Etymology: Lattimore's **decorated head**

Taxonomy: Ornithischia: Thyreophora:
Pachycephalosauria: Pachycephalosauridae
cf. Pachycephalosauria: Homalocephalidae (*1)

[References: (*1) Perle A, Maryańska T, Osmólska H (1982). *Goyocephale lattimorei* gen. *et* sp. n., a new flat-headed pachycephalosaur (Ornithischia, Dinosauria) from the Upper Cretaceous of Mongolia. *Acta Palaeontologica Polonica* 27(1–4): 115–127.]

Graciliceratops mongoliensis

> *Graciliceratops mongoliensis* Sereno, 2000 [Cenomanian–Santonian; Bayan Shireh Formation, Shireegiin Gashuun locality, Ömnögovi, Mongolia]

Generic name: *Graciliceratops* ← {(Lat.) gracilis: slender + *Ceratops* [(Gr.) κέρας/keras: horn + ὤψ/ōps: face]}

This specimen is "characterized by the very slender median and posterior parietal frill margins and high tibiofemoral ratio" (*1).

Specific name: *mongoliensis* ← {(place-name) Mongolia + -ensis}

Etymology: **slender-horned face (ceratopsian)** from Mongolia

Taxonomy: Ornithischia: Cerapoda: Ceratopsia

Notes: The specimen of *Graciliceratops* represents an immature form which is estimated to be 0.8 m in length.

[References: (*1) Sereno PC (2000). The fossil record, systematics and evolution of pachycephalosaurs and ceratopsians from Asia. *The age of dinosaurs in Russia and Mongolia*: 480–516.]

Graciliraptor lujiatunensis

> *Graciliraptor lujiatunensis* Xu & Wang, 2004 [Barremian–Aptian; Yixian Formation, Beipiao, Liaoning, China]

Generic name: *Graciliraptor* ← {(Lat.) gracilis: slender + raptor: robber}; "derived from the slender limbs and tail of the animal, and raptor, commonly used for dromaeosaurid dinosaur names" (*1).

Specific name: *lujiatunensis* ← {(Chin. place-name) Lùjiātún陸家屯 + -ensis}

Etymology: **slender robber (raptor)** from Lujiatun village

Taxonomy: Theropoda: Dromaeosauridae:
Microraptoria

[References: (*1) Xu X, Wang X-L (2004). A new dromaeosaur (Dinosauria: Theropoda) from the Early Cretaceous Yixian Formation of western Liaoning. *Vertebrata PalAsiatica* 42 (2): 11–119.]

Gravitholus albertae

> *Gravitholus albertae* Wall & Galton, 1979 [late Campanian; Dinosaur Park Formation, Alberta, Canada]

Generic name: *Gravitholus* ← {(Lat.) gravis: heavy + tholus: dome}; "referring to the enlarged dome" (*1).

Specific name: *albertae* ← {(place-name) Alberta + -ae}

Etymology: Alberta's **heavy dome**

Taxonomy: Ornithischia: Marginocephalia:
Pachycephalosauridae

[References: (*1) Wall WP, Galton PM (1979). Notes on pachycephalosaurid dinosaurs (Reptilia: Ornithischia) from North America, with comments on their status as ornithopods. *Canadian Journal of Earth Sciences* 16: 1176–1186.]

Gryphoceratops morrisoni

> *Gryphoceratops morrisoni* Ryan *et al.*, 2012 [Santonian; Milk River Formation, Alberta, Canada]

Generic name: *Gryphoceratops* ← {(Eng.) Gryphon: mythological figure [(Gr.) γρύψ/gryps: griffin] + *Ceratops*: horned face [κέρας/keras: horn + ὤψ/ōps: face]}; referring to the legendary gryphon that had the body of a lion and the head of an eagle; a reference to the beaked-face (*1).

Specific name: *morrisoni* ← {(person's name) Morrison + -ī}; in honor of Royal Ontario Museum technician Ian Morrison for his expert preparation of the holotype jaw and his contributions to vertebrate paleontology (*1).

Etymology: Morrison's **Griffin horned face (ceratopsian)**

Taxonomy: Ornithischia: Ceratopsia: Neoceratopsia: Leptoceratopsidae (*1)

Notes: *Gryphoceratops* is a small leptoceratopsid.

[References: (*1) Ryan MJ, Evans DC, Currie PJ, Brown CM, Brinkman D (2012). New leptoceratopsids from the Upper Cretaceous of Alberta, Canada. *Cretaceous Research* (35): 69–80.]

Gryponyx africanus

> *Gryponyx africanus* Broom, 1911 [Hettangian–Sinemurian; upper Elliot Formation, Free State (= Orange Free State), South Africa]

Generic name: *Gryponyx* ← {(Gr.) γρυπός/grypos: hooked + ὄνυξ/onyx: claw}

Specific name: *africanus* ← {(Lat.) Āfricānus: African}; referring to Africa.

Etymology: African **hooked claw**

Taxonomy: Saurischia: Sauropodomorpha: Gryponychidae

Notes: *Gryponyx* was originally described as a large carnivorous dinosaur (theropod), by Broom (*1). According to *The Dinosauria 2nd edition*, *Gryponyx* is considered to be a *nomen dubium*.

[References: (*1) Broom R (1911). On the dinosaurs of the Stormberg, South Africa. *Annals of the South African Museum* 7(4): 291–308.]

Gryposaurus notabilis

> *Gryposaurus notabilis* Lambe, 1914 ⇒ *Kritosaurus notabilis* (Lambe, 1914) Lull & Wright, 1942 ⇒ *Hadrosaurus notabilis* (Lambe, 1914) Horner, 1979 [Campanian; Dinosaur Park Formation, Alberta, Canada]

Generic name: *Gryposaurus* ← {(Gr.) γρυπός/grypos: hooked beak + σαῦρος/sauros: lizard}; "referring to the arch on the dorsal surface of the nasal, resembling a gryphin" (*1).

Specific name: *notabilis* ← {(Lat.) notābilis: notable}.

Etymology: notable **hooked-nosed lizard**

Taxonomy: Ornithischia: Ornithopoda: Iguanodontia: Hadrosauridae: Saurolophinae: Kritosaurini
cf. Ornithischia: Hadrosauridae: Hadrosaurinae (*1)

Other species:

> *Kritosaurus incurvimanus* Parks, 1920 ⇒ *Gryposaurus incurvimanus* (Parks, 1920) Gates & Sampson, 2007 [Campanian; Dinosaur Park Formation, Alberta, Canada]

> *Gryposaurus latidens* Horner, 1992 [late Santonian; Two Medicine Formation, Montana, USA]

> *Gryposaurus monumentensis* Gates & Sampson, 2007 [late Campanian; Kaiparowits Formation, Utah, USA]

> *Gryposaurus alsatei* Lehman *et al.*, 2016 [late Maastrichtian; Javelina Formation, Texas, USA]

[References: (*1) Gates TA, Sampson SD (2007). A new species of *Gryposaurus* (Dinosauria: Hadrosauridae) from the late Campanian Kaiparowits Formation, southern Utah, USA. *Zoological Journal of the Linnean Society* 151: 351–376.; Lambe LM (1914). On *Gryposaurus notabilis*, a new genus and species of trachodont dinosaur from the Belly River Formation of Alberta, with a description of the skull of *Chasmosaurus belli*. *The Ottawa Naturalist* 27(11): 145–155.]

Guaibasaurus candelariensis

> *Guaibasaurus candelariensis* Bonaparte *et al.*, 1999 [Carnian–Rhaetian?; Caturrita Formation, Rio Grande do Sul, Brazil]

Generic name: *Guaibasaurus* ← {(place-name) Guaíba + (Gr.) σαῦρος/sauros: lizard}; named after Rio Guaíba Hydrographic Basin where the holotype was collected.

Specific name: *candelariensis* ← {(place-name) Candelária + -ensis}

Etymology: **Rio Guaiba lizard** from Candelária

Taxonomy: Saurischia: Sauropodomorpha: Guaibasauridae

[References: Bonaparte JF, Ferigolo J, Ribeiro AM (1999). A new early Late Triassic saurischian dinosaur from Rio Grande do Sul State, Brazil. In: Tomida Y, Rich TH, Vickers-Rich P (eds), *Proceedings of the Second Gondwana Dinosaur Symposium. Nationa Science Museum Monographs* 15: 89–109.]

Gualicho shinyae

> *Gualicho shinyae* Apesteguía *et al.*, 2016 [late Cenomanian–early Turonian; Huincul Formation, Neuquén Basin, northern Patagonia, Argentina]

Generic name: *Gualicho* (Spanish) ←"(Gennaken

[günün-a-künna] or northern Tehuelche language) *watsiltsüm*; for a goddess who was considered the owner of animals and later, following the introduction of Christianity, reinterpreted as a demonic entity. She is now considered a source of misfortune by rural settlers (gauchos) of the Southern Cone. The name was chosen to reflect the difficult circumstances surrounding the discovery and study of the specimen, and its contentious history following excavation" (*1).

Specific name: *shinyae* ← {(person's name) Shinya + -e}; "in honor of Ms. Akiko Shinya, chief fossil preparator at the Field Museum, Chicago, for her many contributions to paleontology including discovery of the holotype of *Gualicho* on February 13th, 2007" (*1).

Etymology: Shinya's **Gualicho**

Taxonomy: Theropoda: Carcharodontosauria: Neovenatoridae
cf. Theropoda: Tetanurae: Avetheropoda (*1)

[References: (*1) Apesteguía S, Smith ND, Juárez Valieri R, Makovicky PJ (2016). An unusual new theropod with a didactyl manus from the Upper Cretaceous of Patagonia, Argentina. *PLoS ONE* 11(7): e0157793.]

Guanlong wucaii

> *Guanlong wucaii* Xu *et al.*, 2006 [Oxfordian; Shishugou Formation, Xingjiang, China]

Generic name: *Guanlong* ← {(Chin.) Guān冠: crown + lóng龍: dragon}; in reference to its headcrest.

Specific name: *wucaii* ← {(Chin.) wǔcǎi五彩: meaning "five colours" + -ī}; "referring to the rich colours of rocks that produced the specimens" (*1). *Guanlong* was found in "Wucaiwan五彩灣 area, Junggar Bassin, Xinjiang".

Etymology: **crowned dragon** of five colours (from Wukaiwan)

Taxonomy: Theropoda: Tyrannosauroidea: Proceratosauridae

Notes: *Guanlong* is a basal tyrannosauroid dinosaur (*1).

[References: (*1) Xu X, Clark JM, Forster CA, Norell MA, Erickson GM, Eberth DA, Jia C, Zhao Q (2006). A basal tyrannosauroid dinosaur from Late Jurassic of China. *Nature* 439: 715–718.]

Gurilynia nessovi [Avialae]

> *Gurilynia nessovi* Kurochkin, 1999 [Maastrichtian; Nemegt Formation, south Gobi, Mongolia]

Generic name: *Gurilynia* ← {(place-name) Gurilyn: Gurilyn Tsav locality + -ia}

Specific name: *nessovi* ← {(person's name) Nessov + -ī}; in honor of Lev Nesov.

Etymology: **Nessov's one from Gurilyn Tsav**

Taxonomy: Theropoda: Avialae: Enatiornithes

Gyposaurus capensis

> *Gyposaurus capensis* Broom, 1911 ⇒ *Anchisaurus capensis* (Broom, 1911) Galton & Culver, 1976 [Hettangian; upper Elliot Formation, Free State (= Orange Free State), South Africa]

Generic name: *Gyposaurus* ← {gypo- [(Gr.) γύψ/gyps: vulture] + (Gr.) σαῦρος/sauros: lizard}

Specific name: *capensis* ← {(place-name) Cape + -ensis}

Etymology: **vulture lizard** from Cape

Taxonomy: Saurischia: Sauropodomorpha: Plateosauria

Other species:

> *Lufengosaurus huenei* Young, 1941 ⇒ *Massospondylus huenei* (Young, 1941) Cooper, 1981 [Hettangian; Lufeng Formation, Yunnan, China]

?= *Gyposaurus sinensis* Young, 1941 ⇒ *Anchisaurus sinensis* (Young, 1941) Dong *vide* Olshevsky, 1991 [Hettangian; Lufeng Formation, Yunnan, China]

Notes: According to *The Dinosauria 2nd edition*, *Gyposaurus capensis* was regarded as a *nomen dubium* and *G. sinensis* as an unnamed prosauropod. Wang *et al.* (2017) indicate that *Gyposaurus sinensis* is a junior synonym of *Lufengosaurus huenei*.

H

Hadrosaurus foulkii

> *Hadrosaurus foulkii* Leidy, 1858 ⇒ *Trachodon foulkii* (Leidy, 1858) Lydekker, 1888 [Campanian; Woodbury Formation, Haddonfield, New Jersey, USA]

= *Ornithotarsus immanis* Cope, 1869 [Campanian; Woodbury Formation, New Jersey, USA]

= *Hadrosaurus cavatus* Cope, 1871 (*nomen dubium*) ⇒ *Trachodon cavatus* (Cope, 1871) Hay, 1902 (*nomen dubium*) [Maastrichtian, Navesink Formation, New Jersey, USA]

Generic name: *Hadrosaurus* ← {(Gr.) ἁδρός/hadros: stout, bulky, sturdy + σαῦρος/sauros: lizard}

Specific name: *foulkii* ← {(person's name) Foulke + -iī}; in honor of William Parker Foulke.

The skeleton of *Hadrosaurus* was recovered by John Estaugh Hopkins from a marl pit in Haddonfield in 1838, and Foulke excavated the site in 1858.

Etymology: Foulke's **bulky lizard**

Taxonomy: Ornithischia: Ornithopoda: Hadrosauridae

Other species:

> *Trachodon mirabilis* Leidy, 1856 (*nomen dubium*) ⇒ *Hadrosaurus mirabilis* (Leidy, 1856) Leidy, 1868 ⇒ *Diclonus mirabilis* (Leidy, 1856) Cope, 1883 [Campanian; Judith River Formation, Montana, USA]
cf. Diclonius mirabilis Cope, 1876 (*nomen dubium*, in part) [Lance Formation] (See *Edmontosaurus*.)

> *Thespesius occidentalis* Leidy, 1856 (*nomen dubium*) ⇒ *Hadrosaurus occidentalis* (Leidy, 1856) Cope, 1874 (*nomen dubium*) ⇒ *Trachodon occidentalis* (Leidy, 1856) Kuhn, 1936 (*nomen dubium*) [Maastrichtian; Lance Formation, South Dakota, USA]

= *Agathaumas milo* Cope, 1874 ⇒ *Hadrosaurus milo* (Cope, 1874) Hay, 1901 [Maastrichtian; Denver Formation, Colorado, USA]

> *Hadrosaurus tripos* Cope, 1869 (*nomen dubium*)* [Pliocene transition zone; Duplin Formation, North Carolina, USA]
*not a dinosaur, Balaenopteridae

> *Hadrosaurus minor* Marsh, 1870 (*nomen dubium*) ⇒ *Edmontosaurus* ("*Hadrosaurus*") *minor* Baird & Horner, 1977 [Maastrichtian; Navesink Formation, New Jersey, USA]

> *Hadrosaurus agilis* Marsh, 1872 ⇒ *Claosaurus agilis* (Marsh, 1872) Marsh, 1890 ⇒*Trachodon agilis* (Marsh, 1872) Kuhn, 1936 [Coniacian–Campanian; Niobrara Formation, Kansas, USA]

> *Trachodon cantabrigiensis* Lydekker, 1888 (*nomen dubium*) ⇒ *Hadrosaurus cantabrigiensis* (Lydekker, 1888) Newton, 1892 (*nomen dubium*) ⇒ *Telmatosaurus cantabrigiensis* (Lydekker, 1888) Olshevesky, 1978 (*nomen dubium*) [Cenomanian; Cambridge Greensand, Cambridgeshire, England, UK]

> *Hadrosaurus paucidens* Marsh, 1889 (*nomen dubium*) ⇒ *Ceratops paucidens* (Marsh, 1889) Hay, 1901 [Campanian; Judith River Formation, Montana, USA]

> *Hadrosaurus breviceps* Marsh, 1889 (*nomen dubium*) ⇒ *Kritosaurus breviceps* (Marsh, 1889) Lull & Wright, 1942 (*nomen dubum*) ⇒ *Trachodon breviceps* (Marsh, 1889) Hay, 1902 (*nomen dubium*) [Campanian; Judith River Formation, Montana, USA]

= *Diclonius pentagonus* Cope, 1876 (*nomen dubium*) [Campanian; Judith River Formation, Montana, USA]

> *Claosaurus annectens* Marsh, 1892 ⇒ *Trachodon annectens* (Marsh, 1892) Hatcher, 1902 ⇒ *Thespesius annectens* (Marsh, 1892) Gilmore, 1924 ⇒ *Anatosaurus annectens* (Marsh, 1892) Lull & Wright, 1942 ⇒ *Edmontosaurus annectens* (Marsh, 1892) Brett-Surman *vide* Chapman & Brett-Surman, 1990 [Maastrichtian; Lance Formation, Wyoming, USA]

= *Trachodon longiceps* Marsh, 1890 ⇒ *Hadrosaurus longiceps* (Marsh, 1890) Nopcsa, 1900 ⇒ *Anatotitan longiceps* (Marsh, 1890) Olshevsky, 1991 [Maastrichtian; Lance Formation, Wyoming, USA]

> *Kritosaurus navajovius* Brown, 1910 ⇒ *Hadrosaurus navajovius* (Brown, 1910) Baird & Horner, 1977 [Campanian; Kirtland Formation, New Mexico, USA]

> *Gryposaurus notabilis* Lambe, 1914 ⇒ *Kritosaurus notabilis* (Lambe, 1914) Lull & Wright, 1942 ⇒ *Hadrosaurus notabilis* (Lambe, 1914) Baird & Horner, 1977, [Campanian; Dinosaur Park Formation, Alberta, Canada]

Notes: Hadrosaurids are known as the duck-billed dinosaurs. The mouth of a hadrosaur had dental batteries which were continually replaced with new teeth. According to *The Dinosauria 2nd edition*, *Hadrosaurus foulkii* was regarded to be valid, and *Hadrosaurus breviceps* Marsh, 1889, *H. cavatus* Cope, 1871, *H. minor* Marsh, 1870, *H. paucidens* Marsh, 1889 were considered as *nomina dubia*.

[References: Prieto-Márquez A, Weishampel DB, Horner JR (2006). The dinosaur *Hadrosaurus foulkii*, from the Campanian of the East Coast of North America, with a reevaluation of the genus. *Acta Palaeontologica Polonica* 51(1): 77–98.]

Haestasaurus becklesii

> *Pelorosaurus becklesii* Mantell, 1852 (misassigned species) ⇒ *Morosaurus becklesi* (Mantell, 1852) Marsh, 1889 ⇒ *Haestasaurus becklesii* (Mantell, 1852) Upchurch *et al.*, 2015 [late Berriasian–Valanginian; Wealden Group, Hastings (exact locality unknown), East Sussex, UK]

Generic name: *Haestasaurus* ← {(person's name) Haesta + (Gr.) σαῦρος/sauros: reptile}; referring to "the name of the putative pre-Roman chieftain whose people apparently settled the area of Hastings and gave the town its name"[*1].

Specific name: *becklesii* ← {(person's name) Beckles + -iī}; in honor of Samuel H. Beckles (1814–1890).

Etymology: Beckles' **Haesta reptile**

Taxonomy: Saurischia: Sauropodomorpha: Sauropoda
cf. Sauropoda: Neosauropoda: Macronaria[*1]

[References: (*1) Upchurch P, Mannion PD, Taylor MP (2015). The anatomy and phylogenetic relationships of "*Pelorosaurus*" *becklesii* (Neosauropoda, Macronaria) from the Early Cretaceous of England. *PLoS ONE* 10(6): e0125819.; Mantell GA (1852). On the structure of the *Iguanodon* and on the fauna and flora of the Wealden Formation, *Proceedings of the Royal Institute of Great Britain* 1: 141–146.]

Hagryphus giganteus

> *Hagryphus giganteus* Zanno & Sampson, 2005 [late Campanian; Kaiparowits Formation, Grand Staircase-Escalante National Monuments, Utah, USA]

Generic name: *Hagryphus* ← {(myth.) "Ha: the ancient Egyptian God of the western desert + gryphus: a fabulous four-footed bird, gender masculine."(*1)}

Specific name: *giganteus* ← {(Lat.) gigantēus: huge}

Etymology: huge **Ha's griffin**

Taxonomy: Theropoda: Oviraptorosauria: Caenagnathidae

Notes: The left carpus and manus are completely preserved. *Hagryphus giganteus* is a relatively large oviraptorosaur. (*1)

[References: (*1) Zanno LE, Sampson SD (2005). A new oviraptorosaur (Theropoda; Maniraptora) from the Late Cretaceous (Campanian) of Utah. *Journal of Vertebrate Paleontology* 25(4): 897–904.]

Halimornis thompsoni [Avialae]

> *Halimornis thompsoni* Chiappe *et al.*, 2002 [early–middle Campanian; Mooreville Chalk Formation, Greene, Alabama, USA]

Generic name: *Halimornis* ← {(Gr) ἅλῐμος/halimos: [ἅλς/hals: the sea]: "belonging to the sea" + (Gr.) ὄρνις/ornis: bird}; "in allusion to the marine deposits in which the specimen was found" (*1).

Specific name: *thompsoni* ← {(person's name) Thompson + -ī}; "after Mrs. W. Thompson, the landlord of the area in which the specimen was found, in recognition of her many years of support of fossil collecting on her property" (*1).

Etymology: Thompson's **bird of the sea**

Taxonomy: Theropoda: Avialae: Enantiornithes: Avisauridae

cf. Aves: Ornithothoraces: Enantiornithes: Euenantiornithes (*1)

[References: (*1) Chiappe LM, Lamb JP, Ericson PGP (2002). New enantiornithine bird from the marine Upper Cretaceous of Alabama. *Journal of Vertebrate Paleontology* 22(1): 170–174.]

Halszkaraptor escuilliei

> *Halszkaraptor escuilliei* Cau *et al.*, 2017 [Campanian; Bayn Dzak Member, Djadochta Formation, Ömnögovi, Mongolia]

Generic name: *Halszkaraptor* ← {(Latinized person's name) Halszka = (Polish) Halżka + (Lat.) raptor: robber}; in honor of "Halszka Osmólska (1930–2008) for her contributions to theropod paleontology, which include the description of the first halszkaraptorine species found (*Hulsanpes perlei*)" (*1).

Specific name: *escuilliei* ← {(person's name) Escuillié + -ī}; referring to "François Escuillié, who returned the poached holotype", which had been owned by several collectors, "to Mongolia" (*1).

Etymology: Escuillié's **Halszka's seizer**

Taxonomy: Theropoda: Maniraptora: Dromaeosauridae: Halszkaraptorinae

[References: (*1) Cau A, Beyrand V, Voeten DFAE, Fernandez V, Tafforeau P, Stein K, Barsbold R, Tsogtbaatar K, Currie PJ, Godefroit P (2017). Synchrotron scanning reveals amphibious ecomorphology in a new clade of bird-like dinosaurs. *Nature* 552(7685): 395–399.]

Halticosaurus longotarsus

> *Halticosaurus longotarsus* von Huene, 1908 [Norian; Löwenstein Formation, Baden-Württemberg, Germany]

Generic name: *Halticosaurus* ← {(Gr.) ἁλτικός/haltikos: good at jumping, nimble + σαῦρος/sauros: lizard}

Specific name: *longotarsus* ← {(Lat.) longus: long + 《anat.》tarsus}

Etymology: **nimble lizard** with long tarsi

Taxonomy: Theropoda: Ceratosauria: Coelophysoidea

Other species:

> *Halticosaurus orbitoangulatus* von Huene, 1932 ⇒ *Apatosuchus orbitoangulatus* (von Huene, 1932) Sues & Schoch, 2013 [Alaunian; Löwenstein Formation, Baden-Württemberg, Germany]

> *Halticosaurus liliensterni* von Huene, 1934 ⇒ *Liliensternus liliensterni* (von Huene, 1934) Welles, 1984 [Norian; Trossingen Formation, Thuringia, Germany]

Notes: According to *The Dinosauria 2nd edition*, *Halticosaurus longotarsus* is considered to be *nomen dubium*. The second species *H. orbitoangulatus* represented a primitive archosaur, and it was given a new generic name *Apatosuchus*.

Hanssuesia sternbergi

> *Troodon sternbergi* Brown & Schlaikjer, 1943 ⇒ *Stegoceras sternbergi* (Brown & Schlaikjer, 1943) Sternberg, 1945 ⇒ *Hanssuesia sternbergi* (Brown & Schlaikjer, 1943) Sullivan, 2003

[late Campanian; Oldman and Dinosaur Park formations, Alberta, Canada]

Generic name: *Hanssuesia* ← {(person's name) Hans Sues + -ia}; "in honor of paleontologist Hans-Dieter Sues for his contributions on dinosaurs, and in particular, for his work on pachycephalosaurs" (*1).

Specific name: *sternbergi* ← {(person's name) Sternberg + -ī}; in honor of paleontologist Charles M. Sternberg (1885–1981), discoverer of the type. (*2)

Etymology: Sternberg and **Hans Sues' one**

Taxonomy: Ornithischia: Pachycephalosauria: Pachycephalosauridae

Notes: According to *The Dinosauria 2nd edition*, *Hanssuesia sternbergi* was considered as a synonym of *Stegoceras validum*.

[References: (*1) Sullivan RM (2003). Revision of the dinosaur *Stegoceras* Lambe (Ornithischia, Pachycephalosauridae). *Journal of Vertebrate Paleontology* 23(1): 181–207.]; (*2) Brown B, Schlaikjer EM (1943). A study of the troödont dinosaurs with the description of a new genus and four new species. *Bulletin of the American Museum of Natural History* 82(5): 115–150.]

Haplocanthosaurus priscus

> *Haplocanthus priscus* Hatcher, 1903 (*nomen rejectum*) ⇒ *Haplocanthosaurus priscus* (Hatcher, 1903) Hatcher, 1903 (*nomen conservandum*) [Kimmeridgian; Brushy Basin Member, Morrison Formation, Colorado, USA]

= *Haplocanthosaurus utterbacki* Hatcher, 1903 [Kimmeridgian; Brushy Basin Member, Morrison Formation, Colorado, USA]

Generic name: *Haplocanthosaurus* ← {(Gr.) ἁπλόος/haploos: single, simple + ἄκανθα/akantha: spine + σαῦρος/sauros: lizard}; in reference to the dorsal neural spines which are non-bifurcated. (*1)

Specific name: *priscus* ← {(Lat.) prīscus: ancient}

Etymology: ancient **single-spined lizard**

Taxonomy: Saurischia: Sauropodomorpha: Sauropoda

Other species:

> *Haplocanthosaurus delfsi* McIntosh & Williams, 1988 [Kimmeridgian; Salt Wash Member, Morrison Formation, Colorado, USA]

Notes: This specimen was originally named *Haplocanthus* by Hatcher (1903) but was thought to be preoccupied by a spiny shark, and changed to *Haplocanthosaurus* (*2). However, the name of a spiny shark was found to be spelled *Haplacanthus* (*3). The original name *Haplocanthus* became a *nomen oblitum*. Therefore, the latest name *Haplocantosaurus* became a *nomen conservandum* (conserved name).

[References: (*1) Forster JR, Wedel MJ (2014). *Haplocanthosaurus* (Saurischia: Sauropoda) from the lower Morrison Formation (Upper Jurassic) near Snowmass, Colorado. *Volumina Jurassica* 12(2): 197–210.; (*2) Hatcher JB (1903). A new name for the dinosaur *Haplocanthus* Hatcher. *Proceedings of the Biological Society of Washington* 16: 100.; (*3) Derycke C, Goujet D (2011). Multicuspidate shark teeth associated with chondrichthyan and acanthodian scales from the Emsian (Devonian) of southern Algeria. *Geodiversitas* 33(2): 209–226.]

Haplocheirus sollers

> *Haplocheirus sollers* Choiniere *et al.*, 2010 [Oxfordian; Shishugou Formation, Junggar Basin, Xinjiang, China]

Generic name: *Haplocheirus* ← {(Gr.) ἁπλόος/haploos: simple + χειρός/cheiros (χείρ/cheir: hand)}; referring to the lack of the specialized manus of derived alvarezsauroids. (*1)

Specific name: *sollers* ← {(Lat.) sollers: skillful}; referring to the presumed ability of this taxon to perform digital actions that would be impossible for derived alvarezsaurids. (*1)

Etymology: skillful **simple-handed one**

Taxonomy: Theropoda: Alvarezsauroidea

Notes: According to Chen & Song (2010), *Haplocheirus* could be a transitional feature from dinosaurs to birds. (*2)

[References: (*1) Choiniere JN, Xu X, Clark JM, Forster CA, Guo Y, Han F (2010). A basal alvarezsauroid theropod from the early Late Jurassic of Xinjiang, China. *Science* 327(5965): 571–574.; (*2) Chen P, Song J (2010). From dinosaurs to birds: puzzles unraveled while evidence building up. *Bulletin of the Chinese Academy of Science* 24(2): 98–102.]

Harpymimus okladnikovi

> *Harpymimus okladnikovi* Barsbold & Perle, 1984 [Albian; Khuren Dukh Formation, Dundgovi Aimag, Gobi Desert, Mongolia]

Generic name: *Harpymimus* ← {(Gr. myth.) Ἅρπυια/Harpyia: Harpy* + μῖμος/mīmos: mimic}

*a cruel creature with a woman's head and body and a bird's wings.

Specific name: *okladnikovi* ← {(person's name) Okladnikov + -ī}; in honor of the late Soviet archeologist Alexey Pavlovich Okladnikov.

Etymology: Okladnikov's **Harpy mimic**

Taxonomy: Theropoda: Coelurosauria: Ornithomimosauria

[References: Barsbold R, Perle A (1984). On first new find of a primitive ornithomimosaur from the Cretaceous of the MPR. *Paleontologicheskii Zhurnal* 2: 121–123.; Kobayashi Y (2004). *Asian Ornithomimosaurs. PhD Thesis*, Southern Methodist University, Dallas: 340pp.; Kobayashi Y, Barsbold R

(2005). Anatomy of *Harpymimus okladnikovi* Barsbold and Perle 1984 (Dinosauria; Theropoda) of Mongolia. In: Carpenter K (ed), *The Carnivorous Dinosaurs*. Indiana University Press, Bloomington: 97–126.]

Haya griva

> *Haya Griva* Makovicky *et al*., 2011 [Santonian; Javkhlant Formation, Khugenetslavkant, Dornogovi, Mongolia]

Generic name: *Haya* ← {(Sanskrit) haya: horse}

Specific name: *griva* ← {(Sanskrit) grīva: head, neck}

Haya griva (Sanskrit): for "the Hindu deity Hayagriva, an avatar of Vishnu characterized by a horse head, in reference to the elongate and faintly horse-like skull of this dinosaur and the common depiction of the deity in the Buddhist art of Mongolia" (*1).

Etymology: **horse** head

Taxonomy: Ornithischia: Ornithopoda: Jeholosauridae

cf. Ornithischia: Neornithischia: Cerapoda: Ornithopoda (*1)

[References: (*1) Makovicky PJ, Kilbourne BM, Sadleir RW, Norell MA (2011). A new basal ornithopod (Dinosauria, Ornithischia) from the Late Cretaceous of Mongolia. *Journal of Vertebrate Paleontology* 31(3): 626–640.]

Helioceratops brachygnathus

> *Helioceratops brachygnathus* Jin L. *et al*., 2009 [Albian; Quantou Formation, Jilin, China]

Generic name: *Helioceratops* ← {(Gr. myth.) Ἥλιος/Hēlios: the god of the sun + *Ceratops* [κέρας/ keras: horn + ὤψ/ōps: face]}; in reference to Helios, "who drove his chariot across the sky from the east to the west every day. He was also the brother of Eos ("dawn"). This generic name suggests an oriental origin for this neoceratopsian, which is closely related to *Auroraceratops*" (*1).

Specific name: *brachygnathus* ← {(Gr.) βραχύς/ brachys: short + γνάθος/gnathos: jaw}

Etymology: **Helios horned-face (ceratopsian)** with short jaw

Taxonomy: Ornithischia: Marginocephalia: Ceratopsia

cf. Ornithischia: Ceratopsia: Neoceratopsia (*1)

[References: (*1) Jin L, Chen J, Zan, S, Godefroit P (2009). A new basal neoceratopsian dinosaur from the middle Cretaceous of Jilin Province, China. *Acta Geologica Sinica* 83(2): 200–206.]

Heptasteornis andrewsi

> *Heptasteornis andrewsi* Harrison & Halker, 1975 [Maastrichtian; Sânpetru Formation, Szèntpeterfalva, Hátszeg, Transylvania, Romania]

Generic name: *Heptasteornis* ← {(GR.) ἑπτά/hepta: seven + ἄστεος/asteos [ἄστυ/asty: city, town] + ὄρνις/ornis: a bird}; "in reference to the name of the area of origin, and is feminine" (*1).

The type was found in the historical region, Transylvania, the German *Siebenbürgen*, which means "seven fortresses". The type was described as a gigantic owl in 1975.

Specific name: *andrewsi* ← {(person's name) Andrews + -ī}; in honor of C. W. Andrews.

Etymology: Andrews' **Transylvanian (seven cities) bird**

Taxonomy: Theropoda: Coelurosauria: Alvarezsauridae

Notes: *Heptasteornis* is a small dinosaur. According to *The Dinosauria 2nd edition* (2004), it is considered a *nomen dubium*.

[References: (*1) Harrison CJO, Walker CA (1975). The Bradycnemidae, a new family of owls from the Upper Cretaceous of Romania. *Palaeontology* 18(3): 563–570.]

Herrerasaurus ischigualastensis

> *Herrerasaurus ischigualastensis* Reig, 1963 [Carnian; Ischigualasto Formation, San Juan, Argentina]

= *Ischisaurus cattoi* Reig, 1963 [Carnian; Ischigualasto Formation, San Juan, Argentina]

= *Frenguellisaurus ischigualastensis* Novas, 1986 [Norian; Ischigualasto Formation, San Juan, Argentina]

Generic name: *Herrerasaurus* ← {(person's name) Herrera + (Gr.) σαῦρος/sauros: lizard}; "in honor of Don Victorino Herrera, settler of the Ischigualasto zone, guide and experienced collector, who was converted into an irreplaceable collaborator of the scientific undertakings developed in these pages" (*1).

Specific name: *ischigualastensis* ← {(place-name) Ischigualasto + -ensis}; referring to the Ischigualasto Valley where fossils were discovered by Victorino Herrera in 1961.

Etymology: **Herrera's lizard** from Ischigualasto

Taxonomy: Dinosauria: Saurischia: Herrerasauridae

[References: (*1) Reig OA (1963). The presence of saurischian dinosaurs in the "Ischigualasto Beds" (upper Middle Triassic) of the provinces of San Juan and La Rioja (Argentine Republic). *Ameghiniana Revista de la Asociación Paleontológia Argentina* 3(1): 3–20.]

Hesperonychus elizabethae

> *Hesperonychus elizabethae* Longrich & Currie, 2009 [Campanian; Dinosaur Park

Formation, Alberta, Canada]

Generic name: *Hesperonychus* ← {(Gr.) ἕσπερος/hesperos: of evening, western + (Gr.) ὄνῠχος/onychos [ὄνυξ/onyx = (Lat.) unguis: claw, talon, hoof]}

Specific name: *elizabethae* ← {(person's name) Elizabeth + -ae}; in honor of "the late Dr. Elizabeth Nicholls, who discovered the holotype"(*1).

Etymology: Elizabeth's **western claw**

Taxonomy: Theropoda: Coelurosauria: Maniraptora: Dromaeosauridae: Microraptorinae

Notes: *Hesperonychus* is a tiny, bird-like predator.

[References: (*1) Longrich NR, Currie PJ (2009). A microraptorine (Dinosauria-Dromaeosauridae) from the Late Cretaceous of North America. *Proceedings of the National Academy of Sciences of the United States of America* 106(13): 5002–5007.]

Hesperornis regalis [Avialae]

> *Hesperornis regalis* Marsh, 1872 [Campanian; Smoky Hill Chalk Member, Niobrara Formation, Kansas, USA]

Generic name: *Hesperornis* ← {(Gr.) ἕσπερος/hesperos: western + ὄρνις/ornis: bird}

In 1871, when Othniel Charles Marsh made an expedition to western Kansas, he discovered the fossils. (according to Discovery of a remarkable fossil bird: by Professor O. C. Marsh, 1871)

Specific name: *regalis* ← {(Lat.) rēgālis: regal}

Etymology: regal **western bird**

Taxonomy: Theropoda: Avialae: Hesperornithiformes: Hesperornithidae

Other species:

> *Lestornis crassipes* Marsh, 1876 ⇒ *Hesperornis crassipes* (Marsh, 1876) Brodkorb, 1963 [Coniacian–early Campanian; Niobrara Chalk Formation, USA]

> *Hesperornis gracilis* Marsh, 1876 ⇒ *Hargeria gracilis* (Marsh, 1876) Lucas, 1903 [Coniacian–early Campanian; Niobrara Chalk Formation, Kansas, USA]

> *Coniornis altus* Marsh, 1893 ⇒ *Hesperonis altus* (Marsh, 1893) [Late Cretaceous; Claggett Shale Formation, Montana, USA]

> *Hesperornis montana* Shufeldt, 1915 [late Campanian; Claggett Formation, Montana, USA] (According to *The Dinosauria 2nd edition*, *H. montanus* was considered a synonym of *Coniornis altus*)

> *Hesperornis rossicus* Nesov & Yarkov, 1993 [early Campanian; Rybushka Formation, Saratov, Russia]

> *Hesperornis bairdi* Martin & Lim, 2002 [Campanian; Pierre Shale Formation, South Dakota, USA]

> *Hesperornis chowi* Martin & Lim, 2002 [Campanian; Pierre Shale Formation, South Dakota, USA]

> *Hesperornis macdonaldi* Martin & Lim, 2002 [Campanian; Pierre Shale Formation, South Dakota, USA]

> *Hesperornis mengeli* Martin & Lim, 2002 [Campanian; Pierre Shale Formation, Manitoba, Canada]

> *Hesperornis lumgairi* Aotsuka & Sato, 2016 [early Campanian; Pierre Shale Formation, Manitoba, Canada]

Notes: Most of the species have been recovered in North America and one in Russia. *Hesperornis* is an extinct genus of large aquatic birds with no wings. It had teeth.

Hesperornithoides miessleri

> *Hesperornithoides miessleri* Hartman *et al.*, 2019 [Oxfordian–Tithonian; Morrison Formation, Wyoming, USA]

Generic name: *Hesperornithoides* ← {(Gr.) hesper- [ἕσπερος/hesperos: of evening, western] + ornith- [ὄρνις/ornis: bird] + oeides: similar]}; "referring to the discovery in the American West" and "to the avian-like form of derived paravians" (*1).

Specific name: *miessleri* ← {(person's name) Miessler + -ī}; in honor of "the Miessler family, who have been avid supporters of the project" (*1).

Etymology: Miessler family's **western bird-like**

Taxonomy: Theropoda: Maniraptora: Paraves: Deinonychosauria: Troodontidae

[References: (*1) Hartman S, Mortimer M, Wahl WR, Lomax DR, Lippincott J, Lovelace DM (2019). A new paravian dinosaur from the Late Jurassic of North America supports a late acquisition of avian flight. *Peer J* 7: e7247.]

Hesperosaurus mjosi

> *Hesperosaurus mjosi* Carpenter *et al.*, 2001 [Kimmeridgian; Morrison Formation, Wyoming, USA]

Generic name: *Hesperosaurus* ← {(Gr.) ἕσπερος/hesperos: western + σαῦρος/sauros: reptile}; "in reference to its discovery in the western United States" (*1).

Specific name: *mjosi* ← {(person's name) Mjos + -ī}; in honor of "Ronald G. Mjos ("mūs"), who was responsible for collecting and preparing the specimen and for mounting a cast of the holotype skeleton" (*1).

Etymology: Mjos' **western lizard**

Taxonomy: Ornithischia: Thyreophora: Stegosauria:

Stegosauridae

[References: (*1) Carpenter K, Miles CA, Cloward K (2001). New primitive stegosaur from the Morrison Formation, Wyoming. In: Carpenter K (ed), *The Armored Dinosaurs*, Indiana University Press, Bloomington: 55–75.]

Heterodontosaurus tucki

> *Heterodontosaurus tucki* Crompton & Charig, 1962 ⇒ *Lycorhinus tucki* (Crompton & Charig, 1962) Thulborn, 1970 [Sinemurian; Clarens Formation, Eastern Cape, South Africa]

According to the authors, *Heterodontosaurus tucki* was thought to be from the Late Triassic period[*1].

Generic name: *Heterodontosaurus* ← {(Gr.) ἕτερος/heteros: different + ὀδοντο-/odonto- [ὀδούς/odūs: tooth] + σαῦρος/sauros: lizard}; "in reference to the differentiated dentition, unexpected in an ornithischian"[*1].

Specific name: *tucki* ← {(person's name) Tuck + -ī}; "in recognition of the generous help afforded to the Expedition by Mr. G. C. Tuck, managing director of the Austin Motor Co. of South Africa (Proprietary), Ltd."[*1]

Etymology: Tuck's **different toothed lizard**

Taxonomy: Ornithischia: Ornithopoda: Heterodontosauridae

[References: (*1) Crompton AW, Charig AJ (1962). A new ornithischian from the Upper Triassic of South Africa. *Nature* 196 (4859): 1074–1077.]

Hexing qingyi

> *Hexing qingyi* Jin *et al.*, 2012 [lower Valanginian–lower Barremian; Yixian Formation, Liaoning, China]

Generic name: *Hexing* ← {(Chin.) hèxíng 鶴形: "like a crane"[*1]}

Specific name: *qingyi* ← {(Chin.) qīngyì 軽翼: "with thin wings"[*1]}

Etymology: **crane-like one** with light wings

Taxonomy: Theropoda: Tetanurae: Coelurosauria: Ornithomimosauria

[References: (*1) Jin L, Chen J, Godefroit P (2012). A new basal Ornithomimosaur (Dinosauria: Theropoda) from the Early Cretaceous Yixian Formation, Northeast China. In: Godefroit P (ed), *Bernissart Dinosaurs and Early Cretaceous Terrestial Ecosystems.* Indiana University Press: 467–487.]

Hexinlusaurus multidens

> *Yandusaurus multidens* He & Cai, 1983 ⇒ *Agilisaurus multidens* (He & Cai, 1983) Peng, 1990 ⇒ *Othnielia multidens* (He & Cai, 1983) Paul, 1996 ⇒ *Hexinlusaurus multidens* (He & Cai, 1983) Barret *et al.*, 2005 [Bathonian–Callovian; lower Shaximiao Formation, Dashanpu, Sichuan, China]

Generic name: *Hexinlusaurus* ← {(Chin. person's name) Hé Xīn-Lù何信禄 + (Gr.) σαῦρος/sauros: lizard}; in honor of Professor He Xin-Lu, who originally named the specimen as *Yandusaurus multidens*.

Specific name: *multidens* ← {(Lat.) multus: many + dēns: tooth}

Etymology: **He Xin-Lu's lizard** with many teeth

Taxonomy: Ornithischia: Neornithischia

[References: Barrett PM, Butler RJ, Knoll F (2005). Small-bodied ornithischian dinosaurs from the Middle Jurassic of Sichuan, China. *Journal of Vertebrate Paleontology* 25(4): 823–834. Abstract.]

Heyuannia huangi

> *Heyuannia huangi* Lü, 2002 [Maastrichitian; Dalangshan Formation, Heyuan, Guangdong, China]

Generic name: *Heyuannia* ← {(Chin. place-name) Héyuàn河源 + -ia}; referring to "the fossil locality, Heyuan City, Guangdong Province"[*1].

Specific name: *huangi* ← {(Chin. person's name) Huáng + -ī}; in honor of "Dong Huang, the director of Heyuan Museum, who made great contributions in the excavation and preservation of these fossils"[*1].

Etymology: Huang's **Heyuan one**

Taxonomy: Theropoda: Oviraptorosauria: Oviraptoridae

Other species:

> *Ingenia yanshini*? Barsbold, 1981 (preoccupied) ⇒ *Ajancingenia yanshini?* (Barsbold, 1981) Easter, 2013 ⇒ *Heyuannia yanshini*? (Barsbold, 1981) Funston *et al.*, 2017 [Campanian; Barun Goyot Formation, Ömnögovi, Mongolia]

Notes: According to Wiemann *et al.* (2017), "eggshell parataxon *Macroolithus yaotunensis* can be assigned to *Heyuannia huangi*". The color of the egg that was reconstructed by researchers is blue-green. [*2]

[References: (*1) Lü J (2002). A new oviraptorosaurid (Theropoda: Oviraptorosauria) from the Late Cretaceous of southern China. *Journal of Vertebrate Paleontology* 22(4): 871–875.; (*2) Wiemann J, Yang T-R, Sander PN, *et al.* (2017). Dinosaur origin of egg color: oviraptors laid blue-green eggs. *PeerJ* 5: e3706.]

Hierosaurus sternbergii

> *Hierosaurus sternbergii* Wieland, 1909 [Coniacian–Campanian; Niobrara Formation, Kansas, USA]

Generic name: *Hierosaurus* ← {(Gr.) ἱερός/hieros:

sacred + σαῦρος/sauros: lizard}
Specific name: *sternbergii* ← {(person's name) Sternberg + iī}; in honor of Charles Hazelius Sternberg who discovered the specimen.
Etymology: Sternberg's **sacred lizard**
Taxonomy: Ornithischia: Ankylosauria: Nodosauridae
Other species:
> *Hierosaurus coleii* Mehl, 1936 ⇒ *Nodosaurus coleii* (Mehl, 1936) Coombs, 1978 ⇒ *Niobrarasaurus coleii* (Mehl, 1936) Carpenter *et al.*, 1995 [Coniacian–Campanian; Niobrara Formation, Kansas, USA]
Notes: According to *The Dinosauria 2nd edition*, *Hierosaurus sternbergi* is considered to be a *nomen dubium*. The second species, *H. coleii* was redescribed as a new genus *Niobrarasaurus* in 1995.

Hippodraco scutodens

> *Hippodraco scutodens* McDonald *et al.*, 2010 [upper Barremian–lowermost Aptian; Yellow Cat Member, Cedar Mountain Formation, Utah, USA]
Generic name: *Hippodraco* ← {(Gr.) ἵππος/hippos: horse + (Lat.) draco: dragon}; "in reference to the long and low overall shape of the skull, grossly resembling that of a horse" (*1). The gender is masculine.
Specific name: *scutodens* ← {(Lat.) scūtum: oblong shield + dēns: tooth}; "in reference to the shape of the dentary tooth crowns" (*1).
Etymology: "shield-toothed **horse-dragon**" (*1)
Taxonomy: Ornithischia: Ornithopoda: Styracosterna

[References: (*1) McDonald AT, Kirkland JI, DeBlieux DD, Madsen SK, Cavin J, Milner ARC, Panzarin L (2010). New basal iguanodonts from the Cedar Mountain Formation of Utah and the evolution of thumb-spiked dinosaurs. *PLoS ONE* 5(11): e14075.]

Histriasaurus boscarollii

> *Histriasaurus boscarollii* Dalla Vecchia, 1998 [late Hauterivian; unnamed formation, Istarska Zupanija, Croatia]
Generic name: *Histriasaurus* ← {(place-name) Histria: an old name for Istria + (Gr.) σαῦρος/sauros: lizard}
Specific name: *boscarollii* ← {(person's name) Boscarolli + -ī}; in honor of Darío Boscarolli who discovered the site.
Etymology: Boscarolli's **Istria (Histria) lizard**
Taxonomy: Saurischia: Sauropodomorpha: Diplodocoidea: Rebbachisauridae
Notes: According to *The Dinosauria 2nd edition*, *Histriasaurus* was described as *nomen dubium*. Then it was included in Rebbachisauridae, according to the cladogram after Caballido *et al.* (2012) and Fanti *et al.* (2013). In Istria, many dinosaur footprints and tracks have also been found (*1).

[References: (*1) Dalla Veccia FM (2001). Terrestrial ecosystems on the Mesozoic peri-Adriatic carbonate platforms: the vertebrate evidence. *Asociació Paleontológica Argentina, Publicación Especial 7, VII International Symposium on Mesozoic Terrestrial Ecosystems*: 77–83.]

Hollanda luceria [Avialae]

> *Hollanda luceria* Bell *et al.*, 2010 [Campanian; Barun Goyot Formation, Ömnögovi, Mongolia]
Generic name: *Hollanda* ← {(person's name) Holland + -a}; "in honor of the Holland family (Janice, Charles, Carl, and J.-P), whose generous donations have supported a great deal of paleontological research and field work at the Dinosaur Institute of the Natural History Museum of Los Angeles County" (*1).
Specific name: *luceria*; "inspired by the band of *Lucero* of Memphis, Tennessee, and coming from the Latin, 'to shine' " (*1).
Etymology: shining, Lucero and **Holland's one**
Taxonomy: Theropoda: Avialae: Ornithothoraces: Euornithes *cf.* Aves: Ornithuromorpha (*1)

[References: (*1) Bell AK, Chiappe LM, Erickson GM, Suzuki S, Watabe M, Barsbold R, Tsogtbaatar K (2010). Description and ecologic analysis of *Hollanda luceria*, a Late Cretaceous bird from the Gobi Desert (Mongolia). *Cretaceous Reserch* 31(1): 16–26.]

Homalocephale calathocercos

> *Homalocephale calathocercos* Maryańska & Osmólska, 1974 [Campanian; Nemegt Formation, Ömnögovi, Mongolia]
Generic name: *Homalocephale* ← {(Gr.) ὁμᾰλος/homalos: even + κεφαλή/kephalē: head}; referring to "the flat skull roof" (*1).
Specific name: *calathocercos* ← {(Gr.) κάλαθος/kalathos: basket + κέρκος/kerkos: tail}; referring to "the caudal tendons arranged in a kind of a basket" (*1).
Etymology: **level head** with a basket tail
Taxonomy: Ornithischia: Marginocephalia: Pachycephalosauria: Homalocephaloidea

[References: (*1) Maryañska T, Osmólska H (1974). Pachycephalosauria, a new suborder of ornithischian dinosaurs. *Palaeontologia Polonica* 30: 45–102.]

Hongshanornis longicresta [Avialae]

> *Hongshanornis longicresta* Zhou & Zhang, 2005 [Aptian; Yixian Formation, Ningcheng, Inner Mongolia, China]

Generic name: *Hongshanornis* ← {(Chin.) Hóngshān + (Gr.) ὄρνις/ornis: bird}; in reference "to the Hongshan culture, one of earliest Chinese cultures mainly recorded in this region"(*1).

Specific name: *longicresta* ← {(Lat.) longus: long, great + crista: crest}; referring to "the raised crest of this bird"(*1).

Etymology: **Hongshan bird** with a raised crest

Taxonomy: Avialae: Euornithes: Ornithuromorpha: Hongshanornithidae

[References: (*1) Zhou Z, Zhang F (2005). Discover of an ornithurine bird and its implication for Early Cretaceous avian radiation. *Proceedings of the National Academy of sciences of the United States of America* 102(52): 18998–19002.]

Hongshanosaurus houi

> *Psittacosaurus lujiatunensis* Zhou *et al*., 2006 [Aptian; Yixian Formation, Liaoning, China]

= *Hongshanosaurus houi* You *et al*., 2003 ⇒ *Psittacosaurus houi* (You *et al*., 2003) Sereno, 2010 [Aptian; Lujiatun Member, Yixian Formation, Liaoning, China]

See also *Psittacosaurus*.

Generic name: *Hongshanosaurus* ← {(Chin.) Hóngshān* 紅山 + (Gr.) σαῦρος/sauros: lizard}; referring to the ancient Hongshan culture.

*"Red Hill Culture' which existed about 6,000 years ago in western Liaoning, northeastern China, where this specimen was found"(*1).

Specific name: *houi* ← {(Chin. person's name) Hóu + -ī}; in honor of Prof. Hou Lianhai, 侯連海, a professor at the Institute of Vertebrate Paleontology and Paleoanthropology in Beijing. (*1)

Etymology: Hou's **Hongshan lizard** [侯氏紅山龍]

Taxonomy: Ornithischia: Ceratopsia: Psittacosauridae

Notes: *Hongshanosaurus* was proposed to be a synonym of *Psittacosaurus* Osborn, 1923 (acc. Sereno, 2010). (*2)

[References: (*1) You H-L, Xu X, Wang X (2003). A new genus of Psittacosauridae (Dinosauria: Ornithopoda) and the origin and early evolution of marginocephalian dinosaurs. *Acta Geologica Sinica* 77(1): 15–20.; (*2) Hedrick BP, Dodson P (2013). Lujiatun Psittacosaurids: understanding individual and taphonomic variation using 3D geometric morphometrics. *PLoS ONE* 8(8): e69295.]

Hoplitosaurus marshi

> *Stegosaurus marshi* Lucas, 1901 ⇒ *Hoplitosaurus marshi* (Lucas, 1901) Lucas, 1902 ⇒ *Polacanthus marshi* (Lucas, 1901) Pereda-Suberbiola, 1991 [Barremian; Lakota Formation, South Dakota, USA]

Generic name: *Hoplitosaurus* ← {(Gr.) ὁπλίτης/hoplites: armed foot-soldier + σαῦρος/sauros: lizard}

Specific name: *marshi* ← {(person's name) Marsh + -ī}; in honor of paleontologist Othniel Charles Marsh.

Etymology: Marsh's **armored lizard**

Taxonomy: Ornithischia: Thyreophora: Ankylosauria: Ankylosauridae / Nodosauridae: Polacanthinae

[References: Lucas FA (1901). A new dinosaur, *Stegosaurus marshi*, from the Lower Cretaceous of South Dakota. *Proceedings of the United States National Museum* 23(1224): 591–592.]

Horshamosaurus rudgwickensis

> *Polacanthus rudgwickensis* Blows, 1996 ⇒ *Horshamosaurus rudgwickensis* (Blows, 1996) Blows, 2015 [Barremian; Weald Clay, Rudgwick, West Sussex, UK]

Generic name: *Horshamosaurus* ← {(place-name) Horsham + (Gr.) σαῦρος/sauros: lizard}

Specific name: *rudgwickensis* ← {(place-name) Rudgwick + -ensis}

Etymology: **Horsham lizard** from Rudgwick

Taxonomy: Ornithischia: Thyreophora: Ankylosauria

Notes: Blows (2015) suggested *Horshamosaurus* was a member of the Nodosauridae. The *Horshamosaurus* fossils can be seen in Horsham District Council's Horsham Museum.

[References: Blows WT (1996). A new species of *Polacanthus* (Ornithischia; Ankylosauria) from the Lower Cretaceous of Sussex, England. *Geological Magazine* 133(6): 671–682.]

Hortalotarsus skirtopodus

> *Hortalotarsus skirtopodus* Seeley, 1894 [Sinemurian; Clarens Formation, Eastern Cape, South Africa]

Generic name: *Hortalotarsus* ← {Hortalo- + (Lat.) tarsus: tarsus}

Specific name: *skirtopodus*

Etymology: uncertain

Taxonomy: Saurischia: Sauropodomorpha: Plateosauria: Massospodylidae

Notes: "A few remains of a skeleton, known locally as the Bushman Fossil was discovered by Mr. William Horner Wallace at 'Eagle's Crag', Barkly East, Cape of Good Hope, 11th June 1888"

[*1]. According to *The Dinosauria 2nd edition*, *Hortalotarsus skirtopodus* is regarded as a *nomen dubium*.

[References: (*1) Seeley HG (1894). LIII.–On *Hortalotarsus skirtopodus*, a new saurischian fossil from Barkly East, Cape Colony. *Annals and Magazine of Natural History* 6 (14):411–419.]

Houornis caudatus [Avialae]

> *Cathayornis caudatus* Hou, 1997 ⇒ "*Similicaudipteryx caudatus*" (Hou, 1997) Wang & Liu, 2015 ⇒ *Houronis caudatus* (Hou, 1997) Wang & Liu, 2015 [Aptian; Jiufotang Formation, Liaoning, China]

Generic name: *Houornis* ← {(person's name) Hou + (Gr.) ὄρνις/ornis: bird}; "in honour of Dr Hou Lianhai for his contribution to the research of Mesozoic birds of China" [*1].

Specific name: *caudatus* ← {(Lat.) cauda: tail + -ātus}

Etymology: **Hou's bird** with tail

Taxonomy: Theropoda: Enantiornithes

Notes: A nearly complete specimen is preserved as an impression [*1].

[References: (*1) Wang M, Liu D (2015). Taxonomical reappraisal of Cathayornithidae (Aves: Enantiornithes). *Journal of Systematic Palaeontology* 14 (1): 29–47.]

Huabeisaurus allocotus

> *Huabeisaurus allocotus* Pang & Cheng, 2000 [Cenomanian; upper member, Huiquanpu Formation, Shanxi, China]

Generic name: *Huabeisaurus* ← {(Chin. place-name) Huáběi華北"'North China': an administrative mega-region" [*1] + (Gr.) σαῦρος/sauros: lizard}

Specific name: *allocotus* ← {(Gr.) ἀλλό-κοτος/allo-kotos; unusual}; meaning "that the new species has special significance in its characteristics and geological horizon" [*1].

Locality: unusual **North China lizard** [不尋常華北龍]

Taxonomy: Saurischia: Sauropodomorpha: Sauropoda: Titanosauria

Notes: *Huabeisaurus allocotus* is a gigantic sauropod, about 20 m in total length [*1].

[References: (*1) Pang Q, Cheng Z (2000). A new family of sauropod dinosaur from the Upper Cretaceous of Tianzhen, Shanxi Province, China. *Acta Geologica Sinica* 74(2): 117–125.]

Hualianceratops wucaiwanensis

> *Hualianceratops wucaiwanensis* Han *et al.*, 2015 [Oxfordian; Shishugou Formation, Wucaiwan, Junggar Basin, Xinjiang, China]

Generic name: *Hualianceratops* ← {(Chin.) huāliǎn 花臉*: ornamental face + ceratops (horned face) [(Gr.) κέρας/keras: horn + ὤψ/ōps: face]; "referring to the texture found on most part of the skull" [*1]. *liǎn臉: face. *cf.* jiǎn瞼: eyelid.

Specific name: *wucaiwanensis* ← {(Chin. place-name) Wǔcǎiwān 五彩灣 'five color bay' + -ensis}; referring to "the area where the specimen was discovered" [*1].

Etymology: **ornamental-faced ceratopsian** from Wukaiwan

Taxonomy: Ornithischia: Ceratopsia: Chaoyangsauridae

[References: (*1) Han F, Forster CA, Clark JM, Xu X (2015). A new taxon of basal ceratopsian from China and the early evolution of Ceratopsia. *PLoS ONE* 10(12): e0143369.]

Huanansaurus ganzhouensis

> *Huanansaurus ganzhouensis* Lü *et al.*, 2015 [Campanian–Maastrichtian; Nanxiong Formation, Ganzhou, Jiangxi, China]

Generic name: *Huanansaurus* ← {(Chin. place-name) Huánán 華南: 'southern China' + (Gr.) σαῦρος/sauros: lizard}; "because the dinosaur was discovered in Ganzhou of Jiangxi Province". [*1]

Specific name: *ganzhouensis* ← {(Chin. place-name) Gànzhōu贛州 + -ensis}; "referring to the locality of Ganzhou" [*1].

The skeleton was uncovered in the vicinity of the Ganzhou Railway Station. [*1]

Etymology: **Huanan lizard** from Ganzhou

Taxonomy: Saurischia: Theropoda: Oviraptorosauria: Oviraptoridae: Oviraptorinae

[References: (*1) Lü J, Pu H, Kobayashi Y, Xu L, Chang H, Shang Y, Liu D, Lee Y-N, Kundrát M, Shen C (2015). A new oviraptorid dinosaur (Dinosauria: Oviraptorosauria) from the Late Cretaceous of southern China and its paleobiogeographical implications. *Scientific Reports* 5: 11490.]

Huanghetitan liujiaxiaensis

> *Huanghetitan liujiaxiaensis* You *et al.*, 2006 [Barremian–Aptian; Hekou Group, Lanzhou Basin, Gansu, China]

Generic name: *Huanghetitan* ← {(Chin. place-name) Huáng Hé黄河: 'Yellow River' + (Gr. myth.) Τῑτάν/Tītan}; referring to "the Yellow River, which flows along the Lanzhou Basin where the fossils were discovered" [*1], and Titan, referring to its "symbolic of great size" [*1].

Specific name: *liujiaxiaensis* ← {(Chin. place-name) Liújiāxiá劉家峡 + -ensis}; referring to "Liujia Gorge, which is part of the Yellow River in Lanzhou Basin, where the Liujiaxia National Dinosaur Geopark is located nearby" [*1].

Etymology: **Yellow River titan** from Liujiaxia
Taxonomy: Saurischia: Sauropodomorpha: Camarasauromorpha: Titanosauriformes
Other species:

> *Huanghetitan ruyangensis* Lu *et al.*, 2007 [Aptian–Albian; Haoling Formation, Henan, China]

[References: (*1) You H-L, Zhou L, Li D-Q, Ji Q (2006). *Huanghetitan liujiaxiaensis*, a new sauropod dinosaur from Lower Cretaceous Hekou Group of Lanzhou Basin, Gansu Province, China. *Geological Review* 52(5): 668–674.]

Huangshanlong anhuiensis

> *Huangshanlong anhuiensis* Huang *et al.*, 2014 [Middle Jurassic; Hongqin Formation, Huangshan, Anhui, China]

Generic name: *Huangshanlong* ← {(Chin. place-name) Huángshān黄山 + lóng龍: dragon} (*1)
Specific name: *anhuiensis* ← {(Chin. place-name) Ānhuī安徽 + -ensis} (*1)
Etymology: **Huangshan dragon** from Anhui Province [安徽黄山龍]
Taxonomy: Saurischia: Sauropodomorpha: Sauropoda: Eusauropoda: Mamechisauridae
Notes: The Hongqin Formation contains also *Anhuilong diboensis*.

[References: (*1) Huang J-D, You H-L, Yang J-T, Ren X-X (2014). A new sauropod dinosaur from the Middle Jurassic of Huangshan, Anhui Province. *Vertebrata PalAsiatica* 52(4): 390–400.]

Huaxiagnathus orientalis

> *Huaxiagnathus orientalis* Hwang *et al.*, 2004 [Aptian; Yixian Formation, Beipiao, Liaoning, China]

Generic name: *Huaxiagnathus* ← {(Chin.) Huáxià華夏: an ancient Mandarin name for China + (Gr.) γνάθος/gnathos: jaw} (*1)
Specific name: *orientalis* ← {(Lat.) orientālis: of or from the east, oriental} (*1)
Etymology: **Huaxia (China) jaw** from the east [東方華夏顎龍]
Taxonomy: Theropoda: Coelurosauria: Maniraptora: Compsognathidae
Notes: The holotype specimen is nearly complete, lacking only the end of the tail.

[References: (*1) Hwang SH, Norell MA, Ji Q, Gao K (2004). A large compsognathid from the Early Cretaceous Yixian Formation of China. *Journal of Systematic Palaeontology* 2(1): 13–30.]

Huaxiaosaurus aigahtens

> *Shantungosaurus giganteus* Hu, 1973 [Campanian; Xingezhuang Formation, Wangshi Group, Shandong, China]
= *Zhuchengosaurus maximus* Zhao *et al.*, 2007 [Campanian; Xingezhuang Formation, Shandong, China]
= *Huaxiaosaurus aigahtens* Zhao *et al.*, 2011 [Campanian; Xingezhuang Formation, Shandong, China]

Generic name: *Huaxiaosaurus* ← {(Chin.) Huáxià 華夏: the ancient name for China + (Gr.) σαῦρος/sauros: lizard}
Specific name: *aigahtens* [*sic*] ← {(Lat.) gigantēus: gigantic}
Etymology: gigantic **Huaxia lizard**
Taxonomy: Ornithischia: Ornithopoda: Iguanodontia: Hadrosauridae
Notes: *Huaxiaosaurus* is seemed to be almost the same as *Shantungosaurus*. *Huaxiaosaurus aigahtens* and *Zhuchengosaurus maximus* are now considered to be junior synonyms of *Shantungosaurus giganteus*.

Huayangosaurus taibaii

> *Huayangosaurus taibaii* Dong *et al.*, 1982 [Bajocian; lower Shaximiao Formation, Sichuan, China]

Generic name: *Huayangosaurus* ← {(Chin. place-name) Huáyáng華陽: an old name for Sichuan + (Gr.) σαῦρος/sauros: lizard} (*1)
Specific name: *taibaii* ← {(Chin. person's name) Tài Bái太白 + -ī}; in honor of the great Chinese poet Li Bai whose courtesy name is Tai Bai. (*1)
Etymology: Tai Bai's **Huayang (Sichuan) lizard** [太白華陽龍]
Taxonomy: Ornithischia: Thyreophora: Stegosauria: Huayangosauridae

[References: (*1) Dong Z, Tang Z, Zhou SW (1982). [Note on the new mid-Jurassic stegosaur from Sichuan Basin, China] (in Chinese). *Vertebrate PalAsiatica* 20(1): 83–87.]

Hudiesaurus sinojaponorum

> *Hudiesaurus sinojaponorum* Dong, 1997 [Late Jurassic; Kalazha Formation, Turpan Basin, Xinjiang, China]

Generic name: *Hudiesaurus* ← {(Chin.) húdié胡蝶 / 蝴蝶: butterfly + (Gr.) σαῦρος/sauros: lizard}; referring to a flat butterfly-shape of neural spine of the anterior dorsal vertebra.
Specific name: *sinojaponorum* ← {Sinojapon: China and Japan + -ōrum}; referring to the Chinese-Japanese expedition, and also referring to Japanese *Chunichi Shinbun* press group.
Etymology: Chinese and Japanese **butterfly (vertebra) lizard** [中日蝴蝶龍]

Taxonomy: Saurischia: Sauropodomorpha: Sauropoda: Mamenchisauridae

[References: Dong Z (1997). A gigantic sauropod (*Hudiesaurus sinojaponorum* gen. *et* sp. nov.) from the Turpan Basin, China. *Sino-Japanese Silk Road Dinosaur Expedition.* China Ocean Press, Beijing: 102–110.]

Huehuecanauhtlus tiquichensis

> *Huehuecanauhtlus tiquichensis* Ramírez-Velasco *et al.*, 2012 [Santonian; unnamed formation, Michoacán, Mexico]

Generic name: *Huehuecanauhtlus* ← {(Náhuatl) huehuetl: ancient + canauhtli: duck}; "in reference to its hadrosauroian affinities" (*1).

Specific name: *tiquichensis* ← {(place-name) Tiquicheo + -ensis}; referring to "the town of Tiquicheo, honoring the generosity and hospitality of its people during the fieldwork season" (*1).

Etymology: **ancient duck** from Tiquicheo

Taxonomy: Ornithischia: Ornithopoda: Hadrosauroidea: Hadrosauromorpha

[References: (*1) Ramírez-Velasco AA, Benammi M, Prieto-Márquez, Ortega JA, Hernández-Rivera R (2012). *Huehuecanauhtlus tiquichenensis*, a new hadrosauroid dinosaur (Ornithischia: Ornithopoda) from the Santonian (Late Cretaceous) of Michoacán Mexico. *Canadian Journal of Earth Sciences* 49(2): 379–395.]

Huinculsaurus montesi

> *Huinculsaurus montesi* Baiano *et al.*, 2020 [Cenomanian–Turonian; Huincul Formation, Neuquén, Argentina]

Generic name: *Huinculsaurus* ← {(place-name) Huincul Formation + σαῦρος/sauros: lizard}

Specific name: *montesi* ← {(person's name) Montes + -ī}; in honor of Eduardo Montes.

Etymology: Montes' **lizard from Huincul Formation**

Taxonomy: Theropoda: Ceratosauria: ?Abelisauroidea: ?Noasauridae: Elaphrosaurinae

[References: Baiano MA, Coria RA, Cau A (2020). A new abelisauroid (Dinosauria: Theropoda) from the Huincul Formation (lower Upper Cretaceous, Neuquén Basin) of Patagonia, Argentina. *Cretaceous Research* 110: 104408.]

Hulsanpes perlei

> *Hulsanpes perlei* Osmólska, 1982 [Campanian; Barun Goyot Formation, Khulsan, Ömnögovi, Mongolia]

Generic name: *Hulsanpes* ← {(place-name) Khulsan + (Lat.) pēs: foot}

Specific name: *perlei* ← {(person's name) Perle + -ī}; in honor of Mongolian paleontologist Altangerel Perle.

Etymology: Perle's **Khulsan foot**

Taxonomy: Theropoda: Maniraptora: Dromaeosauridae: Halszkaraptorinae (*1)

Notes: According to Cau *et al.* (2017), *Hulsanpes* is placed in the Halszkaraptorinae with *Halszkaraptor* and *Mahakala*.

[References: Osmólska, H. (1982). *Hulsanpes perlei* n. g. n. sp. (Deinonychosauria, Saurischia, Dinosauria) from the Upper Cretaceous Barun Goyot Formation of Mongolia. *Neues Jahrbuch fur Geologie und Palaeontologie, Monatsheffte* 1982(7): 440–448.; (*1) Cau A, Madzia D (2018). Redescription and affinities of *Hulsanpes perlei* (Dinosauria, Theropoda) from the Upper Cretaceous of Mongolia. *Peer J* 6: e4868.]

Hungarosaurus tormai

> *Hungarosaurus tormai* Ösi, 2005 [Santonian; Csehbánya Formation, Bakony Mountains, Veszprém, Hungary]

Generic name: *Hungarosaurus* ← {(place-name) Hungary + (Gr.) σαῦρος/sauros: lizard}

Specific name: *tormai* ← {(person's name) Torma + -ī}; in honor of András Torma, one of the discoverers of the Iharkút locality.

Etymology: Torma's **Hungary lizard**

Taxonomy: Ornithischia: Ankylosauria: Nodosauridae

[References: Ösi A (2005). *Hungarosaurus tormai*, a new ankylosaur (Dinosauria) from the Upper Cretaceous of Hungary. *Journal of Vertebrate Paleontology* 25(2): 370–383.]

Huxleysaurus hollingtoniensis

> *Iguanodon fittoni* Lydekker, 1889 ⇒ *Hypselospinus fittoni* (Lydekker, 1889) Norman, 2010 ⇒ *Wadhurstia fittoni* (Lydekker, 1889) Carpenter & Ishida, 2010 [late Valanginian; Wadhurst Clay Formation, East Sussex, UK]

= *Iguanodon hollingtoniensis* Lydekker, 1889 ⇒ *Huxleysaurus hollingtoniensis* (Lydekker, 1889) Paul, 2012 [early Valanginian; Wadhurst Clay Formation, Hollington, Hastings, East Sussex, UK]

Generic name: *Huxleysaurus* ← {(person's name) Huxley + (Gr.) σαῦρος/sauros}; "for Darwin's 'Bulldog' and coiner of the term agnostic, Thomas Huxley" (*1).

Specific name: *hollingtoniensis* ← {(place-name) Hollington + -iensis}

Etymology: **Huxley's lizard** from Hollington

Taxonomy: Ornithischia: Ornithopoda: Ankylopollexia: Styracosterna

Notes: *Huxleysaurus* is regarded as a junior synonym of *Hypselospinus*.

[References: (*1) Paul G (2012). Notes on the rising diversity of Iguanodont taxa, and Iguanodonts named after Darwin, Huxley, and evolutionary science. *Actas de V Jornadas Internacionales sobre Paleontologia de Dinosaurios y su Entorno, Salas de los Infantes, Burgos*. Colectivo de Arquelogico-Paleontologico de Salas de los Infantes (Burgos).pp. 121–131.]

Hylaeosaurus armatus

> *Hylaeosaurus armatus* Mantell, 1833 [Valanginian; Tunbridge Wells Sand Formation, UK]
= *Hylosaurus mantelli* Fitzinger, 1843 (acc. *The Dinosauria 2nd edition*)
= *Hylaeosaurus oweni* Mantell, 1844 [Valanginian; Tunbridge Wells Sand Formation, UK]

Generic name: *Hylaeosaurus* ← {(Gr.) ὑλαῖος/hylaios: of the wood, of the forest [ὕλη/hylē: forest] + σαῦρος/sauros: lizard}; indicating "its locality, the Forest of Tilgate" (*1), or in reference to the Wealden Group. Weald means "forest, woodland".

Specific name: *armatus* ← {(Lat.) armātus: armoured, armed}.

Etymology: armoured **forest** (Wealden) **lizard**

Taxonomy: Ornithischia: Ankylosauria: Nodosauridae

Other species:

> *Polacanthus foxii* Hulke, 1881 ⇒ *Hylaeosaurus foxii* (Hulke, 1881) Coombs, 1971 (*nomen ex dissertatione*) [Barremian; Wessex Formation, UK]

[References: (*1) Mantell GA (1833). Observations on the remains of *Iguanodon*, and other fossil reptiles, of the strata of Tilgate Forest in Sussex. *Proceedings of the geological Society of London* 1: 410–411.]

Hypacrosaurus altispinus

> *Hypacrosaurus altispinus* Brown, 1913 [Maastrichtian; Horseshoe Canyon Formation, Alberta, Canada]
= *Cheneosaurus tolmanensis* Lambe, 1971 [Maastrichtian; Horseshoe Canyon Formation, Alberta, Canada]

Generic name: *Hypacrosaurus* ← {(Gr.) ὕπακρος/hypakros*: nearly the highest + σαῦρα/saura: lizard}

Brown (1913) described the size which is "approaching the great carnivorous dinosaur *Tyrannosaurus* of the later Lance Formation" (*1).

**cf.* [(Gr.) ὑπό-/ hypo-: less + ἄκρος/ akros: highest, topmost]

Specific name: *altispinus* ← {(Lat.) altus: high + spīna: spine}

The mid-dorsal spines are "three times the height of respective centra." "The dorsal vertebrae are characterized by extremely high, massive spines and comparatively small centra" (*1)

Etymology: high-spined **near-topmost lizard**

Taxonomy: Ornithischia: Ornithopoda: Iguanodontia: Hadrosauridae: Lambeosaurinae

Other species:

> *Hypacrosaurus stebingeri* Horner & Currie, 1994 [Campanian; upper Two Medicine Formation, Montana, USA]

[References: (*1) Brown B (1913). A new trachodont dinosaur, *Hypacrosaurus*, from the Edmonton Cretaceous of Alberta. *Bulletin of the American Museum of Natural History* 32(20): 395–406.]

Hypselosaurus priscus

> *Hypselosaurus priscus* Matheron, 1869 [Maastrichtian; Provence-Alpes-Côte d'Azur, France]

Generic name: *Hypselosaurus* ← {(Gr.) ὑψηλός/hypsēlos: high, lofty + σαῦρος/sauros: lizard}

Specific name: *priscus* ← {(Lat.) priscus: ancient, primitive}

Etymology: ancient, **high lizard**

Taxonomy: Saurischia: Sauropodomorpha: Sauropoda: Titanosauria

Notes: Matheron thought *Hypselosaurus* was an aquatic crocodile. Eggs with abnormally thin shells have been found. According to *The Dinosauria 2nd edition, Hypselosaurus priscus* is considered a *nomen dubium*.

[References: Matheron P (1869). Note sur les reptiles fossiles des dépôts fluvio-lacustres Crétacés du basin à lignite de Fuveau. *Bulletin de la Société géologique de France, série 2* 26: 781–795. : Hirsch KF (2001). Pathological amniote eggshell – fossil and modern. In: Tanke DH, Carpenter K (eds), *Mesozoic Vertebrate Life*, Indiana University Press, Bloomington: 378–392.]

Hypselospinus fittoni

> *Iguanodon fittoni* Lydekker, 1889 ⇒ *Hypselospinus fittoni* (Lydekker, 1889) Norman, 2010 ⇒ *Wadhurstia fittoni* (Lydekker, 1889) Carpenter & Ishida, 2010 [late Valanginian; Wadhurst Clay Formation, East Sussex, UK]
= *Iguanodon hollingtoniensis* Lydekker, 1889 ⇒ *Huxleysaurus hollingtoniensis* (Lydekker, 1889) Paul, 2012 [early Valanginian; Wadhurst Clay Formation, East Sussex, UK]
= ?*Darwinsaurus evolutionis* Paul, 2012 [Valanginian; Wadhurst Clay Formation, Hastings, East Sussex, UK]

Generic name: *Hypselospinus* ← {(Gr.) ὕψηλός/hypsēlos: high + (Lat.) spina: thorn, spine}; in reference to the high vertebral spines.

Specific name: *fittoni* ← {(person's name) Fitton + -ī}; in honor of William Henry Fitton.
Etymology: Fitton's **high-spined one**
Taxonomy: Ornithischia: Ornithopoda: Iguanodontia: Ankylopollexia: Styracosterna

[References: Lydekker R (1889). On the remains and affinities of five genera of Mesozoic reptiles. *Quarterly Journal of the Geological Society of London* 45: 41–59.; Norman DB (2010). A taxonomy of iguanodontians (Dinosauria: Ornithopoda) from the lower Wealden Group (Cretaceous: Valanginian) of southern England. *Zootaxa* 2489: 47–66.]

Hypsibema crassicauda

> *Hypsibema crassicauda* Cope, 1869 [Campanian; Black Creek Group, North Carolina, USA]

Generic name: *Hypsibema* ← {(Gr.) ὕψι-/hypsi-: high + βῆμα/bēma: step}
Specific name: *crassicauda* ← {(Lat.) crassus: fat, solid + cauda: tail}
Etymology: solid-tailed **high step**
Taxonomy: Ornithischia: Ornithopoda: Hadrosauroidea
Other species:

> *Neosaurus missouriensis* Gilmore & Stewart, 1945 (preoccupied) (*nomen dubium*) ⇒ *Parrosaurus missouriensis* (Gilmore & Stewart, 1945) Gilmore, 1945 (*nomen dubium*) ⇒ *Hypsibema missouriensis* (Gilmore & Stewart, 1945) Baird & Horner, 1979 (*nomen dubium*) [Maastrichtian; Ripley Formation, Missouri, USA]

Notes: According to *The Dinosauria 2nd edition*, *Hypsibema crassicauda* and *Neosaurus missouriensis*, respectively, were considered to be *nomina dubia*.

Hypsilophodon foxii

> *Hypsilophodon foxii* Huxley, 1869 [Barremian–early Aptian; Wessex Formation, Isle of Wight, UK]

Generic name: *Hypsilophodon* ← {*Hypsilophus* [(Gr.) ὑψί-/hypsi-: high + λόφος/lophos: crest] + ὀδών/odōn [←ὀδούς/odūs: tooth]}
Specific name: *foxii* ← {(person's name) Foxius←Fox}; in honor of paleontologist William Fox (1813–1881).
Etymology: Fox's ***Hypsilophus* (high-crested) tooth**
Taxonomy: Ornithischia: Ornithopoda

I

Iaceornis marshi [Avialae]

> *Iaceornis marshi* Clarke, 2004* [Santonian; Niobrara Formation, Kansas, USA]

*The second specimen of *Apatornis celer* (Marsh, 1873) Marsh, 1873 was given its own genus name *Iaceornis marshi*, different from the type specimen of *Apatornis celer*. *cf. Ichthyornis*.

Generic name: *Iaceornis* ← {(Lat.) iaceō: figurative for "to be neglected" + (Gr.) ὄρνις/ornis: bird} (*1)
Specific name: *marshi* ← {(person's name) Marsh + -ī}; for "O. C. Marsh, who originally described the holotype specimen (Marsh, 1880)" (*1).
Etymology: Marsh's **neglected bird**
Taxonomy: Theropoda: Avialae: Ornithuromorpha: Ornithurae

[References: (*1) Clarke JA (2004). Morphology, phylogenetic taxonomy, and systematics of *Ichthyornis* and *Apatornis* (Avialae: Ornithurae). *Bulletin of the American Museum of Natural History* 286: 1–179.]

Ichthyornis dispar [Avialae]

> *Ichthyornis dispar* Marsh, 1872 [Coniacian; Niobrara Formation, Kansas, USA]
= *Colonosaurus mudgei* Marsh, 1872 [Coniacian; Niobrara Formation, Kansas, USA]
= *Graculavus anceps* Marsh, 1872 ⇒ *Ichthyornis anceps* (Marsh, 1872) Marsh, 1872 [Coniacian, Niobrara Formation, Kansas, USA]
= *Graculavus agilis* Marsh, 1873 ⇒ *Ichthyornis agilis* (Marsh, 1873) Marsh, 1880 [Coniacian; Niobrara Formation, Kansas, USA]
= *Ichthyornis victor* Marsh, 1876 [Coniacian; Niobrara Formation, Kansas, USA]
= *Ichthyornis validus* Marsh, 1880 [Coniacian; Niobrara Formation, Kansas, USA]
= *Plegadornis antecessor* Wetmore, 1962 ⇒ *Angelinornis antecessor* (Wetmore, 1962) Kashin, 1972 ⇒ *Ichthyornis antecessor* (Wetmore, 1962) Olson, 1975 [Santonian; Mooreville Chalk Formation, Alabama, USA]

Generic name: *Ichthyornis* ← {(Gr.) ἰχθύς/ichthys: fish + ὄρνις/ornis: bird}; in reference to "having biconcave vertebrae" (*1).
Specific name: *dispar* ← {(Lat.) dispār: unlike, different}; "differing widely from all known birds in having biconcave vertebrae" (*1)
Etymology: different, **fish-like bird**
Taxonomy: Theropoda: Avialae: Ornithurae
Other species:

> *Ichthyornis celer* Marsh, 1873 ⇒ *Apatornis celer* (Marsh, 1873) Marsh, 1873 [Coniacian; Niobrara Formation, Kansas, USA]
> *Ichthyornis maltshevskyi* Nessov, 1986 ⇒ *Lenesornis maltshevskyi* (Nessov, 1986) Kurochkin, 1996 [Turonian; Bissekty Formation, Navoi, Uzbekistan]

> *Ichthyornis minusculus* Nessov, 1990 (*nomen dubium*) [Turonian; Bissekty Formation, Navoi, Uzbekistan]

Notes: *Ichthyornis* is a genus of toothed seabirds having binocave vertebrae.

[References: (*1) Marsh OC (1872). Notice of a new and remarkable fossil bird. *American Journal of Science*, series 3, 4 (22): 344.]

Ichthyovenator laosensis

> *Ichthyovenator laosensis* Allain *et al.*, 2012 [Aptian; Grés Supérieurs Formation, Savannakhet, Laos]

Generic name: *Ichthyovenator* ← {(Gr.) ἰχθύς/ichthys: fish + (Lat.) vēnātor: hunter}

Specific name: *laosensis* ← {(place-name) Laos + -ensis}

Etymology: **fish hunter** from Laos

Taxonomy: Theropoda: Spinosauridae: Spinosaurinae / Baryonychinae

[References: Allain R, Xaisanavong T, Richir P, Khentavong B (2012). The first definitive Asian spinosaurid (Dinosauria: Theropoda) from the Early Cretaceous of Laos. *Naturwissenschaften* 99 (5): 369–377.]

Ignavusaurus rachelis

> *Ignavusaurus rachelis* Knoll, 2010 [Hettangian; upper Elliot Formation, Ha Ralekoala, Qacha's Nek District, Lesotho]

Generic name: *Ignavusaurus* ← {(Lat.) īgnāvus: coward + (Gr.) σαῦρος/sauros: lizard}; referring to the place of discovery, Ha Ralekoala which means "the place of the father of the coward".

Specific name: *rachelis* ← {(person's name) Rachel + -is}; in honor of paleontologist Raquel López-Antoñanzas.

Etymology: Rachel's **coward lizard**

Taxonomy: Saurischia: Sauropodomorpha: Massopoda

[References: Knoll F (2010). A primitive sauropodomorph from the upper Elliot Formation of Lesotho. *Geological Magazine* 147 (6): 814–829.; McPhee BW, Bordy EM, Sciscio L, Choiniere JN (2017). The sauropodomorph biostratigraphy of the Elliot Formation of southern Africa: Tracking the evolution of Sauropodomorpha across the Triassic-Jurassic boundary. *Acta Palaeontologica Polonica* 62(3): 441–465.]

Ignotosaurus fragilis [Silesauridae]

> *Ignotosaurus fragilis* Martínez *et al.*, 2013 [late Carnian; Ischigualasto Formation, San Juan, Argentina]

Generic name: *Ignotosaurus* ← {(Lat.) īgnōtus: unknown + (Gr.) σαῦρος/sauros: lizard}; "in reference to the previously unknown presence of silesaurids in the well-sampled Ischigualasto Formation".

Specific name: *fragilis* ← {(Lat.) fragilis: fragile}; "in reference to the extremely thin central portion of the blade of the ilium and gracile proportions of the femur".

Etymology: fragile, **unknown lizard**

Taxonomy: Dinosauriformes: Silesauridae

Iguanacolossus fortis

> *Iguanacolossus fortis* MacDonald *et al.*, 2010 [Barremian; Yellow Cat Member, Cedar Mountain Formation, Utah, USA]

Generic name: *Iguanacolossus* ← {(genus) *Iguana* + (Lat.) colossus: colossal, colossus}; "in reference to the herbivorous lizards of the genus *Iguana*, the teeth of which have been historically compared to those of basal iguanodonts, and to the large size of the holotype skeleton" (*1). The gender of this genus is masculine.

Specific name: *fortis* ← {(Lat.) fortis: mighty}

Etymology: "mighty **Iguana colossus**" (*1)

Taxonomy: Ornithischia: Ornithopoda: Iguanodontia: Styracosterna

[References: (*1) McDonald AT, Kirkland JI, DeBlieux DD, Madsen SK, Cavin J, Milner ARC, Panzarin L (2010). New basal iguanodonts from the Cedar Mountain Formation of Utah and the evolution of thumb-spiked dinosaurs. *PLoS ONE* 5(11): e14075.]

Iguanodon bernissartensis

> *Iguanodon bernissartensis* Boulenger, 1881 [Barremian–Aptian; Sainte-Barbe Clay Formation, Bernissart, Hainaut, Belgium]

?= *Iguanodon seelyi* Hulke, 1882 ⇒ *Dollodon seeley* (Hulke, 1882) Carpenter & Ishida, 2010 [Hauterivian; Wealden Group, England, UK]

= *Delapparentia turolensis* Ruiz-Omeñaca, 2011 [early Barremian; Camarillas Formation, Aragón, Spain]

Generic name: *Iguanodon* Mantell, 1825 ← {*Iguana* + (Gr.) ὀδών/odōn [ὀδούς/odūs]: tooth}; referring to the fossil teeth resembling those of an iguana.

Specific name: *bernissartensis* ← {(place-name) Bernissart + -ensis}; referring to the place in which the skeletons were discovered in a coal mine.

Etymology: **Iguana's tooth** from Bernissart

n.b. Mantell named the teeth discovered in 1822 in England *Iguanodon* in 1825, but he did not add a specific name. In 1829, it was supplied by Friedrich Holl and formed a proper binomial, *Iguanodon angulicus*.

Taxonomy: Ornithischia: Ornithopoda: Iguanodontia: Iguanodontoidea

Other species:

[valid]

> *Iguanodon galvensis* Verdú *et al.*, 2015 [Barremian; Camarillas Formation, Aragon, Spain]

[*nomen dubium*]

> *Iguanodon anglicum* / *anglicus* Holl, 1829 (*nomen dubium*) ⇒ *Therosaurus anglicus* (Holl, 1829) Fitzingerf, 1843 [Barremian; lower Wealden Formation, Kent, England, UK]

= *Iguanodon mantelli* von Meyer, 1832 ⇒ *Therosaurus mantelli* (von Meyer, 1832) Fitzingerf, 1843 [Early Cretaceous; Wealden Formation, UK]

= *Iguanodon mantelli* Hulke, 1882 (Type of *I. mantelli*) [Valanginian–Hauterivian; Wessex Formation, Isle of Wight, UK]

= *Streptospondylus major* Owen, 1842 (*Streptospondylus recentior* is a museum label for syntype specimens) ⇒*Iguanodon major* (Owen, 1842) Delair, 1966 (*nomen dubium*) [Early Cretaceous, Wealden Formation, West Sussex, England, UK]

> *Iguanodon praecursor* Sauvage, 1876 (*nomen dubium*) ⇒ *Pelorosaurus praecursor* (Sauvage, 1876) Sauvage, 1895 (?*nomen dubium*) [Tithonian; Nord-Pas-de-Calais, France]

> *Sphenospondylus gracilis* Lydekker, 1888 ⇒ *Iguanodon gracilis* (Lydekker, 1888) Steel, 1969 [Valanginian–Hauterivian; Wessex Formation, Isle of Wight, England, UK]

> *Iguanodon hilli* Newton, 1892 (*nomen dubium*) [Coniacian; Zig Zag Chalk Formation, UK]

> *Iguanodon ottingeri* Galton & Jensen, 1979 (*nomen dubium*) [Aptian; Cedar Mountain Formation, Utah, USA]

> *Iguanodon mongoliensis* Whitfield, 1992 (*nomen nudum*) ⇒ *Altirhinus kurzanovi* Norman, 1998 [middle Albian; Khuren Dukh Formation, Dornogov, Mongolia]

[species transferred to other genera]

> *Hypsilophodon foxii* Huxley, 1869 ⇒ *Iguanodon foxii* (Huxley, 1869) Owen 1873 or 1874 [late Barremian; Wessex Formation, England, UK]

> *Iguanodon phillipsi* Seeley, 1869 ⇒ *Priodontognathus phillipsi* (Seeley, 1869) Seeley, 1875 [Oxfordian; Calcareous Grit Formation, England, UK]

> *Iguanodon hoggi* Owen, 1874 ⇒ *Camptosaurus hoggi* (Owen, 1874) Norman & Barrett, 2002 ⇒ *Owenodon hoggi* (Owen, 1874) Galton, 2009 [Berriasian; Lulworth Formation, Dorset, UK]

> *Iguanodon prestwichii* Hulke, 1880 ⇒ *Cumnoria prestwichii* (Hulke, 1880) Seeley, 1888 ⇒ *Camptosaurus prestwichii* (Hulke, 1880) Lydekker, 1889 [late Kimmeridgian; Kimmeridge Clay Formation, England, UK]

> *Iguanodon dawsoni* Lydekker, 1888 ⇒ *Barilium dawsoni* (Lydekker, 1889) Norman, 2010 ⇒ *Torilion dawsoni* (Lydekker, 1888) Carpenter & Ishida, 2010 [Valanginian; Wadhurst Clay Formation, UK]

> *Iguanodon fittoni* Lydekker, 1889 ⇒ *Hypselospinus fittoni* (Lydekker, 1889) Norman, 2010 ⇒ *Wadhurstia fittoni* (Lydekker, 1889) Carpenter & Ishida, 2010 [late Valanginian; Wadhurst Clay Formation, England, UK]

= *Iguanodon hollingtoniensis* Lydekker, 1889 ⇒ *Huxleysaurus hollingtoniensis* (Lydekker, 1889) Paul, 2012 [Valanginian; Wadhurst Clay Formation, UK]

> *Iguanodon atherfieldensis* Hooley, 1925 ⇒ *Mantellisaurus atherfieldensis* (Hooley, 1925) Paul, 2006 [late Hauterivian; Wealden Group, England, UK]

?= *Vectisaurus valdensis* Hulke, 1879 (*nomen dubium*) ⇒ *Iguanodon valdensis* (Hulke, 1879) van der Broeck, 1900 [late Hauterivian; Wessex Formation, England, UK]

> *Iguanodon suessi* Bunzel, 1871 ⇒ *Mochlodon suessi* (Bunzel, 1871) Seeley, 1881 [early Campanian; Grünbach Formation, Niederosterreich, Austria]

> *Iguanodon exogirarum* Fritsch, 1878 ⇒ *Procerosaurus exogirarum* (Fritsch, 1878) Fritsch, 1905 (preoccupied) ⇒ *Ponerosteus exogryrarum* (Fritsch, 1878) Olshevsky, 2000 (species name amended) [Cenomanian; Bohemia, Czech]

> *Iguanodon albinus* Fritsch, 1893 (*nomen dubium*) ⇒ *Albisaurus albinus* (Fritsch, 1893) Fritsch, 1905 [Turonian–Santonian; Pardubice, Czech]

= *Albisaurus scutifer* Fritsch, 1905 [Coniacian; Priesener Formation, Vychodocesky, Czech] (based on the same specimen as *A. albinus*, which has priority to *A. scutifer.*)

> *Iguanodon lakotaensis* Weishampel & Bjork, 1989 ⇒ *Dakotadon lakotaensis* (Weishampel & Bjork, 1989) Paul, 2008 [Valanginian; Lakota Formation, South Dakota, USA]

Notes: In 1878, over 30 skeletons were discovered in a coal mine in Bernissart [*1]. According to *The*

Dinosauria 2nd edition, 6 species (*Iguanodon anglicus*, *I. atherfieldensis*, *I. bernissartensis*, *I. dawsoni*, *I. fittoni* and *I. lakotaensis*) belonged to the genus *Iguanodon*. But, now only 2 species (*I. bernissartensis* and a new species *I. galvensis*) are seemed to be valid species of *Iguanodon*.

[References: (*1) Dennis R. Dean (1999). *Gideon Mantell and the Discovery of Dinosaurs*. 〈https://books.google.co.jp/books/about/Gideon_Mantell_and_the_Discovery_of_Dino.html?id=37AT5l1DnaQC&redir_esc=y〉]

Iliosuchus incognitus

> *Iliosuchus incognitus* von Huene, 1932 ⇒ *Megalosaurus incognitus* (von Huene, 1932) Romer, 1966 [middle Bathonian; Stonesfield Slate, Oxfordshire, UK]

Generic name: *Iliosuchus* ← {(Lat.) ilium + (Gr.) Σοῦχος/Sūchos: the crocodile god}; referring to the ilia which were found in England. (acc.*The Dinosauria 2nd edition*)

Specific name: *incognitus* ← {(Lat.) incōgnitus: unknown}

Etymology: unknown **ilium crocodile god**

Taxonomy: Theropoda: Neotheropoda: Tetanurae

Notes: According to Benson (2009), *Iliosuchus incognitus* is considered a *nomen dubium*.

[References: von Huene F (1932). Die fossile Reptil-Ordnung Saurischia, ihre Entwicklung und Geschichte. *Monographien zur Geologie und Palaeontologie, serie 1* 4(1-2): 1–361.]

Ilokelesia aguadagrandensis

> *Ilokelesia aguadagrandensis* Coria & Salgado, 1998 [late Cenomanian; Huincul Formation, Neuquén, Argentina]

Generic name: *Ilokelesia* ← {(Mapuche) ilo: flesh + kelesio: lizard}; "a flesh-eating reptile". (*1)

Specific name: *aguadagrandensis* ← {(place-name) Aguada Grande + -ensis}; referring to the place where the specimen was found. (*1)

Etymology: **flesh-eating reptile** from Aguada Grande

Taxonomy: Theropoda: Ceratosauria: Neoceratosauria: Abelisauria: Abelisauridae

Notes: Medium-sized theropod.

[References: (*1) Coria RA, Salgado L (1998). A basal Abelisauria Novas, 1992 (Theropoda–Ceratosauria) from the Cretaceous of Patagonia, Argentina. *Gaia* 15: 89–102.]

Imperobator antarcticus

> *Imperobator antarcticus* Ely & Case, 2019 [Maastrichtian; Snow Hill Island Formation, James Ross Basin, Antarctic Peninsula]

Generic name: *Imperobator* ← {(Lat.) impero-* [imperō: to command] + (Mong.) bator: warrior} (*1) *imperiōsis: powerful.

Specific name: *antarcticus* ← {(Lat.) antarcticus: Antarctic]}; for its location of discovery. This specimen was collected from James Ross Island, Antarctic Peninsula (*1).

Etymology: Antarctic, **powerful warrior**

Taxonomy: Theropoda: Maniraptora: Paraves

[References: (*1) Ely RC, Case JA (2019). Phylogeny of a new gigantic paravian (Theropoda; Coelurosauria; Maniraptora) from the Upper Cretaceous of James Ross Island, Antarctica. *Cretaceous Research* 101: 1–16.]

Incisivosaurus gauthieri

> *Incisivosaurus gauthieri* Xu *et al.*, 2002 [Barremian; Yixian Formation, Liaoning, China]

Generic name: *Incisivosaurus* ← {(Lat.) incisivus: incisor + (Gr.) σαῦρος/sauros: lizard}; referring to "the presence of incisor-like premaxillary teeth" (*1).

Specific name: *gauthieri* ← {(person's name) Gauthier + -ī}; "in honor of Jacques Gauthier for his contributions to theropod systematics" (*1).

Etymology: Gauthier's **incisor lizard**

Taxonomy: Theropoda: Maniraptoriformes: Oviraptorosauria

[References: (*1) Xu X, Cheng Y-N, Wang X-L, Chang C-H (2002). An unusual oviraptorosaurian dinosaur from China. *Nature* 419: 291–293.]

Indosaurus matleyi

> *Indosaurus matleyi* von Huene & Matley, 1933 [Maastrichtian; Lameta Formation, Madhya Pradesh, India]

Generic name: *Indosaurus* ← {Indo- [(Gr.) Ἰνδός/Indos: an Indian, the river Indus] + σαῦρος/sauros: lizard}

Specific name: *matleyi* ← {(person's name) Matley + -ī}; in honor of Charles Alfred Matley who discovered the remains.

Etymology: Matley's **Indian lizard**

Taxonomy: Theropoda: Ceratosauria: Abelisauridae: Majungasaurinae

Notes: According to *The Dinosauria 2nd edition*, *Indosaurus matleyi* was considered to be Neoceratosauria *incertae sedis*.

Indosuchus raptorius

> *Indosuchus raptorius* von Huene & Matley, 1933 [Maastrichtian; Lameta Formation, Madhya Pladesh, India]

Generic name: *Indosuchus* ← {Indo- [(Gr.) Ἰνδός/

Indos: an Indian, the river Indus] + Σοῦχος/Sūchos: Egyptian crocodile god}

Specific name: *raptorius* ← {(Lat.) raptōrius: raptorial}

Etymology: raptorial, **Indian crocodile god**

Taxonomy: Theropoda: Ceratosauria: Abelisauridae

Notes: According to *The Dinosauria 2nd edition*, *Indosuchus raptorius* was considered to be Neoceratosauria *incertae sedis*.

[References: Chatterjee S (1978). *Indosuchus* and *Indosaurus*, Cretaceous carnosaurs from India. *Journal of Paleontology* 52(3): 570–580.]

Indrasaurus wangi [Squamata]

> *Indrasaurus wangi* O'Connor *et al.*, 2019 [Early Cretaceous; Liaoning, China]

Generic name: *Indrasaurus* ← {(Vedic Myth.) Indra: the god Indra + (Gr.) σαῦρος/sauros: lizard}

According to a Vedic legend, the God Indra was swallowed by the dragon Vritra. On the other hand, *Indrasaurus* was found in the stomach of a *Microraptor* specimen.

Specific name: *wangi* ← {(person's name) Wang + -ī}; after the paleoherpetologist Wang Yuan.

Etymology: Wang's **Indra lizard**

Taxonomy: Squamata: Scleroglossa

[References: O'Connor J, Zheng X, Dong L, Wang X, Wang Y, Zhang X, Zhou Z (2019). *Microraptor* with ingested lizard suggests non-specialized digestive function. *Current Biology* 29(14): 2423–2429. e2.]

Ingenia yanshini

> *Ingenia yanshini* Barsbold, 1981 (preoccupied) (*1) ⇒ *Oviraptor yanshini* (Barsbold, 1981) Paul 1988 ⇒ *Ajancingenia yanshini* (Barsbold, 1981) Easter, 2013 ⇒ *Heyuannia yanshini* (Barsbold, 1981) Funston *et al.*, 2018 [Campanian; Barun Goyot Formation, Ömnögovi, Mongolia]

Generic name: *Ingenia* ← {(place-name) Ingen + -ia}; derived from the Ingen Khoboor Depression.

Specific name: *yanshini* ← {(person's name) Yanshin + -ī}; in honor of Aleksandr Leonidovich Yanshin.

Etymology: Yanshin's **one from Ingen Khoboor Depression**

Taxonomy: Theropoda: Coelurosaria: Oviraptorosauria: Oviraptoridae

Notes: The generic name *Ingenia* had been preoccupied by the generic name of a nematode, therefore, *Ajancingenia* was proposed by Easter in 2013. (*1)

[References: (*1) Easter J (2013). A new name for the oviraptorid dinosaur "*Ingenia*" *yanshini* (Barsbold, 1981; preoccupied by Gerlach, 1957) *Zootaxa* 3737(2): 184–190.]

Ingentia prima

> *Ingentia prima* Apaldetti *et al.*, 2018 [Norian-Rhaetian; Quebrada del Barro Formation, San Juan, Argentina]

Generic name: *Ingentia* ← {(Lat.) ingēns: huge} (*1)

Specific name: *prima* ← {(Lat.) prima, *f.*: first}; "referring to the large body size acquired during the early evolution of Dinosauria" (*1).

Etymology: first, **huge one**

Taxonomy: Saurischia: Sauropodomorpha: Lessemsauridae

[References: (*1) Apaldetti C, Martinez RN, Cerda IA, Pol D, Alcober O (2018). An early trend towards gigantism in Triassic sauropodomorph dinosaurs. *Nature Ecology & Evolution* 2: 1227–1232.]

Inosaurus tedreftensis

> *Inosaurus tedreftensis* de Lapparent, 1960 [early Berriasian; Irhazer Group, In Tedreft, Agadez, Niger]

Generic name: *Inosaurus* ← {(place-name) In Tedreft + (Gr.) σαῦρος/sauros}; referring to the In Tedreft locality which is situated lower in the Continental Intercalaire. (*1)

Specific name: *tedreftensis* ← {(place-name) In Tedreft + -ensis} (*1)

Etymology: **In Tedreft lizard**

Taxonomy: Theropoda

Notes: According to *The Dinosauria 2nd edition*, *Inosaurus* is considered to be a *nomen dubium*. From the In Tedreft locality of the central Sahara, *Carcharodontosaurus saharicus*, *Elaphrosaurus iguidiensis*, *E. gautieri* and *Rebbachisaurus tamesnensis* are also found. (*1)

[References: (*1) de Lapparent AF (1960). Les dinosauriens du "Continental intercalaire" du Sahara central. *Mémoires de la Société géologique de France, nouvelle série* 39(88A): 1–57.]

Intiornis inexpectatus [Avialae]

> *Intiornis inexpectatus* Novas *et al.*, 2010 [Campanian; Morales Member, Las Curtiembres Formation, Salta, Argentina]

Generic name: *Intiornis* ← {(Quechua) Inti: sun + (Gr.) ὄρνις/ornis: bird} (*1)

Specific name: *inexpectatus* ← {(Lat.) inexpectātus = inexspectātus: unlooked for}; "in reference to the casual and weird technical situation of the finding of the holotype specimen" (*1).

Etymology: unexpected, **Sun bird**

Taxonomy: Theropoda: Avialae: Enantiornithes: Avisauridae

cf. Aves: Ornithothoraces: Enantiornithes: Avisauridae (*1)

[References: (*1) Novas FE, Agnolín FL, Scanferla CA (2010). New enantiornithine bird (Aves, Ornithothoraces) from the Late Cretaceous of NW Argentina. *Comptes Rendus Palevol* 9(8): 499–503.]

Invictarx zephyri

> *Invictarx zephyri* McDonald & Wolfe, 2018 [Campanian: Menefee Formation, San Juan, New Mexico, USA]

Generic name: *Invictarx* ← {(Lat.) invīctus: invincible, unconquerable + arx: fortress}; "in reference to the well-armored nature of ankylosaurian dinosaurs".

Specific name: *zephyri* ← {(Lat.) zephyrus (masculine): west wind}; "in reference to the blustery conditions that prevail among the outcrops where the specimens were discovered".

Etymology: **unconquerable fortness** of the western wind

Taxonomy: Ornithischia: Thyreophora: Ankylosauria: Nodosauridae

[References: McDonald AT, Wolfe DG (2018). A new nodosaurid ankylosaur (Dinosauria: Thyreophora) from the Upper Cretaceous Menefee Formation of New Mexico. *PeerJ* 6: e5435.]

Irisosaurus yimenensis

> *Irisosaurus yimenensis* Peyre de Fabrègues *et al.*, 2020 [Early Jurassic; Fengjiahe Formation, Yunnan; China]

Generic name: *Irisosaurus* ←{(Lat.) Īris = (Gr.) Ἶρις/Iris: the goddess of the rainbow + σαῦρος/sauros: lizard}; "referring to the famous iridescent clouds of Yunnan Province (彩雲之南*)".

*[Pinyin: cǎiyùn zhī nán]

Specific name: *yimenensis* ← {(place-name) Yimen 易門 + -ensis}; "referring to Yimen County, where the type locality is located".

Etymology: **iridescent lizard** from Yimen [易門彩雲龍]

Taxonomy: Saurischia: Sauropodomorpha: Sauropodiformes

[References: Peyre de Fabrègues C, Bi S, Li H, Li G, Yang L, Xu X (2020). A new species of early-diverging Sauropodiformes from the Lower Jurassic Fengjiahe Formation of Yunnan Province, China. *Scientific Reports* 10(1): 10961.]

Irritator challengeri

> *Irritator challenger* Martill *et al.*, 1996 [Albian; Romualdo Member, Santana Formation, Ceará, Brazil]

?= *Angaturama limai* Kellner & Campos, 1996 [Albian; Romualdo Member, Santana Formation, Ceará, Brazil]

Generic name: *Irritator* ← {(Lat.) Irrītātor: irritator}; "the feeling the authors felt (understated here) when discovering that the snout had been artificially elongated" (*1).

Specific name: *challengeri* ← {(novel) Challenger + -ī}; "from Professor Challenger, the ficticious hero and dinosaur discoverer of Sir Artur Conan-Doyle's *Lost World*" (*1).

Etymology: Challenger's **irritator**

Taxonomy: Theropoda: Spinosauridae

Notes: According to *The Dinosauria 2nd edition*, *Angaturama limai* was valid. If *Angaturama* is a synonym of *Irritator*, *Irritator* has priority because it was named one month earlier than *Angaturama*.

[References: (*1) Martill DM, Cruickshank ARI, Frey E, Small PG, Clarke M (1996). A new crested maniraptoran dinosaur from the Santana Formation (Lower Cretaceous) of Brazil. *Journal of the Geological Society*, London 153(1): 5–8.]

Isaberrysaura mollensis

> *Isaberrysaura mollensis* Salgado *et al.*, 2017 [Toarcian–Bajocian; Los Molles Formation, Neuquén, Argentina]

Generic name: *Isaberrysaura* ← {(person's name) Isabel Berry + (Gr.) saura [the feminine form of the masculine "sauros"]: lizard)}; "in honor of Isabel Valdivia Berry, who reported the finding of the holotype material" (*1).

Specific name: *mollensis* ← {(place-name) Los Molles + -ensis} (*1)

Etymology: **Isabel Berry's lizard** from Los Molles

Taxonomy: Ornithischia: Genasauria: Neornithischia

Notes: The specimen was found in the marine-deltic deposits of the Los Molles Formation (Toarcian–Bajocian).

[References: (*1) Salgado L, Canudo JI, Garrido AC, Moreno-Azanza M, Martínez LCA, Coria RA, Gasca JM (2017). A new primitive neornithischian dinosaur from the Jurassic of Patagonia with gut contents. *Scientific Reports* 7: 42778.]

Isanosaurus attavipachi

> *Isanosaurus attavipachi* Buffetaut *et al.*, 2000 [Norian–Rhaetian; upper Member, Nam Phong Formation, Khonkaen, Thailand]

Generic name: *Isanosaurus* ← {(place-name) Isan: "the local name for north-eastern Thailand" (*1) | (Pali) īsāna: northeast + (Gr.) σαῦρος/sauros: lizard}

Specific name: *attavipachi* ← {(person's name) Attavipach + -ī}; "in honour of Pricha Attavipach, former Director General of the Thai Department of Mineral Resources, a long-time supporter of palaeontological research" (*1).

Etymology: Attavipach's **Isan lizard**

Taxonomy: Saurischia: Sauropodomorpha: Sauropoda

[References: (*1) Buffetaut E, Suteethorn V, Cuny G, Tong H, Le Loeuff J, Khansubha S, Jongautchariyakul S (2000). The earliest known sauropod dinosaur. *Nature* 407 (6800): 72–74.]

Isasicursor santacrucensis

> *Isasicursor santacrucensis* Novas *et al.*, 2019 [Campanian–Maastrichtian; Chorrillo Formation, Santa Cruz, Argentina]

Generic name: *Isasicursor* ← {(person's name) Isasi + (Lat.) cursor: runner} ; in honor of "the skilled technician Marcelo P. Isasi (conicet – MACN*), who discovered the remains of this new iguanodontian", and cursor, meaning runner in Latin. (*1)

*Museo Argentina de Ciencias Naturales

Specific name: *santacrucensis* ← {(place-name) Santa Cruz + -ensis}; in regard to "Santa Cruz, the Argentine province where fossils were found" (*1).

Etymology: **Isasi's runner** from Santa Cruz

Taxonomy: Ornithischia: Elasmaria

[References: (*1) Novas FE, Agnolin F, Rozadilla S, *et al.* (2019). Paleontological discoveries in the Chorrillo Formation (upper Campanian–lower Maastrichtian, Upper Cretaceous), Santa Cruz Province, Patagonia, Argentina. *Revista del Museo Argentino de Ciencias Naturales, Nueva Serie* 21(2): 217–293.]

Ischioceratops zhuchengensis

> *Ischioceratops zhuchengensis* He *et al.*, 2015 [Campanian; Wangshi Group, Kugou, Zhucheng, Shandong, China]

Generic name: *Ischioceratops* ← {《anat.》ischium + ceratops [(κέρας/keras: horn +ὤψ /ōps: face)]}; "in reference to the unique morphology of the ischium" (*1).

Specific name: *zhuchengensis* ← {(Chin. place-name) Zhūchéng諸城 + -ensis} (*1)

Etymology: **ischium horned face (ceratopsian)** from Zhucheng City

Taxonomy: Ornithischia: Ceratopsia: Leptoceratopsidae

[References: (*1) He Y, Makovicky PJ, Wang K, Chen S, Sullivan C, Han F, Xu X (2015). A new leptoceratopsid (Ornithischia, Ceratopsia) with a unique ischium from the Upper Cretaceous of Shandong Province, China. *PLoS ONE* 10(12): e0144148.]

Ischisaurus cattoi

> *Herrerasaurus ischigualastensis* Reig, 1963 [Carnian; Ischigualasto Formation, San Juan, Argentina]

= *Ischisaurus cattoi* Reig, 1963 [Carnian; Ischigualasto Formation, San Juan, Argentina]

= *Frenguellisaurus ischigualastensis* Novas, 1986 [Norian; Ischigualasto Formation, San Juan, Argentina]

Generic name: *Ischisaurus* ← {(place-name) Ischi-: Ischigualasto + (Gr.) σαῦρος/sauros: lizard} (*1)

Specific name: *cattoi* ← {(person's name) Cattoi}; "dedicated to Dr. Noemi V. Cattoi, in recognition of his repeated collaboration in the development of our activities in the Museo Argentino de Ciencias Naturales" (*1).

Etymology: Cattoi's **Ischigualasto lizard**

Taxonomy: Saurischia: Herrerasauridae

Notes: *Ischisaurus cattoi* is considered as a junior synonym of *Herrerasaurus ischigualastensis* (*2).

[References: (*1) Reig OA (1963). La presencia de dinosaurios saurisquios en los "Estratos de Ischigualasto" (Mesotriásico Superior) de las provincias de San Juan y La Rioja (República Argentina) [The presence of saurischian dinosaurs in the "Ischigualasto beds" (upper Middle Triassic) of the provinces of San Juan and La Rioja (Argentine Republic)]. *Ameghiniana* 3(1): 3–20.; (*2) Novas FE (1994). New information on the systematic and postcranial skeleton of *Herrerasaurus ischigualastensis* (Theropoda: Herrerasauridae) from the Ischigualasto Formation (Upper Triassic) of Argentina. *Journal of Vertebrate Paleontology* 13(4): 400–423.]

Ischyrosaurus manseli

> *Ischyrosaurus manseli* Hulke, 1874 *vide* Lydekker, 1888 (*nomen dubium*) ⇒ *Ornithopsis manseli* (Hulke, 1874 *vide* Lydekker, 1888) Lydekker, 1888 ⇒ *Pelorosaurus manseli* (Hulke, 1874 *vide* Lydekker, 1888) Sauvage, 1900 [early Tithonian; Kimmeridge Clay, Dorset, UK]

?= *Morinosaurus typus* Sauvage, 1874 (*nomen dubium*) [Kimmeridgian; Aspidoceras caletanus ammonoid zone, Nord-Pas-de-Calais, France]

Generic name: *Ischyrosaurus* ← {(Gr.) ισχυρός/ischyros: strong + σαῦρος/sauros: lizard}; referring to its large humerus.

Specific name: *manseli* ← {(person's name) Mansel + -ī}; in honor of J. C. Mansel- Pleydell, one of the founder members of Dorset County Museum.

Etymology: Mansel's **strong lizard**

Taxonomy: Saurischia: Sauropodomorpha: Sauropoda

Notes: According to *The Dinosauria 2nd edition*, *Ischyrosaurus manseli* is considered to be a *nomen dubium*.

[References: Hulke JW (1874). Note on a very large saurian limb-bone adapted for progression upon land, from the Kimmeridge Clay of Weymouth, Dorset. *Quarterly Journal of the Geological Society* 30: 16–17.]

Isisaurus colberti

> *Titanosaurus colberti* Jain & Bandyopadhyay, 1997 ⇒ *Isisaurus colberti* (Jain & Bandyopadhyay, 1997) Wilson & Upchurch, 2003 [Maastrichtian; Lameta Formation, Maharashtra, India]

Generic name: *Isisaurus* ← {(acronym) ISI=Indian Statistical Institute + (Gr.) σαῦρος/sauros: reptile}; referring to "the Indian Statiscal Institute (ISI), which houses India's foremost collection of Mesozoic fossil vertebrates and whose scholars discovered and described the holotype skeleton" [*1].

Specific name: *colberti* ← {(person's name) Colbert + -ī}; in honor of American vertebrate paleontologist Edwin Harris Colbert (1905–2001).

Etymology: Colbert's **Indian Statistical Institute's lizard**

Taxonomy: Saurischia: Sauropodomorpha: Titanosauria: Lithostrotia: Saltasauridae: Saltasaurinae

[References: (*1) Wilson JA, Upchurch P (2003). A revision of Titanosaurus Lydekker (Dinosauria–Sauropoda), the first dinosaur genus with a 'Gondwanan' distribution. *Journal of Systematic Palaeontology* 1(3): 125–160.]

Itapeuasaurus cajapioensis

> *Itapeausaurus cajapioensis* Lindoso *et al.*, 2019 [Cenomanian; Alcântara Formation, Maranhão, Brazil]

Generic name: *Itapeuasaurus* ← {(place-name) Itapéua + (Gr.) σαῦρος/sauros: lizard}

Specific name: *cajapioensis* ← {(place-name) Cajapió + -ensis}

Etymology: **Itapeua lizard** from Cajapió

Taxonomy: Saurischia: Sauropodomorpha: Rebbachisauridae

Notes: A humerus was discovered on the beach at Itapeua, Cajapió, Brazil by a fisherman Carlos Wagner Silva in 2014, according to Lindoso *et al.* (2019).

Itemirus medullaris

> *Itemirus medullaris* Kurzanov, 1976 [Turonian; Bissekty Formation, Navoi (= Navoiy), central Kyzylkum Desert, Uzbekistan]

Generic name: *Itemirus* ← {(place-name) Itemir + -us}

According to Sues and Averianov (2014), the type locality is Dzharakuduk II (not Itemir, contra Kurzanov, 1976). [*1]

Specific name: *medullaris* ← {(Lat.) medullāris: of medulla}; referring to a partial braincase.

Etymology: **Itemir one** with medulla

Taxonomy: Theropoda: Dromaeosauridae: Eudromaeosauria: Dromaeosaurinae [*1]
cf. Theropoda: Coelurosauria: Tyrannosauroidea (acc. *The Dinosauria 2nd edition*)

Notes: *Itemirus* was thought to be possible Tyrannosauroidea, according to *The Dinosauria 2nd edition*, however, it is now considered as Dromaeosauridae [*1].

[References: (*1) Sues H-D, Averianov A (2014). Dromaeosauridae (Dinosauria: Theropoda) from the Bissekty Formation (Upper Cretaceous: Turonian) of Uzbekistan and the phylogenetic position of *Itemirus medullaris* Kurzanov, 1976. *Cretaceous Research* 51: 225–240.]

Iteravis huchzermeyeri [Avialae]

> *Iteravis huchzermeyeri* Zhou *et al.*, 2014 [Early Cretaceous; Yixian Formation, Liaoning, China]

Generic name: *Iteravis* ← {(Lat.) iter: journey + avis: bird} [*1].

Specific name: *huchzermeyeri* ← {(person's name) Huchzermeyer + -ī}; "in honor of the late archosaur biologist, Dr. Fritz Huchzermeyer, and his endless quest for knowledge" [*1].

Etymology: Huchzermeyer's **journey bird**

Taxonomy: Aves: Ornithothoraces: Ornithuromorpha [*1]

Notes: The specimen preserves feather impressions. *Iteravis* seems to have a short fork-like tail. [*1]

[References: (*1) Zhou S, O'Connor JK (2014). A new species from an ornithuromorph dominated locality of the Jehol Group. *Chinese Science Bulletin* 59(36): 5366–5378.]

Iuticosaurus valdensis

> *Titanosaurus valdensis* von Huene, 1929 ⇒ *Iuticosaurus valdensis* (von Huene, 1929) le Loeuff *et al.*, 1993 [Barremian; Wessex Formation, Wealden Group, Isle of Wight, UK]

Generic name: *Iuticosaurus* ← {(demonym) Iuti: the Jutes + (Gr.) σαῦρος/sauros: lizard}

Specific name: *valdensis* ← {(place-name) Wealden + -ensis}

Etymology: **Jute lizard** from Wealden

Taxonomy: Saurischia: Sauropoda: Titanosauria

Notes: According to *The Dinosauria 2nd edition* (2004), this genus is described *Titanosaurus valdensis* von Huene, 1929b (type of *Iuticosaurus valdensis* Le Loeuff, 1993) and a *nomen dubium*.

J

Jainosaurus septentrionalis

> *Antarctosaurus septentrionalis* von Huene & Matley, 1993⇒ *Jainosaurus septentrionalis* (von Huene & Matley, 1933) Hunt *et al*., 1995 [Maastrichtian; Lameta Formation, Madhya Pradesh, India]

Generic name: *Jainosaurus* ← {(person's name) Jain + (Gr.) σαῦρος/sauros: lizard}; in honor of Indian paleontologist Sohan Lal Jain.

Specific name: *septentrionalis* ← {(Lat.) septentriōnālis*; northern}

*[septem: seven + trio: plough-ox]; referring to the seven stars of Ursa Major (*i.e.*, the big dipper or the plough), which in Roman times was a convenient way to refer to "northern". (e.g., used in Caesar's "Gallic Wars"; see Hower, 1951).

Etymology: northern **Jain's lizard**

Taxonomy: Saurischia: Sauropoda: Titanosauria

Notes: *Antarctosaurus septentrionalis* von Huene & Matley, 1933 from India, whose fragment is very closely comparable with and similar to the posterior part of the skull of *Antarctosaurus wichmannianus* von Huene, 1929 from Patagonia, was renamed *Jainosaurus septentrionalis.* (*1)

[References: (*1) Wilson JA, D'Emic MD, Curry Rogers CA, Mohabey DM, Sen S (2009). Reassessment of the sauropod dinosaur *Jainosaurus* (="*Antarctosaurus*") *septentrionalis* from the Upper Cretaceous of India. *Contributions from the Museum of Paleontology, University of Michigan* 32(2): 17–40.]

Jaklapallisaurus asymmetrica

> *Jaklapallisaurus asymmetrica* Novas *et al*., 2011 [Norian; upper Maleri Formation, Andhra Pradesh, India]

Generic name: *Jaklapallisaurus* ← {(place-name) Jaklapalli: the Indian town + (Gr.) σαῦρος/sauros: lizard} (*1)

Specific name: *asymmetrica* ← {(Gr.) ἀ-σύμμετρος/a-symmetros: unsymmetrical}; "in allusion to the strongly asymmetrical astragalus of this species in distal view" (*1).

Etymology: asymmetrical, **Jaklapalli lizard**

Taxonomy: Saurischia: Sauropodomorpha: Unaysauridae*

*Unaysauridae Müller *et al*., 2018 includes *Jaklapallisaurus*, *Macrocollum* and *Unaysaurus*.

[References: (*1) Novas FE, Ezcurra MD, Chatterjee S, Kutty TS (2011). New dinosaur species from the Upper Triassic Maleri and Lower Dharmaram formations of central India. *Earth and Environmental Science Transactions of the Royal Society of Edinburgh* 101: 333–349.]

Janenschia robusta

> *Gigantosaurus robustus* Fraas, 1908 ⇒ *Tornieria robusta* (Fraas, 1908) Sternfeld, 1911 ⇒ *Barosaurus robustus* (Fraas, 1908) Haughton, 1928 ⇒ *Janenschia robusta* (Fraas, 1908) Wild, 1991 [Tithonian; Tendaguru Formation, Mtwara, Tanzania]

Generic name: *Janenschia* ← {(person's name) Janensch + -ia}; in honor of German paleontologist and geologist Werner Janensch (1878–1969). He contributed to the expeditions to Tanzania, collecting fossils from Tendaguru Beds.

Specific name: *robusta*, *f.* ← {(Lat.) rōbustus, *m.*: solid, robust}; referring to robust limb bones.

Etymology: robust **one for Janensch**

Taxonomy: Saurischia: Sauropodomorpha: Sauropoda

cf. Saurischia: Sauropodomorpha: Sauropoda: Titanosauriformes: Titanosauria (acc. *The Dinosauria 2nd edition*)

Jaxartosaurus aralensis

> *Jaxartosaurus aralensis* Riabinin, 1938 [Coniacian–Santonian; Dabrazhin Formation, Asht, Tajikistan]

= *Procheneosaurus convincens* Rozhdestvensky, 1968 ⇒ *Kazaklambia convincens* (Rozhdestvensky, 1968) Bell & Brink, 2013 [Coniacian; Dabrazhin Formation, Ongtustik Qazaqstan, Kazakhstan]

Generic name: *Jaxartosaurus* ← {(Gr. place-name) ὁ Ιαξάρτης/ho Iaxartēs (= ancient name for the Syr Darya) + (Gr.) σαῦρος/sauros: lizard}

Specific name: *aralensis* ← {(place-name) Aral + -ensis}

Etymology: **Jaxartes River (the Syr Darya) lizard** from Aral

Taxonomy: Ornithischia: Ornithopoda: Hadrosauridae: Lambeosaurinae

Other species:

> *Jaxartosaurus fuyunensis* Wu, 1984 (*nomen dubium*) [Late Cretaceous; Ulungurhe Formation, Xinjiang, China]

Notes: *Jaxartosaurus* had a large helmet-like crest. According to *The Dinosauria 2nd edition, Procheneosaurus convincens* was considered to be a synonym of *Jaxartosaurus aralensis*, but *P. convincens* was proposed its own genus *Kazaklambia* by Bell & Brink (2013).

[References: Riabinin AN (1938). [Some results of the studies of the Upper Cretaceous dinosaurian fauna from the vicinity of the station Sary-Agach, South Kazakhstan. *Problems of Paleontology*] 4: 125–135.]

Jeholornis prima [Avialae]

> *Jeholornis prima* Zhou & Zhang, 2002 [Aptian; Jiufotang Formation, Liaoning, China]

= *Shenzhouraptor sinensis* Ji *et al.*, 2002 (the same month as *Jeholornis*) [early Aptian; Dakangpu Member, Yixian Formation, Liaoning, China]

Generic name: *Jeholornis* ← {(place-name) Jehol + (Gr.) ὄρνις/ornis: bird}; referring to the " 'Jehol Group', which contains the Jehol Biota" (*1).

Specific name: *prima* ← {(Lat.) prīmus: the first, first}; in reference to "its primitive appearance in the tail" (*1).

Etymology: primitive, **Jehol bird** [原始熱河鳥]

Taxonomy: Theropoda: Maniraptora: Avialae
cf. Aves (*1)

Other species:

> *Jeholornis palmapenis* O'Connor *et al.*, 2012 [early Aptian; Jiufotang Formation, Liaoning, China]

> *Jeholornis curvipes* Lefèvre *et al.*, 2014 [early Aptian; Dakangpu Member, Yixian Formation, Liaoning, China]

Notes: According to Zou & Zhang (2002) (2003), the seeds which were found in the stomach of *Jeholornis* represent the evidence for seed-eating adaptation in the Mesozoic.

[References: (*1) Zhou Z, Zhang F (2002). A long-tailed, seed-eating bird from the Early Cretaceous of China. *Nature* 418: 405–409.]

Jeholosaurus shangyuanensis

> *Jeholosaurus shangyuanensis* Xu *et al.*, 2000 [early Aptian; Lujiatun Member, Yixian Formation, Shangyuan, Beipiao, Liaoning, China]

Generic name: *Jeholosaurus* ← {place-name) Jehol (= Rehol) 熱河 + (Gr.) σαῦρος/sauros: lizard}; referring to " 'Jehol', old geographic name for western Liaoning and northern Hebei" (*1).

Specific name: *shangyuanensis* ← {(Chin. place-name) Shàngyuán上園 + -ensis}; referring to "the larger geographical area including the type locality" (*1).

Etymology: **Jehol lizard** from Shangyuan

Taxonomy: Ornithischia: Ornithopoda:
Jeholosauridae

Notes: The specimens represent juvenile.

[References: (*1) Xu X, Wang X-L, You H-L (2000). A primitive ornithopod from the Early Cretaceous Yixian Formation of Liaoning. *Vertebrata PalAsiatica* 38: 318–325.]

Jeyawati rugoculus

> *Jeyawati rugoculus* McDonald *et al.*, 2010 [Turonian; Moreno Hill Formation, New Mexico, USA]

Generic name: *Jeyawati* ← {(Zuni*) jeyawati [pron.: HEY-a-WHAT-ee] "grinding mouth"}; referring to "the sophisticated chewing mechanism evolved by the herbivorous lineage to which *Jeyawati* belongs" (*1).

*"the language of Zuni people, a Native American tribe located around the Zuni River in western New Mexico" (*1).

Specific name: *rugoculus* ← {(Lat.) rūga: wrinkle + oculus: eye}; referring to "a rugose texture that covers the entire lateral surface of the postorbital" (*2).

Etymology: wrinkle-eyed **grinding mouth**

Taxonomy: Ornithischia: Ornithopoda:
Hadrosauroidea: Hadrosauromorpha

Notes: Several of the rib fragments indicate that the animal suffered broken ribs in its life and the injuries healed (*1).

[References: (*1) University of Pennsylvania "New species of plant-eating dinosaur named for 'grinding mouth and wrinkle eye'." Science Daily, 1 June 2010. 〈https://www.sciencedaily.com/releases/2010/05/100526111330.htm〉; (*2) McDonald AT, Wolfe DG, Kirkland JI (2010). A new basal hadrosauroid (Dinosauria: Ornithopoda) from the Turonian of New Mexico. *Journal of vertebrate Paleontology* 30(3): 799–812.]

Jianchangornis microdonta [Avialae]

> *Jianchangornis microdonta* Zhou *et al.*, 2009 [early Aptian; Jiufotang Formation, Gehol Group, Jiangchang, Huludao, Liaoning, China]

Generic name: *Jianchangornis* ← {(Chin. place-name) Jiànchāng 建昌: Jianchang County + (Gr.) ὄρνις/ornis: bird} (*1)

Specific name: *microdonta* ← {(Gr.) μῑκρός/mīkros: small + ὀδών/odōn: tooth}; "indicating small teeth on the dentary" (*1).

Etymology: **Jianchang bird** with small teeth

Taxonomy: Theropoda: Avialae: Ornithothoraces:
Euornithes
cf. Aves: Pygostylia: Ornithurae (*1)

[References: (*1) Zhou Z-H, Zhang F-C, Li Z-H (2009). A new basal ornithurine bird (*Jianchangornis microdonta* gen. *et* sp. nov.) from the Lower Cretaceous of China. *Vertebrata PalAsiatica* 47(4): 299–310.]

Jianchangosaurus yixianensis

> *Jianchagosaurus yixianensis* Pu *et al.*, 2013 [early Aptian; Yixian Formation, Jehol Group, Jianchang, Liaoning, China]

Generic name: *Jianchangosaurus* ← {(Chin. place-name) Jiànchāng建昌: the county of Liaoning Province + (Gr.) σαῦρος/sauros: lizard}

Specific name: *yixianensis* ← {(Chin. place-name) Yìxiàn義縣 + -ensis}; "referring to the formation which yielded this specimen".

Etymology: **Jianchang lizard** from Yixian Formation

Taxonomy: Theropoda: Coelurosauria: Therizinosauria

[References: Pu H, Kobayashi Y, Lü J, Xu L, Wu Y, Chang H, Zhang J, Jia S (2013). An unusual basal Therizinosaur dinosaur with an Ornithischian dental arrangement from northeastern China. *PLoS ONE* 8(5): e63423.]

Jiangjunosaurus junggarensis

> *Jiangjunosaurus junggarensis* Jia *et al.*, 2007 [Oxfordian; Shishugou Formation, Jiangjunmiao, Junggar Basin, Xinjiang, China]

Generic name: *Jiangjunosaurus* ← {(Chin.) Jiāngjūn将軍: 'General' + (Gr.) σαῦρος/sauros: lizard}; in reference to the deserted town of Jiangjunmiao near the holotype locality, named for an ancient general who died nearby (*1).

Specific name: *junggarensis* ← {(place-name) Junggar + -ensis}; derived from the Junggar Basin, the large geological area that includes the type locality (*1).

Etymology: **General lizard** from Junggar Basin

Taxonomy: Ornithischia: Stegosauria: Stegosauridae

[References: (*1) Jia C, Forster CA, Xu X, Clark JM (2007). The First stegosaur (Dinosauria, Ornithischia) from the Upper Jurassic Shishugou Formation of Xinjiang, *China. Acta Geologica Sinica* 81(3): 351–356.]

Jiangshanosaurus lixianensis

> *Jiangshanosaurus lixianensis* Tang *et al.*, 2001 [Albian; Jinhua Formation, Jiangshan, Quzhou, Zhejiang, China]

Generic name: *Jiangshanosaurus* ← {(Chin. place-name) Jiāngshān江山 + (Gr.) σαῦρος/sauros: dinosaur}; after the fossil-producing Lixian Village, Jiangshan County.

Specific name: *lixianensis* ← {(Chin. place-name) Lǐxián禮賢 + -ensis}

Etymology: **Jiangshan County lizard** from Lixian Village

Taxonomy: Saurischia: Sauropodomorpha: Sauropoda: Titanosauria

[References: Tang F, Kang X-M, Jin X-S, Wei F, Wu W-T (2001). A new sauropod dinosaur of Cretaceous from Jiangshan, Zhejiang Province. *Vertebrata PalAsiatica* 39(4): 272–281.]

Jiangxisaurus ganzhouensis

> *Jiangxisaurus ganzhouensis* Wei *et al.*, 2013 [Maastrichtian; Nanxiong Formation, Ganzhou, Jiangxi, China]

Generic name: *Jiangxisaurus* ← {(Chin. place-name) Jiāngxī江西: the Jiangxi Province of southern China + (Gr.) σαῦρος/sauros}; referring to "the Chinese administrative unit Jiangxi Province of southern China" (*1).

Specific name: *ganzhouensis* ← {(Chin. place-name) Gànzhōu贛州 + -ensis}

Etymology: **Jiangxi lizard** from Ganzhou

Taxonomy: Theropoda: Oviraptorosauria: Oviraptoridae

[References: (*1) Wei X, Pu H, Xu L, Liu D, Lü J (2013). A new oviraptorid dinosaur (Theropoda: Oviraptorosauria) from the Late Cretaceous of Jiangxi Province, southern China. *Acta Geologica Sinica (English Edition)* 87(4): 899–904.]

Jianianhualong tengi

> *Jianianhualong tengi* Xu *et al.*, 2017 [early Aptian; Dakangpu Member, Yixian Formation, Liaoning, China]

Generic name: *Jianianhualong* ← {(Chin. company) Jiāniánhuā嘉年華 + (Chin.) lóng龍: dragon}; referring to "the company that supported this study" (*1).

Specific name: *tengi* ← {(Chin. person's name) Teng + -ī}; in honor of "Ms Fangfang Teng, who secured the specimen for study" (*1).

Etymology: Ms Teng's **Jianianhua dragon**

Taxonomy: Theropoda: Coelurosauria: Maniraptora: Troodontidae

[References: (*1) Xu X, Currie P, Pittman M, Xing L, Meng Q, Lü J, Hu D, Yu C (2017). Mosaic evolution in an asymmetrically feathered troodontid dinosaur with transitional features. *Nature Communications* 8: 14972.]

Jinbeisaurus wangi

> *Jinbeisaurus wangi* Wu *et al.*, 2019 [Late Cretaceous; Huiquanpu Formation, Shanxi, China]

Generic name: *Jinbeisaurus* ← {(Chin. place-name) Jìnběi晉北 + (Gr.) σαῦρος/sauros: lizard}; "derived from the Chinese 'jin'晉: the abbreviation of Shanxi Province and 'bei'北: north, from which the material was recovered, with the suffix 'saurus' from the Latin for 'lizard'" (*1).

Specific name: *wangi* ← {(Chin. person's name) Wáng王 + -ī}; "in honour of Mr. Suozhu Wang for his great contribution to the vertebrate paleontology of Shanxi Province, including organizing field explorations and discovering

many Mesozoic fossils of reptiles within the province" (*1).
Etymology: Wang's **Jinbei lizard**
Taxonomy: Theropoda: Tyrannosauroidea

[References: (*1) Wu X-C, Shi J-R, Dong L-Y, Carr TD, Yi J, Xu S-C (2019). A new tyrannosauroid from the Upper Cretaceous of Shanxi, China. *Cretaceous Research* 108: 104357.]

Jinfengopteryx elegans

> *Jinfengopteryx elegans* Ji *et al.*, 2005 [Barremian; Qiaotou Member, Huajiying Formation, Hebei, China]

Generic name: *Jinfengopteryx* ← {(Chin. folklore) Jīnfèng金鳳: 'golden phoenix' + (Gr.) πτέρυξ/pteryx: feather}
Specific name: *elegans* ← {(Lat.) ēlegāns: delicate, elegant}
Etymology: elegant, **Jinfeng (golden phoenix) feather** [華美金鳳鳥]
Taxonomy: Theropoda: Coelurosauria: Deinonychosauria: Troodontidae
Notes: According to the authors, *Jinfengopteryx* was considered "a new well-preserved avialian bird" (*1).

[References: (*1) Ji Q, Ji S, Lu J, You H, Chen W, Liu Y, Liu Y (2005). First avialian bird from China (*Jinfengopteryx elegans* gen. *et* sp. nov.) *Geological Bulletin of China* 24(3): 197–205.]

Jingshanosaurus xinwaensis

> *Jingshanosaurus xinwaensis* Zhang and Yang, 1995 [Hettangian; Shawan Member, Lufeng Formation, Xinwa, Jingshan, Yunnan, China]
= *Chuxiongosaurus lufengensis* Lü *et al.*, 2010 [Hettangian; Shawan Member, Lufeng Formation, Zhongcun, Chuxiong, Yunnan, China] (according to Zhang *et al.* 2019)

Generic name: *Jingshanosaurus* ← {(Chin. place-name) Jīngshān金山 'Golden Hill': name of the town + (Gr.) σαῦρος/sauros: lizard}
Specific name: *xinwaensis* ← {(Chin. place-name) Xīnwā新窪: name of the village + -ensis}
Etymology: **Jingshan lizard** from Xinwa
Taxonomy: Saurischia: Sauropodomorpha: Sauropodiformes

Jinguofortis perplexus [Avialae]

> *Jinguofortis perplexus* Wang *et al.*, 2018 [Early Cretaceous; Dabeigou Formation, Hebei, China]

Generic name: *Jinguofortis* ← {(Chin.) jīnguó 巾幗: referring to female warrior (*1) + (Lat.) fortis: brave, strong}; "in honor of woman scientists around the world" (*2).
Specific name: *perplexus* ← {(Lat.) perplexus: interwoven, tangled}; "referring to the combination of plesiomorphic and derived characters present in the holotype specimen" (*1).
Etymology: tangled, **brave female warrior**
Taxonomy: Theropoda: Avialae: Pygostilia: Jinguofortisidae
cf. Aves: Pygostylia: Jinguofortisidae (*1)

[References: (*1) Wang M, Stidham TA, Zhou Z (2018). A new clade of basal Early Cretaceous pygostylian birds and developmental plasticity of the avian shoulder girdle. *Proceedings of the National Academy of Sciences of the United States of America* 115(42): 10708–10713.; (*2) Chinese Academy of Scienses (September 24, 2018). Chinese Cretaceous fossil highlights avian evolution.]

Jintasaurus meniscus

> *Jintasaurus meniscus* You & Li, 2009 [Aptian; Xiagou Formation, Xinminpu Group, Jinta, Gansu, China]

Generic name: *Jintasaurus* ← {(Chin. place-name) Jīntǎ金塔: 'Golden temple' + (Gr.) σαῦρος/sauros: lizard}; referring to "the county where the fossil was discovered" (*1).
Specific name: *meniscus* ← {(Gr.) μηνίσκος/mēniskos: crescent}; referring to "the crescentic paroccipital processes and occipital condyle of the specimen" (*1).
Etymology: crescent, **Jinta County lizard**
Taxonomy: Ornithischia: Ornithopoda: Iguanodontia: Hadrosauriformes: Hadrosauroidea: Hadrosauromorpha

[References: (*1) You H-L, Li D-Q (2009). A new basal hadrosauriform dinosaur (Ornithischia: Iguanodontia) from the Early Cretaceous of northwestern China. *Canadian Journal of Earth Sciences* 46(12): 949–957.]

Jinyunpelta sinensis

> *Jinyunpelta sinensis* Zheng *et al.*, 2018 [Albian–Cenomanian; Liangtoutang Formation, Jinyun, Zhejiang, China] (*1)

Generic name: *Jinyunpelta* ← {(Chin. place-name) Jìnyún縉雲: Jinyun County + (Lat.) pelta: a small shield}; "in reference to the osteoderms found on all ankylosaurians" (*1).
Specific name: *sinensis* ← {(Gr.) 'sin': China + -ensis}; referring to China, the country of origin (*1).
Etymology: **Jinyun shield (ankylosaur)** from China
Taxonomy: Ornithischia: Thyreophora: Eurypoda: Ankylosauria: Ankylosauridae: Ankylosaurinae

[References: (*1) Zheng W, Jin X, Azuma Y, Wang Q, Miyata K, Xu X (2018). The most basal ankylosaurine dinosaur from the Albian–Cenomanian of China, with implications for the evolution of the tail club. *Scientific Reports* 8(3711).]

Jinzhousaurus yangi

> *Jinzhousaurus yangi* Wang & Xu, 2001 [early Aptian; Dakangpu Member, Yixian Formation, Jinzhou, Liaoning, China]

Generic name: *Jinzhousaurus* ← {(Chin. place-name) Jǐnzhōu錦州 + (Gr.) σαῦρος/sauros: lizard}

Specific name: *yangi* ← {(Chin. person's name) Yáng楊 + -ī}; "in honor of Yáng Zhōngjián 楊鍾健 who is the founder of the vertebrate paleontology in China"(*1).

Etymology: Yang's **Jinzhou lizard**

Taxonomy: Ornithischia: Ornithopoda: Hadrosauroidea

[References: (*1) Wang X-L, Xu X (2001). A new iguandontid (*Jinzhousaurus yangi* gen. *et* sp. nov.) from the Yixian Formation of western Liaoning, China. *Chinese Science Bulltin* 46(19): 1669–1672.]

Jiuquanornis niui [Avialae]

> *Jiuquanornis niui* Wang *et al*., 2013 [Aptian; Xiagou Formation, Xinminpu Group, Changma Basin, Gansu, China]

Generic name: *Jiuquanornis* ← {(Chin. place-name) Jiǔquán酒泉: the name of the city + (Gr.) ὄρνις/ornis: bird} (*1)

Specific name: *niui* ← {(Chin. person's name) Niu + -ī}; "dedicated to Professor Shao-Wu Niu, for his contribution to geological research in the Changma Basin"(*1).

Etymology: Prof. Niu's **Jiuquan bird**

Taxonomy: Theropoda: Avialae: Ornithuromorpha

[References: (*1) Wang Y-M, O'Connor JK, Li D-Q, You H-L (2013). Previously unrecognized Ornithuromorph bird diversity in the Early Cretaceous Changma Basin, Gansu Province, Northwestern China. *PLoS ONE* 8(10): e77693.]

Jiutaisaurus xidiensis

> *Jiutaisaurus xidiensis* Wu *et al*., 2006 [Aptian–Cenomanian; Quantou Formation, Jiutai, Jilin, China]

Generic name: *Jiutaisaurus* ← {(Chin. place-name) Jiǔtái九台 + (Gr.) σαῦρος/sauros: lizard}

Specific name: *xidiensis* ← {(Chin. place-name) Xīdì西地 + -ensis}

Etymology: **Jiutai lizard** from Xidi [西地九台龍]

Taxonomy: Saurischia: Sauropodomorpha: Sauropoda: Titanosauriformes

[References: Wu W-H, Dong Z-M, Sun Y-W, Li C-T, Li T (2006). A new sauropod dinosaur from the Cretaceous of Jiutai, Jilin, China. *Global Geology* 25(1): 6–9.]

Jixiangornis orientalis [Avialae]

> *Jixiangornis orientalis* Ji *et al*., 2002 [early Aptian; Dakangpu Member, Yixian Formation, Liaoning, China]

Generic name: *Jixiangornis* ← {(Chin.) Jíxiáng 吉祥: lucky + (Gr.) ὄρνις/ornis: bird}

Specific name: *orientalis* ← {(Lat.) orientālis: oriental, eastern}

Etymology: oriental, **lucky bird** [東方吉祥鳥]

Taxonomy: Theropoda: Avialae: Euavialae

[References: Ji Q, Ji S, Zhang H, You H, Zhang J, Wang L, Yuan C, Ji X (2002). A new avialian bird—*Jixiangornis orientalis* gen. *et* sp. nov. from the Lower Cretaceous of Western Liaoning. NE China. *Journal of Nanjing University (Natural Sciences)* 38(6): 723–736 [in Chinese with English abstract.]]

Jobaria tiguidensis

> *Jobaria tiguidensis* Sereno *et al*., 1999 [Bathonian; Tiourarén Formation, Agadez, Niger]

Generic name: *Jobaria* ← {(Tuareg folklore) Jobar* + -ia: pertaining to}; "named after the mythical creature 'Jobar', to whom local Tuaregs had attributed the exposed bones"(*1).

*According to Sereno, "In Western cultures, the equivalent to Jobar would be the bogeyman, the evil, unseen spirit lurking in the imaginations of children." (by Mullen W, Chicago Tribune, Nov. 12, 1999)

Specific name: *tiguidensis* ← {(Tuareg, place-name) Tiguidi + -ensis: from}; "after the Falaise de Tiguidi, a cliff near the base of which lie the horizons yielding all of its remains" (*1).

Etymology: **Jobar's one** from Tiguidi cliff

Taxonomy: Saurischia: Sauropodomorpha: Sauropoda: Diplodocoidea

[References: (*1) Sereno PC, Beck AL, Dutheil DB, Larsson HCE, Lyon GH, Moussa B, Sadleir RW, Sidor CA, Varricchio DJ, Wilson GP, Wilson JA (1999). Cretaceous sauropods from the Sahara and the uneven rate of skeletal evolution among dinosaurs. *Science* 286: 1342–1347.]

Jubbulpuria tenuis

> *Jubbulpuria tenuis* von Huene & Matley, 1933 [Maastrichtian; Lameta Formation, Madhya Pradesh, India]

Generic name: *Jubbulpuria* ← {(place-name) Jubbulpore [= Jabalpur] + -ia}

Specific name: *tenuis* ← {(Lat.) tenuis: slender}.

Etymology: slender **Jubbulpore (Jabalpur) one**

Taxonomy: Theropoda: Ceratosauria

Notes: According to *The Dinosauria 2nd edition*, *Jubbulpuria* is considered to be a *nomen dubium*.

Judiceratops tigris

> *Judiceratops tigris* Longrich, 2013 [middle Campanian; McClelland Ferry Member, Judith River Formation, Montana, USA]

Generic name: *Judiceratops* ← {(place-name) Judith River Formation + (Gr.) κέρας/keras: horn + ὤψ/ōps: face}

Specific name: *tigris* ← {(Lat.) tigris: a tiger}; referring to the mascot of Prinston University. The researchers are from Prinston.

Etymology: tiger's **Judith horned face (ceratopsian)**

Taxonomy: Ornithischia: Ceratopsia: Ceratopsidae: Chasmosaurinae

[References: Longrich NR (2013). *Judiceratops tigris*, a new horned dinosaur from the middle Campanian Judith River Formation of Montana. *Bulletin of the Peabody Museum of Natural History* 54(1): 51–65.]

Judinornis nogontsavensis [Avialae]

> *Judinornis nogontsavensis* Nessov & Borkin, 1983 [early Maastrichtian; Nemegt Svita, Bayankhongor, Mongolia]

Generic name: *Judinornis* ← {(person's name) Judin + (Gr.) ὄρνις/ornis: bird}; in honor of Konstantin Alekseyevich Yudin (= Judin).

Specific name: *nogontsavensis* ← {(place-name) Nogon Tsav: locality + -ensis}

Etymology: **Judin's bird** from Nogon Tsav

Taxonomy: Theropoda: Avialae: Ornithurae: Hesperornithes

Notes: *Judinornis* is a flightless bird genus.

Juehuaornis zhangi [Avialae]

> *Juehuaornis zhangi* Wang *et al*., 2015 (*1) [Aptian; Jiufotang Formation, Liaoning, China]

Generic name: *Juehuaornis* ← {(Chin. place-name) the island of Juéhuā覚華: the "Chrysanthemum island" off the coast of Liaoning + (Gr.) ὄρνις/ornis: bird}

Specific name: *zhangi* ← {(Chin. person's name) Zhang + -ī}; in honor of curator Zhāng Dàyǒng 張大勇.

Etymology: Zhang's **Juehua bird** [張氏覚華鳥] (*1)

Taxonomy: Theropoda: Avialae: Ornithothoraces: Ornithuromorpha

Notes: *Juehuaornis, Dingavis* and *Changzuiornis* were found in the same Jiufotang Formation.

[References: (*1) Wang R-F, Wang Y, Hu D-Y (2015). Discovery of a new ornithuromorph genus, *Juehuaornis* gen. nov. from Lower Cretaceous of western Liaoning, China. *Global Geology* (1): 7–11.]

Junornis houi [Avialae]

> *Junornis houi* Liu *et al*., 2017 [Early Cretaceous; Yixian Formation, Ningcheng, Inner Mongolia, China]

Generic name: *Junornis* ← {(Chin.) jùn 俊: beautiful + (Gr.) ὄρνις/ornis: bird} (*1)

Specific name: *houi* ← {(Chin. person's name) Hou + -ī}; in honor of "Dr. Hóu Liánhǎi侯連海 and his important contributions to Chinese paleornithology" (*1).

Etymology: Dr. Hou's **beautiful bird**

Taxonomy: Theropoda: Avialae: Ornithothoraces: Enantiornithes

cf. Aves: Pygostylia: Ornithothoraces: Enantiornithes (*1)

[References: (*1) Liu D, Chiappe LM, Serrano F, Habib M, Zhang Y, Meng Q (2017). Flight aerodynamics in enantiornithines: Information from a new Chinese Early Cretaceous bird. *PLoS ONE* 12(10): e0184637.]

Juratyrant langhami

> *Stokesosaurus langhami* Benson, 2008 ⇒ *Juratyrant langhami* (Benson, 2008) Brusatte & Benson, 2013 [Tithonian; Kimmeridge Clay, Dorset, UK]

Generic name: *Juratyrant* ← {[geol.] Jura: the Jurassic age + (Eng.) tyrant [← (Gr.) τύραννος/tyrannos | (Lat.) tyrannus]}; "in reference to the vernacular characterization of tyrannosauroids as 'tyrant dinosaurs' (based on the original etymology of *Tyrannosaurus rex*)" (*1).

Specific name: *langhami* ← {(person's name) Langham + -ī}; in honor of Peter Langham, who uncovered the specimen.

Etymology: Langham's **Jurassic tyrant**

Taxonomy: Theropoda: Coelurosauria: Tyrannosauroidea

Notes: A second species, *Stokesosaurus langhami* was given its own new genus, *Juratyrant*, in 2013 (*1). According to Brusatte & Carr (2016), *Stokesosaurus* and *Juratyrant* were analized more advanced than the Proceratosauridae.

[References: (*1) Brusatte SL, Benson RBJ (2013). The systematic of Late Jurassic tyrannosauroid theropods from Europe and North America. *Acta Palaeontologica Polonica* 58(1): 47–54.]

Juravenator starki

> *Juravenator starki* Göhlich & Chiappe, 2006 [late Kimmeridgian; Painten Formation, Bayern (=Bavaria), Germany]

Generic name: *Juravenator* ← {(German, place-name) Jura | (Lat.) Jūra + (Lat.) vēnātor: hunter}; "referring to the Bavarian Jura Mountains" (*1).

Specific name: *starki* ← {(person's name) Stark + -ī}; in honor of "the family Stark, owners of the Quarry Stark" (*1).

Etymology: Stark Family's **Jura hunter**

Taxonomy: Theropoda: Coelurosauria: Tyrannoraptora

cf. Theropoda: Tetanurae: Coelurosauria: Compsognathidae (*1)

Notes: The holotype is a juvenile, which is estimated to be 75–80 cm in length. *Juravenator*, a small bipedal predator, was originally classified as a member of the Compsognathidae (*1). *Juravenator* may have been nocturnal (*2). Nodes on epidermal scales on the tail of *Juravenator* suggest that the tail had a sensory function (*3). According to Foth *et al.* (March, 2020), the tubercle-like structures had been reinterpreted as remains of adipocere.

[References: (*1) Göhlich UB, Chiappe LM (2006). A new carnivorous dinosaur from the Late Jurassic Solnhofen archipelago. *Nature* 440: 329–332.; (*2) Schmitz L, Motani R (2011). Nocturnality in dinosaurs inferred from scleral ring and orbit morphology. *Science* 332(6030): 705–708.; (*3) Bell PR, Hendrickx C (2020). Crocodile-like sensory scales in a Late Jurassic theropod dinosaur. *Current Biology* 30: R1068–R1070.]

K

Kaatedocus siberi

> *Kaatedocus siberi* Tschopp & Mateus, 2012 [Kimmeridgian; Morrison Formation, Wyoming, USA]

Generic name: *Kaatedocus* ← {(Crow / Absaroka) kaate: small + (Gr.) δοκός/dokos: beam, alluding to *Diplodocus*} (*1)

The term *docus* is the common suffix to denote members of the diplodocid sauropod.

Specific name: *siberi* ← {(person's name) Siber + -ī}; in honor of "Hans-Jakob 'Kirby' Siber, b. 1942, doctor honoris causa of the University of Zurich, Switzerland. Siber is the founder and director of the Sauriermuseum Aathal, Switzerland, and organized and funded the excavation, preparation and curation of the holotype specimen of *Kaatedocus siberi*" (*1).

Etymology: Siber's **small beam (*Diplodocus*)**

Taxonomy: Saurischia: Sauropodomorpha: Sauropoda: Diplodocidae: Diplodocinae

cf. Sauropoda: Eusauropoda: Neosauropoda: Diplodocoidea: Flagellicaudata: Diplodocidae (*1)

Notes: *Kaatedocus* had a long neck and a long tail.

[References: (*1) Tschopp E, Mateus OV (2012). The skull and neck of a new flagellicaudatan sauropod from the Morrison Formation and its implication for the evolution and ontogeny of diplodocid dinosaurs. *Journal of Systematic Palaeontology* 11(7): 853–888.]

Kaijiangosaurus lini

> *Kaijiangosaurus lini* He, 1984 [Bathonian–Callovian; lower Shaximiao Formation, Sichuan, China]

Generic name: *Kaijiangosaurus* ← {(Chin. place-name) Kāijiāng開江+ (Gr.) σαῦρος/sauros: lizard}

Specific name: *lini* ← {(Chin. person's name) Lín林 + -ī}; in honor of the paleontologist Lin Wenqiu.

Etymology: Lin's **Kaijiang lizard** [林氏開江龍]

Taxonomy: Theropoda: Tetanurae

Kaijutitan maui

> *Kaijutitan maui* Filippi *et al.*, 2019 [Coniacian; Sierra Barrosa Formation, Neuquén Basin, Argentina]

Generic name: *Kaijutitan* ← {(Jpn.) kaijū怪獣: monster + (Gr.) Τῑτάν/tītan: giant}

Specific name: *maui* ← {(acronym) MAU: Museo Municipal Argentino Urquiza, Rincón de los Sauces, Neuquén, Argentina + -ī}

Etymology: Museo Municipal Argentino Urquiza's **monster giant**

Taxonomy: Saurischia: Sauropoda: Titanosauriformes: Somphospondyli: Titanosauria

[References: Filippi LS, Salgado L, Garrido AC (2019). A new giant basal titanosaur sauropod in the Upper Cretaceous (Coniacian) of the Neuquén Basin, Argentina. *Cretaceous Research* 100: 61–81.]

Kakuru kujani

> *Kakuru kujani* Molnar & Pledge, 1980 [Aptian; Bulldog Shale, Andamooka, South Australia, Australia]

Generic name: *Kakuru* ← {(Guyani, legend) "Kakuru"}: referring to "the ancestral serpent" (*1). The gender is masculine. (*1)

The tibia of *Kakuru* was opalized like the colour of an ancestral Rainbow serpent.

Specific name: *kujani* ← {(demonym) Kujani or Guyani}; "the tribe that inhabited the Andamooka area" (*1).

Etymology: Kujani's **Kakuru**

Taxonomy: Theropoda: Coelurosauria: Maniraptoriformes: Maniraptora

cf. Theropoda: Coelurosauria *incertae sedis* (acc. *The Dinosauria 2nd edition*)

Notes: *Kakuru* is seemed to have been dug up in the opal fields of Andamooka.

[References: (*1) Molnar RE, Pledge NS (1980). A new theropod dinosaur from South Australia. *Alcheringa* 4: 281–287.]

Kamuysaurus japonicus

> *Kamuysaurus japonicus* Kobayashi *et al.*, 2019 [early Maastrichitian; Hakobuchi Formation, Mukawa, Hokkaido, Japan]

Generic name: *Kamuysaurus* ← {(Ainu) Kamuy: deity + (Gr.) σαῦρος/sauros: lizard (reptile)}

Specific name: *japonicus* ← {(Lat.) japonicus: of Japan}

Etymology: **God-like lizard** of Japan

Taxonomy: Ornithischia: Ornithopoda: Hadrosauridae: Hadrosaurinae

Notes: The tail was discovered by Mr. Yoshiyuki Horita in 2004, and noticed to be a possibly dinosaur by Dr. Tamaki Sato. A nearly complete skeleton was unearthed in Mukawa-cho, Hokaido. Its nickname is Mukawaryu, which means "dragon of Mukawa" in Japanese.

[References: Kobayashi Y, Nishimura T, Takasaki R, Chiba K, Fiorillo AR, Tanaka K, Chinzorig T, Sato T, Sakurai K (2019). A new hadrosaurine (Dinosauria: Hadrosauridae) from the marine deposits of the Late Cretaceous Hakobuchi Formation, Yezo Group, Japan. *Scientific Reports* 9(1): 1–14.]

Kangnasaurus coetzeei

> *Kangnasaurus coetzeei* Haughton, 1915 [Early Cretaceous; Bushmanland, Cape, South Africa]

Generic name: *Kangnasaurus* ← {(place-name) Kangnas Farm + (Gr.) σαῦρος/sauros: lizard} (*1)

Specific name: *coetzeei* ← {(person's name) Coetzee + -ī}; "in honor of Mr. Coetzee, the owner of the farm Kangnas, by whom the remains were first brought to Dr. Rogers" (*1).

Etymology: Coetzee's **Kangnas Farm lizard**

Taxonomy: Ornithischia: Ornithopoda: Dryosauridae

Notes: A tooth was found in a well at Farm Kangnas. According to *The Dinosauria 2nd edition*, *Kangnasaurus* was considered a *nomen dubium*.

[References: (*1) Haughton SH (1915). On some dinosaur remains from Bushmanland. *Transactions of the Royal Society of South Africa* 5: 259–264.]

Karongasaurus gittelmani

> *Karongasaurus gittelmani* Gomani, 2005 [Aptian; Dinosaur Beds Formation, Karonga, Northern, Malawi]

Generic name: *Karongasaurus* ← {(place-name) Karonga District + (Gr.) σαῦρος/sauros: lizard}

Specific name: *gittelmani* ← {(person's name) Steve Gittelman + -ī}; "in honor of Steve Gittelman, friend of science, for his work as president of the Dinosaur Society" (*1).

Etymology: Gittelman's **Karonga lizard**

Taxonomy: Saurischia: Sauropodomorpha: Sauropoda: Titanosauria

[References: (*1) Gomani EM (2005). Sauropod dinosaurs from the Early Cretaceous of Malawi, Africa. *Palaeontologia Electronica* 8(1): 1–37.]

Katepensaurus goicoecheai

> *Katepensaurus goicoecheai* Ibiricu *et al.*, 2013 [Cenomanian–Turonian; Bajo Barreal Formation, Chubut, Patagonia, Argentina]

Generic name: *Katepensaurus* ← {(Tehuelche) katepenk: hole + (Gr.) σαῦρος/sauros: lizard}; "in reference to one of the most distinctive features on the new taxon, the fenestrae in the dorsal vertebral transverse processes" (*1).

Specific name: *goicoecheai* ← {(person's name) Goicoechea + -ī}; "in honor of Mr. Alejandro Goicoechea, owner of the Estancia Laguna Palacios" (*1).

Etymology: Goicoechea's **hole lizard**

Taxonomy: Saurischia: Sauropodomorpha: Sauropoda: Diplodocoidea: Rebbachisauridae

[References: (*1) Ibiricu LM, Casal GA, Martinez RD, Lamanna MC, Luna M, Salgado L (2013). *Katepensaurus goicoecheai*, gen. *et* sp. nov., a Late Cretaceous rebbachisaurid (Sauropoda, Diplodocoidea) from central Patagonia, Argentina. *Journal of Vertebrate Paleontology* 33(6): 1351–1366.]

Kayentavenator elysiae

> *Kayentavenator elysiae* Gay, 2010 [Sinemurian–Pliensbachian; Kayenta Formation, Arizona, USA]

Generic name: *Kayentavenator* ← {(place-name) Kayenta Formation + (Lat.) vēnātor: hunter}

Specific name: *elysiae* ← {(person's name) Elysia + -ae}

Etymology: Elysia's **hunter of Kayenta Formation**

Taxonomy: Theropoda: Tetanurae

Notes: The holotype is seemed to be a juvenile because of its unfused neural spines (*1). This specimen was originally assigned to *Syntarsus kayentakatae* Rowe, 1989.

[References: (*1) Gay R (2010). *Kayentavenator elysiae,* a new tetanuran from the Early Jurassic of Arizona. In: Gay R (ed), *Notes on Early Mesozoic Theropods.* Lulu Press, Morrisville: 27–43.]

Kazaklambia convincens

> *Procheneosaurus convincens* Rozhdestvensky, 1968 ⇒ *Kazaklambia convincens* (Rozhdestvensky, 1968) Bell & Brink, 2013 [Coniacian–Santonian; Dabrazhin Formation, Syuk-Syuk, South Kazakhstan, Kazakhstan]

Generic name: *Kazaklambia* ← {(place-name) Kazak: Kazakhstan + lambia: "denoting the

affiliation to Lambeosaurinae"[*1]}
Specific name: *convincens* ← {(Lat.) convincens [convincō: convince]}; "referring to Rozhdestvensky's (1968) conviction that this specimen proved a Cretaceous age for the Dabrazinskaya Svita"[*1].
Etymology: convincing **Kazakh lambeosaurine**
Taxonomy: Ornithischia: Ornithopoda: Hadrosauridae: Lambeosaurinae
Notes: Rozhdestvensky (1968) named *"Procheneosaurus" convincens*, but it had not been admitted formally. Bell & Brink (2013) gave it a new name *Kazaklambia*. According to *The Dinosauria 2nd edition*, however, *Procheneosaurus convincens* had been included in *Jaxartosaurus aralensis* Riabinin, 1939.
[References: (*1) Bell PR, Brink KS (2013). *Kazaklambia convincens* comb. nov., a primitive juvenile lambeosaurine from the Santonian of Kazakhstan. *Cretaceous Research* 45: 265–274.]

Keichousaurus hui [Sauropterygia]
> *Keichousaurus hui* Young, 1958 [Spathian; Huixia Formation, Guizhou, China]
Generic name: *Keichousaurus* ← {(Chin. place-name) Keichou貴州 [Pinyin: Guìzhōu] + (Gr.) σαῦρος/sauros}
Specific name: *hui* ← {(Chin. person's name) Hú胡 + -ī}; "dedicated to Mr. C. C. Hu, the leader of the field party, who first discovered the interesting fossils"[*1].
Etymology: Hu's **Keichou (= Guizhou) lizard**
Taxonomy: Lepidosauromorpha: Sauropterygia: Nothosauroidea: Keichousauridae
Other species:
> *Keichousaurus yuananensis* Young, 1965 [Spathian; Jialingjiang Formation, Hubei, China]
> *Keichousaurus lusiensis* Gonzui, 1978 [Ladinian; Falang Formation, Guizhou, China]
[References: (*1) Young C-C (1958). On the new Pachypleurosauroidea from Keichow, South-West China. *Vertebrata PalAsiatica* 2: 72–81.; Liu GB, Yixi GZ, Wang X-H, Luo Y-M, Wang SY (2003). New discovered fishes from *Keichousaurus* bearing horizon of Late Triassic in Xingyi of Guizhou. *Acta Palaeontologica Sinica* 42: 346–366.]

Kelmayisaurus petrolicus
> *Kelmayisaurus petrolicus* Dong, 1973 [?Valanginian–Albian; Lianmuqin Formation, Karamay, Xingjiang, China]
Generic name: *Kelmayisaurus* ← {(place-name) Karamay = Kelmayi* + (Gr.) σαῦρος/sauros: lizard}
*(Uyghur) Qaramay [qara: black + may: oil]
Specific name: *petrolicus* ← {petrolicus: petrolic}; referring to the discovery area.
Etymology: petrolic, **Kelmay lizard**
Taxonomy: Theropoda: Tetanuae: Allosauroidea: Carcharodontosauria: Carcharodontosauridae[*1]
Notes: According to *The Dinosauria 2nd edition*, *Kelmayisaurus* was considered Tetanurae *incertae sedis*, but, it was classified into Carcharodontosauridae (Brusatte *et al*., 2012).
[References: (*1) Brusatte SL, Benson RBJ, Xu X (2012). A reassessment of *Kelmayisaurus petrolicus*, a large theropod dinosaur from the Early Cretaceous of China. *Acta Palaeontologica Polonica* 57(1): 65–72.]

Kentrosaurus aethiopicus
>*Kentrosaurus aethiopicus* Hennig, 1915 ⇒ *Kentrurosaurus aethiopicus* (Hennig, 1915) Hennig 1916 ⇒ *Doryphorosaurus aethiopicus* (Hennig, 1915) Nopcsa, 1916 [Kimmeridgian; Tendaguru Formation, Mtwara, German East Africa (now Tanzania)]
Generic name: *Kentrosaurus* ← {(Gr.) κέντρον/kentron: sharp point, prickle + σαῦρος/sauros: lizard}
The spiked-tail called "thagomizer" was probably used for defence, according to Mallison (2011).[*1]
Specific name: *aethiopicus* ← {(Gr. place-name) Αἰθιοπία/Aithiopia + -icus}; denoting the provenance from Africa.
Etymology: Ethiopian (African) **spike lizard**
Taxonomy: Ornithischia: Thyreophora: Stegosauria: Stegosauridae
Notes: *Kentrosaurus* is the valid name, because it did not have to be renamed to either *Kentrurosaurus* or *Doryphorosaurus*.
[References: (*1) Mallison H (2011). Defense capabilities of *Kentrosaurus aethiopicus* Hennig, 1915. *Palaeontologia Electronica* 14(2): 1–25.]

Kerberosaurus manakini
> *Kerberosaurus manakini* Bolotsky & Godefroit, 2004 [Maastrichtian; Tsagayan Formation, Blagoveschensk, Amur, Russia]
Generic name: *Kerberosaurus* ← {(Gr. myth.) Κέρβερος/Kerberos*: Cerberus + σαῦρος/sauros: lizard}
*"this monstrous dog was the guardian of the Tartarus; it had a dragon tail and bore snake heads along its back"[*1].
Specific name: *manakini* ← {(person's name) Manakin + -ī}; "in honor of the Russian Colonel Manakin, pioneer of the palaeontological research in the Amur Region, who first obtained dinosaur

bones from the banks of the Amur River" [*1].

Etymology: Manakin's **Cerberus lizard**

Taxonomy: Ornithischia: Ornithopoda: Hadrosauridae: Saurolophinae

[References: (*1) Bolotsky YL, Godefroit P (2004). A new hadrosaurine dinosaur from the Late Cretaceous of Far Eastern Russia. *Journal of Vertebrate Paleontology* 24(2): 351–365.]

Khaan mckennai

> *Khaan mckennai* Clark *et al.*, 2001 [middle Campanian; Djadokhta Formation, Ömnögovi, Mongolia]

Generic name: *Khaan* ← {(Mong.) Khaan: ruler} [*1]

Specific name: *mckennai* ← {(place-name) McKenna + -ī}; "for Malcolm McKenna, in recognition of his passionate interest in and efforts towards the exploration for fossils in Cretaceous deposits of Mongolia" [*1].

Etymology: Mckenna's **lord**

Taxonomy: Theropoda: Oviraptorosauria: Oviraptoridae

Notes: The holotype consists of an almost complete skeleton. It was found together with another specimen.

[References: (*1) Clark JM, Norell MA, Barsbold R (2001). Two new oviraptorids (Theropoda: Oviraptorosauria), Upper Cretaceous Djadokhta Formation, Ukhaa Tolgod, Mongolia. *Journal of Vertebrate Paleontology* 21(2): 209–213.]

Khetranisaurus barkhani

> *Khetranisaurus barkhani* Malkani, 2006 (*nomen dubium*) [Maastrichtian; Pab Formation, Balochistan, Pakistan]

Generic name: *Khetranisaurus* ← {(demonym) the Khetran + (Gr.) σαῦρος/sauros: reptiles}; "in honor of the Khetran tribe of Barkhan District" [*1].

Specific name: *barkhani* ← {(place-name) Barkhan District + -ī}; "honoring the Barkhan which is the host District of dinosaurs" [*1].

Etymology: **Khetran reptile** from Barkhan

Taxonomy: Saurischia: Sauropodomorpha: Sauropoda: Titanosauria: Pakisauridae (= Titanosauridae) [*1]

[References: (*1) Malkani MS (2006). Biodiversity of saurischian dinosaurs from the Latest Cretaceous Park of Pakistan. *Journal of Applied and Emerging Sciences* 1(3): 108–140.]

Kholumolumo ellenbergerorum

> *Thotobolosaurus mabeatae* Ellenberger, 1970 (*nomen nudum*) ⇒*Kholumolumo ellenbergerorum* Peyre de Fabrègues & Allain, 2020 [Norian; lower Elliot Formation, Maphutseng, Lesotho]

Generic name: *Kholumolumo* ← {(Sotho folklore) Kholumolumo: the name of a type of reptilian monster}

Specific name: *ellenbergerorum* ← {(person's name) Ellenberger + -orum}; in honour of the Ellenberger family.

Etymology: Ellenberger brothers' **folkloric reptilian monster**

Taxonomy: Saurischia: Sauropodomorpha: Massopoda

Kileskus aristotocus

> *Kileskus aristotocus* Averianov *et al.*, 2010 [Bathonian; Itat Formation, Krasnoyarsk Territory, West Siberia, Russia]

Generic name: *Kileskus* ← {(Khakas) Kileskus: a lizard} [*1]

Specific name: *aristotocus* ← {(Gr.) ἀριστοτόκος/aristotokos: "of noble origin"}; "in allusion to the phylogenetic position of *Kileskus* gen. nov. within the clade of derived theropods (Coelurosauria)" [*1].

Etymology: **lizard** of noble origin

Taxonomy: Theropoda: Coelurosauria: Tyrannosauroidea: Proceratosauridae

[References: (*1) Averianov AO, Krasnolutskii SA, Ivantsov SV (2010). A new basal coelurosaur (Dinosauria: Theropoda) from the Middle Jurassic of Siberia. *Proceedings of the Zoological Institute of the Russian Academy of Sciences* 314(1): 42–57.]

Kinnareemimus khonkaenensis

> *Kinnareemimus khonkaenesis* Buffetaut *et al.*, 2009 [Barremian; Sao Khua Formation, Khon Kaen, Thailand]

Generic name: *Kinnareemimus* ← {(Buddhist myth.) Kinnaree/Kinnari + (Gr.) μῖμος/mīmos: mimic}; "in reference to graceful beings of Thai mythology, with the body of a woman and the legs of a bird, said to inhabit the depths of the legendary Himmapan Forest, by allusion to the bird-like feet of this dinosaur" [*1].

Specific name: *khonkaenensis* ← {(place-name) Khon Kaen Province in NE Thailand + -ensis} [*1]

Etymology: **Kinnaree mimic** from Khon Kaen

Taxonomy: Theropoda: Coelurosauria: Ornithomimosauria

[References: (*1) Buffetaut E, Suteethorn V, Tong H (2009). An early 'ostrich dinosaur' (Theropoda: Ornithomimosauria) from the Early Cretaceous Sao Khua Formation of NE Thailand. In: Buffetaut E, Cuny G, Le Loeuff J, Suteethorn V (eds), *Late Palaezoic and Mesozoic Ecosystems in SE Asia, Geological Society, London, Special Publications* 315(1): 229–243.]

Kizylkumavis cretacea [Avialae]

> *Kizylkumavis cretacea* Nessov, 1984 [Turonan–Coniacian; Bissekty Formation, Bukhoro/Bukhara, Uzbekistan]

Generic name: *Kizylkumavis* ← {(place-name) Kizylkum or Kyzylkum Desert + (Lat.) avis: bird}; referring to the Bissekty Formation of the Kizylkum Desert, Uzbekistan.

Specific name: *cretacea* ← {cretacea: of Cretaceous}

Etymology: **Kizylkum Desert bird** of Cretaceous

Taxonomy: Theropoda: Avialae: Enantiornithes

Klamelisaurus gobiensis

> *Klamelisaurus gobiensis* Zhao, 1993 [Callovian–Oxfordian; Wucaiwan Member, Shishugou Formation, Junggar Basin, Xinjiang, China]

Generic name: *Klamelisaurus* ← {(place-name) Klameli: 克拉美麗 [Pinyin: Kèlāměilì] + (Gr.) σαῦρος/sauros: lizard}; referring to "the fossil locality near the Kelameilishan Mts" [*1].

Specific name: *gobiensis* ← {(place-name) Gobi + -ensis}; "referring to the town of Jiàngjūn将軍 being in the Gobi desert" [*1].

cf. "Locality: 35 km north of the town of Jiangjunmiao (将軍廟), south of the Kelameilishan Mts., Jiangjun Gebi (将軍戈壁), eastern Junggar Basin,[*1]"

Etymology: **Klameli lizard** from Jiangjun Gobi

Taxonomy: Saurischia: Sauropodomorpha: Sauropoda: Eusauropoda

cf. Saurischia: Sauropoda: Bothrosauropodea: Klamelisaurinae [*1]

Notes: The Gurbantünggüt Desert's southeastern corner is known locally as Jiangjun Gobi (将軍戈壁 'General's Desert').

[References: (*1) Zhao X (1993). A new Middle Jurassic sauropod subfamily (Klamelisaurinae subfam. Nov.) from Xinjiang Autonomous Region, China. *Vertebrata PalAsiatica* 31(2): 132–138.]

Kol ghuva

> *Kol ghuva* Turner *et al*., 2009 [Campanian; Djadochta Formation, Ukhaa Tolgod, Ömnögovi, Mongolia]

Generic name: *Kol* ← {(Mong.) köl: foot [*1]}; referring to a well-preserved pes.

Specific name: *ghuva* ← {(Mong.) ghuv-a: beautiful [*1]}

Etymology: beautiful **foot**

Taxonomy: Theropoda: Coelurosauria: Maniraptora: Alvarezsauridae

[References: (*1) Turner AH, Nesbitt SJ, Norell MA (2009). A large alvalezsaurid from the Cretaceous of Mongolia. *American Museum Novitates* 3648: 1–14.]

Koparion douglassi

> *Koparion Douglassi* Chure, 1994 [Kimmeridgian; Morrison Formation, Utah, USA]

Generic name: *Koparion* ← {(Gr.) κοπάριον/koparion: a small surgical knife}; "an allusion to the small size of the serrated tooth that is the type specimen" [*1].

Specific name: *douglassi* ← {(person's name) Douglass + -ī}; "in honor of Earl Douglass, who discovered and excavated the great dinosaur quarry for which Dinosaur National Monument was created" [*1].

Etymology: Douglass' **knife**

Taxonomy: Theropoda: Coelurosauria: Troodontidae

Notes: According to *The Dinosauria 2nd edition*, *K. douglassi* was regarded as a *nomen dubium*.

[References: (*1) Chure DJ (1994). *Koparion douglassi*, a new dinosaur from the Morrison Formation (Upper Jurassic) of Dinosaur National Monument; the oldest troodontid (Theropoda: Maniraptora). *Brigham Young University Geology Studies* 40: 11–15.]

Koreaceratops hwaseongensis

> *Koreaceratops hwaseongensis* Lee & Kobayashi, 2011 [Albian; Tando Beds, Hwaseong, South Korea]

Generic name: *Koreaceratops* ← {(place-name) Korea + (Gr.) κερατ-/kerat- [κέρας/ keras: horn] + ὤψ/ōps: face} [*1]

Specific name: *hwaseongensis* ← {(place-name) Hwaseong華城 + -ensis}

Etymology: **Korean horned face (ceratopsian)** from Hwaseong

Taxonomy: Ornithischia: Marginocephalia: Ceratopsia: Neoceratopsia

Notes: *Koreaceratops* had the tall neural spines on its caudal vertebrae [*1].

[References: (*1) Lee Y-N, Ryan MJ, Kobayashi Y (2011). The first ceratopsian dinosaur from South Korea. *Naturwissenschaften* 98(1): 39–49.]

Koreanosaurus boseongensis

> *Koreanosaurus boseongensis* Huh *et al*., 2011 [Santonian–Campanian; Seonso Conglomerate, Boseong, Jeonnam-do, South Korea]

Generic name: *Koreanosaurus* ← {(place-name) Korean + (Gr.) σαῦρος/sauros: lizard}; referring to the fact that "this is the first relatively complete dinosaur discovered in Korea" [*1].

Specific name: *boseongensis* ← {(place-name) Boseong宝城 + -ensis}; "in reference to Boseong County where the holotype and paratype were discovered" [*1].

Etymology: **Korean lizard** from Boseong

Taxonomy: Ornithischia: Cerapoda: Ornithopoda: Parksosauridae (= Thescelosauridae)
cf. Ornithischia: Neornithischia: Ornithopoda [*1]

[References: (*1) Huh M, Lee D-G, Kim J-K, Lim J-D, Godefroit P (2011). A new basal ornithopod dinosaur from the Upper Cretaceous of South Korea. *Neues Jahrbuch für Geologie und Paläontologie, Abhandlungen* 259(1): 1–24.]

Koshisaurus katsuyama

> *Koshisaurus katsuyama* Shibata & Azuma, 2015 [Barremian–Aptian; Kitadani Formation, Katsuyama, Fukui, Japan]

Generic name: *Koshisaurus* ← {(Jpn. place-name) Koshi 越*: an old regional name + (Gr.) σαῦρος/sauros: lizard}
*The southern part of Koshi is now Fukui Prefecture.

Specific name: *katsuyama* ← {(Jpn. place-name) Katsuyama勝山}; in reference to the Katsuyama City where the described material was found.

Etymology: **Koshi lizard** from Katsuyama

Taxonomy: Ornithischia: Ornithopoda: Hadrosauroidea

[References: Shibata M, Azuma Y (2015). New basal hadrosauroid (Dinosauria: Ornithopoda) from the Lower Cretaceous Kitadani Formation, Fukui, central Japan. *Zootaxa* 3914(4): 421–440.]

Kosmoceratops richardsoni

> *Kosmoceratops richardsoni* Sampson *et al.*, 2010 [Campanian; Kaiparowits Formation, Utah, USA]

Generic name: *Kosmoceratops* ← {(Gr.) κόσμος/kosmos: ornamented [κοσμέω/kosmeō: order, adorn] + ceratops [κέρας/keras: horn + ὤψ/ōps: face]}

Specific name: *richardsoni* ← {(person's name) Richardson + -ī}; "in honor of Scott Richardson, who discovered the holotype and many other significant fossils" [*1].

Etymology: Richardson's **ornamented horned face (ceratopsian)**

Taxonomy: Ornithischia: Ceratopsia: Ceratopsidae: Chasmosaurinae

Notes: *Kosmoceratops* possesses the most ornate skull, with 15 well developed horns or horn-like structures [*1].

[References: (*1) Sampson SD, Loewen MA, Farke AA, Roberts EM, Forster CA, Smith JA, Titus AL (2010). New horned dinosaurs from Utah provide evidence for intracontinental dinosaur endemism. *PLoS ONE* 5(9): e12292.]

Kotasaurus yamanpalliensis

> *Kotasaurus yamanpalliensis* Yadagiri, 1988 [Hettangian–Pliensbachian; Kota Formation, Yamanpalli, Telangana, Andhra Pradesh, India]

Generic name: *Kotasaurus* ← {(place-name) Kota + (Gr.) σαῦρος/sauros: lizard}; referring to the Kota Formation where it was found.

Specific name: *yamanpalliensis* ← {(place-name) Yamanpalli: village + -ensis}

Etymology: **Kota lizard** from Yamanpalli

Taxonomy: Saurischia: Sauropodomorpha: Sauropoda

Notes: It had a long neck and a long tail. Nearly complete skeleton without skull was found [*1].

[References: (*1) Yadagiri P (2001). The osteology of *Kotasaurus yamanapalliensis*, a sauropod dinosaur from the Early Jurassic Kota Formation of India. *Journal of Vertebrate Paleontology* 21 (2): 242–252.]

Koutalisaurus kohlerorum

> *Pararhabdodon isonensis* Casanovas Cladellas, 1993 [Maastrichtian; Conques Formation, Tremp Group, Sant Romà D'Abella, Cataluña, Spain]

= *Koutalisaurus kohlerorum* Prieto-Márquez *et al.*, 2006 [Maastrichtian; Conques Formation, Tremp Group, Abella de la Conca, Cataluña, Spain]

Generic name: *Koutalisaurus* ← {(Gr.) κουτάλι/koutali: spoon + σαῦρος/sauros: lizard}; "in reference to the extreme medial extension of the edentulous region of the dentary, which would have given a 'spoon-like' appearance to the jaws of the animal when both dentaries were articulated" [*1].

Specific name: *kohlerorum* ← {(person's name) Kohler + -ōrum}; "in honor of Terry and Mary Kohler, for their support of research in vertebrate paleontology" [*1].

Etymology: Kohler's **spoon lizard**

Taxonomy: Ornithischia: Ornithopoda: Hadrosauridae

Notes: According to Prieto-Márquez & Wagner (2009), *Koutalisaurus kohlerorum* is "most probably the junior synonym of *Pararhabdodon isonensis*" [*2].

[References: (*1) Prieto-Marquéz A, Gaete R, Rivas G, Galobart Á, Boada M (2006). Hadrosauroid dinosaurs from the Cretaceous of Spain: *Pararhabdodon isonensis* revisited and *Koutalisaurus kohlerorum*, gen. *et* sp. nov. *Journal of Vertebrate Paleontology* 26(4): 929–943.; (*2) Prieto-Márquez A, Wagner JR (2009). *Pararhabdodon isonensis* and *Tsintaosaurus spinorhinus*: a new clade of lambeosaurine hadrosaurids from Eurasia. *Cretaceous Research* 30(5): 1238–1246.]

Kritosaurus navajovius

> *Kritosaurus navajovius* Brown, 1910 ⇒ *Hadrosaurus navajovius* (Brown, 1910) Baird & Horner, 1977 [Campanian; Kirtland Formation, New Mexico, USA]

Generic name: *Kritosaurus* ← {(Gr.) κρῐτός/kritos: separated, divided, chosen [κρίνω/krinō: separate, divine] + σαῦρος/sauros: lizard}

Specific name: *navajovius* ← {(demonym) Navajo + vius}

Etymology: **separated lizard** of the Navajo

Taxonomy: Ornithischia: Ornithopoda: Hadrosauridae: Saurolophinae: Kritosaurini

Other species:

> *Hadrosaurus breviceps* Marsh, 1889 (*nomen dubium*) ⇒ *Kritosaurus breviceps* (Marsh, 1889) Lull & Wright, 1942 (*nomen dubium*) ⇒ *Trachodon breviceps* (Marsh, 1889) Hay, 1902 (*nomen dubium*) [Campanian; Judith River Formation, Montana, USA]

> *Gryposaurus notabilis* Lambe, 1914 ⇒ *Kritosaurus notabilis* (Lambe, 1914) Lull & Wright, 1942 ⇒ *Hadrosaurus notabilis* (Lambe, 1914) Horner, 1979 [Campanian; Dinosaur Park Formation, Alberta, Canada]

= *Kritosaurus incurvimanus* Parks, 1919 ⇒ *Gryposaurus incurvimanus* (Parks, 1919) Gates & Samson, 2007 ⇒ *Hadrosaurus incurvimanus* (Parks, 1919) Norman, 1985 [Campanian; Dinosaur Park Formation, Alberta, Canada]

> *Secernosaurus koerneri* Brett-Surman, 1979 [Campanian–Maastrichtian; Lago Colhué Huapi Formation, Río Chico, Argentina]

= *Kritosaurus australis* Bonaparte *et al.*, 1984 [Campanian–Maastrichtian; Los Alamitos Formation, Rio Negro, Argentina]

> *Anasazisaurus horneri* Hunt & Lucas, 1993 ⇒ *Kritosaurus horneri* (Hunt & Lucas, 1993) Prieto-Márquez, 2016 [Campanian; Kirtland Formation, New Mexico, USA].

Notes: According to *The Dinosauria 2nd edition*, *Kritosaurus navajovius* is considered a *nomen dubium*.

[References: (*1) Brown B (1910). The Cretaceous Ojo Alamo beds of New Mexico with description of the new dinosaur genus *Kritosaurus*. *Bulletin of the American Museum of Natural History* 28(24): 267–274.]

Kronosaurus queenslandicus [Plesiosauria]

> *Kronosaurus queenslandicus* Longman, 1924 [Albian; Toolebuc Formation, Queensland, Australia]

Generic name: *Kronosaurus* ← {(Gr. myth.) Κρόνος/Kronos: Cronos, Cronus + σαῦρος/sauros: lizard}; referring to the Greek titan Kronos, who devoured his own children.

Specific name: *queenslandicus* ← {(place-name) Queensland + -icus}

Etymology: **Kronos lizard** from Queensland

Taxonomy: Plesiosauria: Pliosauroidea: Pliosauridae: Brachaucheninae

Other species:

> *Kronosaurus boyacensis* Hampe, 1992 [Aptian; Paja Formation, Boyacá, Columbia]

Notes: *Kronosaurus* is a short necked pliosaur.

Kryptops palaios

> *Kryptops palaios* Sereno & Brusatte, 2008 [Aptian–Albian; Elrhaz Formation, Ténéré Desert, Niger]

Generic name: *Kryptops* ← {(Gr.)κρυπτός/kryptos: covered, hidden + ὤψ/ōps: face}; "in reference to the pitted surface and impressed vessel tracks on the maxilla, which is indicative of a firmly attached, possibly keratinous, integument or covering" (*1).

Specific name: *palaios* ← {(Gr.) πᾰλαιός/palaios: old; "in reference to its Early Cretaceous age" (*1).}

Etymology: old, **covered face**

Taxonomy: Theropoda: Ceratosauria: Abelisauroidea: Abelisauridae

[References: (*1) Sereno PC, Brusatte SL (2008). Basal abelisaurid and charcharodontosaurid theropods from the Lower Cretaceous Elrhaz Formation of Niger. *Acta Palaeontologica Polonica* 53 (1): 15–46.]

Kukufeldia tilgatensis

> *Iguanodon dawsoni* Lydekker, 1888 ⇒ *Barilium dawsoni* (Lydekker, 1888) Norman, 2010 ⇒ *Torilion dawsoni* (Lydekker, 1888) Carpenter & Ishida, 2010 [Valanginian; Wadhurst Clay, East Sussex, UK]

= *Kukufeldia tilgatensis* McDonald *et al.*, 2010 [Valanginian; Tunbridge Wells Sand, West Sussex, UK]

Generic name: *Kukufeldia* ← {(place-name) Kukufeld: "an Old English name for the village of Cuckfield" + -ia} (feminine) (*1)

Specific name: *tilgatensis* ← {(place-name) Tilgate + (Lat.) -ensis: from}; referring to Tilgate Forest, the region in which many of Mantell's fossils were obtained. (*1)

Etymology: **Cuckfield's one** from Tilgate Forest

Taxonomy: Ornithischia: Ornithopoda: Iguanodontia: Ankylopollexia: Styracosterna

Notes: According to Norman (2014), *Kukufeldia tilgatensis* is considered to be synonymous with

Barilium dawsoni.

[References: (*1) Mcdonald AT, Chapman SD, Barrett PM (2010). A new basal iguanodont (Dinosauria: Ornithischia) from the Wealden (Lower Cretaceous) of England. *Zootaxa* 2569: 1–43.]

Kulceratops kulensis

> *Kulceratops kulensis* Nescov, 1995 [Albian; Khodzhakul Formation, Kizylkum Desert, Uzbekistan]

Generic name: *Kulceratops* ← {kul [(Uzbek) ko`l: lake] + ceratops [(Gr.) κέρας/ceras: horn + ὤψ/ōps: face]}

Specific name: *kulensis* ← {(place-name) Kul: Khodzhakul Formation + -ensis}

Etymology: **lake horned face (ceratopsian)** from Khodzhakul Formation

Taxonomy: Ornithischia: Cerapoda: Ceratopsia

Notes: According to *The Dinosauria 2nd edition*, *Kulceratops* is considered as a *nomen dubium*, because the fossils have been sparse.

Kulindadromeus zabaikalicus

> *Kulindadromeus zabaikalicus* Godefroit *et al*., 2014 [Bathonian; Ukureyskaya Formation, Transbaikalia, Russia]

= *Kulindapteryx ukureica* Alifanov & Saveliev, 2014 (invalid species) [Bathonian; Ukureyskaya Formation, Transbaikalia, Russia]

= *Daurosaurus olovus* Alifanov & Saveliev, 2014 (invalid species) [Bathonian; Ukureyskaya Formation, Transbaikalia, Russia]

= "*Lepidocheirosaurus natatilis*" Alifanov & Saveliev, 2015 (invalid species) [Bathonian; Ukureyskaya Formation, Transbaikalia, Russia]

Generic name: *Kulindadromeus* ← {(place-name) Kulinda* + (Gr.) δρομεύς/ dromeus: runner}

*Kulinda is the first Russian and northernmost Asiatic locality. (according to Alifanov & Saveliev, 2014)

Specific name: *zabaikalicus* ← {(place-name) the Zabaykalsky Krai + -icus}

Etymology: **Kulinda runner** of Zabaykalsky Krai

Taxonomy: Ornithischia: Neornithischia

[References: Godefroit P, Sinitsa SM, Dhouailly D, Bolotsky YL, Sizov AV, McNamara ME, Benton MJ, Spagna P (2014). A Jurassic ornithischian dinosaur from Siberia with both feathers and scales. *Science* 345 (6195): 451–455.; Alifanov VR, Saveliev SV (2014). Two new ornithischian dinosaurs (Hypsilophodontia, Ornithopoda) from the Late Jurassic of Russia. Paleontological Journal 48(4): 414–425.]

Kulindapteryx ukureica

> *Kulindadromeus zabaikalicus* Godefroit *et al*., 2014 [Bathonian; Ukureyskaya Formation, Transbaikalia, Russia]

= *Kulindapteryx ukureica* Alifanov & Saveliev, 2014 [Bathonian; Ukureyskaya Formation, Transbaikalia, Russia]

Generic name: *Kulindapteryx* ← {(place-name) Kulinda locality + (Gr.) πτέρυξ/pteryx: wing}

Specific name: *ukureica* ← {(place-name) ukureica: Ukureyskaya Formation}

Etymology: **Kulinda wing** from Ukureyskaya Formation

Taxonomy: Ornithischia: Neornithischia: Jeholosauridae

cf. Ornithischia: Ornithopoda: Hypsilophodontia (*1)

Notes: *Kulindapteryx* is considered to be a synonym of *Kulindadromeus*.

[References: (*1) Alifanov VR, Saveliev SV (2014). Two new ornithischian dinosaurs (Hypsilophodontia, Ornithopoda) from the Late Jurassic of Russia. *Paleontological Journal* 48(4): 414–425.]

Kunbarrasaurus ieversi

> *Kunbarrasaurus ieversi* Leahey *et al*., 2015 [Albian–Cenomanian; Allaru Formation, Marathon Station, Queensland, Australia]

Generic name: *Kunbarrasaurus* ← {(Mayi: the language of the Wunumara People) Kunbarra [Kunbara]: shield + (Gr.) σαῦρος/sauros: lizard}; "referring the animal's heavily ossified skin" (*1).

Specific name: *ieversi* ← {(person's name) Ievers + -ī}; "in honor of Mr. Ian Ievers, discoverer of the holotype" (*1).

Etymology: Ievers' **shield-lizard**

Taxonomy: Ornithischia: Eurypoda: Ankylosauria

Notes: A more complete skeleton discovered between 1989 and 1996, which was referred to a *Minmi* sp., was named as a separate genus, *Kunbarrasaurus* in 2015.

[References: (*1) Leahey LG, Molnar RE, Carpenter K, Witmer LM, Salisbury SW (2015). Cranial osteology of the ankylosaurian dinosaur formerly known as *Minmi* sp. (Ornithischia: Thyreophora) from the Lower Cretaceous Allaru Mudstone of Richmond, Queensland, Australia. *PeerJ* 3: e1475.]

Kundurosaurus nagornyi

> *Kundurosaurus nagornyi* Godefroit *et al*., 2012 [Maastrichtian; Udurchukan Formation, Kundur, Amur, Russia]

Generic name: *Kundurosaurus* ← {(place-name) Кундур/Kundur: the type-locality + (Gr.) σαῦρος/sauros: lizard} (*1)

Specific name: *nagornyi* ← {(person's name) Nagorny + -ī}; "in honor of V. A. Nagorny (Far Eastern

Institute of Mineral Resources, FEBRAS*), who discovered the Kundur locality" (*1).

*Far East Branch of Russian Academy of Sciences

Etymology: Nagorny's **Kundur lizard**

Taxonomy: Ornithischia: Ornithopoda: Hadrosauridae/Saurolophidae: Saurolophinae

[References: (*1) Godefroit P, Yuri L, Bolotsky YL, Lauters P (2012). A new saurolophine dinosaur from the Latest Cretaceous of Far Eastern Russia. *PLoS ONE* 7(5): e36849.]

Kuszholia mengi [Avialae]

> *Kuszholia mengi* Nesov, 1992 [Turonian; Bissekty Formation, Kyzyl Kum, Bukhara, Uzbekistan]

Generic name: *Kuszholia* ← {(Kazakh) kus zholi: 'the Milky Way' + -ia}

According to the Kazakh beliefs the Milky Way showed the direction of the bird's migration, that's where it got its name 'Bird's path' (Kus zholy).

Specific name: *mengi* ← {(person's name) Meng + -ī}

Etymology: Meng's **Milky Way bird**

Taxonomy: Theropoda: Avialae: Ornithothoraces: Enantiornithes

Kwanasaurus williamparkeri [Silesauridae]

> *Kwanasaurus williamparkeri* Martz & Small, 2019 [Norian–Rhaetian; Chinle Formation, Colorado, USA]

Generic name: *Kwanasaurus* ← {(Ute) kwana-: eagle + (Gr.) σαῦρος/sauros: lizard}; in honour of "the town and county of Eagle in Colorado, located near the fossil localities that produced the type and referred specimens, as well as the Ute people. The town and county of Eagle are named for the Eagle River (Río Águila in Spanish), said to be translated from a local Ute name for the river or from the name of a Ute chief" (*1).

Specific name: *williamparkeri* ← {(person's name) William Parker + -ī}; in honour of "friend and colleague Bill Parker, whose research has helped to greatly clarify our understanding of Late Triassic dinosauromorph diversity in the western United States" (*1).

Etymology: William Parker's **Eagle lizard**

Taxonomy: Dinosauriformes: Drachors: Silesauridae

[References: (*1) Martz JW, Small BJ (2019). Non-dinosaurian dinosauromorphs from the Chinle Formation (Upper Triassic) of the Eagle Basin, northern Colorado: *Dromomeron romeri* (Lagerpetidae) and a new taxon, *Kwanasaurus williamparkeri* (Silesauridae). *PeerJ* 7: e7551.]

L

Labocania anomala

> *Labocania anomala* Molnar, 1974 [Santonian–Campanian; La Bocana Roja Formation, Baja California, Mexico]

Generic name: *Labocania* ← {(place-name) La Bocana Roja Formation: 'the red estuary' + -ia}

Specific name: *anomala* ← {(Gr.) ἀνώμαλος/anōmalos: anomalous}; in reference to the distinctive build.

Etymology: anomalous **one from La Bocana Roja Formation**

Taxonomy: Theropoda: Coelurosauria: Tyrannosauroidea

[References: Molnar RE (1974). A distinctive theropod dinosaur from the Upper Cretaceous of Baja California (Mexico). *Journal of Paleontology* 48(5): 1009–1017.]

Laevisuchus indicus

> *Laevisuchus indicus* von Huene & Matley, 1933 [Maastrichtian; Lameta Formation, Madhya Pradesh, India]

Generic name: *Laevisuchus* ← {(Lat.) levis: light, swift (*1) + (Gr.) Σοῦυχος/Souchos [Sūchos]: Egyptian crocodile god}

cf. lēvis (= laevis): smooth.

Specific name: *indicus* ← {(Lat.) indicus: of India, Indian} (*1)

Etymology: **light crocodile** of India

Taxonomy: Theropoda: Abelisauria: Noasauridae (*1)

[References: (*1) Novas FE, Agnolin FL, Bandyopadhyay S (2004). Cretaceous theropods from India: a review of specimens described by von Huene and Matley (1933). *Revista del Museo Argentino de Ciencias Naturales, nuevo serie* 6(1): 67–103.]

Lagosuchus talampayensis [Dinosauromorpha]

Lagosuchus talampayensis Romer, 1971 [Carnian*; Chañares Formation, La Rioja, Argentina]

*The Chañares Formation where the fossils were found was originally thought to be formed during Ladinian age (lower Chanarian), of the Middle Triassic (*1). Marsicano *et al.* (2016) dated it to early Carnian (235–234 Ma), Late Triassic.

Generic name: *Lagosuchus* ← {(Gr.) λᾰγός/lagos: hare, rabbit + Suchus [(Gr.) Σοῦχος/Souchos [Sūchos]: a crocodile god, an Egyptian deity]}

Specific name: *talampayensis* ← {(place-name) Talampaya + -ensis}

Etymology: **hare-like crocodile** from Talampaya

Taxonomy: Archosauria: Ornithodira: Dinosauromorpha: Lagosuchia: Lagosuchidae

Other species:

> *Lagosuchus lilloensis* Romer, 1972 ⇒ *Marasuchus lillocensis* (Romer, 1972) Sereno & Arcucci, 1994 (*2) [Carnian; Chañares Formation, La Rioja, Argentina]

Notes: The genus and species of *Lagosuchus talampayensis* are regarded as *nomina dubia* by some. For the species "*Lagosuchus*" *lilloensis*, a new genus *Marasuchus* was designated (*2).

[References: (*1) Bonaparte JF (1975). New materials of *Lagosuchus talampayensis* Romer (Thecodontia-Pseudosuchia) and its significance on the origin of the Saurischia Lower Chanarian, Middle Triassic of Argentina. *Acta Geologica Lilloana* 13(1): 5–90.]; (*2) Sereno PC, Arcucci A (1994). Dinosaurian precursors from the Middle Triassic of Argentina: *Marasuchus lilloensis*, gen. nov. *Journal of Vertebrate Paleontology* 14(1): 53–73.]

Laiyangosaurus youngi

> *Laiyangosaurus youngi* Zhang, 2017 [Campanian, Jingangkou Formation, Wangshi Group, Laiyang, Shandong, China]

Generic name: *Laiyangosaurus* ← {(Chin. place-name) Láiyáng萊陽 + (Gr.) σαῦρος/sauros: lizard}; "referring to the city where the fossils discovered" (*1).

Specific name: *youngi* ← {(Chin. person's name) Young Chungchien + -ī}; "in commemoration of the 120th Anniversary of Dr. Chungchien Young's Birth. Dr. Young is the pioneer of vertebrate paleontological research in Laiyang, and who has discovered many dinosaurs at this locality" (*1).

Etymology: Young's **Laiyang lizard** [楊氏萊陽龍]

Taxonomy: Ornithischia: Ornithopoda:
Hadrosauroidea: Hadrosauridae: Saurolophinae

[References: (*1) Zhang J, Wang X, Wang Q, Jiang S, Cheng X, Li N, Qiu R (2017). A new saurolophine hadrosaurid (Dinosauria: Ornithopoda) from the Upper Cretaceous of Shandong, China. *Anais da Academia Brasileira de Ciências* 91: e20160920.]

Lajasvenator ascheriae

> *Lajasvenator ascheriae* Coria *et al.*, 2019 [Valanginian; Mulichinco Formation, Neuquén, Argentima]

Generic name: *Lajasvenator* ← {(place-name) Las Lajas + (Lat.) venator: hunter}; "referring to the city of Las Lajas, within the jurisdiction of which the specimen was found" (*1).

Specific name: *ascheriae* ← {(person's name) Ascheri + -ae}; "after Susana Ascheri, for her kindness in allowing us to work on her land" (*1).

Etymology: Ascheri's **hunter from Las Lajas**

Taxonomy: Saurischia: Theropoda:
Carcharodontosauridae

[References: Coria RA, Currie PJ, Ortega F, Baiano MA (2019). An Early Cretaceous, medium-sized carcharodontosaurid theopod (Dinosauria, Saurischia) from the Mulichinco Formation] (upper Valanginian), Neuquén Province, Patagonia, Argentina. *Cretaceous Research* (in press). 104319.]

Lamaceratops tereschenkoi

> *Bagaceratops rozhdestvenskyi* Maryańska & Osmólska, 1975 [Campanian; Barun Goyot Formation, Mongolia]

= *Lamaceratops tereschenckoi* Alifanov, 2003 [Campanian; Barun Goyot Formation, Khulsan, Ömnögovi, Mongolia]

Generic name: *Lamaceratops* ← {(Tibetan) lama←[blama: chief or high priest]}: Buddhistic monk + (Gr.) κέρας/keras: horn + ὤψ/ōps: face}

Specific name: *tereschenkoi* ← {(person's name) Tereschenko + -ī}; in honor of paleontologist V. S. Tereschenko.

Etymology: Tereschenko's **Lama horned face (ceratopsian)**

Taxonomy: Ornithischia: Ceratopsia: Coronosauria:
Protoceratopsidae*: Bagaceratopidae / Bagaceratopsidae
*Due to Sereno's phylogenetic definition, Alifanov's Bagaceratopidae appears to be a subclade of Protoceratopsidae.

Notes: According to Czepiński (2019), *Lamaceratops tereschenkoi*, *Gobiceratops minutus*, *Platyceratops tatarinovi*, and *Magnirostris dodsoni* are considered to be junior synonyms of *Bagaceratops rozhdestvenskyi*.

[References: Alifanov VR (2003). Two new dinosaurs of the infraorder Neoceratopsia (Ornithischia) from the Upper Cretaceous of the Nemegt Depression, Mongolian People's Republic. *Paleontological Journal* 37(5): 524–534.; Czepiński Ł (2019). Ontogeny and variation of a protoceratopsid dinosaur *Bagaceratops rozhdestvenskyi* from the Late Cretaceous of the Gobi Desert. *Historical Biology* 32(10): 1394–1421.]

Lambeosaurus lambei

> *Lambeosaurus lambei* Parks, 1923 [Campanian; Dinosaur Park Formation, Alberta, Canada]

= *Trachodon marginatus* Lambe, 1902 (*partim*) (*nomen dubium*) ⇒ *Stephanosaurus marginatus* (Lambe, 1902) Lambe, 1914 [Campanian; Judith River Formation, Alberta, Canada]

= *Tetragonosaurus praeceps* Parks, 1931 ⇒ *Procheneosaurus praeceps* (Parks, 1931) Lull & Wright, 1942 [Campanian; Dinosaur Park Formation, Alberta, Canada]

= *Corythosaurus frontalis* Parks, 1935 [Campanian; Dinosaur Park Formation,

Alberta, Canada]
= *Lambeosaurus clavinitialis* Sternberg, 1935 [Campanian; Dinosaur Park Formation, Alberta, Canada]

Generic name: *Lambeosaurus* ← {(person's name) Lambe + (Gr.) σαῦρος/sauros: lizard}; in honor of Canadian geologist and palaeontologist Lawrence Lambe.

Specific name: *lambei* ← {(person's name) Lambe + -ī}; in honor of Lawrence Lambe.

Etymology: Lambe's **Lambeosaurus (Lambe's lizard)**

Taxonomy: Ornithischia: Ornithopoda: Hadrosauridae: Lambeosaurinae Lambeosaurini

Other species:
> *Lambeosaurus magnicristatum* Sternberg, 1935 ⇒*Lambeosaurus magnicristatus* (Sternberg, 1935) Sternberg, 1935 [late Campanian; Dinosaur Park Formation, Alberta, Canada]
> *Hadrosaurus paucidens* Marsh, 1889 (*nomen dubium*) ⇒ *Ceratops paucidens* (Marsh, 1889) Hay, 1901 ⇒ *Lambeosaurus paucidens*? (Marsh, 1889) Ostrom, 1964(*1) (*nomen dubium*) [Campanian; Judith River Formation, Montana, USA]
> *Trachodon altidens* Lambe, 1902 ⇒ *Didanodon altidens* (Lambe, 1902) Osborn, 1902 ⇒ *Procheneosaurus altidens* (Lambe, 1902) Lull & Wright, 1942 ⇒ *Lambeosaurus* sp.? Lund & Gates, 2006 [Campanian; Judith River Formation, Montana, USA]
> ? *Lambeosaurus laticaudus* Morris, 1981 ⇒ *Magnapaulia laticaudus* (Morris, 1981) Prieto-Márquez *et al.*, 2012 [Campanian; El Gallo Formation, Baja California Norte, Mexico]

Notes: *Lambeosaurus* is known for its hollow cranial crest resembling a hatchet.

[References: (*1) Ostrom JH (1964). The systematic position of *Hadrosaurus* (Ceratops) *paucidens* Marsh. *Journal of Paleontology* 38(1): 130–134.]

Lametasaurus indicus
> *Lametasaurus indicus* Matley, 1923 [Maastrichtian; Lameta Formation, Jabalpur Madhya Pradesh, India]

Generic name: *Lametasaurus* ← {(place-name) Lameta Formation + (Gr.) σαῦρος/sauros: lizard}

Specific name: *indicus* ← {(Lat.) indicus: of India, Indian}

Etymology: **lizard of Lameta Formation** of India

Taxonomy: Theropoda: Abelisauridae: Carnotaurinae

Notes: According to *The Dinosauria 2nd edition*, *Lametasaurus* is regarded as a *nomen dubium*.

Lamplughsaura dharmaramensis
> *Lamplughsaura dharmaramensis* Kutty *et al.*, 2007 [Sinemurian; Dharmaram Formation, Andhra Pradesh, India]

Generic name: *Lamplughsaura* ← {(person's name) Lamplugh [pron.: Lam-plo] + (Gr.) σαύρα/saura (the femine form of σαῦρος/sauros: lizard)}; "in honor of the late Pamela Lamplugh Robinson of University College, University of London, England, who founded and guided the ISI (Indian Statistical Institute)" (*1).

Specific name: *dharmaramensis* ← {(palce-name) the Dharmaram Formation + -ensis}

Etymology: **Lamplugh's lizard** from Dharmaram Formation

Taxonomy: Saurischia: Sauropodomorpha

[References: (*1) Kutty TS, Chatterjee S, Galton PM, Upchurch P (2007). Basal sauropodomorphs (Dinosauria: Saurischia) from the Lower Jurassic of India: their anatomy and relationships. *Journal of Paleontology* 81(6): 1218–1240.]

Lanasaurus scalpridens
> *Lycorhinus angustidens* Haughton, 1924 [Hettangian–Sinemurian; upper Elliot Formation, Eastern Cape, South Africa]
= *Lanasaurus scalpridens* Gow, 1975 [Hettangian–Sinemurian; upper Elliot Formation, Free State, South Africa]

Generic name: *Lanasaurus* ← {(Lat.) lāna: wool + (Gr.) σαῦρος/sauros: lizard}; in honor of Professor Alfred Walter Crompton, whose nickname was Fuzz because he had woolly hair.

Specific name: *scalpridens* ← {(Lat.) scalprum: chisel + dēns: tooth}

Etymology: **wool lizard** with chisel-like tooth

Taxonomy: Ornithischia: Heterodontosauridae: Heterodontosaurinae

Notes: According to *The Dinosauria 2nd edition*, *Lanasaurus scalpridens* is considered as a synonym of *Lycorhinus angustidens*.

Lanzhousaurus magnidens
> *Lanzhousaurus magnidens* You *et al.*, 2005 [Barremian; Hekou Group, Gansu, China]

Generic name: *Lanzhousaurus* ← {(Chin. place-name) Lánzhōu蘭州: the capital city of Gansu Province + (Gr.) σαῦρος/sauros: lizard}

Specific name: *magnidens* ← {(Lat.) magnus: big + dēns: tooth}

Etymology: **Lanzhou lizard** with large tooth

Taxonomy: Ornithischia: Ornithopoda: Iguanodontia: Styracosterna (*1)

[References: (*1) You H, Ji Q, Li D (2005).

Lanzhousaurus magnidens gen. *et* sp. nov. from Gansu Province, China: the largest-toothed herbivorous dinosaur in the world. *Geological Bulletin of China* 24(9): 785–794.]

Laosaurus celer

> *Laosaurus celer* Marsh, 1878 (*nomen dubium*) [Kimmeridgian–Tithonian*; Morrison Formation, Wyoming, USA]
>
> *Marsh (1878) first described remains from the Oxfordian–Tithonian age.

Generic name: *Laosaurus* ← {(Gr.) λᾶος/lao- [= λᾶας/laas: stone] + σαῦρος/sauros: lizard}

Specific name: *celer* ← {(Lat.) celer: swift, fleet}

Etymology: swift, **stone lizard**

Taxonomy: Ornithischia: Neornithischia

Other species:

> *Nanosaurus rex* Marsh, 1877 ⇒ *Othnielia rex* (Marsh, 1877) Galton, 1977 (*nomen dubium*) [Kimmeridgian–Tithonian; Morrison Formation, Colorado, USA]

= *Laosaurus gracilis* Marsh, 1878 (*nomen dubium*) [Kimmeridgian–Tithonian; Morrison Formation, Wyoming, USA]

= *Laosaurus consors* Marsh, 1894 ⇒ *Othnielia consors* (Marsh, 1894) Galton, 1983 ⇒ *Othnielosaurus consors* (Marsh, 1894) Galton, 2007 [Kimmerdgian–Tithonian, Morrison Formation, Wyomimg, USA]

> *Laosaurus altus* Marsh, 1878 ⇒ *Dryosaurus altus* (Marsh, 1878) Galton, 1983 [Late Jurassic; Morrison Formation, Wyoming, USA]
>
> *Laosaurus minimus* Gilmore, 1924 (*nomen dubium*) [Campanian; Allison Formation, Alberta, Canada]

Notes: According to *The Dinosauria 2nd edition*, *Laosaurus gracilis and L. consors* were considered to be synonyms of *Othnielia rex*, but now three species: *L. celer*, *L. gracilis* and *L. minimus* are thought to be *nomina dubia*.

Lapampasaurus cholinoi

>*Lapampasaurus cholinoi* Coria *et al.*, 2013 [Campanian–Maastrichtian; Allen Formation, La Pampa, Argentina]

Generic name: *Lapampasaurus* ← {(place-name) La Pampa Province + (Gr.) σαῦρος/sauros: lizard}; "en referencia a la provincia de La Pampa de donde procede el ejemplar tipo" (*1). [in reference to the province of La Pampa where the type specimen was found.]

Specific name: *cholinoi* ← {(person's name) Cholino + -ī}; "in honor of José Cholino, por alertar sobre la existencia del material y apoyar su divulgación científica" (*1). [for notifying of the existence of the material and supporting its scientific dissemination].

Etymology: Cholino's **La Pampa lizard**

Taxonomy: Ornithischia: Ornithopoda: Hadrosauridae

[References: (*1) Coria RA, González Riga B, Casadío S (2012). Un nuevo hadrosáurido (Dinosauria, Ornithopoda) de la Formación Allen, provincia de La Pampa, Argentina [A new hadrosaurid (Dinosauria, Ornithopoda) from Allen Formation, La Pampa, Argentina]. *Ameghiniana* 49(4): 552–572.]

Laplatasaurus araukanicus

> *Laplatasaurus araukanicus* von Huene, 1929 ⇒ *Titanosaurus araukanicus* (von Huene, 1929) Powell, 1992 [Campanian; Allen Formation, Río Negro, Argentina] / [Campanian; Anacleto Formation, Río Negro (Provence), Argentina] (lectotype, according to Bonaparte, 1979) / [Campanian–Maastrichtian; Asencio Formation, Palmitas, Río Negro (Department), Uruguay] (according to von Huene, 1929)

Generic name: *Laplatasaurus* ← {(place-name) La Plata* + (Gr.) σαῦρος/sauros: lizard}

*Von Huene based *Laplatasaurus* on materials from three locations as described above. The etymology of La Plata is not clear, but it can be thought possibly either Río de la Plata which flows between Argentina and Uruguay or Museo de La Plata where the lectotype has been housed.

Specific name: *araukanicus* ← {(demonym) Araucanos or Mapuche + -icus}

Etymology: **La Plata lizard** of Araucanos

Taxonomy: Saurischia: Sauropodomorpha: Sauropoda: Titanosauria: Lithostrotia

Other species:

> *Titanosaurus madagascariensis* Depéret, 1896 ⇒ *Laplatasaurus madagascariensis* Depéret, 1896) von Huene & Matley, 1933 (*nomen dubium*) [Coniacian; Ankazomihaboka Formation, Mahajanga Basin, Madagascar]

Notes: According to *The Dinosauria 2nd edition*, *Laplatasaurus* is regarded as valid genus. Gallina and Otero (2015) argued that the type specimen of *Laplatasaurus araukanicus* comes from the Anacleto Formation, although it had been regarded as a lectotype by Bonaparte (1979).

[References: von Huene F (1929). Terrestrische Oberkreide in Uruguay [The terrestrial Upper Cretaceous in Uruguay]. *Centralblatt fül Mineralogie, Geologie und Paläontologie Abteilung B* 1929: 107–112 [M. Carrano]; Wilson JA, Upchurch

P (2003). A revision of *Titanosaurus* Lydekker (Dinosauria-Sauropoda), the first dinosaur genus with a 'Gondwanan' distribution. *Journal of Systematic Palaentology* 1(3): 125–160.]

Lapparentosaurus madagascariensis

> *Bothriospondylus madagascariensis* Lydekker, 1895 ⇒ *Lapparentosaurus madagascariensis* (Lydekker, 1895) Bonaparte, 1986 [Bathonian; Isalo III Formation, Majunga, Madagascar]

Generic name: *Lapparentosaurus* ← {(person's name) Lapparent + (Gr.) σαῦρος/sauros: lizard}; in honor of French palaeontologist Albert-Félix de Lapparent.

Specific name: *madagascariensis* ← {(place-name) Madagascar + -iensis}

Etymology: **Lapparent's lizard** from Madagascar

Taxonomy: Saurischia: Sauropodomorpha: Sauropoda: Titanosauriformes

Laquintasaura venezuelae

> *Laquintasaura venezuelae* Barrett *et al.*, 2014 [Hettangian; La Quinta Formation, Táchira, Venezuela]

Generic name: *Laquitasaura* ← {(place-name) La Quinta: the type horizon + (Gr.) σαῦρα/saura, *f.*: lizard}

Specific name: *venezuelae* ← {(place-name) Venezuela + -ae}; "for the country and people of Venezuela" (*1).

Etymology: Venezuela's **La Quinta lizard**

Taxonomy: Ornithischia

[References: (*1) Barrett PM, Butler RJ, Mundil R, Scheyer TM, Irmis RB, Sánchez-Villagra MR (2014). A palaeoequatorial ornithischian and new constraints on early dinosaur diversification. *Proceedings of the Royal Society B: Biological Sciences* 281(1791): 1–7.]

Largirostrornis sexdentoris [Avialae]

> *Largirostrornis sexdentoris* Hou, 1997 [Aptian; Jiufotang Formation, Chaoyang, Liaoning, China]

Generic name: *Largirostrornis* ← {(Lat.) largus: plentiful or bountiful (*1), large + rostrum: bill or mouth + (Gr.) ὄρνις/ornis}; referring to the large beak on this genus.

Specific name: *sexdentoris* ← {(Lat.) sex: six + dent-: teeth} (*1)

Etymology: six-toothed, **large beak bird**

Taxonomy: Theropoda: Avialae: Enantiornithes

[References: (*1) Hou L (1997). *Mesozoic Birds of China.* Phoenix Valley Provincial Aviary, Taipei [In Chinese].]

Latenivenatrix mcmasterae

> *Latenivenatrix mcmasterae* Reest & Currie, 2017 [Campanian; upper Dinosaur Park Formation, Alberta, Canada]

Generic name: *Latenivenatrix* ← {(Lat.) latēns: latent and hiding + venatrix: feminine form of "venator": hunter}; "referring to the taxon having been in multiple collections for nealy 100 years but unrecognized until now", "also referring to a predatory animal hiding in cover until a suitable time to attack its prey" (*1).

Specific name: *mcmasterae* ← {(person's name) McMaster + -ae}; "in honor of the late mother of the first author, Lynne (McMaster) van der Reest" (*1).

Etymology: McMaster's **hidden hunter**

Taxonomy: Theropoda: Maniraptora: Troodontidae: Troodontinae

[References: (*1) van der Reest AJ, Currie PJ (2017). Troodontids (Theropoda) from the Dinosaur Park Formation, Alberta, with a description of a unique new taxon: implications for deinonychosaur diversity in North America. *Canadian Journal of Earth Sciences* 54(9): 919–935.]

Latirhinus uitstlani

> *Latirhinus uitstlani* Prieto-Márquez & Brañas, 2012 [Campanian; Cerro del Puerbo Formation, Coahuila, Mexico]

Generic name: *Latirhinus* ← {(Lat.) lātus: broad + rhinus [(Gr.) ῥις/rhis: nose]}; "referring to the great width of the narial foramen in this animal" (*1).

Specific name: *uitstlani* ← {(Náhuatl) uitstlani: southern}; "in reference to the southern occurrence of this taxon within North America and the Late Cretaceous landmass Laramidia" (*1).

Etymology: southern, **broad nose**

Taxonomy: Ornithischia: Ornithopoda: Hadrosauridae: Saurolophinae

[References: (*1) Prieto-Márquez A, Serrano Brañas CI (2012). *Latirhinus uitstlani*, a 'broad-nosed' saurolophine hadrosaurid (Dinosauria, Ornithopoda) from the late Campanian (Cretaceous) of northern Mexico. *Historical Biology* 24(6): 607–619.]

Lavocatisaurus agrioensis

> *Lavocatisaurus agrioensis* Canudo *et al.*, 2018 [Aptian–Albian; Rayoso Formation, Neuquén Basin, Argentina]

Generic name: *Lavocatisaurus* ← {(person's name) Lavocat + (Gr.) σαῦρος/sauros: lizard}; "in honor of the French researcher René Lavocat (1909–2007), who described *Rebbachisaurus*, the first known representative of Rebbachisauridae" (*1).

Specific name: *agrioensis* ← {(place-name) Agrio del Medio: the locality + -ensis} (*1)

Etymology: **Lavocat's lizard** from Agrio del Medio

Taxonomy: Saurischia: Sauropoda: Diplodocoidea:

Diplodocimorpha: Rebbachisauridae

[References: (*1) Canudo JI, Carballido JL, Garrido A, Salgado L (2018). A new rebbachisaurid sauropod from the Aptian–Albian, Lower Cretaceous Rayoso Formation, Neuquén, Argentina. *Acta Palaeontologica Polonica* 63(4): 679–691.]

Leaellynasaura amicagraphica

> *Leaellynasaura amicagraphica* Rich & Rich, 1989 [Aptian–Albian; Eumeralla Formation, Victoria, Australia]

Generic name: *Leaellynasaura* ← {(person's name) Leaellyn Rich: daughter of the authors + (Gr.) σαύρα/saura (feminine form of sauros): lizard} (*1)

Specific name: *amicagraphica* ← {(Lat.) amicus: friend + (Gr.) γραφικός/graphikos: painting}; for "the friends of Museum of Victoria, and for the National Geographic Society" (*1).

Etymology: **Leaellyn's lizard** for friends of Museum of Victoria and of National Geographic Society

Taxonomy: Ornithischia: Ornithopoda

[References: Rich TH, Rich PV (1989). Polar dinosaurs and biotas of the Early Cretaceous of southeastern Australia. *National Geographic Research* 5(1): 15–53.; (*1) Vickers-Rich P, Rich TH (1993). Australia's Polar Dinosaurs. Scientific American July 1993.]

Lectavis bretincola [Avialae]

> *Lectavis bretincola* Chiappe, 1993 [Campanian–Maastrichtian; Lecho Formation, Salta, Argentina]

Generic name: *Lectavis* ← {(Lat.) lectus = (Spanish) lecho: bed + avis: bird}; in reference to the Lecho Formation.

Specific name: *bretincola* ← {(place-name) El Brete + (Lat.) incola: an inhabitant}

Etymology: **bird from the Lecho Formation**, living in El Brete

Taxonomy: Theropoda: Avialae: Pygostylia: Enantiornithes

Ledumahadi mafube

> *Ledumahadi mafube* McPhee *et al.*, 2018 [Hettangian–Sinemurian; upper Elliot Formation, Free State, on the border of South Africa and Lesotho]

Generic name: *Ledumahadi* ← {(Southern Sotho / Sesotho) Ledumahadi: a giant thunderclap}; "in recognition of the tremendous size of this taxon" (*1).

Specific name: *mafube* ← {(Southern Sotho / Sesotho) mafube: dawn}; "in the sense of the stratigraphically early position of this taxon" (*1).

Etymology: **giant thunderclap** at dawn

Taxonomy: Saurischia: Sauropodomorpha: Sauropodiformes

Notes: The limb bones indicate the specimen was about fourteen years old at death. (*1)

[References: (*1) McPhee BW, Benson RBJ, Botha-Brink J, Bordy EM, Choiniere JN (2018). A giant dinosaur from the earliest Jurassic of South Africa and the transition to quadrupedality in early sauropodomorphs. *Current Biology* 28(19): 3143–3151.e7.]

Leinkupal laticauda

> *Leinkupal laticauda* Gallina *et al.* 2014 [late Berriasian–early Valanginian; Bajada Colorada Formation, Neuquén Basin, Argentina]

Generic name: *Leinkupal* ← {(Mapudungun) lein: vanishing + kupal: family}; "referring to the record of the last known representative of the family Diplodocidae" (*1).

Specific name: *laticauda* ← {(Lat.) lātus: wide + cauda: tail}; "referring to the broad tail evidenced by the lateral extension of the transverse processes in proximal caudal vertebrae" (*1).

Etymology: wide-tailed, **vanishing family**

Taxonomy: Saurischia: Sauropoda: Diplodocoidea: Flagellicaudata: Diplodocidae: Diplodocinae

[References: (*1) Gallina PA, Apesteguía S, Haluza A, Canale JI (2014). A diplodocid sauropod survivor from the Early Cretaceous of South America. *PLoS ONE* 9(5): e97128.]

Lenesornis maltshevskyi [Avialae]

> *Ichthyornis maltshevskyi* Nesov, 1986 ⇒ *Lenesornis maltshevskyi* (Nesov, 1986) Kurochkin, 1996 [Turonian; Bissekty Formation, Bukhoro*, Uzbekistan]

* = (Uzbek Latin) Buxoro | (Rus) Bukhara

Generic name: *Lenesornis* ← {(person's name) Lev. A. Nesov + (Gr.) ὄρνις/ornis: bird}; in honor of the late Lev. A. Nesov.

Specific name: *maltshevskyi* ← {(person's name) Maltshevsky + -ī}

Etymology: Maltshevsky and **Lev. A. Nesov's bird**

Taxonomy: Theropoda: Avialae: Enantiornithes

Leonerasaurus taquetrensis

> *Leonerasaurus taquetrensis* Pol *et al.*, 2011 [Sinemurian–Toarcian; Las Leoneras Formation, Chubut, Argentina]

Generic name: *Leonerasaurus* ← {(place-name) Leoneras + (Gr.) σαῦρος/sauros: lizard}; "in reference to the lithostratigraphic unit where this taxon was found" (*1).

Specific name: *taquetrensis* ← {(place-name) Taquetrén + -ensis}; "referring to the Sierras de Taquetrén, where Las Leoneras Formation crops out in Central Patagonia" (*1).

Etymology: **Las Leoneras lizard** from Sierras de

Taquetrén
Taxonomy: Saurischia: Sauropodomorpha: Anchisauria

[References: (*1) Pol D, Garrido A, Cerda IA (2011). A new Sauropodomorph dinosaur from the Early Jurassic of Patagonia and the origin and evolution of the sauropod–type sacrum. *PLoS ONE* 6(1): e14572.]

Lepidus praecisio

> *Lepidus praecisio* Nesbitt & Ezcurra, 2015 [Norian; Dockum Group, Texas, USA]

Generic name: *Lepidus* ← {(Lat.) lepidus: fascinating} (gender masculine) [*1]

Specific name: *praecisio* ← {(Lat.) praecīsio: fragment or scrap}; "in referece to the common preservation of early dinosaurs from North America as bony fragments" [*1].

Etymology: **fascinating one** of fragment

Taxonomy: Theropoda: Neotheropoda: Coelophysoidea

[References: (*1) Nesbitt SJ, Ezcurra MD (2015). The early fossil record of dinosaurs in North America: A new neotheropod from the base of the Upper Triassic Dockum Group of Texas. *Acta Palaeontologica Polonica* 60(3): 513–526.]

Leptoceratops gracilis

> *Leptoceratops gracilis* Brown, 1914 [Maastrichtian; Scollard Formation, Alberta, Canada]

Generic name: *Leptoceratops* ← {(Gr.) λεπτός/leptos: small + κερατ-/kerat- [κέρας/ keras: horn] + ὤψ/ōps: face}

Specific name: *gracilis* ← {(Lat.) gracilis: slender, thin}

Etymology: slender, **small horned face (ceratopsian)**

Taxonomy: Ornithischia: Marginocephalia: Ceratopsia: Neoceratopsia: Leptoceratopsidae

Other species:

> *Leptoceratops cerorhynchus* Brown & Schlailjer, 1942 ⇒ *Montanoceratops cerorynchus* (Brown & Schlailjer, 1942) Sternberg, 1951 [Maastrichtian; St. Mary River Formation, Montana, USA]

Notes: The skeleton of *Leptoceratops gracilis* was small. The skull was short and deep, without a nasal horn. [*1].

[References: (*1) Brown B (1914). *Leptoceratops*, a new genus of Ceratopsia from the Edmonton Cretaceous of Alberta. *Bulletin of the American Museum of Natural History* 33(36): 567–580.]

Leptorhynchos gaddisi

> *Leptorhynchos gaddisi* Longrich *et al.*, 2013 [Campanian; Aguja Formation, Texas, USA]

Generic name: *Leptorhynchos* ← {(Gr.) λεπτός/leptos: slim, small + ῥύγχος/rhynchos: snout, muzzle}

Specific name: *gaddisi* ← {(person's name) Gaddis + -ī}; in honor of Gaddis family, owner of the land where the holotype was found.

Etymology: Gaddis' **little snout**

Taxonomy: Theropoda: Oviraptorosauria: Caenagnathidae

Other species:

> *Ornithomimus elegans* Parks, 1933 ⇒ *Elmisaurus elegans* (Parks, 1933) Currie, 1989 ⇒ *Chirostenotes elegans* (Parks, 1933) Sues, 1997 ⇒ *Leptorhynchos elegans* (Parks, 1933) Longrich *et al.*, 2013 ⇒ *Citipes elegans* (Parks, 1933) Funston, 2020 [Campanian; Dinosaur Park Formation, Alberta, Canada]

[References: Longrich NR, Barnes K, Clark S, Millar L (2013). Caenagnathidae from the Upper Campanian Aguja Formation of West Texas, and a revision of the Caenagnathinae. *Bulletin of the Peabody Museum of Natural History* 54(1): 23–49.]

Leshansaurus qianweiensis

> *Leshansaurus qianweiensis* Li *et al.*, 2009 [Oxfordian; upper Shaximiao Formation, Qianwei, Sichuan, China]

Generic name: *Leshansaurus* ← {(Chin. place-name) Lèshān楽山 + (Gr.) σαῦρος/sauros: lizard}

Specific name: *qianweiensis* ← {(Chin. place-name) Qiánwéi犍為 + -ensis}

Etymology: **Leshan lizard** from Qianwei

Taxonomy: Theropoda: Megalosauridae: Afrovenatorinae

[References: Li F, Peng G, Ye Y, Jiang S, Huang D (2009). A new carnosaur from the Late Jurassic of Qianwei, Sichuan, China. *Acta Geologica Sinica* 83(9): 1203–1213.]

Lesothosaurus diagnosticus

> *Lesothosaurus diagnosticus* Galton, 1978 [Hettangian; upper Elliot Formation, Mafeteng, Lesotho]

= *Stormbergia dangershoeki* Butler, 2005 [Hettangian; upper Elliot Formation, Eastern Cape, South Africa]

Generic name: *Lesothosaurus* ← {(place-name) Lesotho + (Gr.) σαῦρος/sauros: lizard}

Specific name: *diagnosticus* ← {(Gr.) διαγνωστικός/diagnōstikos: able to distinguish}

Etymology: distinguishable **Lesotho lizard**

Taxonomy: Ornithischia: Genasauria

Notes: According to Baron *et al.* (2016), *Stormbergia* is a junior subjective synonym of *Lesothosaurus*. *Stormbergia* represents the adult form of *Lesothosaurus*.

Lessemsaurus sauropoides

> *Lessemsaurus sauropoides* Bonaparte, 1999 [Norian; Los Colorados Formation, La Rioja, Argentina]

Generic name: *Lessemsaurus* ← {(person's name) Donald Lessem: founder of the Dinosaur Society, from USA. + (Gr.) σαῦρος/sauros: lizard} (*1)

Specific name: *sauropoides* ; "like sauropods, to stress that the morphology of its presacral vertebrae recalls that of primitive sauropods" (*1).

Notes: **Lessem's lizard** like sauropods

Taxonomy: Saurischia: Sauropodomorpha: Sauropoda: Lessemsauridae

[References: (*1) Bonaparte J (1999). Evolución de las vértebras presacras en Sauropodomorpha [Notes on the evolution of vertebrae in the Sauropodomorpha]. *Ameghiniana* 36(2): 115–187.]

Levnesovia transoxiana

> *Levnesovia transoxiana* Sues & Averianov, 2009 [Turonian; Bissekty Formation, Kyzylkum Desert, Uzbekistan]

Generic name: *Levnesovia* ← {(person's name) Lev Nesov* + -ia}

*"named for Lev Nessov (1947–1995), using the most accurate English-language spelling of his surname as transliterated from the Cyrillic" (*1).

Specific name: *transoxiana* ← {(Lat. ancient place-name) Transoxiana [trans: beyond + Oxus: the Amu Darya River]}; "corresponding to present-day Uzbekistan" (*1).

Etymology: **for Lev Nessov** from Uzbekistan (Transoxiana)

Taxonomy: Ornithischia: Ornithopoda: Iguanodontia: Hadrosauroidea (*1)

[References: (*1) Sues H-D, Averianov A (2009). A new basal hadrosaurid dinosaur from the Late Cretaceous of Uzbekistan and the early radiation of duck-billed dinosaurs. *Proceedings of the Royal Society B. Biological Sciences* 276(1667): 2549–2555.]

Lexovisaurus durobrivensis

> *Omosaurus durobrivensis* Hulke, 1887 ⇒ *Stegosaurus durobrivensis* (Hulke, 1887) Hulke, 1887 ⇒*Dacentrurus durobrivensis* (Hulke, 1887) Hennig, 1915 ⇒ *Dacentrurosaurus durobrivensis* (Hulke, 1887) Hennig, 1924 ⇒*Lexovisaurus durobrivensis* (Hulke, 1887) Hoffstetter, 1957 [Callovian; Oxford Clay, Cambridgeshire, UK]

Generic name: *Lexovisaurus* ← {(Lat. demonym) Lexoviī: a Gallic tribe/an ancient Celtic people of northern France + (Gr.) σαῦρος/sauros: lizard}; referring to the region of Normandy which the Lexovii inhabited.

Specific name: *durobrivensis* ← {(place-name) Durobrivae: an old Roman town in Cambridgeshire + -ensis}

Etymology: **Lexovii lizard**, from Durobrivae

Taxonomy: Ornithischia: Thyreophora: Stegosauria: Stegosauridae

Other species:

> *Omosaurus leedsi* Seeley, 1901 (*partim*) ⇒ *Dacentrurus leedsi* (Seeley, 1901) Hennig, 1915 ⇒ *Lexovisaurus leedsi* (Seeley, 1901) Kuhn, 1964 [Callovian; Oxford Clay Formation, Northamptonshire, UK]

> *Stegosaurus priscus* Nopcsa, 1911 ⇒ *Lexovisaurus priscus* (Nopcsa, 1901) Kuhn, 1964 ⇒ *Loricatosaurus priscus* (Nopcsa, 1911) Maidment *et al.*, 2008 [Callovian; Oxford Clay Formation, Peterborough, Cambridgeshire, UK]

> *Omosaurus vetustus* von Huene, 1910 ⇒ *Dacentrurus vetustus* (von Huene, 1910) Hennig, 1915 ⇒ *Lexovisaurus vetustus* (von Huene, 1910) Galton, 1983 ⇒ *Eoplophysis vetustus* (von Huene, 1910) Ulansky, 2014 (*nomen dubium*) [Bathonian; Cornbrash Formation, UK]

Notes: Remains of *Lexovisaurus durobrivensis* were found in Oxford Clay in England and France.

[References: Maidment SCR, Norman DB, Barrett PM, Upchurch P (2008). Systematics and phylogeny of Stegosauria (Dinosauria: Ornithischia). *Journal of Systematic Palaeontology* 6(4): 367–407.; Galton PM (2016). Notes on plated dinosaurs (Ornithischian: Stegosauria), mostly on dermal armor from Middle and Upper Jurassic of England (also France, Iberia), with a revised diagnosis of *Loricatosaurus priscus* (Callovian, England). *Neues Jahrbuch Für Geologie und Paläontologie*, Abhandlungen 282(1): 1–25.]

Leyesaurus marayensis

> *Leyesaurus marayensis* Apaldetti *et al.*, 2011 [Early Jurassic; Quebrada del Barro Formation, San Juan, Argentina]

Generic name: *Leyesaurus* ← {(person's name) Leyes family + (Gr.) σαῦρος/sauros: lizard}; "in honor of the Leyes family, inhabitants of the small town Balde de Leyes, who made the discovery and notified the paleontologists of the San Juan Museum" (*1).

Specific name: *marayensis* ← {(place-name) Marayes: Marayes-El Carrizal Basin + -ensis} (*1)

Etymology: **Leyes' lizard** from Marayes-El Carrizal Basin

Taxonomy: Saurischia: Sauropodomorpha:

Massospondylidae

[References: (*1) Apaldetti C, Martìnez RN, Alcober OA, Pol D (2011). A new basal sauropodomorph (Dinosauria: Saurischia) from Quebrada del Barro Formation (Marayes-EL Carrizal Basin), Northwestern Argentina. *PLoS ONE* 6(11): e26964.]

Liaoceratops yanzigouensis

> *Liaoceratops yanzigouensis* Xu *et al.*, 2002 [Barremian; Yixian Formation, Yanzigou, Liaoning, China]

Generic name: *Liaoceratops* ← {(Chin. place-name) Liáo遼 (= Liáoníng遼寧) + (Gr.) κερατ- [κέρας/keras: horn] + ὤψ/ōps: face}

Specific name: *yanzigouensis* ← {(Chin. place-name) Yānzǐgōu燕子溝 + -ensis}; "referring to the village near which the holotype was found"(*1).

Etymology: **Liaoning horned face (ceratopsian)** from Yanzigou

Taxonomy: Ornithischia: Marginocephalia: Ceratopsia: Neoceratopsia

[References: (*1) Xu X, Makovicky PJ, Wang X-L, Norell MA, You H-L (2002). A ceratopsian dinosaur from China and the early evolution of Ceratopsia. *Nature* 416: 314–317.]

Liaoningornis longidigitris [Avialae]

> *Liaoningornis longidigitris* Hou, 1996 [Barremian; Yixian Formation, Liaoning, China]

Generic name: *Liaoningornis* ← {(Chin. place-name) Liáoníng Province遼寧省 + (Gr.) ὄρνις/ornis: bird}

Specific name: *longidigitris* ← {(Lat.) longus: long + digitus: digit, finger}

Etymology: long-fingered **bird from Liaoning**

Taxonomy: Theropoda: Avialae: Enantiornithes

Liaoningosaurus paradoxus

> *Liaoningosaurus paradoxus* Xu *et al.*, 2001 [Aptian; Yixian Formation, Liaoning, China]

Generic name: *Liaoningosaurus* ← {(Chin. place-name) Liáoníng遼寧 + (Gr.) σαῦρος/sauros: lizard}; "referring to Liaoning Province"(*1).

Specific name: *paradoxus* ← {(Lat.) paradoxus = (Gr.) παράδοξος/paradoxos: incredible, paradoxical}; "referring to the surprising characteristics of this animal"(*1), because of its fragmentary skeleton of a juvenile.

Etymology: paradoxical **Liaoning lizard**

Taxonomy: Ornithischia: Ankylosauria: Ankylosauridae

[References: (*1) Xu X, Wang X-L, You H-L (2001). A juvenile ankylosaur from China. *Naturwissenschaften* 88: 297–300.]

Liaoningotitan sinensis

> *Lioaningotitan sinensis* Zhou *et al.*, 2018 [Barremian; Yixian Formation, Liaoning, China]

Generic name: *Liaoningotitan* ← {(place-name) Liaoning + (Gr. myth.) Τιτάν/Tītan}

Specific name: *sinensis* ← {(place-name) Sin + -ensis}

Etymology: **Liaoning titan** from China

Taxonomy: Saurischia: Sauropodomorpha: Sauropoda: Somphospondyli

[References: Zhou C-F, Wu W-H, Sekiya T, Dong Z–M (2018). A new Titanosauriformes dinosaur from Jehol Biota of western Liaoning, China. *Global Geology* 37(2): 327–333.]

Liaoningvenator curriei

> *Liaoningvenator curriei* Shen *et al.*, 2017 [Barremian; Yixian Formation, Liaoning, China]

Generic name: *Liaoningvenator* ← {(Chin. place-name) Liáoníng遼寧 + (Lat.) vēnātor: hunter}(*1)

Specific name: *curriei* ← {(person's name) Currie + -ī}; "in honor of Philip J. Currie for his outstanding contribution to the research on small theropod dinosaurs"(*1).

Etymology: Currie's **Liaoning hunter**

Taxonomy: Theropoda: Maniraptora: Troodontidae

[References: (*1) Shen C-Z, Zhao B, Gao C-L, Lü J-C, Kundrát M (2017). A new troodontid dinosaur (*Liaoningvenator curriei* gen. *et* sp. nov.) from the Early Cretaceous Yixian Formation in Western Liaoning Province. *Acta Geoscientica Sinica* 38(3): 359–371.]

Liaoxiornis delicatus [Avialae]

> *Liaoxiornis delicatus* Hou & Chen, 1999* (*nomen dubium*) [Barremian; Yixian Formation, Dawangzhangzi, Liaoning, China]

= *Lingyuanornis parvus* Ji & Ji, 1999* [Barremian; Yixian Formation, Dawangzhangzi, Liaoning, China]

**Liaoxiornis delicatus* was named one month earlier than *Lingyuanornis parvus,* and takes precedence.

Generic name: *Liaoxiornis* ← {(Chin. place-name) Liáoxī遼西: the western part of Liaoning Province (遼寧省) + (Gr.) ὄρνις/ornis: bird}

Specific name: *delicatus* ← {(Lat.) dēlicātus: alluring, small}

Etymology: small, **Liaoxi bird**

Taxonomy: Avialae: Enantiornithes

[References: Hou L & Chen P (1999). *Liaoxiornis delicates* gen. *et* sp. nov., the smallest Mesozoic bird. *Chinese Science Bulletin* 44(9): 834–838.]

Ligabueino andesi

> *Ligabueino andesi* Bonaparte, 1996 [Barremian–Aptian; La Amarga Formation, Patagonia, Argentina]

Generic name: *Ligabueino* ← {(person's name) Ligabue + (Ital. suffix) -ino: little one}; in honor of Dr. Giancarlo Ligabue, who is its discoverer.

Specific name: *andesi* ← {(place-name) Cordillera de los Andes + -ī}

Etymology: **Ligabue's little one** from the Andes

Taxonomy: Theropoda: Abelisauroidea: Noasauridae

Notes: According to the *Dinosauria 2nd edition*, this genus was considered to be Neoceratosauria *incertae sedis*.

Ligabuesaurus leanzai

> *Ligabuesaurus leanzai* Jose Bonaparte *et al.*, 2006 [late Aptian–Albian; Lohan Cura Formation, Neuquén, Argentina]

Generic name: *Ligabuesaurus* ← {(person's name) Giancarlo Rigabue + (Gr.) σαῦρος/sauros: reptile}; "in honor of Italian philanthropist, and friend Dr. Giancarlo Ligabue" (*1).

Specific name: *leanzai* ← {(person's name) Leanza + -ī}; "in honor of Dr. Héctor Leanza, geo-paleontologist, colleague and friend, who informed us about the palaeontological riches of Cerro León, Picún Leufú, Neuquén Province" (*1).

Etymology: Leanza and **Ligabue's lizard**

Taxonomy: Saurischia: Sauropodomorpha: Sauropoda: Titanosauriformes: Titanosauria

[References: (*1) Bonaparte JF, González Riga BJ, Apesteguía S (2006). *Ligabuesaurus leanzai* gen. *et* sp. nov. (Dinosauria, Sauropoda), a new titanosaur from the Lohan Cura Formation (Aptian, Lower Cretaceous) of Neuquén, Patagonia, Argentina. *Cretaceous Research* 27(3): 364–376.]

Liliensternus liliensterni

> *Halticosaurus liliensterni* von Huene, 1934 ⇒ *Liliensternus liliensterni* (von Huene, 1934) Welles, 1984 [late Norian; Trossingen Formation, Knollenmergel Thüringen, Germany]

Generic name: *Liliensternus* ← {(Lat. person's name) Lilienstern + -us}; in honor of Hugo Rühle von Lilienstern who discovered the specimens.

The *Liliensternus* fossils were transferred to the Humboldt Museum in Berlin in 1969 from his castle.

Specific name: *liliensterni* ← {(person's name) Lilienstern + -ī}

Etymology: **Liliensternus** of Lilienstern

Taxonomy: Theropoda: Neotheropoda

Other species:

> *Liliensternus airelensis* Cuny & Galton, 1993 (*nomen dubium*) ⇒ *Lophostropheus airelensis* (Cuny & Galton, 1993) Ezcurra & Cuny, 2007 [Rhaetian–Hettangian; Moon-Airel Formation, Manche, France]

[References: (*1) Mohr BAR, Kustatscher E, Hiller C, Böhme G (2008). Hugo Rühle von Lilienstern and his palaeobotanical collection: an east-west German story. *Earth Sciences History* 27(2): 278–296.]

Limaysaurus tessonei

> *Rebbachisaurus tessonei* Calvo & Salgado, 1995 ⇒ *Rayososaurus tessonei* (Calvo & Salgado, 1995) Bonaparte, 1996 ⇒ *Limaysaurus tessonei* (Calvo & Salgado, 1995) Salgado *et al.*, 2004 [Cenomanian; Candeleros Formation, Neuquén, Patagonia, Argentina]

Generic name: *Limaysaurus* ← {(place-name) Río Limay: Limay River + (Gr.) σαῦρος/sauros: lizard}; "for the Río Limay, the most important waterway in the area in which the holotype of the type species and referred material were found" (*2).

Specific name: *tessonei* ← {(person's name) Tessone + -ī}; "in honor to Mr. Lieto Tessone, discoverer of the holotype" (*1).

Etymology: Tessone's **Río Limay lizard**

Taxonomy: Saurischia: Sauropodomorpha: Sauropoda: Diplodocoidea: Diplodochimorpha: Rebbachisauridae (*2)

Notes: Gastroliths were found in fossil remains of *Limaysaurus*.

[References: (*1) Carvo JO, Salgado L (1995). *Rebbachisaurus tessonei* sp. nov. a new Sauropoda from the Albian–Cenomanian of Argentina; new evidence on the origin of the Diplodocidae. *Gaia* 11: 13–33.]; (*2) Salgado L, Garrido A, Cocca SE, Cocca JR (2004). Lower Cretaceous rebbachisaurid sauropods from Cerro Aguada del León (Lohan Cura Formation), Neuquén Province, northwestern Patagonia, Argentina. *Journal of Vertebrate Paleontology* 24(4): 903–912.]

Limenavis patagonica [Avialae]

> *Limenavis patagonica* Clarke & Chiappe, 2001 [middle Campanian; Allen Formation, Río Negro, Argentina]

Generic name: *Limenavis* ← {(Lat.) līmen: threshold + avis: bird}; "for the window it offers into the origin of the radiation of the avian crown clade" (*1)

Specific name: *patagonica* ← {(place-name) Patagonia + -icus}; "from the provenience of the specimen from northern Patagonia" (*1).

Etymology: **threshold bird** from Patagonia

Taxonomy: Theropoda: Avialae: Ornithuromorpha: Ornithurae

cf. Theropoda: Avialae: Carinatae [*1]

[References: (*1) Clarke JA, Chiappe LM (2001). A new carinate bird from the Late Cretaceous of Patagonia (Argentina). *American Novitates* 3323: 1–23.]

Limusaurus inextricabilis

> *Limusaurus inextricabilis* Xu *et al.*, 2009 [Oxfordian; Shishugou Formation, Xinjiang, China]

Generic name: *Limusaurus* ← {(Lat.) līmus: mud or mire + (Gr.) σαῦρος/sauros: lizard}

Specific name: *inextricabilis* ← {(Lat.) inextrīcābilis: impossible to extricate, inextricable}; "in reference to the specimens' inferred death in a mire" [*1].

Etymology: **lizard** inextricable from **mud**

Taxonomy: Theropoda: Ceratosauria: Noasauridae: Elaphrosaurinae

cf. Theropoda: Ceratosauria [*1]

[References: (*1) Xu X, Clark JM, Mo J, Choiniere J, Forster CA, Erickson GM, Hone DWE, Sullivan C, Eberth DA, Nesbitt S, Zhao Q, Hernandez R, Jia C-K, Han F-L, Guo Y (2009). A Jurassic ceratosaur from China helps clarify avian digital homologies. *Nature* 459: 940–944.]

Lingyuanosaurus sihedangensis

> *Lingyuanosaurus sihedangensis* Yao *et al.*, 2019 [Aptian; Yixian Formation, Gehol Group, Sihedang, Lingyuan, Liaoning, China]

Generic name: *Lingyuanosaurus* ← {(Chin. place-name) Língyuán凌源 + (Gr.) σαῦρος/sauros: lizard}

Specific name: *sihedangensis* ← {(Chin. place-name) Sìhédāng四合當: the type locality + -ensis}

Etymology: **Lingyuan lizard** from Sihedang

Taxonomy: Theropoda: Coelurosauria: Therizinosauria

[References: Yao X, Liao C-C, Sullivan C, Xu X (2019). A new transitional therizinosaurian theropod from the Early Cretaceous Jehol Biota of China. *Scientific Reports* 9(5026): 1–12.]

Lingwulong shenqi

> *Lingwulong shenqi* Xu *et al.*, 2018 [Toarcian–Bajocian; Yanan Formation, Lingwu, Yinchuan, Ningxia Hui, China]

Generic name: *Lingwulong* ← {(Chin. place-name) Língwǔ 霊武 + (Chin.) lóng龍: dragon} [*1]

Specific name: *shenqi* ← {(Chin.) shénqí 神奇: amazing}; "reflecting the unexpected discovery of a dicraeosaurid in the Middle Jurassic of China" [*1].

Etymology: amazing **dragon from Lingwu** [神奇霊武龍]

Taxonomy: Sauropoda: Neosauropoda: Diplodocoidea: Diplodocoidae

[References: (*1) Xu X, Upchurch P, Mannion PD, Barrett PM, Regalado-Fernandez OR, Mo J, Ma J, Liu H (2018). A new Middle Jurassic diplodocoid suggests an earlier dispersal and diversification of sauropod dinosaurs. *Nature Communications* 9 (2700).]

Linhenykus monodactylus

> *Linhenykus monodactylus* Xu *et al.*, 2011 [Campanian; Wulansuhai Formation, Inner Mongolia, China]

Generic name: *Linhenykus* ← {(Chin. place-name) Línhé 臨河: a city in Inner Mongolia + (Gr.) ὄνυξ/onyx: claw}

Specific name: *monodactylus* ← {(Gr.) μόνος/monos: single, one + δάκτυλος/daktylos: finger}; "referring to the presence of a single finger in this animal" [*1].

Etymology: single-fingered **Linhe claw**

Taxonomy: Theropoda: Coelurosauria: Alvarezsauroidea: Alvarezsauridae: Parvicursorinae [*1]

[References: (*1) Xu X, Sullivan C, Pittman M, Choiniere JN, Hone D, Upchurch P, Tan Q, Xiao D, Tan L, Han F (2011). A monodactyl nonavian dinosaur and the complex evolution of the alvarezsauroid hand. *Proceedings of the National Academy of Sciences* 108(6): 2338–2342.]

Linheraptor exquisitus

> *Linheraptor exquisitus* Xu *et al.*, 2010 [Campanian; Wulansuhai (Bayan Mandahu) Formation, Linhe, Inner Mongolia, China]

Generic name: *Linheraptor* ← {(Chin. place-name) Línhé臨河 + (Lat.) raptor}; "referring to the animal's status as a predatory dinosaur ('raptor') from Linhe, Nei Mongol, China (area of origin)" [*1].

Specific name: *exquisitus* ← {(Lat.) exquīsītus: excellent, exquisite}; "referring to the exceptional preservation of the holotype specimen" [*1].

Etymology: excellent, **Linhe raptor**

Taxonomy: Theropoda: Coelurosauria: Maniraptora: Dromaeosauridae

Notes: *Linheraptor* had large toe claws.

[References: (*1) Xu X, Choiniere JN, Pittman MD, Tan Q, Xiao D, Li Z, Sullivan C (2010). A new dromaeosaurid (Dinosauria: Theropoda) from the Upper Cretaceous Wulansuhai Formation of Inner Mongolia, China. *Zootaxa* (2403): 1–9.]

Linhevenator tani

> *Linhevenator tani* Xu *et al.*, 2011 [Campanian; Wulansuhai (Bayan Mandahu) Formation, Bayan Mandahu, Linhe, Inner Mongolia, China]

Generic name: *Linhevenator* ← {(Chin. place-name) Línhé臨河 + (Lat.) vēnātor: hunter}

Specific name: *tani* ← {(person's name) Tan + -ī}; "in honor of Tán Lín for his contributions

to the field of vertebrate paleontology in Inner Mongolia" [*1].
Etymology: Tan's **Linhe hunter**
Taxonomy: Theropoda: Coelurosauria: Maniraptora: Troodontidae
[References: (*1) Xu X, Tan Q, Sullivan C, Han F, Xiao D (2011). A short-armed troodontid dinosaur from the Upper Cretaceous of Inner Mongolia and its implications for troodontid evolution. *PLoS ONE* 6(9): e22916.]

Liopleurodon ferox [Plesiosauria]
> *Liopleurodon ferox* Sauvage, 1873 ⇒ *Polyptychodon ferox* (Sauvage, 1873) Sauvage, 1880 ⇒ *Pliosaurus ferox* (Sauvage, 1873) Lydekker, 1888 [Callovian; Oxford Clay, Le Wast, Pas-de-Calais, France]
= *Thaumatosaurus mosquensis*? Kiprijanoff, 1883 [Oxfordian–Kimmeridgian; Moscow Basin, Russia]
Generic name: *Liopleurodon* ← {(Gr.) λεῖος/leios: smooth + πλευρά/pleura: side, rib + ὀδών/odōn [= ὀδούς/odūs]: tooth}
Specific name: *ferox* ← {(Lat.) ferōx: wild, savage, fierce}
Etymology: savage, **smooth-sided teeth**
Taxonomy: Sauropterygia: Plesiosauria: Pliosauridae
Other species:
> *Liopleurodon grossouvrei* Sauvage, 1873 [Callovian; Charly, France]
= *Pliosaurus andrewsi* Tarlo, 1960 [Callovian; Oxford Clay, Peterborough, UK]
> *Liopleurodon bucklandi*? Sauvage, 1873 [Callovian; Caen, France]
> *Pliosaurus pachydeirus* Seeley, 1869 ⇒ *Liopleurodon pachydeirus* (Seeley, 1869) Tarlo, 1960 [Callovian; England, UK]
> *Pliosaurus rossicus* Novozhilov, 1948 ⇒ *Liopleurodon rossicus* (Novozhilov, 1948) Halstead, 1971 ⇒ *Strongylokroptaphus rossicus* (Novozhilov, 1948) Novozhilov, 1964 [Tithonian; Volga beds, Volgain Russia]
= *Pliosaurus brachydeirus* Owen, 1841 [Kimmeridgian; Kimmeridge Clay, Market Rasen, UK]
Notes: *Liopleurodon*, a short-necked plesiosaur, is seemed to have been the apex predator.

Lirainosaurus astibiae
> *Lirainosaurus astibiae* Sanz *et al.*, 1999 [late Campanian; Unit S3U1*, Laño, Condado de Treviño, Burgos, Spain]
*according to *The Dinosauria 2nd edition.*
Generic name: *Lirainosaurus* ← {(Basque) lirain: slender + (Gr.) σαῦρος/sauros: lizard} [*1]
Specific name: *astibiae* ← {(person's name) Astibia +-ae}; "devoted to Dr. Humberto Astibia, leader of the research at the outcrop of Laño" [*1].
Etymology: Astibia's **slender lizard**
Taxonomy: Saurischia: Sauropoda: Titanosauria: Saltasauridae
cf. Sauropoda: Titanosauria: Eutitanosauria [*1]
[References: (*1) Sanz JL, Powell JE, Le Loeuff J, Martínez R, Pereda-Suberbiola X (1999). Restos de saurópodos del Cretácico superior de Laño (España centroseptentrional). Relaciones filogenéticas de los Titanosaurios. [Sauropod remains from the Upper Cretaceous of Laño (north central Spain). Titanosaur phylogenetic relationships] *Estudios del Museo de Ciencias Naturales de Alava* 14(1): 235–255.]

Liubangosaurus hei
> *Liubangosaurus hei* Mo *et al.*, 2010 [Aptian; Xinlong Formation, Guangxi, China]
Generic name: *Liubangosaurus* ← {(Chin. place-name) Liùbàng六榜 + (Gr.) σαῦρος/sauros: lizard}
Specific name: *hei* ← {(Chin. person's name) Hé何 + -ī}
Etymology: He's **Liubang lizard** [何氏六榜龍]
Taxonomy: Saurischia: Sauropodomorpha: Eusauropoda
[References: Mo J, Xu X, Buffetaut E (2010). A new eusauropod dinosaur from the Lower Cretaceous of Guangxi Province, Southern China. *Acta Geologica Sinica* 84(6): 1328–1335.]

Lohuecotitan pandafilandi
> *Lohuecotitan pandafilandi* Diaz *et al.*, 2016 [Campanian–Maastrichtian; Villalba de la Sierra Formation, Lo Hueco, Cuenca, Spain]
Generic name: *Lohuecotitan* ← {(place-name) Lo Hueco: the type locality + (Gr.) titan: the giants} [*1]
Specific name: *pandafilandi* ← {Pandafilando: a character, in the novel *Don Quixote* + -ī} ; "referring to Pandafilando de la fosca vista, one of the characters in the novel '*The Ingenious Gentleman Don Quixote of La Mancha*' (*El ingenioso hidalgo don Quijote de la Mancha*) written by Miguel de Cervantes and published in the early seventeenth century. Pandafilando is, in the mind of the protagonist, a giant against who he must fight" [*1].
Etymology: Pandaflilando's **Lo Hueco giant**
Taxonomy: Saurischia: Sauropodomorpha: Sauropoda: Titanosauria: Lithostrotia
[References: Díaz VD, Mocho P, Páramo A, Escaso F, Marcos-Fernández F, Sanz JL, Ortega F (2016). A new

titanosaur (Dinosauria, Sauropoda) from the Upper Cretaceous of Lo Hueco (Cuenca, Spain). *Cretaceous Research* 68: 49–60.]

Longicrusavis houi [Avialae]

> *Longicrusavis houi* O'Connor *et al.*, 2010 [Barremian–Aptian; Yixian Formation, Liaoning Province, China]

Generic name: *Longicrusavis* ← {(Lat.) longus: long + crūs: shin, leg + avis: bird}; "referring to the elongate hindlimb and, in particular, tibiotarsus that characterizes the taxon and the Hongshanornithidae in general" (*1).

Specific name: *houi* ← {(person's name) Hou + -ī}; "in honor of Lianhai Hou in recognition of his contribution to avian paleontology of the Jehol Group" (*1).

Etymology: Hou's **long shin bird**

Taxonomy: Theropoda: Avialae: Ornithuromorpha: Hongshanornithidae

[References: (*1) O'Connor JK, Gao K-Q, Chiappe LM (2010). A new ornithuromorph (Aves: Ornithothoraces) bird from the Jehol Group indicative of higher-level diversity. *Journal of Vertebrate Paleontology* 30(2): 311–321.]

Longipteryx chaoyangensis [Avialae]

> *Longipteryx chaoyangensis* Zhang *et al.*, 2001 [Aptian; Jiufotang Formation, Chaoyang, Liaoning, China]

= ? *Camptodontus yangi* Li *et al.*, 2010 (preoccupied) ⇒ *Camptodontornis yangi* (Li *et al.*, 2010) Demirjian, 2019 [Aptian; Jiufotang Formation, Liaoning, China]

Generic name: *Longipteryx* ← {(Lat.) longus: long + (Gr.) πτέρυξ/pteryx: wing, feather, pinion}

Specific name: *chaoyangensis* ← {(Chin. place-name) Cháoyáng City朝陽市 + -ensis}

Etymology: **long-wing bird** from Chaoyang [朝陽長翼鳥]

Taxonomy: Theropoda: Avialae: Enantiornithes: Longipterygiformes: Longipterygidae (*1)

Notes: According to Wang *et al.* (2015), *Camtodontornis yangi* is considered to be a probable synonym of *Longipteryx chaoyangensis*.

[References: (*1) Zhang F, Zhou Z, Hou L, Gu G (2001). Early diversification of bird: evidence from a new opposite bird. *Chinese Science Bulletin* 46(11): 945–949.]

Longirostravis hani [Avialae]

> *Longirostravis hani* Hou *et al.*, 2004 [Barremian–Aptian; Yixian Formation, Yixian, Liaoning, China]

Generic name: *Longirostravis* ← {(Lat.) longus: long + rōstrum: rostrum, beak + avis: bird} (*1)

Specific name: *hani* ← {(person's name) Han + -ī}; in honor of Mr. Han, discoverer. (*1)

Etymology: Han's **bird with a long rostrum**

Taxonomy: Theropoda: Avialae: Enantiornithes: Longipterygidae

[References: (*1) Hou L, Chiappe LM, Zhang F, Chuong C-M (2004). New Early Cretaceous fossil from China documents a novel trophic specialization for Mesozoic birds. *Naturwissenschaften* 91(1): 22–25.]

Longusunguis kurochkini [Avialae]

> *Longusunguis kurochkini* Wang *et al.*, 2014 [Aptian; Jiufotang Formation, Liaoning, China]

Generic name: *Longusunguis* ← {(Lat.) longus: long + unguis: claw}; "referring to the distinctly long pedal unguals that characterize this taxon and other bohaiornithids" (*1).

Specific name: *kurochkini* ← {(person's name) Kurochkin + -ī}; in honor of the late Prof. Evgeny Kurochkin, an eminent paleontologist. (*1)

Etymology: Kurochkin's **long claw**

Taxonomy: Theropoda: Avialae: Enantiornithes: Bohaiornithidae

cf. Aves: Ornithothoraces: Enantiornithes: Bohaiornithidae (*1)

[References: (*1) Wang M, Zhou Z-H, O'Connor JK, Zelenkv NV (2014). A new diverse enantiornithine family (Bohaiornithidae fam. nov.) from the Lower Cretaceous of China with information from two new species. *Vertebrata PalAsiatica* 52(1): 31–76.]

Lophorhothon atopus

> *Lophorhothon atopus* Langston, 1960 [early Campanian; Mooreville Chalk Member, Selma Formation, Alabama, USA]

Generic name: *Lophorhothon* ← {(Gr.) "λόφος/lophos: a ridge or crest on the head + ῥώθων/rhōthōn: nose" (*1)}

Specific name: *atopus* ← {(Gr.) ἄτοπος/atopos: strange}; "out of place, allusion to the form and position of the nasal crest" (*1).

Etymology: strange **crested nose**

Taxonomy: Ornithischia: Ornithopoda: Hadrosauroidea: Hadrosauromorpha

[References: (*1) Langston W (1960). The vertebrate fauna of the Selma Formation of Alabama, part VI: The dinosaurs. *Fieldiana, Geology Memoirs* 3(5): 315–361.]

Lophostropheus airelensis

> *Liliensternus airelensis* Cuny & Galton, 1993 ⇒ *Lophostropheus airelensis* (Cuny & Galton, 1993) Ezcurra & Cuny, 2007 [uppermost Rhaetian–early Hettangian; Moon-Airel

Formation, Manche, Normandy, France]

Generic name: *Lophostropheus* ← {(Gr.) λόφος/lophos: crest + στροφή/strophe: vertebrae}; "in allusion to the prominent dorsal and ventral laminae present in the postaxial cranial cervical vertebrae" (*1).

Specific name: *airelensis* ← {(place-name) Airel Quarry + -ensis}

Etymology: **crested vertebrae** from Airel Quarry

Taxonomy: Theropoda: Coelophysoidea

[References: (*1) Ezcurra MD, Cuny G (2007). The coelophysoid *Lophostropheus airelensis*, gen. nov.: a review of the systematics of "*Liliensternus*" *airelensis* from the Triassic–Jurassic outcrops of Normandy (France). *Journal of Vertebrate Paleontology* 27(1): 73–86.]

Loricatosaurus priscus

> *Stegosaurus priscus* Nopcsa, 1911 ⇒ *Lexovisaurus priscus* (Nopcsa, 1911) Kuhn, 1964 ⇒ *Loricatosaurus priscus* (Nopcsa, 1911) Maidment *et al*., 2008 [Callovian; Oxford Clay, Peterborough, UK]

Generic name: *Loricatosaurus* ← {(Lat.) lōrīcātus: armoured, clothed in mail + (Gr.) σαῦρος/sauros: reptile} (*1)

Specific name: *priscus* ← {(Lat.) prīscus: ancient}

Etymology: ancient, **armoured lizard**

Taxonomy: Ornithischia: Thyreophora: Stegosauria: Stegosauridae

[References: (*1) Maidment SCR, Norman DB, Barrett PM, Upchurch P (2008). Systematics and phylogeny of Stegosauria (Dinosauria: Ornithischia). *Journal of Systematic Palaeontology* 6(4): 367–407.]

Loricosaurus scutatus

> *Titanosaurus australis* Lydekker, 1893 ⇒ *Saltasaurus australis* (Lydekker, 1893) McIntosh, 1990 ⇒ *Neuquensaurus australis*? (Lydekker, 1893) Powell, 1992 [Santonian; Bajo de la Carpa Formation, Argentina]

= *Loricosaurus scutatus* von Huene, 1929 [late Campanian–early Maastrichtian; Allen Formation, Río Negro, Argentina]

For the other synonyms, see *Neuquensaurus australis*.

Generic name: *Loricosaurus* ← {(Lat.) lōrīca: a leather cuirass + (Gr.) σαῦρος/sauros: lizard}

Specific name: *scutatus* ← {(Lat.) scūtātus: armed with a long shield}

Etymology: **cuirass lizard** armed with a long shield

Taxonomy: Saurischia: Sauropodomorpha: Sauropoda: Titanosauria: Saltasauridae

Notes: According to *The Dinosauria 2nd edition*, *Loricosaurus scutatus* is considered possibly a *synonym* of *Neuquensaurus australis*.

Losillasaurus giganteus

> *Losillasaurus giganteus* Casanovas *et al*., 2001 [Kimmeridgian–Tithonian; Villar del Arzobispo Formation, Los Serranos Basin, Valencia, Spain]

Generic name: *Losillasaurus* ← {(place-name) Losilla: "village in the district of Serranos (Valencia, Spain)" + (Gr.) σαῦρος/ sauros: lizard} (*1)

Specific name: *giganteus* ← {(Lat.) gigantēus: giant}; "due to its great size". (*1)

Etymology: giant **lizard of Losilla**

Taxonomy: Saurischia: Sauropodomorpha: Sauropoda: Triasauria

[References: (*1) Casanovas ML, Santafé JV, Sanz JL (2001). *Losillasaurus giganteus*, un nuevo saurópodo del tránsito Jurásico–Cretácico e la Cuenca de "Los Serranos" (Valencia, España)" [*Losillasaurus giganteus*, a new sauropod from the transitional Jurassic–Cretaceous of the Los Serranos basin (Valencia, Spain)]. *Paleontologia i Evolució* 32–33: 99–122.]

Lourinhanosaurus antunesi

> *Lourinhanosaurus antunesi* Mateus, 1998 [Kimmeridgian–Tithonian; Lourinhã Formation, Lourinhã, Portugal]

Generic name: *Lourinhanosaurus* ← {(place-name) Lourinhã + (Gr.) σαῦρος/sauros: lizard} (*1)

Specific name: *antunesi* ← {(person's name) Antunes + -ī}; "after Prof. Miguel Telles Antunes, Portuguese palaeontologist" (*1).

Etymology: Antunes' **Laurinhã lizard**

Taxonomy: Theropoda: Avetheropoda: Carnosauria: Allosauroidea

Notes: The remains of several embryos among the some hundred eggs were recovered at Paimogo. (*2)

[References: (*1) Mateus O (1998). *Lourinhanosaurus antunesi*, a new Upper Jurassic allosauroid (Dinosauria: Theropoda) from Lourinhã (Portugal). *Memórias da Academia de Ciências de Lisboa* 37: 111–24.; (*2) Mateus I, Mateus H, Antunes MT, Mateus O, Taquet P, Ribeiro V, Manuppella G (1998). Upper Jurassic theropod dinosaur embryos from Lourinhã (Portugal). *Memórias da Academia de Ciências de Lisboa* 37: 101–109.]

Lourinhasaurus alenquerensis

> *Apatosaurus alenquerensis* Lapparent & Zbyszewski, 1957⇒ *Atlantosaurus alenquerensis* (Lapparent & Zbyszewski, 1957) Steel, 1970 ⇒*Brontosaurus alenquerensis* (Lapparent & Zbyszewski, 1957) Olshevsky, 1978 ⇒ *Camarasaurus alenquerensis* (Lapparent

& Zbyszewski, 1957) McIntosh, 1990 ⇒ *Lourinhasaurus alenquerensis* (Lapparent & Zbyszewski, 1957) Dantas *et al.*, 1998 [Oxfordian–Kimmeridgian; Sobral Member, Lourinhã Formation, Lourinhã, Portugal]

Generic name: *Lourinhasaurus* ← {(place-name) Lourinhã + (Gr.) σαῦρος/sauros: lizard}; "from Lourinã, an area in west-central Portugal (Lisbon district) rich in remains of sauropod dinosaurs" (*1).

Specific name: *alenquerensis* ← {(place-name) Alenquer: a municipality north of Lisbon + -ensis}

Etymology: **Lourinhã lizard** from Alenquer

Taxonomy: Saurischia: Sauropodomorpha: Sauropoda: Camarasauridae

[References: Dantas P, Sanz JL, Marques Da Silva C, Ortega F, dos Santos VF, Cachão M (1998). *Lourinhasaurus* n. gen. novo dinossáurio saurópode do Jurássico superior (Kimmeridgiano superior–Tithoniano inferior) de Portugal [*Lourinhasaurus* n. gen. A new sauropod dinosaur from the Upper Jurassic (upper Kimmeridgian–lower Tithonian) of Portugal]. *Comunicações do Instituto Geológico e Mineiro* 84 (1A): 91–94.]

Luanchuanraptor henanensis

> *Luanchuanraptor henanensis* Lü *et al.*, 2007 [Late Cretaceous; Qiupa Formation, Luanchuan, Henan, China]

Generic name: *Luanchuanraptor* ← {Chin. place-name) Luánchuān欒川 + (Lat.) raptor: thief}

Specific name: *henanensis* ← {(Chin. place-name) Hénán河南 + -ensis}

Etymology: **Luanchuan raptor (thief)** from Henan

Taxonomy: Theropoda: Dromaeosauridae

Lucianovenator bonoi

> *Lucianovenator bonoi* Martinez & Apaldetti, 2017 [Norian–Rhaetian; Quebrada del Barro Formation, San Juan, Argentina]

Generic name: *Lucianovenator* ← {(person's name) Luciano Leyes + (Lat.) vēnātor: hunter}; "in honor of Don Luciano Leyes, inhabitant of the small town Balde de Leyes, who in 2001 called the attention of RNM* about 'some white bones buried in their lands'" (*1).

*RNM stands for the first author Ricardo N. Martínez.

Specific name: *bonoi* ← {(person's name) Bono + -ī}; in honor of "Tulio del Bono, main authority of the Secretaría de Ciencia, Técnica e Innovación of the Government of San Juan, who is strongly collaborating and helping to develop the paleontology research in San Juan Province" (*1).

Etymology: Bono and **Luciano's hunter**

Taxonomy: Theropoda: Neotheropoda: Coelophysoidea: Coelophysidae

[References: (*1) Martínez RN, Apaldetti C (2017). A late Norian- Rhaetian coelophysid neotheropod (Dinosauria, Saurischia) from the Quebrada del Barro Formation, northwestern Argentina. *Ameghiniana* 54(5): 488–505.]

Lufengosaurus huenei

> *Lufengosaurus huenei* Young, 1941 ⇒ *Massospondylus huenei* (Young, 1941) Cooper, 1981 [Hettangian; Lufeng Formation, Shawan, Lufeng, Yunnan, China]

= ?*Gyposaurus sinensis* Young, 1941 ⇒ *Anchisaurus sinensis* (Young, 1941) Dong & Olshevsky, 1991 [Hettangian; Lufeng Formation, Shawan, Yunnan, China]

= *Lufengosaurus magnus* Young, 1947 [Hettangian; Lufeng Formation, Tachung, Yunnan, China]

= *Tawasaurus minor* Young, 1982 [Early Jurassic; Lufeng Formation, Yunnan, China]

Generic name: *Lufengosaurus* ← {(Chin. place-name) Lùfēng祿豐/禄豊 + (Gr.) σαῦρος/sauros: lizard}

Specific name: *huenei* ← {(person's name) Huene + -ī}; in honor of German paleontologist Friedrich von Huene.

Etymology: Huene's **Lufeng lizard**

Taxonomy: Saurischia: Sauropodomorpha: Massospondylidae

Other species:

> *Lufengosaurus changduensis* Zhao, 1985 [Early Jurassic; Changdu*昌都, Tibet, China] (*nomen dubium*) *a city known Chamdo

Notes: According to Lee *et al.* (2017), they found evidence of protein inside vascular canals in the rib of a *Lufengosaurus* fossil.

Luoyanggia liudianensis

> *Luoyanggia liudianensis* Lü *et al.*, 2009 [Aptian–Albian*, Haoling Formation, Ruyang Basin, Henan, China]

*The type horizon of *Luoyanggia* was thought to date Cenomanian stage (Lü *et al.*, 2009), but the recent research indicates an Aptian–Albian (Xu *et al.*, 2012).

Generic name: *Luoyanggia* ← {(Chin. place-name) Luòyáng洛陽 + -ia}

Specific name: *liudianensis* ← {(Chin. place-name) Liúdiàn劉店 + -ensis}

Etymology: **Ruoyang's one** from Liudian

Taxonomy: Theropoda: Caenagnathoidea: Oviraptoridae

[References: Lü J, Xu L, Jiang X, Jia S, Lì M, Yuan C, Zhang X, Ji Q (2009). A preliminary report on the new dinosaurian fauna from the Cretaceous of the Ruyang Basin, Henan Province of central China. *Journal of the Palaeontological Society of Korea* 25: 43–56.]

Lurdusaurus arenatus

> "*Gravisaurus tenerensis*" Chabli, 1988 (invalid name) ⇒ *Lurdusaurus arenatus* Taquet & Russell, 1999 [late Aptian: Elrhaz Formation, Agadez, Niger]

Generic name: *Lurdusaurus* ← {lurdus: heavy + (Gr.) σαῦρος/sauros: lizard}

Specific name: *arenatus* ← {(Lat.) arēnātus: sandy}; in reference to the Ténéré Desert.

Etymology: sandy, **heavy lizard**

Taxonomy: Ornithischia: Ornithopoda: Hadrosauriformes

Notes: "*Gravisaurus tenerensis*" meaning "heavy lizard from Ténéré desert" was regarded as an invalid name because the name was not published, and later Taquet & Russell (1999) formally gave the remains the name *Lurdusaurus arenatus*.

[References: Taquet P, Russell DA (1999). A massively-constructed iguanodont from Gadoufaoua, Lower Cretaceous of Niger. *Annales de Paléontologie* 85(1): 85–96.]

Lusitanosaurus liasicus

> *Lusitanosaurus liasicus* Lapparent & Zbyszewski, 1957 [Sinemurian?; near São Pedro de Moel, Leiria, Portugal]

Generic name: *Lusitanosaurus* ← {(Lat. place-name) Lusitania: now Portugal and part of Spain + (Gr.) σαῦρος/sauros: lizard}

Specific name: *liasicus* ← {[geol.] Lias: the Early Jurassic period + -icus}

Etymology: **Portuguese lizard** of Early Jurassic

Taxonomy: Ornithischia: Thyreophora

Notes: According to *The Dinosauria 2nd edition*, *Lusitanosaurus liasicus* is regarded as a *nomen dubium*.

Lusotitan atalaiensis

> *Brachiosaurus atalaiensis* Lapparent & Zbyszewski, 1957 ⇒ *Astrodon atalaiensis* (Lapparent & Zbyszewski, 1957) Kinghan, 1962 ⇒ *Lusotitan atalaiensis* (Lapparent & Zbyszewski, 1957) Antunes & Mateus, 2003 (lectotype) [Kimmeridgian–Tithonian; Lourinhã Formation, Lourinhã, Portugal]

Generic name: *Lusotitan* ← {(Lat. demonym) "Luso: an inhabitant of Lusitania*" (*1) + (Gr. myth.) Τῑτάν/Tītan: "a mythological giant" (*1)}

* "an ancient region that partly corresponds to Portugal" (*1)

Specific name: *atalaiensis* ← {(place-name) Atalaia + -ensis}

Etymology: **Portuguese giant** from Atalaia

Taxonomy: Saurischia: Sauropodomorpha: Titanosauriformes: Brachiosauridae

[References: de Lapparent AF, Zbyszewski G (1957). Les dinosauriens du Portugal [Dinosaurs of Portugal]. *Mémoires des Services Géologiques du Portugal, nouvelle série* 2: 1–63.; (*1) Antunes MT, Mateus O (2003). Dinosaurs of Portugal. *Comptes Rendus Palevol* 2(1): 77–95.]

Lusovenator santosi

> *Lusovenator santosi* Malafaia *et al.*, 2020 [upper Kimmeridgian; Praia da Amoreira-Porto Novo Formation, Lourinhã, Portugal]

Generic name: *Lusovenator* ← {(Lat.) Luso: Lusitania + vēnātor: hunter}; "referring to Lusitania, the province in Roman Hispania related to the current Portugal".

Specific name: *santosi* ← {(person's name) Santos + -ī}; "after José Joaquim dos Santos, who found and collected the holotype".

Etymology: Santos' **Portuguese hunter**

Taxonomy: Theropoda: Allosauroidea: Carcharodontosauria

Lutungutali sitwensis [Silesauridae]

> *Lutungutali sitwensis* Peecook *et al.*, 2013 [Middle Triassic; Ntawera Formation, Zambia]

Generic name: *Lutungutali* ← {(Bemba) lutungu: hip + tali: high}

Specific name: *sitwensis* ← {(place-name) Sitwe: the village of Sitwe + -ensis}

Etymology: **high hip** from Sitwe

Taxonomy: Dinosauriformes: Silesauridae

Lycorhinus angustidens

> *Lycorhinus angustidens* Haughton, 1924 [Hettangian; upper Elliot Formation, Paballong, Mount Fletcher, Cape (now, Eastern Cape), South Africa]

= *Lanasaurus scalpridens* Gow, 1975 [Hettangian; upper Elliot Formation, Free State, South Africa]

Generic name: *Lycorhinus* ← {(Gr.) λύκος/lykos: wolf + ῥῑνός/rhinos [ῥίς/rhis: snout]}; named for the "canine" teeth. (*1)

Specific name: *angustidens* ← {(Lat.) angustus: constricted, narrow + dēns: tooth}

Etymology: **wolf snout** with constricted teeth

Taxonomy: Ornithischia: Ornithopoda: Heterodontosauridae

Other species:

> *Lycorhinus consors** Thulborn, 1974 ⇒ *Abrictosaurus consors* (Thulborn, 1974) Hopson, 1975 [Hettangian; upper Elliot Formation, Noosi, Qacha's Nek, Lesotho]
>
> *(Lat.) consors: spouse, in allusion to suspected feminine gender of the holotype (Thulborn, 1974).

Notes: *Lycorhinus* was first thought to be a cynodont. (*1)

[References: (*1) Haughton SH (1924). The fauna and stratigraphy of the Stormberg Series. *Annals of the South African Museum* 12: 323–497.; Thulborn RA (1970). The systematic position of the Triassic ornithischian dinosaur *Lychorhinus angustidens*. *Zoological Journal of the Linnean Society* 49(3): 235–245.]

Lythronax argestes

> *Lythronax argestes* Loewen *et al.*, 2013 [middle Campanian; Wahweap Formation, Kane County, Utah, USA]

Generic name: *Lythronax* ← {(Gr.) λύθρον/lythron: gore + ἄναξ/anax: king} (*1)

Specific name: *argestes* ← {(Gr.) ἀργεστής/argestēs: the Homeric wind* from the southwest}; "in reference to the geographic location of the specimen within North America" (*1).

*Homeric winds;
North Winds: Zephyrus (NW), Boreas (N), Eurus (NE)
South Winds: Argestes (SW), Notos (S), Apeliotes (SE)

Etymology: southwest wind's **gore king**

Taxonomy: Theropoda: Coelurosauria: Tyrannosauroidea: Tyrannosauridae: Tyrannosaurinae (*1)

[References: (*1) Loewen MA, Irmis RB, Sertich JJW, Currie PJ, Sampson SD (2013). Tyrant dinosaur evolution tracks the rise and fall of Late Cretaceous oceans. *PLoS ONE* 8(11): e79420.]

M

Maaqwi cascadensis [Avialae]

> *Maaqwi cascadensis* McLachlan *et al.*, 2017 [Campanian; Northumberland Formation, Hornby Island, British Columbia, Canada]

Generic name: *Maaqwi* ← {(Coast Salish) ma'aqwi: "water bird"} (*1)

Specific name: *cascadensis* ← {(place-name) Cascadia + -ensis}; "reflecting provenance in the Cascadia region of western North America" (*1).

Etymology: **water bird** from Cascadia

Taxonomy: Avialae: Ornithuraces: Ornithuromorpha: Ornithurae: Vegaviidae (*1)

[References: (*1) McLachlan SMS, Kaiser GW, Longrich NR (2017). *Maaqwi cascadensis*: A large, marine diving bird (Avialae: Ornithurae) from the Upper Cretaceous of British Columbia, Canada. *PLoS ONE* 12(12): e0189473.]

Machairasaurus leptonychus

> *Machairasaurus leptonychus* Longrich *et al.*, 2010 [Campanian; Bayan Mandahu Formation, Inner Mongolia, China]

Generic name: *Machairasaurus* ← {(Gr.) μάχαιρα/machaira: a forward-curving sword + σαῦρος/sauros: lizard}; "referring to the shape of the unguals" (*1).

Specific name: *leptonychus* ← {(Gr.) λεπτός/leptos: slender + onychos [ὄνυξ/onyx: claw]}

Etymology: **sword lizard** with slender claw

Taxonomy: Theropoda: Maniraptora: Oviraptorosauria: Oviraptoridae

[References: (*1) Longrich NR, Currie PJ, Dong Z-M (2010). A new oviraptorid (Dinosauria: Theropoda) from the Upper Cretaceous of Bayan Mandahu, Inner Mongolia. *Palaeontology* 53(5): 945–960.]

Machairoceratops cronusi

> *Machairoceratops cronusi* Lund *et al.*, 2016 [Campanian; Wahweap Formation, Utah, USA]

Generic name: *Machairoceratops* ← {(Gr.) μᾰχαιρίς/machairis: bent sword, razor + ceratops [(Gr.) κέρας/keras: horn + ὤψ/ōps: face]}; "in reference to the posterodorsally projecting, anteriorly curved epiparietal ornamentation, and ceratops (Latinized Greek), horned-face" (*1).

Specific name: *cronusi* ← {Cronus [(Gr. myth.) Κρόνος/Kronos] + -ī}; "referring to the Greek god Cronus who, according to mythology, deposes his father Uranus with a sickle or scythe, and as such is depicted carrying a curved bladed weapon" (*1).

Etymology: Cronus' **sword horned face (ceratopsian)**

Taxonomy: Ornithischia: Ceratopsia: Ceratopsidae: Centrosaurinae

[References: (*1) Lund EK, O'Connor PM, Loewen MA, Jinnah ZA (2016). A new centrosaurine ceratopsid, *Machairoceratops cronusi* gen. *et* sp. nov., from the Upper Sand Member of the Wahweap Formation (middle Campanian), southern Utah. *PLoS ONE* 11(5): e0154403.]

Macrocollum itaquii

> *Macrocollum itaquii* Müller *et al.*, 2018 [Norian; Candelária Sequence, Agudo, Rio Grande do Sul, Brazil]

Generic name: *Macrocollum* ← {(Gr.) μακρός/makros: long + (Lat.) collum: neck}; "referring to the elongated neck of the new taxon" (*1).

Specific name: *itaquii* ← {(person's name) Itaqui + -ī}; "in honor of Mr. José Jerundino Machado Itaqui, one of main actors behind the creation of CAPPA/UFSM" (Centro de Apoio à Psequisa Paleontológica da Quarta Colônia / Universidade Federal de Santa Maria) (*1).

Etymology: Itaqui's **long neck**

Taxonomy: Saurischia: Sauropodomorpha: Unaysauridae

[References: (*1) Müler T, Langer MC, Dias-da-Silva S (2018). An exceptionally preserved association of complete dinosaur skeletons reveals the oldest long-necked sauropodomorphs. *Biology Letters* 14: 20180633.]

Macrogryphosaurus gondwanicus

> *Macrogryphosaurus gondwanicus* Calvo *et al.*, 2007 [Coniacian; Portezuelo Formation, Neuquén Group, Patagonia, Argentina]

Generic name: *Macrogryphosaurus* ← {(Gr.) μάκρος/macros: big + γρίφος/griphos: enigmatic + σαῦρος/sauros: lizard} (*1)

Specific name: *gondwanicus* ← {[geol] Gondwana + -icus}; "in reference to the Gondwana continent" (*1).

Etymology: **big enigmatic lizard** from Gondwana

Taxonomy: Ornithischia: Neornithischia: Elasmaria

[References: (*1) Calvo JO, Porfiri JD, Novas FE (2007). Discovery of a new ornithopod dinosaur from the Portezuelo Formation (Upper Cretaceous), Neuquén, Patagonia, Argentina. *Arquivos do Museu Nacional, Rio de Janeiro* 65(4): 471–483.]

Macrurosaurus semnus

> *Macrurosaurus semnus* Seeley, 1869 (*nomen nudum*) ⇒ *Macrurosaurus semnus* Seeley, 1876 (*nomen dubium*) [Cenomanian; West Melbury Marly Chalk Formation, Coldham Common, Barnwell, UK]

Generic name: *Macrurosaurus* ← {(Gr.) μακρός/macros: long, large + οὖρα/ūra: tail + σαῦρος/sauros: lizard}

Specific name: *semnus* ← {(Gr.) σεμνός/semnos: stately, impressive}

Etymology: stately, **long-tailed lizard**

Taxonomy: Saurischia: Sauropodomorpha: Sauropoda: Titanosauriformes: Titanosauria

Other species:

> *Acanthopholis platypus* Seeley, 1869 (*nomen dubium*) ⇒ *Macrurosaurus platypus* (Seeley, 1869) von Huene, 1956 [early Cenomanian; upper Greensand Formation, England, UK]

Notes: According to *The Dinosauria 2nd edition*, *Macrurosaurus semnus* and *Acanthopholis platypus* are considered as *nomina dubia*.

[References: Seeley HG (1876). On *Macrurosaurus semnus* (Seeley), a long tailed animal with procoelous vertebrae from the Cambridge Upper Greensand, preserved in the Woodwardian Museum of the University of Cambridge. *Quarterly Journal of the Geological Society* 32: 440–444.]

Magnamanus soriaensis

> *Magnamanus soriaensis* Vidarte *et al.*, 2016 [Early Cretaceous; Golmayo Formation, Soria, Spain]

Generic name: *Magnamanus* ← {(Lat.) magnus: great, large + manus: hand}

Specific name: *soriaensis* ← {(place-name) Soria + -ensis}

Etymology: **great hand** from Soria

Taxonomy: Ornithischia: Ornithopoda: Iguanodontia: Ankylopollexia: Styracosterna

[References: Vidarte CF, Calvo MM, Fuentes FM, Fuentes MM (2016). Un Nuevo dinosaurio estiracosterno (Ornithopoda: Ankylopollexia) del Cretácico Inferior de españa [A new styracosternan dinosaur (Ornithopoda: Ankylopollexia) from the Lower Cretaceous of Spain]. *Spanish Jounal of Palaeontology* 31(2): 407–446.]

Magnapaulia laticaudus

> *Lambeosaurus laticaudus* Morris, 1981 ⇒ *Magnapaulia laticaudus* (Morris, 1981) Prieto–Márquez *et al.*, 2012 [Campanian; El Gallo Formation, Baja California Norte, Mexico]

Generic name: *Magnapaulia* ← {(Lat.) magna [magnus: large] + (person's name) Paul + -ia}; "referring to the unusually large size reached by at least some specimens of this lambeosaurine, and in honor of Mr. Paul Haaga for his outstanding support to the research and public programs of the Natural History Museum of Los Angeles County and its Dinosaur Institute" (*1).

Specific name: *laticaudus* ← {(Lat.) lati- [latus: wide] + caudus [cauda: tail]}

Etymology: **Paul's big one** with a wide tail

Taxonomy: Ornithischia: Ornithopoda: Hadrosauridae: Lambeosaurini

Notes: *Magnapaulia laticaudus* was first described as a species of *Lambeosaurus* and in 2012, it was given a new own genus.

[References: (*1) Prieto-Márquez A, Chiappe LM, Joshi SH (2012). The lambeosaurine dinosaur *Magnapaulia laticaudus* from the Late Cretaceous of Baja California, Northwestern Mexico. *PLoS ONE* 7(6): e38207.]

Magnirostris dodsoni

> *Bagaceratops rozhdestvenskyi* Maryańska & Osmólska, 1975 [Campanian; Barun Goyot

Formation, Ömnögovi, Mongolia]
= *Magnirostris dodsoni* You & Dong, 2003 [Campanian; Bayan Mandahu Formation, Inner Mongolia, China]

Generic name: *Magnirostris* ← {(Lat.) magnus: big + rostrum: beak, snout}

Specific name: *dodsoni* ← {(person's name) Dodson + -ī}; in honor of paleontologist Peter Dodson.

Etymology: Dodson's **big snout**

Taxonomy: Ornithischia: Marginocephalia: Ceratopsia: Protoceratopsidae

[References: You H-L, Dong Z-M (2003). A new protoceratopsid (Dinosauria: Neoceratopsia) from the Late Cretaceous of Inner Mongolia, China. *Acta Geologica Sinica* 77(3): 299–303.]

Magnosaurus nethercombensis

> *Megalosaurus nethercombensis* von Huene, 1926 ⇒ *Magnosaurus nethercombensis* (von Huene, 1926) von Huene, 1932 [Bajocian; Inferior Oolite, Dorset, UK]

Generic name: *Magnosaurus* ← {(Lat.) magnus: large, great + (Gr.) σαῦρος/sauros: lizard}

Specific name: *nethercombensis* ← {(place-name) Nethercomb + -ensis}

Etymology: **large lizard** from Nethercomb

Taxonomy: Theropoda: Tetanurae: Megalosauridae: Afrovenatorinae
cf. Theropoda: Tetanurae *incertae sedis* (acc. *The Dinosauria 2nd edition*)

Other species:

> *Megalosaurus woodwardi* Lydekker, 1909 (*nomen dubium*) [Hettangian; Blue Lias Formation, Lyme Regis, Warwickshire, UK]
= *Megalosaurus lydekkeri* von Huene, 1926 ⇒ *Magnosaurus lydekkeri* (von Huene, 1926) von Huene, 1932 (*nomen dubium*) [Hettangian; Blue Lias Formation, Lyme Regis, UK]
> *Sarcosaurus woodi* Andrews, 1921 ⇒ *Magnosaurus woodi* (Andrews, 1921) von Huene, 1932 [Hettangian–Sinemurian; England, UK]
= *Sarcosaurus andrewsi* von Huene, 1932 ⇒ *Megalosaurus andrewsi* (von Huene, 1932) Waldman, 1974 (*nomina dubia*) [Hettangian; Blue Lias Formation, Warwickshire, UK]
(=)* *Magnosaurus woodwardi* von Huene, 1932 (*partim*) [Hettangian; Blue Lias Formation, Warwickshire, UK]
*objective synonym of *Sarcosaurus andrewsi*
> *Eustreptospondylus oxoniensis* Walker, 1964 ⇒ *Magnosaurus oxoniensis* (Walker, 1964) Rauhut, 2003 [Callovian; Oxford Clay Formation, Wolvercote, Oxfordshire, UK]

Notes: In 1932, von Huene gave *Megalosaurus nethercombensis* a new name *Magnosaurus nethercombensis*. Until the 1990s, it had been seen as a species of *Megalosaurus*. According to *The Dinosauria 2nd edition*, *Magnosaurus nethercombensis* is considered to be a valid taxon. Concerning to *Megalosaurus lydekkeri* and *Magnosaurus woodwardi,* they are regarded as synonyms of *Sarcosaurus woodi* because *Sarcosaurus* has priority.

[References: Holtz TR, Molnar RE, Currie PJ (2004). Basal Tetanurae. In: Weishampel DB, Dodson P, Osmólska H (eds). *The Dinosauria 2nd edition*. University of California Press, Berkley: 71–110.]

Magyarosaurus dacus

> *Titanosaurus dacus* Nopcsa, 1915 ⇒ *Magyarosaurus dacus* (Nopcsa, 1915) von Huene, 1932 [Campanian–Maastrichtian; Sânpetru Formation, Szentpéterfalva, Romania]
= *Magyarosaurus hungaricus* von Huene, 1932 [Campanian–Maastrichtian; Sânpetru Formation, Szentpéterfalva, Romania]
= *Magyarosaurus transsylvanicus* von Huene, 1932 [Campanian–Maastrichtian; Sânpetru Formation, Szentpéterfalva, Romania]

Generic name: *Magyarosaurus* ← {(Finno-Ugric, demonym*) Magyar: derived from the name of Hungarian tribe + (Gr.) σαῦρος/sauros: lizard}
*demonym [(Gr.) demo-[δῆμος/dēmos: people] + -onym [ὄνομα/onoma or ὄνῦμα/onyma: a name]

Specific name: *dacus* ← {(Lat.) Dacus}; in honor of the Dacians of ancient Romania.

Dacia is located in the area near the Carpathian Mountains and west of the Black Sea.

Etymology: Dacian **Magyars' lizard**

Taxonomy: Saurischia: Sauropodomorpha: Sauropoda: Titanosauria

Notes: *Magyarosaurus* had a long neck and a long tail. Fossil eggs, which have been attributed to *Magyarosaurus*, were discovered in the Maastrichtian Sânpetru Formation, Hunedoara. (*1)

[References: (*1) Grellet-Tinner G, Codrea V, folie A, Higa A, Smith T (2012). First evidence of reproductive adaptation to "island effect" of a dwarf Cretaceous Romanian titanosaur, with embryonic integument in ovo. *PLoS ONE* 7(3): e32051.]

Mahakala omnogovae

> *Mahakala omnogovae* Turner *et al.*, 2007 [Campanian; Djadokhta Formation, Ömnögovi, Mongolia]

Generic name: *Mahakala* ← {(Sanskrit) mahā: great + kāla: black, time}; "one of the eight protector

deities (dharmapalas) in Tibetan Buddhism"(*1).
Specific name: *omnogovae* ← {(place-name) Ömnögov + -ae}; "referring to the southern Gobi provenance of this taxon"(*1).
Etymology: **Mahakala** from Ömnögovi
Taxonomy: Theropoda: Dromaeosauridae: Halszkaraptorinae

[References: (*1) Turner AH, Pol D, Clarke JA, Erickson GM, Norell MA (2007). A basal dromaeosaurid and size evolution preceding avian flight. *Science* 317(5843): 1378–1381.]

Mahuidacursor lipanglef

> *Mahuidacursor lipanglef* Cruzado-Caballero *et al.*, 2019 [Santonian; Bajo de la Carpa Formation, Neuquén Basin, Argentina]

Generic name: *Mahuidacursor* ← {(Mapudungun) mahuida = (Sp.) montaña: mountain + (Lat.) cursor: runner}
Specific name: *lipanglef* ← {(Mapuche) lipanglef = (Sp.) de brazos ligeros: with light arms}
Etymology: **mountain runner** with light arms [**corredor de la montaña** de brazos ligeros]
Taxonomy: Ornithischia: Ornithopoda

[References: Cruzado-Caballero P, Gasca JM, Filippi LS, Cerda I, Garrido AC (2019). A new ornithopod dinosaur from the Santonian of Northern Patagonia (Rincón de los Sauces, Argentina). *Cretaceous Research* 98: 211–229.]

Maiasaura peeblesorum

> *Maiasaura peeblesorum* Horner & Makela, 1979 [Campanian; upper Two medicine Formation, Montana, USA]

Generic name: *Maiasaura* ← {(Gr.) μαῖα/maia: good mother | (Gr. myth.) Μαῖα/Maia: daughter of Atlas, mother of Hermes + σαύρα/saura: female form of the male "sauros": lizard}; referring to the find of nests with eggs, embryos and young animals.
Specific name: *peeblesorum* ← {(person's name) Peebles + -ōrum}; in honor of the Peebles family, owners of the land where the specimens were collected.
Etymology: Peebles' **good mother lizard**
Taxonomy: Ornithischia: Ornithopoda: Hadrosauridae: Saurolophinae: Brachylophosaurini

[References: Horner JR, Makela R (1979). Nest of juveniles provides evidence of family structure among dinosaurs. *Nature* 282: 296–298.]

Majungasaurus crenatissimus

> *Megalosaurus crenatissimus* Depéret, 1896 ⇒ *Dryptosaurus crenatissimus* (Depéret, 1896) Depéret & Savornin, 1928 ⇒ *Majungasaurus crenatissimus* (Depéret, 1896) Lavocat, 1955 [Maastrichtian; Maevarano Formation, Mevarana, Mahajanga, Madagascar]

= *Majungatholus atopus* Sues & Taquet, 1979 [Maastrichtian; Maevarano Formation, Mahajanga, Madagascar]

Generic name: *Majungasaurus* ← {(place-name) Majunga + (Gr.) σαῦρος/sauros: lizard}; "in reference to the province of Mahajanga (formerly Majunga)"(*1).
Specific name: *crenatissimus* ← {(Lat.) crēnātus: "notched or toothed" + -issimus: "most or very much"}; "in reference to the serrations which are extended along the entire length of the two trenchant ridges of the teeth"(*1).
Etymology: most crenated, **Majunga lizard**
Taxonomy: Theropoda: Ceratosauria: Neoceratosauria: Abelisauridae: Majungasaurinae

[References: (*1) Krause DW, Sampson SD, Carrano MT, O'Connor PM (2007). Overview of the history of discovery, taxonomy, phylogeny and biogeography of *Majungasaurus crenatissimus* (Theropoda: Abelisauridae) from the Late Cretaceous of Madagascar. *Journal of Vertebrate Paleontology* 27(2): 1–20.]

Majungatholus atopus

> *Megalosaurus crenatissimus* Depéret, 1896 ⇒ *Dryptosaurus crenatissimus* (Depéret, 1896) Depéret & Savornin, 1928 ⇒ *Majungasaurus crenatissimus* (Depéret, 1896) Lavocat, 1955 [Maastrichtian; Maevarano Formation, Mevarana, Mahajanga, Madagascar]

= *Majungatholus atopus* Sues & Taquet, 1979 [Maastrichtian; Maevarano Formation, Mahajanga (= Majunga), Madagascar]

Generic name: *Majungatholus* ← {(place-name) Majunga + (Gr.) θόλος/tholos: a vaulted chamber}
Specific name: *atopus* ← {(Gr.) ἄτοπος/atopos: strange, extraordinary}
Etymology: strange **Mahajanga (Majunga) dome**
Taxonomy: Theropoda: Ceratosauria: Neoceratosauria (acc. *The Dinosauria 2nd edition*)
Notes: *Majungatholus atopus* was reported as a pachycephalosaurid dinosaur by Sues & Taquet in 1979(*1). However, this dinosaur is now considered to be a synonym of *Majungasaurus crenatissimus*.

[References: (*1) Sues H-D, Taquet P (1979). A pachycephalosaurid dinosaur from Madagascar and a Laurasia-Gondwanaland connection in the Cretaceous. *Nature* 279: 633–635.]

Malarguesaurus florenciae

> *Malarguesaurus florenciae* González Riga *et al.*, 2009 [upper Turonian–lower Coniacian; Portezuelo Formation, Mendoza, Argentina]

Generic name: *Malarguesaurus* ← {(place-name) Malargüe + (Gr.) σαῦρος/sauros: lizard, reptile}

Specific name: *florenciae* ← {(person's name) Florencia + -ae}; "in honor of Florencia Fernández Favarón, who collaborated in our field work for years, and found the first fossil remains of this species" (*1).

Etymology: Florencia's **Malargüe lizard**

Taxonomy: Saurischia: Sauropodomorpha: Sauropoda: Titanosauriformes: Somphospondyli: Titanosauria

[References: (*1) González Riga BJ, Previtera E, Pirrone CA (2009). *Malarguesaurus florenciae* gen. *et* sp. nov., a new titanosauriform (Dinosauria, Sauropoda) from the Upper Cretaceous of Mendoza, Argentina. *Cretaceous Research* 30(1): 135–148.]

Malawisaurus dixeyi

> *Gigantosaurus dixeyi* Haughton, 1928 (preoccupied) ⇒ *Tornieria dixeyi* (Haughton, 1928) Sternfeld, 1911 ⇒ *Malawisaurus dixeyi* (Haughton, 1928) Jacobs *et al.*, 1993 [Aptian; Dinosaur Beds, Northern Province, Malawi]

Generic name: *Malawisaurus* ← {(place-name) Malawi + (Gr.) σαῦρος/sauros: lizard} (*1)

Specific name: *dixeyi* ← {(person's name) Frank Dixey + -ī}; in honor of Dr. Frank Dixey, geological explorer.

Etymology: Dixey's **Malawi lizard**

Taxonomy: Saurischia: Sauropodomorpha: Sauropoda: Titanosauria: Eutitanosauria: Lithostrotia
cf. Saurischia: Sauropodomorpha: Sauropoda: Titanosauridae (*1)

[References: Haughton SH (1928). On some reptilian remains from the Dinosaur Beds of Nyasaland. *Transactions of the Royal Society of South Africa* 16: 67–75.; (*1) Jacobs LL, Winkler DA, Downs WR, Gomani EM (1993). New material of an Early Cretaceous titanosaurid sauropod dinosaur from Malawi. *Palaeontology* 36(3): 523–534.]

Maleevus disparoserratus

> *Syrmosaurus disparoserratus* Maleev, 1952 ⇒ *Talarurus disparoserratus* (Maleev, 1952) Maryańska, 1977 ⇒ *Maleevus disparoserratus* (Maleev, 1952) Tumanova, 1987 (*nomen dubium*) [Cenomanian–Santonian; Baynshire Formation, Shiregin Gashun, Mongolia]

Generic name: *Maleevus* ← {(person's name) Maleev + -us}; in honor of Soviet palaeontologist E. A. Maleev.

Specific name: *disparoserratus* ← {(Lat.) dispār: unequal + serratus: serrated}; referring to the unequal serrations on the teeth.

Etymology: Maleev's unequal serrated-toothed one

Taxonomy: Ornithischia: Thyreophora: Ankylosauria

Notes: According to *The Dinosauria 2nd edition*, *Syrmosaurus disparoserratus* Maleev, 1952 (type of *Maleevus* Tumanova, 1987) is considered a *nomen dubium*, and *Syrmosaurus viminicaudus* Maleev, 1952 is regarded a synonym of *Pinacosaurus granger* Gilmore, 1933. Maryańska renamed *Syrmosaurus disparoserratus* into a second species: *Talarurus disparoserratus*, and Tumanova gave it a new genus *Maleevus* in 1987.

Mamenchisaurus constructus

> *Mamenchisaurus constructus* Young, 1954 [Oxfordian; upper Shaximiao Formation, Yibin, Sichuan, China]

Generic name: *Mamenchisaurus* ← {(place-name) Mǎménchī馬門溪 + (Gr.) σαῦρος/sauros: lizard}
The site of discovery mentioned is in the vicinity of Mamenchi (馬門溪), according to Young (1954) (*1). However, the site is actually called "Mǎmíngxī" (馬鳴溪), due to an accentual mix-up by Young (*2).

Specific name: *constructus* ← {(Lat.) constructus: constructed}
Fossils were discovered during the construction of the Yitang Highway in 1952.

Etymology: **Mamenchi (Mamingxi) lizard** from the construction site

Taxonomy: Saurischia: Sauropodomorpha: Eusauropoda

Other species:

> *Mamenchisaurus hochuanensis* Young & Zhao, 1972 [Oxfordian; upper Shaximiao Formation, Xinjiang, China]

> *Mamenchisaurus sinocanadorum* Russell & Zheng, 1994 [Oxfordian; Shishugou Formation, Sichuan, China]

> *Mamenchisaurus youngi* Pi *et al.*, 1996 [Oxfordian; upper Shaximiao Formation, Sichuan, China]

> *Mamenchisaurus anyuensis* He *et al.*, 1996 [Late Jurassic; Penglaizhen Formation, Sichuan, China]

> *Mamenchisaurus jingyanensis* Zhang *et al.*, 1998 [Late Jurassic; upper Shaximiao Formation, Sichuan, China]

> *Omeisaurus changshouensis* Young, 1958 ⇒ *Mamenchisaurus changshouensis* (Young, 1958) Zhang & Chen, 1996 [Oxfordian; Shangshaximiao Formation, Sichuan, China]

> *Zigongosaurus fuxiensis* Hou *et al.*, 1976 ⇒ *Omeisaurus fuxiensis* (Hou *et al.*, 1976) Dong *et al.*, 1983 ⇒ *Mamenchisaurus fuxiensis* (Hou *et al*, 1976) Zhang & Chen, 1996 [Bathonian–Callovian; lower Shaximiao Formation, Sichuan, China]

> *Mamenchisaurus yunnanensis* Fang *et al.*, 2004 [Late Jurassic; Lufeng Basin, Lufeng, China]

[References: (*1) Young C-C [楊鍾健] (1954). On a new sauropod from Yiping, Szechuan, China. *Acta Palaeontologica Sinica* 2(4): 355–369 (First published in Chinese).; (*2) 它為什麼叫做馬門溪龍？ 每日頭條 2017-04-12由 集郵 發表于收藏 〈https://kknews.cc/collect/z6yygkl.html〉; (*3) Don Z, Zhou S, Zhang Y (1983). Dinosaurs from the Jurassic of Sichuan. *Palaeontologica Sinica* 162 C, 23: 1–136.]

Mandschurosaurus amurensis

> *Trachodon amurense* Riabinin, 1925 ⇒ *Mandschurosaurus amurensis* (Riabinin, 1925) Riabinin, 1930 [Maastrichtian; Yuliangze Formation, Jiayin, Heilongjiang, China]

Generic name: *Mandschurosaurus* ← {(Eng. place-name) Manchuria満洲 [Pinyin: Mǎnzhōu] + (Gr.) σαῦρος/sauros: lizard}; referring to Manchuria where the skeleton was found. (*1)

Specific name: *amurensis* ← {(place-name) Amur + -ensis}; referring to the bank of the Amur River in Manchuria where an incomplete skeleton was found. (*1)

Etymology: **Manchurian lizard** from Amur

Taxonomy: Ornithischia: Ornithopoda: Hadrosauridae

Other species:

> *Mandschurosaurus laosensis* Hoffet, 1943 (*nomen dubium*) [Aptian–Albian; Grès supérieurs Formation, Laos]

> *Mandschurosaurus mongoliensis* Gilmore, 1933 ⇒ *Gilmoreosaurus mongoliensis* (Gilmore, 1933) Brett-Surman, 1979 [Campanian; Iren Dabasu Formation, Inner Mongolia, China]

Notes: Riabinin first named the skeleton *Trachodon amurense*, but its incomplete preservation makes it impossible to characterize the genus (*1). Horner *et al.* (2004) listed *Mandschurosaurus laosensis* as a *nomen dubium*.

[References: (*1) Gilmore CW (1933). On the Dinosaurian fauna of the Iren Dabasu Formation. *Bulletin of the American Museum of Natural History* 67: 23–78.]

Manidens condorensis

> *Manidens condorensis* Pol *et al.*, 2011 [Aalenian–early Bathonian; Cañadon Asfalto Formation, Cerro Cóndor, Chubut, Argentina]

Generic name: *Manidens* ← {(Lat.) manus: hand + dēns: tooth}; "referring to the similarity of the posteriormost tooth to the human hand" (*1).

Specific name: *condorensis* ← {(place-name) Cerro Cóndor + -ensis}; "referring to the nearby village of Cerro Cóndor, Chubut Province, Argentina" (*1).

Etymology: **hand-like tooth** from Cerro Cóndor

Taxonomy: Ornithischia: Ornithopoda: Heterodontosauridae

[References: (*1) Pol D, Rauhut OWM, Becerra M (2011). A Middle Jurassic heterodontosaurid dinosaur from Patagonia and the evolution of heterodontosaurids. *Naturwissenschaften* 98 (5): 369–379.]

Mansourasaurus shahinae

> *Mansourasaurus shahinae* Sallam *et al.*, 2018 [Campanian; Quseir Formation, Dakhla Oasis, New Valley, Western Desert, Egypt]

Generic name: *Mansourasaurus* ← {(place-name / university) Mansoura: Mansoura University, Egypt + (Gr.) sauros: lizard}; referring to "Mansoura University in Mansoura, Egypt, home institution of the research collaborative that undertook the field and laboratory work" (*1)

Specific name: *shahinae* ← {(person's name) Shahin + -ae}; "in honor of Ms. Mona Shahin for her contributions to the foundation of the Mansoura University Vertebrate Paleontology Center (MUVP)" (*1).

Etymology: Shahin's **lizard of Mansoura University**

Taxonomy: Saurischia: Sauropodomorpha: Sauropoda: Titanosauria: Lithostrotia

[References: Sallam HM, Gorscak E, O'Connor PM, El-Dawoudi IA, El-Sayed S, Saber S, Kora MA, Sertich JJW, Seiffert ER, Lamanna MC (2018). New Egyptian sauropod reveals Late Cretaceous dinosaur dispersal between Europe and Africa. *Nature Ecology and Evolution* 2: 445–451.]

Mantellisaurus atherfieldensis

> *Iguanodon atherfieldensis* Hooley, 1925 ⇒ *Mantellisaurus atherfieldensis* (Hooley, 1925) Paul, 2007 [Hauterivian–Aptian; Wealden Group, Isle of Wight, UK]

= *Heterosaurus neocombiensis* Cornuel 1850 (*partim*) [Hauterivian; Calcaires à Spatangues Formation, France]

= *Vectisaurus valdensis* Hulke, 1879 [Hauterivian–Barremian; Wessex Formation, UK]

= *Sphenospondylus gracilis* Lydekker, 1888 (*nomen dubium*) [Valanginian–Hauterivian; Wessex Formation, Isle of Wight, UK]

= *Dollodon bampingi* Paul, 2008 (*nomen dubium*) [Barremian; Sainte-Barbe Clays, Belgium]

= *Proplanicoxa galtoni* Carpenter & Ishida,

2010 [Barremian; Wessex Formation, Isle of Wight, UK]
= *Mantellodon carpenteri* Paul, 2012 [Aptian; Hythe Formation, England, UK]

Generic name: *Mantellisaurus* ← {(person's name) Mantell + (Gr.) σαῦρος/sauros: lizard}; (emended by Paul, 2008) "For Mary and Gideon Mantell, who discovered and described the first Wealden iguanodonts". (*1)

Specific name: *atherfieldensis* ← {(place-name) Atherfield + -ensis}

Etymology: **Mantell's lizard** from Atherfield

Taxonomy: Ornithischia: Ornithopoda: Hadrosauriformes

Notes: *Mantellisaurus* was more lightly built than *Iguanodon*.

[References: (*1) Paul GS (2008). A revised taxonomy of the iguanodont dinosaur genera and species. *Cretaceous Research* 29(2): 192–216.]

Mantellodon carpenteri

> *Iguanodon atherfieldensis* Hooley, 1925 ⇒ *Mantellisaurus atherfieldensis* (Hooley, 1925) Paul, 2007 [Hauterivian–Aptian; Isle of Wight, UK]
= *Mantellodon carpenteri* Paul, 2012 [early Aptian; Hythe Formation, UK]

Generic name: *Mantellodon* ← {(person's name) Gideon Mantell + (Gr.) ὀδών/odōn: tooth}; referring to the holotype "known as Gideon Mantell's 'Mantel-piece' " (*1).

Specific name: *carpenteri* ← {(person's name) Carpenter + -ī}; "in recognition of Kenneth Carpenter's work on dinosaurs including iguanodonts" (*1).

Etymology: Carpenter's **Mantel-piece's tooth**

Taxonomy: Ornithischia: Ornithopoda: Styracosterna

[References: (*1) Paul G (2012). Notes on the rising diversity of Iguanodont taxa, and Iguanodonts named after Darwin, Huxley and evolutionary science. In: Hurtado PH, Fernandez-Baldro FT, Sanagustin JIC (eds). *Actas de V Jornadas Internacionales sobre Paleontologia de Dinosaurios y su Entorno,* Salas de Los Infantes, Burgos: 121–131.]

Mapusaurus roseae

> *Mapusaurus roseae* Coria & Currie, 2006 [Cenomanian; Huincul Formation, Río Limay Group, Neuquén, Argentina]

Generic name: *Mapusaurus* ← {(Mapuche*) Mapu: Earth + (Gr.) σαῦρος/sauros: reptile} (*1)

*Mapuche [mapu: land + če: people]; an American Indian people of Southern Chile.

Specific name: *roseae* ← {Rose + -ae}; "referring to the rose-coloured rocks" and "to Rose Letwin (Seattle) who sponsored of the expeditions in 1999, 2000 and 2001" (*1).

Etymology: Rose's rose-coloured **Earth reptile**

Taxonomy: Theropoda: Carcharodontosauridae: Gigantosaurinae (*1)

Notes: *Mapusaurus* is considered to be a giant carnosaurian dinosaur.

[References: (*1) Coria RA, Currie PJ (2006). A new carcharodontosaurid (Dinosauria, Theropoda) from the Upper Cretaceous of Argentina. *Geodiversitas* 28(1): 71–118.]

Maraapunisaurus fragillimus

> *Amphicoelias fragillimus* Cope 1878 ⇒ *Maraapunisaurus fragillimus* (Cope, 1878) Carpenter, 2018 [Tithonian; Morrison Formation, Colorado, USA]

Generic name: *Maraapunisaurus* ← {(Southern Ute) Ma-ra-pu-ni [pron: mah-rah-poo-nee]: huge + (Gr.) σαῦρος/sauros: reptile}; in reference to the huge size of the animal. This name was recommended by the Southern Ute Cultural Department, Ignacio, Colorado, because the Garden Park area where the specimen was found was traditionally Ute tribal territory. (*1)

Specific name: *fragillimus** ← {(Lat.) fragilimus; very fragile}

*Hay (1902) corrected *fragillimus* to *fragilimus*, but such emendations are not allowed by the ICZN.

Etymology: very fragile **huge reptile**

Taxonomy: Sauropoda: Diplodocoidae: Rebbachisauridae

Notes: According to *The Dinosauria 2nd edition*, *Amphicoelias fragillimus* was regarded as a synonym of *A. altus*, but it was given a new generic name *Maraapunisaurus*.

[References: (*1) Carpenter K (2018). *Maraapunisaurus fragillimus*, N.G. (formerly *Amphicoelias fragillimus*), a basal rebbachisaurid from the Morrison Formation (Upper Jurassic) of Colorado. *Geology of the Intermountain West* 5: 227–244.]

Marasuchus lilloensis [Dinosauriformes]

> *Lagosuchus lilloensis* Romer, 1972 ⇒ *Marasuchus lilloensis* (Romer, 1972) Sereno & Arcucci, 1994 [Ladinian–Carnian; Chañares Formation, La Rioja, Argentina]

Generic name: *Marasuchus* ← {(mammal) mara: *Dolichotis*, a large rodent genus dwelling in Patagonia + (Gr.) Σοῦχος/Souchos [Sūchos]: crocodile}

Specific name: *lilloensis* ← {Lillo + -ensis}

Etymology: **mara crocodile** from Lillo

Taxonomy: Archosauria: Ornithodira:

Dinosauromorpha: Dinosauriformes

Notes: *Marasuchus lilloensis* was originally described as a second species of *Lagosuchus*.

[References: Sereno PC, Arcucci AB (1994). Dinosaurian precursors from the Middle Triassic of Argentina: *Marasuchus lilloensis* gen. nov. *Journal of Vertebrate Paleontology* 14: 53–73.]

Marisaurus jeffi

> *Marisaurus jeffi* Malkani, 2006 (*nomen dubium*) [Maastrichtian; Pab Formation, Barkhan, Balochistan, Pakistan]

Generic name: *Marisaurus* ← {(demonym & place-name) Mari tribes + (Gr.) σαῦρος/sauros: reptiles}; "in honor of the Mari tribes Bohri locality from Central Sulaiman Range" (*1).

Specific name: *jeffi* ← {(person's name) Jeff + -ī}; "in honor of Dr. Jeffery A. Wilson, Museum of Paleontology, University of Michigan, USA, for verifying the ideas of author regarding first biconvex caudal of Saltasaurids dinosaur from Pakistan" (*1).

Etymology: Jeffery's **Mari lizard**

Taxonomy: Saurischia: Sauropodomorpha: Sauropoda: Somphospondyli: Titanosauria
cf. Saurischia: Sauropodomorpha: Sauropoda: Titanosauria: Balochisauridae (= Saltasauridae) (*1).

[References: (*1) Malkani MS (2006). Biodiversity of saurischian dinosaurs from the Latest Cretaceous Park of Pakistan. *Journal of Applied and Emerging Sciences* 1(3): 108–140.]

Marshosaurus bicentesimus

> *Marshosaurus bicentesimus* Madsen, 1976 [Kimmeridgian; Morrison Formation, Utah, USA]

Generic name: *Marshosaurus* ← {(person's name) Marsh + (Gr.) σαῦρος/sauros: lizard}; in honor of the paleontologist, Professor Othniel Charles Marsh (1831–1899).

Marsh is known for so-called "Bone Wars" against Edward Drinker Cope.

Specific name: *bicentesimus* ← {(Lat.) bi- [bis: 2] + centēsimus: hundredth}; commemorating the United States Bicentennial.

Etymology: **Marsh's lizard** of the Bicentennial

Taxonomy: Theropoda: Tetanurae: Megalosauroidea: Piatnitzkysauridae

Notes: Originally, Madsen marked *Marshosaurus bicentesimus* as *incertae sedis*. Holtz (2004) classified it in Coelurosauria. *Marshosaurus* is now thought to be Piatnitzkysauridae Carrano *et al*, 2012, which consists of *Condorraptor*, *Piatnitzkysaurus* and *Marshosaurus*.

Martharaptor greenriverensis

> *Martharaptor greenriverensis* Senter *et al.*, 2012 [? Barremian; Yellow Cat Member, Cedar Mountain Formation, Emery, Utah, USA]

Generic name: *Martharaptor* ← {(person's name) Martha Hayden + raptor: plunderer, robber}; "in honor of Martha Hayden, who co-discovered the site and has served as the assistant to three successive state paleontologists of Utah over a period of about 25 years" (*1).

Specific name: *greenriverensis* ← {(place-name) Green River + -ensis}; "referring to the city of Green River in Emery County, Utah" (*1).

Etymology: **Martha's robber** from Green River

Taxonomy: Theropoda: Coelurosauria: Therizinosauroidea

[References: (*1) Senter P, Kirkland JI, DeBlieux DD (2012). *Martharaptor greenriverensis*, a new theropod dinosaur from the Lower Cretaceous of Utah. *PLoS ONE* 7(8): e43911.]

Martinavis cruzyensis [Avialae]

> *Martinavis cruzyensis* Walker *et al.*, 2007 [Campanian; Grès à Reptiles Formation, Cruzy, France]

Generic name: *Martinavis* ← {(peson's name) Martin + avis: bird}; "in honor of Larry D. Martin, in recognition of his contributions to the study of Mesozoic birds and for his support of Cyril Walker in the 1980's. Many of the original illustrations of the El Brete collection were rendered by KU-NM* artists in the 1980s (Chiappe & Walker, 2002)" (*1).

*University of Kansas, Museum of Natural History, Lawrence, Kansas, USA.

Specific name: *cruzyensis* ← {(place-name) Cruzy + -ensis}; "for the village of Cruzy, Hérault, southern France, where this specimen was collected (Massecaps locality)" (*1).

Etymology: **Martin's bird** from Cruzy

Taxonomy: Theropoda: Avialae: Ornithothoraces: Enantiornithes: Euenantiornithes
cf. Aves: Ornithothoraces: Enantiornithes (*1)

Other species:

> *Martinavis vincei* Walker *et al.*, 2007 [Maastrichtian; Lecho Formation, Argentina]
> *Martinavis minor* Walker & Dyke, 2009 [Maastrichtian; Lecho Formation, Argentina]
> *Martinavis saltariensis* Walker & Dyke, 2009 [Maastrichtian; Lecho Formation, Argentina]
> *Martinavis whetstonei* Walker & Dyke, 2009 [Maastrichtian; Lecho Formation, Argentina]

[References: (*1) Walker CA, Buffetaut E, Dyke GJ (2007). Large euenantiornithine birds from the Cretaceous of southern France, North America and Argentina. *Geological Magazine* 144(6): 977–986.]

Masiakasaurus knopfleri

> *Masiakasaurus knopfleri* Sampson *et al*., 2001 [Maastrichtian; Maevarano Formation, Mahajanga Basin, Madagascar]

Generic name: *Masiakasaurus* ← {(Malagasy) masiaka: vicious + (Gr.) σαῦρος/sauros: lizard} (*1) *Masiakasaurus* was seemed to be a unique predatory "with a highly procumbent and distinctly heterodont lower dentition" (*1).

Specific name: *knopfleri* ← {(person's name) Knopfler + -ī}; "after singer/songwriter Mark Knopfler, whose music inspired expedition crews" (*1).

Etymology: Knopfler's **vicious lizard**

Taxonomy: Theropoda: Ceratosauria: Abelisauria (= Abelisauroidea): Noasauridae

[References: (*1) Sampson SD, Carrono MT, Forster CA (2001). A bizarre predatory dinosaur from the Late Cretaceous of Madagascar. *Nature* 409(6819): 504–506.]

Massospondylus carinatus

> *Massospodylus carinatus* Owen, 1854 [Hettangian–Sinemurian; upper Elliot Formation, Harrismith, South Africa]

Generic name: *Massospondylus* ← {(Gr.) μάσσων/massōn: longer, greater + σπόνδυλος/spondylos: vertebra}

Specific name: *carinatus* ← {(Lat.) carinātus: keeled} *cf.* carīna: a keel (of a ship)

Etymology: keeled, **longer vertebra**

Taxonomy: Saurischia: Sauropodomorpha: Plateosauria: Massospondylidae

cf. Saurischia: Sauropodomorpha: Prosauropoda (acc. *The Dinosauria 2nd edition*)

Other species:

> *Massospondylus browni* Seeley, 1895 ⇒ *Thecodontosaurus browni* (Seeley, 1895) von Huene, 1914 (*nomen dubium*) [Hettangian–early Sinemurian; upper Elliot Formation, Eastern Cape, South Africa]

> *Massospondylus harriesi* Broom, 1911 (*nomen dubium*) [Hettangian; upper Elliot Formation, Free State, South Africa]

> *Massospondylus hislopi* Lydekker, 1890 (*nomen dubium*) [Norian; Maleri Formation, Andhra Pradesh, India]

> *Massospondylus rawesi* Lydekker, 1890 (*nomen dubium*) ⇒ *Megalosaurus rawesi* (Lydekker, 1890) Vianey-Liaud *et al*., 1988 (*nomen dubium*) [Maastrichtian; Takli Formation, Nagpur, Maharashtra, India]

> *Massospondylus schwarzi* Haughton, 1924 (*nomen dubium*) [Hettangian–Sinemurian; upper Elliot Formation, Cape, South Africa]

> *Lufengosaurus huenei* Young, 1941 ⇒ *Massospondylus huenei* (Young, 1941) Cooper, 1981 [Hettangian; Dull Purplish Beds, Lufeng, Yunnan, China]

> *Massospondylus kaalae* Barrett, 2009 [Hettangian–Sinemurian, upper Elliot Formation, Eastern Cape, South Africa]

Notes: According to Chapelle *et al.* (2019), a skull and partial skeleton previously assigned to *Massospondylus carinatus* became the holotype of *Ngwevu intloko* in 2019.

Matheronodon provincialis

> *Matheronodon provincialis* Godefroit *et al*., 2017 [late Campanian; Aix-en-Provence Basin, Bouches-du-Rhône, France]

Generic name: *Matheronodon* ← {(person's name) Matheron + (Gr.) ὀδών/odōn (= ὀδούς/odous [odūs]): tooth}; "in honor of Philippe Matheron, who was the first to describe dinosaur remains in Provence" (*1).

Specific name: *provincialis* ← {(Lat.): "from Provence (southern France)"} (*1).

Etymology: **Matheron's tooth** from Provence

Taxonomy: Ornithischia: Ornithopoda: Iguanodontia: Rhabdodontidae

[References: (*1) Godefroit P, Garcia G, Gomez B, Stein K, Cincotta A, Lefèvre U, Valentin X (2017). Extreme tooth enlargement in a new Late Cretaceous rhabdodontid dinosaur from Southern France. *Scientific Reports* 7: 13098.]

Maxakalisaurus topai

> *Maxakalisaurus topai* Kellner *et al*., 2006 [Turonian–Santonian; Adamantina Formation, Minas Gerais, Brazil]

Generic name: *Maxakalisaurus* ← {(demonym, Maxakalí*) Maxakali/Mashakalí + (Gr.) σαῦρος/sauros: lizard}; in homage to the Maxakali ethic group, that inhabits in the Minas Gerais State where this dinosaur was found.

*one of the Macro-jê stock language. (*1)

Specific name: *topai* ← {(Maxakalí, myth.) Topa + -ī}; in reference to Topa (= Topar) (*2), which is a tribal god worshiped by the Maxakali ethic group (*1).

Etymology: Topa's **Maxakalian lizard**

Taxonomy: Saurischia: Sauropodomorpha: Sauropoda: Macronaria: Titanosauria:

Lithostrotia: Aeolosaurini [*3]

[References: (*1) Kellner AWA, Campos DdA, Azevedo SAKd , Trotta MNF, Henriques DDR, Craik MMT, Silva HdP (2006). On a new titanosaur sauropod from the Bauru Group, Late Cretaceous of Brazil. *Boletim do Museu Nacional Nova Série Rio de Janeiro-Brasil Geologia* 74: 1–31.; (*2) Warren JW (2001). *Antiracism and Indian Resurgence in Brazil.* Durham: Duke University Press. 392p.; (*3) França MAG, Marsola JCdA, Riff D, Hsiou AS, Langer MC (2016). New lower jaw and teeth referred to *Maxakalisaurus topai* (Titanosauria: Aeolosaurini) and their implications for the phylogeny of titanosaurid sauropods. *PeerJ* 4: e2054]

Medusaceratops lokii

> *Medusaceratops lokii* Ryan *et al.*, 2010 [Campanian; Judith River Formation, Havre, Montana, USA]

Generic name: *Medusaceratops* ← {(Gr. myth.) Μέδουσα/Medūsa: "a monster with 'hair' comprised of snakes and a gaze that could turn men to stone" [*1] + ceratops [(Gr.) κερατ-/kerat-(κέρας/keras: horn) + ὤψ/ōps: face]}; "alluding to the large, thick snake-like spikes that extend from the lateral margins of the posterior portion of the parietal" [*1].

Specific name: *lokii* ← {(Norse myth.) Loki : a troublemaking god + -ī}; referring to "a Norse god, Loki, who contrived mischief for his fellow gods, the name thus alluding to the confusion experienced in trying to assign taxonomic designations to the material collected from bonebed" [*1].

Etymology: Loki's **Medusa horned face (ceratopsian)**

Taxonomy: Ornithischia: Ceratopsia: Neoceratopsia: Ceratopsidae: Chasmosaurinae [*1]

[References: (*1) Ryan MJ, Russell AP, Hartman S (2010). A new chasmosaurine ceratopsid from Judith River Formation, Montana. In: Ryan MJ, Chinnery-Allgeier BJ, Eberth DA (eds), *New Perspectives on Horned Dinosaurs: The Royall Tyrrell Museum Ceratopsian Symposium*, Indiana University Press, Bloomington: 181–188.]

Megalosaurus bucklandi

> *Scrotum humanum* Brookes, 1763 (*nomen oblitum*) ⇒ *Megalosaurus bucklandi* Mantell, 1827 [Bathonian; Taynton Limestone Formation, Stonesfield, Oxfordshire, UK]

= *Megalosaurus conybeari* von Ritgen, 1826 (*nomen oblitum*)

Generic name: *Megalosaurus* Buckland, 1824 ← {(Gr.) μεγάλος/megalos [μέγας/megas]: big, great + σαῦρος/sauros: lizard}

Megalosaurus is the first genus which was named in 1824 by William Buckland (1784–1856), though it was thought to be a great lizard literally in those days.

Specific name: *bucklandi* ← {(person's name) Buckland + -ī}; in honor of William Buckland, who was an English theologian, and also geologist and palaeontologist. He described on *Megalosaurus* but he didn't give it a specific name in a paper. Later, in 1826, Ritgen named it *M. conybeari*, but it was a *nomen oblitum* (forgotten name). In 1827, Mantell named it formally *M. bucklandii*.

The latinized name of Buckland's is 'Bucklandus' or 'Bucklandius'. The genitive form of Bucklandus is 'bucklandi', and that of Bucklandius is 'bucklandii'.

Etymology: Buckland's **great lizard**

Taxonomy: Theropoda: Tetanurae: Megalosauroidea: Megalosauridae

Other species:

> *Deinodon horridus* Leidy, 1856 ⇒ *Megalosaurus horridus* (leidy, 1856) Leidy, 1857 (*nomen dubium*) [Campanian: Judith River Formation, Montana, USA]

> *Megalosaurus cloacinus* Quenstedt, 1858 ⇒ *Plateosaurus cloacinus* (Quenstedt, 1858) von Huene, 1905 ⇒ *Gresslyosaurus cloacinus* (Quenstedt, 1858) Steel, 1970 (*nomen dubium*) [Rhaetian; Exter Formation, Bebenhause, Germany]

> *Megalosaurus insignis* Eudes-Deslongchamps & Lennier vide Lennier, 1870 ⇒ *Erectops insignis* (Eudes-Deslongchamps & Lennier vide Lennier, 1870) Stromer, 1931 (*nomen dubium*) [Kimmeridgian: Marnes à Deltoideum delta Formation, Normandy, France]

> *Megalosaurus meriani* Greppin, 1870 ⇒ *Labrosaurus meriani* (Greppin, 1870) Janensh, 1920 ⇒ *Antrodemus meriani* (Greppin, 1870) Steel, 1970 ⇒ *Allosaurus meriani* (Greppin, 1870) Olshevsky, 1991 ⇒ *Ceratosaurus meriani* (Greppin, 1870) emend Madsen & Welles, 2000 (*nomen dubium*) [Kimmeridgian; Reuchenette Formation, Moutier, Switzerland]

> *Megalosaurus schnaitheimi* Bunzel, 1871 (*nomen nudum*) [Kimmeridgian; Mergelstätten Formation, Schnaitheim, Germany]

> *Megalosaurus obtusus* Henry, 1876 ⇒ *Plateosaurus obtusus* (Henry, 1876) von Huene 1908 (*nomen dubium*) [Rhaetian; Moissey, France]

> *Megalosaurus pannoniensis* Seeley, 1881 (*nomen dubium*) [Campanian; Grünbach Formation, Muthumannsdorf, Austria]

> *Megalosaurus bredai* Seeley, 1883 ⇒ *Betasuchus bredai* (Seeley, 1883) von Huene, 1932 ⇒ *Ornithomimidorum gen. b. bredai* (Seeley, 1883) von Huene, 1932 (*nomen oblitum*) [Maastrichtian; Maastricht Formation, Maastricht, Netherland]

> *Megalosaurus superbus* Sauvage, 1882 ⇒ *Erectopus superbus* (Sauvage, 1882) von Huene, 1923 (*nomen dubium*) [early Albian; La Penthiève, Beds, Louppy-le-Château, France]

> *Megalosaurus dunkeri* Dames, 1884 [Berriasian; Bückeberg Formation, Deisters (locality), Niedesachsen, Germany]

> *Megalosaurus dunkeri* von Huene, 1923 ⇒ *Altispinax dunkeri* (von Huene, 1923) von Huene, 1923, Kuhn, 1939 [Valanginian; Hastings Bed, Sussex, England, UK]

> *Dakosaurus gracilis* Quenstedt, 1885 ⇒ *Megalosaurus gracilis* (Quenstedt, 1885) Douville, 1885 [uncertain]

> *Megalosaurus oweni* Lydekker, 1889 ⇒ *Altispinax oweni* (Lydekker, 1889) von Huene, 1923 ⇒ *Valdosaurus oweni* (Lydekker, 1889) Olshevsky, 1991 [late Valanginian; Tunbridge Wells Sand Formation, Cuckfield, UK]

> *Ceratosaurus nasicornis* Marsh, 1884 ⇒ *Megalosaurus nasicornis* (Marsh, 1884) Cope, 1892 [Kimmeridgian–Tithonian; Morrison Formation, Colorado, USA]

> *Megalosaurus crenatissimus* Depéret, 1896 ⇒ *Dryptosaurus crenatissimus* (Depéret, 1896) Depéret & Savornin, 1928 ⇒ *Majungasaurus crenatissimus* (Depéret, 1896) Lavocat, 1955 [Maastrichtian; Maevarano Formation, Maevarana, Madagascar]

> *Laelaps aquilunguis* Cope, 1866 (preoccupied) ⇒ *Dryptosaurus aquilunguis* (Cope, 1866) Marsh, 1877 ⇒ *Megalosaurus aquilunguis* (Cope, 1866) Osborn, 1898 [Maastrichtian; New Egypt Formation, New Jersey, USA]

> *Laelaps trihedrodon* Cope, 1877 ⇒ *Megalosaurus trihedrodon* (Cope, 1877) Nopcsa, 1901 ⇒ *Antrodemus trihedrodon* (Cope, 1877) Kuhn, 1939 ⇒ *Dryptosaurus trihedrodon* (Cope, 1877) Kuhn, 1939 (*nomen dubium*) [Kimmeridgian–Tithonian; Morrison Formation, Colorado, USA]

> *Poicilopleuron valens* Leidy, 1870 ⇒ *Allosaurus valens* (Leidy, 1870) Gilmore, 1920 ⇒ *Megalosaurus valens* (Leidy, 1870) Nopcsa, 1901 [Kimmeridgian–Tithonian; Morrison Formation, Colorado, USA]

> *Megalosaurus hungaricus* Nopcsa, 1901 [Maastrichtian; Bihor, Romania (former Kingdom of Hungary)]

> *Megalosaurus lonzeensis* Dollo, 1903 [Santonian-Campanian; Glauconie de Lonzée Formation, Lonzée, Namur, Belgium]

> *Streptospondylus cuvieri* Owen, 1842 ⇒ *Megalosaurus cuvieri* (Owen, 1842) von Huene, 1907/1908 (*nomen dubium*) [Bajocian; Chipping Norton, Oxfordshire, UK]

> *Megalosaurus woodwardi* Lydekker, 1909 (*nomen dubium*) [Hettangian; Blue Lias Formation, Warwickshire, UK]

= *Megalosaurus lydekkeri* von Huene, 1926 ⇒ *Magnosaurus lydekkeri* (von Huene, 1926) von Huene 1932 (*nomen dubium*) [Hettangian; Blue Lias Formation, England, UK]

> *Megalosaurus bradleyi* Woodward, 1910 ⇒ *Proceratosaurus bradleyi* (Woodward, 1910) von Huene, 1926 [Bathonian; White Limestone, Groucestershire, UK]

> *Megalosaurus ingens* Janensch, 1920 ⇒ *Ceratosaurus ingens* (Janensch, 1920) Paul, 1988 [Tithonian; Tendaguru Formation, Mandawa Basin, Tanzania]

> *Poekilopleuron bucklandii* Eudes-Deslongchamps, 1838 ⇒ *Megalosaurus poikilopleuron* (Eudes-Deslongchamps, 1838) von Huene, 1923 [Bathonian; Calcaire de Caen Formation, France]

> *Megalosaurus parkeri* von Huene, 1923 ⇒ *Metriacanthosaurus parkeri* (von Huene, 1923) Walker, 1964 [Oxfordian; Oxford Clay Formation, Dorset, UK]

> *Megalosaurus nethercombensis* von Huene, 1926 ⇒ *Magnosaurus nethercombensis* (von Huene, 1926) von Huene, 1932 [Bajocian; Middle Inferior Oolite Formation, Nethercomb, UK]

> *Megalosaurus saharicus* Depéret & Savornin, 1925 ⇒ *Carcharodontosaurus saharicus* (Depéret & Savornin, 1925) Stromer, 1931 [Cenomanian; Kem Kem Beds, Algeria]

= *Megalosaurus africanus* von Huene, 1956 [Albian; Kem Kem Beds, Algeria]

> *Megalosaurus terquemi* von Huene, 1926 (*nomen dubium*) [Hettangian, Lorraine, France]

> *Megalosaurus wetherilli* Welles, 1954 ⇒ *Dilophosaurus wetherilli* (Welles, 1954) Welles, 1970 [Sinemurian-Priesbachian; Kayenta Formation, Arizona, USA]

> *Megalosaurus mersensis* de Lapparent, 1955 [Bathonian–Callovian; El Mers Formation, High Atlas, Morocco]

> *Aggiosaurus nicaeensis* Ambayrac, 1913 ⇒

Megalosaurus nicaeensis (Ambayrac, 1913) Romer, 1956 [Oxdordian; Cap d'Aggio-La Turbie, France]
> *Eutynichnium pombali* de Lapparent *et al.*, 1951 ⇒ *Megalosaurus pombali* de Lapparent & Zbyszewski, 1957 (*nomen dubium*) [Callovian–Oxfordian; Pombal, Portugal]
> *Zanclodon silesiacus* Jaekel, 1910 ⇒ *Megalosaurus silesiacus* (Jaekel, 1910) Kuhn, 1965 (*nomen dubium*) [Anisian; Chorzow Formation, Opole, Poland]
> *Megalosaurus inexpectatus* Corro, 1966 [Albian–Cenomanian; Cerro Barcino Formation, Chubut, Argentina]
> *Megalosaurus dapukaensis* Zhao, 1985 (*nomen nudum*) [Middle Jurassic; Dabuka Formation, Xizang, China]
> *Megalosaurus tibetensis* Zhao, 1985 (*nomen nudum*) [Early Jurassic; Duogaila Member, Xizang, China]
> *Massospondylus rawesi* Lydekker, 1890 ⇒ *Megalosaurus rawesi* (Lydekker, 1890) Viaey-Liaud *et al.*, 1988 (*nomen dubium*) [Maastrichtian?; Takli Formation?, Nagpur, India]
> *Megalosaurus chubutensis* Corro, 1974 (*nomen dubium*) [Cenomanian; Castillo Formation, Chubut, Argentina]
> *Megalosaurus hesperis* Waldman, 1974 ⇒ *Walkersaurus hesperis* (Waldman, 1974) Welles *et al.*, 1995 ⇒ *Duriavenator hesperis* (Waldman, 1974) Benson, 2008 [Bajocian; Upper Inferior Oolite Formation, Sherborne, UK]
> *Torvosaurus tanneri* Galton & Jansen, 1979 ⇒ *Megalosaurus tanneri* (Galton & Jensen, 1979) Paul, 1988 [Kimmeridgian–Tithonian; Morrison Formation, Colorado, USA]
> *Poekilopleuron schmidti* Kiprianow, 1883 ⇒ *Megalosaurus schmidti* (Kiprianow, 1883) Olshevsky, 1991 [Albian–Cenomanian; Sekmenevsk Formation, Tuskar, Kursk, Russia]
> *Plateosaurus ornatus* von Huene, 1905 ⇒ *Megalosaurus ornatus* (von Huene, 1905) Probst & Windolf, 1993 (*nomen vanum*) [Rhaetian; Exter Formation, Baden-Württemberg, Germany]
> *Megalosauripus teutonicus* Kaever & Lapparent, 1974 ⇒ *Megalosaurus teutonicus* (Kaever & Lapparent, 1974) Probst & Windolf, 1933 (ichinospecies) [Kimmeridgian; Niedersachsen, Germany]
> *Saurocephalus monasterii* Münster, 1846 ⇒ *Megalosaurus monasterii* (Münster, 1846) Windolf, 1997 (*nomen dubium*) [Oxfordian: Korallenkalk Formation, Niedersachsen, Germany]
> *Zanclodon cambrensis* Newton, 1899 ⇒ *Megalosaurus cambrensis* (Newton, 1899) Galton, 1998 [Rhatian; Bridgend, Wales, UK]

Notes: According to *Zoological Journal of the Linnean Society*, 2010, Megalosauroidea [= Spinosauroidea] includes Megalosauridae and Spinosauridae.

[References: Buckland W (1824). Notice on the *Megalosaurus* or great fossil lizard of Stonedfield. *Transactions of the Geological Society of London series 2* 1: 390–396.; Benson RBJ (2010). A description of *Megalosaurus bucklandii* (Dinosauria: Theropoda) from the Bathonian of the UK and the relationships of Middle Jurassic theropods. *Zoological Journal of the Linnean Society* 158(4): 882–935.]

Megapnosaurus rhodesiensis

> *Syntarsus rhodesiensis* Raath, 1969 (preoccupied) ⇒ *Coelophysis rhodesiensis* (Raath, 1969) Paul, 1988 ⇒ *Megapnosaurus rhodesiensis* (Raath, 1969) Ivie *et al.*, 2001 [Hettangian–Sinemurian; Forest Sandstone Formation, Southern Rhodesia, Matabeleland North, Zimbabwe]

Generic name: *Megapnosaurus* ← {(Gr.) μέγᾰς/megas: big + ἄπνοος/apnoos: dead + σαῦρος/sauros: lizard}

Specific name: *rhodesiensis* ← {(place-name) Rhodesia + -ensis}

Etymology: **big dead lizard** from Rhodesia

Taxonomy: Theropoda: Ceratosauria: Coelophysoidea

Other species:

> *Syntarsus kayentakatae* Rowe, 1989 ⇒ *Megapnosaurus kayentakatae* (Rowe, 1989) Ivie *et al.*, 2001 ⇒ *Coelophysis kayentakatae* (Rowe, 1989) Bristowe & Raath, 2004 [Sinemurian–Pliensbachian; Kayenta Formation, Arizona, USA]

Megaraptor namunhuaiquii

> *Megaraptor namunhuaiquii* Novas, 1998 [Turonian–Coniacian; Portezuelo Formation, Neuquén, Patagonia, Argentina]

Generic name: *Megaraptor* ← {(Gr.) μέγας/megas: large + (Lat.) raptor: thief} [*1]

Specific name: *namunhuaiquii* ← {(Mapuche) namun: foot + huaiqui: lance + -ī}; "referring to the enormous and sharp pedal ungual" [*1].

Etymology: **giant thief** with foot lance

Taxonomy: Theropoda: Megaraptora: Megaraptoridae

cf. Theropoda: ?Coelurosauria [*1]

Notes: *Megaraptor* had sickle-shaped foot claws. It

was considered as a problematic taxa, according to *The Dinosauria 2nd edition*.

[References: (*1) Novas FE (1998). *Megaraptor namunhuaiquii*, gen. *et* sp. nov., a large-clawed, Late Cretaceous theropod from Patagonia. *Journal of Vertebrate Paleontology* 18(1): 4–9.]

Mei long

> *Mei long* Xu & Norell, 2004 [Aptian; Lujiatun Member, Yixian Formation, Beipiao, Liaoning, China]

Generic name: *Mei* ← {(Chin.) mèi寐: to sleep soundly}; in reference to "the stereotypical sleeping or resting posture found in extant Aves" (*1).

Specific name: *long* ← {(Chin.) lóng龍: dragon} (*1)

Etymology: **sleeping** dragon

Taxonomy: Theropoda: Maniraptora: Troodontidae: Sinovenatorinae

cf. Theropoda: Maniraptora: Troodontidae (*1)

Notes: The type fossil is a juvenile about 53 centimeters long.

[References: (*1) Xu X, Norell MA (2004). A new troodontid dinosaur from China with avian-like sleeping posture. *Nature* 431(7010): 838–841.]

Melanorosaurus readi

> *Melanorosaurus readi* Haughton, 1924 [Norian–Rhaetian; lower Elliot Formation, Herschel, Eastern Cape, South Africa]

Generic name: *Melanorosaurus* ← {(Gr. place-name) μελαν-/melan- [μέλας/melas: black] + ὄρος/oros: mountain + σαῦρος/sauros: lizard}; in reference to its discovery on Thaba Nyama (= Black Mountain) (*1).

Specific name: *readi* ← {(person's name) Read + -ī}; "in honor of Mr. B. Read, former Principal of the Bensonvale Training School, of whose kindness, display of interest, and hospitality I have a lively recollection" (*1).

Etymology: Read's **Black Mountain lizard**

Taxonomy: Saurischia: Sauropodomorpha: Melanorosauridae

Other species:

> *Melanorosaurus thabanensis* Gauffre, 1993 ⇒ *Meroktenos thabanensis* (Gauffre, 1993) Peyre de Fabrègues & Allain, 2016 [Late Triassic; lower Elliot Formation, Thabana-Morena, Mafeteng, Lesotho]

[References: (*1) Haughton SH (1924). The fauna and stratigraphy of the Stormberg Series. *Annals of the South African Museum* 12: 323–497.]

Mendozasaurus neguyelap

> *Mendozasaurus neguyelap* González-Riga, 2003 [late Turonian–late Coniacian; Sierra Barrosa Formation, Mendoza, Argentina]

Generic name: *Mendozasaurus* ← {(place-name) Mendoza Province + σαῦρος/sauros: lizard}

Specific name: *neguyelap* ← {(Huarpes indigenous terms [Millcayac language, Márquez Miranda, 1943]) neguy: first + yelap: beast}; "referring to the first species of dinosaur discovered in Mendoza Province. The ending of the term yelap has not been modified, since is not a Latin or latinized word (Art.31.2.3. Comisión Internacional de Nomenclatura Zoológica, 2000)". (*1)

Etymology: **Mendoza lizard**, first beast

Taxonomy: Saurischia: Sauropodomorpha: Sauropoda: Titanosauria: Longkosauria

cf. Saurischia: Sauropoda: Titanosauria: Titanosauridae (*1)

Notes: *Mendozasaurus* is estimated to be 18–25 m long (*1).

[References: (*1) González Riga BJ (2003). A new titanosaur (Dinosauria, Sauropoda) from the Upper Cretaceous of Mendoza Province, Argentina. *Ameghiniana* 40(2): 155–172.]

Mercuriceratops gemini

> *Mercuriceratops gemini* Ryan *et al.*, 2014 [Campanian; Judith River Formation, Montana, USA.]

Generic name: *Mercuriceratops* ← {(Roman God) Mercury: Roman messenger god + (Gr.) ceratops [(Gr.) κέρας/keras: horn + ὤψ/ōps: face]}; referring to the butterfly-shaped frill which looked like wings Mercury wore on his helmet.

Specific name: *gemini* ← {(Lat.) Gemini: the Twins*}; referring to the two specimens which were found from Judith River Formation, Montana, USA (holotype) and Dinosaur Park Formation, Alberta, Canada

*Gemini is one of the constellations of the zodiac. In Greek mythology, the twins are Castor and Pollux.

Etymology: twin **Mercury horned-face (ceratopsian)**

Taxonomy: Ornithischia: Ceratopsia: Ceratopsidae: Chasmosaurinae

[References: Ryan MJ, Evans DC, Currie PJ, Loewen MA (2014). A new chasmosaurine from northern Laramidia expands frill disparity in ceratopsid dinosaurs. *Naturwissenschaften* 101(6): 505–512.]

Meroktenos thabanensis

> *Melanorosaurus thabanensis* Gauffre, 1993 ⇒ *Meroktenos thabanensis* (Gauffre, 1993) Peyre de Fabrègues & Allain, 2016 [Late Triassic; lower Elliot Formation, Mafeteng, Lesotho]

Generic name: *Meroktenos* ← {(Gr.) μηρός/mēros: femur (thigh bone) + κτῆνος/ktēnos: animal, beast}; "because the species was first described based only on its femur"(*1).
Specific name: *thabanensis* ← {(place-name) Thabana + -ensis}; referring to "the area of Thabana Morena, Lesotho" where "this material was originally collected in 1959 by a team led by François Ellenberger"(*1).
Etymology: **femur beast** from Thabana Morena
Taxonomy: Saurischia: Sauropodomorpha: Sauropodiformes (*1)

[References: (*1) Fabrègues CP, Allain R (2016). New material and revision of *Melanorosaurus thabanensis*, a basal sauropodomorph from the Upper Triassic of Lesotho. *PeerJ* 4: e1639.]

Metriacanthosaurus parkeri

> *Megalosaurus parkeri* von Huene, 1923 ⇒ *Metriacanthosaurus parkeri* (von Huene, 1923) Walker, 1964 [Oxfordian, Oxford Clay, Weymouth, Dorset, UK]

Generic name: *Metriacanthosaurus* ← {(Gr.) metri- [μέτριος/metrios: modelate] + ἄκανθο- [ἄκανθα/akantha: spine] + σαῦρος/sauros: lizard}
Specific name: *parkeri* ← {(person's name) W. Parker + -ī}
Etymology: Parker's **moderately spined lizard**
Taxonomy: Theropoda: Allosauroidea: Metriacanthosauridae: Metriacanthosaurinae

[References: Walker AD (1964). Triassic reptiles from the Elgin area: *Ornithosuchus* and the origin of carnosaurs. *Philosophical Transactions of the Royal Society B, Biological Sciences* 248(744): 53–134.]

Microceratops gobiensis

> *Microceratops gobiensis* Bohlin, 1953 (preoccupied)* ⇒ *Microceratus gobiensis* (Bohlin, 1953) Mateus, 2008 [Campanian–Maastrichtian; Minhe Formation, Inner Monglia, China]

**Microceratops* had been used as a genus name of an insect by Seyrig, 1952.

Generic name: *Microceratops* ← {(Gr.) μῑκρός/mīcros: small + ceratops [κέρας/keras: horn + ὤψ/ōps: face]}
Specific name: *gobiensis* ← {(place-name) Gobi + -ensis}
Etymology: **small-horned face (ceratopsian)** from Gobi
Taxonomy: Ornithischia: Marginocephalia: Ceratopsia
Other species:

> *Microceratops sulcidens* Bohlin, 1953 ⇒ *Asiaceratops sulcidens* (Bohlin, 1953) Nessov *et al.*, 1989 [Barremian–Aptian; Xinminbao Group, Gansu, China]

Notes: According to *The Dinosauria 2nd edition*, *Microceratops gobiensis* and *M. sulcidens* were *nomina dubia*. *Microceratops sulcidens* was renamed into a second species of *Asiaceratops*.

Microceratus gobiensis

> *Microceratops gobiensis* Bohlin, 1953 (preoccupied by Seyrig, 1952) ⇒ *Microceratus gobiensis* (Bohlin, 1953) Mateus, 2008 (*1) [Campanian–Maastrichtian; Minhe Formation, Inner Mongolia, China]

Generic name: *Microceratus* ← {(Gr.) μῑκρός/mīkros: small + ceratos: horned} (*1)
Specific name: *gobiensis* ← {(place-name) Gobi Desert + -ensis}
Etymology: **small horned one** from Gobi Desert
Taxonomy: Ornithischia: Marginocephalia: Ceratopsia: Neoceratopsia

[References: (*1) Mateus O (2008). Two ornithischian dinosaurs renamed: *Microceratops* Bohlin, 1953 and *Diceratops* Lull, 1905. *Journal of Paleontology* 82(2): 423.]

Microcoelus patagonicus

> *Titanosaurus australis* Lydekker, 1893 ⇒ *Saltasaurus australis* (Lydekker, 1893) McIntosh, 1990 ⇒ *Neuquensaurus australis* (Lydekker, 1893) Powell, 1992 [Santonian; Bajo de la Carpa Formation, Neuquén, Argentina]

= *Microcoelus patagonicus* Lydekker, 1893 [Santonian; Bajo de la Carpa Formation, Sierra Roca, Río Neuquén, Neuquén, Argentina]

Generic name: *Microcoelus* ← {(Gr.) μῑκρός/mikros: small + κοῖλος/koilos: hollow}
Specific name: *patagonicus* ← {(place-name) Patagonia + -icus}; in reference to the province of Patagonia in Argentina.
Etymology: **small hollow** from Patagonia
Taxonomy: Saurischia: Sauropodomorpha: Sauropoda: Titanosauria: Saltasauridae: Saltasaurinae
Notes: According to *The Dinosauria 2nd edition*, *Microcoelus patagonicus* is considered to be a synonym of *Neuquensaurus australis*.

Microhadrosaurus nanshiungensis

> *Microhadrosaurus nanshiungensis* Dong, 1979 [Campanian– Maastrichtian; Nanxiong Formation, Guangdong, China]

Generic name: *Microhadrosaurus* ← {(Gr.) μῑκρός/mīkros: small + *Hadrosaurus* [ἁδρός/hadros:

sturdy + σαῦρος/sauros: lizard]}
Specific name: *nanshiungensis* ← {(place-name) Nanshiung 南雄 [Pinyin:Nánxióng] + -ensis}
Etymology: **small sturdy lizard (hadrosaur)** from Nanshiong
Taxonomy: Ornithischia: Ornithopoda:
Hadrosauridae
Notes: *Microhadrosaurus* is based on juvenile remains. According to *The Dinosauria 2nd edition, M. nanshiungensis* is regarded as a *nomen dubium*.
[References: Dong Z (1979). The Cretaceous dinosaur fossils in south China. In: Institute of Vertebrate Paleontology and Paleoanthropology and Nanjing Institute of Paleontology (eds), *Mesozoic and Cenozoic Red Beds of South China. Nanxiong, China*: 342–350.]

Micropachycephalosaurus hongtuyanensis
> *Micropachycephalosaurus hongtuyanensis* Dong, 1978 [Campanian; Wangshi Group, Laiyang, Shandong, China]
Generic name: *Micropachycephalosaurus* ← {(Gr.) μῑκρός/mīkros: small + παχύς/pachys: thick + κεφαλή/kephalē: head + σαῦρος/sauros: lizard}
The specimen is "a small member of the Pachycephalosauria".(*1)
Specific name: *hongtuyanensis* ← {(Chin.) hóngtǔyán紅土岩 + -ensis}
This specimen was collected "in a cliff southwest of the Laiyang train station"(*1).
Etymology: **small, thick-headed lizard** from hongtuyán
Taxonomy: Ornithischia: Marginocephalia: Ceratopsia
cf. Pachycephalosauria: Homalocephaleridae (*1)
Notes: *Micropachycephalosaurus hongtuyanensis* was originally described as a member of the Pachycephalosauria, however, it was classified as a basal member of the Ceratopsia by Butler *et al.* (2011).
[References: (*1) Dong Z (1977). A pachycephalosaur from the Wangshi Fm. of Laiyang Co., Shandong Province, *Micropachycephalosaurus hongtuyanensis* gen. *et* sp. nov. *Vertebrata PalAsiatica* 16(4): 225–228.]

Microraptor zhaoianus
> *Microraptor zhaoianus* Xu *et al.*, 2000 [Aptian; Jiufotang Formation, Xiasanjiazi, Liaoning, China]
(=)* "*Archaeoraptor liaoningensis*" Sloan, 1999 (*partim*) [Aptian; Jiufotang Formation, Xiasanjiazi, Liaoning, China]
*objective synonym
= *Cryptovolans pauli* Czerkas *et al.*, 2002 [Aptian; Jiufotang Formation, Shangheshou, Liaoning, China]
Generic name: *Microraptor* ← {(Gr.) μῑκρός/mīkros: small + (Lat.) raptor: robber, thief}; "referring to the small size of this new dromaeosaurid dinosaur"(*1).
Specific name: *zhaoianus* ← {(person's name) Zhào趙 + -ianus}; "in honor of Zhào Xǐjìn, a distinguished dinosaurologist who introduced the first author to the field of vertebrate paleontology"(*1).
Etymology: Zhao's **small plunderer** [趙氏小盜龍]
Taxonomy: Theropoda: Maniraptora:
Dromaeosauridae: Microraptoria
cf. Theropoda: Maniraptora: Dromaeosauridae (*1)
Other species:
> *Microraptor gui* Xu *et al*, 2003 [Barremian; Jiufotang Formation, Dapingfang, Liaoning, China]
> *Microraptor hanqingi* Gong *et al.*, 2012 [Aptian; Jiufotang Formation, Liaoning, China]
Notes: The plumage of *Microraptor* is thought to be predominantly iridescent(*2). A nearly complete lizard (*Indrasaurus*) was found in the stomach of a *Microraptor,* according to Zhou *et al.* (2019).
[References: (*1) Xu X, Zhou Z, Wang X (2000).The smallest known non-avian theropod dinosaur. *Nature* 408: 705–708.; (*2) Li Q, Gao K-Q, Meng Q, Clarke JA, Shawkey MD, D'Alba L, Pei R, Ellison M, Norell MA, Vinther J (2012). Reconstruction of *Microraptor* and the evolution of iridescent plumage. *Science* 335(6073): 1215–1219.]

Microvenator celer
> *Microvenator celer* Ostrom, 1970 [late Aptian; Cloverly Formation, Montana, USA]
Generic name: *Microvenator* ← {(Gr.) μῑκρός/mīkros: small + (Lat.) vēnātor, *m.*: hunter} (*1)
Specific name: *celer* ← {(Lat.) celer: swift}; "in reference to the probable rapid-running capabilities indicated by the tibia-femur ratio" (*1).
Etymology: swift, **small hunter**
Taxonomy: Theropoda: Coelurosauria:
Oviraptorosauria: Caenagnathoidea:
Caenagnathidae
Notes: The holotype fossil is likely to be a juvenile.
[References: (*1) Ostrom JH (1970). Stratigraphy and paleontology of the Cloverly Formation (Lower Cretaceous) of the Bighorn Basin area, Wyoming and Montana. *The Peabody Museum of Natural History Yale University, Bulletin* 35: 1–234.]

Mierasaurus bobyoungi
> *Mierasaurus bobyoungi* Royo-Torres *et al.*, 2017 [late Barremian–early Aptian; lower Yellow Cat Member, Cedar Mountain Formation, Utah, USA]
Generic name: *Mierasaurus* ← {(person's name)

Miera + (Gr.) σαῦρος/sauros: lizard}; in honor of "Bernardo de Miera y Pacheco, Spanish cartographer and chief scientist for the 1776 Domínguez-Escalante Expedition: the first European scientist to enter what is now Utah" (*1).
Specific name: *bobyoungi* ← {(person's name) Bob Young + -ī}; acknowledging "the importance of the underappreciated research by Robert Young on the Early Cretaceous of Utah" (*1).
Etymology: Robert Young and **Miera's lizard**
Taxonomy: Saurischia: Sauropoda: Eusauropoda: Turiasauria

[References: (*1) Royo-Torres R, Upchurch P, Kirkland JI, DeBlieux DD, Foster JR, Cobos A, Alcalá L (2017). Descendants of the Jurassic turiasaurs from Iberia found refuge in the Early Cretaceous of western USA. *Scientific Reports* 7: 14311.]

Minmi paravertebra

> *Minmi paravertebra* Molnar, 1980 [Aptian; Bungil Formation, Queensland, Australia]

Generic name: *Minmi* ← {(place-name) Minmi}; "from the Minmi Crossing, near the site of discovery. 'Minmi' seems to be of aboriginal origin, but uncertain meaning, either being a corruption of Min Min (a kind of will-o-the-wisp light), or referring to a large lily (Reed 1967)" (*1).

The holotype was discovered in the Minmi Member of the Bungil Formation.

Specific name: *paravertebra* ← {(Gr.) παρά/para + (Lat.) vertebra}; "referring to the unique paravertebral elements" (*1).
Etymology: paravertebral **Minmi**
Taxonomy: Ornithischia: Thyreophora: Ankylosauria: Ankylosauridae

[References: (*1) Molnar RE (1980). An ankylosaur (Ornithischia: Reptilia) from the Lower Cretaceous of southern Queensland. *Memoirs of the Queensland Museum* 20: 65–75.]

Minotaurasaurus ramachandrani

> *Minotaurasaurus ramachandrani* Miles & Miles, 2009 [Late Cretaceous; Gobi Desert of either Mongolia or China]

Generic name: *Minotaurasaurus* ← {(Gr. myth.) Μινώταυρος/Minōtauros: Minotaur, man-bull + (Gr.) σαῦρος/sauros: reptile}; "in reference to the bull-like appearance of the skull, similar to the Minotaur of Greek mythology" (*1).
Specific name: *ramachandrani* ← {(person's name) Ramachandran + -ī}; "for Vilayanur S. Ramachandran: paleontology patron who made sure that this skull was described and made available to science" (*1).
Etymology: Ramachandra's **Minotaur (man-bull) reptile**
Taxonomy: Ornithischia: Thyreophora: Ankylosauria: Ankylosauridae: Ankylosaurinae
Notes: The holotype specimen was originally purchased by V. S. Ramachandran and displayed at the Victor Valley Museum, California, USA. The stratigraphic information for the specimen was missing (*1). In 2013, a new specimen of *Minotaurasaurus* was reported from Djadokhta Formation, Mongolia by Alicea & Loewen (*2). According to Arbour *et al.* (2014), *Minotaurasaurus ramachandrani* is considered a junior synonym of *Tarchia kielanae* (*3). But, Penkalski & Tumanova (2016) concluded that *Minotaurasaurus* is a valid taxon (*4).

[References: (*1) Miles CA, Miles CJ (2009). Skull of *Minotaurasaurus ramachandrani*, a new Cretaceous ankylosaur from the Gobi Desert. *Current Science* 96(1): 65–70.; (*2) Alicea J, Loewen M (2013). New *Minosaurasaurus* material from the Djodokta Formation establishes new taxonomic and stratigraphic criteria for the taxon. *Journal of Vertebrate Paleontology.* Program and Abstracts: 76.; (*3) Arbour VM, Currie PJ, Badamgarav D (2014). The ankylosaurid dinosaurs of the Upper Cretaceous Baruungoyot and Nemegt formations of Mongolia. *Zoological Journal of the Linnean Society* 172(3): 631–652.; (*4) Penkalski P, Tumanova T (2017). The cranial morphology and taxonomic status of Tarchia (Dinosauria: Ankylosauridae) from the Upper Cretaceous of Mongolia. *Cretaceous Research* 70: 117–127.]

Miragaia longicollum

> *Miragaia longicollum* Mateus *et al.*, 2009 [upper Kimmeridgian–lower Tithonian; Sobral Unit, Lourinhã Formation, Lourinhã, Portugal]

Generic name: *Miragaia* ← {(place-name) Miragaia [Mira [(Lat.) mīrus: wonderful] + (Gr. myth.) Γαῖα/Gaia: the Greek goddess of the Earth]}; "after the locality and the geological unit of the same name" (*1).
Specific name: *longicollum* ← {(Lat.) longus: long + collum: neck} (*1)
Etymology: **wonderful goddess of the Earth (Miragaia)** with a long neck
Taxonomy: Ornithischia: Stegosauria: Stegosauridae: Dacentrurinae
Notes: *Miragaia* was proposed to be a synonym of *Dacentrurus*, however, Costa & Mateus (2019) affirmed the validity of *Miragaia longicollum*.

[References: (*1) Mateus O, Maidment SCR, Christiansen NA (2009). A new long-necked 'sauropod-mimic' stegosaur and the evolution of the plated dinosaurs. *Proceedings of the Royal Society B. Biological Sciences* 276 (1663): 1815–1821.]

Mirarce eatoni [Avialae]

> *Mirarce eatoni* Atterholt *et al.*, 2018 [Campanian; Kaiparowits Formation, Utah, USA]

Generic name: *Mirarce* ← {(Lat.) mīrus: wonderful + (Gr. myth) Arce: winged messenger of the titans}; "named for its spectacular preservation and level of morphological detail", and "for the evidence suggesting a refined flight apparatus in this species" (*1).

Specific name: *eatoni* ← {(person's name) Eaton + -ī}; "in honor of Dr. Jeffrey Eaton, for his decades of work contributing to our understanding of the Kaiparowits Formation and the fossils recovered from it" (*1).

Etymology: Eaton's **wonderful Arce (winged messenger)**

Taxonomy: Theropoda: Ornithothoraces: Enantiornithes: Avisauridae
cf. Aves: Ornithothoraces: Enantiornithes: Avisauridae (*1)

[References: (*1) Atterholt J, Hutchison JH, O'Connor JK (2018). The most complete enantiornithine from North America and a phylogenetic analysis of the Avisauridae. *PeerJ* 6: e5910.]

Mirischia asymmetrica

> *Mirischia asymmetrica* Naish *et al.*, 2004 [Albian; Romualdo member, Santana Formation, Chapada do Araripe, Pernambuco, Brazil]

Generic name: *Mirischia* ← {(Lat.) mīrus: wonderful + (Gr.) ischia*: "pertaining to the pelvis (and not the ischia alone)"} (*1)
* [(Gr.) ἰσχίον/ischion: hip joint]

Specific name: *asymmetrica* ← {(Gr.) ἀ-σύμμετρος: asymmetric}; referring to the fact that "the ischia of *Mirischia* are asymmetrical" (*1).

Etymology: asymmetric, **wonderful hip joints**

Taxonomy: Theropoda: Compsognathidae

[References: (*1) Naish D, Martill DM, Frey E (2004). Ecology, systematics and biogeographical relationships of dinosaurs, including a new theropod, from the Santana Formation (? Albian, Early Cretaceous) of Brazi. *Historical Biology* 16(2–4): 57–70.]

Mnyamawamtuka moyowamkia

> *Mnyamawamtuka moyowamkia* Gorscak & O'Connor, 2019 [Aptian–Cenomanian; Mtuka Member, Galula Formation, Rukwa Rift Basin, Tanzania]

Generic name: *Mnyamawamtuka* [Mm-nya-ma-wah-mm-too-ka (*1)] ← {(Kiswahili) mnyama: animal, beast + wa Mtuka: of the Mtuka}; 'mnyama' acting as "a conceptual proxy to the titans in Titanosauria", and 'wa mkia' referring "to the river drainage that yielded the type specimen" (*1).

Specific name: *moyowamkia* ← {(Kiswahili) moyo: heart + wa mkia: of the tail}; "in reference to the posterolateral expansion of the posterior centrum on the middle caudal vertebrae that gives the posterior centrum surface a heart-shape outline" (*1).

Etymology: **beast from Mtuka** with heart of the tail

Taxonomy: Saurischia: Sauropoda: Titanosauria: Lithostrotia

[References: (*1) Gorscak E, O'Connor PM (2019). A new African titanosaurian sauropod dinosaur from the middle Cretaceous Galula Formation (Mtuka Member), Rukwa Rift Basin, southwestern Tanzania. *PLoS ONE* 14(2): e211412.]

Moabosaurus utahensis

> *Moabosaurus utahensis* Britt *et al.*, 2017 [Aptian; Cedar Mountain Formation, Utah, USA]

Generic name: *Moabosaurus* ← {(place-name) Moab: the city of Moab + (Gr.) σαῦρος/sauros: lizard}

Specific name: *utahensis* ← {(place-name) Utah: the state + -ensis}

Etymology: **Moab lizard** from Utah

Taxonomy: Saurischia: Sauropoda: Turiasauria
cf. Saurischia: Sauropoda: Neosauropoda: Macronaria (*1)

[References: (*1) Britt BB, Scheetz RD, Whiting MF, Wilhite DR (2017). *Moabosaurus utahensis*, n. gen., n. sp., a new sauropod from the early Cretaceous (Aptian) of North America. *Contributions from the Museum of Paleontology, University of Michigan* 32(11): 189–243.]

Mochlodon suessi

> *Iguanodon suessi* Bunzel, 1871 ⇒ *Mochlodon suessi* (Bunzel, 1871) Seeley, 1881 ⇒ *Rhabdodon suessi* (Bunzel, 1871) Steel, 1969 [Campanian; Grünbach Formation, Gosau Group, Niederösterreich, Austria]

Generic name: *Mochlodon* ← {(Gr.) μοχλός/mochlos: bar, lever + ὀδών/odōn: tooth}; referring to the ridge on the tooth.

Specific name: *suessi* ← {(person's name) Suess + -ī}; in honor of geologist Eduard Suess.

Etymology: Suess' **barred tooth**

Taxonomy: Ornithischia: Ornithopoda: Rhabdodontidae

Other species:

> *Mochlodon inkeyi* Nopcsa, 1899 ⇒ *Rhabdodon inkeyi* (Nopcsa, 1899) Nopcsa, 1899 [Campanian–Maastrichtian; Sânpetru Formation, Comitat Hunyad, Transylvania, Romania]

> *Mochlodon robustum* Nopcsa, 1899 ⇒ *Rhabdodon robustum* (Nopcsa, 1899) Nopcsa, 1915 ⇒ *Zalmoxes robustus* (Nopcsa, 1899)

Weishampel *et al.*, 2003 [Maastrichtian; Transylvania, Romania]
> *Mochlodon vorosi* Ősi *et al.*, 2012 [Santonian; Csehbánya Formation, Veszprém, Hungary]

Notes: According to *The Dinosauria 2nd edition*, *Mochlodon suessi* was considered as a synonym of *Rhabdodon priscus*. According to Ősi *et al.* (2012), *M. suessi* is resurrected as a valid species. *M. robustus* was renamed *Zalmoxes robustus*.

Moganopterus zhuiana [Pterosauria]

> *Moganopterus zhuiana* Lü *et al.*, 2012 [Berremian–Aptian; Yixian Formation, Liaoning, China]

Generic name: *Moganopterus* ← {(person's name/ sword name) Mò Yé莫邪 + Gān Jiāng 干将 + (Gr.) πτέρον/pteron: wing}; referring to the very long jaws which look like a famous pair of swords.

Specific name: *zhuiana* ← {(person's name) Zhu + -iana}; in honor of Ms. Zhu Haifen, who offered the specimen for scientific research.

Etymology: **Mo and Gan's wing** of Zhu

Taxonomy: Archosauromorpha: Pterosauria: Ctenochasmatidae

[References: Lü J, Pu H, Xu L, Wu Y, Wei X (2012). Largest toothed pterosaur skull from the Early Cretaceous Yixian Formation of western Liaoning, China, with coments on the family Boreopteridae. *Acta Geologica Sinica* 86(2): 387–293.]

Mojoceratops perifania

> *Mojoceratops perifania* Longrich, 2010 [late Campanian; Dinosaur Park Formation, Alberta, Canada]

Generic name: *Mojoceratops* ← {(an early 20th-century African-American term) mojo: "a magic-charm or talisman, often used to attract members of the opposite sex" (*1) + (Gr.) ceras: horn + ops: face}

Specific name: *perifania* ← {(Gr.) περιφάνεια/ perifaneia: conspicuousness}; in reference to the skull frill which is heart-shaped. (*1)

Etymology: ostentatious, **mojo horned-face (ceratopsian)**

Taxonomy: Ornithischia: Marginocephalia: Ceratopsia: Ceratopsidae: Chasmosaurinae

[References: Longrich NR (2010). *Mojoceratops perifania*, a new chasmosaurinae ceratopsid from the Late Campanian of western Canada. *Journal of Paleontology* 84(4): 681–694.; (*1) July 8, 2010. *Mojoceratops*: New dinosaur species named for flamboyant frill. ScienceDaily. Source: Yale University.]

Mongolosaurus haplodon

> *Mongolosaurus haplodon* Gilmore, 1933 [Aptian–Albian; On Gong Formation, Hu Khung Ulan, Inner Mongolia, China (*1)]

Generic name: *Mongolosaurus* ← {(place-name) Mongolia + (Gr.) σαῦρος/sauros: lizard}

Specific name: *haplodon* ← {(Gr.) ἁπλόος /haploos: single, simple, plain + ὀδών/odōn: tooth}

Etymology: **Mongolian lizard** with simple teeth

Taxonomy: Saurischia: Sauropodomorpha: Sauropoda: Titanosauriformes: Somphospondyli

Notes: According to *The Dinosauria 2nd edition*, *Mongolosaurus haplodon* was considered to be a *nomen dubium*.

[References: (*1) Gilmore CW (1933). Two new dinosaurian reptiles from Mongolia with notes on some fragmentary specimens. *American Museum Novitates* 679: 1–20.]

Mongolostegus exspectabilis

> *Wuerhosaurus* "mongoliensis" Ulansky, 2014 *vide* Galton & Carpenter, 2016 (*nomen nudum*) ⇒ *Mongolostegus exspectabilis* Tumanova & Alifanov, 2018 [Aptian–Albian; Dzunbain Formation, Dornogovi, Mongolia]

Generic name: *Mongolostegus* ← {(place-name) Mongolia + stegus: roof}

Specific name: *exspectabilis* ← {exspectabilis: long expected}

Etymology: expected **Mongolian roof (stegosaurian)**

Taxonomy: Ornithischia: Stegosauria: Stegosauridae

Notes: Ulansky (2014) informally dubbed the material, which had been reported as an indeterminate stegosaur in 2005 and 2012, *Wuerhosaurus* "mongoliensis" (*nomen nudum*). It was formally named *Mongolostegus expectabilis* in 2018.

[References: Tumanova TA & Alifanov VR (2018). First record of stegosaur (Ornithischian, Dinosauria) from the Aptian-Albian of Monglia. *Paleontological Journal* 52(14): 1771–1779.]

Monkonosaurus lawulacus

> *Monkonosaurus lawulacus* Zhao, 1986 (*nomen nudum*) ⇒ *Monkonosaurus lawulacus* Zhao, 1986 *vide* Dong, 1990 [Kimmeridgian; Loe-ein Formation, Monko, Xizang, China]

Generic name: *Monkonosaurus* ← {(Chin. place-name) Monko: Markam County + (Gr.) σαῦρος/ sauros: lizard}; referring to Markam County, also known as Monko.

Specific name: *lawulacus* ← {(place-name) Lawula: the name of a mountain in Markam, Tibet (Xixang)}

Etymology: **Monkon lizard** from Lawula
Taxonomy: Ornithischia: Stegosauria
Notes: According to *The Dinosauria 2nd edition*, *Monkonosaurus lawulacus* was a valid taxon, but Maidement & Wei (2006) regard it as a *nomen dubium* because it is "based on fragmentary and undiagnostic material".

Monoclonius crassus

> *Monoclonius crassus* Cope, 1876 (*nomen dubium*) [late Campanian; Judith River Formation, Montana, USA]

Generic name: *Monoclonius* ← {(Gr.) μόνος/monos: single + κλών/klōn: sprout + -ius}
According to Creisler (1992), "The generic name *Monoclonius* was paired with *Diclonius* (double sprout) in allusion to different modes of tooth replacement in each genus" (*1), not in reference with a single nose horn.
Specific name: *crassus* ← {(Lat.) crassus: fat, solid}.
Etymology: fat, **single sprout**
Taxonomy: Ornithischia: Marginocephalia: Ceratopsia: Ceratopsidae: Centrosaurinae
Other species:

> *Monoclonius fissus* Cope, 1876 (*nomen dubium*) [Campanian; Judith River Formation, Montana, USA]

> *Monoclonius recurvicornis* Cope, 1889 (*nomen dubium*) ⇒ *Ceratops recurvicornis* (Cope, 1889) Hatcher *vide* Stanton & Hatcher, 1905 (*nomen dubium*) ⇒ *Centrosaurus recurvicornis* (Cope, 1889) Sternberg, 1940 (*nomen dubium*) ⇒ *Eucentrosaurus recurvicornis* (Cope, 1889) Chure & McIntosh, 1989 (*nomen dubium*) [Campanian; Judith River Formation, Montana, USA]

> *Monoclonius sphenocerus* Cope, 1889 ⇒ *Agathaumas sphenocerus* (Cope, 1889) Ballou, 1897 ⇒ *Styracosaurus sphenocerus* (Cope, 1889) Lambe, 1915 (*nomen dubium*) [Campanian; Judith River Formation, Montana, USA]

> *Monoclonius canadensis* Lambe, 1902 ⇒ *Ceratops canadensis* (Lambe, 1902) Hatcher *vide* Stanton & Hatcher, 1905 ⇒ *Eoceratops canadensis* (Lambe, 1902) Lambe, 1915 ⇒ *Chasmosaurus canadensis* (Lambe, 1902) Lehman, 1990 [Campanian; Belly River Group, Alberta, Canada]

> *Centrosaurus apertus* Lambe, 1904 (*nomen dubium*) ⇒ *Monoclonius apertus* (Lambe, 1904) Lull, 1933 (*nomen dubium*) [Campanian; Dinosaur Park Formation, Alberta, Canada]

> *Monoclonius dawsoni* Lambe, 1902 ⇒ *Brachyceratops dawsoni* (Lambe, 1902) Lambe, 1915 ⇒ *Centrosaurus dawsoni* (Lambe, 1902) Dodson, 1990 [Campanian; Dinosaur Park Formation, Alberta, Canada]

> *Monoclonius flexus* Lambe, 1914 ⇒ *Centrosaurus flexus* (Lambe, 1914) Lull, 1933 ⇒ *Eucentrosaurus flexus* (Lambe, 1914) Chure & McIntosh, 1989 [Campanian; Dinosaur Park Formation, Alberta, Canada]

> *Monoclonius nasicornus* Brown, 1917 ⇒ *Centrosaurus nasicornus* (Brown, 1917) Lull, 1933 [Campanian; Dinosaur Park Formation, Alberta, Canada]

> *Monoclonius cutleri* Brown, 1917 ⇒ *Centrosaurus cutleri* (Brown, 1917) Russell, 1933 [Campanian; Oldman Formation, Alberta, Canada]

> *Monoclonius lowei* Sterberg, 1940 [late Campanian; Dinosaur Park Formation, Alberta, Canada]

> *Centrosaurus longirostris* Sternberg, 1940 ⇒ *Monoclonius longirostris* (Sternberg, 1940) Kuhn, 1964 [Campanian; Dinosaur Park Formation, Alberta, Canada]

> *Monoclonius belli* Lambe, 1902 ⇒ *Ceratops belli* (Lambe, 1902) Hatcher *vide* Stanton & Hatcher, 1905 ⇒ *Chasmosaurus belli* (Lambe, 1902) Lambe, 1914 ⇒ *Protorosaurus belli* (Lambe, 1902) Lambe, 1914 [Campanian; Dinosaur Park Formation, Alberta, Canada]

Notes: According to *The Dinosauria 2nd edition*, *Monoclonius crassus* (type) was a valid taxon and *M. lowei* was regarded as a junior synonym of *M. crassus.*

[References: (*1) Creisler BS (1992). Why Monoclonius Cope was not named for its horn: the etymologies of Cope's dinosaurs. *Journal of Vertebrate Paleontology* 12(3): 313–317.; Cope ED (1876). Descriptions of some vertebrate remains from the Fort Union Beds of Montana. *Proceedings of the Academy of Natural Sciences of Philadelphia* 28: 248–261.]

Monolophosaurus jiangi

> *Monolophosaurus jiangjunmiaoi* Dong, 1992 (*nomen nudum*) ⇒ *Monolophosaurus dongi* Grady, 1993 (*nomen nudum*) ⇒ *Monolophosaurus jiangi* Zhao & Currie, 1993 [Callovian; Wucaiwan Member, Shishugou Formation, Xinjiang, China]

Generic name: *Monolophosaurus* ← {(Gr.) μόνος/monos: single + λόφος/lophos: crest + σαῦρος/sauros: lizard}; "referring to the single crest on the midline of the skull roof"(*1).
Specific name: *jiangi* ← {(Chin.) jiang 将 [= Jiàngjūnmiào "(translated as General Jiang's Temple)" (*1) 将軍廟] + -ī}; referring to "the site

of an abandoned desert inn in the Gurbantunggut (previous transliteration is Kurban Tangut) Desert of the Junggar Basin of Xinjiang" (*1).

Etymology: **single crested lizard** from Jiangjunmiao

Taxonomy: Theropoda: Tetanurae

cf. Theropoda: Carnosauria (acc. *The Dinosauria 2nd edition*)

[References: (*1) Zhao X-J, Currie PJ (1993). A large crested theropod from the Jurassic of Xinjiang, People's Republic of China. *Canadian Journal of Earth Sciences* 30: 2027–2036.]

Mononykus olecranus

> *Mononychus olecranus* Perle *et al.*, 1993 (preoccupied) ⇒ *Mononykus olecranus* (Perle *et al.*, 1993) Perle *et al.*, 1993 [Maastrichtian; Nemegt Formation, Bugin Tsav, Ömnögovi, Mongolia]

Generic name: *Mononykus* ← {(Gr.) μόνος/monos: one + ὄνυχ/onyx: claw} (*1)

Mononychus was seemed to have "a short, robust forelimb with a single stout claw" (*1).

Specific name: *olecranus* ← {(Gr.) ὠλέκρανον/ōlecranon: elbow head} (*1)

Etymology: **single claw** with elbow head

Taxonomy: Theropoda: Coelurosauria: Alvarezsauridae

[References: (*1) Perle A, Norell MA, Chiappe LM, Clark JM (1993). Flightless bird from the Cretaceous of Mongolia. *Nature* 362(6421): 623–626.]

Montanoceratops cerorhynchus

> *Leptoceratops cerorhynchus* Brown & Schlaikjer, 1942 ⇒ *Montanoceratops cerorhynchus* (Brown & Schlaikjer, 1942) Sternberg, 1951 [early Maastrichtian; St. Mary River Formation, Montana, USA]

Generic name: *Montanoceratops* ← {(place-name) Montana + (Gr.) κέρας/keras: horn + ὦψ/ōps: face}

Specific name: *cerorhynchus* ← {(Gr.) κερο- [κέρας/keras: horn] + ῥύγχος/rhynchos: snout, muzzle}

According to the authors, this specimen had "its large nasal and very pronounced nasal horn-core" (*1).

Etymology: **Montana horned face (ceratopsian)** with nasal horn-core

Taxonomy: Ornithischia: Marginocephalia: Ceratopsia: Leptoceratopsidae

According to Brown (1942), *Leptoceratops* was thought to be close to *Protoceratops* (*1), and later, these two genera were placed in the family Protoceratopsidae.

Notes: *Leptoceratops cerorhynchos* was once thought to have a horn on its nose (*1), but it was actually a cheek horn. Sternberg reassigned *L. cerorhynchos* collected in Montana a new genus *Montanoceratops* in 1951.

[References: (*1) Brown B, Schlaikjer EM (1942). The skeleton of *Leptoceratops* with the description of a new species. *American Museum Novitates* 1169: 1–15.]

Morelladon beltrani

> *Morelladon beltrani* Gasulla *et al.*, 2015 [early Aptian; Arcillas de Morella Formation, Morella, Spain]

Generic name: *Morelladon* ← {(place-name) Morella: the name of the type locality + (Gr.) ὀδών/odōn: tooth} (*1)

Specific name: *beltrani* ← {(person's name) Beltrán + -ī}; in honor of "Victor Beltrán and Vega del Moll S. A. Company, for his involvement and collaboration in the localization of the different fossil sites at the Mas de la Parreta Quarry" (*1).

Etymology: Beltrán's **Morella tooth**

Taxonomy: Ornithischia: Ornithopoda: Iguanodontia: Ankylopollexia: Styracosterna

[References: (*1) Gasulla JM, Escaso F, Narváez I, Ortega F, Sanz JL (2015). A new sail-backed styracosternan (Dinosauria: Ornithopoda) from the Early Cretaceous of Morella, Spain. *PLoS ONE* 10(12): e0144167.]

Moros intrepidus

> *Moros intrepidus* Zanno *et al.*, 2019 [Cenomanian; lower Mussentuchit Member, upper Cedar Mountain Formation, Utah, USA]

Generic name: *Moros* ← {(Gr.) μόρος/moros: fate, doom, "the embodiment of impending doom" (*1)}; "in reference to the establishment of the Cretaceous tyrannosauroid lineage in North America" (*1).

Specific name: *intrepidus* ← {(Lat.) intrepidus: intrepid}; "in reference to the hypothesized intracontinental dispersal of tyrannosaurs during this interval" (*1).

Etymology: intrepid **fate**

Taxonomy: Theropoda: Coelurosauria: Tyrannosauroidea

Notes: Moros was a small-bodied tyrannosauroid and seemed to have been a fast runner.

[References: Zanno LE, Tucker RT, Canoville A, Avrahami HM, Gates TA, Makovicky PJ (2019). Diminutive fleet-footed tyrannosauroid narrows the 70-million-year gap in the North American fossil record. *Communications Biology* 2(64): 1–12.]

Morosaurus agilis

> *Morosaurus agilis* Marsh, 1889 ⇒ *Camarasaurus agilis* (Marsh, 1889) Kuhn, 1939 ⇒ *Smitanosaurus agilis* (Marsh, 1889) Whitlock

& Wilson, 2020 [Kimmeridgian; Brushy Basin Member, Morrison Formation, Colorado, USA]
Generic name: *Morosaurus* Marsh, 1878 ← {(Gr.) μωρός/mōros: dull, sluggish, stupid + σαῦρός/sauros: lizard}
Specific name: *agilis* ← {(Lat.) agilis: nimble, quick}
Etymology: agile, **dull lizard**
Taxonomy: Sauropoda: Camarasauromorpha: Camarasauridae
Other species:
> *Apatosaurus grandis* Marsh, 1877 ⇒*Morosaurus grandis* (Marsh, 1877) Williston, 1898 ⇒ *Camarasaurus grandis* (Marsh, 1877) Gilmore, 1925 [Kimmeridgian; Brushy Basin Member, Morrison Formation, Wyoming, USA]
= *Morosaurus impar* Marsh, 1878 ⇒ *Camarasaurus impar* (Marsh, 1878) Steel, 1970 [Kimmeridgian–Tithonian; Brushy Basin Member, Morrison Formation, Wyoming, USA]
= *Morosaurus robustus* Marsh, 1878 ⇒ *Camarasaurus robustus* (Marsh, 1878) White, 1958 [Kimmeridgian–Tithonian; Brushy Basin Member, Morrison Formation, USA]
> *Morosaurus lentus* Marsh, 1889 ⇒ *Camarasaurus lentus* (Marsh, 1889) Mook, 1914 [Kimmeridgian; Lake Como Member, Morrison Formation, Wyoming, USA]
Notes: According to *The Dinosauria 2nd edition*, *Morosaurus* is considered to be a junior synonym of *Camarasaurus*.

[References: Marsh OC (1889). Notice of new American Dinosauria. *The American Journal of Science and Arts, series 3* 38: 331–336.]

Morrosaurus antarcticus

> *Morrosaurus antarcticus* Rozadilla *et al*., 2016 [Maastrichtian; Snow Hill Island Formation, El Morro (The Naze) Peninsula, James Ross Island, Antarctica]
Generic name: *Morrosaurus* ← {(place-name) El Morro + (Gr.) σαῦρος/sauros: lizard}
Specific name: *antarcticus* ← {(Gr.) ἀντ/ant-: oppsite + ἄρκτος/arctos: a bear, the North}; in reference to the Antarctic continent.
Etymology: **El Morro lizard** of Antarctica
Taxonomy: Ornithischia: Ornithopoda: Euiguanodontia: Elasmaria

[References: Rozadilla S, Agnolin FL, Novas FE, Aranciaga Roland AM, Motta MJ, Lirio JM, Isasi MP (2016). A new ornithopod (Dinosauria, Ornthischia) from the Upper Cretaceous of Antarctica and its palaeobiogeographical implications. *Cretaceous Research* 57: 311–324.]

Mosaiceratops azumai

> *Mosaiceratops azumai* Zheng *et al*., 2015 [Turonian–Campanian; Xiaguan Formation, Henan, China]
Generic name: *Mosaiceratops* ← {(Lat.) mosaicus: mosaic + ceratops: ceratopsian}; "in reference to the specimen's unique (mosaic) combination of characters that were previously considered diagnostic of basal ceratopsians, psittacosaurids, or basal neoceratopsians" (*1).
Specific name: *azumai* ← {(person's name) Azuma + -ī}; "in honor of Dr. Yoichi Azuma from Fukui Prefectual Dinosaur Museum, who co-organized and participated in several dinosaur expeditions in China. One of those expeditions led to the discovery of the basal neoceratopsian *Archaeoceratops*" (*1).
Etymology: Azuma's **mosaic ceratopsian**
Taxonomy: Ornithischia: Ceratopsia: Neoceratopsia

[References: (*1) Zheng W, Jin X, Xu X (2015). A psittacosaurid-like basal neoceratopsian from the Upper Cretaceous of central China and its implications for basal ceratopsian evolution. *Scientific Reports* 5: 14190.]

Murusraptor barrosaensis

> *Murusraptor barrosaensis* Novas & Currie, 2016 [Coniacian; Sierra Barrosa Formation, northeast of Plaza Huincul, Neuquén, Argentina]
Generic name: *Murusraptor* ← {(Lat.) mūrus: wall + raptor: robber}; "referring to the discovery of the specimen in the wall of a canyon" (*1).
Specific name: *barrosaensis* ← {(place-name) Barrosa + -ensis}; "alluding to Sierra Barrosa, the locality where it was collected" (*1).
Etymology: **wall robber (raptor)** from Sierra Barrosa
Taxonomy: Theropoda: Tetanurae: Megaraptora: Megaraotoridae

[References: (*1) Coria RA, Currie PJ (2016). A new megaraptoran dinosaur (Dinosauria, Theropoda, Megaraptoridae) from the Late Cretaceous of Patagonia. *PLoS ONE* 11(7): e157973.]

Mussaurus patagonicus

> *Mussaurus patagonicus* Bonaparte & Vince, 1979 [Norian; Laguna Colorada Formation, Santa Cruz, Patagonia, Argentina]
Generic name: *Mussaurus* ← {(Lat.) mūs: mouse, rat + saurus [σαῦρος/sauros: lizard]}; "in reference to the rat size of this dinosaur" (*1).
Specific name: *patagonicus* ← {patagonicus: from Patagonia} (*1)
Etymology: **rat lizard** from Patagonia
Taxonomy: Saurischia: Sauropodomorpha: Mussauridae

Notes: Seven juvenile individuals of prosauropod dinosaurs and two fossil eggs were found. (*1)

[References: (*1) Bonaparte JF, Vince M (1979). El hallazgo del primer nido de Dinosaurios Triásicos (Saurischia, Prosauropoda), Triásico Superior de Patagonia, Argentina [The Discovery of the first nest of Triassic dinosaurs (Saurischia, Prosauropoda) from the Upper Triassic of Patagonia, Argentina]. *Ameghiniana* 16(1–2): 173–182.]

Muttaburrasaurus langdoni

> *Muttaburrasaurus langdoni* Bartholomai & Molnar, 1981 [Albian; Mackunda Formation, Muttaburra, Queensland, Australia]

Generic name: *Muttaburrasaurus* ← {(place-name) Muttaburra + (Gr.) σαῦρος/sauros: lizard}

Specific name: *langdoni* ← {(person's name) Langdon + -ī}; in honor of M. D. Langdon, of Muttaburra, who discovered the specimen, and reported it to the Queensland Museum (*1).

Etymology: Langdon's **Muttaburra lizard**

Taxonomy: Ornithischia: Ornithopoda: Iguanodontia: Rhabdodontomorpha

Notes: *Muttaburrasaurus* had a big, hollow, upward-bulging nasal muzzle.

[References: (*1) Bartholomai A, Molnar RE (1981). *Muttaburrasaurus*: a new iguanodontid (Ornithischia: Ornithopoda) dinosaur from the Lower Cretaceous of Queensland. *Memoirs of the Queensland Museum* 20(2): 310–349.]

Muyelensaurus pecheni

> *Muyelensaurus pecheni* Calvo *et al.*, 2007 [Coniacian; Portezuelo Formation, Rincón de los Sauces, Neuquén, Argentina]

Generic name: *Muyelensaurus* ← {(Mapuche) Muyelen: one of the names of the Colorado River + (Gr.) σαῦρος/sauros: lizard} (*1)

Specific name: *pecheni* ← {(person's name) Pechen + -ī}; "in honor of Dra. Ana María Pechén, main head of the National University of Comahue (2002–2006), who supported the study of dinosaur fossils in Neuquén Province, Patagonia" (*1).

Etymology: Pechén's **Muyelen lizard**

Taxonomy: Saurischia: Sauropoda: Titanosauria: Titanosauridae: Rinconsauria (*1)

[References: (*1) Calvo JO, González-Riga BJ, Porfiri JD (2007). A new titanosaur sauropod from the Late Cretaceous of Neuquén, Patagonia, Argentina. *Arquivos do Museu Nacional*, Rio do Janeiro 65(4): 485–504.]

Mymoorapelta maysi

> *Mymoorapelta maysi* Kirkland & Carpenter, 1994 [Kimmeridgian–Tithonian; Morrison Formation, Colorado, USA]

Generic name: *Mymoorapelta* ← {(person's name) Mymoor [←Mygatt-Moore] + (Lat.) pelta= (Gr.) πέλτη/pelte: shield}; "in honor of Peter and Marilyn Mygatt and John D. and Vanetta Moore, who discovered the Mygatt-Moore Quarry" (*1).

Specific name: *maysi* ← {(person's name) Mays + -ī}; "for Chris Mays, president of Dinamation International Corporation and founder of the Dinamation International Society (a nonprofit organization to promote paleontology), who with DIS funded this research" (*1).

Etymology: Mays' **shield of Mygatt-Moore**

Taxonomy: Ornithischia: Thyreophora: Ankylosauria: Nodosauridae

[References: (*1) Kirkland JI, Carpenter K (1994). North America's first pre-Cretaceous ankylosaur (Dinosauria) from the Upper Jurassic Morrison Formation of western Colorado. *Brigham Young University Geology Studies* 40: 25–42.]

N

Naashoibitosaurus ostromi

> *Naashoibitosaurus ostromi* Hunt & Lucas, 1993 [late Campanian; lower Kirtland Formation, San Juan Basin, New Mexico, USA]

Generic name: *Naashoibitosaurus* ← {(Navajo, place-name) Naashoibito Member (*1) + (Gr.) σαῦρος/sauros: lizard}; referring to "the Naashoibito Member of the Kirtland Formation that yielded the holotype" (*2).

This specimen was thought to come from the Naashoibito Member of the Kirtland Formation, but in fact it was discovered from an older member, De-na-zin Member (upper Shale Member).

Specific name: *ostromi* ← {(person's name) Ostrom + -ī}; "in honor of John H. Ostrom for his work on the cranial morphology of hadrosaurs and for describing one of the taxa from the San Juan Basin" (*2).

Etymology: Ostrom's **Naashoibito lizard**

Taxonomy: Ornithischia: Ornithopoda: Hadrosauridae: Saurolophinae: Kritosaurini

[References: (*1) Hunt AP, Lucas SG (1992). Stratigraphy, paleontology, and age of the Fruitland and Kirtland formations (Upper Cretaceous) San Juan Basin, New Mexico. In: Lucas SG, Kues BS, Williamson TE, Hunt AP (eds), *New Mexico Geological Society 43rd Annual Fall Field Conference Guidebook*, New Mexico Geological Society, Socorro, USA: 217–239.; (*2) Hunt AP, Lucas SG (1993). Cretaceous vertebrates of New Mexico. In: Lucas SG, Zidek J. (eds), *Dinosaurs of New Mexico. New Mexico Museum of Natural History and Science Bulletin* 2: 77–91.]

Nambalia roychowdhurii

> *Nambalia roychowdhurii* Novas *et al.*, 2011 [Norian; upper Maleri Formation, Nambal Village, Andhra Pradesh, India]

Generic name: *Nambalia* ← {(place-name) Nambal: Indian town of Nambal + -ia} [*1]

Specific name: *roychowdhurii* ← {(person's name) Roy Chowdhuri + -ī}; "in honor of Dr. Roy Chowdhuri, for his outstanding research on the Triassic vertebrate faunas of India" [*1].

Etymology: Roy Chowdhuri's **one from Nambal**

Taxonomy: Saurischia: Sauropodomorpha

[References: (*1) Novas FE, Ezcurre MD, Chatterjee S, Kutty TS (2011). New dinosaur species from the Upper Triassic upper Maleri and lower Dharmaram formations of Central India. *Earth and Environmental Science Transactions of the Royal Society of Edinburgh* 101: 333–349.]

Nanantius eos [Avialae]

> *Nanantius eos* Molnar 1986 [Albian; Toolebuc Formation, Queensland, Australia]

Generic name: *Nanantius* ← {(Gr.) νᾶνος/nānos: dwarf + ἀντίος/antios: opposite}

Specific name: *eos* ← {(Gr.) ἠώς/ēōs: dawn}

Etymology: **dwarf enantiornithine** of dawn

Taxonomy: Theropoda: Avialae: Enantiornithes

Other species:

> *Gobipteryx minuta* Elżanowski, 1974 [Campanian; Barun Goyot Formation, Ömnögovi, Mongolia]

= *Nanantius valifanovi* Kurochkin, 1996 [Campanian; Barun Goyot Formation, Ömnögivi, Mongolia]

Nankangia jiangxiensis

> *Nankangia jiangxiensis* Lü *et al.*, 2013 [Maastrichtian; Nanxiong Formation, Jiangxi, China]

Generic name: *Nankangia* ← {(Chin. place-name) Nánkāng南康 + -ia}; "referring to the Chinese administrative unit Nankang City in Jiangxi Province" [*1].

Specific name: *jiangxiensis* ← {(Chin. place-name) Jiāngxī 江西 + -ensis}; "referring to the Jiangxi Province, where the holotype site in Nankang City is located" [*1].

Etymology: **Nankang's one** from Jiangxi

Taxonomy: Theropoda: Oviraptorosauria: Caenagnathoidea

cf. Theropoda: Oviraptorosauria [*1]

[References: (*1) Lü J, Yi L, Zhong H, Wei X (2013). A new oviraptorosaur (Dinosauria: Oviraptorosauria) from the Late Cretaceous of southern China and its paleoecological implications. *PLoS ONE* 8(11): e80557.]

Nanningosaurus dashiensis

> *Nanningosaurus dashiensis* Mo *et al.*, 2007 [Late Cretaceous; Dashi site, Nanning, Guangxi, China]

Generic name: *Nanningosaurus* ← {(Chin. place-name) Nánnìng南寧 + (Gr.) σαῦρος/sauros: lizard}

Specific name: *dashiensis* ← {(Chin. place-name) Dàshí大石 + -ensis}

Etymology: **Nanning lizard** from Dashi

Taxonomy: Ornithischia: Ornithopoda: Hadrosauridae: Lambeosaurinae

[References: Mo J, Zhao Z, Wang W, Xu X (2007). The first hadrosaurid dinosaur from southern China. *Acta Geologica Sinica* 81(4): 550–554.]

Nanosaurus agilis

> *Nanosaurus agilis* Marsh, 1877 [Kimmeridgian–Tithonian; Morrison Formation, Colorado, USA]

= *Nanosaurus rex* Marsh, 1877 ⇒ *Laosaurus rex* (Marsh, 1877) Marsh, 1896 ⇒ *Othnielia rex* (Marsh, 1877) Galton, 1977 [Kimmeridgian; Morrison Formation, Wyoming, USA]

= *Laosaurus consors* Marsh, 1894 ⇒ *Othnielia consors* (Marsh, 1894) Galton, 1977 ⇒ *Othnielosaurus consors* (Marsh, 1894) Galton, 2006 [Kimmeridgian; Morrison Formation, Wyoming, USA]

= *Drinker nisti* Bakker *et al.*, 1990 [Tithonian; Morrison Formation, Wyoming, USA]

Generic name: *Nanosaurus* ← {(Gr.) νᾶνος/nānos: dwarf + σαῦρος/sauros: lizard}

Marsh (1877) calls this small dinosaur "pigmy dinosaur" [*1].

Specific name: *agilis* ← {(Lat.) agilis: nimble, quick}

Etymology: nimble, **dwarf lizard**

Taxonomy: Ornithischia: Neornithischia

Other species:

> *Nanosaurus victor* Marsh, 1877 ⇒ *Hallopus victor* (Marsh, 1877) Marsh, 1881 [Kimmeridgian–Tithonian; Morrison Formation, Colorado, USA]

Notes: According to the *Dinosauria 2nd edition*, *Nanosaurus agilis* was considered to be a *nomen dubium* and *N. rex* was a synonym of *Othnielia rex*. *N. victor* was named by Marsh as a dinosaur [*2]. Now it is thought to be a pseudosuchian.

[References: (*1) Marsh OC (1877). Notice of some new vertebrate fossils. *American Journal of Science (Series 3)* 14: 249–256.; (*2) Marsh OC (1877) Notice of new dinosaurian reptiles from the Jurassic formations. *American Journal of Science (Series 3)* 14: 514–516.]

Nanotyrannus lancensis

> *Tyrannosaurus rex* Osborn, 1905 [Maastrichtian; Hell Creek Formation, Montana, USA]

= *Gorgosaurus lancensis* Gilmore, 1946 ⇒ *Deinodon lancensis* (Gilmore, 1946) Kuhn, 1965 ⇒ *Aublysodon lancensis* (Gilmore, 1946) Charig, 1967 ⇒ *Albertosaurus lancensis* (Gilmore, 1946) Russell, 1970 ⇒ *Nanotyrannus lancensis* (Gilmore, 1946) Bakker *et al.*, 1988 [late Maastrichtian; Hell Creek Formation, Montana, USA]

= *Aublysodon molneri*? Paul, 1988 ⇒ *Stygivenator molnari* ? (Paul, 1988) Olshevsky, 1955 [Maastrichtian; Hell Creek Formation, Montana, USA]

See also *Tyrannosaurus rex*.

Generic name: *Nanotyrannus* ← {(Gr.) νᾶνος/nanos: dwarf + τύραννος/tyrannos: tyrant, king}

Specific name: *lancensis* ← {(place-name) Lance Formation + -ensis}

Etymology: **pygmy tyrant** from Lance Formation [*1]

Taxonomy: Theropoda: Coelurosauria: Tyrannosauroidea: Tyrannosauridae

Notes: Rozhdestvensky (1965) first suggested that *Nanotyrannus* might be a juvenile *Tyrannosaurus rex*. According to Bakker *et al.* (1988), the specimen is estimated to be 5.2 m in total length when it died. Carr considered *Nanotyrannus* to be juvenile specimens of *Tyrannosaurus rex* [*1]. Woodward *et al.* (2020) indicate that *Tyrannosaurus* and *Nanotyrannus* are synonymous [*2].

[References: (*1) Carr TD (1999). Craniofacial ontogeny in Tyrannosauridae (Dinosauria, Coelurosauria). *Journal of Vertebrate Paleontology* 19: 497–520.; (*2) Woodward HN, Tremaine K, Williams SA, Zanno LE, Horner JR, Myhrvold N (2020). Growing up *Tyrannosaurus rex*: Osteohistology refutes the pygmy "*Nanotyrannus*" and supports ontogenetic niche partitioning in juvenile *Tyrannosaurus*. *Science Advances* 6(1): eaax6250.]

Nanshiungosaurus brevispinus

> *Nanshiungosaurus brevispinus* Dong, 1979 [Campanian; Yuanpu Formation, Nanxiong, Guangdong, China]

Generic name: *Nanshiungosaurus* ← {(place-name) Nanshiung南雄 [Pinyin: Nánxióng] + σαῦρος/sauros: lizard}

Specific name: *brevispinus* ← {(Lat.) brevis: short + spinus [spina: thorn]} short-spined; referring to the short spines on its vertebrae.

Etymology: short-spined **Nanxiong lizard**

Taxonomy: Theropoda: Coelurosauria: Therizinosauridae

Other species:

?> *Nanshiungosaurus bohlini* Dong & Yu, 1997 [Barremian; Chijinbao Formation, Xinminbao Group, Gansu, China]

Notes: Zanno (2010) considered that *Nanshiungosaurus bohlini* to be unrelated to *Nanshiungosaurus* [*1].

[References: (*1) Zanno L (2010). A taxonomic and phylogenetic re-evaluation of Therizinosauria (Dinosauria: Maniraptora). *Journal of Systematic Palaeontology* 8(4): 503–543.]

Nanuqsaurus hoglundi

> *Nanuqsaurus hoglundi* Fiorillo & Tykoshi, 2014 [Maastrichtian; Prince Creek Formation, Alaska, USA]

Generic name: *Nanuqsaurus* ← {(Iñupiat or Inupiaq) nanuq: polar bear + (Gr.) sauros: lizard} [*1]

Specific name: *hoglundi* ← {(person's name) Hoglund + -ī}; "in recognition of Forrest Hoglund for his career in earth sciences and his philanthropic efforts in furthering cultural institutions" [*1].

Etymology: Hoglund's **polar bear lizard**

Taxonomy: Theropoda: Tyrannosauroidea: Tyrannosauridae: Tyrannosaurinae

[References: (*1) Fiorillo AR, Tykoski RS (2014). A diminutive new tyrannosaur from the top of the world. *PLoS ONE* 9(3): e91287.]

Nanyangosaurus zhugeii

> *Nanyangosaurus zhugeii* Xu *et al.*, 2000 [? Albian; Sangping Formation, Henan, China]

Generic name: *Nanyangosaurus* ← {(Chin. place-name) Nányáng南陽+ (Gr.) σαῦρος/sauros: lizard}

Specific name: *zhugeii* ← {(person's name) Zhūgě Liàng 諸葛亮 + -iī}; in memory of the famous ancient strategist and politician, Zhuge Liang, who lived in Nanyang for a long time. [*1]

Etymology: Zhuge's **Nanyang lizard**

Taxonomy: Ornithischa: Ornithopoda: Hadrosauroidea: Hadrosauromorpha

[References: (*1) Xu X, Zhao JC, Huang WB, Li Z-Y, Dong Z-M (2000). A new iguanodontian from Sangping Formation of Neixiang, Henan and its stratigraphical implication. *Vertebrata PalAsiatica* 38(3): 176–191.]

Narambuenatitan palomoi

> *Narambuenatitan palomoi* Filippi *et al.*, 2011 [Campanian; Anacleto Formation, Neuquén, Argentina]

Generic name: *Narambuenatitan* ← {(place-name) Puesto Narambuena* + (Gr.) Τῑτάν/Tītan: "to correspond to a titanosaur specimen" [*1]} *situated about 20 km west from Rincón de los Sauces, Neuquén Province. [(*1)]

Specific name: *palomoi* ← {(person's name) Palomo + -ī}; "in reference to Salvador Palomo, technician of the Museo Municipal "Argentino Urquiza", Rincón de los Sauces, Neuquén, who found the specimen, and in acknowledgement of his permanent contribution to the local paleontology" [(*1)].

Etymology: Palomo's **titanosaur from Narambuena**

Taxonomy: Sauropoda: Titanosauriformes: Titanosauria: Lithostrotia

[References: (*1) Filippi LS, García RA, Garrido AC (2011). A new titanosaur sauropod dinosaur from the Upper Cretaceous of north Patagonia, Argentina. *Acta Palaeontologica Polonica* 56(3): 505–520.]

Narindasaurus thevenini

> *Narindasaurus thevenini* Royo-Torres *et al.*, 2020 [Bathonian; Isalo III Formation, Ankinganivalaka, Madagascar]

Generic name: *Narindasaurus* ← {(place-name) Narinda Bay: close to the Ankinganivalaka site + (Gr.) σαῦρος/sauros: lizard} [(*1)]

Specific name: *thevenini* ← {(person's name) Thevenin + -ī}; in honor of "Armand Thevenin, French palaeontologist who was interested in the Jurassic dinosaurs of Madagascar at the 20th century" [(*1)].

Etymology: Thevenin's **lizard from Narinda Bay**

Taxonomy: Saurischia: Sauropodomorpha: Sauropoda: Turiasauria

[References: Royo-Torres R, Cobos A, Mocho P, Alcalá L (2020). Origin and evolution of turiasaur dinosaurs set by means of a new 'rosetta' specimen from Spain. *Zoological Journal of the Linnean Society* 191(1):201–227.]

Nasutoceratops titusi

> *Nasutoceratops titusi* Sampson *et al.*, 2013 [Campanan; Kaiparowits Formation, Utah, USA]

Generic name: *Nasutoceratops* ← {(Lat.) nāsūtus: large-nosed + ceratops: horned-face [(Gr.) κέρας/keras: horn + ὤψ/ōps: face]}

Specific name: *titusi* ← {(person's name) Titus + -ī}; "in honor of Alan Titus, palaeontologist at Grand Staircase-Escalante National Monument, for his exemplary efforts assisting palaeontological fieldwork in the Monument" [(*1)].

Etymology: Titus' **large-nosed horned-face (ceratopsian)**

Taxonomy: Ornithischia: Ceratopsia: Ceratopsidae: Centrosaurinae

Notes: *Nasutoceratops* had unique rounded horns above its eyes.

[References: (*1) Sampson SD, Lund EK, Loewen MA, Farke AA, Clayton KE (2013). A remarkable short-snouted horned dinosaur from the Late Cretaceous (late Campanian) of southern Laramidia. *Proceedings of the Royal Society B: Biological Sciences* 280(1766): 20131186.]

"Natronasaurus" longispinus

> *Stegosaurus longispinus* Gilmore, 1914 ⇒ *Natronasaurus longispinus* (Gilmore, 1914) Ulansky, 2014 (invalid name) ⇒ *Alcovasaurus longispinus* (Gilmore, 1914) Galton & Carpenter, 2016 [Kimmeridgian–Tithonian; Alcova Quarry, Morrison Formation, Natrona, Wyoming, USA]

Generic name: *Natronasaurus* ← {(place-name) Natrona + (Gr.) σαῦρος/sauros: lizard}

Specific name: *longispinus* ← {(Lat.) longus + spina}; referring to the long tail spines.

Etymology: long-spined **Natrona lizard**

Taxonomy: Ornithischia: Thyreophora: Stegosauria: Stegosauridae

Notes: Ulansky gave *Stegosaurus longispinus* a new genus, *Natronasaurus*, in a self-published electronic publication. But this name was not admitted.

Navajoceratops sullivani

> *Navajoceratops sullivani* Fowler & Freedman Fowler, 2020 [Campanian; Kirtland Formation, New Mexico, USA]

Generic name: *Navajoceratops* ← {(demonym) Navajo + ceratops [(Gr.) κέρας/keras + ὤψ/ōps: face]}; "after the Navajo people indigenous to the San Juan Basin".

Specific name: *sullivani* ← {(person's name) Sullivan + -ī}; "after Dr. Robert Sullivan, leader of the SMP* expeditions to the San Juan Basin".

*State Museum of Pennsylvania, Harrisburg, Pennsylvania, USA.

Etymology: Sullivan's **Navajo horned face (ceratopsian)**

Taxonomy: Ornithischia: Ceratopsia: Ceratopsidae: Chasmosauridae

[References: Fowler DW, Freedman Fowler EA (2020). Transitional evolutionary forms in chasmosaurine ceratopsid dinosaurs: evidence from the Campanian of New Mexico. *PeerJ* 8: e9251.]

Nebulasaurus taito

> *Nebulasaurus taito* Xing *et al.*, 2013 [Aalenian–Bajocian; Zhanghe Formation, Yuanmou, Yunnan, China]

Generic name: *Nebulasaurus* ← {(Lat.) nebulae: misty, cloud + (Gr.) σαῦρος/sauros: lizard}; "after the alpine province of Yunnan (= southern cloudy

province, Chinese)"[*1]
Specific name: *taito* ← {(Jpn. company) Taito}; "in honor of the Taito Corporation of Japan, which funded the field project in and near the type locality"[*1].
Etymology: Taito's **misty cloud lizard (from Yunnan)**
Taxonomy: Saurischia: Sauropodomorpha:
Eusauropoda

[References: (*1) Xing L, Miyashita T, Currie PJ, You H, Dong Z (2015). A new basal eusauropod from the Middle Jurassic of Yunnan, China, and faunal compositions and transitions of Asian sauropodomorph dinosaurs. *Acta Palaeontologica Polonica* 60(1): 145–154.]

Nedcolbertia justinhofmanni

> *Nedcolbertia justinhofmanni* Kirkland *et al.*, 1998 [Barremian; Cedar Mountain Formation, Utah, USA]

Generic name: *Nedcolbertia* ← {(person's name) Ned (= Edwin) + -ia}; "named for Edwin Harris Colbert, one of the foremost dinosaur paleontologists of the twentieth century. Dr. Colbert is known as Ned to his friends and has conducted much significant research on small theropod dinosaurs"[*1].
Specific name: *justinhofmanni* ← {(person's name) Justin Hofmann + -ī}; "named for Justin Hofmann, a young boy from New Jersey, for his love of dinosaurs"[*1].
Etymology: Justin Hofmann and **Ned Colbert's one**
Taxonomy: Theropoda: Ornithomimosauria
cf. Theropoda: Tetanurae: Coelurosauria[*1]

[References: (*1) Kirkland JI, Britt BB, Whittle CH, Madsen SK, Burge DL (1998). A small coelurosaurian theropod from the Yellow Cat Member of the Cedar Mountain Formation (Lower Cretaceous, Barremian) of eastern Utah. *New Mexico Museum of Natural History and Science Bulletin* 14: 239–248.]

Nedoceratops hatcheri

> *Diceratops hatcheri* Lull *vide* Hatcher, 1905 (preoccupied by Foerster, 1868) ⇒ *Triceratops (Diceratops) hatcheri* (Lull *vide* Hatcher, 1905) Lull, 1933 ⇒ *Nedoceratops hatcheri* (Lull *vide* Hatcher, 1905) Ukrainsky, 2007 ⇒ *Diceratus hatcheri* (Lull *vide* Hatcher, 1905) Mateus, 2008 [Maastrichtian; Lance Formation, Wyoming, USA]

Generic name: *Nedoceratops* ← {(Russian prefix) nedo-: insufficience + (generic name) *Ceratops* Marsh, 1888[*1]}; in reference to its lack of a nasal horn.
Specific name: *hatcheri* ← {(person's name) Hatcher + -ī}; in honor of John Bell Hatcher who made the discovery.
Etymology: Hatcher's **insufficient horned-face (ceratopsian)**
Taxonomy: Ornithischia: Ceratopsia: Ceratopsidae:
Chasmosaurinae
Notes: *Diceratops* means "two horned face" and therefore it was named *Nedoceratops* meaning "insufficient horned face".

[References: Scannella JB, Horner JR (2011). 'Nedoceratops'; an example of a transitional morphology. *PLoS ONE* 6(12): e28705.; (*1) Ukrainsky AS (2007). A new replacement name for *Diceratops* Lull, 1905 (Reptilia; Ornithischia: Ceratopsidae). *Zoosystematica Rossica* 16(2): 292.]

Neimongosaurus yangi

> *Neimongosaurus yangi* Zhang *et al.*, 2001 [Campanian (Senonian[*1])*; Iren Dabasu Formation, Sanhangobi, Inner Mongolia, China]
> *Senonian comprises the Coniacian, Santonian, Campanian and Maastrichtian ages.

Generic name: *Neimongosaurus* ← {(place-name) Nei Mongol: Inner Mongolia + (Gr.) σαῦρος/sauros: reptile}
Specific name: *yangi* ← {(person's name) Yáng楊 + -ī}; "in memory of the founder of vertebrate paleontology in China, Yang Zhongjian (C. C. Young)"[*1].
Etymology: Yang's **Inner Mongolian reptile**
Taxonomy: Theropoda: Coelurosauria:
Therizinosauroidea: Therizinosauridae

[References: (*1) Zhang X-H, Xu X, Zhao Z-J, Sereno P, Kuang X-W, Tan L (2001). A long-necked therizinosauroid dinosaur from the Upper Cretaceous Iren Dabasu Formation of Nei Mongol, People's Republic of China. *Vertebrata PalAsiatica* 39(4): 282–290.]

Nemegtia barsboldi

> *Nemegtia barsboldi* Lü *et al.*, 2004 (preoccupied) ⇒ *Nemegtomaia barsboldi* (Lü *et al.*, 2004) Lü *et al.*, 2005 [Maastrichtian; Nemegt Formation, Nemegt Basin, Mongolia]

Generic name: *Nemegtia* ← {(place-name) Nemegt + -ia}; "referring to the locality, the Nemegt Basin of southwestern Mongolia"[*1].
Specific name: *Barsboldi* ← {(person's name) Barsbold + -ī}: "in honor of Dr. R. Barsbold, the Mongolian vertebrate paleontologist, one of the leaders of the Mongolian Highland International Dinosaur Project"[*1].
Etymology: Barsbold's **one of Nemegt Basin**
Taxonomy: Theropoda: Oviraptorosauria:
Oviraptoridae

[References: (*1) Lü J, Tomida Y, Azuma Y, Dong Z, Lee Y-N (2004). New oviraptorid dinosaur (Dinosauria: Oviraptorosauria) from the Nemegt Formation of southwestern Mongolia. *Bulletin of the National Science Museum, Tokyo, Series C* 30: 95–130.]

Nemegtomaia barsboldi

> *Nemegtia barsboldi* Lü *et al*., 2004 (preoccupied) ⇒ *Nemegtomaia barsboldi* (Lü *et al*., 2004) Lü *et al*., 2005 [Maastrichtian; Nemegt Formation, Nemegt Basin, Mongolia]

Generic name: *Nemegtomaia* ← {(place-name) Nemegt Basin*: geographic name + (Gr.) μαῖα/maia: good mother}; in reference to the fact that oviraptorids are thought to have brooded their eggs. "The word 'maia' stands for the fact that recent idea of oviraptorids is brooding eggs (e.g. Norell *et al*., 1995) rather than stealing eggs (Osborn, 1924)" (*1).

*the Nemegt Formation where the specimen was found is a geological formation in the Nemegt Basin in the northwestern Gobi Desert.

Specific name: *barsboldi* ← {(person's name) Barsbold + -ī}; in honor of Mongolian paleontologist Rinchen Barsbold.

Etymology: Barsbold's **good mother of Nemegt**

Taxonomy: Theropoda: Oviraptoridae

[References: Lü J, Tomida Y, Azuma Y, Dong Z, Lee Y-N (2005). *Nemegtomaia* gen. nov., a replacement name for the oviraptorosaurian dinosaur Nemegtia Lü *et al.* 2004, a preoccupied name. *Bulletin of the National Science Museum, Tokyo, Series C.* 31: 51.]

Nemegtonykus citus

> *Nemegtonykus citus* Lee *et al*., 2019 [Late Cretaceous; Nemegt Formation, Ömnögovi, Mongolia]

Generic name: *Nemegtonykus* ← {(stratal name) Nemegt Formation + (Gr.) ὄνυξ/onyx: claw}

Specific name: *citus* ← {(Lat.) citus: swift}

Etymology: swift **Nemegt claw**

Taxonomy: Theropoda: Alvarezsauria: Alvarezsauridae: Parvicursorinae

Nemegtosaurus mongoliensis

> *Nemegtosaurus mongoliensis* Nowinski, 1971 [early Maastrichtian; Nemegt Formation, Ömnögovi, Mongolia]

Generic name: *Nemegtosaurus* ← {(place-name) Nemegt + (Gr.) σαῦρος/sauros: lizard}; named after the Nemegt Bassin (= Basin) where the skull was found.

Specific name: *mongoliensis* ← {(place-name) Mongolia + -ensis}

Etymology: **Nemegt lizard** from Mongolia

Taxonomy: Saurischia: Sauropodomorpha: Sauropoda: Titanosauria: Nemegtosauridae

Other species:

> *Nemegtosaurus pachi* Dong, 1977 (*nomen dubium*) [Campanian–Maastrichtian; Subashi Formation, Turpan, Xinjiang, China]

Notes: *Nemegtosaurus* had a long neck and a long tail.

Neosaurus missouriensis

> *Neosaurus missouriensis* Gilmore & Stewart, 1945 (preoccupied) ⇒ *Parrosaurus missouriensis* (Gimore & Stewart, 1945) Gilmore, 1945 (*nomen dubium*) ⇒ *Hypsibema missouriensis* (Gilmore & Stewart, 1945) Baird & Horner, 1979 (*nomen dubium*) [Maastrichtian; Ripley Formation, Missouri, USA]

Generic name: *Neosaurus* ← {(Gr.) νέος/neos: new + σαῦρος/sauros: lizard}

Specific name: *missouriensis* ←{(place-name) Missouri + -ensis}

Etymology: **new lizard** from Missouri

Taxonomy: Ornithischia: Ornithopoda: Iguanodontia: Hadrosauroidea

Neovenator salerii

> *Neovenator salerii* Hutt *et al*., 1996 [Barremian; Wessex Formation, Isle of Wight, UK]

Generic name: *Neovenator* ← {(Gr.) νέος/neos: new + (Lat.) vēnātor: hunter}

Specific name: *salerii* ← {(person's name) Salero + -ī}; in honor of the Salero family, owner of the land.

Etymology: Salero family's **new hunter**

Taxonomy: Theropoda: Allosauroidea: Carcharodontosauria: Neovenatoridae

Notes: *Neovenator* is seemed perhaps the most complete theropod from the European Cretaceous (*1).

[References: (*1) Hutt S, Martill DM, Barker MJ (1996). The first European allosauroid dinosaur (Lower Cretaceous, Wealden Group, England). *Neues Jahrbuch für Geologie und Paläontologie Monatshefte* 1996(10): 635–644.]

Neuquenornis volans [Avialae]

> *Neuquenornis volans* Chiappe & Calvo, 1994 [Santonian; Bajo de la Carpa Formation, Neuquén, Argentina] (*1)

Generic name: *Neuquenornis* ← {(place-name) Neuquén Province + (Gr.) ὄρνις/ornis: bird}; "referring to the place at which the holotype was found" (*1).

Specific name: *volans* ← {(Lat.) volāns: flying}; in reference to the flying capability inferred for the new species.

Etymology: flying, **Neuquén bird**

Taxonomy: Theropoda: Avialae: Enantiornithes: Avisauridae

[References: (*1) Chiappe LM, Calvo JO (1994). *Neuquenornis volans*, a new Late Cretaceous bird (Enantiornithes: Avisauridae) from Patagonia, Argentina. *Journal of Vertebrate Paleontology* 14(2): 230–246.]

Neuquenraptor argentinus

> *Neuquenraptor argentinus* Novas & Pol, 2005 [Coniacian; Portezuelo Formation, Neuquén, Argentina]

Generic name: *Neuquenraptor* ← {(place-name) Neuquén: a province of northwest Patagonia + (Lat.) raptor: robber} (*1)

Specific name: *argentinus* ←{(place-name) Argentina: Argentine}; in reference to Argentina (*1).

Etymology: **Neuquén robber** from Argentina

Taxonomy: Theropoda: Dromaeosauridae: Unenlagiinae

[References: (*1) Novas FE, Pol D (2005). New evidence on deinonychosaurian dinosaurs from the Late Cretaceous of Patagonia. *Nature* 433(7028): 858–861.]

Neuquensaurus australis

> *Titanosaurus australis* Lydekker, 1893 ⇒ *Neuquensaurus australis* (Lydekker, 1893) Powell, 1986 (*nomina ex dissertatione*) ⇒ *Saltasaurus australis* (Lydekker, 1893) McIntosh, 1990 ⇒ *Neuquensaurus australis* (Lydekker, 1893) Powell, 1992 [Santonian; Bajo de la Carpa Formation, Sierra Roca, Río Neuquén, Argentina]

= *Titanosaurus nanus* Lydekker, 1893 [Santonian; Bajo de la Carpa Formaton, Sierra Roca, Río Neuquén, Argentina]

= ?*Microcoelus patagonicus* Lydekker, 1893 [Santonian; Bajo de la Carpa Formation, Sierra Roca, Río Neuquén, Argentina]

= *Loricosaurus scutatus* von Huene, 1929 [late Campanian–late Maastrichtian; Allen Formation, Río Negro, Argentina]

Generic name: *Neuquensaurus* ← {(place-name) Neuquén + (Gr.) σαῦρος/sauros: lizard}

Specific name: *australis* ← {(Lat.) austrālis: southern}

Etymology: southern, **Neuquén lizard**

Taxonomy: Saurischia: Titanosauria: Saltasauridae

Other species:

> *Titanosaurus robustus* von Huene, 1929 ⇒ *Saltasaurus robustus* (von Huene, 1929) McIntosh, 1990 ⇒*Neuquensaurus robustus* (von Huene, 1929) Powell, 1992 (*nomen dubium*) [Campanian–Maastrichtian; Allen Formation, Río Negro, Argentina]

Ngwevu intloko

> *Ngwevu intloko* Chapelle *et al.*, 2019 [Early Jurassic; upper Elliot Formation, Free State, South Africa]

Generic name: *Ngwevu* ← {(Xhosa) ngwevu: grey}; "in reference to the affectionate nickname, 'grey skull', that had been given to BP/1/4779 by many of the scientists who worked on it previously" (*1).

Specific name: *intloko* ← {(Xhosa) intloko [pron.: in-tloh-koh]: head}

Etymology: **grey** skull

Taxonomy: Saurischia: Sauropodomorpha: Massospondylidae

Notes: A partially complete skeleton including skull previously assigned to *Massospondylus carinatus* was given the new genus and species *Ngwevu intloko* in 2019.

[References: (*1) Chapelle KEJ, Barrett PM, Botha J, Choiniere JN (2019). *Ngwevu intloko*: a new early sauropodomorph dinosaur from the Lower Jurassic Elliot Formation of South Africa and comments on cranial ontogeny in *Massospondylus carinatus*. *PeerJ* 7: e7240.]

Nhandumirim waldsangae

> *Nhandumirim waldsangae* Marsola *et al.*, 2019 [Carnian; Alemoa Member, Santa Maria Formation, Paraná Basin, Brazil]

Generic name: *Nhandumirim* ← {(Tupi-Guarani) "Nhandu: running bird, common rhea + Mirim: small"}; "in reference to the size and inferred cursorial habits of a new dinosaur" (*1).

Specific name: *waldsangae* ← {(place-name) Waldsanga + -ae}; "referring to the Waldsanga site, the historic outcrop (Langer, 2005a) that yielded this new species" (*1).

Etymology: **small rhea** from Waldsanga

Taxonomy: Saurischia: Theropoda

[References: (*1) Marsola JCA, Bittencourt JS, Butler RJ, Da Rosa AA, Sayão JCA, Langer MC (2019). A new dinosaur with theropod affinities from the Late Triassic Santa Maria Formation, South Brazil. *Journal of Vertebrate Paleontology*: e1531878.]

Nigersaurus taqueti

> *Nigersaurus taqueti* Sereno *et al.*, 1999 [Aptian–Albian; Elrhaz Formation, Agadez, Niger]

?= *Rebbachisaurus tamesnensis* Lapparent, 1960 [Albian–Cenomanian; Echkar Formation, Agadez, Niger]

Generic name: *Nigersaurus* ← {(place-name) Niger + (Gr.) σαῦρος/sauros: lizard}

Specific name: *taqueti* ← {(person's name) Taquet + -ī}; in honor of French paleontologist Philippe

Taquet, who discovered the first remains.

Etymology: Taquet's **Niger lizard**

Taxonomy: Saurischia: Sauropodomorpha: Diplodocoidea: Rebbachisauridae: Rebbachisaurinae

Notes: *Nigersaurus* had a wide muzzle filled with more than 500 teeth, according to Sereno *et al.* (2009). In *The Dinosauria 2nd edition*, *Rebbachisaurus tamesnensis* is considered a synonym of *Nigersaurus taqueti*, although Sereno (1999) considers *R. tamesnensis* to be a synonym of *Jobaria tiguidensis*.

[References: Sereno PC, Beck AL, Dutheil DB, Larsson HCE, Lyon GH, Moussa B, Sadleir RW, Sidor CA, Varricchio DJ, Wilson GP, Wilson JA (1999). Cretaceous sauropods from the Sahara and the uneven rate of skeletal evolution among dinosaurs. *Science* 286(5443): 1342–1347.]

Ningyuansaurus wangi

> *Ningyuansaurus wangi* Ji *et al.*, 2012 [Aptian; Yixian Formation, Jianchang, Liaoning, China]

Generic name: *Ningyuansaurus* ← {(Chin. place-name) Ningyuan寧遠: the ancient name of Xingcheng City興城市+ (Gr.) σαῦρος/sauros: lizard}

Specific name: *wangi* ← {(Chin. person's name) Wáng王 + -ī}; in honor of Mr. Wang Qiuwu, the private owner of the specimen who donated the specimen for scientific study.

Etymology: Wang's **Ningyuan lizard**

Taxonomy: Theropoda: Oviraptorosauria

[References: Ji Q, Lü J-C, Wei X-F, Wang X-R (2012). A new oviraptorosaur from the Yixian Formation of Jianchang, western Liaoning Province, China. *Geological Bulletin of China* 31(12): 2102–2107.]

Niobrarasaurus coleii

> *Hierosaurus coleii* Mehl, 1936 ⇒ *Nodosaurus coleii* (Mehl, 1936) Coombs, 1978 ⇒ *Niobrarasaurus coleii* (Mehl, 1936) Carpenter *et al.*, 1995 [Coniacian; Smoky Hill Chalk Member, Niobrara Formation, Kansas, USA]

Generic name: *Niobrarasaurus* ← {(place-name) Niobrara + (Gr.) sauros: lizard}; in reference to the Smoky Hill Chalk Member of the Niobrara Formation. (*1)

Specific name: *coleii* ← {(person's name) Cole + -iī}; in honor of Virgil Cole, a geologist working for the Gulf Oil Company, who discovered the holotype specimen. (*1)

Etymology: Cole's **Niobrara lizard**

Taxonomy: Ornithischia: Ankylosauria: Nodosauridae

[References: (*1) Everhart MJ (2004). Notice of the transfer of the holotype specimen of *Niobrarasaurus coleii* (Ankylosauria; Nodosauridae) to the Sternberg Museum of Natural History. *Transactions of the Kansas Academy of Science* 107: 173–174.; Coombs WP (1978). The families of the ornithischian dinosaur order Ankylosauria. *Palaeontology* 21(1): 143–170.]

Nipponosaurus sachalinensis

> *Nipponosaurus sachalinensis* Nagao, 1936 [Santonian–early Campanian; Sachalinskaya Oblast, Russia (former Kawakami, Karafuto, where Imperial Japan once ruled)]

Generic name: *Nipponosaurus* ← {(Jpn. place-name) Nippon日本: Japan + (Gr.) σαῦρος/sauros: lizard}

Specific name: *sachalinensis* ← {(place-name) Sachalin* + -ensis}

*Sachalin (Karafuto) where the specimen was discovered was part of Japan in 1934.

Etymology: **Nippon (Japanese) lizard** from Sachalin

Taxonomy: Ornithischia: Ornithopoda: Iguanodontia: Hadrosauridae: Lambeosaurinae

Notes: The validity of *Nipponosaurus sachalinensis* has been douted, but redescriptions from 2004 and 2017 concluded it was a valid taxon, according to Takasaki, R. *et al.* (*1).

[References: Nagao T (1936). *Nipponosaurus sachalinensis*: a new genus and species of trachodont dinosaur from Japanese Saghalien. *Journal of the Faculty of Science, Hokkaido Imperial University, Series 4, Geology and Mineralogy* 3(2): 185–220.; (*1) Takasaki R, Chiba K, Kobayashi Y, Currie PJ, Fiorillo AR (2017). Reanalysis of the phylogenetic status of *Nipponosaurus sachalinensis* (Ornithopoda: Dinosauria) from the Late Cretaceous of Southern Sakhalin. *Historical Biology* 30(5): 1–18.]

Noasaurus leali

> *Noasaurus leali* Bonaparte & Powell, 1980 [Maastrichtian; Lecho Formation, Salta, Argentina]

Generic name: *Noasaurus* ← {(acronym, place-name) NOA (= Noroeste Argentino): Northwestern Argentina + (Gr.) σαῦρος/sauros: lizard}

Specific name: *leali* ← {(person's name) Leal + -ī}

Etymology: Leal's **Northwestern Argentine lizard**

Taxonomy: Theropoda: Ceratosauria: Abelisauria: Noasauridae

Notes: *Noasaurus* was a small theropod.

Nodocephalosaurus kirtlandensis

> *Nodocephalosaurus kirtlandensis* Sullivan, 1999 [late Campanian; De-na-zin Member, lower Kirtland Formation, New Mexico, USA]

Generic name: *Nodocephalosaurus* ← {(Lat.) nōdus: knob + (Gr.) κεφαλή/kephalē: head + σαῦρος/

sauros: lizard}; referring to "bulbous, polygonal, cranial osteoderms"(*1).

Specific name: *kirtlandensis* ← {(place-name) the Kirtland Formation + -ensis}

Etymology: **knob-headed lizard** from the Kirtland Formation

Taxonomy: Ornithischia: Thyreophora: Ankylosauria: Ankylosauridae: Ankylosaurinae

[References: (*1) Sullivan RM (1999). *Nodocephalosaurus kirtlandensis*, gen. *et* sp. nov., a new ankylosaurid dinosaur (Ornithischia: Ankylosauria) from the Upper Cretaceous Kirtland Formation (upper Campanian), San Juan Basin, New Mexico. *Journal of Vertebrate Paleontology* 19(1): 126–139.]

Nodosaurus textilis

> *Nodosaurus textilis* Marsh, 1889 [middle Cenomanian; Frontier Formation, Wyoming, USA]

Generic name: *Nodosaurus* ← {(Lat.) nōdus: knob, node + (Gr.) σαῦρος/sauros: lizard}; referring to "a series of rounded knobs in rows and these protuberances have suggested the generic name"(*1).

Specific name: *textilis* ← {(Lat.) woven, texile}; referring to the dermal ossifications whose surface is "marked by a texture that appears interwoven, like a coarse cloth"(*1).

Etymology: interwoven, **knobbed lizard**

Taxonomy: Ornithischia: Thyreophora: Ankylosauria: Nodosauridae

[References: (*1) Marsh OC (1889). Notice of gigantic horned Dinosauria from the Cretaceous. *American Journal of Science* 38: 173–175.]

Nomingia gobiensis

> *Nomingia gobiensis* Barsbold *et al.*, 2000 [early Maastrichtian; Nemegt Formation, Ömnögovi, Gobi Desert, Mongolia]

Generic name: *Nomingia* ← {(place-name) Nomingiin + -ia}; "after Nomingiin Gobi, a part of the Gobi Desert close to the type locality"(*1).

Specific name: *gobiensis* ← {(place-name) the Gobi Desert + -ensis}

Etymology: **Nomingiin Gobi's one** from Gobi Desert [戈壁天青石龍*]

cf. 天青石: celestite (mineral)

Taxonomy: Theropoda: Coelurosauria: Oviraptorosauria: Oviratoridae

Notes: *Nomingia* has a pygostyle, the structure known only in birds(*1).

[References: (*1) Barsbold R, Osmólska H, Watabe M, Currie PJ, Tsogtbaatar K (2000). A new oviraptorosaur (Dinosauria, Theropoda) from Mongolia: The first dinosaur with a pygostyle. *Acta Palaeontologica Polonica* 45(2): 97–106.]

Nopcsaspondylus alarconensis

> *Nopcsaspondylus alarconensis* Apesteguía, 2007 [Cenomanian; Candeleros Formation, Neuquén, Argentina]

Generic name: *Nopcsaspondylus* ← {(person's name) Franz Nopcsa + (Gr.) σπόνδῦλος/spondylos: vertebra}

Specific name: *alarconensis* ← {(place-name) Barda Alarcón + -ensis}

Etymology: **Nopcsa's vertebra** from Barda Alarcón

Taxonomy: Saurischia: Sauropodomorpha: Sauropoda: Rebbachisauridae

Normanniasaurus genceyi

> *Normanniasaurus genceyi* Le Loeuff *et al.*, 2013 [Albian; Poudingue Ferrugineux Formation, Le Havre, Seine-Maritime, France]

Generic name: *Normanniasaurus* ← {(Lat. place-name) Normannia: Normandy + σαῦρος/sauros: lizard}; referring to "the region where the bones were discovered"(*1)

Specific name: *genceyi* ← {(person's name) Gencey + -ī}; "dedicated to the discoverer, Mr. Pierre Gencey"(*1).

Etymology: Gencey's **lizard from Normandy**

Taxonomy: Saurischia: Sauropodomorpha: Sauropoda: Titanosauria

[References: (*1) Le Loeuff J, Suteethorn S, Buffetaut E (2013). A new sauropod from the Albian of Le Havre (Normandy, France). *Oryctos* 10: 23–30.]

Notatesseraeraptor frickensis

> *Notatesseraeraptor frickensis* Zahner & Brinkmann, 2019 [Norian; Klettgau Formation, Aargau, Switzerland]

Generic name: *Notatesseraeraptor* ← {"(Lat.) nota: feature + tesserae: individually shaped tiles used to create a mosaic + raptor: predator"}; "in reference to the intermixture of features typically known from eather dilophosaurid or coelophysoid neotheropods"(*1).

Specific name: *frickensis* ← {(place-name) Frick: a village + -ensis} (*1)

Etymology: **feature mosaic tile predator** from Frick

Taxonomy: Saurischia: Theropoda: Neotheropoda

[References: (*1) Zahner M, Brinkmann W (2019). A Triassic averostran-line theropod from Switzerland and the early evolution of dinosaurs. *Nature Ecology & Evolution* 13; 3(8): 1146–1152.]

Nothronychus mckinleyi

> *Nothronychus mckinleyi* Kirkland & Wolfe, 2001 [middle Turonian; Moreno Hill Formation, New Mexico, USA]

Generic name: *Nothronychus* ← {(Gr.) νωθρός/nōthros: slothful + ὄνυξ/onyx: claw} (*1)
Specific name: *mckinleyi* ← {(person's name) McKinley + -ī}; in honor of "Bobby McKinley for his support of this research" (*1).
Etymology: McKinley's **slothful claw**
Taxonomy: Theropoda: Coelurosauria: Therizinosauridae
Other species:
> *Nothronychus graffami* Zanno *et al.*, 2009 [Cenomanian; Tropic Shale Formation, Utah, USA]

[References: (*1) Kirkland JI, Wolfe DG (2001). First definitive therizinosaurid (Dinosauria; Theropoda) from North America. *Journal of Vertebrate Paleontology* 21(3): 410–414.]

Notoceratops bonarellii

> *Notoceratops bonarellii* Tapia, 1918 [Late Cretaceous; near Lago Colhué Huapí, Chubut, Argentina]

Generic name: *Notoceratops* ← {(Gr.) νότος/notos: the south + ceratops [κέρας/ceras: horn + ὤψ/ōps: face]}; referring to the Southern Hemisphere where it was found.
Specific name: *Bonarellii** ← {(person's name) Bonarelli + -ī}; in honor of Guido Bonarelli who advised Tapia. *genitive form of Bonarellius
Etymology: Bonarelli's **southern horned face (ceratopsian)**
Taxonomy: Ornithischia: Marginocephalia: Ceratopsia
Notes: According to *The Dinosauria 2nd edition*, *Notoceratops* is regarded as a *nomen dubium*. But, an analysis by Tom Rich *et al.* (2014) focused on the validity of possible ceratopsian *Serendipaceratops*, and they concluded *Notoceratops* would be probably valid, because it had ceratopsian features.

Notocolossus gonzalezparejasi

> *Notocolossus gonzalezparejasi* González Riga *et al.* 2016 [Coniacian–Santonian; Plottier Formation, Mendoza, Argentina]

Generic name: *Notocolossus* ← {(Gr.) νότος/notos: the south, south-west wind + (Lat.) colossus = (Gr.) κολοσσός: a gigantic statue, colossus}; "in reference to the gigantic size and Gondwanan provenance of the new taxon" (*1).
Specific name: *gonzalezparejasi* ← {(person's name) González Parejas + -ī}; "in honor of Dr. Jorge González Parejas, who has collaborated and provided legal guidance on the research, protection, and preservation of dinosaur fossils from Mendoza Province for nearly two decades. In so doing, he has advised researchers on the creation of a natural park that serves to protect dinosaur footprints in Mendoza" (*1).
Etymology: González Parejas' **southern colossus**
Taxonomy: Saurischia: Sauropoda: Titanosauriformes: Somphospondyli: Titanosauria: Lithostrotia

[References: (*1) González Riga BJ, Lamanna MC, Ortiz David LD, Calvo JO, Coria JP (2016). A gigantic new dinosaur from Argentina and the evolution of the sauropod hind foot. *Scientific Reports* 6: 19165.]

Notohypsilophodon comodorensis

> *Notohypsilophodon comodorensis* Martinez, 1998 [Cenomanian–Turonian; Bajo Barreal Formation, Chubut, Argentina]

Generic name: *Notohypsilophodon* ← {(Gr.) νότο-/noto-: southern- + ὑψίλοφος/hypsilophus + ὀδών/odōn: tooth}
Specific name: *comodorensis* ← {(place-name) Comodoro Rivadavia + -ensis}
Etymology: **southern Hypsilophodon** from Comodoro Rivadavia
Taxonomy: Ornithischia

[References: Martínez RD (1998). *Notohypsilophodon comodorensis* gen. *et* sp. nov. un Hypsilophodontidae (Ornitischia: Ornithopoda) del Cretacico Superior de Chubut, Patagonia central, Argentina. *Acta Geologica Leopoldensia* 21 (46/47):119–135.]

Nqwebasaurus thwazi

> *Nqwebasaurus thwazi* de Klerk *et al.*, 2000 [Berriasian–Valanginian; upper Kirkwood Formation, Cape, South Africa]

Generic name: *Nqwebasaurus* ← {(Xhosa, place-name) Nqweba [pron.: n-KWE-bah (*1)]: Kirkwood region + (Gr.) σαῦρος/sauros: lizard}
Specific name: *thwazi* ← {(Xhosa) thwazi [pron.: TWAH-zee]: fast runner} (*1)
Etymology: **Nqweba lizard**, fast runner
Taxonomy: Theropoda: Ornithomimosauria
cf. Theropoda: Tetanurae: Coelurosauria (*1)
Notes: Gastroliths were found in the abdominal cavity of the specimen.

[References: (*1) De Klerk WJ, Forster CA, Sampson SD, Chinsamy A, Ross CF (2000). A new coelurosaurian dinosaur from the Early Cretaceous of South Africa. *Journal of Vertebrate Paleontology* 20(2): 324–332.]

Nullotitan glaciaris

> *Nullotitan glaciaris* Novas *et al.*, 2019 [late Campanian; Chorrillo Formation, Santa Cruz, Argentina]

Generic name: *Nullotitan* ← {(person's name) Nullo + (Gr. myth) Τῑτάν/Tītan: giant}; "in honor of geologist Francisco E. Nullo, discoverer of the

holotype specimen, and titan, powerful giant" [*1].
Specific name: *glaciaris* ← {(Lat.) glaciālis: icy, frozen, glacial}; "referring to the majestic Perito Moreno Glacier, observable from excavation site" [*1].
Etymology: **Nullo's giant** from Perito Moreno Glacier
Taxonomy: Saurischia: Sauropodomorpha:
Sauropoda: Titanosauria: Colossosauria

[References: (*1) Novas F, Agnolin F, Rozadilla S, *et al.*, (2019). Paleontological discoveries in the Chorrillo Formation (upper Campanian–lower Maastrichtian, Upper Cretaceous), Santa Cruz Province, Patagonia, Argentina. *Revista del Museo Argentino de Ciencias Naturales, Nueva Serie* 21 21(2): 217–293.]

"*Nurosaurus qaganensis*"

> *Nuoerosaurus chaganensis* Dong & Li, 1991 (*nomen nudum*) ⇒ *Nurosaurus qaganensis* Dong, 1992 (*nomen nudum*) [Early Cretaceous; Qagannur Formation, Inner Mongolia, China]

Generic name: *Nurosaurus* ← {(Mong.) нуур: lake + (Gr.) σαῦρος/sauros: lizard}
Specific name: *qaganensis* or *chaganensis* ← {(place-name) Qagan*: Qaggannur/Qagan Nur + -ensis} *(Mong.) цагаан: white.
Etymology: **lake lizard** from Qagan Nur Formation
Taxonomy: Saurischia: Sauropoda
Notes: "Nurosaurus" has not been officially described.

Nuthetes destructor

> *Nuthetes destructor* Owen, 1854 (*partim*) ⇒ *Megalosaurus destructor* (Owen, 1854) Swinton, 1934* , Steel, 1970 (*nomen dubium*) [Berriasian; Lulworth Formation, Dorset, UK]
> *In 1854, Owen named it as a lizard, later classified it as a crocodilian [*1] [*2]. In 1888, Lydekker thought it was a dinosaur. In 1934, Swinton thought it was a member of Megalosauridae and in 1970, Steel renamed it *Megalosaurus destructor*. According to Sweetman (2004) [*3], this specimen is considered as a member of Dromaeosauridae.

Generic name: *Nuthetes* ← {(Gr.) νουθέτης/nūthetēs (abbreviated from νουθέτητης: monitor}; "in reference to the affinities of the fossil to the modern lizards so called" [*1].
Specific name: *destructor* ← {(Lat.) destructor: destroyer}
Etymology: **monitor lizard**, destroyer
Taxonomy: Theropoda: Dromaeosauridae
Notes: According to *The Dinosauria 2nd edition* (2004), *Nuthetes* is considered to be a *nomen dubium*.

[References: (*1) Owen R (1854). On some fossil Reptilian and mammalian remains from the Purbecks. *Quarterly Journal of the Geological Society* 10: 420–433.; (*2) Owen R (1879). *Monograph on the Fossil Reptilia of the Wealden and Purbeck Formations, Supplement no IX. Crocodilia (Goniopholis, Brachydectes, Nannosuchus, Theriosuchus, and Nuthetes)*. The Palaeontographical Society, London: 1–19.; (*3) Sweetman SC (2004). The first record of velociraptorine dinosaurs (Saurischia, Theropoda) from the Wealden (Early Cretaceous, Barremian) of southern England. *Cretaceous Research* 25(3): 353–364.]

Nyasasaurus parringtoni [Dinosauriformes]

> *Nyasasaurus parringtoni* Nesbitt *et al.*, 2013 [Anisian; Manda Formation, Ruhuhu Basin, Ruvuma, Tanzania]
= *Thecodontosaurus alophos* Haughton, 1932 [Anisian; Manda Formation, Ruvuma, Tanzania]

Generic name: *Nyasasaurus* ← {(place-name) Lake Nyasa + (Gr.) σαῦρος/sauros: lizard}
Specific name: *parringtoni* ← {(person's name) Parrington + -ī}; in honour of Francis Rex Parrington, collector of the holotype [*1].
Etymology: Parrington's **lizard from Lake Nyasa**
Taxonomy: Archosauria: Ornithodira:
Dinosauromorpha: Dinosauriformes
cf. Dinosauria: Saurischia: Sauropodomorpha [*2]
cf. Archosauria: Dinosauriformes: Dinosauria [*1]
Notes: Estimated to be 2–3 m in total length. *Niasasaurus* comes from a deposite that dates back to the middle Triassic Anisian stage, near Lake Nyasa (= Lake Malawi). It may be the earliest known dinosaur. [*1]

[References: (*1) Nesbitt SJ, Barrett PM, Werning S, Sidor CA, Charig AJ (2013). The oldest dinosaur? A Middle Triassic dinosauriform from Tanzania. *Biology Letters* 9(1): 20120949.; (*2) Baron MG, Norman DB, Barrett PM (2017). A new hypothesis of dinosaur relationships and early dinosaur evolution. *Nature* 543: 501–506.]

O

Oceanotitan dantasi

> *Oceanotitan dantasi* Mocho *et al.*, 2019 [Late Jurassic; Praia da Amoreira-Porto Novo Formation, Lourinhã, Portugal]

Generic name: *Oceanotitan* ← {(Gr.) Ὠκεᾰνός/Ōkeanos: Oceanus + Τῑτάν/Tītan: giant}; referring to the Atlantic Ocean.
Specific name: *dantasi* ← {(person's name) Dantas + -ī}; in honor of Portuguese paleontologist Pedro Dantas.
Etymology: Dantas' **Oceanus giant**
Taxonomy: Saurischia: Sauropodomorpha:
Sauropoda: Camarasauromorpha:
Titanosauriformes

Oculudentavis khaungraae [Squamata]

> *Oculudentavis khaungraae* Xing *et al.*, 2020 [Cenomanian; Angbamo site, Hukawng Valley, Kachin, Myanmar]

Generic name: *Oculudentavis* ← {(Lat.) oculus: eye + dēns: tooth + avis: bird}

Specific name: *khaungraae* ← {(person's name) Khaung Ra + -ae}; "from Khang Ra, who donated the specimen to the Hupoge Amber Museum" (*1).

Etymology: Khaung Ra's **eye-toothed bird**

Taxonomy: Lepidosauria: Squamata
cf. Saurischia: Theropoda: Avialae
cf. Aves (*1) (*2)

[References: (*1) Xing L, O'Connor JK, Schmitz L, Chiappe LM, McKellar RC, Yi Q, Li G (2020). Hummingbird-sized dinosaur from the Cretaceous period of Myanmar. *Nature* 579(7798): 245–249.]; (*2) Xing L, O'Connor JK, Schmitz L, Chiappe LM, McKellar RC, Yi Q, Li G (2020). Retraction Note: Hummingbird-sized dinosaur from the Cretaceous period of Myanmar. *Nature* 584 (7822): 652.]

Ohmdenosaurus liasicus

> *Ohmdenosaurus liasicus* Wild, 1978 [middle Toarcian; Posidonienschiefer (Posidonia Shale), Baden-Württemberg, Germany]

Generic name: *Ohmdenosaurus* ← {(place-name) Ohmden + (Gr.) σαῦρος/sauros: lizard}; referring to Ohmden, a town in Baden-Württemberg.

Specific name: *liasicus* ← {[geol.] lias + -icus}; referring to the Lias, an old name for the Early Jurassic.

Etymology: **Ohmden lizard** of Lias (Early Jurassic)

Taxonomy: Saurischia: Sauropodomorpha:
Sauropoda: ?Vulcanodontidae

Notes: *Ohmdenosaurus* was first mistaken for a plesiosaur. But, Wild (1978) recognized the fossil to be a dinosaur bone.

[References: Wild R (1978). Ein Sauropoden-Rest (Reptilia, Saurischia) aus dem Posidonienschiefer (Lias, Toarcium) von Holzmaden. *Stuttgarter Beiträge zur Naturkunde, Serie B (Geologie und Paläontologie)* 41: 1–15.]

Ojoceratops fowleri

> *Ojoceratops fowleri* Sullivan & Lucas, 2010 [Maastrichtian; Ojo Alamo Formation, New Mexico, USA]

Generic name: *Ojoceratops* ← {(place-name) Ojo Alamo: name of the formation + ceratops [(Gr.) κέρας/ceras: horn + ὤψ/ōps: face]}

Specific name: *fowleri* ← {(person's name) Fowler + -ī}; "in honor of Denver Fowler, who discovered the holotype and who also discovered, and collected, a number of the specimens" (*1).

Etymology: Fowler's **ceratops from Ojo Alamo Formation**

Taxonomy: Ornithischia: Ceratopsia: Ceratopsidae:
Chasmosaurinae (*1)

[References: (*1) Sullivan RM, Lucas SG (2010). A new chasmosaurine (Ceratopsidae, Dinosauria) from the Upper Cretaceous Ojo Alamo Formation (Naashoibito Member), San Juan Basin, New Mexico. In: Ryan MJ, Chinnery-Allgeier BJ, Eberth DA (eds), *New Perspectives on Horned Dinosaurs: The Roal Tyrrell Museum Ceratopsian Symposium*, Indiana University Press, Bloomington: 169–180.]

Ojoraptorsaurus boerei

> *Ojoraptorsaurus boerei* Sullivan *et al.*, 2011 [early Maastrichtian; Naashoibito Member, Ojo Alamo Formation, New Mexico, USA]

Generic name: *Ojoraptorsaurus* ← {(place-name) Ojo Alamo Formation + (Lat.) raptor: plunderer + (Gr.) σαῦρος/sauros: lizard} (*1)

Specific name: *boerei* ← {(person's name) Boere + -i}; "in honor of Arjan C. Boere, who discovered and collected the specimen in 2002" (*1).

Etymology: Boere's **plunderer lizard from Ojo Alamo Formation**

Taxonomy: Theropoda: Coelurosauria:
Oviraptorosauria: Caenagnathidae

[References: (*1) Sullivan RM, Jasinski SE, van Tomme MPA (2011). A new caenagnathid *Ojoraptorsaurus boerei*, n. gen., n. sp. (Dinosauria, Oviraptorosauria), from the Upper Cretaceous Ojo Alamo Formation (Naashoibito Member), San Juan Basin, New Mexico. In: Sullivan *et al.* (eds), *The Fossil Record 3. New Mexico Museum of Natural History and Science, Bulletin* 53: 418–428.]

Oligosaurus adelus

> *Rhabdodon priscus* Matheron, 1869 [Maastrichtian; Provence-Alpes-Côte d'Azur, France]

= *Oligosaurus adelus* Seeley, 1881 [Campanian; Grünbach Formation, Niederösterreich, Austria]

Generic name: *Oligosaurus* ← {(Gr.) ὀλίγος/oligos: few + σαῦρος/sauros: lizard}

Specific name: *adelus* ← {(Gr.) ἄδηλος/adēlos: unknown, obscure}

Etymology: obscure, **small lizard**

Taxonomy: Ornithischia: Ornithopoda: Iguanodontia

Notes: According to *The Dinosauria 2nd edition* (2004), *Oligosaurus adelus* is considered as a synonym of *Rhabdodon priscus*.

Olorotitan arharensis

> *Olorotitan arharensis* Godefroit *et al.*, 2003 [Maastrichtian; Udurchukan Formation, Arhara,

Amur, Russia]
Generic name: *Olorotitan* ← {(Lat.) olor: swan + (Gr.) Τιτάν/Tītan: gigantic, giant}
Specific name: *arharensis* ← {(place-name) Arhara County + -ensis}
Etymology: **gigantic swan** from Arhara
Taxonomy: Ornithischia: Ornithopoda: Hadrosauridae: Lambeosaurinae: Lambeosaurini
Notes: *Olorotitan* had a helmet-like hollow crest.

[References: Godefroit P, Bolotsky Y, Alifanov V (2003). A remarkable hollow-crested hadrosaur from Russia: an Asian origin for lambeosaurines. *Comptes Rendus Palevol* 2(2): 143–151.]

Omeisaurus junghsiensis

> *Omeisaurus junghsiensis* Young, 1939 [Bathonian–Callovian; lower Shaximiao* Formation, Sichuan, China]
> *Lower Shaximiao = Xiashaximiao, lower = [Pinyin: xià 下]

Generic name: *Omeisaurus* ← {(place-name) Omei 峨眉 [Pinyin: Éméi]: Mount Emei峨眉山 + (Gr.) σαῦρος/sauros: lizard}
Specific name: *junghsiensis* ← {(place-name) Junghsien榮縣 [Pinyin: Róngxiàn] + -ensis}
Etymology: **Emei lizard** from Junghsien
Taxonomy: Saurischia: Sauropodomorpha: Eusauropoda: Mamenchisauridae
Other species:

> *Omeisaurus changshouensis* Young, 1958 ⇒ *Mamenchisaurus changshouensis* (Young, 1958) Zhang & Chen, 1996 [Oxfordian; upper Shaximiao Formation, Sichuan, China]
> *Zigongosaurus fuxiensis* Hou *et al.*, 1976 ⇒ *Omeisaurus fuxiensis* (Hou *et al.*, 1976) Dong *et al.*, 1983 ⇒ *Mamenchisaurus fuxiensis* (Hou *et al.*, 1976) Zhang & Chen, 1996 [Oxfordian; upper Shaximiao Formation, Zigong, Sichuan, China]
> *Omeisaurus tianfuensis* He *et al.*, 1984 [Bathonian–Callovian* or Oxfordian; lower Shaximiao Formation, Sichuan, China]
> *(acc. *The Dinosauria 2nd edition*)
> *Omeisaurus luoquanensis* He *et al.*, 1988 [Middle Jurassic; lower Shaximiao Formation, Luoquan, Sichuan, China]
> *Omeisaurus maoianus* Tang *et al.*, 2001 [Late Jurassic; upper Shaximiao Formation, Sichuan, China]
> *Omeisaurus jiaoi* Jiang *et al.*, 2011 [Bajocian–Callovian; Middle Jurassic; lower Shaximiao Formation, Sichuan, China]
> *Omeisaurus puxiani* Tan *et al.*, 2020 [Oxfordian; lower Shaximiao Formation, Chongqing, China]

Notes: *Omeisaurus* had a long neck. According to Dong *et al.* (1988), tail clubs have been referred to *Omeisaurus tianfuensis*. But Upchurch in 2004 considered them possible *Shunosaurus* clubs.

According to *The Dinosauria 2nd edition*, three species (*Omeisaurus changshouensis, O. fuxiensis and O. luoquanensis*) were regarded to be *nomina dubia*, but now to be species of *Omeisaurus*.

[References: Young C (1939). On a new Sauropoda, with notes on other fragmentary reptiles from Szechuan. *Bulletin of the Geological Society of China* 19(3): 279–315.]

Omnivoropteryx sinousaorum [Avialae]

> *Omnivoropteryx sinousaorum* Czerkas & Ji, 2002 [Aptian; Jiufotang Formation, Liaoning, China]

Generic name: *Omnivoropteryx* ← {(Lat.) omnis: all + voro-: eating [vorō: devour] + (Gr.) πτέρυξ/pteryx: wing}
Specific name: *sinousaorum* ← {(place-name) Sino-: China + USA + -ōrum}
Etymology: Chinese and American **omnivorous wing**
Taxonomy: Theropoda: Avialae: Euavialae: Omnivoropterygidae
Notes: *Omnivoropteryx* may be a junior synonym of *Sapeornis* Zhou & Zhang, 2002.

[References: Czerkas SA, Ji Q (2002). A preliminary report on an omnivorous volant bird from northeast China. In: Czerkas SJ (ed), *Feathered Dinosaurs and the origin of flight. The Dinosaur Museum Journal* 1: 127–135.]

Omosaurus armatus

> *Omosaurus armatus* Owen, 1875 (preoccupied) ⇒ *Stegosaurus armatus* (Owen, 1875) Lydekker, 1890 ⇒ *Dacentrurus armatus* (Owen, 1875) Lucas, 1902 ⇒ *Dacentrurosaurus armatus* (Owen, 1875) Hennig, 1925 [Oxfordian–Kimmeridgian; lower Kimmeridge Clay, Wiltshire, UK]
= *Omosaurus lennieri* von Nopcsa, 1911 [Kimmeridgian; Argiles d'Octeville Formation, France]

Generic name: *Omosaurus* ← {(Gr.) ὦμος/ōmos: humerus, shoulder + (Gr.) σαῦρος/sauros: lizard}
Specific name: *armatus* ← {(Lat.) armātus: armed}
Etymology: **armed humerus** lizard
Taxonomy: Ornithischia: Stegosauria: Stegosauridae: Dacentrurinae
Other species:

> *Omosaurus phillipsi* Seeley, 1893 ⇒ *Dacentrurus phillipsi* (Seeley, 1893) Hennig,

1915 (*nomen dubium*) [Oxfordian; Calcareous Grit Formation, UK]

> *Omosaurus vetustus* von Huene, 1910 (*nomen dubium*) ⇒ *Dacentrurus vetustus* (von Huene, 1910) Hennig, 1915 (*nomen dubium*) ⇒ *Omosaurus (Dacentrurus) vetustus* (von Huene, 1910) Hoffstetter, 1957 ⇒ *Lexovisaurus vetustus* (von Huene, 1910) Galton & Powell, 1983 ⇒ *Eoplophysis vestustus* (von Huene, 1910) Ulansky, 2014 [Bathonian, Cornbrash Formation, UK]

> *Omosaurus durobrivensis* Hulke, 1887 ⇒ *Stegosaurus durobrivensis* (Hulke, 1887) Hulke, 1887 ⇒ *Dacentrurus durobrivensis* (Hulke, 1887) Hennig, 1915 ⇒ *Lexovisaurus durobrivensis* (Hulke, 1887) Hoffstetter, 1957 [Callovian; Oxford Clay Formation, UK]

= *Omosaurus leedsi* Seeley, 1901 (*partim*) ⇒ *Dacentrurus leedsi* (Seeley, 1901) Hennig, 1915 [Callovian; Oxford Clay, Northamptonshire, UK]

> *Omosaurus hastiger* Owen, 1877 ⇒ *Stegosaurus hastiger* (Owen, 1877) Lydekker, 1890 ⇒ *Dacentrurus hastiger* (Owen, 1877) Hennig, 1915 [Kimmeridge Clay, Wiltshire, UK]

Notes: *Omosaurus* had been preoccupied by a phytosaur, *Omosaurus perplexus* Leidy, 1856.

Omphalosaurus nevadanus [Ichthyosauria]

> *Omphalosaurus nevadanus* Merriam, 1906 [Middle Triassic; Prida Formation, Nevada, USA]

Generic name: *Omphalosaurus* ← {(Gr.) ὀμφᾰλός/omphalos: the navel, a button or knob on the horse's yoke, the knob or boss in the middle of the shield + σαῦρος/sauros: lizard}; referring to button-like teeth.

Specific name: *nevadanus* ← {(place-name) Nevada}

Etymology: **button lizard** from Nevada

Taxonomy: Ichthyosauria: Omphalosauridae

Other species:

> *Omphalosaurus merriami* Maisch, 2010 [Anisian; Sticky Keep Formation, Svalbard, Norway]

> *Omphalosaurus nettarhynchus* Mazin & Bucher, 1987 [Spathian; Prida Formation, Nevada, USA]

> *Omphalosaurus peyeri* Maisch & Lehmann, 2002 [Anisian; Schaumkalk Bed, Rüdersdorf, Germany]

> *Omphalosaurus wolfi* Tichy, 1995 [Ladinian; Lercheck Limestone, Dürrnberg Mountain, Germany]

Oohkotokia horneri

> *Oohkotokia horneri* Penkalski, 2013 [Campanian; upper Two Medicine Formation, Montana, USA]

Generic name: *Oohkotokia* ← "{(Blackfoot) animate noun *ooh'kotoka*: large stone or rock + (Lat.) -ia: indicating made of or derived from}; literally 'child of stone', an allusion to the all-encompassing armour, and in honor of the Blackfeet people, on whose land the specimen was found" (*1).

Specific name: *horneri* ← {(person's name) Horner + -ī}; "in honor of John R. Horner for his work on dinosaurs from Montana" (*1).

Etymology: Horner's **child of stone**

Taxonomy: Ornithischia: Ankylosauria: Ankylosauridae: Ankylosaurinae

Notes: According to Arbour & Currie (2013), they concluded that *Oohkotokia* is a junior synonym of *Scolosaurus*.

[References: (*1) Penkalski P (2014). A new ankylosaurid from the Late Cretaceous Two Medicine Formation of Montana, USA. *Acta Palaeontologica Polonica* 59(3): 617–634.]

Opallionectes andamookaensis [Plesiosauria]

> *Opallionectes andamookaensis* Kear, 2006 [Aptian–Albian; Bulldog Shale Formation, Andamooka, South Australia, Australia]

Generic name: *Opallionectes* ← {(Gr.) ὀπάλλιος/opallios: opal + νηχης/nēktēs: swimmer}; "alluding to the holotype specimen's discovery in an opal mine and replacement of the fossil bone by opal" (*1).

Specific name: *andamookaensis* ← {(place-name) Andamooka + -ensis}; "referring to derivation of the holotype specimen from near the opal-mining township of Andamooka" (*1).

Etymology: **opal swimmer** from Andamooka

Taxonomy: Diapsida: Sauropterygia: Plesiosauria: Plesiosauroidea

[References: (*1) Kear BP (2006). Marine reptiles from the Lower Cretaceous of South Australia: elements of a high-latitude cold-water assemblage. *Palaeontology* 49(4): 837–856.]

Opisthocoelicaudia skarzynskii

> *Opisthocoelicaudia skarzynskii* Borsuk-Bialynicka, 1977 [early Maastrichtian; Nemegt Formation, Ömnögovi, Mongolia]

Generic name: *Opisthocoelicaudia* ← {(Gr.) ὀπισθο-/opistho-: back + κοῖλος/koilos: hollow, hollowed + (Lat.) cauda: tail + -ia}; "because of the opisthocoelian structure of the anterior caudals" (*1).

Specific name: *skarzynskii* ← {(person's name)

Skarzynski + -ī}; in honor of "Mr. Wojciech Skarżyński who prepared the specimen" (*1).
Etymology: Skarzynski's **opisthocoelian tail**
Taxonomy: Saurischia: Sauropodomorpha:
Sauropoda: Titanosauria
cf. Sauropoda: Camarasauridae:
Euhelopodinae (*1)
Notes: According to Borsuk-Bialynicka (1977), the specimen indicates that the skull and neck must have been separated before burial.

[References: (*1) Borsuk-Bialynicka MM (1977). A new camarasaurid sauropod *Opisthocoelicaudia skarzynskii* gen., sp. n. from the Upper Cretaceous of Mongolia. *Palaeontologia Polonica* 37(5): 5–64.]

Oplosaurus armatus

> *Oplosaurus armatus* Gervais, 1852 ⇒ *Hoplosaurus armatus* (Gervais, 1852) Lydekker, 1888 ⇒ *Pelorosaurus armatus* (Gervais, 1852) Lydekker, 1889 [Valanginian–Barremian; Wessex Formation, Wealden Group, Isle of Wight, UK]

Generic name: *Oplosaurus* ← {(Gr.) ὅπλον/hoplon: weapon, shield + σαῦρος/sauros: lizard}; based one tooth which "belonged to a carnivorous reptile" (*1). It may have been used as the weapon of a carnivore.
Specific name: *armatus* ← {(Lat.) armātus: armed, in arms}
Etymology: armed, **weapon lizard**
Taxonomy: Saurischia: Sauropodomorpha:
Sauropoda

[References: (*1) Wright T (1852). Contributions to the palaeontology of the Isle of Wight. *Annals and Magazine of Natural History* 2: 87–93.]

Orienantius ritteri [Avialae]

> *Orienantius ritteri* Liu *et al.*, 2019 [Hauterivian; Huajiying Formation, Hebei, China]

Generic name: *Orienantius* ← {(Lat.) oriēns: 'dawn' + enantius: Enantiornithes} (*1)
Specific name: *ritteri* ← {(person's name) Ritter + -ī}; "in honor of Polish-born physicist Johann Wilhelm Ritter whose discovery of ultraviolet light has allowed visualization of the soft tissues that are extraordinarily well-preserved in these new fossils (*1).
Etymology: Ritter's **Orient Enantiornithes** [里氏黎明鳥]
Taxonomy: Theropoda: Enantiornithes
cf. Aves: Pygostylia: Ornithothoraces:
Enantiornithes (*1)

[References: (*1) Liu D, Chiappe LM, Zhang Y, Serrano FJ, Meng Q (2019). Soft tissue preservation in two new enatiornithine specimens (Aves) from the Lower Cretaceous Huajiying Formation of Hebei Province, China. *Cretaceous Research* 95: 191–207.]

Orkoraptor burkei

> *Orkoraptor burkei* Novas *et al.*, 2008 [Campanian–Maastrichtian; Cerro Fortaleza Formation* (=Pari Aike Formation), Santa Cruz, Argentina]
> *according to Sickman *et al.* (2018)

Generic name: *Orkoraptor* ← {(Tehuelche, place-name) Orr-Korr: 'Toothed River' + (Lat.) raptor: thief}
"Orr-Korr, name applied by Aoniken Patagonian native people to the La Leona River, which runs not far from the fossil site" (*1).
Specific name: *burkei* ← {(person's name) Burke + -ī}; "in honour of Coleman Burke, an American amateur geologist and paleontologist, who loves Patagonia and kindly supported our explorations in Argentina" (*1).
Etymology: Burke's **Orr-Korr thief**
Taxonomy: Theropoda: Megaraptora

[References: (*1) Novas FE, Ezcurra MD, Lecuona A (2008). *Orkoraptor burkei* nov. gen. *et* sp., a large theropod from the Maastrichtian Pari Aike Formation, Southern Patagonia, Argentina. *Cretaceous Research* 29(3): 468–480.]

Ornatotholus browni

> *Stegoceras validum* Lambe, 1902 [Campanian; Dinosaur Park Formation, Alberta, Canada]
> = *Troodon validus* Gilmore, 1924 (paratype) [Campanian; Dinosaur Park Formation]
> = *Stegoceras browni* Wall & Galton, 1979 ⇒ *Ornatotholus browni* (Wall & Galton, 1979) Galton & Sues, 1983 [Campanian; Dinosaur Park Formation, Alberta, Canada]

Generic name: *Ornatotholus* ← {(Lat.) ornātus: adorned, decorated + tholus: dome}
Specific name: *browni* ← {(person's name) Brown + -ī}; in honor of Barnum Brown, who found the holotype.
Etymology: Brown's **adorned dome**
Taxonomy: Ornithischia: Marginocephalia:
Pachycephalosauria
Notes: According to *The Dinosauria 2nd edition* (2004), *Ornatotholus browni* was valid, however, Schott *et al.* (2011) think that *O. browni* is a synonym of *Stegoceras validum*.

Ornithocheirus simus [Pterosauria]

> *Pterodactylus simus* Owen, 1861⇒*Ornithocheirus simus* (Owen, 1861) Seeley, 1869 ⇒ *Criorhynchus simus* (Owen, 1861) Owen, 1874 [late Albian; Cambridge Greensand Member, West Melbury

Marly Chalk Formation, UK]
= *Ornithocheirus platyrhinus* Seeley, 1870 [late Albian; Cambridge Greensand Member, West Melbury Marly Chalk Formation, UK]

Generic name: *Ornithocheirus* ← {(Gr.) ὀρνιθο-/ornitho- [ὄρνις/ornis: bird] + χειρός/cheiros [χείρ/cheir: hand]}
Specific name: *simus* ← {(Lat.) sīmus = (Gr.) σιμός/simos: flat-nosed}
Etymology: **bird hand** with flat nose
Taxonomy: Pterosauria: Pterodactyloidea: Ornithocheiridae
Other species: *Ornithocheirus* is thought to be monotypic and most other species are seemed to be placed in other genera. (*1)
> *Ornithocheirus wiedenrothi* Wild 1990 ⇒ *Targaryendraco wiedendrothi* (Wild, 1990) Pêgas *et al.*, 2019 [Hauterivian; Clay Pit, Engelbostel, Hannover, Germany] *etc.*

[References: (*1) Rodrigues T, Kellner AWA (2013). Taxonomic review of the *Ornithocheirus* complex (Pterosauria) from the Cretaceous of England. *Zookeys* 308: 1–112.]

Ornitholestes hermanni

> *Ornitholestes hermanni* Osborn, 1903 ⇒ *Coelurus hermanni* (Osborn, 1903) Hay, 1930 [Kimmeridgian; Morrison Formation, Bone Cabin Quarry, Wyoming, USA]

Generic name: *Ornitholestes* ← {(Gr.) ὀρνιθο-/ornitho- [ὄρνις/ornis: bird] + ληστής/lēstēs: robber}; "suggesting the hypothesis that the animal may have been adapted to the pursuit of the Jurassic birds" (*1)
The genus was named as suggested by Dr. Theodore Gill (*1).
Specific name: *hermanni* ← {(person's name) Hermann + -ī}; in honor of Mr. Adam Hermann, who was the head preparatory. (*1)
Etymology: Hermann's **bird robber**
Taxonomy: Theropoda: Coelurosauria
Notes: *Ornitholestes hermanni* was renamed *Coelurus hermanni* by O. P. Hay in 1930. In 1980, J. Ostrom revived the genus.

[References: (*1) Osborn HF (1903). *Ornitholestes hermanni*, a new compsognathoid dinosaur from the Upper Jurassic. *Bulletin of the American Museum of Natural History* 19(12): 459–464.]

Ornithomerus gracilis

> *Rhabdodon priscus* Matheron, 1869 [Maastrichtian; Bouches-du-Rhône, Provence-Alpes-Côte d'Azur, France]
= *Ornithomerus gracilis* Seeley, 1881 [Campanian; Grünbach Formation, Niederosterreich, Austria]

Generic name: *Ornithomerus* ← {(Gr.) ὄρνις/ornis: bird + μηρός/mēros: shin}
Specific name: *gracilis* ← {(Lat.) gracilis: thin, slender}
Etymology: slender **bird's shin**
Taxonomy: Ornithischia; Ornithopoda: Iguanodontia
Notes: According to *The Dinosauria 2nd edition*, *Ornithomerus gracilis* was regarded to be a synonym of *Rhabdodon priscus*.

Ornithomimoides barasimlensis

> *Ornithomimoides barasimlensis* von Huene & Matley, 1933 (*nomen dubium*) [Maastrichtian; Lameta Formation, Madhya Pradesh, India]

Generic name: *Ornithomimoides* ← {(Gr.) ὄρνις/ornis: bird + μῖμος/mimos: mimic + -oides}
Specific name: *barasimlensis* ← {(place-name) Bara Simla Hill + -ensis}
Etymology: ***Ornithomimus*-like one** from Bara Simla Hill
Taxonomy: Theropoda: Ceratosauria: Abelisauridae
Other species:
> *Ornithomimoides mobilis* von Huene & Matley, 1933 (*nomen dubium*) [Maastrichtian; Lameta Formation, Madhya Pradesh, India]

Notes: According to *The Dinosauria 2nd edition*, *O. brasimlensis* and *O. mobilis* are regarded as *nomina dubia*.

Ornithomimus velox

> *Ornithomimus velox* Marsh, 1890 [Maastrichtian; Denver Formation, Colorado, USA]

Generic name: *Ornithomimus* ← {(Gr.) ὀρνιθο-/ornitho- [ὄρνις/ornis: bird] + μῖμος/mīmos: mimic}; in reference to the bird-like foot.
Specific name: *velox* ← {(Lat.) vēlōx: swift, speedy}
Etymology: swift **bird mimic**
Taxonomy: Theropoda: Coelurosauria: Maniraptoriformes: Ornithomimosauria: Ornithomimidae
Other species:
> *Struthiomimus brevetertius* Parks, 1926 ⇒ *Ornithomimus brevitertius* (Parks, 1926) Russell, 1930 ⇒ *Dromiceiomimus brevitertius* (Parks, 1926) Russell, 1972 [Campanian; Horseshoe Canyon Formation, Alberta, Canada]
= *Ornithomimus edmontonicus* Sternberg, 1933 [Campanian; Horseshoe Canyon Formation, Alberta, Canada]
= *Struthiomimus currellii* Parks, 1933 ⇒ *Ornithomimus currei* (Parks, 1933) Russell, 1967 [Campanian; Horseshoe Canyon Formation, Alberta, Canada]
= *Struthiomimus ingens* Parks, 1933 ⇒

Ornithomimus ingens (Parks, 1933) Russell, 1967 [Campanian; Horseshoe Canyon Formation, Alberta, Canada]
> *Ornithomimus minutus* Marsh, 1892 (*nomen dubium*) ⇒ *Dromaeosaurus minutus* (Marsh, 1892) Russell, 1972 [Maastrichtian; Lance Formation, Niobrara, Wyomimg, USA]
> *Ornithomimus sedens* Marsh, 1892 ⇒ *Struthiomimus sedens* (Marsh, 1892) Farlow, 2001 [Maastrichtian; Lance Formation, Wyoming, USA]
> *Ornithomimus tenuis* Marsh, 1890 ⇒ *Struthiomimus tenuis* (Marsh, 1890) Osborn, 1916 [Campanian; Judith River Formation, Montana, USA]
> *Struthiomimus samueli* Parks, 1928 ⇒ *Ornithomimus samueli*? (Parks, 1928) Russell, 1930 [late Campanian; Dinosaur Park Formation, Alberta, Canada]

[References: Marsh OC (1890). Description of new dinosaurian reptiles. *The American Journal of Science, series 3* 39: 81–86.; Zelenitsky DK, Therrien K, Erickson GM, DeBuhr CL, Kobayashi Y, Eberth DA, Hadfield F (2012). Fethered non-avian dinosaurs from North America provide insight into wing origins. *Science* 338(6106): 510–514.]

Ornithopsis hulkei

> *Ornithopsis hulkei* Seeley, 1870 (Lectotype) ⇒ *Bothriospondylus magnus* (Seeley, 1870) Owen, 1875 ⇒ *Chondrosteosaurus magnus* (Seeley, 1870) Owen,1876 ⇒ *Pelorosaurus hulkei* (Seeley, 1870) Lydekker, 1893 ⇒ *Eucamerotus hulkei* (Seeley, 1870) Sebaschan, 2005 [Barremian; Wessex Formation, Isle of Wight, UK]

Generic name: *Ornithopsis* ← {(Gr.) ὄρνις/ornis: bird + ὄψις/opsis: appearance (←ὤψ/ōps: face)}
Specific name: *hulkei* ← {(person's name) Hulke + -ī}; in honor of Dr. John Whitaker Hulke, who was a British surgeon, fossil collector.
Etymology: Hulke's **bird-likeness**
Taxonomy: Saurischia: Sauropodomorpha: Sauropoda: Titanosauriformes
cf. Saurischia: Sauropodomorpha: Sauropoda: Brachiosauridae [*1]
Other species:
> *Ornithopsis leedsii* Hulke, 1887 ⇒ *Pelorosaurus*? *leedsi* (Hulke, 1887) Lydekker, 1895 ⇒ *Cetiosaurus leedsi* (Hulke, 1887) Woodward, 1905* ⇒ *Cetiosauriscus leedsi* (Hulke, 1887) von Huene, 1927* [Callovian-Oxfordian; Oxford Clay, Peterborough, UK] *referred specimen
> *Ornithopsis eucamerotus* Hulke, 1882 (*nomen dubium*)
> *Cetiosaurus conybearei* Melville, 1849 ⇒ *Pelorosaurus conybearei* (Merville, 1849) Mantell, 1850 ⇒ *Ornithopsis conybearei* (Merville, 1849) von Huene, 1929 [Valanginian; Tunbridge Wells Sand Formation, England, UK]
= *Cetiosaurus conybearei* Melville, 1849 (*nomen dubium*) (syntype) [Early Cretaceous; upper Wealden Formation, West Sussex, UK]
> *Bothriospondylus suffossus* Hulke, 1875 (*nomen dubium*) ⇒ *Astrodon suffossus* (Owen, 1875) Hatcher, 1903 ⇒ *Ornithopsis suffossa* (Owen, 1875) von Huene, 1922 [Kimmeridgian; Kimmeridge Clay Formation, Wiltshire, UK]
> *Cetiosaurus humerocristatus* Hulke, 1874 ⇒ *Pelorosaurus humerocristatus* (Hulke, 1874) Sauvage, 1887 ⇒ *Ornithopsis humerocristatus* (Hulke, 1874) Lydekker, 1888 ⇒ *Duriatitan humerocristatus* (Hulke, 1874) Barrett *et al.*, 2010 [late Kimmeridgian; Kimmeridge Clay Formation, Dorset, UK]
> *Megalosaurus meriani* Greppin, 1870 (*partim*) ⇒ *Ornithopsis greppini* von Huene, 1922 ⇒ *Cetiosauriscus greppini* (von Huene, 1922) von Huene, 1927 ⇒ *Amanzia greppini* (von Huene, 1922) Schwatz *et al.*, 2020 [early Kimmeridgian; Reuchenette Formation, Moutier, Bern, Swizerland]

Notes: Medium-sized brachiosaurid sauropod.

[References: Seeley HG (1870). On *Ornithopsis*, a gigantic animal of the pterodactyle kind from the Wealden. *Annals and Magazine of Natural History, 4th series* 4(5): 305–318.; (*1) Blows WT (1995). The early Cretaceous brachiosaurid dinosaurs *Ornithopsis* and *Eucamerotus* from the Isle of Wight, England. *Palaeontology* 38(1): 187–197.]

Orodromeus makelai

> *Orodromeus makelai* Horner & Weishampel, 1988 [?late Campanian; upper Two Medicine Formaton, Montana, USA]

Generic name: *Orodromeus* ← {(Gr.) ὄρος/oros: mountain + δρομεύς/dromeus: runner}; "alluding to the Egg Mountain, as well as the state of Montana, and to the animals, presumed cursorial habits" [*1].
Specific name: *makelai* ← {(person's name) Makela + -ī}; "in honor of the late Robert Makela for his many dinosaur discoveries including the holotype" [*1].
Etymology: Makela's **mountain runner** of Montana
Taxonomy: Ornithischia: Ornithopoda: Parksosauridae: Orodrominae

[References: (*1) Horner JR, Weishampel DB (1988). A comparative embryological study of two ornithischian dinosaurs. *Nature* 332(6161): 256–257.]

Orthomerus dolloi

> *Orthomerus dolloi* Seeley, 1883 (*nomen dubium*) [late Maastrichtian, Maastricht Formation, Limburg, Belgium]

Generic name: *Orthomerus* ← {(Gr.) ὀρθῶς/orthōs: straight + μηρός/mēros: femur}

Specific name: *dolloi* ← {(person's name) Dollo + -ī}; in honor of a French-born Belgian palaeontologist, Louis Dollo.

Etymology: Dollo's **straight femur**

Taxonomy: Ornithischia: Ornithopoda: Hadrosauridae

Other species:

> *Orthomerus weberi* Riabinin, 1945 (*nomen dubium*) ⇒ *Orthomerus weberae* (Riabinin, 1945) Nessov, 1995 ⇒ *Riabininohadros weberi* (Riabinin, 1945) Ulansky, 2015 ⇒ *Riabininohadros weberae* (Riabinin, 1945) Lopatin & Averianov, 2020 (formally) [Maastrichtian; Mt. Besh-Kosh, Crimea, Ukraine (former Soviet Union)]

Notes: According to *The Dinosauria 2nd edition* (2004), *Orthomerus dolloi* and *O. weberi* were considered to be *nomina dubia*.

[References: Seeley HG (1883). On the dinosaurs from the Maastricht beds. *Quarterly Journal of the Geological Society of London* 39: 246–253.; Ulansky R (2015). *Riabininohadros* [*Riabininohadros*, a new genus for hadrosaur from Maastrichtian of Crimea, Russia]. Dinologia (2015): 1-10 [Russian].]

Oryctodromeus cubicularis

> *Oryctodromeus cubicularis* Varricchio *et al.*, 2007 [Cenomanian; Blackleaf Formation, Montana, USA]

Generic name: *Oryctodromeus* ← {(Gr.) ὀρυκτό-/orykto-: digging + δρομεύς/dromeus: runner} "digging runner"(*1).

Specific name: *cubicularis* ← {(Lat.) cubiculāris: of the lair}; "referring to the inferred denning habit of the taxon" (*1).

Etymology: **digging runner** of the lair

Taxonomy: Ornithischia: Neornithischia: Parksosauridae: Orodrominae

cf. Ornithischia: Ornithopoda: Euornithopoda(*1)

[References: (*1) Varricchio DJ, Martin AJ, Katsura Y (2007). First trace and body fossil evidence of a burrowing, denning dinosaur. *Proceedings of the Royal Society B: Biological Sciences* 274(1616): 1361–1368.]

Osmakasaurus depressus

> *Camptosaurus depressus* Gilmore, 1909 ⇒ *Planicoxa depressa* (Gilmore, 1909) Carpenter & Wilson, 2008 ⇒ *Osmakasaurus depressus* (Gilmore, 1909) McDonald, 2011 [Barremian; Lakota Formation, South Dakota, USA]

Generic name: *Osmakasaurus* ← {(Lakota) ósmaka: canyon + (Gr.) σαῦρος/sauros: lizard}

Specific name: *depressus* ← {(Lat.) dēpressus: depressed, sunken, low}; referring to the narrowness or depressed nature of the ilia. (according to Gilmore, 1909)

Etymology: depressed, **canyon lizard**

Taxonomy: Ornithischia: Ornithopoda: Ankylopollexia: Styracosterna

[References: McDonald AT (2011). The taxonomy of species assigned to *Camptosaurus* (Dinosauria: Ornithopoda). *Zootaxa* 2783: 52–68.]

Ostafrikasaurus crassiserratus

> *Ostafrikasaurus crassiserratus* Buffetaut, 2012 [Tithonian; upper Dinosaur Member, Tendaguru Formation, Lindi, Tanzania]

Generic name: *Ostafrikasaurus* ← {(German, place-name) Deutsch-Ostafrika (German colony in East Africa) + (Gr.) σαῦρος/sauros: lizard}; "(*[Deutsch]-Ostafrika*) for the part of East Africa including Tendaguru at the time of the expeditions of the Berlin Museum"(*1).

Specific name: *crassiserratus* ← {(Lat.) crassus: thick + serrātus: serrated}; "in reference to the large size of the serrations on the carinae*"(*1).

*carinae: the sharp edges of the teeth

Etymology: **East Africa lizard** with thick-serrated teeth

Taxonomy: Theropoda: Spinosauridae

[References: (*1) Buffetaut E (2012). An early spinosaurid dinosaur from the Late Jurassic of Tendaguru (Tanzania) and the evolution of the spinosaurid dentition. *Oryctos* 10: 1–8.]

Ostromia crassipes [Anchiornithidae]

> *Pterodactylus crassipes* von Meyer, 1857 (*nomen oblitum*) [Pterosauria] (incorrectly classified at the time) ⇒ *Pterodactylus* (*Rhamphorhynchus*) *crassipes* (von Meyer, 1857) (*nomen oblitum*) von Meyer, 1857 ⇒ *Scaphognathus crassipes* (von Meyer, 1857) Wagner 1861 (*nomen oblitum*) ⇒*Archaeopteryx crassipes* (von Meyer, 1857) Ostrom, 1970 ⇒ *Ostromia crassipes* (von Meyer, 1857) Foth & Rauhut, 2017 [Tithonian; Painten Formation, Bavaria, Germany]

Generic name: *Ostromia* ← {(person's name) Ostrom + -ia}; "in honor of the late John Ostrom, who identified the Haarlem specimen as a theropod"(*1).

Specific name: *crassipes* ← {(Lat.) crassus: thick +

pēs: foot}
Etymology: thick-footed one **for Ostrom**
Taxonomy: Theropoda: Maniraptora: Avialae (?): Anchiornithidae
cf. Theropoda: Maniraptora: Anchiornithidae [*1]
Notes: According to *The Dinosauria 2nd edition*, *Pterodactylus crassipes* was regarded as a synonym of *Archaeopteryx lithographica*. The new generic name *Ostromia* was proposed in 2017 [*1].

[References: (*1) Foth C, Rauhut OWM (2017). Re-evaluation of the Haarlem *Archaeopteryx* and the radiation of maniraptoran theropod dinosaurs. *BMC Evolutionary Biology* 17: 236.]

Othnielia rex

> *Nanosaurus rex* Marsh, 1877 ⇒ *Othnielia rex* (Marsh, 1877) Galton, 1977 [Kimmeridgian–Tithonian; Morrison Formation, Wyoming, USA]
= *Laosaurus gracilis* Marsh, 1878 [Kimmeridgian; Morrison Formation, Wyoming, USA]
= *Laosaurus consors* Marsh, 1894 ⇒ *Othnielia consors* (Marsh, 1894) Galton, 1977 ⇒*Othnielosaurus consors* (Marsh, 1894) Galton, 2007 [Kimmeridgian–Tithonian; Morrison Formation, Wyoming, USA]
See also *Nanosaurus agilis*.
Generic name: *Othnielia* ← {(person's name) Othniel + -ia}; in honor of paleontologist Othniel Charles Marsh. [*1]
Specific name: *rex* ← {(Lat.) rēx: king}
Etymology: **Othniel's** king
Taxonomy: Ornithischia: Neornithischia
Notes: *Othnielia rex* is regarded as a synonym of *Nanosaurus agilis*, according to Carpenter *et al.* (2018).

[References: (*1) Galton PM (1977). The ornithopod dinosaur *Dryosaurus* and a Laurasia-Gondwanaland connection in the Upper Jurassic. *Nature* 268: 230–232.]

Othnielosaurus consors

> *Nanosaurus agilis* Marsh, 1877 [Kimmeridgian; Morrison Formation, Colorado, USA]
= *Laosaurus consors* Marsh, 1894 ⇒ *Othnielia consors* (Marsh, 1894) Galton, 1977 ⇒*Othnielosaurus consors* (Marsh, 1894) Galton, 2007 [Kimmeridgian–Tithonian; Morrison Formation, Wyoming, USA]
See also *Nanosaurus agilis*.
Generic name: *Othnielosaurus* ← {(person's name) Othniel Charles Marsh + (Gr.) σαῦρος/sauros: lizard}
Specific name: *consors* ← {(Lat.) cōnsors: partner}.
Etymology: partner's **Othniel's lizard**
Taxonomy: Ornithischia: Neornithischia

Otogosaurus sarulai

> *Otogosaurus sarulai* Zhao, 2004 [Campanian–Maastrichtian; Inner Mongolia, China]
Generic name: *Otogosaurus* ← {(place-name) Otog 鄂托克 + (Gr.) σαῦρος/sauros: lizard}
Specific name: *sarulai* ← {(demonym) Sarula薩茹拉 + -ī} *cf.* Sarulai (female), discoverer [*1]
Etymology: **Otog lizard** of Sarula (or Sarulai)
Taxonomy: Saurischia: Sauropodomorpha: Sauropoda

[References: (*1) updated May 19, 2005. Largest shank fossil of dinosaur recovered in Inner Mongolia. ⟨http://en.people.cn/200505/19/eng20050519_185853.html⟩]

Ouranosaurus nigeriensis

> *Ouranosaurus nigeriensis* Taquet, 1976 [late Aptian; Elrhaz Formation, Agadez, Niger]
Generic name: *Ouranosaurus* ← {(Arabic) ourane: valour, courage, recklessness | (by the Touareg of Niger and by the Berbers of Algeria) ourane: sand monitor + (Gr.) σαῦρος/sauros: lizard} [*1].
Specific name: *nigeriensis* ← {(place-name) Niger + -ensis} [*1]
Etymology: **brave, big lizard** from Niger
Taxonomy: Ornithischia: Ornithopoda: Iguanodontidae [*1]

[References: (*1) Taquet P (1976). Géologie et paléontologie du gisement de Gadoufaoua (Aptien du Niger). *Cahiers de Paléontologie, CNRS*, Paris: 1–191.]

Overoraptor chimentoi

> *Overoraptor chimentoi* Motta *et al.*, 2020 [Cenomania; Huincul Formation, Río Negro, Argentina]
Generic name: *Overoraptor* ← {(Sp.) overo: piebald + (Lat.) raptor: thief}; "in reference to the coloration of the *O. chimentoi* bones, which consists of a pattern of light and dark spots" [*1].
Specific name: *chimentoi* ← {(person's name) Chimento + -ī}; "in honor of its discoverer, the paleontologist Dr. Roberto Nicolás Chimento" [*1].
Etymology: Chimento's **piebald thief**
Taxonomy: Saurischia: Theropoda: Paraves: Maniraptora

[References: (*1) Motta MJ, Agnolín FL, Egli FB, Novas FE (2020). New theropod dinosaur from the Upper Cretaceous of Patagonia sheds light on the paravian radiation in Gondwana. *The Science of Nature* 107(3): 24.]

Overosaurus paradasorum

> *Overosaurus paradasorum* Coria *et al.*, 2013 [Campanian; Anacleto Formation, Cerro Overo, Neuquén, Argentina]
Generic name: *Overosaurus* ← {(place-name) Cerro Overo + (Gr.) σαῦρος/sauros: lizard}
Specific name: *paradasorum* ← {(person's name) Parada and his family + -ōrum}; "in recognition

of Carlos Parada and his family, who have been always generous, helpful and supportive of the work in Rincón de los Sauces area" [*1].

Etymology: Parada family's **lizard from Cerro Overo**

Taxonomy: Saurischia: Sauropodomorpha: Sauropoda: Titanosauria: Lithostrotia: Aeolosaurini

Notes: According to Coria (2013), "in November 2002, a joint expedition of the Museo Carmen Funes (Plaza Huincul), the Museo Argentino Urquiza (Rincón de los Sauces) and the Natural History Museum of Los Angeles County uncovered the partial skeleton of a titanosaur sauropod" [*1].

[References: (*1) Coria RA, Filippi LS, Chiappe LM, Garcia R, Arcucci AB (2013). *Overosaurus paradasorum* gen. *et* sp. nov., a new sauropod dinosaur (Titanosauria: Lithostrotia) from the Late Cretaceous of Neuquén, Patagonia, Argentina. *Zootaxa* 3683(4): 357–376.]

Oviraptor philoceratops

> *Oviraptor philoceratops* Osborn, 1924 [? middle Campanian; Djadokhta Formation, Ömnögovi, Mongolia]

Generic name: *Oviraptor* ← {(Lat.) ovi- [ovum: egg] + raptor: plunderer, robber} "egg seizer"; "named because the type skull (Amer. Mus. 6517) was found lying directly over a nest of dinosaur eggs" [*1].

Specific name: *philoceratops* ← {(Gr.) φίλος/ philos: fond of + κέρας/ceras: horn + ὤψ/ōps: face}; "fondness for ceratopsian eggs" [*1]

Etymology: **egg seizer**, fondness of ceratopsian eggs

Taxonomy: Theropoda: Oviraptoridae

Other species:

> *Oviraptor mongoliensis* Barsbold, 1986 ⇒ *Rinchenia mongoliensis* (Barsbold, 1986) Osmólska *et al.*, 2004 [Maastrichtian; Nemegt Formation, Ömnögovi, Mongolia]

[References: (*1) Osborn HF (1924). Three new Theropoda, Protoceratops zone, central Mongolia. *American Museum Novitates* 144: 1–12.]

Owenodon hoggii

> *Iguanodon hoggii* Owen, 1874 ⇒ *Camptosaurus hoggi* (Owen, 1874) Norman & Barrett, 2002 ⇒ *Owenodon hoggii* (Own, 1874) Galton, 2009 [Berriasian; Lulworth Formation, Dorset, UK]

Generic name: *Owenodon* ← {(person's name) Richard Owen + (Gr.) ὀδών/odōn: tooth}; in honor of Sir Richard Owen, who first described.

Specific name: *hoggii* ← {(person's name) Hogg + -iī}; in honor of A. J. Hogg.

Etymology: Hogg and **Owen's tooth**

Taxonomy: Ornithischia: Ornithopoda: Iguanodontia: Ankylopollexia

Oxalaia quilombensis

> *Oxalaia quilombensis* Kellner *et al.*, 2011 [Cenomanian; Alcântara Formation, Maranhão, Brazil]

Generic name: *Oxalaia* ← {(myth.) Oxalá + -ia}; referring to Oxalá, "the most respected masculine deity in the African pantheon, introduced in Brazil during slavery" [*1].

Specific name: *quilombensis* ← {(Portuguese expression, place-name) quilombo + -ensis}; in reference to "the place where the Quilombola (the descendants of former Brazilian slaves) live. The Cajual Island, where the specimens were collected, is one of these places" [*1].

Etymology: **Oxala's** one from Quilombo

Taxonomy: Theropoda: Spinosauroidea: Spinosauridae

[References: (*1) Kellner AWA, Azevedo SAK, Machado EB, de Carvalho LB, Henriques DDR (2011). A new dinosaur (Theropoda, Spinosauridae) from the Cretaceous (Cenomanian) Alcântara Formation, Cajual Island, Brazil. *Anais da Academia Brasileira de Ciências* 83(1): 99–108.]

Ozraptor subotaii

> *Ozraptor subotaii* Long & Molnar, 1998 [Aalenian–Bajocian; Colalura Sandstone Formation, Western Australia, Australia]

Generic name: *Ozraptor* ← {(place-name) Oz (= Ozzie) + (Lat.) raptor: thief}

Specific name: *subotaii* ← {(movie) Subotai + -ī}; referring to the character, the swift-running thief and archer Subotai from the movie"Conan the Barbarian" [*1].

Etymology: Subotai's **Australian thief**

Taxonomy: Theropoda: Abelisauroidea

[References: Long JA, Molnar RE (1998). A new Jurassic theropod dinosaur from Western Australia. *Records of the Western Australian Museum* 19(1): 221–229.; (*1)"Conan the Barbarian"(1982, Universal Pictures), based on the Robert E. Howard books.]

P

Pachycephalosaurus wyomingensis

> *Troodon wyomingensis* Gilmore, 1931 [*1] ⇒ *Pachycephalosaurus wyomingensis* (Gilmore, 1931) Brown & Schlaijer, 1943 (*nomen conservandum*) [Maastrichtian; Lance Formation, Wyoming, USA]

= *Tylosteus ornatus* Leidy, 1872 (*nomen rejectum*) [Maastrichtian; Lance Formation?, Wyoming, USA]

= *Pachycephalosaurus grangeri* Brown & Schlaijer, 1943 [Maastrichtian; Hell Creek Formation, Montana, USA]

= *Pachycephalosaurus reinheimeri* Brown & Schlaikjer, 1943 [Maastrichtian; Lance

Formation, South Dakota, USA]
= *Stygimoloch spinifer* Galton & Sues, 1983 [Maastrichtian; Hell Creek Formation, Montana, USA][*2]
= *Stenotholus kohleri* Giffin *et al.*, 1988 [Maastrichtian; Hell Creek Formation, Montana, USA]
= *Dracorex hogwartsia* Bakker *et al.*, 2006 [Maastrichitian; Hell Creek Formation, South Dakota, USA][*2]

Generic name: *Pachycephalosaurus* ← {(Gr.) πᾰχύς/pachys: thick + κεφᾰλή/kephale: head + σαῦρος/sauros: lizard}
Specific name: *wyomingensis* ← {(place-name) Wyoming + -ensis}
Etymology: **thick-headed lizard** from Wyoming
Taxonomy: Ornithischia: Marginocephalia: Pachycephalosauria: Pachycephalosauridae
Notes: According to *The Dinosauria 2nd edition*, *Stygimoloch* was thought to be a valid taxon and *Stenotholus kohleri* was regarded as a synonym of *Stygimoloch spinifer*. According to Horner & Goodwin (2009), *Dracorex hogwartsia* (juvenile), *Stygimoloch spinifer* (subadult), *Pachycephalosaurus wyomingensis* (adult) are proposed to be the same taxon.

[References: (*1) Gilmore CW (1931). A new species of troödont dinosaur from the Lance Formation of Wyoming. *Proceedings of the United States National Museum* 79(2875): 1–6.]; (*2) Goodwin MB, Evans DC (2016). The early expression of squamosal horns and parietal ornamentation confirmed by new end-stage juvenile *Pachycephalosaurus* fossils from the upper Cretaceous Hell Creek Formation, Montana. *Journal of Vertebrate Paleontology* 36(2): e1078343.]

Pachyrhinosaurus canadensis

> *Pachyryinosaurus canadensis* Sternberg, 1950 ⇒ *Centrosaurus canadensis* (Sternberg, 1950) Paul, 2010 [Maastrichtian; lower Horseshoe Canyon Formation, Alberta, Canada]

Generic name: *Pachyrhinosaurus* ← {(Gr.) πᾰχύς/pachys: thick + ῥινό-/rhino- [ῥίς/rhis: nose] + σαῦρος/sauros: lizard}
Specific name: *canadensis* ← {(place-name) Canada + -ensis}
Etymology: **thick-nosed lizard** from Canada
Taxonomy: Ornithischia: Marginocephalia: Ceratopsia: Ceratopsidae: Centrosaurinae: Pachyrhinosaurini: Pachyrostra
Other species:

> *Pachyrhinosaurus lakustai* Currie *et al.*, 2008 ⇒ *Centrosaurus lakustai* (Currie *et al.*, 2008) Paul, 2010 [Campanian; Wapiti Formation, Alberta, Canada]

> *Pachyrhinosaurus perotorum* Fiorillo & Tykoshi, 2012 [Maastrichtian; Prince Creek Formation, Alaska, USA.]

Pachysuchus imperfectus

> *Pachysuchus imperfectus* Young, 1951 (*nomen dubium*) [Sinemurian; lower Lufeng Formation, Lufeng, Yunnan, China]

Generic name: *Pachysuchus* ← {(Gr.) παχύς/pachys: thick + suchos: crocodile [Σουχος/Souchos [Sūchos]: Egyptian crocodile god Sobek]}
Specific name: *imperfectus* ← {(Lat.) imperfectus: imperfect}
Etymology: imperfect, **thick crocodile**
Taxonomy: Saurischia: Sauropodomorpha[*1]
Notes: Young identified the rostrum as that of a phytosaur. But, later, Xu & Barrett show that "the holotype of *Pachysuchus imperfectus* is not a phytosaur, but an indeterminate sauropodomorph dinosaur"[*1].

[References: (*1) Barrett PM, Xu X (2012). The enigmatic reptile *Pachysuchus imperfectus* Young, 1951 from the Lower Lufeng Formation (Lower Jurassic) of Yunnan, China. *Vertebrate PalAsiatica* 50: 151–159.]

Padillasaurus leivaensis

> *Padillasaurus leivaensis* Carballido *et al.*, 2015 [Barremian–Aptian; Paja Formation, Boyacá, Ricaurte, Colombia]

Generic name: *Padillasaurus* ← {(person's name) Padilla + saurus [(Gr.) σαῦρος/sauros: lizard]}; "honor of Dr. Carlos Bernardo Padilla Bernal (1957–2013), a paleontological enthusiast who led the creation of the Centro de Investigaciones Paleontológicas of Villa de Leiva (Colombia)"[*1], promoted paleontological collection and research on the Colombian fossil record, and encouraged the study of this specimen by combining 'Padilla' and 'saurus'.
Specific name: *leivaensis* ← {(place-name) Leiva + -ensis}; "referring to the locality of Villa de Leiva, from which the specimen derives"[*1].
Etymology: **Padilla's lizard** from Leiva
Taxonomy: Saurischia: Sauropodomorpha: Sauropoda: Brachiosauridae

[References: (*1) Carballido JL, Pol D, Parra Ruge ML, Bernal SP, Páramo-Fonseca ME, Etayo-Serna F (2015). A new Early Cretaceous brachiosaurid (Dinosauria, Neosauropoda) from northwestern Gondwana (Villa de Leiva, Colombia). *Jornal of Vertebrate Paleontology* 35(5): e980505.]

Pakisaurus balochistani

> *Pakisaurus balochistani* Malkani, 2006 [Maastrichtian; Pab Formation, Balochistan,

Pakistan]

Generic name: *Pakisaurus* ← {(place-name) Pakistan + (Gr.) σαῦρος/sauros: reptiles} [*1]

Specific name: *balochistani* ← {(place-name) Balochistan + -ī}; "referring to the province of origin as Balochistan" [*1].

Etymology: **Pakistan reptile** of Balochistan

Taxonomy: Saurischia: Sauropoda: Titanosauria: Pakisauridae (= Titanosauridae) [*1]

[References: (*1) Malkani MS (2006). Biodiversity of saurischian dinosaurs from the Latest Cretaceous Park of Pakistan. *Journal of Applied and Emerging Sciences* 1(3): 108–140.]

Palaeopteryx thomsoni

> *Palaeopteryx thomsoni* Jensen, 1981 (*nomen dubium*) [Tithonian–Kimmeridgian; Morrison Formation, Colorado, USA]

Generic name: *Palaeopteryx* ← {(Gr.) παλαιός/palaios: ancient + πτέρυξ/pteryx: wing}; referring to the bone end which was first identified as an avian proximal tibia [*1].

Specific name: *thomsoni* ← {(person's name) Thomson + -ī}

Etymology: Thomson's **ancient wing**

Taxonomy: Theropoda: Maniraptora: Paraves

Notes: According to *The Dinosauria 2nd edition*, *Palaeopteryx* is regarded to be a *nomen dubium*.

[References: (*1) Jensen JA, Padian K (1989). Small pterosaurs and dinosaurs from the Uncompahgre fauna (Brushy Basin Member, Morrison Formation: ?Tithonian), Late Jurassic, western Colorado. *Journal of Paleontology* 63(3): 364–373.]

Palaeoscincus costatus

> *Palaeoscincus costatus* Leidy, 1856 (*nomen dubium*) [Campanian; Judith River Formation, Montana, USA]

Generic name: *Palaeoscincus* ← {(Gr.) παλαιός/palaios: ancient + σκίγγος/skinkos: skink, a kind of lizard}

Specific name: *costatus* ← {(Lat.) costa: rib}; referring to a tooth.

Etymology: ribbed, **ancient skink**

Taxonomy: Ornithischia: Nodosauridae

Other species:

> *Palaeoscincus africanus* Broom, 1912 ⇒ *Paranthodon africanus* (Broom, 1912) Nopcsa, 1929 [Berriasian; Kirkwood Formation, Eastern Cape, South Africa]

> *Palaeoscincus asper* Lambe, 1902 (*nomen dubium*) [Campanian; Oldman Formation or Dinosaur Park Formation, Alberta, Canada]

> *Palaeoscincus latus* Marsh, 1892 [Maastrichtian; Lance Formation, Wyoming, USA]

> *Palaeoscincus rugosidens* Gilmore, 1930 ⇒ *Edmontonia rugosidens* (Gilmore, 1930) [Campanian; Two Medicine Formation, Montana, USA]

> *Stereocephalus tutus* Lambe, 1902 ⇒ *Euoplocephalus tutus* (Lambe, 1902) Lambe, 1910 ⇒ *Palaeoscincus tutus* (Lambe, 1902) Hennig, 1915 [Campanian; Dinosaur Park Formation, Alberta, Canada]

> *Palaeoscincus magoder* Hennig, 1914 (*nomen nudum*)

Notes: According to *The Dinosauria 2nd edition*, *Palaeoscincus costatus*, *P. asper* and *P. latus* are thought to be *nomina dubia*.

Paludititan nalatzensis

> *Paludititan nalatzensis* Csiki *et al.*, 2010 [Maastrichtian; Sânpetru Formation, Hunedoara, Romania]

Generic name: *Paludititan* ← {(Lat.) palūs: marsh + (Gr.) Τῑτάν/Tītan}; referring to a silty mudstone layer where the holotype was found.

Specific name: *nalatzensis* ← {(place-name) Nălaţ-Vad + -ensis}

Etymology: **marsh Titan** from Nălaţ-Vad

Taxonomy: Saurischia: Sauropodomorpha: Sauropoda: Titanosauria

[References: Csiki Z, Codrea V, Jipa-Murzea C, Godefroit P (2010). A partial titanosaur (Sauropoda, Dinosauria) skeleton from the Maastrichtian of Nălaţ-Vad, Haţeg Basin, Romania. *Neues Jahrbuch für Geologie und Paläontologie-Abhandlungen* 258(3): 297–324(28).]

Paluxysaurus jonesi

> *Sauroposeidon proteles* Wedel *et al.*, 2000 [Aptian–Albian; Antlers Formation, Oklahoma, USA]

= *Paluxysaurus jonesi* Rose, 2007 [late Aptian; Twin Mountains Formation, Texas, USA]

Generic name: *Paluxysaurus* ← {(place-name) Paluxy + (Gr.) σαῦρος/sauros: lizard}; "referring to the nearby town of Paluxy, Texas, and the Paluxy River, which flows through this region" [*1].

Specific name: *jonesi* ← {(person's name) Jones + -ī}; "in honor of William R. (Bill) Jones, who for nearly two decades has graciously allowed the excavation of these important fossils on his land" [*1].

Etymology: Jones' **Paluxy lizard**

Taxonomy: Saurischia: Sauropodomorpha: Sauropoda: Somphospondyli

cf. Saurischia: Sauropodomorpha: Titanosauriformes: Brachiosauridae [*1]

Notes: D'Emic & Foreman (2012) concluded that *Paluxysaurus* was a junior synonym of

Sauroposeidon.

[References: (*1) Rose PJ (2007). A new titanosauriform sauropod (Dinosauria: Saurischia) from the Early Cretaceous of central Texas and its phylogenetic relationships. *Palaeontologia Electronica* 10(2); 8A: 65.]

Pampadromaeus barberenai

> *Pampadromaeus barberenai* Cabreira *et al.*, 2011 [Carnian; Santa Maria Formation, Rio Grande do Sul, Brazil]

Generic name: *Pampadromaeus* ← {(Quechua) pampa: plain + dromaeus [(Gr.) δρομεύς/dromeus: runner]}; "*pampa*, in reference to the grassland landscape that covers parts of Rio Grande do Sul, and *dromaeus*, in reference to the probable cursoriality of the animal" (*1).

Specific name: *barberenai* ← {(person's name) Barberena + -ī}; "in honor of the Brazilian palaeontologist Mário C. Barberena" (*1).

Etymology: Barberena's **plain runner**

Taxonomy: Saurischia: Sauropodomorpha

[References: (*1) Cabreira SF, Schultz CL, Bittencourt JS, Soares MB, Fortier DC, Silva LR, Langer MC (2011). New stem-sauropodomorph (Dinosauria, Saurischia) from the Triassic of Brazil. *Naturwissenschaften* 98: 1035–1040.]

Pamparaptor micros

> *Pamparaptor micros* Porfiri *et al.*, 2011 [Turonian–Coniacian; Portezuelo Formation, Neuquén, Patagonia, Argentina]

Generic name: *Pamparaptor* ← {(Quechua, demonym) Pampa + (Lat.) raptor: robber}; "in honor of the Indian Pampas that lived in the central plain of Argentina" (*1).

Specific name: *micros* ← {(Gr.) μῑκρός/micros: small}; "for the small size of specimen" (*1).

Etymology: small **Pampa robber**

Taxonomy: Theropoda: Tetanurae: Dromaeosauridae
cf. Theropoda: Deinonychosauria (*1)

Notes: *Pamparaptor* had a troodontid-like pes (*1).

[References: (*1) Porfiri JD, Calvo JO, dos Santos D (2011). A new small deinonychosaur (Dinosauria: Theropoda) from the Late Cretaceous of Patagônia, Argentina. *Anais da Academia Brasileira de Ciências* 83(1): 109–116.]

Panamericansaurus schroederi

> *Panamericansaurus schroederi* Calvo & Porfiri, 2010 [Campanian–Maastrichtian; Allen Formation, Neuquén, Argentina]

Generic name: *Panamericansaurus* ← {(company) Pan American Energy + (Gr.) σαῦρος/sauros: lizard}; "in honor of the Pan American Energy company which financially supported Proyecto Dino" (*1).

Specific name: *schroederi* ← {(person's name) Schroeder + -ī}; "honor of the Schroeder family on whose land the remains were found" (*1).

Etymology: Schroeder's **Pan American lizard**

Taxonomy: Saurischia: Sauropodomorpha: Titanosauria: Titanosauridae: Aeolosaurini

[References: (*1) Calvo JO, Porfiri LJD (2010). *Panamericansaurus schroederi* gen. nov. sp. nov. un nuevo Sauropoda (Titanosauridae– Aeolosaurini) de la Provincia del Neuquén, Cretácico Superior de Patagonia, Argentina. *Brazilian Geographical Journal: Geosciences and Humanities Research Medium* 1: 100–115.]

Pandoravenator fernandezorum

> *Pandoravenator fernandezorum* Rauhut & Pol, 2017 [Oxfordian–Kimmeridgian; Cañadón Calcáreo Formation, Chubut, Argentina]

Generic name: *Pandoravenator* ← {(place-name) Pandora*: referring to the type locality 'Caja de Pandora'**+ (Lat.) vēnātor: hunter} (*1)

*(Gr.) παν-δώρα/pandōra: giver of all. Pandora.

**Caja de Pandora locality, about 1km west of the fish locality of Puesto Almada, Chubut Province, Argentina. (*1) *cf.* (Sp.) Caja de Pandora = (Eng.) Pandora's box.

Specific name: *fernandezorum* ← {(person's name) Fernandez + -ōrum}; "in honor of the Fernández family, including Daniel Fernández and the late Victoriano Fernández and his daughters and sons (especially Abel). The family has helped in many ways the exploration of the Museo Paleontológico Egidio Feruglio on their land in the Upper Jurassic rocks of central Chubut for more than twenty years" (*1).

Etymology: Fernandez family's **hunter from Pandora's box**

Taxonomy: Theropoda: Tetanurae

[References: (*1) Rauhut OWM, Pol D (2017). A theropod dinosaur from the Late Jurassic Cañadón Calcáreo Formation of Central Patagonia, and the evolution of the theropod tarsus. *Ameghiniana* 54(5): 539–566.]

Panguraptor lufengensis

> *Panguraptor lufengensis* You *et al.*, 2014 [Sinemurian–? Toarcian; Shawan Member, Lufeng Formation, Lufeng, Yunnan, China]

Generic name: *Panguraptor* ← {(Chin. myth) Pángŭ 盤古: "the first living being and the creator of all reality" (*1) + (Lat.) raptor: thief, robber}

Specific name: *lufengensis* ← {(place-name) Lufeng + -ensis}; "referring to 'Lufeng County', one of the world's richest sources of Early Jurassic terrestrial vertebrate fossils" (*1). Type horizon is Shawan Member of the Lufeng Formation (Fang *et al.* 2000).

Etymology: **Pangu thief** from Lufeng County

Taxonomy: Theropoda: Neotheropoda:

Coelophysoidea: Coelophysidae

[References: (*1) You H-L, Azuma Y, Wang T, Wang Y-M, Dong Z-M (2014). The first well-preserved coelophysoid theropod dinosaur from Asia. *Zootaxa* 3873(3): 233–249.]

Panoplosaurus mirus

> *Panoplosaurus mirus* Lambe, 1919 [Campanian; Dinosaur Park Formation, Alberta, Canada]

Generic name: *Panoplosaurus* ← {(Gr.) πάν-οπλος/pan-oplos: completely armoured + σαῦρος/sauros: lizard}

Specific name: *mirus* ← {(Lat.) mīrus: wonderful, amazing}

Etymology: wonderful, **completely armoured lizard**

Taxonomy: Ornithischia: Thyreophora: Ankylosauria: Nodosauridae

Other species:

> *Edmontonia longiceps* Sternberg, 1928 ⇒ *Panoplosaurus longiceps* (Sternberg, 1928) Coombs, 1979 [Maastrichtian; Horseshoe Canyon Formation, Alberta, Canada]

> *Palaeoscincus rugosidens* Gilmore, 1930 ⇒ *Edmontonia rugosidens* (Gilmore, 1930) Russell, 1939 ⇒ *Panoplosaurus rugosidens* (Gilmore, 1930) Coombs, 1979 ⇒ *Chassternbergia rugosidens* (Gilmore, 1930) Olshevsky, 1991 (*partim*) [Campanian; Two Medicine Formation, Montana, USA]

Notes: Originally *Panoplosaurus* was assigned to the Ankylosauridae, but, according to *The Dinosauria 2nd edition*, it is considered a member of the Nodosauridae.

[References: Carpenter K (1990). Ankylosaur systematics: example using *Panoplosaurus* and *Edmontonia* (Ankylosauria: Nodosauridae). In: Carpenter K, Currie PJ (eds), *Dinosaur Systematics: Approaches and Perspectives,* Cambridge University Press, Cambridge: 281–298.]

Panphagia protos

> *Panphagia protos* Martínez & Alcober, 2009 [Carnian; Ischigualasto Formation, San Juan, Argentina]

Generic name: *Panphagia* ← {(Gr.) πάν-/pan-: all + φαγειν/phagein: to eat + (Gr.) -ia: pertaining to [*1]}; "in reference to the inferred omnivorous diet of the new taxon, which appears to be transitional between carnivory and herbivory" [*1].

Specific name: *protos* ← {(Gr.) πρῶτος/prōtos: first}; "in reference to the basal position of the new taxon within Sauropodomorpha" [*1].

Etymology: first, **omnivorous one**

Taxonomy: Saurischia: Sauropodomorpha

[References: (*1) Martinez RN, Alcober OA (2009). A basal sauropodomorph (Dinosauria: Saurischia) from the Ischigualasto Formation (Triassic, Carnian) and the early evolution of Sauropodomorpha. *PLoS ONE* 4(2): e4397.]

Pantydraco caducus

> *Thecodontosaurus antiquus* Morris, 1843 [Rhaetian; Magnesian Conglomerate Formation, Avon, UK]

?= *Thecodontosaurus caducus* Yates, 2003 [*1] ⇒ *Pantydraco caducus* (Yates, 2003) Galton *et al.*, 2007 [Rhaetian; Pant-y-ffynnon Quarry, Bonvilston, South Glamorgan, Wales, UK]

Generic name: *Pantydraco* ← {(Welsh, place-name) Pant-y: Pant-y-ffynnon* Quarry + (Lat.) draco: a fabulous lizard-like animal} [*1]

*(Welsh) pant: a hollow place, a valley + ffynnon: a well or spring [*2]

Specific name: *caducus* ← {(Lat.) cadūcus: fallen}; referring to the fact that the holotype is an articulated specimen presereved in a fissure fill, indicating the animal may have fallen into the fissure and died there [*1] [*3].

Etymology: fallen, **Pant-y-ffynnon dragon**

Taxonomy: Saurischia: Sauropodomorpha

Notes: Ballell *et al.* (2020) considered *Pantydraco caducus* a possible juvenile of *Thecodontosaurus anticuus*.

[References: (*1) Galton PM, Yates AM, Kermack D (2007). *Pantydraco* n. gen. for *Thecodontosaurus caducus* Yates, 2003, a basal sauropodomorph dinosaur from the Upper Triassic or Lower Jurassic of South Wales, UK. *Neues Jahrbuch für Geologie und Paläontologie - Abhandlungen* 243(1): 119–125.; (*2) Welsh Place Names ⟨https://www.welshholidaycottages.com/welsh-history/place-names/⟩; (*3) Yates AM (2003). A new species of the primitive dinosaur *Thecodontosaurus* (Saurischia: Sauropodomorpha) and its implications for the systematics of early dinosaurs. *Journal of Systematic Palaeontology* 1(1): 1–42.]

Parabohaiornis martini [Avialae]

> *Parabohaiornis martini* Wang *et al.*, 2014 [Aptian; Jiufotang Formation, Liaoning, China]

Generic name: *Parabohaiornis* ← {(Lat.) para + *Bohaiornis* [(Chin. place-name) Bóhǎi 渤海 + (Gr.) ὄρνις/ornis: bird}; "derived from the Latin prefix 'para' to indicate similar morphology with bohaiornithids" [*1].

Specific name: *martini* ← {(person's name) Martin + -ī}; "in honor of the late Prof. Larry D. Martin, a paleontologist who made great contributions to the study of the evolution of birds during the course of his life" [*1].

Etymology: Martin's ***Bohaiornis*-like**

Taxonomy: Theropoda: Avialae: Ornithothoraces: Enantiornithes: Bohaiornithidae

[References: (*1) Wang M, Zhou Z-H, O'Connor JK, Zelenkov NV (2014). A new diverse enantiornithine family (Bohaiornithidae fam. nov.) from the Lower Cretaceous of China with information from two new species. *Vertebrata PalAsiatica* 52(1): 31–76.]

Parahesperornis alexi [Avialae]

> *Parahesperornis alexi* Martin, 1984 [Coniacian–Santonian; Niobrara Chalk Formation, Kansas, USA]

Generic name: *Parahesperornis* ← {(Lat.) para: near + *Hesperornis* [(Gr.) ἕσπερος/hesperos: western + (Gr.) ὄρνις/ornis: bird]}

Specific name: *alexi* ← {(person's name) Alex + -ī}; in honor of Alexander Wetmore, a former University of Kansas student and most distinguished avian paleontologist.

Etymology: Alex's **near *Hesperornis***

Taxonomy: Avialae: Ornithurae: Hespeornithes: Hesperornithidae

[References: Martin LD (1984). A new hesperornithid and the relationships of the Mesozoic birds. *Transactions of the Kansas Academy of Science* 87(3–4): 141–150.]

Parahongshanornis chaoyangensis [Avialae]

> *Parahongshanornis chaoyangensis* Li *et al.*, 2011 [Aptian; Jiufotang Formation, Liaoning, China]

Generic name: *Parahongshanornis* ← {(Lat.) para: near + *Hongshanornis* [(Chin.) Hóngshān 紅山 + (Gr.) ὄρνις/ornis: bird]} (*1)

Specific name: *chaoyangensis* ← {(Chin. place-name) Cháoyáng朝陽 + -ensis}

Etymology: **near *Hongshanornis*** from Chaoyang [朝陽副紅山鳥]

Taxonomy: Theropoda: Avialae: Ornithuromorpha: Hongshanornithidae

cf. Aves: Ornithurae: Hongshanornithidae (*1)

[References: (*1) Li L, Wang J-Q, Hou S-L (2011). A new ornithurine bird (Hongshanornithidae) from the Jiufotang Formation of Chaoyang, Liaoning, China. *Vertebrata PalAsiatica* 49(2): 195–200.]

Paralititan stromeri

> *Paralititan stromeri* Smith *et al.*, 2001 [Cenomanian; Baharîje Formation, Egypt]

Generic name: *Paralititan* ← {(Gr.) parali- [παρά/para: near + ἅλς/hals: sea] + Τιτάν/Tītan: giant}; "referring to tidal environments and Titan, an offspring of Uranus and Gaea, symbolic of brute strength and large size" (*1).

Specific name: *stromeri* ← {(person's name) Stromer + -ī}; "in honor of Ernst Stromer" (*1).

Etymology: Stromer's **tidal giant**

Taxonomy: Saurischia: Sauropodomorpha: Sauropoda: Titanosauria

Notes: The Bavarian geologist Ernst Stromer described a diverse biota from the upper Cretaceous Bahariya Formation of the Bahariya Oasis, Egypt in the early 20th century (*1).

[References: (*1) Smith JB, Lamanna MC, Lacovara KJ, Dodson P, Smith JR, Pool JC, Giegengack R, Attia Y (2001). A giant sauropod dinosaur from an Upper Cretaceous mangrove deposit in Egypt. *Science* 292(5522): 1704–1706.]

Paranthodon africanus

> *Palaeoscincus africanus* Broom, 1912 ⇒ *Paranthodon africanus* (Broom, 1912) Nopcsa, 1929 [Berriasian–Valanginian; Kirkwood Formation, Eastern Cape, South Africa]

= *Paranthodon owenii* Nopcsa, 1929 (renamed) [South Africa]

= *Anthodon serrarius* Owen, 1876 (*partim*) [South Africa]

cf. Anthodon serrarius Owen, 1876 (Pareiasauria) [Wuchiapingian; Fort Beaufort, South Africa]

Generic name: *Paranthodon* ← {(Gr.) πᾰρά/para: near + *Anthodon* [ἄνθος/anthos: flower + ὀδών/odon: tooth]}; representing the initial referral of the remains.

The dinosaur material was separated out by Broom in 1912 and was renamed *Paranthodon* by Nopcsa in 1929.

Specific name: *africanus* ← {(Lat.) Āfricānus: African}

Etymology: African, **near-*Anthodon***

Taxonomy: Ornithischia: Thyreophora: Stegosauria: Stegosauridae

Notes: Owen initially identified the fragments as those of *Anthodon* (Owen, 1876).

[References: Raven TJ, Maidment SCR (2018). The systematic position of the enigmatic thyreophoran dinosaur *Paranthodon africanus*, and the use of basal exemplifiers in phylogenetic analysis. *PeerJ* 6: e4529.]

Parapengornis eurycaudatus [Avialae]

> *Parapengornis eurycaudatus* Hu *et al.* 2015 [Aptian; Jiufotang Formation, Liaoning, China]

Generic name: *Parapengornis* ← {(Lat.) para-: prefix + *Pengornis* [(Chin. myth.) Peng鵬: a mythological giant bird] + (Gr.) ὄρνις/ornis: bird}; indicating the close relationship between the new taxon and *Pengornis* (*1).

Specific name: *eurycaudatus* ← {eury- [(Gr.) εὐρυς/eurus: wide, broad] + (Lat.) caudatus [cauda: tail]}; "indicating the unique broad and laterally expanded pygostyle of the new taxon" (*1).

Etymology: **close *Pengornis*** with broad tail

Taxonomy: Enantiornithes

cf. Aves: Ornithothoraces: Enatiornithes:

Pengornithidae [*1]

[References: (*1) Hu H, O'Connor JK, Zhou, Z, Farke, AA (2015). A new species of Pengonithidae (Aves: Enantiornithes) from the Lower Cretaceous of China suggests a specialized scansorial habitat previously unknown in early birds. *PLoS ONE* 10(6): e0126791.]

Paraprotopteryx gracilis [Avialae]

> *Paraprotopteryx gracilis* Zheng *et al*., 2007 [Barremian–Aptian; Huajiying Formation, Fengning, Hebei, China]

Generic name: *Paraprotopteryx* ← {(Lat.) para-: beside + *Protopteryx*: 'primitive feather'}; in reference to the new bird similar to *Protopteryx*.

Specific name: *gracilis* ← {(Lat.) gracilis: pretty, slender}

Etymology: pretty, **near *Protopteryx***

Taxonomy: Theropoda: Avialae: Enantiornithes

[References: Zheng X, Zhang Z, Hou L (2007). A new enantiornithine bird with four long rectrices from the Early Cretaceous of northern Hebei, China. *Acta Geologica Sinica* 81(5): 703–708.]

Pararhabdodon isonensis

> *Pararhabdodon isonensis* Casanovas-Cladellas *et al*., 1993 [Maastrichtian; Conques Formation, Sant Romà d'Abella, Cataluña, Spain]

= *Koutalisaurus kohlerorum* Prieto-Márquez *et al*., 2006 [Maastrichtian; Conques Formation, Abella de la Conca, Cataluña, Spain]

Generic name: *Pararhabdodon* ← {(Gr.) πᾰρά/para: near + *Rhabdodon* [ῥάβδος/rhabdos: rod, stripe + ὀδών/odōn: tooth]}

Specific name: *isonensis* ← {(place-name) Isona + -ensis}

Etymology: **similar to *Rhabdodon*** from Isona

Taxonomy: Ornithischia: Ornithopoda: Hadrosauridae: Tsintaosaurini

cf. Ornithischia: Ornithopoda: Hadrosauroidea [*1]

[References: (*1) Prieto-Marquez A, Gaete R, Rivas G, Galobart A, Boada M (2006). Hadrosauroid dinosaurs from the Late Cretaceous of Spain: *Pararhabdodon isonensis* revisited and *Koutalisaurus kohlerorum*, gen. *et* sp. nov. *Journal of Vertebrate Paleontology* 26(4): 929–943.]

Parasaurolophus walkeri

> *Parasaurolophus walkeri* Parks, 1922 [Campanian; Dinosaur Park Formation, Alberta, Canada]

Generic name: *Parasaurolophus* ← {(Gr.) πᾰρά/para: near + *Saurolophus* [σαῦρος/sauros: lizard + λόφος/lophos: crest]}

Specific name: *walkeri* ← {(person's name) Walker + -ī}; in honor of Sir Byron Edmund Walker.

Etymology: Walker's **near *Saurolophus***

Taxonomy: Ornthischia: Ornithopoda: Hadrosauridae: Lambeosaurinae: Parasaurolophini

Other species:

> *Parasaurolophus tubicen* Wiman, 1931 [late Campanian; lower Kirtland Formation, New Mexico, USA]

> *Parasaurolophus cyrtocristatus* Ostrom, 1961 [late Campanian; Kirtland Formation, New Mexico, USA]

Notes: *Parasaurolophus* is characterized by a remarkable crest. *Parasaurolophus walkeri* is smaller than *Saurolophus osboni*. [*1]

[References: (*1) Parks WA (1922). *Parasaurolophus walkeri*, a new genus and species of crested trachodont dinosaur discovery. *University of Toronto Studies: Geological Series* 13: 5–32.]

Paraxenisaurus normalensis

> *Paraxenisaurus normalensis* Serrano-Brañas *et al*., 2020 [Campanian; Cerro del Pueblo Formation, Coahuila, Mexico]

Generic name: *Paraxenisaurus* ← {(Gr.) πᾰρά/para: near + ξένος/xenos: strange + σαῦρος/sauros: lizard}

Specific name: *normalensis* ← {(institution) Normal + -ensis}; "after the Benemérita Escuela Normal de Coahuila, a teacher training institution".

Etymology: Benemérita Escuela Normal de Coahuila's **strange lizard**

Taxonomy: Theropoda: Ornithomimosauria: Deinocheiridae

Notes: *Paraxenisaurus normalensis* is the first deinocheirid dinosaur found in the Campanian of North America, according to Serrano-Brañas *et al.* (2020).

Pareisactus evrostos

> *Pareisactus evrostos* Párraga & Prieto-Márqez, 2019 [Maastrichtian; Conquès Member, Tremp Formation, Spain]

Generic name: *Pareisactus* ← {(Mod. Gr.) παρείσακτος/pareisaktos: intruder}

Specific name: *evrostos* (*sic*) ← {(Gr.) εύρωστος/eurostos: robust}

Etymology: robust **intruder**

Taxonomy: Ornithischia: Ornithopoda: Rhabdodontidae

[References: Párraga J, Prieto-Márquez A (2019). *Pareisactus evrostos*, a new basal iguanodontian (Dinosauria: Ornithopoda) from the Upper Cretaceous of southwestern Europe. *Zootaxa* 4555(2): 247–258.]

Parksosaurus warreni

>*Thescelosaurus warreni* Parks, 1926 ⇒ *Parksosaurus warreni* (Parks, 1926) Sternberg,

1937 [Maastrichtian; Horseshoe Canyon Formation, Alberta, Canada]
Generic name: *Parksosaurus* ← {(person's name) Parks + (Gr.) σαῦρος/sauros}; in honor of Canadian paleontologist William Arthur Parks.
Specific name: *warreni* ← {(person's name) Warren + -ī}; in honor of Mrs. H. D. Warren, who financially supported the research.

The species epithet, *warreni* (masculine) was emended to *warrenae* (feminine) by Olshevsky (1992), because Warren is a woman. However, in *The Dinosauria 2nd edition*, the original spelling, *warreni*, is used.
Etymology: Warren and **Parks' lizard**
Taxonomy: Ornithischia: Ornithopoda: Thescelosauridae

Paronychodon lacustris
> *Paronychodon lacustris* Cope, 1876 [Campanian; Judith River Formation, Montana, USA]
Generic name: *Paronychodon* ← {(Gr.) παρά/para: near, beside + ὄνυχος/onychos [ὄνυξ/onyx: claw] + ὀδών/odōn: tooth}
Specific name: *lacustris* ← {lacustris: of lake}
cf. (Lat.) lacus, ūs, *m.*: an opening, hollow, lake, pond, pool.
Etymology: **near claw tooth** of lake
Taxonomy: Theropoda: Troodontidae
Other species:
> *Tripriodon caperatus* Marsh, 1889 ⇒ *Paronychodon caperatus* (Marsh, 1889) Olshevsky, 1991 (*nomen dubium*) [Lancian (= latest Maastrichtian); Lance Formation, Wyoming, USA]
Notes: At first, Cope thought the teeth of *Paronychodon* have the general character of those of *Plesiosaurus*, *Elasmosaurus*, *etc.* [*1]. But the same year he realized the teeth represented a carnivorous dinosaur. According to *The Dinosauria 2nd edition* (2004), *Paronychodon lacustris* and *Tripriodon caperatus* Marsh, 1889 are regarded as *nomina dubia*.

[References: (*1) Cope ED (1876). Descriptions of some vertebrate remains from the Fort Union Beds of Montana. *Proceedings of the Academy of Natural Sciences of Philadelphia* 28: 248–261.]

Parrosaurus missouriensis
> *Neosaurus missouriensis* Gilmore & Stewart, 1945 (preoccupied) ⇒*Parrosaurus missouriensis* (Gilmore & Stewart, 1945) Gilmore, 1945 (*nomen dubium*) ⇒ *Hypsibema missouriensis* (Gilmore & Stewart, 1945) Baird & Horner, 1979 (*nomen dubium*) [Maastrichtian; Ripley Formation, Missouri, USA]
Generic name: *Parrosaurus* ← {(person's name) Parr + (Gr.) σαῦρος/sauros: lizard}; in honor of Albert Eide Parr, American zoologist.
Specific name: *missouriensis* ← {(place-name) Missouri + -ensis}
Etymology: **Parr's lizard** from Missouri
Taxonomy: Ornithischia; Ornithopoda; Iguanodontia; Hadrosauroidea
Notes: Brownstein (2018) considers *Parrosaurus* valid, based on new discoveries.

[References: Brownstein CD (2018). The biogeography and ecology of the Cretaceous non-avian dinosaurs of Appalachia. *Palaeontologia Electronica* 21.1.5A: 1–56.]

Parvicursor remotus
> *Parvicursor remotus* Karhu & Rautian, 1996 [Campanian; Barun Goyot Formation, Khulsan, Ömnögovi, Mongolia]
Generic name: *Parvicursor* ← {(Lat.) parvus: small + cursor: runner}
Specific name: *remotus* ← {(Lat.) remōtus: deviating, distant}
Etymology: deviating, **small runner**
Taxonomy: Theropoda: Alvarezsauridae

[References: Karhu AA, Rautian AS (1996). A new family of Maniraptora (Dinosauria: Saurischia) from the Late Cretaceous of Mongolia. *Paleontological Journal Russian Academy of Sciences* 30(5): 583–592.]

Pasquiaornis hardiei [Avialae]
> *Pasquiaornis hardiei* Tokaryk *et al.*, 1997 [Cenomanian; Belle Fourche Formation, Saskatchewan, Canada]
Generic name: *Pasquiaornis* ← {(place-name) Pasquia + (Gr.) ὄρνις/ornis: bird}; "referring to the Pasquia Hills region" [*1].
Specific name: *hardiei* ← {(person's name) Hardie + -ī}; "in honor of Dickson Hardie of Arborfield, Saskatchewan, who donated his collection from the type locality, which included some of the bird specimens" [*1].
Etymology: Hardie's **bird from Pasquia Hills**
Taxonomy: Theropoda: Avialae: Ornithurae: Hesperornithes
cf. Aves: Hesperornithiformes: Baptornithidae [*1]
Other species:
> *Pasquiaornis tankei* Tokaryk *et al.*, 1997 [Cenomanian; Belle Fourche Formation, Saskatchewan, Canada]

[References: (*1) Tokaryk TT, Cumbaa SL, Storer JE (1997). Early Late Cretaceous birds from Saskatchewan, Canada: the oldest diverse avifauna known from North America. *Journal of Vertebrate Paleontology* 17(1): 172–176.]

Patagonykus puertai [Avialae [*1]]
> *Patagonykus puertai* Novas, 1996 [Turonian;

Portezuelo Formation, Neuquén, Argentina]
Generic name: *Patagonykus* ← {(place-name) Patagonia + (Gr.) ὄνυξ/onyx: claw}
Specific name: *puertai* ← {(person's name) Puerta + -ī}; in honor of Pablo Puerta, who is fossil hunter and preparator.
Etymology: Puerta's **claw from Patagonia**
Taxonomy: Theropoda: Alvarezsauridae
cf. Coelurosauria: Maniraptora: Avialae: Metornithes: Alvarezsauridae [(*1)]

[References: (*1) Novas FE, Molnar RE (1996). Alvarezsauridae, Cretaceous basal birds from Patagonia and Mongolia. *Proceedings of the Gondwanan Dinosaur Symposium. Memoirs of the Queensland Museum* 39(3): iv + 489–731; 675–702.]

Patagopteryx deferrariisi [Avialae]
> *Patagopteryx deferrariisi* Alvarega & Bonaparte, 1992 [Santonian; Bajo de la Carpa Formation, Neuquén, Argentina]

Generic name: *Patagopteryx* ← {(place-name) Patagonia + (Gr.) πτέρυξ/pteryx: wing}
Specific name: *deferrariisi* ← {(person's name) de Ferrariis + -ī}; in homage to Professor Oscar de Ferrariis.
Etymology: de Ferrariis' **wing from Patagonia**
Taxonomy: Theropda: Avialae: Ornithothoraces: Euornithes: Patagopterygiformes: Patagopterygidae

Patagosaurus fariasi
> *Patagosaurus fariasi* Bonaparte, 1979 [Callovian; Cañadón Asfalto Formation, Chubut, Argentina]

Generic name: *Patagosaurus* ← {(place-name) Patagonia + (Gr.) σαῦρος/sauros: lizard}
Specific name: *fariasi* ← {(person's name) Farias + -ī}; in honor of villager Ricardo Farias for his information. He found the site near the Farias farm, in Cerro Cóndor North. [(*1)]
Etymology: Farias' **Patagonian lizard**
Taxonomy: Saurischia: Sauropodomorpha: Sauropoda: Cetiosauridae

[References: (*1) Bonaparte JF (1986). The dinosaurs (carnosaurs, allosaurids, sauropods, cetiosaurids) of the Middle Jurassic of Cerro Cóndor (Chubut, Argentina). *Annales de Paléontologie* (*Vert.-Invert.*) 72(4): 325–386.]

Patagotitan mayorum
> *Patagotitan mayorum* Carballido *et al.*, 2017 [Albian; Cerro Barcino Formation, Chubut, Argentina]

Generic name: *Patagotitan* ← {(place-name) Patagonia: southern South America + (Gr. myth.) titan: "symbolic of strength and large size" [(*1)]}
Specific name: *mayorum* ← {(person's name) Mayo + -ōrum}; "in honor of the Mayo family, for their hospitality during fieldwork at the 'La Flecha' ranch" [(*1)].
Etymology: Mayo family's **Patagonian giant**
Taxonomy: Saurischia: Sauropoda: Titanosauria: Eutitanosauria: Lognkosauria

[References: (*1) Calballido JL, Pol D, Otero A, Cerda IA, Salgado L, Garrido AC, Ramezani J, Cúneo NR, Krause JM (2017). A new giant titanosaur sheds light on body mass evolution among sauropod dinosaurs. *Proceedings of the Royal Society B: Biological Sciences* 284(1860): 20171219.]

Pawpawsaurus campbelli
> *Pawpawsaurus campbelli* Lee, 1996 [Albian; Paw Paw Formation, Texas, USA]

Generic name: *Pawpawsaurus* ← {(place-name) Paw Paw Formation + (Gr.) σαῦρος/sauros: lizard}
Specific name: *campbelli* ← {(person's name) Campbell + -ī}; in honor of Mr. Cameron Campbell, discoverer.
Etymology: Campbell's **lizard from Paw Paw Formation**
Taxonomy: Ornithischia: Thyreophora: Ankylosauria: Nodosauridae

[References: Paulina-Carabajal A, Lee Y-N, Jacobs LL (2016). Endocranial morphology of the primitive nodosaurid dinosaur *Pawpawsaurus campbelli* from the Early Cretaceous of North America. *PLoS ONE* 11(3): e0150845.; Lee Y-N (1996). A new nodosaurid ankylosaur (Dinosauria: Ornithischia) from the Paw Paw Formation (late Albian) of Texas. *Journal of Vertebrate Paleontology* 16(2): 232–245.]

Pectinodon bakkeri
> *Pectinodon bakkeri* Carpenter, 1982 ⇒ *Troodon bakkeri* (Carpenter, 1982) Olshevsky, 1991 [late Maastrichtian; Lance Formation, Wyoming, USA]

Generic name: *Pectinodon* ← {(Lat.) pectin- [pecten: comb] + (Gr.) odon: tooth}; referring to "posterior margin with large serrations" of the tooth. [(*1)]
Specific name: *bakkeri* ← {(person's name) Bakker + -ī}; "in honor of Dr. Robert Thomas Bakker, who has contributed considerably to the study of dinosaurs" [(*1)].
Etymology: Bakker's **comb-like tooth**
Taxonomy: Theropoda: Troodontidae
Other species:
> *Pectinodon asiamericanus* Nesov, 1985 (*nomen dubium*) [Cenomanian; Khodzhakul Formation, Navoi, Uzbekistan]

Notes: According to *The Dinosauria 2nd edition*, *Pectinodon bakkeri* was regarded as a synonym of *Troodon formosus*. However, *Pectinodon* was concluded to be valid by Currie & Larson (2013).

[References: (*1) Carpenter K (1982). Baby dinosaurs from the Late Cretaceous Lance and Hell Creek formations and a description of a new species of theropod. *Contributions to Geology, University of Wyoming* 20(2): 123–134.]

Pedopenna daohugouensis [Anchiornithidae]

> *Pedopenna daohugouensis* Xu & Zhang, 2005 [early Callovian; Tiaojishan Formation, Inner Mongolia, China]

Generic name: *Pedopenna* ← {(Lat.) pedo- [pēs: foot] + penna: feather}; referring to the long pennaceous feathers on the feet[*1].

Specific name: *daohugouensis* ← {(Chin. place-name) Dàohŭgōu道虎溝 + -ensis}

Etymology: **foot feather** from Daohugou

Taxonomy: Theropoda: Eumaniraptora: Avialae (?): Anchiornithidae

[References: (*1) Xu X, Zhang F-C (2005). A new maniraptoran dinosaur from China with long feathers on the metatarsus. *Naturwissenschaften* 92(4): 173–177.]

Pegomastax africana

> *Pegomastax africana* Sereno, 2012 [Hettangian–Sinemurian; Elliot Formation, Eastern Cape, South Africa]

Generic name: *Pegomastax* ← {(Gr.) πηγός/pēgos: strong + μάσταξ/mastax, *f.*: jaw}

Specific name: *africana* ← {Africa + (Lat.) -ana (*f.*): pertaining to}: of Africa, African.

"*Pegomastax africanus*" is corrected to *Pegomastax africana*. (Article 34.2)[*1]

Etymology: **strong jaw** of Africa

Taxonomy: Ornithischia: Ornithopoda: Heterodontosauridae

Notes: *Pegomastax* had a short, parrot-shaped beak.

[References: (*1) Sereno PC (2012). Corrigenda: Taxonomy, morphology, masticatory function and phylogeny of heterodontosaurid dinosaurs. *Zookeys* 226: 1–225.]

Peishansaurus philemys

> *Peishansaurus philemys* Bohlin, 1953 (*nomen dubium*) [Campanian; Minhe Formation, Gansu, China[*1]]

Generic name: *Peishansaurus* ← {(Chin. place-name) Peishan北山 [Pinyin: Běishān] + (Gr.) σαῦρος/sauros: lizard}; named after Peishan 'North Mountain'.

Specific name: *philemys* ← {(Gr.) φιλέω: to love + ἐμύς/emys: water turtle}; in reference to the fact that the turtle *Peishanemys latipons* was found at the same site.

Etymology: **Peishan lizard**, fond of water turtle

Taxonomy: Ornithischia

Notes: According to *The Dinosauria 2nd edition*, *Peishanosaurus philemys* (doubtfully ankylosaurian) is considered to be a *nomen dubium*.

[References: (*1) Arbour VM (2014). *Systematics, Evolution, and Biogeography of the ankylosaurid dinosaurs*. Ph.D thesis, University of Alberta, Edmonton.; Bohlin B (1953). Fossil reptiles from Mongolia and Kansu: Reports from the Scientific Expedition to the North-western Provinces of China under the leadership of Dr. Sven Hedin. Vl. Vertebrate Palaeontology. *The Sino-Swedish Expedition Publications* 37: 1–113.]

Pelecanimimus polyodon

> *Pelecanimimus polyodon* Pérez-Moreno *et al.*, 1994 [Barremian; La Huérguina Formation, Las Hoyas, Cuenca, Castilla-La Mancha, Spain]

Generic name: *Pelecanimimus* ← {(Lat.) pelecānus: pelican + (Gr.) μῖμος/mīmos: mimic}; "because of the very long facial part of the skull and the integumentary impressions below the skull, which resemble the gular pouch in the pelican"[*1].

Specific name: *polyodon* ← {(Gr.) πολύς/polys: many + ὀδών/odōn [=ὀδούς/odūs]: tooth}; referring to "its large number of teeth" [*1].

Etymology: **Pelican mimic** with many teeth

Taxonomy: Theropoda: Tetanurae: Ornithomimosauria

Notes: *Pelecanimimus* had about 220 very small teeth in total.[*1]

[References: (*1) Pérez-Moreno BP, Sanz JL, Buscalioni AD, Moratalla JJ, Ortega F, Rasskin-Gutman D (1994). A unique multitoothed ornithomimosaur from the Lower Cretaceous of Spain. *Nature* 370 : 363–367.]

Pellegrinisaurus powelli

> *Pellegrinisaurus powelli* Salgado, 1996 [Campanian; Anacleto Formation (= lower part of Allen Formation), Río Negro, Argentina]

Generic name: *Pellegrinisaurus* ← {(place-name) Lago Pellegrini: Pellegrini Lake + (Gr.) σαῦρος/sauros}; "in reference to lago Pellegrini, the locality where the holotype was collected"[*1].

Specific name: *powelli* ← {(person's name) Powell+ -ī}; "in honor of Jaime E. Powell, in recognition to his work on titanosaurids" [*1].

Etymology: Powell's **reptile from Pellegrini Lake**

Taxonomy: Saurischia: Sauropodomorpha: Sauropoda: Titanosauria: Saltasauridae
cf. Saurischia: Sauropodomorpha: Sauropoda: Titanosauridae[*1]

[References: (*1) Salgado L (1996). *Pellegrinisaurus powelli* nov. gen. *et* sp. (Sauropoda, Titanosauridae) from the Upper Cretaceous of Lago Pellegrini, northwestern Patagonia, Argentina. *Ameghiniana* 33(4): 355–365.]

Peloroplites cedrimontanus

> *Peloroplites cedrimontanus* Carpenter *et al.*, 2008 [Aptian–Albian; Cedar Mountain Formation, Utah, USA]

Generic name: *Peloroplites* ← {(Gr.) πέλωρος/pelōros: monstrous, gigantic + ὁπλίτης/hoplitēs: heavily armed soldier}; "monstrous heavy one"(*1).

Specific name: *cedrimontanus* ← {(Lat.) cedri- [cedrus: cedar] + montānus: mountain}; referring to the Cedar Mountain Formation in eastern Utah.

Etymology: **monstrous heavy armed soldier** from Cedar Mountain Formation

Taxonomy: Ornithischia: Thyreophora: Ankylosauria: Nodosauridae (*1)

[References: (*1) Carpenter K, Bartlett J, Bird J, Barrick R (2008). Ankylosaurs from the Price River Quarries, Cedar Mountain Formation (Lower Cretaceous), east-central Utah. *Journal of Vertebrate Paleontology* 28(4): 1089–1101.]

Pelorosaurus brevis

> *Cetiosaurus brevis* Owen, 1842 ⇒ *Morosaurus brevis* (Owen, 1842) Lydekker, 1889 ⇒*Pelorosaurus brevis* (Owen, 1842) von Huene, 1927 [Valanginian–Hauterivian; Tunbridge Wells Sand Formation, Cuckfield Quarry, West Sussex, UK]

= *Cetiosaurus conybeari* Melville, 1849 ⇒ *Pelorosaurus conybearei* (Melville, 1849) Mantell, 1850 ⇒ *Ornithopsis conybearei* (Melville, 1849) von Huene, 1929 [Valanginian–Hauterivian; Tunbridge Wells Sand Formation, Cuckfield Quarry, West Sussex, UK]

Generic name: *Pelorosaurus* ← {(Gr.) πέλωρος/pelōros: monstrous + σαῦρος/sauros: lizard}

Specific name: *brevis* ← {(Lat.) brevis: short, small}

Etymology: short, **monstrous lizard**

Taxonomy: Saurischia: Sauropodomorpha: Sauropoda: Titanosauriformes

Other species: (misassigned species)

> *Pelorosaurus becklesii* Mantell, 1852 ⇒ *Morosaurus becklesi* (Mantell, 1852) Marsh, 1889 ⇒ *Haestasaurus becklesii* (Mantell, 1852) Upchurch *et al.*, 2015 [Berriasian–Valanginian; Hastings, UK]

> *Ischyrosaurus manseli* Hulke, 1874 ⇒ *Ornithopsis manseli* (Hulke, 1874) Lydekker, 1888 ⇒ *Pelorosaurus manseli* (Hulke, 1874) von Huene, 1909 (*nomen dubium*) [Tithonian; Kimmeridgian Clay Formation, Kimmeridge Bay, UK]

> *Cetiosaurus humerocristatus* Hulke, 1874 ⇒ *Pelorosaurus humerocristatus* (Hulke, 1874) Sauvage, 1887 ⇒ *Ornithopsis humerocristatus* (Hulke, 1874) Lydekker, 1888 ⇒ *Duriatitan humerocristatus* (Hulke, 1874) Barrett *et al.*, 2010 [Kimmeridgian; Kimmeridge Clay Formation, Dorset, UK]

> *Oplosaurus armatus* Gervais, 1852 (*nomen dubium*) ⇒ *Hoplosaurus armatus* (Gervais, 1852) Lydekker, 1888 ⇒ *Pelorosaurus armatus* (Gervais, 1852) Lydekker, 1889 [Valanginian–Barremian; Wealden Group, UK]

> *Ornithopsis hulkei* Seeley, 1870 (Lectotype) ⇒ *Bothriospondylus magnus* (Seeley, 1870) Owen, 1875 ⇒ *Chondrosteosaurus magnus* (Seeley, 1870) Owen, 1876 ⇒ *Pelorosaurus hulkei* (Seeley, 1870) Lydekker, 1889 ⇒ *Hoplosaurus hulkei* (Seeley, 1870) Lydekker, 1893 ⇒ *Eucamerotus hulkei* (Seeley, 1870) Sebaschan, 2005 [Barremian; Wessex Formation, Isle of Wight, England, UK]

= *Ornithopsis hulkei* Seeley, 1870 (*partim*) (syntype) ⇒ *Bothriospondylus elongatus* (Seeley, 1870) Owen, 1875 [Valanginian; Tunbridge Wells Formation, East Sussex, England, UK]

= *Ornithopsis eucamerotus* Hulke, 1882 [Barremian; Isle of Wight, UK] (according to Newton, 1889, Sauvage, 1895)

> *Ornithopsis leedsii* Hulke, 1887 ⇒ *Pelorosaurus? leedsi* (Hulke, 1887) Lydekker, 1895 ⇒ *Cetiosaurus leedsi* (Hulke, 1887) Woodward, 1905* ⇒ *Cetiosauriscus leedsii* (Hulke, 1887) von Huene, 1927* [Callovian-Oxfordian; Oxford Clay, Peterborough, UK] *referred specimen

> *Neosodon* Sauvage, 1876 (*nomen dubium*) ⇒ *Iguanodon praecursor* (Sauvage, 1876) Sauvage, 1888 ⇒ *Pelorosaurus praecursor* (Sauvage, 1876) Romer, 1956 [Tithonian; Wimereux, Pas-de-Calais, France]

> *Dinodocus mackesoni* Owen, 1884 ⇒ *Pelorosaurus mackesoni* (Owen, 1884) Steel, 1970 (*nomen dubium*) [Aptian; Atherfield Clay Formation, UK]

> *Gigantosaurus megalonyx* Seeley, 1869 ⇒ *Pelorosaurus megalonyx* (Seeley, 1869) von Huene, 1909 (*nomen dubium*) [Kimmeridgian–Tithonian; Kimmeridge Clay Formation, UK]

Notes: According to *The Dinosauria 2nd edition*, *Cetiosaurus brevis* Owen, 1842 (*partim*) was considered as a *nomen dubium*, but now *C. brevis* and *C. conybeari* are considered to be synonyms of *Pelorosaurus brevis* (Owen, 1842).

[References: (*1) Mantell GA (1850). XVI.On the *Pelorosaurus*: an undescribed gigantic terrestrial reptile whose remains are associated with those of the iguanodon and other saurians in the strata of Tilgate Forest, in Sussex. *Philosophical Transactions of the Royal Society of London* 140: 379–390.]

Penelopognathus weishampeli

> *Penelopognathus weishampeli* Godefroit *et al.*, 2005 [Aptian; Bayan Gobi Formation, Inner Mongolia, China]

Generic name: *Penelopognathus* ← {penelopo- [(Gr.) πηνέλοψ/penelops: wild duck] + (Gr.) γνάθος/gnathos: jaw} (*1)

"Penelope is also the name of Odysseus' wife, forced to fend off suitors while her husband is away fighting at Troy" (*1).

Specific name: *weishampeli* ← {(person's name) Weishampel + -ī}; "in honor of D. B. Weishampel, for his important contribution in the knowledge of the duck-billed dinosaurs" (*1).

Weishampel is an American palaeontologist, and one of the authors of *The Dinosauria 2nd edition* (2004).

Etymology: Weishampel's **wild duck jaw**

Taxonomy: Ornithischia: Ornithopoda: Iguanodontia: Hadrosauroidea

[References: (*1) Godefroit P, Li H, Shang Ch-Y (2005). A new primitive hadrosauroid dinosaur from the Early Cretaceous of Inner Mongolia (P. R. China). *Comptes Rendus Palevol* 4(8): 697–705.]

Pengornis houi [Avialae]

> *Pengornis houi* Zhou, *et al.*, 2008 [Aptian; Jiufotang Formation, Liaoning, China]

Generic name: *Pengornis* ← {(Chin. myth) Péng鵬: Chinese mythological bird + (Gr.) ὄρνις/ornis: bird}

Specific name: *houi* ← {(Chin. person's name) Hóu 侯 + -ī}; "in honor of Lianhai Hou, a pioneering palaeo-ornithologist" (*1).

Etymology: Hou's **Peng bird**

Taxonomy: Avialae: Enantiornithes: Pengornithidae

[References: (*1) Zhou Z, Clarke J, Zhang F (2008). Insight into diversity, body size and morphological evolution from the largest Early Cretaceous enantiornithine bird. *Journal of Anatomy* 212: 565–577.]

Pentaceratops sternbergii

> *Pentaceratops sternbergii* Osborn, 1923 [Campanian; Fruitland Formation, San Juan, New Mexico, USA]

= *Pentaceratops fenestratus* Wiman, 1930 [Campanian; Kirtland Formation, San Juan, New Mexico, USA]

Generic name: *Pentaceratops* ← {(Gr.) πέντε/pente: five, 5 + κερατ-/kerat- [κέρας/keras: horn] + ὤψ/ōps: face}; referring to its five horns: "one nasal horn, two prominent anteroverted postorbital horns, two lateral jugal osseous horns" (*1).

Specific name: *sternbergii* ← {(person's name) Sternberg + -iī}; "in honor of Charles Hazelius Sternberg, veteran explorer and discoverer in the fossil beds of western America" (*1).

Etymology: Sternberg's **five-horned ceratopsians**

Taxonomy: Ornithischia: Ceratopsia: Ceratopsidae: Chasmosaurinae

Other species:

> *Pentaceratops aquilonius* Longrich, 2014 (*?nomen dubium**) [Campanian; Dinosaur Park Formation, Alberta, Canada]
> *according to Jordan *et al.* (2016).

[References: (*1) Osborn HF (1923). A new genus and species of Ceratopsia from New Mexico, *Pentaceratops sternbergii*. *American Museum Novitates* 93: 1–3.]

Peteinosaurus zambellii [Pterosauria]

> *Peteinosaurus zambellii* Wild, 1978 [Alaunian (= middle Norian); Zorzino Limestone Formation, Cene, Bergamo, Lombardia, Italy]

Generic name: *Peteinosaurus* ← {(Gr.) πετεινός/peteinos: winged + σαῦρος/sauros: lizard}

Specific name: *zambellii* ← {(person's name) Zambelli + -ī}; in honor of Rocco Zambelli, the curator of the Bergamo natural history museum.

Etymology: Zambelli's **winged lizard**

Taxonomy: Archosauria: Ornithodira: Pterosauria: Eopterosauria

Petrobrasaurus puestohernandezi

> *Petrobrasaurus puestohernandezi* Filippi *et al.*, 2011 [Coniacian–Santonian; Plottier Formation, Neuquén, Argentina]

Generic name: *Petrobrasaurus* ← {(company) Petrobras: Brazilian oil company + (Gr.) σαῦρος/sauros: lizard}; "in recognition of the Petrobras oil company for its constant collaboration in the maintenance and preservation of the palaeontological heritage in the area of Rincón de los Sauces" (*1).

Specific name: *puestohernandezi* ← {(place-name) the Puesto Hernández + -ī}; "referring to the Puesto Hernández oil field, where the fossil remains were found" (*1).

Etymology: **Petrobras' lizard** from Puesto Hernández oil field

Taxonomy: Saurischia: Sauropodomorpha:

Sauropoda: Titanosauriformes: Titanosauria

[References: (*1) Filippi LS, Canudo JI, Salgado JL, Garrido A, García R, Cerda I, Otero A (2011). A new sauropod titanosaur from the Plottier Formation (Upper Cretaceous) of Patagonia (Argentina). *Geologica Acta* 9(1): 1–12.]

Philovenator curriei

> *Philovenator curriei* Xu *et al.*, 2012 [Campanian; Wulansuhai Formation, Inner Mongolia, China]

Generic name: *Philovenator* ← {(person's name) Dr. Philip J. ("Phil") Currie | (Gr.) φιλειν/philein: to love + (Lat.) vēnātor: hunter}; "in honor of Dr. Philip J. Currie for his contributions to the study of maniraptoran dinosaurs, including the initial description of IVPP V 10597 (holotype)"(*1).

Specific name: *curriei* ← {(person's name) Currie + -ī}; in honor of Phillip J. Currie.

Etymology: Philip Currie's **lover of the hunt**

Taxonomy: Theropoda: Coelurosauria: Maniraptora: Troodontidae (*1)

[References: (*1) Xu X, Zhao Q, Sullivan C, Tan Q-W, Sander M, Ma Q-Y (2012). The taxonomy of the troodontid IVPP V 10597 reconsidered. *Vertebrata PalAsiatica* 50(2): 140–150.]

Phuwiangosaurus sirindhornae

> *Phuwiangosaurus sirindhornae* Martin *et al.*, 1994 [Barremian–Aptian; Sao Khua Formation, Khon Kaen, Thailand]

Generic name: *Phuwiangosaurus* ← {(place-name) Phu Wiang + (Gr.) σαῦρος/sauros: lizard}

Specific name: *sirindhornae* ← {(person's name) Sirindhorn + -ae}; in honor of Princess Maha Chakri Sirindhorn of Thailand.

Etymology: **Phu Wiang lizard** of Princess Sirindhorn

Taxonomy: Saurischia: Sauropodomorpha: Sauropoda: Euhelopodidae

Notes: *Phuwiangosaurus* was originally assigned to Titanosauria.

[References: Suteethorn S, Le Loeuff J, Buffetaut E, Suteethorn V (2010). Description of topotypes of *Phuwiangosaurus sirindhornae*, a sauropod from the Sao Khua Formation (Early Cretaceous) of Thailand, and their phylogenetic implications. *Neues Jahrbuch für Geologie und Paläontologie - Abhandlungen* 256(1): 109–121.]

Phuwiangvenator yaemniyomi

> *Phuwiangvenator yaemniyomi* Samathi *et al.*, 2019 [Barremian; Sao Khua Formation, Khon Kaen, Thailand]

Generic name: *Phuwiangvenator* ← {(place-name) Phu Wiang Mountain + (Lat.) venator: hunter}

Specific name: *yaemniyomi* ← {(person's name) Yaemniyom + -ī}; "in honor of Sudham Yaemniyom, former geologist of the Department of Mineral Resources, Bangkok, who found the first dinosaur bone of Thailand in 1976 at Phu Wiang Mountain".

Etymology: Yaemniyom's **hunter of Phu Wiang**

Taxonomy: Theropoda: Tetanurae: Coelurosauria: Megaraptora

[References: Samathi A, Chanthasit P, Martin Sander P (2019). Two new basal coelurosaurian theropod dinosaurs from the Lower Cretaceous Sao Khua Formation of Thailand. *Acta Palaeontologica Polonica* 64(2): 239–260.]

Phyllodon henkeli

> *Phyllodon henkeli* Thulborn, 1973 (*nomen dubium**) [Kimmeridgian; unnamed unit, Guimarota lignite mine, Leiria, Portugal]
> *according to Sues & Norman (1990).

Generic name: *Phyllodon* ← {(Gr.) φύλλον/phyllon: leaf + ὀδών/odōn: tooth}

Specific name: *henkeli* ← {(person's name) Henkel + -ī}; in honor of Dr. Siegfried Henkel, paleontologist at Freie Universität Berlin.

Etymology: Henkel's **leaf tooth**

Taxonomy: Ornithischia: Neornithischia

Notes: According to *The Dinosauria 2nd edition* (2004), *Phyllodon* is considered a *nomen dubium,* although Rauhut (2001) indicated that *Phyllodon* could be a valid genus.

Piatnitzkysaurus floresi

> *Piatnitzkysaurus floresi* Bonaparte, 1979 [Toarcian–Bajocian; Cañadón Asfalto Formation, Chubut, Argentina]

Generic name: *Piatnitzkysaurus* ← {(person's name) Piatnitzky + σαῦρος/sauros: lizard}; in honor of Alejandro Mateievich Piatnitzky, a Russian-born Argentine geologist.

Specific name: *floresi* ← {(person's name) Flores + -ī}; in honor of Miguel Flores, geologist.

Etymology: Flores and **Piatnitzky's lizard**

Taxonomy: Theropoda: Megalosauroidea: Piatnitzkysauridae

Notes: The holotype is a subadult, estimated to be about 4.5 m long. Piatnitzkysauridae Carrano *et al.*, 2012 consists of *Piatnitzkysaurus*, *Marshosaurus* and *Condorraptor*. Spinosauroidea, Torvosauroidea, and Spinosauria are synonyms of Megalosauroidea.

[References: Bonaparte JF (1979). Dinosaurs: a Jurassic assemblage from Patagonia. *Science* 205(4413): 1377–1379.]

Pilmatueia faundezi

> *Pilmatueia faundezi* Coria *et al.*, 2018 [Valanginian; Mulichinco Formation, Pilmatué, Las Lajas, Neuquén, Argentina]

Generic name: *Pilmatueia* ← {(place-name) Pilmatué + -ia}; "referring to the Pilmatué locality, where the material was collected" (*1).

Specific name: *faundezi* ← {(person's name) Faúndez + -ī}; "in recognition to Mr. Ramón Faúndez – manager of the Museo Municipal de Las Lajas – for supporting the project since 2009" (*1).

Etymology: Faúndez's **one from Pilmatué**

Taxonomy: Saurischia: Sauropodomorpha: Sauropoda: Diplodocoidea: Dicraeosauridae

[References: (*1) Coria RA, Windholz GJ, Ortega F, Currie PJ (2019). A new dicraeosaurid sauropod from the Lower Cretaceous (Mulichinco Formation, Valanginian, Neuquén Basin) of Argentina. *Cretaceous Research* 93: 33–48.]

Pinacosaurus grangeri

> *Pinacosaurus grangeri* Gilmore, 1933 [Campanian; Djadokhta Formation, Ömnögovi, Mongolia]

= *Pinacosaurus ninghsiensis* Young, 1935 [Late Cretaceous; North Alashan Desert, Ningxia, China]

= *Syrmosaurus viminicaudus* Maleev, 1952 [Campanian; Bayn Dzak Member, Djadokhta Formation, Ömnögovi, Mongolia]

?= *Heishansaurus pachycephalus* Bohlin, 1953 (*nomen dubium*) [Campanian–Maastrichtian; Minhe Formation, Gansu, China]

Generic name: *Pinacosaurus* ← {(Gr.) πινακο-/pinako- [πίναξ/pinax: plank] + σαῦρος/sauros: lizard}; in reference to "numerous osseous scutes" covering the head. (*1)

Specific name: *grangeri* ← {(person's name) Granger + -ī}; in honor of Walter Willis Granger, who collected "a badly crushed skull and jaws and a few scattered dermal bones", 1923. (*1)

Etymology: Granger's **plank lizard**

Taxonomy: Ornithischia: Thyreophora: Ankylosauria: Ankylosauridae: Ankylosaurinae (*2)

Other species:

> *Pinacosaurus mephistocephalus* Godefroit *et al.*, 1999 [Campanian; Djadokhta Formation, Inner Mongolia, China]

[References: (*1) Gilmore CW (1933). Two new dinosaurian reptiles from Mongolia with notes on some fragmentary specimens. *American Museum Novitates* (679): 1–20.; (*2) Maryañska T (1971). New data on the skull of *Pinacosaurus grangeri* (Ankylosauria) *Palaeontologia Polonica* 25: 45–53.]

Pisanosaurus mertii [Silesauridae]

> *Pisanosaurus mertii* Casamiquela, 1967 [Carnian; Ischigualasto Formation, La Rioja, Argentina]

Generic name: *Pisanosaurus* ← {(person's name) Pisano + (Gr.) σαῦρος/sauros: lizard}; "in honor of Argentine paleontologist Juan A. Pisano, work companion in the Facultad de Ciencias Naturales and the Museo de La Plata, recently deceased" (*1).

Specific name: *mertii* ← {(person's name) Merti + -ī}; "in honor of Araucanian naturalist Carlos Merti, recently departed" (*1).

Etymology: Merti and **Pisano's lizard**

Taxonomy: Dinosauriformes: Dracohors: Silesauridae (*2)

cf. Ornithischia: Ornithopoda: Pisanosauridae (*1)

Notes: *Pisanosaurus mertii* has been known as a dinosaur until 2017, but recently it is considered to be a member of the non-dinosaurian Silesauridae, according to Agnolín & Rozadilla (2017) (*2), Baron *et al.* (2017) and Baron (2018).

[References: (*1) Casamiquela RM (1967). Un Nuevo dinosaurio ornitisquio triásico (*Pisanosaurus mertii*; Ornithopoda) de la Formación Ischigualasto, Argentina. *Ameghiniana* 4(2): 47–64.; (*2) Agnolín FL, Rozadilla S (2017). Phylogenetic reassessment of *Pisanosaurus mertii* Casamiquela, 1967, a basal dinosauriform from the Late Triassic of Argentina. *Journal of Systematic Palaeontology* 16(10): 853–879.]

Pitekunsaurus macayai

> *Pitekunsaurus macayai* Filippi & Garrido, 2008 [Campanian; Anacleto Formation, Neuquén, Argentina]

Generic name: *Pitekunsaurus* ← {(Mapuche) pitëkun: to discover + (Gr.) σαῦρος/sauros: reptile}

Specific name: *macayai* ← {(person's name) Macaya + -ī}; in honor of oil company explorer Sr. Luis Macaya, who found the fossil in 2004.

Etymology: **reptile discovered** by Macaya

Taxonomy: Saurischia: Sauropodomorpha: Titanosauria: Aeolosauridae

[References: Filippi LS, Garrido AC (2008). *Pitekunsaurus macayai* gen. *et* sp. nov., nuevo titanosaurio (Saurischia, Sauropoda) del Cretácico Superior de la Cuenca Neuquina, Argentina [new titanosaur (Saurischia, Sauropoda) from Upper Cretaceous Neuquén Basin, Argentina]. *Ameghiniana* 45(3): 575–590.]

Piveteausaurus divesensis

> *Eustreptospondylus divesensis* Walker, 1964 ⇒ *Piveteausaurus divesensis* (Walker, 1964) Taquet & Welles, 1977 ⇒ *Proceratosaurus divesensis* (Walker, 1964) Paul, 1988

[Callovian; Marnes de Dives, Calvados, France]

Generic name: *Piveteausaurus* ← {(person's name) Piveteau + (Gr.) σαῦρος/sauros: lizard}; in honor of paleontologist Jean Piveteau.

Specific name: *divesensis* ← {(place-name) Dives + -ensis}

According to Buffetaut & Enos (1992), "Piveteau clearly indicated that it was found in the Vaches Noires 'near Dives', whereas Taquet & Welles said it came from Dives, some kilometers to the west".

Etymology: **Piveteau's lizard** from Dives

Taxonomy: Theropoda: Tetanurae *incertae sedis* (*1)

[References: Taquet P, Welles SP (1977). Redescription du crâne de dinosaur théropode de Dives (Normandie) [Redescription of a theropod dinosaur skull from Dives (Normandy). *Annales de Paléontologie (Vertébrés)* (in French)]. 63(2): 191–206.; (*1) Holtz TR Jr, Molnar RE, Currie PJ (2004). Basal Tetanurae. In: Weishampel DB, Dodson P, Osmólska H (eds), *The Dinosauria 2nd edition*, University of California Press, Berkeley: 71–110.]

Planicoxa venenica

> *Planicoxa venenica* DiCroce & Carpenter, 2001 [Barremian; Poison Strip Sandstone Member, Cedar Mountain Formation, Utah, USA]

Generic name: *Planicoxa* ← {(Lat.) plani- [planus: flat, level] + coxa, *f.*: hip}; referring to the flat appearance of the ilium, the defining characteristic (*1).

Specific name: *venenica* ← {(Lat.) venēnum: poison}; referring to the Poison Strip Sandstone Member of the Cedar Mountain Formation, Utah (*1).

Etymology: **flat hip** from Poison Strip Member

Taxonomy: Ornithischia: Ornithopoda: Iguanodontia

Other species:

> *Camptosaurus depressus* Gilmore, 1909 ⇒ *Planicoxa depressa* (Gilmore, 1909) Carpenter & Wilson, 2008 ⇒ *Osmakasaurus depressus* (Gilmore, 1909) McDonald, 2011 [Barremian; Lakota Formation, South Dakota, USA]

Notes: A second species, *Planicoxa depressa* Carpenter & Wilson (2008) from South Dakota was assigned to its own genus, *Osmakasaurus*.

[References: (*1) DiCroce K, Carpenter K (2001). New ornithopod from the Cedar Mountain Formation (Lower Cretaceous) of Eastern Utah. In: Tanke D, Carpenter K (eds), *Mesozoic Vertebrate Life*. Indiana University Press, Bloomington: 183–196.]

Plateosauravus cullingworthi

> *Plateosaurus cullingworthi* Haughton, 1924 (*nomen dubium*) ⇒*Plateosauravus cullingworthi* (Haughton, 1924) von Huene, 1932 [Norian–Rhaetian; Elliot Formation, Eastern Cape, South Africa]

Generic name: *Plateosauravus* ← {*Plateosaurus* + (Lat.) avus: grandfather}

Specific name: *cullingworthi* ← {(person's name) Cullingworth + -ī}; in honor of collector T. L. Cullingworth.

Etymology: Cullingworth's, **grandfather of *Plateosaurus***

Taxonomy: Saurischia: Sauropodomorpha: Prosauropoda

Other species:

> *Plateosaurus stormbergensis* Broom, 1915 (*nomen dubium*) ⇒ *Plateosauravus stormbergensis* (Broom, 1915) Haughton & Brink, 1954 [Norian–Rhaetian; Elliot Formation, South Africa]

Plateosaurus engelhardti

> *Plateosaurus engelhardti* Meyer, 1837 [Norian; Feuerletten Formation, Bayern, Germany]

= *Dinosaurus gresslyi* Rütimeyer, 1856 (*nomen nudum*) ⇒ *Gresslyosaurus ingens* (Rütimeyer, 1856) Rütimeyer, 1857 ⇒ *Plateosaurus ingens* (Rütimeyer, 1856) Galton, 1986 [Sevatian; Knollenmergel Formation, Liestal, Switzerland]

= *Megalosaurus cloacinus* Quenstedt, 1858 ⇒ *Plateosaurus cloacinus* (Quenstedt, 1858) von Huene, 1907–1908 ⇒ *Gresslyosaurus cloacinus* (Quenstedt, 1858) von Huene, 1932 [Rhaetian; Exter Formation, Bebenhausen, Germany]

= *Dimodosaurus poligniensis* Pidancet & Chopard, 1862 ⇒ *Plateosaurus poligniensis* (Pidancet & Chopard, 1862) von Huene, 1907–1908 [Norian; Marnes Irisées Supérieures Formation, Poligny, France]

= *Megalosaurus obtusus* Henry, 1846 ⇒ *Plateosaurus obtusus* (Henry, 1846) von Huene, 1907–1908 [Rhaetian; Moissey, France]

= *Plateosaurus erlenbergiensis* von Huene, 1905 (*nomen dubium*) [Rhaetian; Trossingen Formation, Germany]

= *Gresslyosaurus plieningeri* von Huene, 1905 ⇒ *Plateosaurus plieningeri* (von Huene, 1905) von Huene, 1932 [Norian; Trossingen Formation, Baden-Württemberg, Germany]

= *Plateosaurus quenstedti* von Huene, 1905 (*nomen dubium*) [Rhaetian; Trossingen Formation, Germany]

= *Plateosaurus reiningeri* von Huene, 1905 ⇒ *Gresslyosaurus reiningeri* (von Huene, 1905) von Huene, 1926 ⇒ *Pachysaurus reiningeri* (von Huene, 1905) von Huene, 1932 [Norian; Trossingen Formation, Germany]

= *Plateosaurus ornatus* von Huene, 1905 [Rhaetian; Exter Formation, Baden-Württemberg, Germany]
= *Pachysaurus magnus* von Huene, 1905 ⇒ *Pachysauriscus magnus* (von Huene, 1905) Kuhn, 1959 ⇒ *Gresslyosaurus magnus* (von Huene, 1905) Steel, 1970 [Rhaetian; Trossingen Formation, Baden-Württemberg, Germany]
= *Pachysaurus ajax* von Huene, 1907–1908 ⇒ *Pachysauriscus ajax* (von Huene, 1907–1908) Kuhn, 1959 ⇒ *Gresslyosaurus ajax* (von Huene, 1907–1908) Steel, 1970 [Norian; Trossingen Formation, Baden-Württemberg, Germany]
= *Gresslyosaurus robustus* von Huene, 1907–1908 ⇒ *Plateosaurus robustus* (von Huene, 1907–1908) von Huene, 1932 [Norian; Trossingen Formation, Rhothen Graben, Germany]
= *Gresslyosaurus torgeri* Jaekel, 1911 [Rhaetian; Trossingen Formation, Sachsen-Anhalt, Germany]

Generic name: *Plateosaurus* ← {(Gr.) πλᾰτέως/plateōs* + σαῦρος/sauros: lizard}**

The original auther didn't provide a meaning.

*The genitive-case form plateos of Greek platys 'broad'.

**According to Agassiz (1844:34), [πλατη: pala + σαυρος: lacerta]. [*1]

Specific name: *engelhardti* ← {(person's name) Engelhardt + -ī}; in honor of Johann Friedrich Engelhardt, who discovered the first fossils of *Plateosaurus*.

Etymology: Engelhardt's **broad-built lizard**

Taxonomy: Saurischia: Sauropodomorpha: Prosauropoda: Plateosauria: Plateosauridae

Other species:

> *Plateosaurus bavaricus* Fraas *vide* Sandberger, 1894 ⇒ *Zanclodon bavaricus* (Fraas, *vide* Sandberger, 1894) (*nomen dubium*) [Rhaetian; Keuper Group, Bayern, Germany]
> *Massospondylus carinatus* Owen, 1854 ⇒ *Plateosaurus carinatus* (Owen, 1854) Paul, 1988 [Hettangian–Sinemurian; Elliot Formation, Free State,, South Africa]
> *Thecodontosaurus elisae* Sauvage, 1907 ⇒ *Plateosaurus elizae* (sauvage, 1907) von Huene, 1908 ⇒ *Thecodontosaurus elizae* (Sauvage, 1907) von Huene, 1914 [Rhaetian; Grès de Infralias Formation, France]
> *Sellosaurus gracilis* Meyer; von Huene, 1907–1908 ⇒ *Plateosaurus gracilis* (von Huene, 1907–1908) von Huene, 1926 [Norian; Löwenstein Formation, Stuttgart, Germany]
= *Thecodontosaurus hermannianus* von Huene, 1905 ⇒ *Sellosaurus hermannianus* (von Huene, 1905) von Huene, 1915 [Norian; Löwenstein Formation, Stuttgart, Germany]
> *Teratosaurus minor* von Huene, 1908 ⇒ *Efraasia minor* (von Huene, 1908) Yates, 2003 [Alaunian; Löwenstein Formation, Baden-Württemberg, Germany]
= *Plateosaurus fraasi* von Huene, 1907–1908 ⇒ *Sellosaurus fraasi* (von Huene, 1907–1908) von Huene, 1914 [Alaunian; Löwestein Formation, Baden-Württemberg, Germany]
= *Thecodontosaurus diagnosticus* Fraas, 1912 ⇒ *Palaeosaurus diagnosticus* (Fraas, 1912) von Huene, 1932 (preoccupied) ⇒ *Palaeosauriscus diagnosticus* (Fraas, 1912) Kuhn, 1959 [Alaunian; Löwenstein Formation, Baden-Württemberg, Germany]
> *Plateosaurus longiceps* Jackel, 1913 [Rhaetian; Trossingen Formation, Sachsen-Anhalt, Germany]
= *Plateosaurus trossingensis* Fraas, 1913 [Rhaetian; Trossingen Formation, Baden-Württemberg, Germany]
= *Plateosaurus integer* Fraas *vide* von Huene, 1915 [Rhaetian; Trossingen Formation, Germany]
= *Plateosaurus fraasianius* von Huene, 1932 [Rhaetian or Norian; Trossingen Formation, Germany]
> *Plateosaurus cullingworthi* Haughton, 1924 (*nomen dubium*) ⇒ *Plateosauravus cullingworthi* (Haughton, 1924) von Huene, 1932 [Norian–Rhaetian; Elliot Formation, Eastern Cape, South Africa]
> *Plateosaurus stormbergensis* Broom, 1915 (*nomen dubium*) ⇒ *Plateosauravus stormbergensis* (Broom, 1915) Haughton & Brink, 1954 (*nomen dubium*) [Norian–Rhaetian; Elliot Formation, South Africa]

Notes: Its nickname is Schwäbischer Lindwurm (Swabian lindworm), because the abundance of fossils of *Plateosaurus* has been found in Swabia, Germany. In 2019, *Plateosaurus trossingensis* Fraas, 1913 was designated as the type species [*2], and, *P. longiceps* and *P. gracilis* are accepted as valid.

[References: (*1) Moser M (2003). *Plateosaurus engerhardti* Meyer, 1837 (Dinosauria: Sauropodomorpha) aus dem Feuerletten (Mittelkeuper; Obertrias) von Bayern. *Zitteliana Reihe B, Abhandlungen der Bayerischen Staatssammlung für Paläontologie und Geologie* 24: 1–186.; (*2) Opinion 2435 (Case 3560) –*Plateosaurus* Meyer, 1837 (Dinosauria, Sauropodomorpha): new type species designated. *The Bulletin of Zoological Nomenclature* 76(1): 144-145. 2019.]

Platyceratops tatarinovi

> ?*Bagaceratops rozhdestvenskyi* Maryańska & Osmólska, 1975 [Campanian; Barun Goyot Formation, Mongolia]

= *Platyceratops tatarinovi* Alifanov, 2003 [Campanian; Barun Goyot Formation, Ömnögovi, Mongolia]

Generic name: *Platyceratops* ← {(Gr.) πλατύς/platys: flat + ceratops [κερατ- (κεράς/ceras: horn) + ὤψ/ōps: face]}

Specific name: *tatarinovi* ← {(person's name) Tatarinov + -ī}; in honor of paleontologist Tatarinov.

Etymology: Tatarinov's **flat Ceratops**

Taxonomy: Ornithischia: Marginocephalia: Ceratopsia: Bagaceratopidae

Notes: *Platyceratops tatarinovi* is possibly a junior synonym of *Bagaceratops rozhdestvenski* Maryańska & Osmólska, 1975. (*1)

[References: (*1) Czepiński L (2019). Ontogeny and variation of a protoceratopsid dinosaur *Bagaceratops rozhdestvenskyi* from the Late Cretaceous of the Gobi Desert. *Historical Biology* 32: 1394–1421.]

Platypelta coombsi

> *Platypelta coombsi* Penkalski, 2018 [Campanian; Dinosaur Park Formation, Alberta, Canada]

Generic name: *Platypelta* ← {(Gr.) πλατύς/platys: wide + (Lat.) pelta [= (Gr.) πέλτη/platē]: small shield}; in reference to the broad osteoderms.

Specific name: *coombsi* ← {(person's name) Coombs + -ī}; in honor of Walter P. Coombs Jr.

Etymology: Coombs' **wide small shield**

Taxonomy: Ornithischia: Ankylosauridae: Ankylosaurinae

[References: Penkalski P (2018). Revised systematics of the armoured dinosaur *Euoplocephalus* and its allies. *Neues Jahrbuch für Geologie und Paläontologie – Abhandlungen* 287(3): 261–306.]

Plesiohadros djadokhtaensis

> *Plesiohadros djadokhtaensis* Tsogtbaatar *et al*., 2014 [late Santonian; Djadokhta Formation, Alag Teg, Ömnögovi, Mongolia]

Generic name: *Plesiohadros* ← {(Gr.) πλησίος/plēsios: near, close to + Hadros}; in reference to its close proximity to Hadrosauridae (the heavy lizards).

Specific name: *djadokhtaensis* ← {(stratal name) the Djadokhta Formation + -ensis}

Etymology: **near-hadrosaurid** from Djadokhta

Taxonomy: Ornithischia: Ornithopoda: Hadrosauriformes: Hadrosauroidea

[References: Tsogtbaatar K, Weishampel DB, Evans DC, Watabe M (2014). A new hadrosauroid (*Plesiohadros djadokhtaensis*) from the Late Cretaceous Djadokhtan fauna of southern Mongolia. In: Eberth DA, Evans DC (eds), *Hadrosaurs: Proceedings of the International Hadrosaur Symposium*. Indiana University Press, Bloomington: 108–135.]

Pleurocoelus nanus

> *Astrodon johnstoni* Leidy, 1865 [Aptian; Arundel Clay Formation, Maryland, USA]

= *Pleurocoelus nanus* Marsh, 1888 ⇒ *Astrodon nanus* (Marsh, 1888) Gilmore, 1921 (*nomina dubia*) [Aptian; Arundel Clay Formation, Maryland, USA]

= *Pleurocoelus altus* Marsh, 1888 ⇒ *Astrodon altus* (Marsh, 1888) Gilmore, 1921 (*nomina dubia*) [Aptian; Arundel Clay Formation, Maryland, USA]

Generic name: *Pleurocoelus* ← {(Gr.) πλευρά = πλευρόν/pleuron: rib, side + κοῖλος/koilos: hollow, concave, hollowed}; referring to the vertebra. (*1)

Specific name: *nanus* ← {(Lat.) nānus, *m*.: a dwarf}

Etymology: small, **hollow-sided**

Taxonomy: Saurischia: Sauropodomorpha: Sauropoda: Titanosauriformes

Other species:

> *Pleurocoelus valdensis* Lydekker, 1889 (*nomen dubium*) ⇒ *Astrodon valdensis* (Lydekker, 1889) Swinton, 1936 [Isle of Wight, UK]

Notes: According to *The Dinosauria 2nd edition* (2004), *Pleurocoelus nanus* and *P. altus* were valid, and *Astrodon johnstoni* was *nomen dubium*. But now *P. nanus* and *P. altus* are considered synonyms of *Astrodon johnstoni*. D'Emic (2013) considered *Astrodon johnstoni*, *Pleurocoelus nanus* and *P. altus* to be *nomina dubia* (*2).

[References: (*1) Marsh OC (1888). Notice of a new genus of Sauropoda and other new dinosaurs from the Potomac Formation. *America Journal of Science* 35: 89–94.; (*2) D'Emic MD (2013). Revision of the sauropod dinosaurs of the Lower Cretaceous Trinity Group, southern USA, with the description of a new genus. *Journal of Systematic Palaeontology* 11(6): 707–726.]

Pneumatoraptor fodori

> *Pneumatoraptor fodori* Ősi *et al*., 2010 [Santonian; Csehbánya Formation, Veszprém, Hungary]

Generic name: *Pneumatoraptor* ← {(Gr.) pneumato- [πνεῦμα/pneuma: air] + (Lat.) raptor: plunderer, thief}; referring to "the pneumatic construction of the scapulocoracoid" (*1).

Specific name: *fodori* ← {(person's name) Fodor + -ī}; in honor of "Géza Fodor, who provided a generous support during the early stages of the Iharkút field works" (*1).

Etymology: Fodor's **pneumatic plunderer**
Taxonomy: Theropoda: Maniraptora: Paraves

[References: (*1) Ösi A, Apesteguía S, Kowalewsli M (2010). Non-avian theropod dinosaurs from the early Late Cretaceous of central Europe. *Cretaceous Research* 31: 304–320.]

Podokesaurus holyokensis

> *Podokesaurus holyokensis* Talbot, 1911 ⇒ *Coelophysis holyokensis* (Talbot, 1911) Colbert, 1964 [Hettangian–Sinemurian; Portland Formation, Massachusetts, USA]

Generic name: *Podokesaurus* ← {(Gr.) ποδώκης/podōkēs: swift-footed + σαῦρα/saura: lizard} (*1)
Specific name: *holyokensis* ← {(place-name) Holyoke + -ensis}; referring to Holyoke, a city in Massachusetts, in the Connecticut River Valley.
Etymology: **swift-footed lizard** from Holyoke
Taxonomy: Theropoda: Coelophysoidea: Podokesauridae
Notes: Original fossil material was destroyed in a fire in 1917.

[References: (*1) Talbot M (1911). *Podokesaurus holyokensis*, a new dinosaur of the Connecticut Valley: *American Journal of Science* 31: 469–479.]

Poekilopleuron bucklandii

> *Poekilopleuron bucklandii* Eudes-Deslongchamps, 1838 ⇒ *Megalosaurus poikilopleuron* (Eudes-Deslongchamps, 1838) von Huene, 1923 [Bathonian; Calcaire de Caen, Calvados, France]

Generic name: *Poekilopleuron* ← {(Gr.) ποικίλος/poikilos: many-coloured, various, varied + πλευρόν/pleuron, *n.* [= πλευρά/pleura, *f.*: a rib, the side of things and places]}
Specific name: *bucklandi / bucklandii* ← {(person's name) Buckland + -ī}; in honor of William Buckland.
Etymology: Buckland's **varied ribs**
Taxonomy: Theropoda: Megalosauridae
cf. Theropoda: Tetanurae: ?Spinosauroidea: family *incertae sedis* (*1)
Other species:

> *Streptospondylus altdorfensis* Meyer, 1832 [upper Callovian; Basse-Normandie, France]
= *Laelaps gallicus* Cope, 1867 ⇒ *Poekilopleuron gallicum* (Cope, 1867) Cope, 1869 [?Callovian–Oxfordian; unnamed unit, Calvados, France]
> *Poikilopleuron pusillus* Owen, 1876 ⇒ *Poecilopleuron minor* (Owen, 1876) Owen *vide* Cope, 1878 ⇒ *Aristosuchus pusillus* (Owen, 1876) Seeley, 1887 [Barremian; Wessex Formation, Isle of Wight, UK]
> *Poekilopleuron schmidti* Kiprijanov, 1883 (*nomen dubium*) ⇒ *Megalosaurus schmidti* (Kiprijanov, 1883) Olshevsky, 1991 [Albian–Cenomanian; Sekmenevsk Formation, Tuskar, Kursk, Russia]
> *Poicilopleuron valens* Leidy, 1870 ⇒ *Antrodemus valens* (Leidy, 1870) Leidy, 1870 (*nomen dubium*) ⇒ *Megalosaurus valens* (Leidy, 1870) Nopcsa, 1901 ⇒ *Allosaurus valens* (Leidy, 1870) Gilmore, 1920 [Kimmeridgian–Tithonian; Morrison Formation, Colorado, USA]
> *Poekilopleuron valesdunensis* Allain, 2002 ⇒ *Dubreuillosaurus valesdunensis* (Allain, 2002) Allain, 2005 [Bathonian; Calcaires de Caen Formation, Basse-Normandie, France]

Notes: The only specimen of *P. bucklandii*, housed in the Musée de la Faculté des Sciences de Caen, was destroyed during World War II (Bigot, 1945). However, casts of some parts of the type skeleton have been found. (*1)

[References: (*1) Allain R, Chure DJ (2002). *Poekilopleuron bucklandii*, the theropod dinosaur from the Middle Jurassic (Bathonian) of Normandy. *Palaeontology* 45(6): 1107–1121.]

Polacanthus foxii

> *Polacanthus foxii* Hulke, 1881 ⇒ *Hylaeosaurus foxii* (Hulke, 1881) Coombs, 1971 (partial skeleton) [Barremian; Wessex Formation, Isle of Wight, England, UK]
= *Euacanthus vectianus* Owen, 1897 (partial skeleton) [UK]
= *Polacanthus foxi* Seeley, 1891, non Hulke, 1881 ⇒ *Polacanthus becklesi* Hennig, 1924 [Valanginian–Hauterivian; Wessex Formation, Isle of Wight, UK]

Generic name: *Polacanthus* Huxley, 1867* or Fox, 1865 (Owen *vide* Hulke, 1881) ← {(Gr.) πολύς/polys: many + ἄκανθα/ akantha: thorn, spine, prickle + -us}; referring to the many spines of the armour. *according to *The Dinosauria 2nd edition.*
Specific name: *foxii* ← {(person's name) Fox + -iī}; in honor of the Reverend William Fox, who discovered the specimen in 1865. (*1)
Etymology: Fox's **many spined one**
Taxonomy: Ornithischia: Thyreophora: Ankylosauria: Nodosauridae: Polacanthinae
Other species:

> *Polacanthus rudgwickensis* Blows, 1996 ⇒ *Horshamosaurus rudgwickensis* (Blows, 1996) Blows, 2015 [Hauterivian–Barremian; Weald Clay, Wessex, England, UK]

Notes: In 2015, Blows made *Polacanthus rudgwickensis* a separate genus *Horshamosaurus*. *Polacanthus* was a medium-sized ankylosaur. Its body was covered with armour plates and spikes.

[References: (*1) Hulke JW (1881). XV. *Polacanthus foxii*, a large undescribed dinosaur from the Wealden Formation in the Isle of Wight. *Philosophical Transactions of the Royal Society of London* 172: 653–662.]

Ponerosteus exogyrarum [Archosauromorpha]

> *Iguanodon exogyrarum* Fritsch, 1878 ⇒ *Procerosaurus* (renamed in 1905 but, preoccupied) ⇒ *Ponerosteus exogyrarum* (Fritsch, 1878) Olshevsky, 2000 (*nomen dubium*) [Cenomanian; Exogyrenkalk, Bohemia, Czech]

Generic name: *Ponerosteus* ← {(Gr.) πονηρός/poneros: bad, worthless, useless + ὀστέον/osteon: bone}; referring to the condition of the type specimen.

Specific name: *exogyrarum* ← {Exogyra: a genus of fossil shell + -arum}

The fossil shell was found in the same deposit as *Ponerosteus* specimen.

Etymology: **useless bone** of Exogyra

Taxonomy: Sauropsida: Diapsida: Archosauromorpha: *incertae sedis*

Notes: The type material is seemed to be too poor to classify within Dinosauria.

Potamornis skutchi [Avialae]

> *Potamornis skutchi* Elżanowski *et al.*, 2001 [late Maastrichtian; Lance Formation, Wyoming, USA]

Generic name: *Potamornis* ← {(Gr.) ποτἄμός/potamos: a river, streem + ὄρνις/ornis: bird} (*1) The locality is near Buck Creek, a tributary of Lance Creek, Niobrara Co., Wyoming.

Specific name: *skutchi* ← {(person's name) Skutch + -ī}; "in honor of Dr. Alexander F. Skutch, an eminent ornithologist, in recognition of his respect for birds' lives" (*1).

Etymology: Skutch's **river bird**

Taxonomy: Theropoda: Avialae: Ornithurae: Hesperornithes

[References: (*1) Elżanowski A, Paul GS, Stidham TA (2000). An avian quadrate from the Late Cretaceous Lance Formation of Wyoming. *Journal of Vertebrate Paleontology* 20(4): 712–719.]

Powellvenator podocitus

> *Powellvenator podocitus* Ezcurra, 2017 [Norian; Los Colorados Formation, La Rioja, Argentina]

Generic name: *Powellvenator* ← {(person's name) Jaime Eduardo Powell + (Lat.) vēnātor: hunter}

Specific name: *podocitus* ← {(Gr.) ποδο-/podo- [πούς/pūs: foot] + (Lat.) citus: fast, swift, rapid}

Etymology: fast, **Powell's hunter**

Taxonomy: Theropoda: Neotheropoda: Coelophysoidea

[References: Ezcurra MD (2017). A new early coelophysoid neotheropod from the Late Triassic of northwestern Argentina. *Ameghiniana* 54(5): 506–538.]

Pradhania gracilis

> *Pradhania gracilis* Kutty *et al.*, 2007 [Sinemurian; Dharmaram Formation, Andhra Pradesh, India]

Generic name: *Pradhania* {(person's name) Pradhan + -ia}; in honor of Indian fossil collector Dhuiya Pradhan.

Specific name: *gracilis* ← {(Lat.) gracilis: slender, thin}

Etymology: slender Pradhan's

Taxonomy: Saurischia: Sauropodomorpha: Massospondylidae

[References: Kutty TS, Chatterjee S, Galton PM, Upchurch P (2007). Basal sauropodomorphs (Dinosauria: Saurischia) from the Lower Jurassic of India: their anatomy and relationships. *Journal of Paleontology* 81(6): 1218–1240.]

Prenocephale prenes

> *Prenocephale prenes* Maryańska & Osmólska, 1974 [Campanian; Nemegt Formation, Ömnögovi, Mongolia]

Generic name: *Prenocephale* ← {(Gr.) πρηνής/prēnēs: inclined, sloping + κεφαλή/kephale: head}; "because of the anterior sloping profile of the head" (*1).

Specific name: *prenes* ← {(Gr.) πρηνής/prēnēs: inclined, sloping (*1)}.

Etymology: sloping, **sloping head**

Taxonomy: Ornithischia: Marginocephalia: Pachycephalosauridae

Other species:

> *Stegoceras brevis* Lambe, 1918 ⇒ *Stegoceras breve* (Lambe, 1918) Sues & Galton, 1987 ⇒ *Prenocephale brevis* (Lambe, 1918) Sullivan, 2000 ⇒ *Foraminacephale brevis* (Lambe, 1918) Schott & Evans, 2016 [Campanian; Dinosaur Park Formation, Alberta, Canada]

> *Troodon edmontonense* Brown & Schlaikjer, 1943 ⇒ *Stegoceras edmontonense* (Brown & Schlaikjer, 1943) Sternberg, 1945 ⇒ *Prenocephale edmontonensis* (Brown & Schlaikjer, 1943) Sullivan, 2000 [Campanian; Edmonton Group, Alberta, Canada]

> *Sphaerotholus goodwini* Williamson & Carr, 2002 ⇒ *Prenocephale goodwini* (Williamson

& Carr, 2002) Sullivan, 2003 [Campanian; Kirtland Formation, New Mexico, USA]

[References: (*1) Maryańska T, Osmólska H (1974). Pachycephalosauria, a new suborder of ornithischian dinosaurs. *Palaeontologia Polonica* 30: 45–102.]

Prenoceratops pieganensis

> *Prenoceratops pieganensis* Chinnery, 2004 [Campanian; Two Medicine Formation, Montana, USA]

Generic name: *Prenoceratops* ← {preno-: sloping [(Gr.) πρηνής/prēnēs: with the face downwards] + ceratops [κερατ-/kerat-: horn + ὤψ/ōps: face]}; "referring to the collection of facial features that distinguish this genus and provide it with a long, low head shape" (*1).

Specific name: *pieganensis* ← {(demonym) Piegan + -ensis}; "in honor of the Piegan tribe of the Blackfeet Indian Nation that resides in Montana, where the specimens were discovered (also known as Piikani)" (*1).

Etymology: **sloping ceratops** of the Piegan Blackfeet

Taxonomy: Ornithischia: Ceratopsia: Neoceratopsia: Leptoceratopsidae

[References: (*1) Chinnery B (2004). Description of *Prenoceratops pieganensis* gen. *et* sp. nov. (Dinosauria: Neoceratopsia) from the Two Medicine Formation of Montana. *Journal of Vertebrate Paleontology* 24(3): 572–590.]

Preondactylus buffarinii [Pterosauria]

> *Preondactylus buffarinii* Wild, 1983 [Norian; Preone Valley, Italy]

Generic name: *Preondactylus* ← {(place-name) Preone valley + (Gr.) δάκτυλος/daktylos: finger}

Specific name: *buffarinii* ← {(person's name) Buffarini + -ī}; in honor of the finder, Mr. Nando Buffarini.

Etymology: Buffarini's **Preone finger**

Taxonomy: Archosauria: Ornithodira: Pterosauria: Preondactylia

Priconodon crassus

> *Priconodon crassus* Marsh, 1888 (*nomen dubium*) [Aptian; Arundel Clay Formation, Prince George's County, Maryland, USA]

Generic name: *Priconodon* ← {(Gr.) πρίων/priōn: saw + κῶνος/kōnos: cone + ὀδών/odōn: tooth}

Specific name: *crassus* ← {(Lat.) crassus: solid, thick, fat, dense}

Etymology: thick **saw-cone tooth** (*2)

Marsh did not mention about the etymology.

Taxonomy: Ornithischia: Thyreophora: Ankylosauria: Nodosauridae

Notes: A large worn tooth, the type specimen, was not identified as an ankylosaurian until Coombs assigned it to Nodosauridae in 1978. Carpenter and Kirkland (1998) considered it as valid. According to *The Dinosauria 2nd edition*, *Priconodon crassus* is regarded as a *nomen dubium*.

[References: (*1) Marsh OC (1888). Notice of a new genus of Sauropoda and other new dinosaurs from the Potomac Formation. *American Journal of Science* 35: 89–94.; (*2) West A, Tibert N (2004). Quantitative analysis for the type material of *Priconodon crassus*: a distinct taxon from the Arundel Formation in southern Maryland. *Geological Society of America Abstracts with Programs* 36(5): 423.]

Priodontognathus phillipsii

> *Iguanodon phillipsi* Seeley, 1869 ⇒ *Priodontognathus phillipsii* (Seeley, 1869) Seeley, 1875 [Oxfordian; Calcareous Grit Formation, England, UK]

= *Omosaurus phillipsi* Seeley, 1893 ⇒ *Dacentrurus phillipsii* (Seeley, 1893) Hennig, 1915 (*nomen dubium*) [Oxfordian; Corallian Oolite Formation, North Yorkshire, UK]

Generic name: *Priodontognathus* ← {(Gr.) πρίων/priōn: saw + odont- [ὀδών/odōn: tooth] + γνάθος/gnathos: jaw}

Specific name: *phillipsii* ← {(person's name) Phillips + -iī}; in honor of geology professor John Phillips.

Etymology: Phillips' **saw-toothed jaw**

Taxonomy: Ornithischia: Thyreophora: Ankylosauria

Notes: *Priodontognathus* has been erroneously mixed up with iguanodonts and stegosaurs. According to *The Dinosauria 2nd edition*, it is considred a *nomen dubium*.

[References: Galton PM (1980). *Priodontognathus phillipsii* (Seeley), an ankylosaurian dinosaur from the Upper Jurassic (or possibly Lower Cretaceous) of England. *Neues Jahrbuch für Geologie und Paläontologie Monatshefte* (8): 477–489.]

Proa valdearinnoensis

> *Proa valdearinnoensis* McDonald *et al*., 2012 [Albian; Escucha Formation, Teruel, Aragón, Spain]

Generic name: *Proa* ← {(Spain) proa: prow}; "in reference to the pointed shape of the predentary" (*1).

Specific name: *valdearinnoensis* ← {(place-name) Val de Ariño + (Latin ending) -ensis: from}; "in reference to Val de Ariño, the traditional name of the coal mining area around the municipality of Ariño, near which the fossils were discovered" (*1).

Etymology: **prow** from Val de Ariño

Taxonomy: Ornithischia: Ornithopoda: Iguanodontia: Hadrosauriformes

[References: (*1) McDonald AT, Espílez E, Mampel L, Kirkland JI, Alcalá L (2012). An unusual new basal iguanodont (Dinosauria; Ornithopoda) from the Lower Cretaceous of Teruel, Spain. *Zootaxa* 3595: 61–76.]

Probactrosaurus gobiensis

> *Probactrosaurus gobiensis* Rozhdestvensky, 1966 [Barremian–Albian; Dashuigou Formation, Inner Mongolia, China]

= *Probactrosaurus alashanicus* Rozhdestvensky, 1966 [Barremian–Albian; Dashuigou Formation, Alashan, Inner Mongolia, China]

Generic name: *Probactrosaurus* ← {(Gr.) πρό-/pro- [πρός/pros: before] + *Bactrosaurus* [βάκτρον/bactron: staff, club + σαῦρος/sauros: lizard]}; indicating that this genus may have been an ancestor of the later *Bactrosaurus*.

Specific name: *gobiensis* ← {(place-name) Gobi + -ensis}

Etymology: **Before *Bactrosaurus*** from Gobi

Taxonomy: Ornithischia: Ornithopoda: Iguanodontia: Hadrosauroidea

Other species:

> *Probactrosaurus mazongshanensis* Lü, 1997 ⇒ *Gongpoquansaurus mazongshanensis* (Lü, 1997) You *et al.*, 2014 [Barremian–Albian; Zhonggou Formation, Xinminbao Group, Gansu, China]

Notes: According to *The Dinosauria 2nd edition*, *Probactrosaurus alashanicus* and ? *P. mazongshanensis* were regarded as species of *Probactrosaurus*. But, *P. alashanicus* is now regarded as a synonym of *P. gobiensis*, and *P. mazongshanensis* was given a new genus name, *Gongpoquansaurus*.

[References: Rozhdestvensky AK (1966). [New iguanodonts from Central Asia. Phylogenetic and taxonomic interrelationships of late Iguanodontidae and early Hadrosauridae] *Paleontologicheskii Zhurnal* 3: 103–116.; Norman DB (2002). On Asian ornithopods (Dinosauria: Ornithischia). 4. *Probactrosaurus* Rozhdestvensky, 1966. *Zoological Journal of the Linnean Society* 136(1): 113–144.]

Probrachylophosaurus bergei

> *Probrachylophosaurus bergei* Fowler & Horner, 2015 [Campanian; Judith River Formation, Montana, USA]

Generic name: *Probrachylophosaurus* ← {(Gr.) πρό-/pro-: before + *Brachylophosaurus* [(Gr.) βρᾰχύς/brachys: short + λόφος/lophos: crest + σαῦρος/sauros: lizard]} (*1)

Specific name: *bergei* [berg-ee-i] ← {(person's name) Berge + -ī}; "in memory of Sam Berge, co-owner of the land where the specimen was discovered, and friend and relative of many members of the Rudyard, Montana community, who have supported paleontologic research for decades" (*1).

Etymology: Berge's **before-*Brachylophosaurus***

Taxonomy: Ornithischia: Ornithopoda: Hadrosauridae: Hadrosaurinae: Brachylophosaurini

[References: (*1) Freedman Fowler EA, Horner JR (2015). A new Brachylophosaurin hadrosaur (Dinosauria: Ornithischia) with an intermediate nasal crest from the Campanian Judith River Formation of northcentral Montana. *PLoS ONE*: e0141304.]

Proceratosaurus bradleyi

> *Megalosaurus bradleyi* Woodward, 1910 ⇒ *Proceratosaurus bradleyi* (Woodward, 1910) von Huene, 1926 [Bathonian; White Limestone Formation, Gloucestershire, UK]

Generic name: *Proceratosaurus* ← {(Gr.) πρό-/pro-: before + *Ceratosaurus* [κερατο- (←κέρας/keras: horn)] + σαῦρος/sauros: lizard}

Specific name: *bradleyi* ← {(person's name) Bradley + -ī}; in honor of F. Lewis Bradley, who discovered the first specimen in the early 1900s. (*1)

Etymology: Bradley's **before *Ceratosaurus***

Taxonomy: Theropoda: Coelurosauria: Tyrannosauroidea: Proceratosauridae

Other species:

> *Eustreptospondylus divesensis* Walker, 1964 ⇒ *Piveteausaurus divesensis* (Walker, 1964) Taquet & Welles, 1977 ⇒ *Proceratosaurus divesensis* (Walker, 1964) Paul, 1988 [Callovian; Marnes de Dives, Calvados, France]

Notes: *Proceratosaurus* had a small crest on the snout, which was similar to that of *Ceratosaurus*.

[References: (*1) Woodward AS (1910). On a skull of *Megalosaurus* from the Great Oolite of Minchinhampton (Gloucestershire). *Quarterly Journal of the Geological Society* 66: 111–115.]

Procompsognathus triassicus

> *Procompsognathus triassicus* Fraas, 1913 [Alaunian (= mid Norian); Löwenstein Formation, Baden-Württemberg, Germany]

Generic name: *Procompsognathus* ← {(Gr.) πρό-/pro-: before + *Compsognathus* [κομψός/kompsos: elegant + γνάθος/gnathos: jaw}

Specific name: *triassicus* ← {triassicus: triassic}; referring to the Triassic age.

Etymology: before-*Compsognathus* of Triassic

Taxonomy: Theropoda: Coelophysoidea: Coelophysidae

Prodeinodon mongoliensis

> *Prodeinodon mongoliensis* Osborn, 1924 (*nomen dubium*) [Cretaceous; Öösh Formation,

Ovorkhangai, Mongolia]
Generic name: *Prodeinodon* ← {(Gr.) πρό-/pro-: before + *Deinodon* [δεινός/deinos: fearful, terrible + ὀδών/odōn: tooth]}
Specific name: *mongoliensis* ← {(place-name) Mongolia + -ensis}
Etymology: **before-*Deinodon*** from Mongolia
Taxonomy: Theropoda
Other species:

> *Prodeinodon kwangshiensis* Hou *et al*., 1975 [Aptian; Xinlong Formation, Guangxi, China]

Notes: According to *The Dinosauria 2nd edition*, *Prodeinodon mongoliensis* and *P. kwangshiensis* are considered to be *nomina dubia*.

Propanoplosaurus marylandicus

> *Propanoplosaurus marylandicus* Stanford *et al*., 2011 [Aptian; Arundel Clay Formation, Maryland, USA]

Generic name: *Propanoplosaurus* ← {(Gr.) πρό-/pro-: before + *Panoplosaurus* [πάν-οπλος/pan-oplos: in full armour, full-armed + σαῦρος/sauros: lizard]}
Specific name: *marylandicus* ← {(place-name) Maryland + -icus}
Etymology: before-*Panoplosaurus* from Maryland
Taxonomy: Ornithischia: Thyreophora: Ankylosauria: Nodosauridae
Notes: Tiny fossilized footprints, which were discovered in the vicinity to the holotype, proved they were those of a hatchling specimen. (*1)

[References: (*1) Stanford R, Weishampel DB, DeLeon VB (2011). The first hatchling dinosaur reported from the eastern United States: *Propanoplosaurus marylandicus* (Dinosauria: Ankylosauria) from the Early Cretaceous of Maryland, U.S.A. *Journal of Paleontology* 85(5): 916–924.]

Proplanicoxa galtoni

> *Iguanodon atherfieldensis* Hooley, 1925 ⇒ *Mantellisaurus atherfieldensis* (Hooley, 1925) Paul, 2006 [late Hauterivian; Wealden Group, UK]

= *Proplanicoxa galtoni* Carpenter & Ishida, 2010 (*nomen dubium*) [Barremian; Wessex Formation, UK]

Generic name: *Proplanicoxa* ← {(Gr.) προ-/pro-: before + *Planicoxa* [(Lat.) plānus: level, flat + coxa: hip]}; "in reference to the ilium trending towards the horizontal postacetabular process seen in *Planicoxa*" (*1).
Specific name: *galtoni* ← {(person's name) Galton + -ī}; "in honor of Peter Galton for his work on European ornithischians" (*1).
Etymology: Galton's **before-*Planicoxa***
Taxonomy: Ornithischia: Ornithopoda
Notes: *Proplanicoxa galtoni* is considered a *nomen dubium* and may be synonymous with *Mantellisaurus atherfieldensis,* according to McDonald (2011).

[References: (*1) Carpenter K, Ishida Y (2010). Early and "middle" Cretaceous iguanodonts in time and spece. *Journal of Iberian Geology* 36(2): 145–164.]

Prosaurolophus maximus

> *Prosaurolophus maximus* Brown, 1916 [late Campanian; Dinosaur Park Formation, Alberta, Canada] (*1)

=*Prosaurolophus blackfeetensis* Horner, 1992 [Campanian; Two Medicine Formation, Montana, USA] (*2)

Generic name: *Prosaurolophus* ← {(Gr.) πρό-/pro-: first, before + *Saurolophus* [σαῦρος/sauros: lizard + λόφος/lophos: crest]}
Specific name: *maximus* ← {(Lat.) maximus: largest}; referring to its skull larger than that of *Saurolophus osborni* (*1).
Etymology: largest, **ancestral to *Saurolophus***
Taxonomy: Ornithischia: Ornithopoda: Hadrosauridae: Saurolophinae
Notes: According to *The Dinosauria 2nd edition*, *P. blackfeetensis* was considered as one of the two species of *Prosaurolophus*. But, according to McGarrity *et al*. (2013), *P. blackfeetensis* is regarded as a junior synonym of *P. maximus*.

[References: (*1) Brown B (1916). A new crested trachodont dinosaur, *Prosaurolophus maximus. Bulletin of the American Museum of Natural History* 35(37): 701–708.; (*2) McGarrity CT, Campione NE, Evans DC (2013). Cranial anatomy and variation in *Prosaurolophus maximus* (Dinosauria: Hadrosauridae). *Zoological Journal of the Linnean Society* 167(4): 531–568.]

Protarchaeopteryx robusta [Avialae]

> *Protarchaeopteryx robusta* Ji & Ji, 1997 [Aptian; Yixian Formation, Liaoning, China]

Generic name: *Protarchaeopteryx* ← {(Gr.) πρῶτος/prōtos: first + *Archaeopteryx* [ἀρχαῖος/archaios: ancient + πτέρυξ/pteryx, *f.*: wing]}; referring to the "acknowledges that the specimen possesses characters more primitive than those of *Archaeopteryx*" (*1).
Specific name: *robusta* ← {(Lat.) robustus: strong, vigorous}; "in reference to the long and powerful hind limb on the specimen" (*1).
Etymology: solid, **primitive *Archaeopteryx***
Taxonomy: Theropoda
cf. Aves: Sauriurae: Archaeopterygiformes: Archaeopterygidae (*1)
Notes: According to Ji & Ji (1997), *Protarchaeopteryx*

had "claviform and unserrated dentition" (*1). But, according to Ji *et al.* (1998), it had four serrated premaxillary teeth (*2).

[References: (*1) Ji Q, Ji S (1997). A Chinese archaeopterygian, *Protarchaeopteryx* gen. nov. *Geological Science and Technology (Di Zhi Ke Ji)* 238: 38–41.; (*2) Ji Q, Currie PJ, Norell MA, Ji S (1998). Two feathered dinosaurs from northeastern China. *Nature* 393(6687): 753–761.]

Protoavis texensis [Avialae?]

> *Protoavis texensis* Chatterjee, 1991 [Norian; Cooper Canyon Formation, Texas, USA]

Generic name: *Protoavis* ← {(Gr.) πρῶτος/prōtos: first + (Lat.) avis: bird} (*1)

Specific name: *texensis* ← {(place-name) Texas + -ensis} (*1)

Etymology: **first bird** from Texas

Taxonomy: Theropoda *incertae sedis*
cf. Aves: Protoaviformes: Protoavidae (*1)

Notes: According to *The Dinosauria 2nd edition*, *Protoavis texensis* is regarded a *nomen dubium*, and thought to be doubtfully avialan.

[References: (*1) Chatterjee S (1991). Cranial anatomy and relationship of a new Triassic bird from Texas. *Phylosophical Transactions of the Roal Society B: Biological Sciences* 332: 277–342.]

Protoceratops andrewsi

> *Protoceratops andrewsi* Granger & Gregory, 1923 [Campanian; Djadokhta Formation, Ömnögovi, Mongolia]

Generic name: *Protoceratops* ← {(Gr.) πρῶτος/prōtos: first + κερατ-/kerat- [κέρας/keras: horn] + ὤψ/ōps: face}

Specific name: *andrewsi* ← {(person's name) Andrews + -ī}; in honor of Roy Chapman Andrews who led the American expedition and collected many specimens of the *Protoceratops* genus.

Etymology: Andrews' **early horned face (ceratopsian)**

Taxonomy: Ornithischia: Marginocephalia:
Ceratopsia: Protoceratopsidae

Other species:

> *Protoceratops hellenikorhinus* Lambert *et al.*, 2001 [late Santonian or early Campanian; Djadokhta Formation, Inner Mongolia, China]

> *Protoceratops kozlowskii* Marianska & Osmólska, 1975 ⇒ *Breviceratops kozlowskii* (Mariańska & Osmólska, 1975) Kurzanov, 1990 [Campanian; Barun Goyot Formation, Mongolia]

Notes: *Protoceratops* had a large neck frill and lacked horns. Skeletons of *Protoceratops andrewsi* and *Velociraptor mongoliensis* in combat were found in Mongolia in 1971.]

Protognathosaurus oxyodon

> *Protognathus oxyodon* Zhang, 1988 (preoccupied) ⇒ *Protognathosaurus oxyodon* (Zhang, 1988) Olshevsky, 1991 [Bajocian–Callovian; Xiashaximiao Formation, Sichuan, China]

Generic name: *Protognathosaurus* ← {(Gr.) πρῶτος/prōtos: first + γνάθος/ gnathos: jaw + σαῦρος/ sauros: lizard}

Specific name: *oxyodon* ← {(Gr.) ὀξύς/oxys: sharp + ὀδών/odōn: tooth}

Etymology: **first jaw lizard** with sharp teeth

Taxonomy: Saurischia: Sauropodomorpha:
Sauropoda: Euhelopodidae

Notes: According to *The Dinosauria 2nd edition*, *Protognathosaurus* is considered to be a *nomen dubium*.

Protohadros byrdi

> *Protohadros byrdi* Head, 1998 [Cenomanian; Woodbine Formation, Texas, USA]

Generic name: *Protohadros* ← {(Gr.) πρῶτος/ prōtos: first + ἁδρός/hadros: thick, great, sturdy}; referring to the fact that Head considered the species the oldest known hadrosaur.

Specific name: *byrdi* ← {(person's name) Byrd + -ī}; in honor of an amateur paleontologist Gary Byrd, discoverer.

Etymology: Byrd's **first hadrosaur**

Taxonomy: Ornithischia: Ornithopoda:
Iguanodontia: Hadrosauroidea

[References: Head JJ (1998). A new species of basal hadrosaurid (Dinosauria, Ornthopoda) from the Cenomanian of Texas. *Journal of Vertebrate Paleontology* 18(4): 718–738.]

Protopteryx fengningensis [Avialae]

> *Protopteryx fengningensis* Zang & Zhou, 2000 [Barremian; Huajiying Formation, Hebei, China]

Generic name: *Protopteryx* ← {(Gr.) πρῶτος/prōtos: first + πτέρυξ/pteryx: feather, wing}

Specific name: *fengningensis* ← {(Chin. place-name) Fēngníng豐寧 + -ensis}

Etymology: **primitive fether** from Fengning

Taxonomy: Theropoda: Avialae: Enantiornithes

[References: Zhang F, Zhou Z (2000). A primitive enantiornithine bird and the origin of feathers. *Science* 290(5498): 1955–1959.]

Psittacosaurus mongoliensis

> *Psittacosaurus mongoliensis* Osborn, 1923 [?Aptian–Albian; Öösh Formation, Hühteeg Svita, Övörkhangai, Mongolia]

= *Protiguanodon mongoliense* Osborn, 1923 ⇒

Psittacosaurus protiguanodonensis (Osborn, 1923) Young, 1958 [Early Cretaceous; Öösh Formation, Övörkhangai, Mongolia]
= *Psittacosaurus osborni* Young, 1931 [Aptian; Xinpongnaobao Formation, Inner Mongolia, China]
= *Psittacosaurus tingi* Young, 1931 [?Aptian–Albian; Lisangou Formation, Inner Mongolia, China]
= *Psittacosaurus guyangensis* Cheng, 1983 [?Aptian–Albian; Lisangou Formation, Inner Mogolia, China]

Generic name: *Psittacosaurus* ← {(Gr.) ψιττᾰκός/psittakos: parrot + (Gr.) σαῦρος/sauros: lizard}; referring to "the parrot-beaked rostrum" (*1).

Specific name: *mongoliensis* ← {(place-name) Mongolia + -ensis}

Etymology: **parrot lizard** from Mongolia

Taxonomy: Ornithischia: Marginocephalia: Ceratopsia: Psittacosauridae

Other species:

> *Psittacosaurus sinensis* Young, 1958 [?Aptian–Albian; Qingshan Formation, Shandong, China]
= *Psittacosaurus youngi* Chao, 1962 [?Aptian–Albian; Qingshan Formation, Shandong, China]
> *Psittacosaurus meileyingensis* Sereno *et al.*, 1988 [Early Cretaceous, Jiufotang Formation, Liaoning, China]
> *Psittacosaurus xinjiangensis* Sereno & Zhao, 1988 [?Valanginian–Albian; Tuguru Group, Xinjiang, China]
> *?Psittacosaurus sattayaraki* Buffetaut & Suteethorn, 1992 (*nomen dubium*) [Aptian–Albian; Khok Kruat Formation, Chaiyaphun, Thailand]
> *Psittacosaurus neimongoliensis* Russell & Zhao, 1996 [Valanginian–Albian; Ejinhoro Formation, Inner Mongolia, China]
> *Psittacosaurus ordosensis* Russell & Zhao, 1996 [Valanginian–Albian; Ejinhoro Formation, Inner Mongolia, China]
> *Psittacosaurus mazongshanensis* Xu, 1997 [Barremian–Albian; Xinminbao Group, Gansu, China]
> *Psittacosaurus sibiricus* Voronkevich & Averianov, 2000 (*2) [Barremian–Aptian; Ilek Formation, Kemerovo, Russia]
> *Psittacosaurus lujiatunensis* Zhou *et al.*, 2006 [Aptian; Yixian Formation, Liaoning, China]
= *Hongshanosaurus houi* You *et al.*, 2003 ⇒ *Psittacosaurus houi* (You *et al.*, 2003) Sereno, 2010 (*nomen dubium*) [Aptian; Yixian Formation, Liaoning, China]
= *Psittacosaurus major* Sereno *et al.*, 2007 [Barremian-Aptian; Yixian Formation, Beipiao, China]
> *Psittacosaurus gobiensis* Sereno *et al.*, 2010 [Aptian; Bayan Gobi Formation, Suhongtu, China]
> *Psittacosaurus amitabha* Napoli *et al.*, 2019 [Barremian; Andakhuduk Formation, Tsaagan Nor Basin, Mongolia]

Notes: A cluster of 34 juveniles found in Liaoning Province, China proves them to be gregarious, according to Zhao *et al.* (2013).

[References: (*1) Osborn HF (1923). Two Lower Cretaceous dinosaurs of Mongolia. *American Museum Novitates* 95: 1–10.; (*2) Leshchinskiy SV, Fayngertz AV, Voronkevich AV, Maschenko EN, Averianov AO (2000). Preliminary results of the investigation of the Shestakovo localities of Early Cretaceous vertebrates. In: Komarov AV (ed), *Materials of the Regional Conference of the Geologists of Siberia, Far East and North East of Russia* (in Russian). Tomsk: GalaPress: 363–366.]

Pteranodon longiceps [Pterosauria]

> *Pteranodon longiceps* Marsh, 1876 ⇒ *Pteranodon (Longicepia) longiceps* (Marsh, 1876) Miller, 1971 [Santonian–Campanian; Niobrara Formation, Kansas, USA]
= *Pteranodon ingens* Marsh, 1872 (*nomen dubium*) ⇒ *Ornithostoma ingens* (Marsh, 1872) Williston, 1893 (*nomen dubium*) ⇒ *Pteranodon ingens* (Marsh, 1872) Eaton, 1910 (*nomen dubium*) ⇒ *Pteranodon (Pteranodon)* ingens (Marsh, 1872) Olshevsky, 1978 (*nomen dubium*) [late Santonian; Niobrara Formation, Kansas, USA]
= *Pteranodon (Longicepia) marshi* Miler, 1971 (*nomen dubium*) ⇒ *Pteranodon (Pteranodon) marshi* (Miller, 1971) Miller, 1973 (*nomen dubium*) [late Santonian; Niobrara Formation, Kansas, USA]

Generic name: *Pteranodon* ← {(Gr.) πτερόν/pteron: wing + ἀν-/an-: without + ὀδών/odōn (= ὀδούς/odous [odūs]: tooth}

Specific name: *longiceps* ← {(Lat.) longus: long + -ceps [caput: head]}

Etymology: **toothless wing** with long head

Taxonomy: Archosauria: Ornithodira: Pterosauria: Pterodactyloidea: Pteranodontidae

Other species:

> *Pteranodon gracilis* Marsh, 1876 ⇒ *Nyctosaurus gracilis* (Marsh, 1876) Marsh, 1876 ⇒ *Nyctodactylus gracilis* (Marsh, 1876) Marsh, 1881 ⇒ *Pteranodon (Nyctosaurus)*

gracilis (Marsh, 1876) Miller, 1971 [Santonian-Campanian; Niobrara Formation, Kansas, USA]
= *Pteranodon nanus* Marsh, 1881 ⇒ *Nyctosaurus nanus* (Marsh, 1881) Schoch, 1984 [Coniacian–Santonian; Niobrara Formation, Kansas, USA]
> *Pteranodon (Nyctosaurus) bonneri* Miller, 1971 ⇒ *Nyctosaurus bonneri* (Miller, 1971) Wellnhofer, 1978 [Santonian–Campanian; Niobrara Formation, Kansas, USA]
> *Pteranodon sternbergi* Harksen, 1966 ⇒ *Pteranodon (Sternbergia) sternbergi* (Harksen, 1966) Miller, 1971 ⇒ *Pteranodon (Geosternbergia) sternbergi* (Harksen, 1966) Miller, 1978 ⇒ *Geosternbergia sternbergi* (Harksen, 1966) Kellner, 2010 [late Coniacian; Niobrara Formation, Graham, Kansas, USA]
= *Pteranodon (Occidentalia) eatoni* (Marsh, 1872) Miller, 1971 ⇒ *Pteranodon (Occidentalia) occidentalis* (Marsh, 1872) Olshevsky, 1978 [late Coniacian; Niobrara Formation, Kansas, USA]
= *Pteranodon walkeri* Miller, 1971 ⇒ *Pteranodon (Sternbergia) walkeri* (Miller, 1971 ⇒ *Pteranodon (Geostermbergia) walkeri* (Miller, 1971) Miller, 1978 ⇒ *Geosternbergia walkeri* (Miller, 1971) Olshevsky, 1989 [late Santonian; Niobrara Formation, Kansas, USA]
> *Pteranodon oregonensis* Gilmore, 1923 ⇒ *Bennettazhia oregonensis* (Gilmore, 1923) Nessov, 1991 [Albian; Hudspeth Formation, Oregon, USA]
> *Ornithostoma orientalis* Bogolubov, 1914 ⇒ *Pteranodon orientalis* (Bogolubov, 1914) Bramwell & Whitfield, 1974 ⇒ *Ornithostoma (Pteranodon) orientalis* (Bogolubov, 1914) Wellnhofer, 1978 ⇒ *Bogolubovia orientalis* (Bogolubov, 1914) Nesov & Yarkov, 1989 [Campanian; Rybushka Formation, Setrovsk, Russia]
> *Pterodactylus velox* Marsh, 1872 (*nomen dubium*) ⇒ *Pteranodon velox* (Marsh, 1872) Eaton, 1910 (*nomen dubium*) [late Santonian; Niobrara Formation, Kansas, USA]
> *Ornithochirus umbrosus* Cope, 1872 (*nomen dubium*) ⇒ *Pterodactylus umbrosus* (Cope, 1872) Cope, 1874 (*nomen dubium*) ⇒ *Ornithocheirus umbrosus* (Cope, 1872) Newton, 1888 (*nomen dubium*) ⇒ *Pteranodon umbrosus* (Cope, 1872) Williston, 1893 (*nomen dubium*) ⇒ *Ornithostoma umbrosum* (Cope, 1872) Williston, 1897 (*nomen dubium*) [late Coniacian; Niobrara Formation, Kansas, USA]
> *Pteranodon camptus* Marsh, 1876 (*nomen dubium*) [late Santonian; Niobrara Formation, Kansas, USA]
> *Ornithochirus harpyia* Cope, 1872 (*nomen dubium*) ⇒ *Ornithocheirus harpyia* (Cope, 1872) Newton, 1888 ⇒ *Pteranodon harpyia* (Cope 1872) Williston, 1903 [late Santonian; Niobrara Formation, Kansas, USA]

Notes: Over 1,000 specimens have been identified. The specimen of *Pteranodon longiceps* from Niobrara Formation had the largest wingspan measured about six meters, according to Bennett (1994).

Pterodactylus antiquus [Pterosauria]
> *Pterodactylus* Cuvier, 1809 ⇒ *Ornithocephalus antiquus* Sömmerring, 1812 ⇒ *Pterodactylus antiquus* (Sömmerring, 1812) Cuvier, 1819 [Tithonian; Solnhofener Plattenkalk Formation, Bayern, Germany]
= *Pterodactylus longirostris* Cuvier, 1819 ⇒ *Ornithocephalus longirostris* (Cuvier, 1819) Sömmerring, 1820 ⇒ *Macrotrachelus longirostris* (Cuvier, 1819) Gibel, 1850 [Tithonian; Solnhofen Formation, Bayern, Germany]
= *Pterodactylus crocodilocephaloides* Ritgen, 1826 (*nomen dubium*) [Germany]

Generic name: *Pterodactylus* ← {(Gr.) πτερόν/pteron: feather, wing + δάκτυλος/daktylos: finger} *Pterodactylus* had wings formed by a skin and muscle membrance.

Specific name: *antiquus* ← {(Lat.) antīquus: ancient, former, of old time}

Etymology: ancient, **winged finger**

Taxonomy: Pterosauria: Pterodactylidae

Other species:
> *Pterodactylus arningi* Reck, 1931 (*nomen dubium*) [Tithonian; Tendaguru Formation, Lindi, Tanzania] *etc.*

Pterodaustro guinazui [Pterosauria]
> *Pterodaustro guinazui* Bonaparte, 1970 [Albian; Lagarcito Formation, Argentina]
= *Puntanipterus globosus* Bonaparte & Sánchez, 1975 [Aptian; La Cruz Formation, San Luis, Argentina]

Generic name: *Pterodaustro* ← {(Gr.) πτερόν/pteron: wing + (Lat.) d'austro [auster]: from the south}

Specific name: *guinazui* ← {(person's name) Guinazu + -ī}; in honor of paleontologist Román Guiñazú. In 1978, specific name *quiñazui* was emended to *guinazui*, because the tilde are not allowed in species names

Etymology: Guinazu's **wing from the south**

Taxonomy: Pterosauria: Pterodactyloidea:

Ctenochasmatidae

Notes: *Pterodaustro* had a very long snout and a thousand bristle-like modified teeth in its lower jaws.

Puertasaurus reuili

> *Puertasaurus reuili* Novas *et al.*, 2005 [Maastrichtian; Pari Aike Formation, Santa Cruz, Argentina]

Generic name: *Puertasaurus* ← {(person's name) Puerta + (Gr.) σαῦρος/sauros}; "in honor of Pablo Puerta and Santiago Reuil, remarkable fossil-hunters who discovered and prepared the specimen"(*1).

Specific name: *reuili* ← {(person's name) Reul + -ī}

Etymology: Reuil and **Puerta's lizard**

Taxonomy: Saurischia: Sauropodomorpha: Sauropoda: Titanosauria: Lognkosauria

Notes: Gigantic size.

[References: (*1) Novas FE, Salgado L, Calvo J, Agnolin F (2005). Giant titanosaur (Dinosauria, Sauropoda) from the Late Cretaceous of Patagonia. *Revisto del Museo Argentino de Ciencias Naturales, n. s.* 7(1): 37–41.]

Pukyongosaurus millenniumi

> *Pukyongosaurus millenniumi* Dong *et al.*, 2001 [Hauterivian; Hasandong Formation, Gyeongsang Basin, Hadong, South Korea]

Generic name: *Pukyongosaurus* ← {(institution name) Pukyong釜慶: Pukyong National University + (Gr.) σαῦρος/sauros: lizard}

Specific name: *millennium* ← {(New Latin) millennium: millennial}; commemorating the year 2000, a new millennium.

Etymology: millennial, **Pukyong lizard**

Taxonomy: Saurischia: Sauropodomorpha: Sauropoda: Titanosauriformes

Notes: According to *The Dinosauria 2nd edition*, *Pukyongosaurus* was considered to be a *nomen dubium*, but in 2016, Park made a comment on its validity(*1).

[References: (*1) Park J-Y (2016). Comments on the validity of the taxonomic status of "*Pukyongosaurus*" (Dinosauria: Sauropoda). *Memoir of the Fukui Prefectural Dinosaur Museum* 15: 27–32.]

Pulanesaura eocollum

> *Pulanesaura eocollum* McPhee *et al.*, 2015 [Hettangian–Sinemurian; Elliot Formation, Free State, South Africa]

Generic name: *Pulanesaura* ← {(Sesotho) pulane: rain-maker, rain-bringer + (Gr.) σαῦρα/saura: lizard}; "in reference to the rain-soaked conditions under which the dinosaur was excavated" (*1).

Specific name: *eocollum* ← {(Gr.) ἠώς/ēos: dawn + (Lat.) collum: neck}; "in reference to the hypothesized function of the neck presaging the sauropod condition in the new taxon" (*1).

Etymology: **rain-maker lizard** with a basal neck

Taxonomy: Saurischia: Sauropodomorpha: Sauropodiformes: Sauropoda

[References: (*1) McPhee BW, Bonnan MF, Yates AM, Neveling J, Choiniere JN (2015). A new basal sauropod from the pre-Toarcian Jurassic of South Africa: evidence of niche-partitioning at the sauropodomorph-sauropod boundary? *Scientific Reports* 5(13224): 1–12.]

Pycnonemosaurus nevesi

> *Pycnonemosaurus nevesi* Kellner & Campos, 2002 [Campanian–Maastrichtian; Adamantina Formation, Mato Grosso, Brazil]

Generic name: *Pycnonemosaurus* ← {"(Gr.) πυκνός/pycnos: dense, thick + νέμος/nemos: pasture, wood + (Gr.) σαῦρος/sauros: reptile or lizard}; in allusion to Mato Grosso"(*1), meaning "thick forest" in Portuguese.

Specific name: *nevesi* ← {(person's name) Neves + -ī}; "in honor of Dr. Iedo Batista Neves (deceased in 2000), who encouraged paleontological studies, particularly of A.W.A. Kellner"(*1).

Etymology: Neves' **lizard from dense forest (Mato Grosso)**

Taxonomy: Theropoda: Abelisauridae

Notes: *Pycnonemosaurus nevesi* (8.9 ± 0.3 m) is the largest abelisaurids, according to Grillo & Decourt (2017).(*2)

[References: (*1) Kellner AWA, Campos DA (2002). On a theropod dinosaur (Abelisauria) from the continental Cretaceous of Brazil. *Arquivos do Museu Nacional, Rio de Janeiro* 60(3): 163–170.; (*2) Grillo ON, Delcourt R (2017). Allometry and body length of abelisauroid theropods: *Pycnonemosaurus nevasi* is the new king. *Cretaceous Research* 69: 71–89.]

Pyroraptor olympius

> *Pyroraptor olympius* Allain & Taquet, 2000 [Maastrichtian; Grès à Reptiles Formation, Bouches-du-Rhône, Provence, France]

Generic name: *Pyroraptor* ← {(Gr.) πῦρός/pyros [πῦρ/pyr: fire] + (Gr.) raptor: thief}; "alluding to the fact that this new, agile small theropod has been discovered after a forest fire"(*1).

Specific name: *olympius* ← {olympius: from Olympus}; "alluding to the Mont Olympe (Provence) at the foot of which is situated the new locality"(*1).

Etymology: **fire thief** from Mont Olympe

Taxonomy: Theropoda: Dromaeosauridae

[References: (*1) Allain R, Taquet P (2000). A new genus of Dromaeosauridae (Dinosauria, Theropoda) from the Upper Cretaceous of France. *Journal of Vertebrate Paleontology* 20: 404–407.]

Q

Qantassaurus intrepidus

> *Qantassaurus intrepidus* Rich & Vickers-Rich, 1999 [Valanginian–Aptian; Wonthaggi Formation, Victoria, Australia]

Generic name: *Qantassaurus* ← {(acronym) QANTAS: Queensland and Northern Territory Air Service + (Gr.) σαῦρος/sauros: lizard}; "in honor of QANTAS Airlines for their long term support of dinosaur research and exhibitions" (*1).

Specific name: *intrepidus* ← {(Lat.) in-trepidus: "unshaken, undaunted, brave, bold" (*1)}

Etymology: brave, **QANTAS lizard**

Taxonomy: Ornithischia: Ornithopoda: Euornithopoda: Hypsilophodontidae (*1)

[References: (*1) Rich TH, Vickers-Rich P (1999). The hypsilophodontidae from southeastern Australia. In: Tomida Y, Rich TH, Vickers-Rich P (eds), *Proceedings of the Second Gondwanan Dinosaur Symposium, National Science Museum Monographs* 15: 167–180.]

Qianzhousaurus sinensis

> *Qianzhousaurus sinensis* Lü *et al.*, 2014 ⇒ *Alioramus sinensis* (Lü *et al.*, 2014) Carr *et al.*, 2017 [Maastrichtian; Nanxiong Formation, Ganzhou, Jiangxi, China]

Generic name: *Qianzhousaurus* ← {(Chn. place-name) Qianzhou: an ancient name of the city of Ganzhou + (Gr.) σαῦρος/sauros: lizard}

Specific name: *sinensis* ← {(Gr.) Sin (*1): China + -ensis}

Etymology: **Qianzhou lizard** from China

Taxonomy: Theropoda: Tetanurae: Coelurosauria: Tyrannosauridae: Tyrannosaurinae: Alioramini

[References: (*1) Lü J, Yi L, Brusatte SL, Yang L, Li H, Chen L (2014). A new clade of Asian Late Cretaceous long-snouted tyrannosaurids. *Nature Communications* 5(3788).]

Qiaowanlong kangxii

> *Qiaowanlong kangxii* You & Li, 2009 [Aptian–Albian; Digou Formation, Gansu, China]

Generic name: *Qiaowanlong* ← {(Chin. place-name) Qiáowān橋湾*: a culture relic + lóng龍: dragon}

*qiao 橋: 'bridge', wan湾: 'bend in a stream'.(*1)

Specific name: *kangxii* ← {(emperor) Kāngxī康熙 + -ī}; "after 'Kangxi', a famous emperor of the Qīng Dynasty, who once had a dream of the scenic beauty of the Qiaowan area" (*1).

Etymology: Kangxi's **Qiaowan dragon**

Taxonomy: Saurischia: Sauropodomorpha: Sauropoda: Brachiosauridae (*1)

[References: (*1) You H-L, Li D-Q (2009). The first well-preserved Early Cretaceous brachiosaurid dinosaur in Asia. *Proceedings of the Royal Society B: Biological Sciences* 276(1675): 4077–4082.]

Qijianglong guokr

> *Qijianglong guokr* Xing *et al.*, 2015 [Oxfordian–Kimmeridgian; Suining Formation, Chongqing, China]

Generic name: *Qijianglong* ← {(Chin. place-name) Qíjiāng綦江: Qijiang District + lóng龍: dragon}

Specific name: *guokr* ← {Guokr [pron.: gu-OH-ke-r]; "named in honor of Guokr (science social network; 'Nutshell' in Chinese) for their support of paleontology in Qijiang" (*1).

Etymology: Guokr's **Qijiang dragon**

Taxonomy: Saurischia: Sauropodomorpha: Eusauropoda: Mamenchisauridae

[References: (*1) Xing L, Miyashita T, Zhang J, Li D, Ye Y, Sekiya T, Wang F, Currie PJ (2010). A new sauropod dinosaur from the Late Jurassic of China and the diversity, distribution, and relationships of mamechisaurids. *Journal of Vertebrate Paleontology* 35(1): e889701]

Qiliania graffini [Avialae]

> *Qiliania graffini* Ji *et al.*, 2011 [Aptian; Xiagou Formation, Gansu, China]

Generic name: *Qiliania* ← {(Xiongnu*) qilian: heaven + -ia}; "for the Qilian Mountains that lie to the south of the Changma Basin. The generic name is fashioned after that of the only other fossil bird thus far named from Changma, Gansus, in that it refers to a local geographical feature but eschews a traditional avian suffix" (*1).

*Xiōngnú 匈奴, nomadic tribes that inhabited central Asia.

Specific name: *graffini* ← {(person's name) Graffin + -ī}; "in honor of Dr. Gregory Graffin, lecturer at the University of California, Los Angeles and co-founder of the musical group Bad Religion, for his contributions to evolutionary biology, his public outreach through music, and his inspiration to young scientists around the world" (*1).

Etymology: Graffin's **one from heaven (Qilian Mountains)**

Taxonomy: Theropoda: Avialale: Ornithothoraces: Enantiornithes

cf. Aves: Ornithothoraces: Enantiornithes (*1)

[References: (*1) Ji S-A, Atterholt J, O'Connor JK, Lamanna MC, Harris JD, Li D-Q, You H-L, Dodson P (2011). A new, three-dimensionally preserved enantiornithine bird (Aves: Ornithothoraces) from

Gansu Province, north- western China. *Zoological Journal of the Linnean Society* 162(1): 201–219.]

Qingxiusaurus youjiangensis

> *Qingxiusaurus youjiangensis* Mo *et al*., 2008 [Late Cretaceous; Dashi Village, Nanning, Guangxi, China]

Generic name: *Qingxiusaurus* ← {(Chin.) qīngxiù清秀* + (Gr.) σαῦρος/sauros: lizard}

*qingxiu (= shan qing shui xiu山清水秀): "a picturesque scenery of mountains and water in Guangxi" (*1).

Specific name: *youjiangensis* ← {(Chin. place-name) Yòujiāng右江 + -ensis}; "after the name of the river near where the specimen is excavated" (*1).

Etymology: **picturesque lizard** from Youjiang

Taxonomy: Saurischia: Sauropodomorpha: Sauropoda: Titanosauria (*1)

[References: (*1) Mo J-Y, Huang C-L, Zhao Z-R, Wang W, Xu X (2008). A new titanosaur (Dinosauria: Sauropoda) from the Late Cretaceous of Guangxi, China. *Vertebrata PalAsiatica* 46(2): 147–156.]

Qinlingosaurus luonanensis

> *Qinlingosaurus luonanensis* Xue *et al*., 1996 [Maastrichtian; Hongtuling Formation, Shaanxi, China]

Generic name: *Qinlingosaurus* ← {(Chin. place-name) Qínlǐng秦嶺 + (Gr.) σαῦρος/sauros: lizard}

Specific name: *luonanensis* ← {(Chin. place-name) Luònán洛南 + -ensis}

Etymology: **Qinling (Shaanxi) lizard** from Luonan

Taxonomy: Saurischia: Sauropodomorpha: Sauropoda

Notes: According to *The Dinosauria 2nd edition*, *Qinlingosaurus* is considered to be a *nomen dubium*.

Qiupalong henanensis

> *Qiupalong henanensis* Xu *et al*., 2011 [Late Cretaceous; Qiupa Formation, Henan, China]

Generic name: *Qiupalong* ← {(place-name) Qiupa秋扒 [Pinyin: Qiūbā] + lóng龍: dragon}; referring to the Qiupa Formation.

Specific name: *henanensis* ← {(Chin. place-name) Hénán河南 + -ensis}

Etymology: **Qiupa dragon** from Henan

Taxonomy: Theropoda: Ornithomimosauria: Ornithomimoidea: Ornithomimidae

[References: Xu L, Kobayashi Y, Lü J, Lee Y-N, Liu Y, Tanaka K, Zhang X, Jia S, Zhang J (2011). A new ornithomimid dinosaur with North American affinities from the Late Cretaceous Qiupa Formation in Henan Province of China. *Cretaceous Research* 32(2): 213–222.]

Qiupanykus zhangi

> *Qiupanykus zhangi* Lü *et al*., 2018 [Maastrichtian; Qiupa Formation, Henan, China]

Generic name: *Qiupanykus* ← {(place-name) Qiupa 秋扒 [Pinyin: Qiūbā]; the town in Luanchuan County + (Gr.) ὄνυξ/onyx: claw} (*1)

Specific name: *zhangi* ← {(person's name) Zhang + -ī}; "in honor of Shuancheng Zhang for his logistic support with fossil searching and excavations in the field" (*1).

Etymology: Zhang's **Qiupa claw**

Taxonomy: Theropoda: Maniraptora: Alvarezsauridae

[References: (*1) Lü J-C, Xu L, Chang H-L, Jia S-H, Zhang J-M, Gao D-S, Zhang Y-Y, Zhang C-J, Ding F (2018). A new alvarezsaurid dinosaur from the Late Cretaceous Qiupa Formation of Luanchuan, Henan Province, central China. *China Geology* 1: 28–35.]

Quaesitosaurus orientalis

> *Quaesitosaurus orientalis* Kurzanov & Bannikov, 1983 [Campanian–Maastrichtian; Barun Goyot Formation, Ömnögovi, Mongolia]

Generic name: *Quaesitosaurus* ← {(Lat.) quaesitus: abnormal + (Gr.) σαῦρα/saura: lizard} (*1)

Specific name: *orientalis* ← {(Lat.) orientālis: eastern}

Etymology: eastern, **abnormal lizard**

Taxonomy: Saurischia: Sauropodomorpha: Titanosauria: Nemegtosauridae

cf. Cetiosauridae: Dicraeosaurininae (*1)

[References: (*1) Kurzanov SM, Bannikov AF (1983). Nouviy sauropod ys uerchnego mela MNR [A new sauropod from the Upper Cretaceous of Mongolia]. *Paleontologicheskii Zhurnal* (*Paleontological Journal*) 2: 90–96.]

Quetecsaurus rusconii

> *Quetecsaurus rusconii* González Riga & Ortiz David, 2014 [Turonian; Cerro Lisandro Formation, Mendoza, Argentina]

Generic name: *Quetecsaurus* ← {(Milcayac*) quetec: fire + (Gr.) σαῦρος/sauros: lizard}

*the language used by the people who inhabited the region of Mendoza (*1).

Specific name: *rusconii* ← {(person's name) Rusconi + -ī}; "in honor of Carlos Rusconi (1898–1969), a naturalist who worked extensively in Mendoza Province and was Director of the Museum of Natural Sciences 'Juan Cornelio Moyano'" (*1).

Etymology: Rusconi' **fire lizard**

Taxonomy: Saurischia: Sauropoda: Neosauropoda: Titanosauriformes: Somphospondyli: Titanosauria: Lithostrotia

[References: (*1) González Riga BJ, Ortiz David L

(2014). A new titanosaur (Dinosauria, Sauropoda) from the Upper Cretaceous (Cerro Lisandro Formation) of Mendoza Province, Argentina. *Ameghiniana* 51(1): 3–25.]

Quetzalcoatlus northropi [Pterosauria]

> *Quetzalcoatlus northropi* Lawson, 1975 [Maastrichtian; Javelina Formation, Texas, USA]

Generic name: *Quetzalcoatlus* ← {(Nahuatl, myth.) Quetzalcoatl: the Aztec feathered serpent god}

Specific name: *northropi* ← {(person's name) Northrop + -ī}

John Knudsen "Jack" Northrop (1895–1981), founder of Northrop Corporation, developed tailless flying wing aircraft in the 1940s.

Etymology: Northrop's **Quetzalcoatl**

Taxonomy: Archosauria: Ornithodira: Pterosauria: Pterodactyloidea: Azhdarchidae

Notes: *Quetzalcoatlus* is one of the largest-known flying creature. According to Lawson (1975), the estimated wingspan is 15.5 m (*1).

[References: (*1) Lawson DA (1975). Pterosaur from the latest Cretaceous of West Texas: discovery of the largest flying creature. *Science* 187(4180): 947–948.]

Quilmesaurus curriei

> *Quilmesaurus curriei* Coria, 2001 [Campanian–Maastrichtian; Allen Formation, Río Negro, Argentina]

Generic name: *Quilmesaurus* ← {(demonym) Quilmes people* + (Gr.) σαῦρος/sauros: lizard}

*an indigenous tribe of the Diaguita group.

Specific name: *curriei* ← {(person's name) Currie + -ī}; in honor of Canadian paleontologist Philip John Currie.

Etymology: Currie's **Quilmes lizard**

Taxonomy: Theropoda: Abelisauridae: Carnotaurini

cf. Theropoda: Ceratosauria: Abelisauroidea: Abelisauridae: Carnotaurinae? (*1)

Notes: According to Juárez Valieri *et al.* (2007), *Quilmesaurus* is considered a *nomen dubium*.

[References: (*1) Juárez Valieri RD, Fiorelli LE, Cruz LE (2007). *Quilmesaurus curriei* Coria, 2001 (Dinosauria, Theropoda). Su validez taxonómica y relaciones filogenéticas. *Revista del Museo Argentino de Ciencias Naturales, Bernardino Rivadavia, Paleontología* 9(1): 59–66.]

R

Rahiolisaurus gujaratensis

> *Rahiolisaurus gujaratensis Novas et al.*, 2010 [Maastrichtian; Lameta Formation, Rahioli Village, Gujarat, India]

Generic name: *Rahiolisaurus* ← {(place-name) Rahioli village + (Gr.) σαῦρος/sauros: lizard} (*1)

Specific name: *gujaratensis* ← {(place-name) Gujarat + -ensis} (*1)

Etymology: **Rahioli lizard** from Gujarat

Taxonomy: Theropoda: Ceratosauria: Abelisauridae: Majungasaurinae

Notes: *Rahiolisaurus gujaratensis* had gracile and slender limbs. (*1)

[References: (*1) Novas FE, Chatterjee S, Rudra DK, Datta PM (2010). *Rahiolisaurus gujaratensis*, n. gen. n. sp., a new abelisaurid theropod from the Late Cretaceous of India. In: Bandyopadhyay S (ed), *New Aspects of Mesozoic Biodiversity.* Springer-Verlag, Berlin: 45–62.]

Rahonavis ostromi

> *Rahona ostromi* Forster *et al.*, 1998 (preoccupied by a genus of moths) ⇒ *Rahonavis ostromi* (Forster *et al.*, 1998) Forster, 1998 [Maastrichtian; Maevarano Formation, Majunga (=Mahajanga), Madagascar]

Generic name: *Rahonavis* ← {(Malagasy) rahona: cloud, menace + (Lat.) avis: bird}

According to Forster *et al.*, 1998a, "Rahona [RAH-hoo-nah]: meaning menace/threat or cloud; intended interpretation: 'menace bird from the clouds'" (*1)

Specific name: *ostromi* ← {(person's name) Ostrom + -ī}; in honor of Dr. John H. Ostrom.

Etymology: Ostrom's menace **bird from the clouds**

Taxonomy: Theropoda: Dromaeosauridae: Unenlagiinae (*2)

cf. Theropoda; Avialae (acc. *The Dinosauria 2nd edition*)

Notes: *Rahonavis* had such a sickle claw as *Velociraptor* had. It was considered a basal avialan but later considered to be a basal dromaeosaurid (*2).

[References: (*1) Forster CA, Sampson SD, Chiappe LM, Krause DW (1998). The theropod ancestry of birds: new evidence from the Late Cretaceous of Madagascar. *Science* 279(5358): 1915–1919.; (*2) Turner AH, Mackovicky PJ, Norrell MA (2012). A review of dromaeosaurid systematics and paravian phylogeny. Buletin of the American Museum of Natural History 371: 75–84.; Malagasy Dictionary and Madagascar Encyclopedia: drahonao malagasyword.org. 〈http://malagasyword.org/bins/teny2?w=drahonao〉]

Rajasaurus narmadensis

> *Rajasaurus narmadensis* Wilson, 2003 [Maastrichtian; Lameta Formation, Gujarat, India]

Generic name: *Rajasaurus* ← {(Sanskrit) raja: prince or princely + (Gr.) σαῦρος/sauros: lizard}

Specific name: *narmadensis* ← {(place-name) Narmada: the Narmada Valley in central India +

-ensis}
Etymology: **prince lizard** from Narmada
Taxonomy: Theropda: Ceratosauria: Abelisauridae: Majungasaurinae
Notes: *Rajasaurus* had a low horn on its forehead [*1]. According to Mohabey & Samant (2013), dinosaur extinction in India tied to Deccan volcanism [*2].

[References: (*1) Wilson JA, Sereno PC, Srivastava S, Bhatt DK, Khosla A, Sahni A (2003). A new abelisaurid (Dinosauria, Theropoda) from the Lameta Formation (Cretaceous, Maastrichtian) of India. *Contributions from the Museum of Paleontology University of Michigan* 31(1): 1–42.; (*2) Mohabey DM, Samant B (2013). Deccan continental flood basalt eruption terminated Indian dinosaurs before the Cretaceous-Paleogene boundary. *Geological Society of India Special Publication* (1): 260–267.]

Rapator ornitholestoides

> *Rapator ornitholestoides* von Huene, 1932 [Albian; Griman Creek Formation, Lightning Ridge, New South Wales, Australia]

Generic name: *Rapator* ← {(Mediaeval Lat.) rapator: violator}. Von Huene gave no etymology.
Specific name: *ornitholestoides* ← {(Gr.) ὄρνις/ornis: bird + ληστής/lēstēs: robber, plunderer + -oides}; resembling *Ornitholestes*.
Etymology: **plunderer (raptor)** resembling *Ornitholestes* [*1]
Taxonomy: Theropoda: Megaraptora [*2]
Notes: The left first metacarpal of *Rapator* is opalised. *Rapator* is different from *Australovenator* in some small details of the bone [*2].

[References: (*1) Australian age of dinosaurs | *Rapator ornitholestoides*. ⟨https://www.australianageofdinosaurs.com/page/90/australian-age-of-dinosaurs-rapator-ornitholestoides⟩; (*2) White MA, Falkingham PL, Cook AG, Hocknull SA, Elliott DA (2013). Morphological comparisons of metacarpal I for *Australovenator wintonensis* and *Rapator ornitholestoides*: implications for their taxonomic relationships. *Alcheringa: An Australasian Journal of Palaeontology* 37(4): 435–441.]

Rapaxavis pani [Avialae]

> *Rapaxavis pani* Morschhauser *et al.*, 2009 [Early Cretaceous; Jiufotang Formation, Liaoning, China]

Generic name: *Rapaxavis* ← {(Lat.) rapāx: grasping + avis: bird}; "referring to the inferred grasping ability of the pes of this individual" [*1].
Specific name: *pani* ← {(person's name) Pan | (Gr. myth.) Πάν/Pan + -ī}; "in honor of the specimen's discoverer, Mr. Pan Lijun. There is also a parallel in the species name with the Greek god Pan, who, according to some traditions, is synonymous with Silvanus, the Roman god of the woods. The ecosystem of the Jehol Group is interpreted as a series of lakes in a forested region" [*1].
Etymology: Pan's **grasping bird**
Taxonomy: Theropoda: Avialae: Enantiornithes: Longipterygidae
cf. Aves: Ornithothoraces: Enantiornithes: Euenantiornithes [*1]

[References: (*1) Morschhauser EM,Varricchio DJ, Gao C, Liu J, Wang X, Cheng X, Meng Q (2009). Anatomy of the Early Cretaceous bird *Rapaxavis pani*, a new species from Liaoning Province, China. *Journal of Vertebrate Paleontology* 29(2): 545–554.]

Rapetosaurus krausei

> *Rapetosaurus krausei* Curry Rogers & Forster, 2001 [Maastrichtian; Anembalemba Member, Maevarano Formation, Mahajanga Basin, Madagascar]

Generic name: *Rapetosaurus* ← {(Malagasy folklore) Rapeto [pron.: ruh-PAY-tu] [*1]: a mischievous giant + (Gr.) σαῦρος/sauros: lizard} [*1]
Specific name: *krausei* ← {(person's name) Krause + -ī}; "in honor of David W. Krause, for his contributions to palaeontology in Madagascar" [*1].
Etymology: Krause's **Rapeto lizard**
Taxonomy: Saurischia: Sauropodomorpha: Sauropoda: Titanosauria: Saltasauridae: Saltasaurinae
cf. Saurischia: Sauropoda: Titanosauria [*1]
Notes: According to Curry Rogers *et al.* (2016), they found tiny fossils that belong to a baby *Rapetosaurus*. The fossils indicate the baby died several weeks after birth. [*2]

[References: (*1) Curry Rogers K, Forster CA (2001). The last of the dinosaur titans: a new sauropod from Madagascar. *Nature* 412: 530–534.]; (*2) Curry Rogers K, Whitney M, D'Emic M, Bagley B (2016). Precocity in a tiny titanosaur from the Cretaceous of Madagascar. *Science* 352(6284): 450–453.]

Raptorex kriegsteini

> *Raptorex kriegsteini* Sereno *et al.*, 2009 [Barremian; Lujiatun Member, Yixian Formation, Inner Mongolia, China]

Generic name: *Raptorex* ← {(Lat.) raptor: robber + rēx: king}; suggesting a small predator and 'rex' linking it to *Tyrannosaurus rex* [*1].
Specific name: *kriegsteini* ← {(person's name) Kriegstein + -ī}; in honor of Roman Kriegstein and his wife, survivors of the Holocaust, whose son Henry Kriegstein donated the specimen to the University of Chicago for scientific study [*1].
Etymology: Kriegstein's **robber king**
Taxonomy: Theropoda: Tyrannosauroidea:

Tyrannosauridae: Tyrannosaurinae

Notes: *Raptorex* is a new small-bodied theropod [*2]. According to Fowler *et al.* (2011), the type species represents the juvenile of a large tyrannosaurid from the Late Cretaceous of Mongolia. *Raptorex* is considered to be a *nomen dubium* [*3].

[References: (*1) Mullen W, tribune reporter (2009). 125 million-year-old fossil a mini-T. rex. *Chicago Tribune*, Sept 18th 2009.; (*2) Sereno PC, Tan L, Brusatte SL, Kriegstein HJ, Zhao X, Cloward K (2009). Tyrannosaurid skeletal design first evolved at small body size. *Science* 326(5951): 418–422.; (*3) Fowler DW, Woodward HN, Freedman EA, Larson PL, Horner J (2011). Reanalysis of "*Raptorex kriegsteini*": A juvenile tyrannosaurid dinosaur from Mongolia. *PLoS ONE* 6(6): e21376.]

Ratchasimasaurus suranareae

> *Ratchasimasaurus suranareae* Shibata *et al.*, 2011 [Aptian; Khok Kruat Formation, Nakhon Ratchasima, Thailand]

Generic name: *Ratchasimasaurus* ← {(place-name) Nakhon Ratchasima + (Gr.) σαῦρος/sauros: lizard}

Specific name: *suranareae* ← {Suranaree/ Suranari + -ae}; named after Thao Suranari "the brave lady" who saved the city of Nakhon Ratchasima in the 19th century.

Etymology: **Ratchasima lizard** of Thao Suranari

Taxonomy: Ornithischia: Ornithopoda: Iguanodontia

[References: Shibata M, Jintasakul P, Azuma Y (2011). A new iguanodontian dinosaur from the Lower Cretaceous Khok Kruat Formation, Nakhon Ratchasima in northeastern Thailand. *Acta Geologica Sinica* (*English edition*) 85(5): 969–976.]

Rativates evadens

> *Rativates evadens* McFeeters *et al.*, 2016 [Campanian; Oldman Formation, Alberta, Canada]

Generic name: *Rativates* ← {(Lat.) ratis: raft, referring to the ratites, a group of large flightless birds that includes the modern ostrich + vātēs: foreteller}; alluding to "the paradox of an ostrich mimic dinosaur existing before ostriches" [*2].

Specific name: *evadens* ← {(Lat.) evadens: evading}; referring to its ability of the swift-running animal to evade predators, and to evade for eighty years following the discovery of the original fossil to be recognized as a new species. [*2]

Etymology: evasive, **foreteller of ratites**

Taxonomy: Theropoda: Coelurosauria:
Maniraptoriformes: Ornithomimosauria:
Ornithomimidae

Notes: The specimen of *Rativates* is a small size, though it is not a juvenile [*1].

[References: (*1) McFeeters B, Ryan MJ, Schröder-Adams C, Cullen TM (2016). A new ornithomimid theropod from the Dinosaur Park Formation of Alberta, Canada. *Journal of Vertebrate Paleontology* 36(6): e1221415.; (*2) Bogar G (Sept. 21, 2016) New Dinosaur named for ability to evade predators. Cleveland Museum of Natural History. 〈https://www.cmnh.org/rativates〉]

Rayososaurus agrioensis

> *Rayososaurus agrioensis* Bonaparte, 1996 [early Cenomanian; Candeleros Formation, Neuquén Group, Neuquén, Argentina]

Generic name: *Rayososaurus* ← {(place-name) Rayoso Formation + (Gr.) σαῦρος/sauros: lizard}

Specific name: *agrioensis* ← {(place-name) Agrio + -ensis}

Etymology: **Rayoso lizard** from Agrio

Taxonomy: Saurischia: Sauropodomorpha:
Sauropoda: Diplodocoidea: Rebbachisauridae

Other species:

> *Rebbachisaurus tessonei* Calvo & Salgado, 1995 ⇒ *Rayososaurus tessonei* (Calvo & Salgado, 1995) Bonaparte, 1996 ⇒*Limaysaurus tessonei* (Calvo & Salgado, 1995) Salgado *et al.*, 2004 [early Cenomanian; Candeleros Formation, Neuquén, Argentina]

Notes: *Rayososaurus* is considered to be a member of Rebbachisauridae by its racket-shaped scapula.

[References: Carballido JL, Garrido AC, Canudo JI, Salgado L (2010). Redescription of *Rayososaurus agrioensis* Bonaparte (Sauropoda, Diplodocoidea), a rebbachisaurid from the early Late Cretaceous of Neuquén. *Geobios* 43: 493–502.]

Rebbachisaurus garasbae

> *Rebbachisaurus garasbae* Lavocat, 1954 [early Cenomanian; Ifezouane Formation, Kem Kem Beds*, Gara Sbaa, Er Rachida, Morocco] [*1]
> *cf.* Tegama Group. 'Tegana Formation' was cited for the Kem Kem beds (*The Dinosauria 2nd edition*). [*2]

Generic name: *Rebbachisaurus* ← {(demonym) Rebbach + (Gr.) σαῦρος/sauros: lizard}; referring to the fossil locality being within "le territoire des Aït Rebbach" (Lavocat, 1954: 68). 'Aït' is a Berber word meaning 'people of'. According to the Lavocat's transliteration, *Rebbachisaurus* would be pronounced 'khehb-bash-ee-sore-us'. [*1]

Specific name: *garasbae* ← {(place-name) Gara Sba or Gara Sbaa + -ae}; "in reference to the holotypic locality at Gara Sbaa, which means 'Lion Hill' in Arabic (N. Ibrahim, pers. Comm., 2013)" [*1].

According to Lavocat (1954), specific name refers to the layer of Gara Sbaa being on the territory of Aït Rebbach.

Etymology: **Rebbach's lizard** from Gara Sbaa

Taxonomy: Saurischia: Sauropodomorpha: Sauropoda: Diplodocoidea: Rebbachisauridae

Other species:

> *Rebbachisaurus tamesnensis* Lapparent, 1960 (*nomen dubium*) [Albian–Cenomanian; Echkar Formation (= upper part of the Continental Intercalaire), Tedreft, Agadez, Niger]

> *Rebbachisaurus tessonei* Calvo & Salgado, 1995 ⇒ *Rayososaurus tessonei* (Calvo & Salgado, 1995) Upchurch *et al.*, 2004 ⇒ *Limaysaurus tessonei* (Calvo & Salgado, 1995) Salgado *et al.*, 2004 [Cenomanian; Candeleros Formation, Neuquén, Argentina]

Notes: According to *The Dinosauria 2nd edition*, *Rebbachisaurus tamesnensis* was regarded to be a synonym of *Nigersaurus taqueti*. The material of *R. tamesnensis* was collected from multiple localities.

[References: Lavocat R. (1954). Sur les dinosauriens du Continental Intercalaire des Kem-Kem de la Daoura. *Comptes Rendus 19th International Geological Congress*, 1952. 1: 65–68.; (*1) Wilson JA, Allain R (2015). Osteology of *Rebbachisaurus garasbae* Lavocat, 1954, a diplodocoid (Dinosauria, Sauropoda) from the early Late Cretaceous-aged Kem Kem Beds of southeastern Morocco. *Journal of Vertebrate Paleontology*: e1000701.; (*2) Ibrahim N, Sereno PC, Varricchio DJ, *et al.* (2020). Geology and paleontology of the Upper Cretaceous Kem Kem Group of eastern Morocco. *Zookeys* 928: 1–216.]

Regaliceratops peterhewsi

> *Regaliceratops peterhewsi* Brown & Henderson, 2015 [Maastrichtian; St. Mary River Formation, Alberta, Canada]

Generic name: *Regaliceratops* ← {(Lat.) rēgālis: royal + (Gr.) ceratops: horned face}; "referring to the crown-shaped parietosquamosal frill and epiossifications and the Royal Tyrrell Museum of Palaeontology (the "Royal" appellation was bestowed on the museum in 1990 by Her Majesty Queen Elizabeth II)" (*1).

Specific name: *peterhewsi* ← {(person's name) Peter Hews + -ī}; "in honor of Peter Hews, who discovered the holotype" (*1).

Etymology: Peter Hews' **Royal horned face (ceratopsian)**

Taxonomy: Ornithischia: Ceratopsia: Ceratopsidae: Chasmosaurinae: Triceratopsini (*1)

Notes: Researchers nicknamed the dinosaur "Hellboy", a comic book character, because of its stubby horns and because of the hellish time researchers had in extricating it from so-called "evil hard rock", according to CNN News, 'Hellboy' dinosaur discovered, June 5, 2015.

[References: (*1) Brown CM, Henderson DM (2015). A new horned dinosaur reveals convergent evolution in cranial ornamentation in Ceratopsidae. *Current Biology* 25(12): 1641–1648.]

Regnosaurus northamptoni

> *Regnosaurus northamptoni* Mantell, 1848 [Valanginian; Tunbridge Wells Sand Formation, Tilgate Forest, Cuckfield, Sussex, UK]

?= *Craterosaurus pottonensis* Seeley, 1874 [early Aptian; Woburn Sand Formation, UK]

Generic name: *Regnosaurus* ← {(demonym) Regni: a British tribe inhabiting Sussex + (Gr.) σαῦρος/sauros: lizard}

Specific name: *northamptoni* ← {(person's name) Northampton + -ī}; "as attribute of respect to the eminent nobleman whose approaching retirement from the Presidency of the Royal Society is so much to be regretted" (*1).

Etymology: Northampton's **Sussex dinosaurs (lizard)**

Taxonomy: Ornithischia: Thyreophora: Stegosauria: Huayangosauridae

Notes: Estimated to be 4 m in total length. According to *The Dinosauria 2nd edition*, *Regnosaurus* is considered to be a *nomen dubium* because the remains are so limited.

[References: (*1) Mantell GA (1848). On the structure of the jaws and teeth of the Iguanodon. *Philosophical Transactions of the Royal Society of London* 138: 183–202.]

Revueltosaurus callenderi [Suchia]

> *Revueltosaurus callenderi* Hunt, 1989 [Norian; Bull Canyon Formation, New Mexico, USA]

Generic name: *Revueltosaurus* ← {(place-name) Revuelto Creek | (Spanish) revuelta: revolution (revolt) + (Gr.) σαῦρος/sauros}; in reference to " 'revolution' (revolt) which is appropriate for an animal from the Late Triassic period during which there was a revolution in terrestrial vertebrate evolution" (*1).

Specific name: *callenderi* ← {(person's name) Callender + -ī}

Etymology: Callender's **Revuelto lizard**

Taxonomy: Archosauria: Pseudosuchia: Suchia

cf. Archosauria: Dinosauria: Ornithischia (*1)

Other species:

> *Pekinosaurus olseni* Hunt & Lucas, 1994 (*nomen dubium*) ⇒ *Revueltosaurus olseni* (Hunt & Lucas, 1994) Irmis *et al.*, 2007 (*nomen dubium*) [Carnian; Pekin Formation, North Carolina, USA]

> *Revueltosaurus hunti* Heckert, 2002 ⇒ *Krzyzanowskisaurus hunti* (Heckert, 2002)

Heckert, 2005 [Carnian; Santa Rosa Formation, New Mexico, USA]

Notes: All of the three species were originally thought to be basal ornithischian dinosaurs (*2). According to *The Dinosauria 2nd edition*, *Revueltosaurus callenderi* was considered a *nomen dubium*.

[References: (*1) Hunt AP (1989). A new? ornithischian dinosaur from the Bull Canyon Formation (Upper Triassic) of east-central New Mexico. In: Lucas SG, Hunt AP (eds), *Dawn of the Age of Dinosaurs in the American Southwest.* The New Mexico Museum of Natural History & Science, New Mexico: 355–358.; (*2) Heckert AB (2002). A revision of the Upper Triassic ornithischian dinosaur *Revueltosaurus*, with a description of a new species. *New Mexico Museum of Natural History & Science Bulletin* (21): 253–268.]

Rhabdodon priscus

> *Rhabdodon priscus** Matheron, 1869 [Maastrichtian; Bouches-du-Rhône, Provence-Alpes-Côte d'Azur, France]
  **Rhabdodon priscum* (*1)

= *Oligosaurus adelus* Seeley, 1881 [Campanian; Grünbach Formation, Niederösterreich, Austria]

= *Ornithomerus gracilis* Seeley, 1881 [Campanian; Grünbach Formation, Niederösterreich, Austria]

?= *Iguanodon suessi* Bunzel, 1871(?*nomen dubum*) ⇒ *Mochlodon suessi* (Bunzel, 1871) Seeley, 1881 (*nomen dubium*) [Campanian; Grünbach Formation, Niederösterreich, Austria]

Generic name: *Rhabdodon* ← {(Gr.) ῥαβδος/rhabdos: a rod, a stripe or strip, stave, flute + ὀδών/odōn: tooth}

Specific name: *priscus* ← {(Lat.) prīscus: of former times, of old, ancient}

Etymology: ancient, **fluted tooth**

Taxonomy: Ornithischia: Ornithopoda: Iguanodontia: Rhabdodontidae

Other species:

> *Rhabdodon septimanicus* Buffetaut & Le Loeuff, 1991 [Campanian–Maastrichtian; Grès de Saint Chinian Formation, Hérault, Franse]

> *Mochlodon robustum* Nopcsa, 1899 (*partim*) ⇒ *Rhabdodon robustum* (Nopcsa, 1899) Nopcsa, 1915 (*partim*) ⇒ *Zalmoxes robustus* (Nopcsa, 1899) Weishampel *et al.*, 2003 [Maastrichtian; Sanpetru Formation, Hunedoara, Transylvania, Romania]

[References: (*1) Matheron P (1869). Notice sur les reptiles fossiles des dépôts fluvio-lacustres crétacés du bassin à lignite de Fuveau. *Mémoires de l'Académie des Sciences, Belles-Lettres, et Arts de Marseille* 1868–1869: 345–379.]

Rhamphorhynchus muensteri [Pterosauria]

n.b. (type species: *Pterodactylus longicaudus* Münster, 1839)

> *Ornithocephalus muensteri* Goldfuss, 1831 ⇒ *Pterodactylus muensteri* (Goldfuss, 1831) von Meyer, 1845 ⇒*Rhamphorhynchus münsteri* (Goldfuss, 1831) Owen, 1861 ⇒ *Rhamphorhynchus munsteri* (Goldfuss, 1831) Lydekker, 1888 [Tithonian; Solnhofen Limestone, Bayern, Germany]

= *Pterodactylus longicaudus* Münster, 1839 ⇒ *Ornithocephalus longicaudus* (Münster, 1839) Wagner, 1851 ⇒*Pterodactylus (Rhamphorhynchus) longicaudus* (Münster, 1839) von Meyer, 1846 ⇒ *Rhamphorhynchus longicaudus* (Münster, 1839) Ammon, 1884 ⇒ *Rhamphorhynchus (Odontorhychus) longicaudus* (Münster, 1839) Stolley, 1936 ⇒ *Odontorhynchus longicaudus* (Münster, 1839) Olshevsky, 1978 (type species) [late Tithonian; Solnhofen Limstone, Bayern, Germany]

= *Pterodactylus (Rhamphorhynchus) gemmingi* von Meyer, 1846 ⇒ *Ornithocephalus gemmingi* (von Meyer, 1846) Wagner, 1851 ⇒ *Rhamphorhynchus gemmingi* (von Meyer, 1846) von Meyer, 1859 [late Tithonian; Lithographic Shale, Solnhofen Limestone, Bayern, Germany]

= *Pterodactylus lavateri* von Meyer, 1838 ⇒ *Ornithopterus lavateri* (von Meyer, 1838) von Meyer, 1838 (*nomen dubium*) [late Tithonian; Lithographic Shale, Solnhofen Limestone, Bayern, Germany]

= *Rhamphorhynchus suevicus* O. Fraas, 1885 [late Tithonian; Nusplinger Formation, Baden-Wurttemberg, Germany]

= *Pterodactylus hirundinaceus* Wagner, 1857 (*nomen dubium*) [no record]

= *Rhamphorhynchus curtimanus* Wagner, 1858 [no record]

= *Rhamphorhynchus longimanus* Wagner, 1858 [late Tithonian; Solnhofen Limestone, Bayern, Germany]

= *Rhamphorhynchus meyeri* Owen, 1870 [late Tithonian; Solnhofen Limestone, Bayern, Germany]

= *Rhamphorhynchus phyllurus* Marsh, 1882 ⇒ *Pteromonodactylus phyllurus* (Marsh, 1882) Teryaev, 1967 [late Tithonian; Solnhofen Limestone, Bayern, Germany]

= *Rhamphorhynchus longiceps* Woodward, 1902 [late Tithonian; Solnhofen Limestone, Bayern, Germany]

= *Rhamphorhynchus kokeni* F. Plieninger, 1907

[late Tithonian; Nusplinger Formation, Baden-Wurttemberg, Germany]
= *Rhamphorhynchus megadactylus* von Koenigswald, 1931 [late Tithonian; Solnhofen Limestone, Bayern, Germany]
= *Rhamphorhynchus carnegiei* Koh, 1937 [late Tithonian; Solnhofen Limestone, Bayern, Germany]
= *Rhamphorhynchus intermedius* Koh, 1937 [late Tithonian; Solnhofen Limestone, Bayern, Germany]

Generic name: *Rhamphorhynchus* Meyer, 1846 ← {(Gr.) ῥάμφος/rhamphos: a beak, bill + ῥύγχος/rhynchos: snout, muzzle, of swine}
Specific name: *muensteri* ← {(person's name) Münster + -ī}; in honor of paleontologist Georg Graf zu Münster.
Etymology: Münster's **beak-snout**
Taxonomy: Pterosauria: Rhamphorhynchoidea: Rhamphorhynchidae
Other species:
> *Rhamphorhynchus jessoni* Lydekker, 1890 (*nomen dubium*) [Oxfordian; Oxford Clay Formation, UK]
> *Rhamphorhynchus tendagurensis* Reck, 1931 (*nomen dubium*) [Tithonian; Tendaguru Formation, Lindi, Tanzania]
> *Rhamphorhynchus etchesi* O'Sullivan & Martill, 2015 [Tithonian; Kimmeridge Clay, Dorset, UK]

Notes: *Rhamphorhynchus* had a long tail, stiffened with ligaments.

Rhinorex condrupus

> *Rhinorex condrupus* Gates & Scheetz, 2014 [Campanian; Nelson Formation, Utah, USA]

Generic name: *Rhinorex* ← {(Gr.) rhino- [(Gr.) ῥίς/rhis: nose] + (Lat.) rēx: king}; "in reference to the large nose possessed by this taxon" (*1).
Specific name: *condrupus* ← {(Lat.) cond- [condō: bury] + rūpēs, *f.*: a rock, cliff}; "for being buried in rock in the Book Cliffs of Utah" (*1).
Etymology: **nose king**, buried in the Book Cliffs
Taxonomy: Ornithischia: Ornithopoda: Hadrosauridae: Saurolophinae: Kritosaurini

[References: (*1) Gates TA, Scheetz R (2014). A new saurolophine hadrosaurid (Dinosauria: Ornithopoda) from the Campanian of Utah, North America. *Journal of Systematic Palaeontology* 13: 711–725.]

Rhoetosaurus brownei

> *Rhoetosaurus brownei* Longman, 1926 [Oxfordian; Walloon Coal Measures*, Queensland, Australia]

*according to Todd *et al.* (2019)

Generic name: *Rhoetosaurus* ← {(Lat.) Rhoetus: the giant Rhoetos in Greek Mythology + (Gr.) σαῦρος/sauros: reptile} (*1) (*2)
Specific name: *brownei* ← {(person's name) Browne + -ī}; in honor of Arthur Browne, manager of Durham Downs, Taloona Station (farm) in Queensland. (*2)
Etymology: Browne's **Rhoetus lizard**
Taxonomy: Saurischia: Sauropodomorpha: Sauropoda: Gravisauria

[References: (*1) Longman HA (1926). A giant dinosaur from Durham Downs, Queensland. *Memoirs of the Queensland Museum* 8: 183–194.; (*2) Dinosaurs - *Rhoetosaurus brownei* - Australian Museum. ⟨https://australian.museum/learn/dinosaurs/fact-sheets/rhoetosaurus-brownei/⟩

Riabininohadros weberae

> *Orthomerus weberi* Riabinin, 1945 ⇒ *Orthomerus weberae* (Riabinin, 1945) Nessov, 1995 ⇒ "Riabininohadros" *weberae* (Riabinin, 1945) Ulansky, 2015 (*1) ⇒ *Riabininohadros weberae* (Riabinin, 1945) Lopatin & Averianov, 2020 (formally) [Maastrichtian; Mt. Besh-Kosh, Crimea, Ukraine]

Generic name: *Riabininohadros* ← {(person's name) Riabinin + hadros}; in honor of Anatoly Nikolaevich Riabinin who first described *Orthomerus weberi*.
Specific name: *weberae* ← {(person's name) Weber + -ae}; in honor of G. F. Weber (female).
Etymology: Weber and **Riabinin's hadrosaur**
Taxonomy: Ornithischia: Ornithopoda: Ankylopollexia
cf. Ornithischia: Ornithopoda: Hadrosauria: Hadrosauridae (*1)

Notes: Although original genus *Orthomerus weberi* was considered a *nomen dubium*, this was given own genus name, *Riabininohadros*.

[References: (*1) Ulansky RE (2015). *Riabininohadros*, a new genus for hadrosaur from Maastrichtian of Crimea, Russia. *Dinologia*, 10pp. [In Russian]]

Richardoestesia gilmorei* / *Ricardoestesia** gilmorei*

*The spelling of generic name is according to ICZN / **according to *The Dinosauria 2nd edition*.

> *Richardoestesia gilmorei* Currie *et al.*, 1990 [Campanian; Dinosaur Park Formation, Alberta, Canada]

Generic name: *Richardoestesia* ← {(person's name) Richard Estes + -ia}; in honor of paleontologist Richard Estes.
Specific name: *gilmorei* ← {(person's name) Gilmore

+ -ī}; in honor of Charles Whitney Gilmore.
Etymology: **for Richard Estes** and Gilmore
Taxonomy: Theropoda: Coelurosauria
Other species:

> *Richardoestesia isosceles* Sankey, 2001 [Campanian; Aguja Formation, Texas, USA]
> *Asiamericana asiatica* Nessov, 1995 ⇒ ?*Richardoestesia asiatica* (Nessov, 1995) Sues & Averianov, 2013 [Turonian; Bissekty Formation, Navoi, Uzbekistan]

[References: Currie PJ, Rigby KJ, Sloan RE (1990). Theropod teeth from the Judith River Formation of southern Alberta, Canada. In: Curry PJ, Carpenter K (eds), *Dinosaur Systematics: Perspectives and Approaches.* Cambridge University Press, Cambridge: 107–125.]

Rinchenia mongoliensis

> *Oviraptor mongoliensis* Barsbold, 1986 ⇒ *Rinchenia mongoliensis* (Barsbold, 1986) Barsbold, 1997 (*nomen nudum*) ⇒ *Rinchenia mongoliensis* (Barsbold, 1986) Osmólska *et al.*, 2004 [Maastrichtian; Nemegt Formation, Ömnögovi, Mongolia]

Generic name: *Rinchenia* ← {(person's name) Rinchen + -ia}

The genus name became official in 2004 when decribed by Halszka Osmólska *et al.* (*1).

Specific name: *mongoliensis* ← {(place-name) Mongolia + -ensis}
Etymology: **Rinchen's one** from Mongolia
Taxonomy: Theropoda: Oviraptorosauria: Oviraptoridae

[References: Barsbold R (1997). Oviraptorosauria. In: Currie PJ, Padian K (eds), *Encyclopedia of Dinosaurs*, Academic Press, San Diego: 505–509.; (*1) Osmólska H, Currie PJ, Barsbold R (2004). Oviraptorosauria. In: Weishampel DB, Dodson P, Osmólska H (eds), *The Dinosauria 2nd edition*. California University Press, Berkley: 165–183.]

Rinconsaurus caudamirus

> *Rinconsaurus caudamirus* Calvo & González Riga, 2003 [late Turonian–Coniacian; Río Neuquén Formation, Neuquén Group, Neuquén, Argentina]

Generic name: *Rinconsaurus* ← {(place-name) Rincón de los Sauces + (Gr.) σαῦρος/sauros: lizard}
Specific name: *caudamirus* ← {(Lat.) cauda: tail + mīrus: astonishing, amazing}; "in reference to the unusual morphology of posterior caudal vertebrae" (*1).
Etymology: **the dinosaur from Rincón** with an amazing tail (*1)
Taxonomy: Saurischia: Sauropodomorpha: Sauropoda: Titanosauria: Lithostrotia: Aeolosaurini
cf. Saurischia: Sauropodomorpha: Sauropoda: Titanosauria: Titanosauridae (*1)

[References: (*1) Calvo JO, González Riga BJ (2003). *Rinconsaurus caudamirus* gen. *et* sp. nov., a new titanosaurid (Dinosauria, Sauropoda) from the Late Cretaceous of Patagonia, Argentina. *Revista Geológica de Chile* 30(2): 333–353.]

Riojasaurus incertus

> *Riojasaurus incertus* Bonaparte, 1969 [Norian; Los Colorados Formation, La Rioja, Argentina]
= *Strenusaurus procerus* Bonaparte, 1969 [Norian; Los Colorados Formation, La Rioja, Argentina]

Generic name: *Riojasaurus* ← {(place-name) La Rioja Province + (Gr.) σαῦρος/sauros: lizard}
Specific name: *incertus* ← {(Lat.) incertus: uncertain}
Etymology: uncertain, **La Rioja lizard**
Taxonomy: Saurischia: Sauropodomorpha: Riojasauridae
Notes: According to *The Dinosauria 2nd edition*, *Strenusaurus procerus* was a *nomen dubium*, but now it is regarded to be a junior synonym of *Riojasaurus incertus*.

Rocasaurus muniozi

> *Rocasaurus muniozi* Salgado & Azpilicueta, 2000 [late Campanian; lower member, Allen Formation, Río Negro, Argentina]

Generic name: *Rocasaurus* ← {(place-name) Roca; "the rionegrine city of General Roca, nearby the locality of Salitral Moreno" (*1) + (Gr.) σαῦρος/sauros: lizard}
Specific name: *muniozi* ← {(person's name) Munioz + -ī}; "dedecated to Sr. Juan Carlos Muñoz, attendant of the Area Paleontologia of the Museo "Carlos Ameghino" of the city of Cipoletti, for his permanent support of the paleontological investigations in the region" (*1).
Etymology: Munioz' **Roca lizard**
Taxonomy: Saurischia: Sauropodomorpha: Sauropoda: Titanosauria: Saltasauridae: Saltasaurinae: Saltasaurini
cf. Sauropodomorpha: Sauropoda: Titanosauridae: Saltasaurinae (*1)

[References: (*1) Salgado L, Azpilicueta CV (2000). A new saltasaurine (Sauropoda: Titanosauridae) from the province of Río Negro (Allen Formation, Upper Cretaceous), Patagonia, Argentina. *Ameghiniana* 37(3): 259–264.]

Rubeosaurus ovatus

> *Styracosaurus ovatus* Gilmore, 1930 ⇒ *Rubeosaurus ovatus* (Gilmore, 1930) McDonald

& Horner, 2010 [Campanian; upper member, Two Medicine Formation, Montana, USA]

Generic name: *Rubeosaurus* ← {(Lat.) rubeus [rubus: a thornbush] + (Gr.) σαῦρος/sauros: lizard}; referring to the array of spikes on its frill.

Specific name: *ovatus* ← {(Lat. mid 18th century) ovatus: ovate, egg-shaped [*1]}; referring to the ovate character of the horn-like processes [*1].

cf. (Lat.) ovātus: participle of [ovō: exult, rejoice]

Etymology: egg-shaped **thornbush lizard**

Taxonomy: Ornithischia: Ceratopsidae: Centrosaurinae: Centrosaurini

Notes: Some researchers suggest that *Brachyceratops* may be juvenile *Rubeosaurus*.

[References: (*1) Gilmore CW (1930). On dinosaurian reptiles from the Two Medicine Formation of Montana. *Proceedings of the United States National Museum* 77(16): 1–39.]

Ruehleia bedheimensis

> *Ruehleia bedheimensis* Galton, 2001 [Norian; Trossingen Formation, Thüringen, Germany]

Generic name: *Ruehleia* ← {(person' name) Ruehle + -ia}; in honor of German paleontologist Hugo Ruehle von Lilienstern.

Specific name: *bedheimensis* ← {(place-name) Bedheim + -ensis}

Etymology: **for Ruehle** from Bedheim

Taxonomy: Saurischia: Sauropodomorpha: Prosauropoda

Rugocaudia cooneyi

> *Rugocaudia cooneyi* Woodruff, 2012 [Albian; Himes Member, Cloverly Formation, Montana, USA]

Generic name: *Rugocaudia* ← {(Lat.) rūga: wrinkle + cauda: tail}

Specific name: *cooneyi* ← {(person's name) Cooney + -ī}; in honor of J. P. Cooney, owner of the land where the fossils were discovered.

Etymology: Cooney's **wrinkly tail**

Taxonomy: Saurischia: Sauropodomorpha: Sauropoda: Titanosauriformes

[References: Woodruff DC (2012). A new titanosauriform from the Early Cretaceous Cloverly Formation of Montana. *Cretaceous Research* 36: 58–66.]

Rugops primus

> *Rugops primus* Sereno *et al.*, 2004 [Cenomanian; Echkar Formation, Agadez, Niger]

Generic name: *Rugops* ← {(Lat.) ruga: wrinkle + (Gr.) ὤψ/ōps: face}

Specific name: *primus* ← {(Lat.) prīmus: first}; "named for its significance as one of the earliest abelisaurids with textured external skull surfaces" [*1].

Etymology: first, **wrinkle face**

Taxonomy: Theropoda: Ceratosauria: Abelisauroidea: Abelisauridae

[References: (*1) Sereno PC, Wilson JA, Conrad JL (2004). New dinosaurs link southern landmasses in the Mid-Cretaceous. *Proceedings of the Royal Society B: Biological Sciences* 271(1546): 1325–1330.]

Rukwatitan bisepultus

> *Rukwatitan bisepultus* Gorscak *et al.*, 2014 [Cenomanian; Namba Member, Galula Formation, Tanzania]

Generic name: *Rukwatitan* ← {(place-name) Rukwa (masc.): Lake Rukwa and the Rukwa Rift Basin + (Gr. myth.) Τῑτάν/Tītan: giants}; "referring to Lake Rukwa of southwestern Tanzania and structural rift basin of the same name from which the holotype and referred humerus were recovered, and titan, offspring of Uranus and Gaea, symbolic of brute strength and large size" [*1].

Specific name: *bisepultus* ← {(Lat.) bi-: two- + sepultus [sepeliō: to bury]} twice buried; "in reference to dual nature of holotype specimen being initially entombed in an overbank-derived mudstone, with a portion of the same skeleton later being mobilized by a paleochannel and reburied nearby as part of a channel sandstone facies" [*1].

Etymology: twice-buried **Rukwa titan**

Taxonomy: Saurischia: Sauropoda: Titanosauria [*1]

[References: (*1) Gorscak E, O'Connor PM, Stevens NJ, Roberts EM (2014). The basal titanosaurian *Rukwatitan bisepultus* (Dinosauria, Sauropoda) from the middle Cretaceous Galula Formation, Rukwa Rift Basin, southwestern Tanzania. *Journal of Vertebrate Paleontology* 34(5): 1133–1154.]

Ruyangosaurus giganteus

> *Ruyangosaurus giganteus* Lü *et al.*, 2009 [Aptian–Albian; Haoling Formation, Ruyang, Henan, China]

Generic name: *Ruyangosaurus* ← {(Chin. place-name) Rǔyáng County汝陽県 + (Gr.) σαῦρος/sauros: lizard, reptile} [*1]

Specific name: *giganteus* ← {(Lat.) giganteus: very large, huge} [*1]

Etymology: huge **Ruyang lizard**

Taxonomy: Saurischia: Sauropodomorpha: Sauropoda: Titanosauria

Notes: *Ruyangosaurus* was originally described as being of early Late Cretaceous (Lü *et al.*, 2009) [*1], but recent work has assigned it an Aptian–Albian Age (Xu *et al.*, 2012). *Ruyangosaurus giganteus* is estimated to be 24.8 m in total length, according to Molina-Perez and Larramendi (2020).

[References: (*1) Lü J, Xu L, Jia S, Zhang X, Zhang J, Yang L, You H, Ji Q (2009). A new gigantic sauropod dinosaur from the Cretaceous of Ruyang, Henan, China. *Geological Bulletin of China* 28(1): 1–10.]

S

Sacisaurus agudoensis [Silesauridae]

> *Sacisaurus agudoensis* Ferigolo & Langer, 2006 [Norian; Caturrita Formation, Agudo, Rio Grande do Sul, Brazil]

Generic name: *Sacisaurus* ← {(Tupi, Brazilian lore) Saci: a one-legged creature + (Gr.) σαῦρος/sauros: lizard} ; referring to the fact that only right femora have been found.

Specific name: *agudoensis* ← {(place-name) Agudo: + -ensis}; "referring to Agudo, the town where the material was found".

Etymology: **Saci lizard** from Agudo

Taxonomy: Dinosauriformes: Dracohors: Silesauridae

[References: Ferigolo J, Langer MC (2006). A Late Triassic dinosauriform from south Brazil and the origin of the ornithischian predentary bone. *Historical Biology* 19(1): 1–11.]

Sahaliyania elunchunorum

> *Sahaliyania elunchunorum* Godefroit *et al.*, 2008 [Maastrichtian; Yuliangze Formation, Wulaga, Heilongjian, China]

Generic name: *Sahaliyania* ← {(Manchu) Sahaliyan: black | (place-name) Sahaliyan Ula + -ia}; "referring to Amur/Heilongjiang River (Sahaliyan Ula)"(*1).

Specific name: *elunchunorum* ← {(Chin. demonym) Èlùnchūn鄂倫春* + -ōrum}

*"The Elunchun nationality is one of the smallest Chinese minorities. These hunters lived for generations in the Wulaga area"(*1).

Etymology: Elunchun's **one from Sahaliyan Ula**

Taxonomy: Ornithischia: Ornithopoda: Hadrosauridae: Lambeosaurinae

[References: (*1) Godefroit P, Hai S, Yu T, Lauters P (2008). New hadrosaurid dinosaurs from the uppermost Cretaceous of northeastern China. *Acta Palaeontologica Polonica* 53(1): 47–74.]

Saichania chulsanensis

> *Saichania chulsanensis* Maryańska, 1977 [Campanian; Barun Goyot / Baruungoyot Formation, Ömnögovi, Mongolia]

Generic name: *Saichania* ← {(Mong.) Сайхан/saichan: beautiful(*1) + -ia}. Reason for name was not explained.

Specific name: *chulsanensis* ← {(place-name) Chulsan: Khulsan, Nemegt Basin, Gobi Desert + -ensis}

Etymology: **beautiful one** from Khulsan

Taxonomy: Ornithischia: Thyreophora: Ankylosauria: Ankylosauridae

[References: (*1) Maryańska T (1977). Ankylosauridae (Dinosauria) from Mongolia. *Palaeontologia Polonica* 37: 85–151.]; Arbour VM, Lech-Hernes NL, Guldberg TE, Hurum JH, Currie PJ (2013). An ankylosaurid dinosaur from Mongolia with in situ armour and keratinous scale impressions. *Acta Palaeontologica Polonica* 58(1): 55–64.]

Saltasaurus loricatus

> *Saltasaurus loricatus* Bonaparte & Powell, 1980 [early Maastrichtian; Lecho Formation, Salta Group, Salta, Argentina]

Generic name: *Saltasaurus* ← {(place-name) Salta + (Gr.) σαῦρος/sauros: lizard}

Specific name: *loricatus* ← {(Lat.) lōrīcātus: clothed in mail, harnessed(*1)}.

Etymology: **Salta lizard** clothed in armor

Taxonomy: Saurischia: Sauropodomorpha: Sauropoda: Titanosauria: Saltasauridae

Other species:

> *Titanosaurus australis* Lydekker, 1893 ⇒ *Saltasaurus australis* (Lydekker, 1893) McIntosh, 1990 ⇒ *Neuquensaurus australis* (Lydekker, 1893) Powell, 1992 [Santonian; Bajo de la Carpa Formation, Neuquén, Argentina]

> *Titanosaurus robustus* von Huene, 1929 ⇒ *Saltasaurus robustus* (von Huene, 1929) McIntosh, 1990 ⇒ *Neuquensaurus robustus* (von Huene, 1929) Powell, 1992 (*nomen dubium*) [late Campanian; Allen Formation, Río Negro, Argentina]

Notes: *Saltasaurus loricatus* had dermal armor, which is composed of bony plates.

[References: (*1) Powell JE (1992). Osteología de *Saltasaurus loricatus* (Sauropoda Titanosauridae) del Cretácico Superior del noroeste Argentino [Osteology of *Saltasaurus loricatus* (Sauropoda Titanosauridae) of the Upper Cretaceous of Northwest Argentina]. In: Sanz J, Buscalioni A (eds), *Los Dinosaurios y su entorno biótico. Actas del Segundo Curso de Paleontologia in Cuenca*: 165–230.]

Saltopus elginensis [Dinosauriformes]

> *Saltopus elginensis* von Huene, 1910 [Carnian; Lossiemouth Sandstone Formation, Elgin, Grampian, Scotland, UK]

Generic name: *Saltopus* ← {(Lat.) salto: leap, hop + (Gr.) πούς/pous [pūs]: foot}

Specific name: *elginensis* ← {(place-name) Elgin + -ensis}

Etymology: **hopping foot** from Elgin
Taxonomy: Ornithodira: Dinosauromorpha:
Dinosauriformes: Dracohors*
Notes: *Saltopus* was described first as a theropod dinosaur, but is considered to be a dinosauriform (*1). In *The Dinosauria 2nd edition* (2004), *Saltopus* was described to be a possible Dinosauria. Dracohors* includes *Saltopus*, Silesauridae and Dinosauria (according to Cau, 2018).
*Dracohors ← {(Lat.) draco: dragon + cohors: cohort, circle} (*2)

[References: (*1) Benton MJ, Walker AD (2010). *Saltopus*, a dinosauriform from the Upper Triassic of Scotland. *Earth and Environmental Science Transactions of the Royal Society of Edinburgh* 101(3–4): 285–299.]; (*2) Cau A (2018). The assembly of the avian body plan: a 160-million-year long process. *Bollettino della Società Paleontologica Italiana* 57(1):1–25.]

Saltriovenator zanellai

> *Saltriovenator zanellai* Dal Sasso *et al.*, 2018 [early Sinemurian; Saltrio Formation, Saltrio, Varese, Lombardy, Italy]
> *cf.* "Saltriosaurus" Dal Sasso, 2000 (*nomen nudum*)

Generic name: *Saltriovenator* ← {(place-name) Saltrio + (Lat.) venator: hunter}; "referring to a type of Roman gladiator" (*1).
Specific name: *zanellai* ← {(Lat.) Zanella + -ī}; "dedicated to Angelo Zanella, who discovered the fossil" (*1).
Etymology: Zanella's **Saltrio hunter**
Taxonomy: Theropoda: Neotheropoda: Ceratosauria

[References: Dal Sasso C, Maganuco S, Cau A (2018). The oldest ceratosaurian (Dinosauria: Theropoda), from the Lower Jurassic of Italy, sheds light on the evolution of the three-fingered hand of birds. *PeerJ* 6: e5976.]

Sanjuansaurus gordilloi

> *Sanjuansaurus gordilloi* Alcober & Martinez, 2010 [Carnian; Ischigualasto Formation, San Juan, Argentina]

Generic name: *Sanjuansaurus* ← {(place-name) San Juan Province + saurus [(Gr.) σαῦρος/sauros: lizard]} (*1)
Specific name: *gordilloi* ← {(person's name) Gordillo + -ī}; "in honor of Raúl Gordillo, head fossil preparatory and artist in the laboratory of the San Juan Museum and team member during many years of excavation" (*1).
Etymology: Gordillo's **San Juan lizard**
Taxonomy: Saurischia: Herrerasauridae (*1)

[References: (*1) Alcober OA, Martinez RN (2010). A new herrerasaurid (Dinosauria, Saurischia) from the Upper Triassic Ischigualasto Formation of northwestern Argentina. *ZooKeys* 63: 55–81.]

Sanpasaurus yaoi

> *Sanpasaurus yaoi* Young, 1944 [Toarcian; Maanshan Member, Ziliujing Formation, Sichuan, China]

Generic name:: *Sanpasaurus* ← {(Chin. place-name) Sanpa三巴 [Pinyin: Sānbā] + (Gr.) sauros: lizard}
Specific name: *yaoi* ← {(person's name) Yao + -ī}
Etymology: Yao's **Sanba lizard**
Taxonomy: Saurischia: Sauropodomorpha:
Sauropoda (*1)
Notes: According to *The Dinosauria 2nd edition*, *Sanpasaurus yaoi* was regarded as a *nomen dubium*. McPjee *et al.* (2016) refered this specimen to Sauropoda (*1).

[References: Young CC (1944). On the reptilian remains from Weiyuan, Szechuan, China. *Bulletin of the Geological Society of China* 24(3–4): 187–205.; (*1) McPhee BW, Upchurch P, Mannion PD, Sullivan C, Butler RJ, Barrett PM (2016). A revision of *Sanpasaurus yaoi* Young, 1944 from the Early Jurassic of China, and its relevance to the early evolution of Sauropoda (Dinosauria). *PeerJ* 4: e2578.]

Santanaraptor placidus

> *Santanaraptor placidus* Kellner, 1999 [early Albian; Romualdo Member, Santana Formation, Ceará, Brazil]

Generic name: *Santanaraptor* ← {(place-name) Santana Formation + (Lat.) raptor: thief}
Specific name: *placidus* ← {(Lat. person's name) Plácido}; in honor of Plácido Cidade Nuvens, who founded the Museu de Paleontologia de Santana do Cariri.
Etymology: Placido's **Santana thief**
Taxonomy: Theropoda: Coelurosauria:
Tyrannosauroidea
Notes: The holotype is seemed to be a juvenile skeleton. According to *The Dinosauria 2nd edition*, *Santanaraptor* is considered to be possible Tyrannosauroidea.

[References: Kellner AWA (1999). Short note on a new dinosaur (Theropoda, Coelurosauria) from the Santana Formation (Romualdo Member, Albian), northeastern Brazil. *Boletim do Museu Nacional, Nova Série geologia* 49: 1–8.]

Sanxiasaurus modaoxiensis

> *Sanxiasaurus modaoxiensis* Li *et al.*, 2019 [Middle Jurassic; Xintiangou Formation, Chongqing, China]

Generic name: *Sanxiasaurus* ← {(Chin.) Sānxiá三峡: 'Three Gorges' of the Yangtze River揚子江 (=長江Chang Jiang) ＋(Gr.) σαῦρος/sauros: lizard}

Specific name: *modaoxiensis* ← {(place-name) Módāoxī 磨刀溪 + -ensis}
Etymology: **Sanxia lizard** from Modaoxi
Taxonomy: Ornithischia: Neornithischia

Sapeornis chaoyangensis [Avialae]

> *Sapeornis chaoyangensis* Zhou & Zhang, 2002 [Aptian; the second member, Jiufotang Formation, Liaoning, China]
= *Didactylornis jii* Yuan, 2008 [early Aptian; Dakangpu Member, Yixian Formation, Liaoning, China]
= *Sapeornis angustis* Zhou & Zhang, 2009 [Aptian; the third member, Jiufotang Formation, Liaoning, China]
= *Shenshiornis primita* Hu *et al.*, 2010 [Aptian; the third member, Jiufotang Formation, Liaoning, China]

Generic name: *Sapeornis* ← {(acronym) SAPE (= Society for Avian Paleontology and Evolution) + (Gr.) ὄρνις/ornis: bird}
Specific name: *chaoyangensis* ← {(Chin. place-name) Cháoyáng 朝陽 + -ensis}
Etymology: **SAPE's bird** from Chaoyang
Taxonomy: Theropoda: Avialae: Euavialae: Avebrevicauda: Omnivoropterygidae
Notes: *Sapeornis* had dentition (*2).

[References: (*1) Zhou Z, Zhang F (2003). Anatomy of the primitive bird *Sapeornis chaoyangensis* from the Early Cretaceous of Liaoning, China. *Canadian Journal of Earth Sciences* 40(5): 731–747.]; (*2) Wang Y, Hu H, O'Connor J, Wang M, Xu X, Zhou Z, Wang X, Xiaoting Z (2017). A previously undescribed specimen reveals new information on the dentition of *Sapeornis chaoyangensis*. *Cretaceous Reseach* 74: 1–10.; Dececchi TA, Larsson HCE, Habib MB (2016). The wings before the bird: an evaluation of flapping-based locomotory hypotheses in bird antecedents. *PeerJ* 4: e2159.]

Sarahsaurus aurifontanalis

> *Sarahsaurus aurifontanalis* Rowe *et al.*, 2011 [Sinemurian–Pliensbachian; Kayenta Formation, Glen Canyon Group, Arizona, USA]

Generic name: *Sarahsaurus* ← {(person's name) Sarah (Mrs Ernest) Butler + (Gr.) σαῦρος/sauros: lizard} ; "named in honor of Sarah (Mrs Ernest) Butler, whose broad interests in the arts, the sciences and medicine have enriched Texas in so many marvelous ways" (*1).
Specific name: *aurifontanalis* ← {(Lat.) aurum: gold + fontanalis [fōns: a spring]: of the spring}; "in reference to Gold Spring, Arizona, where the holotype was discovered" (*1).
Etymology: **Sarah's lizard** from Gold Spring
Taxonomy: Saurischia: Sauropodomorpha: Massopoda

[References: (*1) Rowe TB, Sues H-D, Reisz RR (2010). Dispersal and diversity in the earliest North American sauropodomorph dinosaurs, with a description of a new taxon. *Proceedings of the Royal Society B: Biological Sciences* 278(1708): 1044–1053.]

Sarcolestes leedsi

> *Sarcolestes leedsi* Lydekker, 1893 [Callovian; Peterborough Member, Oxford Clay Formation, Cambridgeshire, UK]

Generic name: *Sarcolestes* ← {(Gr.) σαρκο-/sarko- [σάρξ/sarx: flesh] + λῃστής/lēstēs: robber}

A single left jaw bone was discovered. It was first thought to belong to a carnivorous dinosaur (*1). But, in 1901, Nopcsa found that *Sarcolestes* was herbivorous.

Specific name: *leedsi* ← {(person's name) Leeds + -ī}; in honor of Alfred Nicholson Leeds.
Etymology: Leeds' **flesh robber**
Taxonomy: Ornithischia: Thyreophora: Ankylosauria: ?Nodosauridae

[References: (*1) Lydekker R (1893). On the jaw of a new carnivorous dinosaur from the Oxford Clay of Peterborough. *Quarterly Journal of the Geological Society of London* 49(1–4): 284–287.; Galton PM (1983). *Sarcolestes leedsi* Lydekker, an ankylosaurian dinosaur from the Middle Jurassic of England. *Neües Jahrbuch fur Geologie und Palaontologie Monatschefte* (3): 141–155.]

Sarcosaurus woodi

> *Sarcosaurus woodi* Andrews, 1921 ⇒ *Magnosaurus woodi* (Andrews, 1921) von Huene, 1932 [Hettangian–Sinemurian; Blue Lias Formation, Leicestershire, UK]

Generic name: *Sarcosaurus* ← {(Gr.) σαρκο-/sarko- [σάρξ/sarx: flesh] + σαῦρος/sauros: lizard}
Specific name: *woodi* ← {(person's name) Wood + -ī}; in honor of S. L. Wood, who discovered the specimen.
Etymology: Wood's **flesh lizard**
Taxonomy: Theropoda: Neotheropoda
Other species:

> *Sarcosaurus andrewsi* von Huene, 1932 ⇒ *Megalosaurus andrewsi* (von Huene, 1932) Waldman, 1974 (*nomina dubia*) [Hettangian; Blue Lias Formation, Warwickshire, UK]
= *Magnosaurus woodwardi* von Huene, 1932 (*partim*) [Hettangian; Blue Lias Formation, Warwickshire, UK]

Notes: Von Huene later made a choice for *Sarcosaurus andrewsi* to be the valid name (von Huene, 1956). According to *The Dinosauria 2nd edition*, *Sarcosaurus andrewsi* is considered to be a *nomen dubium*.

[References: Andrews CW (1921). On some remains of a theropodous dinosaur from the Lower Lias of Barrow-on-Soar. *Annals and Magazine of Natural History, series* 9 (8): 570–576.]

Sarmientosaurus musacchioi

> *Sarmientosaurus musacchioi* Martínez *et al.*, 2016 [Cenomanian–Turonian; Bajo Barreal Formation, Chubut, Argentina]

Generic name: *Sarmientosaurus* ← {(place-name) Sarmiento: the Patagonian town and the administrative department + (Gr.) σαῦρος/sauros: lizard}; referring to "Salmiento, for the Patagonian town and the administrative department in which it is located, the latter of which has yielded numerous Cretaceous dinosaur fossils" (*1).

Specific name: *musacchioi* ← {(person's name) Musacchio + -ī}; in honor of "the late Dr. Eduardo Musacchio, a model scientist and educator at the Universidad Nacional de la Patagonia San Juan Bosco in Comodoro Rivadavia, Argentina" (*1).

Etymology: Dr. Musacchio's **lizard from Salmiento**

Taxonomy: Saurischia: Sauropodomorpha: Sauropoda: Titanosauriformes: Titanosauria: Lithostrotia

[References: (*1) Martínez RDF, Lamanna MC, Novas FE, Ridgely RC, Casal GA, Martínez JE, Vita JR, Witmer LM (2016). A basal lithostrotian titanosaur (Dinosauria: Sauropoda) with a complete skull: implications for the evolution and paleobiology of *Titanosauria. PLoS ONE* 11(4): e0151661.]

Saturnalia tupiniquim

> *Saturnalia tupiniquim* Langer *et al.*, 1999 [Carnian; Santa Maria Formation, Río Grande do Sul, Brazil]

Generic name: *Saturnalia* ← {(Lat.) Sāturnālia: carnival}; "in reference to the feasting period when the paratypes were found" (*1).

Specific name: "*tupiniquim*,* Portuguese word of indigenous–Guarani–origin, an endearing way of referring to native things from Brazil" (*1).

*Tupiniquim: an ethnic group of Brazil

Etymology: **Carnival's one** from Brazil

Taxonomy: Saurischia: Sauropodomorpha: Guaibasauridae

[References: (*1) Langer MC, Abdala F, Richter M, Benton MJ (1999). A sauropodomorph dinosaur from the Upper Triassic (Carnian) of southern Brazil. *Comptes Rendus de l'Académie des Sciences* 329: 511–517.]

Saurolophus osborni

> *Saurolophus osborni* Brown, 1912 [Maastrichtian; Horseshoe Canyon Formation, Edmonton Group, Alberta, Canada]

Generic name: *Saurolophus* ← {(Gr.) σαῦρος/sauros: lizard + λόφος/lophos: crest}; referring to its long median, dorsal crest.

Specific name: *osborni* ← {(person's name) Osborn + -ī}; in honor of Henry Fairfield Osborn.

Etymology: Osborn's **crested lizard**

Taxonomy: Ornithischia: Ornithopoda: Hadrosauridae: Saurolophinae: Saurolophini

Other species:

> *Saurolophus angustirostris* Rozhdestvensky, 1952 [Campanian; Nemegt Formation, Ömnögovi, Mongolia]

> *Saurolophus kryschtofovici* Riabinin, 1930 (*nomen dubium*) [late Maastrichtian; Yuliangze Formation, Heilongjiang, China]

> *Saurolophus yaoi* Young, 1946 (*partim*)

> *Saurolophus morrisi* Prieto-Márquez & Wagner, 2013 ⇒ *Augustynolophus morrisi* (Prieto-Márquez & Wagner, 2013) Prieto-Márquez *et al.*, 2014 [late Maastrichtian; Marca Member, Moreno Formation, California, USA]

Notes: According to *The Dinosauria 2nd edition*, *Saurolophus yaoi* (*partim*) and *S. kryschtofovici* are considered *nomina dubia*. *Saurolophus morrisi* was determined to be a separate genus (*2).

[References: (*1) Brown B (1912). A crested dinosaur from the Edmonton Cretaceous. *Bulletin of the American Museum of Natural History* 31(14): 131–136.; (*2) Prieto-Márquez A, Wagner JR, Bell PR, Chiappe LM (2015). The late-surviving 'duck-billed' dinosaur Augustynolophus from the upper Maastrichtian of western North America and crest evolution in Saurolophini. *Geological Magazine* 152(2): 225–241.]

Sauroniops pachytholus

> *Sauroniops pachytholus* Cau *et al.*, 2012 [Cenomanian; Ifezouane Formation, Meknès-Tafilalet, Morocco]

Generic name: *Sauroniops* ← {(fictional character) Sauron + (Gr.) ὤψ/ōps: eye}; "formed by Sauron, fictional character created by J. R. R. Tolkien (1892–1973), and 'eye'" (*1).

Specific name: *pachytholus* ← {(Gr.) πᾰχύς/pachys: thick + θόλος/tholos: dome}; "in reference to the thickened frontal dome above the orbit, diagnostic of this taxon" (*1).

Etymology: thick-domed, **Sauron's eye**

Taxonomy: Theropoda: Carnosauria: Carcharodontosauridae

Notes: A large-sized theropod dinosaur.

[References: (*1) Cau A, Dalla Vecchia FM, Fabbri M (2012). A thick-skulled theropod (Dinosauria,

Saurischia) from the Upper Cretaceous of Morocco with implications for carcharodontosaurid cranial evolution. *Cretaceous Research* 40: 251–260.]

Sauropelta edwardsorum

> *Sauropelta edwardsorum* Ostrom, 1970 [Aptian–Albian; Cloverly Formation, Montana, USA]

Generic name: *Sauropelta* ← {(Gr.) σαῦρος/sauros (masculine): lizard + (Lat.) pelta: shield = (Gr.) πέλτη/peltē (feminine): small shield}; "in reference to dermal armor"(*1).

Specific name: *edwardsorum* ← {(person's name) Edwards + -ōrum}; "named for Nell and Tom Edwards of Bridger, Montana, in appreciation of the hospitality and assistance they gave to Yale field crews" (*1).

Etymology: Edwards' **lizard shield**

Taxonomy: Ornithischia: Thyreophora: Ankylosauria: Nodosauridae

cf. Ankylosauria: Acanthopholidae (*1)

Notes: Large pointed spines lined the side of the neck.

[References: (*1) Ostrom JH (1970). Stratigraphy and paleontology of the Cloverly Formation (Lower Cretaceous) of the Bighorn Basin area, Wyoming and Montana. *Bulletin of the Peabody Museum of Natural History* 35: 1–234.]

Saurophaganax maximus

> *Saurophaganax maximus* Chure, 1995 ⇒ *Allosaurus maximus* (Chure, 1995) Smith, 1998 [Kimmeridgian; Morrison Formation, Oklahoma, USA]

= *Saurophagus maximus* Stovall, 1941* (preoccupied, *nomen nudum*) [late Kimmeridgian: Morrison Formation, Oklahoma, USA]

*Ray, a journalist, (1941) named a gigantic theropod dinosaur *Saurophagus maximus* in a popular magazine article. (*1)

Generic name: *Saurophaganax* ← {(Gr.) saurophagos [σαῦρος/sauros: reptile + φăγειν/phagein: eat] + ἄναξ/anax: master, lord, ruler, king} 'king of the reptile-eater'. (*1)

Specific name: *maximus* ← {(Lat.) māximus: greatest}.

Etymology: greatest **king of the reptile-eaters**

Taxonomy: Theropoda: Carnosauria: Allosauridae

[References: (*1) Chure DJ (1995). A reassessment of the gigantic theropod *Saurophagus maximus* from the Morrison Formation (Upper Jurassic) of Oklahoma, USA. In: Sun A, Wang Y (eds), *Sixth Symposium of Mesozoic Terrestrial Ecosystems and Biota, Short Papers.* China Ocean Press, Beijing: 103–106.]

Sauroplites scutiger

> *Sauroplites scutiger* Bolin, 1953 [Aptian; Zhidan Group, Inner Mongolia, China]

Generic name: *Sauroplites* ← {(Gr.) σαῦρος/sauros: lizard + ὁπλίτης/hoplitēs: a heavy-armed foot-soldier, armored soldier}

Specific name: *scutiger* ← {(New Latin) scutiger: shield bearer}

Etymology: **Saurian hoplite**, shield bearer

Taxonomy: Ornithischia: Thyreophora: Ankylosauria

Notes: According to *The Dinosauria 2nd edition*, *Sauroplites scutiger* was considered a *nomen dubium*, but in 2014, it was regarded as a valid taxon by V. M. Arbour.

Sauroposeidon proteles

> *Sauroposeidon proteles* Wedel *et al.*, 2000 [Aptian–Albian; Antlers Formation, Atoka County, Oklahoma, USA]

= *Paluxysaurus jonesi* Rose, 2007 [late Aptian; Twin Mountains Formation, Texas, USA]

Generic name: *Sauroposeidon* ← {(Gr.) σαῦρος/sauros: lizard + (Gr. myth.) Ποσειδῶν/Poseidōn: the god of earthquakes} (*1)

Specific name: *proteles* ← {(Gr.) πρό-/pro- + τέλεος/teleos: perfect}; "(perfect before the end), in reference to the species' culmination of brachiosaurid adaptations just before the extinction of North American sauropods"(*1).

Etymology: perfect before the end, **Poseidon lizard**

Taxonomy: Saurischia: Sauropodomorpha: Sauropoda: Somphospondyli

cf. Saurischia: Sauropodomorpha: Sauropoda: Brachiosauridae

[References: (*1) Wedel MJ, Cifelli RL, Sanders RK (2000). *Sauroposeidon proteles*, a new sauropod from the Early Cretaceous of Oklahoma. *Journal of Vertebrate Paleontology* 20(1): 109–114.]

Saurornithoides mongoliensis

> *Saurornithoides mongoliensis* Osborn, 1924 ⇒ *Troodon mongoliensis* (Osborn, 1924) Paul, 1988 [late Campanian; Bayn Dzak Member, Djadokhta Formation, Ömnögovi, Mongolia]

Generic name: *Saurornithoides* ← {(Gr.) σαῦρος/sauros: lizard + ὀρνιθ-/ornith- [ὄρνις/ornis: bird] + -oides: -like}; "signifying 'the saurian with birdlike rostrum' " (*1).

Specific name: *mongoliensis* ← {(place-name) Mongolia + -ensis}

Etymology: **bird-like saurian** from Mongolia (*1)

Taxonomy: Theropoda: Maniraptora: Troodontidae

Other species:

> *Saurornithoides junior* Barsbold, 1974 ⇒ *Zanabazar junior* (Barsbold, 1974) Norell *et al.*, 2009 [Maastrichtian; Nemegt Formation, Ömnögovi, Mongolia]

Notes: In 2009, *Saurornithoides junior* was re-classified in the new genus *Zanabazar.* (*2)

[References: (*1) Osborn HF (1924). Three new Theropoda, *Protoceratops* zone, central Mongolia. *American Museum Novitates* (144): 1–12.; (*2) Norell MA, Makovicky PJ, Bever GS, Balanoff AM, Clark JM, Barsbold R, Rowe T (2009). A review of the Mongolian Cretaceous dinosaur *Saurornithoides* (Troodontidae: Theropoda). *American Museum Novitates* 3654: 1–63.]

Saurornitholestes langstoni

> *Saurornitholestes langstoni* Sues, 1978 [late Campanian; Dinosaur Park Formation, Alberta, Canada]

Generic name: *Saurornitholestes* ← {(Gr.) σαῦρος/sauros: lizard + ὄρνιθο-/ornitho- [ὄρνις/ornis: bird] + λῃστής/lēstēs: robber}; "in reference to its similarity to the saurornithoididae and its carnivorous mode of life" (*1).

Specific name: *langstoni* ← {(person's name) Langston + -ī}; "for Dr Wann Langston Jr., now of Austin, Texas, in recognition of his contributions to Canadian vertebrate palaeontology and his personal interest and generous support of the author's work" (*1).

Etymology: Langston's **lizard-bird robber**

Taxonomy: Theropoda: Coerulosauria: Maniraptora: Dromaeosauridae: Eudromaeosauria

Other species:

> *Saurornitholestes robustus* Sullivan, 2006 (*nomen dubium*) [late Campanian; De-na-zin Member, Kirtland Formation, New Mexico, USA]

> *Saurornitholestes sullivani* Jasinski, 2015 [late Campanian; De-na-zin Member, Kirtland Formation, New Mexico, USA]

Notes: According to Evans *et al.* (2014), *Saurornitholestes robustus* is thought to be assignable to Troodontidae.

[References: (*1) Sues H-D (1978). A new small theropod dinosaur from the Judith River Formation (Campanian) of Alberta Canada. *Zoological Journal of the Linnean Society* 62: 381–400.]

Savannasaurus elliottorum

> *Savannasaurus elliottorum* Poropat *et al.*, 2016 [Cenomanian–lower Turonian; Winton Formation, Queensland, Australia]

Generic name: *Savannasaurus* ← {(Spanish, Taino) zavana: savanna + (Gr.) σαῦρος/sauros: lizard}; "in reference to the countryside in which the specimen was found" (*1).

Specific name: *elliottorum* ← {(person's name) Elliott + -ōrum}; "in honor of the Elliott family for their ongoing contributions to Australian palaeontology" (*1).

Etymology: The Elliott family's **savanna lizard**

Taxonomy: Saurischia: Sauropoda: Titanosauriformes: Titanosauria

[References: (*1) Poropat SF, Mannion PD, Upchurch P, Hocknull SA, Kear BP, Kundrát M, Tischler TR, Sloan T, Sinapius GHK, Elliott JA, Elliott DA (2016). New Australian sauropods shed light on Cretaceous dinosaur palaeobiogeography. *Scientific Reports* 6: 34467.]

Scansoriopteryx heilmanni [Scansoriopterygidae]

> *Scansoriopteryx heilmanni* Czerkas & Yuan, 2002 [uncertain; uncertain, Liaoning, China]

= *Epidendrosaurus ninchengensis* Zhang *et al.*, 2002 [Bathonian–Oxfordian; Tiaojishan Formation, Inner Mongolia, China]

Generic name: *Scansoriopteryx* ← {(Lat.) scansōrius: scansorial + (Gr.) πτέρυξ/pteryx: feather, wing}; meaning "climbing wing" (*1).

cf. scansōrius: destiné à faire monter (*2)

Specific name: *heilmanni* ← {(person's name) Heilmann + -ī}; "in honor of Gerhard Heilmann, the pioneer of avian paleontological studies who championed the concept of birds being derived from an arboreal ancestry" (*1).

Etymology: Heilmann's **climbing wing**

Taxonomy: Saurischia: Theropoda: Maniraptora: Avialae (?): Scansoriopterygidae

Notes: According to *The Dinosauria 2nd edition, Scansoriopteryx* was considered to be a synonym of *Epidendrosaurus*. *Scansoriopteryx* and *Epidendrosaurus* are so similar that they are considered the same genus. Article 21 of the International Code of Zoological Nomenclature (ICZN) would give priority to *Scansoriopteryx*, which was the first to be published.

[References: (*1) Czerkas SA, Yuan C (2002). An arboreal maniraptoran from northeast China. In: Czerkas SJ (ed), *Feathered Dinosaurs and the Origin of Flight. The Dinosaur Museum Journal* 1. The Dinosaur Museum, Blanding: 63–95.; (*2) Gaffiot F (2001). *Le Gaffiot de poche Dictionnaire Latin-Français.* Hachette-Livre, Paris.]

Scaphognathus crassirostris [Pterosauria]

> *Pterodactylus crassirostris* Goldfuss, 1831 ⇒ *Scaphognathus crassirostris* (Goldfuss, 1831) Wagner, 1861 [Tithonian; Plattenkalk Formation, Bayern, Germany]

Generic name: *Scaphognathus* ← {(Gr.) scapho-

[σκάφη/skaphe: anything dug or scooped, a trough or tub, a light boat] + γνάθος/gnathos: the jaw}; in reference to the blunt shape of the lower jaws.
Specific name: *crassirostris* ← {(Lat.) crassus: solid, thick, fat + rostrum: a beak, bill, snout}
Etymology: **boat-like jaws** with a fat snout
Taxonomy: Pterosauria: Rhamphorhynchidae: Scaphognathinae
Other species:
> *Scaphognathus robustus* Cheng, 2012 [Bathonian–Kimmeridgian; Tiaojishan Formation, Liaoning, China]

Scelidosaurus harrisonii
> *Scelidosaurus harrisonii* Owen, 1861 [Sinemurian; Charmouth Mudstone, Charmouth, Dorset, UK]
Generic name: *Scelidosaurus* ← {(Gr.) σκέλος/skelos: leg + σαῦρος/sauros: lizard}; referring to a few fragmentary fossils of limb-bones [(*1)].
Specific name: *harrisonii* ← {(person's name) Harrison + -iī}; in honor of the discoverer James Harrison [(*1)].
Etymology: Harrison's **leg lizard**
Taxonomy: Ornithischia: Thyreophora
Other species:
> *Tatisaurus oehleri* Simmons, 1965 (*nomen dubium*) ⇒ *Scelidosaurus oehleri* (Simmons, 1965) Lucas, 1996 [Sinemurian; Lufeng Formation, Yunnan, China]

[References: (*1) Owen R (1861). A monograph of a fossil dinosaur (*Scelidosaurus harrisonii*, Owen) of the Lower Lias, part 1. *Monographs on the British Fossil Reptilia from the Oolitic Formations* 1: 1–14.]

Schizooura lii [Avialae]
> *Schizooura lii* Zhou *et al.*, 2012 [Aptian; Jiufotang Formation, Liaoning, China]
Generic name: *Schizooura* ← {(Gr.) schizo- [schizo: split] + οὐρά/oura [ūrā]: tail}
Specific name: *lii* ← {(person's name) Li + -ī}; "dedicated to Mr. Li Yutong (IVPP*), who prepared this delicate specimen and many other birds and feathered dinosaurs" [(*1)].
*IVPP stands for Institute of Vertebrate Paleontology and Paleoanthropology.
Etymology: Li's **split tail**
Taxonomy: Theropoda: Avialae: Pygostylia: Euornithes *cf.* Aves: Pygostylia: Ornithurae [(*1)]

[References: (*1) Zhou S, Zhou Z-H, O'Connor JK (2012). A new basal beaked ornithurine bird from the Lower Cretaceous of western Liaoning, China. *Vertebrata PalAsiatica* 50(1): 9–24.]

Schleitheimia schutzi
> *Schleitheimia schutzi* Rauhut *et al.*, 2020 [Norian; Klettgau Formation, Canton Schaffhausen, Switzerland]
Generic name: *Schleitheimia* ← {(place-name) Schleitheim + -ia}; referring to "the type locality at Schleitheim, Canton Schaffhausen, Switzerland".
Specific name: *schutzi* ← {(person's name) Schutz + -ī}; in honor of "the collector of the type material, Emil Schutz (1916–1974)".
Etymology: Schutz's **one from Schleitheim**
Taxonomy: Saurischia: Sauropodomorpha: Sauropodiformes

Scipionyx samniticus
> *Scipionyx samniticus* Dal Sasso & Signore, 1998 [early Albian; Calcari selciferi ed ittiolitiferi di Pietraroia, Salerno, Campania, Italy]
Generic name: *Scipionyx* ← {(Lat. male name) Scipio: "dedicated to Scipione Breislak, who first described the Pietraroja Plattenkalk, and *Publius Cornelius Scipio* (nicknamed *Africanus*), *consul militaris* of the Roman Army, who fought in the Mediterranean area" + "(Gr.) ὄνυξ/onyx: claw"} [(*1)]
Specific name: *samniticus* ← {(Lat.) samnīticus; "of Samnium, the ancient name of the region that includes Pietraroja and the Benevento Province" [(*1)]}.
Etymology: **Scipione and Scipio's claw** from Samnium
Taxonomy: Theropoda: Coelurosauria: Compsognathidae
cf. Theropoda: Tetanurae: Coelurosauria: Maniraptoriformes [(*1)]

[References: (*1) Dal Sasso C, Signore M (1998). Exceptional soft-tissue preservation in a theropod dinosaur from Italy. *Nature* 392: 383–387.]

Sciurumimus albersdoerferi
> *Sciurumimus albersdoerferi* Rauhut *et al.*, 2012 [Kimmerdgian; Torleite Formation, Bayern, Germany]
Generic name: *Sciurumimus* ← {(Lat.) sciūrus: tree squirrel, *Sciurus* + (Gr.) μῖμος/mīmos: mimic}; "in reference to the bushy tail of the animal" [(*1)].
Specific name: *albersdoerferi* ← {(person's name) Albersdoerfer + -ī}; "in honor of Raimund Albersdörfer, who made the specimen available for study" [(*1)].
Etymology: Albersdöerfer's **squirrel mimic**
Taxonomy: Theropoda: Coelurosauria (according to Godefroit *et al.* 2013)
cf. Theropoda: Tetanurae: Megalosauroidea [(*1)]
Notes: A preserved skeleton of a juvenile megalosauroid, *Sciurumimus albersdoerferi,* preserves a filamentous plumage at the tail base

and on parts of the body [*1]. But, *Sciurumimus* was found to be one of the most primitive members of the Coelurosauria, according to Godefroit *et al.* (2013).

[References: (*1) Rauhut OWM, Foth C, Tischlinger H, Norell MA (2012). Exceptionally preserved juvenile megalosauroid theropod dinosaur with filamentous integument from the Late Jurassic of Germany. *Proceedings of the National Academy of Sciences* 109(29): 11746–11751.]

Scolosaurus cutleri

> *Scolosaurus cutleri* Nopcsa, 1928 [Campanian; Dinosaur Park Formation, Alberta, Canada]

Generic name: *Scolosaurus* ← {(Gr.) σκῶλος/skōlos: spine, thorn, pointed stake + σαῦρος/sauros: lizard}

Specific name: *cutleri* ← {(person's name) Cutler + -ī}; in honor of the discoverer William E. Cutler.

Etymology: Cutler's **thorn lizard**

Taxonomy: Ornithischia: Thyreophora: Ankylosauridae: Ankylosaurinae

Other species:

> *Scolosaurus thronus* Penkalski, 2018 [*3] [Campanian; Dinosaur Park Formation, Belly River Group, Alberta, Canada]

Notes: According to *The Dinosauria 2nd edition* (2004), *S. cutleri* was considered to be a synonym of *Euoplocephalus tutus*, since *Scolosaurus cutleri*, *Anodontosaurus lambei* and *Dyoplosaurus acutosquameus* had been combined into the genus *Euoplocephalus tutus* which had no thorns on the tail (Coombs, 1971). [*1] The research, later, on *Scolosaurus* described the differences among them in the pelvis and armour. [*2]

[References: (*1) Arbour VM, Currie, PJ (2013). *Euoplocephalus tutus* and the diversity of ankylosaurid dinosaurs in the Late Cretaceous of Albera, Canada, and Montana, USA. *PLoS ONE* 8(5): e62421.; (*2) Penkalski P, Blows WT (2013). *Scolosaurus cutleri* (Ornithischia: Ankylosauria) from the Upper Cretaceous Dinosaur Park Formation of Alberta, Canada. *Canadian Journal of Earth Sciences* 50(2): 171–182.; (*3) Penkalski P (2018). Revised systematics of the armoured dinosaur *Euoplocephalus* and its allies. *Neues Jahrbuch für Geologie und Paläontologie – Abhandlungen* 287(3): 261–306.]

Scutellosaurus lawleri

> *Scutellosaurus lawleri* Colbert, 1981 [Sinemurian; Kayenta Formation, Arizona, USA]

Generic name: *Scutellosaurus* ← {(Lat.) scutellum: small shield + (Gr.) σαῦρος/sauros: lizard}

Specific name: *lawleri* ← {(person's name) Lawler + -ī}; in honor of D. Lawler who collected the fossil.

Etymology: Lawler's **small-shielded lizard**

Taxonomy: Ornithischia: Thyreophora

Notes: *Scutellosaurus lowleri* was found on the land of the Navajo Nation in Arizona.

Secernosaurus koerneri

> *Secernosaurus koerneri* Brett-Surman, 1979 [Campanian–Maastrichtian; Colhué Huapi Formation, Chubut, Argentina]

= *Kritosaurus australis* Bonaparte *et al.*, 1984 [*2] [Campanian–Maastrichtian; Los Alamitos Formation, Río Negro, Argentina]

Generic name: *Secernosaurus* ← {(Lat.) secerno: sever, separate + (Gr.) σαῦρος/sauros: lizard}; "referring to its non-Laurisian origin" [*1].

Specific name: *koerneri* ← {(person's name) Koerner + -ī}; "after Dr. Harold E. Koerner, professor emeritus, University of Colorado" [*1].

Etymology: Koerner's **severed lizard**

Taxonomy: Ornithischia: Ornithopoda: Iguanodontia: Hadrosauridae: Saurolophinae: Kritosaurini

[References: (*1) Brett-Surman MK (1979). Phylogeny and palaeobiogeography of hadrosaurian dinosaurs. *Nature* 277: 560–562.; (*2) Prieto-Marquez A, Salinas GC (2010). A re-evaluation of *Secernosaurus koerneri* and *Kritosaurus australis* (Dinosauria, Hadrosauridae) from the Late Cretaceous of Argentina. *Journal of Vertebrate Paleontology* 30 (3): 813–837.]

Sefapanosaurus zastronensis

> *Sefapanosaurus zastronensis* Otero *et al.*, 2015 [Norian–Sinemurian; Elliot Formation, Free State, South Africa]

Generic name: *Safapanosaurus* ← {(Sesotho) safapano: cross + (Gr.) σαῦρος/sauros: lizard}; "in reference to the Cross T-shaped ascending process of the astragalus" [*1].

Specific name: *zastronensis* ← {(place-name) Zastron: type locality + -ensis}

Etymology: **cross lizard** from Zastron

Taxonomy: Saurischia: Sauropodomorpha

cf. Saurischia: Sauropodomorpha: Massopoda: Anchisauria: Sauropodiformes [*1]

[References: (*1) Otero A, Krupandan E, Pol D, Chinsamy A, Choiniere J (2015). A new basal sauropodiform from South Africa and the phylogenetic relationships of basal sauropodomorphs. *Zoological Journal of the Linnean Society* 174(3): 589–634.]

Segisaurus halli

> *Segisaurus halli* Camp, 1936 [Pliensbachian–Toarcian; Navajo Sandstone Formation, Coconino, Arizona, USA]

Generic name: *Segisaurus* ← {(place-name) Segi Canyon (= Tsegi Canyon) | (Navajo) Tsegi [pron.: tuh SAY gee] + σαῦρος/sauros: lizard}

Specific name: *halli* ← {(person's name) Hall + -ī}; in honor of Ansel Franklin Hall.
Etymology: Hall's **Segi Canyon lizard**
Taxonomy: Theropoda: Ceratosauria: Coelophysoidea: Coelophysidae

Segnosaurus galbinensis

> *Segnosaurus galbinensis* Perle, 1979 [Cenomanian–Turonian; Baynshire Formation, Ömnögovi, Mongolia]

Generic name: *Segnosaurus* ← {(Lat.) sēgnis: slow + (Gr.) σαῦρος/sauros: lizard} (*1)
Specific name: *galbinensis* ← {(place-name) Galbin + -ensis}; referring to the Galbin region of the Gobi (*1).
Etymology: **slow lizard** from Galbin
Taxonomy: Theropda: Therizinosauridae (*2)
cf. Theropoda: Segnosauridae (*1)
Notes: *Segnosaurus* had an elongated head, large clawed hands, and a broad pelvis. Segnosauridae was distinguished from the families Deinocheiridae and Therizinosauridae (*1).

[References: (*1) Perl A (1979). Segnosauridae—a new family of Theropoda from the Lower Cretaceous of Mongolia. *Trudy – Sovmestnaya Sovetsko-Mongol'skaya Paleontologicheskaya Ekspeditsiya* 8: 45–55.; (*2) Zanno LE (2010). A taxonomic and phylogenetic re-evaluation of Therizinosauria (Dinosauria: Maniraptora). *Journal of Systematic Paleontology* 8(4): 503–543.]

Seismosaurus hallorum

> *Seismosaurus halli* Gillette, 1991 ⇒ *Seismosaurus hallorum* (Gillette, 1991) Gilette, 1994 ⇒ *Diplodocus hallorum* (Gilette, 1991) Lucas *et al.*, 2006 [Kimmeridgian–Tithonian; Morrison Formation, New Mexico, USA]

As the specific name honours two people, Olshevsky suggested to emend *S. halli* to *S. hallorum*.

Generic name: *Seismosaurus* ← {(Gr.) σεισμός/seismos: earthquake + σαῦρος/sauros: lizard}
Specific name: *hallorum* ← {(person's name) Hall + -ōrum}; in honor of James Hall and his wife Ruth for their support of paleontology in northern New Mexico spanning three decades.
Etymology: Hall and his wife's **earthshaker lizard**
Taxonomy: Saurischia: Sauropodomorpha: Sauropoda: Diplodocidae
Notes: According to *The Dinosauria 2nd edition*, *Seismosaurus hallorum* was considered to be valid, but, according to the Geological Society of America (2004), *Seismosaurus* was considered to be a junior subjective synonym of *Diplodocus* (*1) and *S. hallorum* to be the second species of *Diplodocus*. Gillette (1991) first described it may have been up to 52 m, but later it was estimated 32 m.

[References: (*1) Lucas S, Herne M, Heckert A, and Sullivan R (2004). Reappraisal of *Seismosaurus*, a Late Jurassic sauropod dinosaur from New Mexico. *The Geological Society of America*, 2004, Denver Annual Meeting (November 7–10, 2004).]

Seitaad ruessi

> *Seitaad ruessi* Sertich & Loewen, 2010 [Pliensbachian–Toarcian; Navajo Sandstone Formation, San Juan, Utah, USA]

Generic name: *Seitaad* ← {(Diné/Navajo, folklore) séít'áád}; "a mythological sand 'monster' of Diné folklore that buried its victims in dunes" (*1). The specimen was recovered from a bed of massive sandstone at the base of the Navajo Sandstone.
Specific name: *ruessi* ← {(person's name) Everett Ruess + -ī}; "in honor of the young artist, poet, naturalist, and explorer Everett Ruess (1914–1934?), who mysteriously disappeared in 1934 while exploring southern Utah" (*1).
Etymology: Ruess' **sand monster**
Taxonomy: Saurischia: Sauropodomorpha: Prosauropoda

[References: (*1) Sertich JJW, Loewen MA (2010). A new basal sauropodomorph dinosaur from the Lower Jurassic Navajo Sandstone of southern Utah. *PLoS ONE* 5(3): e9789.]

Sektensaurus sanjuanboscoi

> *Sektensaurus sanjuanboscoi* Ibiricu *et al.*, 2019 [Coniacian–Maastrichtian; Lago Colhué Huapí Formation, Chubut, Argentina]

Generic name: *Sektensaurus* ← {(Tehuelche) sekten: island + (Gr.) σαῦρος/sauros: lizard}; "in reference of the site* where the fossils were found" (*1).
*an ephemeral island near the southeastern shore of Lago Colhue Huapí.
Specific name: *sanjuanboscoi* ← {San Juan Bosco + -ī}; "in honor of Universidad Nacional de la Patagonia Sun Juan Bosco, university where most authors (LMI; GAC; RDM; ML; BNA) either work or have completed their university degree" (*1).
Etymology: **island lizard** of National University of Patagonia San Juan Bosco
Taxonomy: Ornithischia: Ornithopoda: Iguanodontia: ? Elasmaria

[References: (*1) Ibiricu LM, Casal GA, Martínez RD, Luna M, Canale JI, Álvarez BN, Riga BG (2019). A new ornithopod dinosaur (Dinosauria: Ornithischia) from the Late Cretaceous of central Patagonia. *Cretaceous Research* 98: 276–291.]

Sellacoxa pauli

> *Iguanodon dawsoni* Lydekker, 1888 ⇒ *Barilium dawsoni* (Lydekker, 1888) Norman, 2010 ⇒ *Torilion dawsoni* (lydekker, 1888) Carpenter & Ishida, 2010 [late Valanginian; Wadhurst Clay Formation, Wealden Group, East Sussex, UK]

= *Sellacoxa pauli* Carpenter & Ishida, 2010 [early Valanginian; Wadhurst Clay Formation, East Sussex, UK]

Generic name: *Sellacoxa* ← {(Lat.) sella: saddle + coxa: hips}; "in reference to the saddle-shaped ilium" (*1).

Specific name: *pauli* ← {(person's name) Paul + -ī}; in honor of "Gregory S. Paul for recognizing that European iguanodon diversity is far greater than formally recognized" (*1).

Etymology: Paul's **saddle-shaped ilium**

Taxonomy: Ornithischia: Ornithopoda: Iguanodontia

Notes: *Sellacoxa* is seemed to be a subjective synonym of *Barilium*, according to Norman (2013).

[References: (*1) Carpenter K, Ishida Y (2010). Early and "middle" Cretaceous iguanodonts in time and space. *Journal of Iberian Geology* 36(2): 145–164.]

Sellosaurus gracilis

> *Sellosaurus gracilis* Meyer; von Huene, 1907–1908 ⇒ *Plateosaurus gracilis* (von Huene, 1907–1908) von Huene, 1926(*1) [Norian; Löwenstein Formation, Stuttgart, Baden-Württemberg, Germany]

= *Thecodontosaurus hermannianus* von Huene, 1905 ⇒ *Sellosaurus hermannianus* (von Huene, 1905) von Huene, 1915 [Norian; Löwenstein Formation, Stuttgart, Germany]

Generic name: *Sellosaurus* ← {(Lat.) sello- [sella: saddle] + (Gr.) σαῦρος/sauros: lizard}; referring to the vertebrae.

Specific name: *gracilis* ← {(Lat.): thin, slight, slender}.

Etymology: slender, **saddle lizard**

Taxonomy: Saurischia: Sauropodomorpha: Plateosauria: Plateosauridae

Other species:

> *Teratosaurus minor* von Huene, 1907–1908 ⇒ *Efraasia minor* (von Huene, 1907–1908) Galton, 1973 [Alaunian (= middle Norian); Löwenstein Formation, Baden-Württemberg, Germany]

= *Sellosaurus fraasi* von Huene, 1907–1908 ⇒ *Plateosaurus fraasi* (von Huene, 1907–1908) [Alaunian (= middle Norian); Löwenstein Formation, Baden-Württemberg, Germany]

Notes: One of the two clusters, which was found as *Sellosaurus* material, included the original *Sellosaurus gracilis*, which is now regarded to be the second species of *Plateosaurus* (*1). The other included "*Sellosaurus*" *fraasi*, "*Teratosaurus*" *minor* & "*Palaeosaurus*" *diagnosticus*. They were combined into a valid species, *Efraasia minor* in 2003.

[References: (*1) Yates AM (2003). The species taxonomy of the sauropodomorph dinosaurs from the Löwenstein Formation (Norian, Late Triassic) of Germany. *Paleontology* 46(2): 317–337.]

Serendipaceratops arthurcclarkei

> *Serendipaceratops arthurcclarkei* Rich and Vickers-Rich, 2003 [Aptian; Wonthaggi Formation, Victoria, Australia]

Generic name: *Serendipaceratops* ← {Sarandīp = (Old Persian) "Sarandib, the name given to Sri Lanka, the home of Sir Arthur C. Clarke, by Muslim traders (Clarke 1979)" (*1) | serendip- [serendipity]: "a fortuitous chance event as in discovery of the holotype of the species *S. arthurcclarkei* at the site that yielded only half a dozen individual fossil bones and bone fragments on a continent where the Neoceratosia were previously unkown" (*1) + (Gr.) κέρας/ keras: horn + ὤψ/ōps: face}

Specific name: *arthurcclarkei* ← {(person's name) Arthur C. Clerke + -ī}; "in honor of Sir Arthur C. Clarke, who inspired the discoverers in their youth with his writings and who in his youth was lured into science by dinosaurs" (*1).

Etymology: Arthur C. Clerke's **serendipitous horned face (ceratopsian)**

Taxonomy: Ornithischia: Genasauria *cf.* Neoceratopsia (*1)

Notes: *Serendipaceratops* is considered a *nomen dubium*, according to Agnolin *et al.* (2010).

[References: (*1) Rich T, Vickers-Rich P (2003). Protoceratopsian? ulnae from Australia. *Records of the Queen Victoria Museum* 113: 1–12.]

Serikornis sungei [Anchiornithidae]

> *Serikornis sungei* Lefèvre *et al.*, 2017 [Oxfordian; Tiaojishan Formation, Liaoning, China]

Generic name: *Serikornis* ← {(Gr.) σηρῐκός/sērikos: silken, silk + ὄρνις/ornis: bird}; in reference to plumulaceous-like feathers which almost covered the body.

Specific name: *sungei* ← {(Chin. person's name) Sun Ge + -ī}; in honor of Sun Ge, curator of the Paleontological Museum of Liaoning.

Etymology: Sun Ge's **silk bird**

Taxonomy: Theropoda: Avialae (?): Anchiornithidae

[References: Lefèvre U, Cau A, Cincotta A, Hu D, Chinsamy A, Escuillié F, Pascal G (2017). A new

Jurassic theropod from China documents a transitional step in the macrostructure of feathers. *The Science of Nature* 104: 74.]

Shamosaurus scutatus

> *Shamosaurus scutatus* Tumanova, 1983 [Aptian; Dzunbain Formation, Dornogovi, Övörkhangai, Mongolia]

Generic name: *Shamosaurus* ← {(Chin. place-name) shāmò沙漠: desert + (Gr.) σαῦρος/sauros: lizard}; in reference to south-east Gobi desert where the remains were discovered.

Specific name: *scutatus* ← {(Lat.) scūtātus: armed with shields [*1]}; in reference to the body armour.

Etymology: **Gobi desert lizard** armed with shields

Taxonomy: Ornithischia: Thyreophora: Ankylosauria: Ankylosauridae

[References: (*1) Tumanova TA (1983). The first ankylosaur from the Lower Cretaceous of Mongolia. *The Joint Soviet-Mongolian Paleontological Expedition. Transaction* 24: 110–128.]

Shanag ashile

> *Shanag ashile* Turner *et al.*, 2007 [Early Cretaceous; Öösh Formation, Övörkhangai, Mongolia]

Generic name: *Shanag* ← {"shanag: black-hatted dancers in the Buddhist Tsam festival" [*1]}

Specific name: *ashile* ← {(place-name) "Ashile: in reference to the old Öösh locality and formation name used by Dr. Henry F. Osborn" [*1]}

Etymology: **Shanag (dancers)** from Ashile (Öösh Formation)

Taxonomy: Theropoda: Coelurosauria: Maniraptora: Dromaeosauridae

Notes: Small dromaeosaurid theropod.

[References: (*1) Turner AH, Hwang SH, Norell MA (2007). A small derived theropod from Öösh, Early Cretaceous, Baykhangor Mongolia. *American Museum Novitates* 3557(1): 1–27.]

Shanshanosaurus huoyanshanensis

> *Tyrannosaurus bataar* Maleev, 1955 ⇒ *Tarbosaurus bataar* (Maleev, 1955) Rozhdestvensky, 1965 ⇒ *Jenghizkhan bataar* (Maleev, 1955) Olshevsky *et al.*, 1995 [late Campanian; Nemegt Formation, Ömnögovi, Mongolia]

= *Shanshanosaurus huoyanshanensis* Dong, 1977 ⇒ *Aublysodon huoyanshanensis* (Dong, 1977) Paul, 1988 [Campanian; Subashi Formation, Xinjiang, China]

Generic name: *Shanshanosaurus* ← {(Chin. place-name) Shànshàn鄯善 + (Gr.) σαῦρος/sauros: lizard}

Specific name: *huoyanshanensis* ← {(Chin. place-name) Huǒyàn Shān 火焔山 'Flaming Mountains' + -ensis}

Etymology: **Shanshan lizard** from Huoyan Shan

Taxonomy: Theropoda: Tyrannosauroidea: Tyrannosauridae: Tyrannosaurinae

Notes: *Shanshanosaurus huoyanshanensis* is a juvenile tyrannosaurine, possibly *Tarbosaurus* [*1]. According to *The Dinosauria 2nd edition*, *Shanshanosaurus* is considered as a synonym of *Tarbosaurus*.

[References: Dong Z (1977). On the dinosaurian remains from Turpan, Xinjiang. *Vertebrata PalAsiatica* (in Chinese) 15: 59–66.; (*1) Currie PJ, Dong Z (2001). New information on *Shanshanosaurus huoyanshanensis*, a juvenile tyrannosaurid (Theropoda, Dinosauria) from the Late Cretaceous of China. *Canadian Journal of Earth Sciences* 38(12):1729–1737.]

Shantungosaurus giganteus

> *Shantungosaurus giganteus* Hu, 1973 [Campanian; Xingezhuang Formation, Wangshi Group, Shandong, China]

= *Zhuchengosaurus maximus* Zhao *et al.*, 2007 [*1] [Campanian; Xingezhuang Formation, Wangshi Group, Shandong, China]

= *Huaxiaosaurus aigahtens* Zhao *et al.*, 2011 [Campanian; Xingezhuang Formation, Wangshi Group, Shandong, China]

Generic name: *Shantungosaurus* ← {(place-name) Shantung山東 [Pinyin: Shāndóng] + (Gr.) σαῦρος/sauros: lizard}

Specific name: *giganteus* ← {(Lat.) Gigantēus = (Gr.) γῐγάντειος: of the giants}

Etymology: gigantic **Shandong lizard**

Taxonomy: Ornithischia: Ornithopoda: Iguanodontia: Hadrosauroidea: Hadrosauridae: Saurolophinae: Edmontosaurini

[References: (*1) Ji Y, Wang X, Liu Y, Ji Q (2011). Systematics, behavior and living environment of *Shantungosaurus giganteus* (Dinosauria: Hadrosauridae). *Acta Geologica Sinica* 85(1): 58–65.; Zhang J-L, Wang Q, Jiang S-X, Cheng X, Li N, Qiu R, Zhang X-J, Wang X-L (2017). Review of historical and current research on the Late Cretaceous dinosaurs and dinosaur eggs from Laiyang, Shandong. *Vertebrata PalAsiatica* 55: 187–200.]

Shanweiniao cooperorum [Avialae]

> *Shanweiniao cooperorum* O'Connor *et al.*, 2009 [early Aptian; Yixian Formation, Liaoning, China]

Generic name: *Shanweiniao* ← {(Chin.) Shàn wěi niǎo 扇尾鳥: 'fan-tail bird'}; "referring to the fact that this specimen preserves the first known occurrence of an enantiornithine fan-shaped feathered tail" (*1).

Specific name: *cooperorum* ← {(person's name) Cooper + -ōrum}; "in honor of Carl and Lynn Cooper for their generous support in the study of Mesozoic birds from China" (*1).

Etymology: Cooper's **fan-tail bird**

Taxonomy: Theropoda: Avialae: Enantiornithes: Longipterygidae

cf. Aves: Pygostylia: Enatiornithes: Longipterygidae (*1)

[References: (*1) O'Connor JK, Wang X, Chiappe LM, Gao C, Meng Q, Cheng X, Liu J (2009). Phylogenetic support for a specialized clade of Cretaceous enantiornithine birds with information from a new species. *Journal of Vertebrate Paleontology* 29(1): 188–204.]

Shanxia tianzhenensis

> *Shanxia tianzhenensis* Barrett *et al.*, 1998 [Late Cretaceous; Huiquanpu Formation, Tianzhen, Shanxi, China]

Generic name: *Shanxia* ← {(Chin. place-name) Shānxī山西: Shanxi Province + -ia}

Specific name: *tianzhenensis* ← {(Chin. place-name) Tiānzhèn天鎮: Tianzhen County, Shanxi Province + -ensis}

Etymology: **for Shanxi**, from Tianzhen

Taxonomy: Ornithischia: Thyreophora: Ankylosauridae: Ankylosaurinae

Notes: *Shanxia* was considered a *nomen dubium* by Sullivan (1999). But Upchurch and Barrett (2000) reaffirmed the validity of *Shanxia*.(*1) According to *The Dinosauria 2nd edition*, *Shanxia* was considered as a *nomen dubium*. Recently Arbour and Currie (2015) treated *Shanxia* as a junior synonym of *Saichania*.

[References: (*1) Upchurch P & Barrett PM (2000). The taxonomic status of *Shanxia tianzhenensis* (Ornithischia, Ankylosauridae); a response to Sullivan (1999). *Journal of Vertebrate Paleontology* 20: 216–217.]

Shanyangosaurus niupanggouensis

> *Shanyangosaurus niupanggouensis* Xue *et al.*, 1996 [Maastrichtian; Shanyang Formation, Shaanxi, China]

Generic name: *Shanyangosaurus* ← {(Chin. place-name) Shānyáng山陽: a county in Shaanxi陝西 + (Gr.) σαῦρος/sauros: lizard}

Specific name: *niupanggouensis* ← {(Chin, place-name) Niupanggou牛蒡溝 [Pinyin: Niúbánggōu] + -ensis}

Etymology: **Shanyang lizard** from Niupanggou

Taxonomy: Theropoda: Coelurosauria

Shaochilong maortuensis

> *Chilantaisaurus maortuensis* Hu, 1964 ⇒ "*Alashansaurus*"*maortunensis* (Hu, 1964) Chure, 2000 ⇒ *Shaochilong maortuensis* (Hu, 1964) Brusatte *et al.*, 2009 [Turonian; Ulansuhai Formation, Inner Mongolia, China]

Generic name: *Shaochilong* ← {(Chin.) shāchǐ鯊齒: shark tooth + lóng龍: dragon}; "for 'shark-toothed' carcharodontosaurid theropods" (*1).

Specific name: *maortuensis* ← {(Chin. place-name) Maortu毛爾圖 + -ensis}

Etymology: **shark-toothed dragon** from Maortu [毛爾圖假鯊齒龍*]

**cf. Carcharodontosaurus* [鯊齒龍]

Taxonomy: Theropoda: Tetanurae: Allosauroidea: Carcharodontosauridae (*1)

Notes: *Shaochilong maortuensis* was originally named *Chilantaisaurus maortuensis*, but reclassified in 2009.

[References: (*1) Brusatte SL, Benson RBJ, Chure DJ, Xu X, Sullivan C, Hone DWE (2009). The first definitive carcharodontosaurid (Dinosauria: Theropoda) from Asia and the delayed ascent of tyrannosaurids. *Naturwissenschaften* 96: 1051–1058.]

Sharovipteryx mirabilis [Archosauromorpha]

> *Podopteryx mirabilis* Sharov, 1971 (preoccupied) ⇒ *Sharovipteryx mirabilis* (Sharov, 1971) Cowen, 1981 [Ladinian; Madygen Formation, Batken, Fergana Valley, Kyrgyzstan]

Generic name: *Sharovipteryx* ← {(person's name) Sharov + (Gr.) πτέρυξ/pteryx: wing}; in honor of Russian paleontologist A. G. Sharov.

Specific name: *mirabilis* ← {(Lat.) mīrābilis: wonderful}.

Etymology: wonderful **Sharov's wing**

Taxonomy: Archosauromorpha: Protorosauria: Sharovipterygidae

Notes: About 20 cm long. According to Dyke *et al.* (2006), the wing membrane would have allowed *Sharovipteryx* to glide.

Shengjingornis yangi [Avialae]

> *Shengjingornis yangi* Li *et al.*, 2012 [Aptian; Jiufotang Formation, Liaoning, China]

Generic name: *Shengjingornis* ← {(Chin. place-name) Shèngjīng盛京: the ancient name of Shěnyáng瀋陽, the capital city of Liaoning Province + (Gr.) ὄρνις/ornis: bird} (*1)

Specific name: *yangi* ← {(person's name) Yang + -ī}; "dedicated to the fossil preparator of Mr. Yang Qiang" (*1).

Etymology: Yang's **Shengjing (Shenyang) bird**

Taxonomy: Theropoda: Avialae: Enantiornithes:

Longipterygidae
cf. Aves: Pygostylia: Enantiornithes: Longipterygidae [*1]

Notes: A pigeon-sized enantiornithine bird.

[References: (*1) Li L, Wang J, Zhang X, Hou S (2012). A new enantiornithine bird from the Lower Cretaceous Jiufotang Formation in Jinzhou Area, western Liaoning Province, China. *Acta Geologica Sinica* (English Edition) 86(5): 1039–1044.]

Shenqiornis mengi [Avialae]

> *Shenqiornis mengi* Wang *et al.*, 2010 [Barremian; Qiaotou Formation, Hebei, China]

Generic name: *Shenqiornis* ← {(Chin.) Shen qi神七: 神舟 7 号 Shénzhōu qī + (Gr.) ὄρνις/ornis: bird}; "celebrating the successful launch of the Shenzhou 7, China's third human mission into space" [*1] in 2008.

Specific name: *mengi* ← {(person's name) Meng + -ī}; "in honor of Meng Qingjin for his contribution to the study and protection of Liaoxi fossils as former Director of the Dalian Natural History Museum" [*1].

Etymology: Meng's **Shenzhou 7's bird**

Taxonomy: Theropoda: Avialae: Enantiornithes: Bohaiornithidae

[References: (*1) Wang X, O'Connor J, Zhao B, Chiappe LM, Gao C, Cheng X (2010). New species of Enantiornithes (Aves: Ornithothoraces) from the Qiaotou Formation in Northern Hebei, China. *Acta Geologica Sinica* 84(2): 247–256.]

Shenzhousaurus orientalis

> *Shenzhousaurus orientalis* Ji e*t al.*, 2003 [Barremian; Yixian Formation, Liaoning, China]

Generic name: *Shenzhousaurus* ← {(Chin. place-name) Shénzhōu神州: an ancient name of China + (Gr.) σαῦρος/sauros: lizard} [*1]

Specific name: *orientalis* ← {(Lat.) orientālis: eastern} [*1]

Etymology: eastern **Chinese lizard**

Taxonomy: Theropoda: Coelurosauria: Ornithomimosauria: Ornithomiminae [*1]

Notes: Numerous pebbles (gastroliths) were found in the thoracic cavity of *Shenzhousaurus orientalis* [*1].

[References: (*1) Ji Q, Norell MA, Makovicky PJ, Gao K-Q, Ji S, Yuan C (2003). An early ostrich dinosaur and implications for ornithomimosaur phylogeny. *American Museum Novitates* 3420: 1–19.]

Shidaisaurus jinae

> *Shidaisaurus jinae* Wu *et al.*, 2009 [Middle Jurassic; Chuanjie Formation, Lufeng, Yunnan, China]

Generic name: *Shidaisaurus* ← {(Chin.) shídài時代 'age' (= the Jin-Shidai Company) + (Gr.) σαῦρος/sauros: lizard}

Specific name: *jinae* ← {(Chin.) jīn金: 'golden' + -ae}; referring to the Jin-Shidai Company.

Etymology: Jin-**Shidai Company's lizard** [金時代龍]

Taxonomy: Theropoda: Tetanurae: Carnosauria: Allosauroidea: Metriacanthosauridae: Metriacanthosaurinae

[References: Wu X-C, Currie PJ, Dong Z, Pan S, Wang T (2009). A new theropod dinosaur from the Middle Jurassic of Lufeng, Yunnan, China. *Acta Geologica Sinica* 83(1): 9–24.]

Shingopana songwensis

> *Shingopana songwensis* Gorscak *et al.*, 2017 [Cenomanian; Galula Formation, Mbeya, Tanzania]

Generic name: *Shingopana* ← {(Swahili / Kiswahili) shingo: neck + pana: wide}; in reference to "the bulbous expansion of the neural spine".

Specific name: *songwensis* ← {(place-name) Songwe area of southwestern Tanzania + -ensis}

Etymology: **wide neck** from Songwe

Taxonomy: Saurischia: Sauropodomorpha: Sauropoda: Titanosauria: Lithostrotia

[References: Gorscak E, O'Connor PM, Roberts EM, Stevens NJ (2017). The second titanosaurian (Dinosauria: Sauropoda) from the middle Cretaceous Galula Formation, southwestern Tanzania, with remarks on African titanosaurian diversity. *Journal of Vertebrate Paleontology* 37(4): e1343250.]

Shishugounykus inexpectus

> *Shishugounykus inexpectus* Qin *et al.*, 2019 [Oxfordian; Shishugou Formation, Xinjiang, China]

Generic name: *Shishugounykus* ← {(place-name) Shíshùgōu Formation石樹溝層* + (Gr.) ὄνυξ/onyx: claw}

*[石: rock + 樹: tree + 溝: wash]; for the abundant petrified wood in the formation [*1].

Specific name: *inexpectus* ← {(Lat.) inexspectātus: unlooked for}; "referring to the unexpected discovery of a new alvarezsaurian species from the Middle-Late Jurassic Shishugou Formation" [*1].

Etymology: unexpected **claw from Shishugou Formation**

Taxonomy: Theropoda: Maniraptora: Alvarezsauria: Alvarezsauroidea

[References: (*1) Qin Z, Clark J, Choiniere J, Xu X (2019). A new alvarezsaurian theropod from the Upper Jurassic Shishugou Formation of western China. *Scientific Reports* 9: 11727.]

Shixinggia oblita

> *Shixinggia oblita* Lu & Zhang, 2005 [Maastrichtian; Pingling Formation, Guangdong, China]

Generic name: *shixinggia* ← {(Chin. place-name) Shǐxīng 始興 + -ia}
Specific name: *oblita* ← {(Lat.) oblīta (feminine form of oblītus); forgetful, not remembering}; in reference to it being undescribed for years.
Etymology: neglected **Shixing's one**
Taxonomy: Theropoda: Oviraptoridae

Shonisaurus popularis [Ichthyosauria]

> *Shonisaurus popularis* Camp, 1976 [Carnian; Luning Formation, Nevada, USA]

Generic name: *Shonisaurus* ← {(place-name) Shoshone Mountains + (Gr.) σαῦρος/ sauros: lizard}
Specific name: *popularis* ← {(Lat.) populāris: of the people, popular, general}
Etymology: popular **lizard from Shoshone Mountains**
Taxonomy: Ichthyosauria: Shastasauria: Shastasauridae
Other species:

> *Shonisaurus mulleri* Camp, 1976 [Lacian; Luning Formation, Nevada, USA]
> *Shonisaurus silberlingi* Camp, 1976 [Lacian; Luning Formation, Nevada, USA]
> *Shonisaurus sikanniensis* Nicholls & Manabe, 2004 ⇒ ?*Shastasaurus sikanniensis* (Nicholls & Manabe, 2004) Sander *et al.*, 2011 [Norian; Pardonet Formation, British Columbia, Canada] (*1)

[References: (*1) Nicholls EL & Manabe M (2004). Giant ichthyosaurs of the Triassic—a new species of *Shonisaurus* from the Pardonet Formation (Norian: Late Triassic) of British Columbia. *Journal of Vertebrate Paleontology* 24(4): 838–849.]

Shringasaurus indicus [Archosauromorpha]

> *Shringasaurus indicus* Sengupta *et al.*, 2017 [Anisian; Denwa Formation, Madhya Pradesh, India]

Generic name: *Shringasaurus* ← {(Sanscrit) Shringa: horn + (Gr.) σαῦρος/sauros: lizard}; "referring to the horned skull" (*1).
Specific name: *indicus* ← {(Latin English) Indicus: Indian}; "referring to the country where such species was discovered" (*1).
Etymology: **horned lizard** from India
Taxonomy: Archosauromorpha: Allokotosauria: Azendohsauridae

[References: (*1) Sengupta S, Ezcurra MD, Bandyopadhyay S (2017). A new horned and long-necked herbivorous stem-archosaur from the Middle Triassic of India. *Scientific Reports* 7: 8366.]

Shuangbaisaurus anlongbaoensis

> *Shuangbaisaurus anlongbaoensis* Wang *et al.*, 2017 [Hettangian; Fengjiahe Formation, Yunnan, China]

Generic name: *Shuangbaisaurus* ← {(Chin. place-name) Shuāngbǎi* 双柏: the name of the county where the holotype was recovered + (Gr.) σαῦρος/ sauros: lizard}
*"Shuangbai was first established in West Han Dynasty (AD 109)" (*1).
Specific name: *anlongbaoensis* ← {(Chin. place-name) Anlongbao*: the name of the town + -ensis}
*Anlongbao [ān安: placing + lóng龍: dragon + bǎo堡: fort]: meaning "dragon-placing fort" (*1)
Etymology: **Shuangbai lizard** from Anlongbao [安龍堡双柏龍]
Taxonomy: Theropoda

[References: (*1) Wang G-F, You H-L, Pan S-G, Wang T (2017). A new crested theropod dinosaur from the Early Jurassic of Yunnan Province, China. *Vertebrata PalAsiatica* 55(2): 177–186.]

Shuangmiaosaurus gilmorei

> *Shuangmiaosaurus gilmorei* You *et al.*, 2003 [Albian; Sunjiawan Formation, Shuangmiao, Beipiao, Chaoyang, Liaoning, China]

Generic name: *Shuangmiaosaurus* ← {(Chin. place-name) Shuāngmiào雙廟 'twin temples': the name of village + σαῦρος/sauros: lizard}
Specific name: *gilmorei* ← {(person's name) Gilmore + -ī}; in honor of Charles Whitney Gilmore.
Etymology: Gilmore's **Shuangmiao lizard**
Taxonomy: Ornithischia: Ornithopoda: Iguanodontia: Hadrosauroidea: Hadrosauromorpha

[References: You H-L, Ji Q, Li J, Li Y (2003). A new hadrosauroid dinosaur from the mid-Cretaceous of Liaoning, China. *Acta Geologica Sinica* 77(2): 148–154.]

Shunosaurus lii

> *Shunosaurus lii* Dong *et al.*, 1983 [Bajocian; lower Shaximiao Formation, Sichuan, China]

Generic name: *Shunosaurus* ← {(Chin. place-name) Shǔ蜀: the ancient abbreviation of Sichuan Province + (Gr.) σαῦρος/sauros: reptile} (*1)
Specific name: *lii* ← {(Chin. person's name) Lǐ 李 + -ī}; "in commemoration of the hydrologist Bin Li, the magistrate who governed what is now Sichuan Province (256–251 BC) for the state of Qin during the Warring States Period. He was particularly celebrated for his flood control measures along the Minjiang River

which included the costruction of the famed Dujiang dike and irrigation system that are still functioning today" (*1).

Etymology: Li's **Shu lizard**

Taxonomy: Saurischia: Sauropodomorpha:
Sauropoda: Eusauropoda

Notes: *Shunosaurus* was a moderate-sized primitive sauropod. It had spoon-shaped teeth (*1) and a bony club at the end of the tail (*2).

[References: (*1) Dong Z, Zhou S, Zhang Y (1983) Dinosaurs from the Jurassic of Sichuan. *Palaeontologica Sinica* 162 New Series C 162(23); 1–136.; (*2) Dong Z, Peng G, Huang D (1989). The discovery of the bony tail club of sauropods. *Vertebrata PalAsiatica* 27: 219–224.]

Shuvosaurus inexpectatus [Suchia]

> *Shuvosaurus inexpectatus* Chatterjee, 1993 [Norian; Cooper Canyon Formation, Texas, USA]

= *Chatterjeea elegans** [Norian; Cooper Canyon Formation, Texas, USA]

**cf. Chatterjeea* Long & Murry, 1995

Generic name: *Shuvosaurus* ← {(person's name) Shuvo + (Gr.) σαῦρος/sauros: lizard}; for the son of Sankar Chatterjee who discovered the remains.

cf. Soumyasaurus aenigmaticus Sangül, Agnolin & Chatterjee, 2018 (Silesauridae) (Soumya is another of Chatterjee's son, Shuvo.)

Specific name: *inexpectatus* ← {(Lat.) inexpectātus: unexpected}.

Etymology: unexpected, **Shuvo's lizard**

Taxonomy: Archosauria: Pseudosuchia: Suchia:
Paracrocodylomorpha: Poposauroidea:
Shuvosauridae

Notes: *Shuvosaurus inexpectus* was described as the earliest ornithomimid by Chatterjee (1993) (*1).

[References: (*1) Makovicky PJ, Kobayashi Y, Currie PJ (2004). Ornithomimosauria. In: Weishampel DB, Dodson P, Osmólska H (eds), *The Dinosauria 2nd edition*. University of California Press, Berkeley: 137–150.]

Shuvuuia deserti

> *Shuvuuia deserti* Chiappe *et al.*, 1998 [Campanian; Djadokhta Formation, Ömnögovi, Mongolia]

Generic name: *Shuvuuia* ← {(Mong.) шувуу/shuvuu: bird + -ia} (*1)

Specific name: *deserti* ← {(Lat.) dēsertus: deserted, desert}; in reference to the semi-arid depositional environment of the Djadokhta Formation (*1).

Etymology: **bird** from Gobi Desert

Taxonomy: Theropoda: Alvarezsauridae
cf. Theropoda: Alvarezsauridae:
Mononykinae (*2)

[References: (*1) Chiappe LM, Norell MA, Clark JM (1998). The skull of a relative of the stem group bird *Mononykus*. *Nature* 392(6673): 275–278.; (*2) Suzuki S, Chiappe LM, Dyke GJ, Watabe M, Barsbold R, Tsogtbaatar K (1998). A new specimen of *Shuvuuia deserti* Chiappe *et al.*, 1998, from the Mongolian Late Cretaceous with a discussion of the relationships of alvarezsaurids to other theropod dinosaurs. *Contributions in Science* (*Los Angeles*) 494: 1–18.]

Siamodon nimngami

> *Siamodon nimngami* Buffetaut & Suteethorn, 2011 [Aptian; Khok Kruat Formation, Khorat Group, Nakhon Ratchasima, Thailand]

Generic name: *Siamodon* ← {(place-name) Siam: the ancient name for Thailand + (Gr.) ὀδών/odōn: tooth}

Specific name: *nimngami* ← {(person's name) Nimngam + -ī}; in honor of W. Nimngam, who kindly donated the specimens.

Etymology: Nimngam's **tooth from Thailand (Siam)**

Taxonomy: Ornithischia: Ornithopoda: Iguanodontia

[References: Buffetaut E, Suteethorn V (2011). A new iguanodontian dinosaur from the Khok Kruat Formation (Early Cretaceous, Aptian) of northeastern Thailand. *Annales de Paléontologie* 97(1–2): 51–62.]

Siamosaurus suteethorni

> *Siamosaurus suteethorni* Buffetaut & Ingavat, 1986 [Barremian; Sao Khua Formation, Khon Kaen, Thailand]

Generic name: *Siamosaurus* ← {(place-name) Siam: an old name for Thailand + (Gr.) σαῦρος/sauros: lizard}

Specific name: *suteethorni* ← {(person's name) Suteethorn + -ī}; in honor of Varavudh Suteethorn.

Etymology: Suteethorn's **lizard from Thailand (Siam)**

Taxonomy: Theropoda: Tetanurae: Spinosauridae:
Spinosaurinae

Notes: According to *The Dinosauria 2nd edition*, *Siamosaurus suteethorni* was regarded as a *nomen dubium*. The teeth of *Siamosaurus* resemble those of *Spinosaurus*.

[References: (*1) Buffetaut E, Ingavat R (1986). Unusual theropod dinosaur teeth from the Upper Jurassic of Phu Wiang, northeastern Thailand. *Revue de Paleobiologie* 5: 217–220.]

Siamotyrannus isanensis

> *Siamotyrannus isanensis* Buffetaut *et al.*, 1996 [Barremian; Sao Khua Formation, Khon Kaen, Thailand]

Generic name: *Siamotyrannus* ← {(place-name) Siam: the old name of Thailand + (Gr.) τύραννος/tyrannos: tyrant}; in reference to a presumed membership of the Tyrannosauridae (*1).

Specific name: *isanensis* ← {(place-name) Isan: northeastern part of Thailand + -ensis}
Etymology: **Thailand (Siam) tyrant** from Isan
Taxonomy: Theropoda: Carnosauria: Allosauroidea: Metriacanthosauridae: Metriacanthosaurinae
Notes: Originally thought to be a tyrannosauroid, according to Carrano *et al.* (2012), *Siamotyrannus isanensis* is thought to be a member of Metriacanthosaurinae.

[References: (*1) Buffetaut E, Suteethorn V, Tong H (1996). The earliest known tyrannosaur from the Lower Cretaceous of Thailand. *Nature* 381: 689–691.]

Siamraptor suwati

> *Siamraptor suwati* Chokchaloemwong *et al.*, 2019 [Aptian; Khok Kruat Formation, Nakhon Ratchasima, Thailand]

Generic name: *Siamraptor* ← {(place-name) Siam = Thailand + (Lat.) raptor: robber}(*1)
Specific name: *suwati* ← {(person's name) Suwat + -ī}; in honor of "Mr. Suwat Liptapanlop, who supports and promotes the work of the Northeastern Research Institute of Petrified Wood and Mineral Resources"(*1).
Etymology: Suwat's **robber of Thailad**
Taxonomy: Saurischia: Theropoda: Tetanurae: Allosauroidea: Carcharodontosauria

[References: Chokchaloemwong D, Hattori S, Cuesta E, Jintasakul P, Shibata M, Azuma Y (2019). A new carcharodontosaurian theropod (Dinosauria: Saurischia) from the Lower Cretaceous of Thailand. *PLoS ONE* 14(10): e0222489.]

Siats meekerorum

> *Siats meekerorum* Zanno & Makovicky, 2013 [Cenomanian; Mussentuchit Member, Cedar Mountain Formation, Emery, Utah, USA]

Generic name: *Siats* ← {(Ute, legend) Siats: "a predatory, man-eating monster from legends of the Ute native tribe of Utah (occasionally spelled as See-atch)" (*1)}.
Specific name: *meekerorum* ← {(person's name) Meeker + -ōrum}; "honouring the Meeker family for their endowment in support of early career palaeontologists at the Field Museum"(*1).
Etymology: Meeker family's **Siats**
Taxonomy: Theropoda: Allosauroidea: Neovenatoridae

[References: (*1) Zanno LE, Makovicky PJ (2013). Neovenatorid theropods are apex predators in the Late Cretaceous of North America. *Nature Communications* 4: 2827.]

Sibirotitan astrosacralis

> *Sibirotitan astrosacralis* Averianov *et al.*, 2018 [Barremian; Ilek Formation, Kemerovo, Western Siberia, Russia]

Generic name: *Sibirotitan* ← {(place-name) Sibiro-: Siberia + (Gr. myth.) Τῑτάν/Tītan: titan}
Specific name: *astrosacralis* ← {(Gr.) ἄστρο-/astro- [ἀστήρ/aster: star] + (Lat.) sacrālis 'os sacrum': sacred bone}; referring to "the unusual configuration of sacral ribs which radiate, in dorsal view, from the middle of the sacrum as the rays of a star".
Etymology: **Siberian titan** with star-like sacrum (sacred bone)
Taxonomy: Saurischia: Sauropodomorpha: Sauropoda: Titanosauriformes: Somphospondyli
Notes: *Sibirotitan astrosacralis* is the second sauropod unearthed and described in Russia(*1), after *Tengrisaurus starkovi* Averianov & Skutschas, 2017.

[References: (*1) Averianov A, Ivantsov S, Skutschas P, Faingertz A, Leshchinskiy S (2018). A new sauropod dinosaur from the Lower Cretaceous Ilek Formation, western Siberia, Russia. *Geobios* 51(1): 1–14.]

Sigilmassasaurus brevicollis

> *Sigilmassasaurus brevicollis* Russell, 1996 [Albian–Cenomanian; Kem Kem Formation, Er Rachidia (formerly Ksar-Es-Souk), Morocco]
> = ? *Spinosaurus maroccanus* Russell, 1996 [Cenomanian; Kem Kem Formation, Er Rachidia (Ksar-es-Souk), Morocco]

Generic name: *Sigilmassasaurus* ← {(place-name) Sigilmassa*: a medieval Moroccan city + (Gr.) σαῦρος/sauros: lizard}
*(Arabic) transliterated Sijilmasa, Sijilmassa.
Specific name: *brevicollis* ← {(Lat.) brevis: short + collum: neck}; alluding to the shortness of the cervical vertebrae.
Etymology: short-necked **Sigilmassa lizard**
Taxonomy: Theropoda: Tetanurae: Megalosauroidea: Megalosauria: Spinosauridae: Spinosaurinae: Spinosaurini
Notes: According to *The Dinosauria 2nd edition*, *Sigilmassasaurus brevicollis* was regarded to be valid. Evans *et al.* (2015) proposed *Spinosaurus marrocanus* as a junior synonym of *Sigilmassasaurus brevicollis.*

[References: Russell DA (1996). Isolated dinosaur bones from the middle Cretaceous of the Tafilalt, Morocco. *Bulletin du Muséum National d'Histoire Naturelle, Paris, Série* 4(18): 349–402.]

Silesaurus opolensis [Silesauridae]

> *Silesaurus opolensis* Dzik, 2003 [Carnian; Drawno Beds Formation, Opole, Poland]

Generic name: *Silesaurus* ← {(place-name) Silesia +

(Gr.) σαῦρος/sauros: lizard} (*1)
Specific name: *opolensis* ← {(place-name) Opole + -ensis} (*1)
Etymology: **Silesia lizard** from Opole
Taxonomy: Archosauria: Ornithodira:
Dinosauromorpha: Dinosauriformes: Silesauridae
[References: (*1) Dzik J (2003). A beaked herbivorous archosaur with dinosaur affinities from the early Late Triassic of Poland. *Journal of Vertebrate Paleontology* 23(3): 556–574.]

Siluosaurus zhangqiani

> *Siluosaurus zhangqiani* Dong, 1997 [Barremian; Zhonggou Formation, Xinminbao Group, Gansu, China]

Generic name: *Siluosaurus* ← {(Chin. place-name) Sīlù絲路: Silk Road + (Gr.) σαῦρος/sauros: lizard}; referring to the recovery during the 1972 Sino-Japanese Silk Road Dinosaur Expedition.
Specific name: *zhangqiani* ← {(Chin. person's name) Zhāng Qiān張騫 + -ī}; in honor of Zhang Qian, who was a Chinese official and diplomat of Han dynasty B. C. 206–A. D. 23.
Etymology: Zhang Qian's **Silk Road lizard**
Taxonomy: Ornithischia: Ornithopoda
Notes: According to *The Dinosauria 2nd edition*, *Siluosaurus* is considered to be a *nomen dubium*.

Silvisaurus condrayi

> *Silvisaurus condrayi* Eaton, 1960 [Albian; Dakota Formation, Ottawa, Kansas, USA]

Generic name: *Silvisaurus* ← {(Lat.) silva: forest, woodland + (Gr.) σαῦρος/sauros: lizard}; referring to fossil leaves indicating that the animal lived in "a warm-temperate deciduous forest" (*1).
Specific name: *condrayi* ← {(person's name) Condray + -ī}; in honor of Mr. Warren H. Condray who found the specimen in his farm in Ottawa County (*1).
Etymology: Condray's **forest lizard**
Taxonomy: Ornithischia: Thyreophora:
Ankylosauria: Nodosauridae: Nodosaurinae
[References: (*1) Eaton, T. H., Jr. (1960). A new armored dinosaur from the Cretaceous of Kansas. *The University of Kansas Paleontological Contributions: Vertebrata* 8: 1–24.]

Similicaudipteryx yixianensis

> *Similicaudipteryx yixianensis* He *et al.*, 2008 [Aptian; Jiufotang Formation, Liaoning, China]

Generic name: *Similicaudipteryx* ← {(Lat.) similis: like, resembling, similar + *Caudipteryx* [(Lat.) cauda: tail + (Gr.) πτέρυξ/pteryx: feather]}; referring to its similarity to *Caudipteryx* (*1).
Specific name: *yixianensis* ← {(Chin. place-name) Yìxiàn 義縣 + -ensis}
Etymology: **Caudipteryx-like** from Yixian [義縣似尾羽龍]
Taxonomy: Theropoda: Oviraptorosauria:
Caudipterigidae
[References: (*1) He T, Wang X-L, Zhou Z-H (2008). A new genus and species of caudipterid dinosaur from the Lower Cretaceous Jiufotang Formation of western Liaoning, China. *Vertebrata PalAsiatica* 46(3): 178–189.]

Sinankylosaurus zhuchengensis

> *Sinankylosaurus zhuchengensis* Wang *et al.*, 2020 [Campanian; Xingezhuang Formation, Shandong, China]

Generic name: *Sinankylosaurus* ← {(place-name) Sin: China, chinese + (Gr.) ankylo-: fused + sauros: lizard}
Specific name: *zhuchengensis* ← {(place-name) Zhūchéng諸城 + -ensis}
Etymology: **Chinese fused lizard (Ankylosaur)** from Zhucheng
Taxonomy: Ornithischia: Thyreophora: Ankylosauria
[References: Wang KB, Zhang YX, Chen J, Chen SQ, Wang PY (2020). A new ankylosaurian from the Late Cretaceous strata of Zhucheng, Shandong Province. *Geological Bulletin of China* 39(7): 958–962.]

Sinocalliopteryx gigas

> *Sinocalliopteryx gigas* Ji *et al.*, 2007 [Barremian–Aptian; Yixian Formation, Liaoning, China]

Generic name: *Sinocalliopteryx* ← {(place-name) Sino-: an ancient name for China + (Gr.) καλλίο-: beautiful [κάλλος/kallos: beauty] + πτέρυξ/pteryx: feather} (*1)
Specific name: *gigas* ← {(Gr.) γίγας/gigas: giant} (*1)
Etymology: giant **Chinese beautiful feather**
Taxonomy: Theropoda: Coelurosauria:
Compsognathidae
Notes: The skeleton of the holotype was very complete and excellently-preserved with a long filamentous integuments (*1). *Sinocalliopteryx gigas* may have been a stealth hunter, because of its abdominal contents which were identified as an individual of *Sinornithosaurus*. A second specimen preserved *Confuciusornis sanctus* (*2).
[References: (*1) Ji S, Ji Q, Lu J, Yuan C (2007). A new giant compsognathid dinosaur with long filamentous integuments from Lower Cretaceous of northeastern China. *Acta Geologica Sinica* 81(1): 8–15.; (*2) Xing L, Bell PR, Scott Persons IV, Ji S, Miyashita T, Burns ME, Ji Q, Currie PJ (2012). Abdominal contents from two large Early Cretaceous compsognathids (Dinosauria: Theropoda) demonstrate feeding on confuciusornithids and dromaeosaurids. *PLoS ONE* 7(8): e44012.]

Sinoceratops zhuchengensis

> *Sinoceratops zhuchengensis* Xu *et al.*, 2010

[Campanian; Xingezhuang Formation, Shandong, China]

Generic name: *Sinoceratops* ← {(place-name) Sino-: China, Chinese + ceratops [(Gr.) κέρας/keras: horn + ωψ/ōps: face]} (*1)

Specific name: *zhuchengensis* ← {(place-name) Zhūchéng諸城 + -ensis}

Etymology: **Chinese horned face (ceratopsian)** from Zhucheng

Taxonomy: Ornithischia: Ceratopsia: Ceratopsidae: Centrosaurinae

[References: (*1) Xu X, Wang K-B, Zhao XJ, Li DJ (2010). First ceratopsid dinosaur from China and its biogeographical implications. *Chinese Science Bulletin* 55(16): 1631–1635.]

Sinocoelurus fragilis

> *Sinocoelurus fragilis* Yang, 1942 [Tithonian; Kuangyuan Formation, Sichuan, China]

Generic name: *Sinocoelurus* ← {(Med. Lat. place-name) Sino-: Chinese + *Coelurus* [(Gr.) κοῖλος/koilos: hollow + οὐρά/oura [ūrā]: tail]}

Specific name: *fragilis* ← {(Lat.) fragilis; fragile}.

Etymology: fragile, **Chinese hollow tail (*Coelurus*)**

Taxonomy: Theropoda

Notes: According to *The Dinosauria 2nd edition*, *Sinocoelurus* is regarded as a *nomen dubium*, because of the small amount of material.

[References: Young C-C (1942). Fossil vertebrates from Kuangyuan, N. Szechuan, China. *Bulletin of the Geological Society of China* 22(3–4): 293–309.]

Sinopterus dongi [Pterosauria]

> *Sinopterus dongi* Wang & Zhou, 2003 [Aptian; Jiufotang Formation, Liaoning, China]

Generic name: *Sinopterus* ← {(place-name) Sino-: China + pterus: wing [(Gr.) πτέρυξ/pteryx: wing]}

Specific name: *dongi* ← {(person's name) Dong + -ī}; dedicated to Chinese dinosaurologist Dong Zhiming (*1).

Etymology: Dong's **Chinese wing**

Taxonomy: Pterosauria: Pterodactyloidea: Tapejaridae

Other species:

> *Sinopterus gui* Li *et al.*, 2003 [Aptian; Jiufotang Formation, Liaoning, China]
> *Sinopterus lingyuanensis* Lü *et al.*, 2016 [Aptian; Jiufotang Formation, Liaoning, China]
> *Sinopterus atavismus* Lü *et al.*, 2016 [Aptian; Jiufotang Formation, Liaoning, China]

[References: Wang X, Zhou Z (2003). A new pterosaur (Pterodactyloidea, Tapejaridae) from the Early Cretaceous Jiufotang Formation of western Liaoning, China and its implications for biostratigraphy. *Chinese Science Bulletin* 48(1): 16–23.]

Sinornis santensis [Avialae]

> *Sinornis santensis* Sereno & Rao, 1992 [Aptian; Jiufotang Formation, Liaoning, China]

Generic name: *Sinornis* ← {(Med. Lat., place-name) Sin-: China + (Gr.) ὄρνις/ornis: bird}

Specific name: *santensis* ← {(place-name) Sāntă三塔: a traditional name of the county + -ensis}

Etymology: **China bird** from Santa

Taxonomy: Theropoda: Avialae: Enantiornithes

Notes: Sereno *et al.* (2001) considered *Cathayornis* to be a junior synonym of *Sinornis*. According to O'Connor & Dyke (2010), there are clear and distinct differences between them (*2).

[References: (*1) Sereno PC, Rao C (1992). Early evolution of avian flight and perching: New evidence from the Lower Cretaceous of China. *Science* 255(5046): 845–848.]; (*2) O'Connor J, Dyke G (2010). A reassessment of *Sinornis santensis* and *Cathayornis yandica* (Aves: Enantiornithes). *Records of the Australian Museum* 62: 7–20.]

Sinornithoides youngi

> *Sinornithoides youngi* Russell & Dong, 1993 [Aptian–Albian; Ejinhoro Formation, Inner Mongolia, China]

Generic name: *Sinornithoides* ← {(Med. Lat, place-name) Sin-: Chinese + (Gr.) ὀρνιθ-/ornith- (ὄρνις/ornis: bird) + -oides: -like, form}

Specific name: *youngi* ← {(person's name) Young + -ī}; in honor of China's vertebrate paleontologist C. C. Young楊鍾健 [Yang Zhongjian].

Notes: Young's **Chinese bird form**

Taxonomy: Theropoda: Troodontidae

Notes: The holotype was in the same roosting position as that of *Mei long*.

[References: Russell D, Dong Z (1993). A nearly complete skeleton of a new troodontid dinosaur from the Early Cretaceous of the Ordos Basin, Inner Mongolia, People's Republic of China. *Canadian Journal of Earth Sciences* 30(10): 2163–2173.]

Sinornithomimus dongi

> *Sinornithomimus dongi* Kobayashi & Lü, 2003 [Turonian; Ulansuhai Formation, Inner Mongolia, China]

Generic name: *Sinornithomimus* ← {(place-name) Sin- [(Med. Lat.) Sinae: China] + (Gr.) ὀρνιθος/ornithos [ὄρνις/ornis: bird] + (Lat.) mimus = (Gr.) μῖμος/mimos: mimic}; referring to occurrence in China (*1).

Specific name: *dongi* ← {(person's name) Dong + -ī}; "named after Professor Zhi-Ming Dong [董枝明Dong Zhiming], who discovered these skeletons and made great contributions to the Mongol Highland International Dinosaur Project" (*1).

Etymology: Dong's **Chinese bird mimic**

Taxonomy: Theropoda: Ornithomimosauria: Ornithomimidae

[References: (*1) Kobayashi Y, Lü J-C (2003). A new ornithomimid dinosaur with gregarious habits from the late Cretaceous of China. *Acta Palaeontologica Polonica* 48(2): 235–259.]

Sinornithosaurus millenii

> *Sinornithosaurus millenii* Xu *et al*., 1999 [early Aptian; Dakangpu Member, Yixian Formation, Liaoning, China]

Generic name: *Sinornithosaurus* ← {(place-name) Sin-: Chinese + (Gr.) ὀρνιθο-/ornitho-[ὄρνις/ornis: bird]} + σαῦρος/sauros: lizard}

Specific name: *millenii* ← {(Lat.) millennium: one-thousand years}; referring to the discovery near the end of the twentieth century.

Etymology: **Chinese bird lizard**, of millennium [千禧中国鳥龍]

Taxonomy: Theropoda: Maniraptora: Dromaeosauridae

Other species:

> *Sinornithosaurus haoiana* Liu *et al*., 2004 [early Aptian; Dakangpu Member, Yixian Formation, Liaoning, China]

Notes: The hypothesis that *Sinornithosaurus* possibly had a venomous bite is controversial (*1).

[References: Xu X, Wang X-L, Wu XC (1999). A dromaeosaurid dinosaur with a filamentous integument from the Yixian Formation of China. *Nature* 401: 262–266.; (*1) Gong E, Martin LD, Burnham DA, Falk AR (2010). The birdlike raptor *Sinornithosaurus* was venomous. *Proceedings of the National Academy of Sciences* 107(2): 766–768.]

Sinosauropteryx prima

> *Sinosauropteryx prima* Ji & Ji, 1996 [Barremian–Aptian; Yixian Formation, Liaoning, China]

Generic name: *Sinosauropteryx* ← {(Lat. place-name) Sino-: China + (Gr.) σαῦρος/sauros: referring to reptilian + πτέρυξ/pteryx: wing}; "indicating the specimen has transitional characters between small Theropoda and Aves" (*1).

Specific name: *prima* ← {(Lat.) prīma (feminine form of prīmus): first} (*1).

Etymology: first, **Chinese lizard (dinosaur) wing** [原始中華龍鳥]

Taxonomy: Theropoda: Coelurosauria: Compsognathidae
cf. Aves: Sauriurae: Sinosauropterygiformes: Sinosauropterygidae (*1)

Notes: Because *Sinosauropteryx* had short and primitive feathers, it was regarded a member of the class Aves (*1), but it was the first discovery of non-avialian dinosaur with feathers (*2).

[References: (*1) Ji Q, Ji S (1996). On the discovery of the earliest bird fossil in China (*Sinosauropteryx* gen. nov.) and the origin of birds. *Chinese Geology* 10(233): 30–33.; (*2) Chen P, Dong Z, Zhen S (1998). An exceptionally well-preserved theropod dinosaur from the Yixian Formation of China. *Nature* 391: 147–152.]

Sinosaurus triassicus

> *Sinosaurus triassicus* Young, 1940 [Sinemurian; Zhangjiawa Member, Lufeng Formation, Yunnan, China]

Generic name: *Sinosaurus* ← {(Med. Lat.) Sinae + (Gr.) σαῦρος/sauros: lizard}

Specific name: *triassicus*; referring to the Triassic.

Later, the Lower Lufeng Formation of Yunnan Province turned out to be Early Jurassic (*1).

Etymology: **Chinese lizard** from Triassic

Taxonomy: Theropoda: Tetanurae

Other species:

> *Dilophosaurus sinensis* Hu, 1993 ⇒ ?*Sinosaurus sinensis* (Hu, 1993) Wang *et al*., 2017 [Hettangian; Lufeng Formation, Yunnan, China]

[References: (*1) Xing L-D, Bell PR, Rothschild BM, Ran H, Zhang JP, Dong ZM, Zhang W, Currie PJ (2013). Tooth loss and alveolar remodeling in *Sinosaurus triassicus* (Dinosauria: Theropoda) from the Lower Jurassic strata of the Lufeng Basin, China. *Chinese Science Bulletin* 58(16): 1931–1935.]

Sinotyrannus kazuoensis

> *Sinotyrannus kazuoensis* Ji *et al*., 2009 [Aptian; Jiufotang Formation, Kazuo, Chaoyang, Liaoning, China]

Generic name: *Sinotyrannus* ← {(Lat. place-name) Sino-: an ancient name of China + (Gr.) τύραννος/tyrannos: tyrant} (*1)

Specific name: *kazuoensis* ← {(place-name) Kazuo: the county + -ensis} (*1)

Etymology: **Chinese tyrant** from Kazuo

Taxonomy: Theropoda: Tyrannosauroidea: Proceratosauridae
cf. Theropoda: Coelurosauria: Tyrannosauroidea: ?Tyrannosauridae (*1)

[References: (*1) Ji Q, Ji S-A, Zhang L-J (2009). First large tyrannosauroid theropod from the Early Cretaceous Jehol biota in northeastern China. *Geological Bulletin of China* 28(10): 1369–1374.]

Sinovenator changii

> *Sinovenator changii* Xu *et al*., 2002 [Early Cretaceous; Yixian Formation, Liaoning, China]

Generic name: *Sinovenator* ← {(place-name) Sino-[(Lat.) Sinae: China] + (Lat.) venator: hunter} (*1)

Specific name: *changii* ← {(person's name) Chang + -iī}; in honor of Chinese paleontologist Meemann Chang of the IVPP* for her significant role in the study of the Jehol fauna [*1].

*IVPP stands for Institute of Vertebrate Paleontology and Paleoanthropology.

Etymology: Chang's **Chinese hunter**

Taxonomy: Theropoda: Deinonychosauria: Troodontidae: Sinovenatorinae

Notes: A small theropod, less than 1 m in length.

[References: (*1) Xu X, Norell MA, Wang X-L, Makovicky PJ, Wu X-C (2002). A basal troodontid from the Early Cretaceous of China. *Nature* 415: 780–784.]

Sinraptor dongi

> *Sinraptor dongi* Currie & Zhao, 1994 [Oxfordian; Shishugou Formation, Xinjiang, China]

Generic name: *Sinraptor* ← {place-name) Sin- [Sinae: China] + (Lat.) raptor: robber}

Specific name: *dongi* ← {(person's name) Dong + -ī}; in honor of董枝明Dong Zhiming.

Etymology: Dong's **Chinese robber**

Taxonomy: Theropoda: Carnosauria: Allosauroidea: Metriacanthosauridae: Metriacanthosaurinae

Other species:

> *Yangchuanosaurus hepingensis* Gao, 1992 ⇒ ?*Sinraptor hepingensis* (Gao, 1992) Currie & Zhao, 1994 [Oxfordian; upper Shaximiao Formation, Sichuan, China]

[References: Currie PJ, Zhao X-J (1993). A new carnosaur (Dinosauria, Theropoda) from the Jurassic of Xinjiang, People's Republic of China. *Canadian Journal of Earth Sciences* 30(10-11): 2037–2081.]

Sinusonasus magnodens

> *Sinusonasus magnodens* Xu & Wang, 2004 ⇒ *Sinucerasaurus magnodens* (Xu & Wang, 2004) Xu & Norell, 2006 [early Aptian; Lujiatun Member, Yixian Formation, Liaoning, China]

Generic name: *Sinusonasus* ← {sinusoid [(Lat.) sinus: curve, hollow] + nāsus: nose}; referring to the sinusoid nasal [*1].

Specific name: *magnodens* ← {(Lat.) māgnus: large + dēns: tooth}; referring to the relatively large teeth [*1].

Etymology: large-toothed, **sinusoid nasal**

Taxonomy: Theropoda: Coelurosauria: Troodontidae: Sinovenatorinae

Notes: Sinovenatorinae Shen *et al*., 2017 includes *Daliansaurus, Mei*, *Sinovenator* and *Sinusonasus.*

[References: (*1) Xu X, Wang X (2004). A new troodontid (Theropoda: Troodontidae) from the Lower Cretaceous Yixian Formation of western Liaoning, China. *Acta Geologica Sinica* 78(1): 22–26.]

Sirindhorna khoratensis

> *Sirindhorna khoratensis* Shibata *et al*., 2015 [Aptian; Khok Kruat Formation, Nakhon Ratchasima, Thailand]

Generic name: *Sirindhorna* ← {(person's name) Sirindhorn + -a}; "dedicated to the Princess Maha Chakri Sirindhorn, Thailand, for her contribution to the support and encouragement of paleontology in Thailand" [*1].

Specific name: *khoratensis* ← {(place-name) Khorat: "the informal name of Nakhon Ratchasima Province" [*1] + -ensis}

Etymology: **Sirindhorn's one** from Khorat

Taxonomy: Ornithischia: Iguanodontia: Hadrosauriformes: Hadrosauroidea

[References: (*1) Shibata M, Jintasakul P, Azuma Y, You H-L (2015). A new basal hadrosauroid dinosaur from the Lower Cretaceous Khok Kruat Formation in Nakhon Ratchasima Province, Northeastern Thailand. *PLoS ONE* 10(12): e0145904.]

Skorpiovenator bustingorryi

> *Skorpiovenator bustingorryi* Canale *et al*., 2009 [Cenomanian–Turonian; Huincul Formation, Neuquén, NW Patagonia, Argentina]

Generic name: *Skorpiovenator* ← {(Lat.) scorpio [= (Gr.) σκορπίος/skorpios]: scorpion + (Lat.) venator: hunter}; "because of the abundance of living scorpions moving around the excavation" [*1].

Specific name: *bustingorryi* ← {(person's name); Bustingorry + -ī}; "in honor of the late Manuel Bustingorry, owner of the farm where the specimen was excavated" [*1].

Etymology: Bustingorry's **scorpion hunter**

Taxonomy: Theropoda: Ceratosauria: Abelisauroidea: Abelisauridae: Carnotaurinae: Brachyrostra [*1]

[References: (*1) Canale JI, Scanferla CA, Agnolin FL, Novas FE (2009). New carnivorous dinosaur from the Late Cretaceous of NW Patagonia and the evolution of abelisaurid theropods. *Naturwissenschaften* 96(3): 409–414.]

Songlingornis linghensis [Avialae]

> *Songlingornis linghensis* Hou, 1997 [Aptian; Jiufotang Formation, Liaoning, China]

Generic name: *Songlingornis* ← {(Chin. place-name) Sōnglǐng* 松嶺: 'pine ridge' + (Gr.) ὄρνις/ornis: bird}

*Songling, "for the name of the northeast-southwest oriented mountain range to the southeast of the fossil quarry, which is the second largest range in Chaoyang Co." [*1]

Specific name: *linghensis* ← {(Chin. place-name) Linhe + -ensis}; referring to "the Linghe River, which is the largest river in the fossiliferous

region of Western Liaoning and traverses the region from the northeast to southwest" (*1).
Etymology: **Songling bird** from Linhe River
Taxonomy: Theropoda: Avialae: Ornithothoraces: Yanornithiformes: Songlingornithidae
[References: (*1) Hou L (2001). *Mesozoic Birds of China*. Phoenix Valley Provincial Aviary of Taiwan: 1–137.]

Sonidosaurus saihangaobiensis
> *Sonidosaurus saihangaobiennsis* Xu *et al.*, 2006 [Campanian; Iren Dabasu Formation, Saihangaobi, Sonid, Inner Mongolia, China]
Generic name: *Sonidosaurus* ← {(place-name) Sonid Zuoqi蘇尼特左旗 + (Gr.) σαῦρος/sauros: lizard}
Specific name: *saihangaobiensis* ← {(place-name) Saihangaobi 賽罕高畢: southwest of Erenhot + -ensis}
Etymology: **Sonid lizard** from Saihangaobi
Taxonomy: Saurischia: Sauropodomorpha: Sauropoda: Titanosauriformes: Titanosauria
[References: Xu X, Zhang X, Tan Q, Zhao X, Tan L (2006). A new titanosaurian sauropod from Late Cretaceous of Nei Mongol, China. *Acta Geologica Sinica* 80(1): 20–26.]

Sonorasaurus thompsoni
> *Sonorasaurus thompsoni* Ratkevich, 1998 [Albian–Cenomanian; Turney Ranch Formation, Arizona, USA]
Generic name: *Sonorasaurus* ← {(place-name) Sonora + (Gr.) σαῦρος/sauros: lizard}; in reference to the Sonoran Desert.
Specific name: *thompsoni* ← {(person's name) Thompson + -ī}; in honor of Richard Thompson who discovered.
Etymology: Thompson's **Sonora lizard**
Taxonomy: Saurischia: Sauropodomorpha: Sauropoda: Brachiosauridae (*1)
Notes: According to *The Dinosauria 2nd edition* (2004), *Sonorasaurus* is considered to be a *nomen dubium*.
[References: (*1) Ratkevich R (1998). New Cretaceous brachiosaurid dinosaur, *Sonorasaurus thompsoni* gen *et* sp. nov, from Arizona. *Journal of the Arizona-Nevada Academy of Science* 31 (1): 71–82.]

Soriatitan golmayensis
> *Soriatitan golmayensis* Royo-Torres *et al.*, 2017 [late Hauterivian; Golmayo Formation, Soria, Castilla y Leon, Spain]
Generic name: *Soriatitan* ← {(place-name) Soria: a province of central Spain + (Gr. myth.) τῑτάν/tītan: giant}
Specific name: *golmayensis* ← {(place-name) Golmayo village + -ensis}
Etymology: **Soria giant** from Golmayo
Taxonomy: Saurischia: Sauropodomorpha: Sauropoda: Titanosauriformes: Brachiosauridae
[References: Royo-Torres R, Fuentes C, Meijide-Fuentes F, Meijide-Fuentes M (2017). A new Brachiosauridae sauropod dinosaur from the Lower Cretaceous of Europe (Soria Province, Spain). *Cretaceous Research* 80: 38–55.]

Soroavisaurus australis [Avialae]
> *Soroavisaurus australis* Chiappe, 1993 [Maastrichtian; Lecho Formation, Salta, Argentina]
Generic name: *Soroavisaurus* ← {(Lat.) soror: sister + *Avisaurus* (Brett-Surman & Paul., 1985)}; "referring to the sister group relationship inferred for these two taxa" (*1).
Specific name: *australis* ← {(Lat.) austrālis: southern}; "referring to the occurrence of this species in the Southern Hemisphere" (*1).
Etymology: southern **sister *Avisaurus***
Taxonomy: Enantiornithes: Avisauridae
[References: (*1) Chiappe LM (1993). Enantiornithine (Aves) tarsometatarsi from the Cretaceous Lecho Formation of northwestern Argentina. *American Museum Novitates* 3083: 1–27.]

Soumyasaurus aenigmaticus [Silesauridae]
> *Soumyasaurus aenigmaticus* Sarigül *et al.*, 2018 [Norian; Tecovas Formation (or Cooper Canyon Formation), Texas, USA]
Generic name: *Soumyasaurus* ← {(person's name) Soumya + (Gr.) σαῦρος/sauros: lizard}; "coined by Sankar Chatterjee to honour his elder son Soumya for his discovery of the specimen".
Specific name: *aenigmaticus* ← {(Lat.) aenigmaticus: enigmatic}; "representing the nature of the specimen".
Etymology: enigmatic, Soumya's lizard
Taxonomy: Archosauriformes: Dinosauriformes: Silesauridae

Sphaerotholus goodwini
> *Sphaerotholus goodwini* Williamson & Carr, 2003 ⇒*Prenocephale goodwini* (Williamson & Carr, 2003) Sullivan, 2003 [Campanian; Kirtland Formation, New Mexico, USA]
Generic name: *Sphaerotholus* ← {(Gr.) σφαῖρα/sphaira: a ball + θόλος/tholos: a dome}; "referring to the sphere-like shape of the dome of subadult and adult specimens in dorsal view" (*1).
Specific name: *goodwini* ← {(person's name) Goodwin + -ī}; "after Mark Goodwin for his contributions to

the study of pachycephalosaurians"(*1).

Etymology: Goodwin's **ball dome**

Taxonomy: Ornithischia: Marginocephalia: Pachycephalosauria: Pachycephalosauridae

Other species:

> *Sphaerotholus buchholtzae* Williamson & Carr, 2002 [Maastrichtian; Hell Creek Formation, Montana, USA]

= *Troodon edmontonense* Brown & Schlaikjer, 1943 ⇒ *Stegoceras edmontonense* (Brown & Schlaikjer, 1943) Sternberg, 1945 ⇒ = *Prenocephale edmontonensis* (Brown & Schlaikjer, 1943) Sullivan, 2000 [late Campanian; Edmonton Group, Alberta, Canada]

Notes: According to Longrich *et al.* (2010), *Troodon edmontonensis* is recombined as *Sphaerotholus edmontonense*.

[References: Williamson TE, Carr TD (2002). A new genus of derived pachycephalosaurian from western North America. *Journal of Vertebrate Paleontology* 22(4): 779–801.]

Spiclypeus shipporum

> *Spiclypeus shipporum* Mallon *et al.*, 2016 [Campanian; lower Coal Ridge Member, Judith River Formation, Montana, USA]

Generic name: *Spiclypeus* ← {(Lat.) spīca: spike + [anat.] (New Lat.) clypeus [(Lat.) clipeus: a round shield]}; "referring to the many large, spike-like epiossifications about the margin of the parietosquamosal frill"(*1).

With reference to its diagnosis, *Spiclypeus* can be distinguished by the large, triangular epiossifications laterally on the parietal and squamosal (*1).

Specific name: *shipporum* ← {(person's name) Shipp + -ōrum}; "in honor of Dr. Bill and Linda Shipp, the original owners of the holotype, and their family" (*1).

Etymology: Shipp's **spiky shield**

Taxonomy: Ornithischia: Ceratopsia: Neoceratopsia: Ceratopsidae: Chasmosaurinae

[References: (*1) Mallon JC, Ott CJ, Larson PL, Iuliano EM, Evans DC (2016). *Spiclypeus shipporum* gen. *et* sp. nov., a boldly audacious new chasmosaurine ceratopsid (Dinosauria: Ornithischia) from the Judith River Formation (Upper Cretaceous: Campanian) of Montana, USA. *PLoS ONE* 11(5): e0154218.]

Spinophorosaurus nigerensis

> *Spinophorosaurus nigerensis* Remes *et al.*, 2009 [Middle Jurassic; Irhazer II Formation, Agadez, Niger]

Generic name: *Spinophorosaurus* ← {(Lat.) spīna: spike + (Gr.) φορός/phoros*: bringing in [φέρω/pherō: to bear, carry] + σαῦρος/sauros: lizard}; "referring to the presence of spike-bearing osteoderms"(*1).

cf. (Lat.) ferō: to bear, carry, support.

Specific name: *nigerensis* ← {(place-name) Niger + -ensis}; "referring to the Republic of Niger, the provenance of this taxon"(*1).

Etymology: **spike-bearing lizard** from Niger

Taxonomy: Saurischia: Sauropodomorpha: Sauropoda: Gravisauria

[References: (*1) Remes K, Ortega F, Fierro I, *et al.* (2009). A new basal sauropod dinosaur from the Middle Jurassic of Niger and the early evolution of Sauropoda. *PLoS ONE* 4(9): e6924.]

Spinops sternbergorum

> *Spinops sternbergorum* Farke *et al.*, 2011 [Campanian; Oldman Formation, Alberta, Canada]

Generic name: *Spinops* ← {(Lat.) spīna: spine + (Gr.) ὤψ/ōps: face}; "referring to the ornamentation on the face"(*1).

Specific name: *sternbergorum* ← {(person's name) Sternberg + -ōrum}; "in honor of Charles H. and Levi Sternberg, collectors of the original specimens"(*1).

Etymology: Sternberg's **spined face**

Taxonomy: Ornithischia: Ceratopsia: Ceratopsidae: Centrosaurinae(*1)

[References: (*1) Farke AA, Ryan MJ, Barrett PM, Tanke DH, Braman DR, Loewen MA, Graham MR (2011). A new centrosaurine from the Late Cretaceous of Alberta, Canada, and the evolution of parietal ornamentation in horned dinosaurs. *Acta Palaeontologica Polonica* 56(4): 691–702.]

Spinosaurus aegyptiacus

> *Spinosaurus aegyptiacus* Stromer, 1915 [Cenomanian; Baharîja Formation, Marsa Matruh, Egypt]

Generic name: *Spinosaurus* ← {(Lat.) spīna: spine + (Gr.) σαῦρος/sauros: lizard}; referring to the neural spines which were high and formed a sail-like structure.

Specific name: *aegyptiacus* ← {(Lat.) aegyptiacus: of Egypt}; after the land of origin (*1).

Etymology: **Spined lizard** of Egypt

Taxonomy: Theropoda: Megalosauroidea: Spinosauridae: Spinosaurinae: Spinosaurini(*3)

Other species:

> *Sigilmassasaurus brevicollis* Russell, 1996 [Albian–Cenomanian; Kem Kem Beds Formation, Er Rachidia (Ksar-es-Souk), Morocco]

= ?*Spinosaurus maroccanus* Russell, 1996 [Albian–Cenomanian; Kem Kem Beds Formation,

Er Rachidia (Ksar-es-Souk), Morocco]

Notes: According to *The Dinosauria 2nd edition*, *Spinosaurus maroccanus* was regarded as a species of *Spinosaurus*, and *Sigilmassasaurus brevicollis* belonged to its own genus. The skull of *Spinosaurus* was similar to that of a modern crocodilian. The original specimens, kept in Munich, were destroyed by an Allied air raid in 1944.

[References: (*1) Stromer E (1915). Ergebnisse der forschungsreisen Prof. E. Stromers in den wüsten Ägyptens. II. Wirbeltier-Reste der Baharîje-Stufe (unterstes Cenoman). 3. Das original des theropoden *Spinosaurus aegyptiacus* nov. gen. nov. spec. *Abhandlungen der Königliche Bayerischen Akademie der Wissenschaften Mathematisch-physikalische Klasse* 28(3): 1–32.; Stromer E (1915). Results of Prof. E. Stromer's Research Expedition in the Deserts of Egypt. II.Translation by R. T. Zanon, 1989.; (*2) Smith JB, Lamanna MC, Mayr H, Lacovara, KJ (2006). New information regarding the holotype of *Spinosaurus aegyptiacus* Stromer, 1915. *Journal of Paleontology* 80(2): 400–406.; (*3) Arden TMS, Klein CG, Zouhri S, Longrich NR (2018). Aquatic adaptation in the skull of carnivorous dinosaurs (Theropoda: Spinosauridae) and the evolution of aquatic habits in spinosaurs. *Cretaceous Research* (93): 275–284.]

Spinostropheus gautieri

> *Elaphrosaurus gautieri* Lapparent, 1960 ⇒ *Spinostropheus gautieri* (Lapparent, 1960) Sereno *et al.*, 2004 [Bathonian–Oxfordian; Tiourarén Formation, Irhazer Group, Agadez, Niger]

Generic name: *Spinostropheus* ← {(Lat.) spina: spine + (Gr.) στροφεύς/stropheus: vertebra}; "named for the prominent epipophyseal processes on the cervical vertebrae and its moderate body size (estimated length 4 m)"(*1).

Specific name: *gautieri* ← {(person's name) Gautier + -ī}; "named after geologist F. Gautier, who discovered the type locality (In Tedreft)"(*1).

Etymology: Gautier's **spined vertebra**

Taxonomy: Theropoda: Ceratosauria

Notes: The first bones were discovered by A. F. de Lapparent in 1959, but incorrectly referred to the Late Jurassic genus *Elaphrosaurus* (Lapparent 1960)(*1).

[References: (*1) Sereno PC, Wilson JA, Conrad JL (2004). New dinosaurs link southern landmasses in the mid-Cretaceous. *Proceedings of the Royal Society B: Biological Sciences* 271: 1325–1330.]

Spondylosoma absconditum [Archosauria]

> *Spondylosoma absconditum* von Huene, 1942 [Ladinian; Santa Maria Formation, Río Grande do Sul, Brazil]

Generic name: *Spodylosoma* ← {(Gr.) σπόνδυλος/spondylos: vertebra + σῶμα/sōma: body}

Specific name: *absconditum* ← {(Lat.) absconditum (neuter form of absconditus, *adj.*): concealed, secret, hidden}

Etymology: concealed, **vertebra body**

Taxonomy: Archosauria: Avemetatarsalia:
Aphanosauria (acc. Nesbitt *et al.* 2017)
cf. Archosauria: Crurotarsi or Saurischia
(acc. *The Dinosauria 2nd edition*)

Notes: According to *The Dinosauria 2nd edition*, *Spondylosoma* was thought to be "possible Dinosauria", but Nesbitt *et al.* (2017) included it into a member of Aphanosauria, which is a sister to Ornithodira.

Staurikosaurus pricei

> *Staurikosaurus pricei* Colbert, 1970 [Carnian; Alemoa Member, Santa Maria Formation, Rio Grande do Sul, Brazil]

= *Teyuwasu barberenai* Kischlat, 1999 [Carnian; Alemoa Member, Santa Maria Formation, Rio Grande do Sul, Brazil] (according to Garcia *et al.* 2019)

Generic name: *Staurikosaurus* ← {(Gr.) σταυρικος/staurikos: of a cross [σταυρός/stauros: the Cross] + (Gr.) σαῦρος/sauros: lizard}; "in allusion of the constellation of the Southern Cross"(*1).

Specific name: *pricei* ← {(person's name) Price + -ī}; "in honor of Llewellyn Ivor Price, who has made extensive collections and studies of fossil reptiles in Brazil" (*1).

Etymology: Price's **(Southern) Cross lizard**

Taxonomy: Dinosauria: Herrerasauridae (*2)

[References: (*1) Colbert EH (1970). A saurischian dinosaur from the Triassic of Brazil. *American Museum Novitates* 2405: 1–39.; (*2) Cau A (2018). The assembly of the avian body plan: a 160-million-year long. *Bollettino della Società Paleontologica Italiana* 57(1): 1–25.]

Stegoceras validum

> *Stegoceras validus* Lambe, 1902 ⇒ *Stegoceras validum* (Lambe, 1902) Lambe, 1918 ⇒ *Troodon validus* (Lambe, 1902) Gilmore, 1924 ⇒ *Stegoceras validum* (Lambe, 1902) emend. Sues & Galton, 1987 [Campanian; Dinosaur Park Formation, Alberta, Canada]

= *Troodon validus* Gilmore, 1924 (paratype) [Campanian; Dinosaur Park Formation, Alberta, Canada]

= *Stegoceras browni* Wall & Galton, 1979 ⇒ *Ornatotholus browni* (Wall & Galton, 1979) Galton & Sues, 1983 [late Campanian; Dinosaur Park Formation, Alberta, Canada]

= *Stegoceras brevis* Lambe, 1918 ⇒ *Stegoceras breve* (Lambe, 1918) Sues & Galton, 1987 ⇒

Prenocephale brevis (Lambe, 1918) Sullivan, 2000 (*nomen dubium*) ⇒ *Sphaerotholus brevis* (Lambe, 1918) Longrich *et al.*, 2010 [Campanian; Dinosaur Park Formation, Alberta, Canada]
= *Stegoceras lambei* Sternberg, 1945 ⇒ *Colepiocephale lambei* (Sternberg, 1945) Sullivan, 2003 [Campanian; Foremost Formation, Alberta, Canada]
= *Troodon sternbergi* Brown & Schlaikjer, 1943 ⇒ *Stegoceras sternbergi* (Brown & Schlaikjer, 1943) Sternberg, 1945 ⇒ *Hanssuesia sternbergi* (Brown & Schlaikjer, 1943) Sullivan, 2003 [Campanian; Belly River Group, Alberta, Canada]

Generic name: *Stegoceras* ← {(Gr.) στέγος/stegos: roof, covered [στέγω/stegō: cover] + κέρας/keras: horn}. "It was originally thought that thickened frontoparietals were part of the nasal region and formed the basis for a horn" (*1).

Specific name: *validum* ← {(Lat.) validum (neuter form of 'validus'): strong}; "presumably in allusion to the thick skull roof" (*1).

Etymology: strong, **roof horn**

Taxonomy: Ornithischia: Marginocephalia: Pachycephalosauridae

Other species:

> *Troodon edmontonense* Brown & Schlaikjer, 1943 ⇒ *Stegoceras edmontonense* (Brown & Schlaikjer, 1943) Sternberg, 1945 ⇒ *Prenocephale edmontonensis* (Brown & Schlaikjer, 1943) Sullivan, 2000 [Campanian–Maastrichtian; Edmonton Group, Alberta, Canada]
> *Stegoceras novomexicanum* Jasinski & Sullivan, 2011 [Campanian; Fossil Forest Member, Fruitland Formation, New Mexico, USA]
> *Troodon bexelli* Bohlin, 1953 ⇒ *Stegoceras bexelli* (Bohlin, 1953) Kuhn, 1964 (*nomen dubium*) [Campanian–Maastrichtian; Minhe Formation, Nei Mongol, China]

[References: Sullivan RM (2003). Revision of the dinosaur *Stegoceras* Lambe (Ornithischia, Pachycephalosauridae). *Journal of Vertebrate Paleontology* 23(1): 181–207.; (*1) Sues H-D, Galton PM (1987). Anatomy and classification of the North American Pachycephalosauria (Dinosauria: Ornithischia). *Palaeontographica Abteilung A*. 198: 1–40.]

Stegopelta landerensis

> *Stegopelta landerensis* Williston, 1905 [Albian–Cenomanian; Belle Fourche Member, Frontier Formation*, Wyoming, USA]
*originally described as the Hailey Shale Formation.

Generic name: *Stegopelta* ← {(Gr.) stego- [στέγω/stegō: cover] + πέλτη/peltē: shield}; referring to its armour.

Specific name: *landerensis* ← {(place-name) Lander + -ensis}

Etymology: **roofed shield** from Lander
cf. covered shield (acc. Holtz, 2008)

Taxonomy: Ornithischia: Ankylosauria: Nodosauridae

Stegosaurus armatus

> *Stegosaurus armatus* Marsh, 1877 [Kimmeridgian–Tithonian; Morrison Formation, Wyoming, USA]
= *Stegosaurus ungulatus* Marsh, 1879 ⇒ *Hypsirhophus ungulatus* (Marsh, 1879) Lydekker, 1893 (*nomen dubium*) [Kimmeridgian–Tithonian; Morrison Formation, Wyoming, USA]
= *Hypsirhophus seeleyanus* Cope, 1879 (*nomen nudum*) ⇒ *Stegosaurus seeleyanus* (Cope, 1879) (*nomen nudum*) [Late Jurassic; Morrison Formation, Colorado, USA]
= *Stegosaurus affinis* Marsh, 1881 (*nomen dubium*) [Kimmeridgian; Lake Como Member, Morrison Formation, Wyoming, USA]
= *Stegosaurus duplex* Marsh, 1887 [Kimmeridgian–Tithonian; Morrison Formation, Wyoming, USA]
= *Stegosaurus sulcatus* Marsh, 1887 [Kimmeridgian–Tithonian; Morrison Formation, Wyoming, USA

Generic name: *Stegosaurus* ← {(Gr.) στέγος/stegos: roof [στέγω/stegō: cover] + σαῦρος/sauros: lizard}; referring to its bony plates. When Marsh described the first fossil of a *Stegosaurus*, he thought that the plates would have lain flat on its back.

It is described in a paper that 'the body was protected by large bony dermal plates, somewhat like those of *Atlantochelys* (*Protostega*)*'. 'One of the large dermal plates was over one meter in length) (*1). *an extinct marine turtle

Specific name: *armatus* ← {(Lat.) armātus: equipped in arms}

Etymology: armored, **roof lizard**

Taxonomy: Ornithischia: Thyreophora: Stegosauria: Stegosauridae

Other species:

> *Stegosaurus stenops* Marsh, 1887 ⇒ *Diracodon stenops* (Marsh, 1887) Bakker, 1986 [Kimmeridgian; Brushy Basin Member, Morrison Formation, Colorado, USA]
= *Diracodon laticeps* Marsh, 1881 [Kimmeridgian–Tithonian; Morrison Formation, Wyoming, USA]
> *Hypsirhophus discurus* Cope, 1878 (*nomen*

dubium) ⇒ *Stegosaurus discurus* (Cope, 1878) Gilmore, 1914 (*nomen dubium*) [Tithonian; Morrison Formation, Colorado, USA]
> *Stegosaurus marshi* Lucas, 1901 ⇒ *Hoplitosaurus marshi* (Lucas, 1901) Lucas, 1902 [?Barremian; Lakota Formation, South Dakota, USA]
> *Stegosaurus priscus* Nopcsa, 1911 ⇒ *Lexovisaurus priscus* (Nopcsa, 1911) Galton & Upchurch, 2004 ⇒ *Loricatosaurus priscus* (Nopcsa, 1911) Maidment *et al.*, 2008 [Callovian; Oxford Clay Formation, Peterborough, UK]
> *Stegosaurus longispinus* Gilmore, 1914 ⇒ *Natronasaurus longispinus* (Gilmore, 1914) Ulansky, 2014 (invalid) ⇒ *Alcovasaurus longispinus* (Gilmore, 1914) Galton & Carpenter, 2016 (*2) [Kimmeridgian–Tithonian; Morrison Formation, Wyoming, USA]
> *Stegosaurus madagascariensis* Piveteau, 1926 (*nomen dubium*) [Maastrichtian; Maevarano Formation, Mahajanga, Madagascar]
> *Wuerhosaurus homheni* Dong, 1973 ⇒ *Stegosaurus homheni* (Dong, 1973) Maidment *et al.*, 2008 [Aptian–Albian; Lianmugin Formation, Xinjiang, China]
> *Priconodon crassus* Marsh, 1888 (*nomen dubium*) ⇒ *Stegosaurus crassus* (Marsh, 1888) Hennig, 1915 [Aptian-Albian; Arundel Clay Formation, Maryland, USA]
> *Hesperosaurus mjosi* Carpenter *et al.*, 2001 ⇒ *Stegosaurus mjosi* (Carpenter, 2001) Maidment *et al.*, 2008 [Kimmeridgian; Morrison Formation, Wyoming, USA]
> *Stegosaurus durobrivensis* (Hulke, 1887) Hulke, 1887 See *Lexovisaurus*.
> *Stegosaurus hastiger* (Owen, 1877) Lydekker, 1890 See *Dacentrurus*.

Notes: Since the type specimen of *Stegosaurus armatus* was very fragmentary, *S. stenops* was replaced as the type species and *S. armatus* was considered a *nomen dubium*. (*3) (*4)

[References: (*1) Marsh OC (1877). A new order of extinct Reptilia (Stegosauria) from the Jurassic of the Rocky Mountains. *American Journal of Sciences* 3–14(84): 513–514.; (*2) Galton PM, Kenneth C (2016). The plated dinosaur *Stegosaurus longispinus* Gilmore, 1914 (Dinosauria: Ornithischia; Upper Jurassic, western USA), type species of *Alcovasaurus* n. gen. *Neues Jahrbuch für Geologie und Paläontologie – Abhandlungen* 279(2): 185–208.; (*3) Galton PM (2010). Species of plated dinosaur *Stegosaurus* (Morrison Formation, Late Jurassic) of western USA: new type pecies designation needed. *Swiss Journal of Geosciences* 103: 187–198.; (*4) International Commission on Zoological Nomenclature (2013). Opinion 2320 (Case 3536): *Stegosaurus* Marsh, 1877 (Dinosauria, Ornithischia): type species replaced with *Stegosaurus stenops* Marsh, 1877. *The Bulletin of Zoological Nomenclature* 70(2): 129–130.]

Stellasaurus ancellae

> *Stellasaurus ancellae* Wilson *et al.*, 2020 [late Campanian; Two Medicine Formation, Montana, USA]

Generic name: *Stellasaurus* ← {(Lat.) stella: star + (Gr.) σαῦρος/sauros: lizard}; "in reference to the oval star-like appearance of the cranial ornamentation and in homage to the song 'Starman' by David Bowie".

Specific name: *ancellae* ← {(person's name) Ancell +-ae}; "in honor of Museum of the Rockies field palaeontologist and fossil preparator Carrie Ancell, who discovered and prepared MOR* 492, the holotype specimen of *Stellasaurus ancellae*, as well as the holotype of *Achelousaurus horneri* and co-discovered the holotype of *Einiosaurus procurvicornis*, and whose decades of extraordinary fossil preparation have furthered vertebrate palaeontology beyond measure".
*Museum of the Rockies

Etymology: Ancell's **star lizard**

Taxonomy: Ornithischia: Ceratopsia: Neoceratopsia: Ceratosauridae: Centrosaurinae

Stenonychosaurus inequalis

> *Troodon formosus* Leidy, 1856 [middle Campanian; Judith River Formation, Montana, USA]
= *Stenonychosaurus inequalis* Sternberg, 1932 ⇒ *Troodon inequalis* (Sternberg, 1932) Currie, 1987 [Campanian; Dinosaur Park Formation, Belly River Group (= Judith River Group), Alberta, Canada]
= *Polyodontosaurus grandis* Gilmore, 1932 [Campanian; Dinosaur Park Formation, Alberta, Canada]

Generic name: *Stenonychosaurus* ← {(Gr.) στενός/stenos: narrow + ὄνυξ/onyx: talon + σαῦρος/sauros: lizard}

Specific name: *inequalis* ← {(Lat.) in-: un- + aequālis: equal}; unequal.

Etymology: unequal, **narrow-clawed lizard**

Taxonomy: Theropoda: Troodontidae: Troodontinae

Notes: According to *The Dinosauria 2nd edition*, *Stenonychosaurus inequalis* was considered to be a synonym of *Troodon formosus*, but *Stenonychosaurus* was thought a separate genus from *Troodon* and reverted (Evans *et al.*, 2017; van der Reest, 2017).

[References: van der Reest AJ, Currie PJ (2017). Troodontids (Theropoda) from the Dinosaur Park Formation, Alberta, with a description of a unique new taxon: implications for deinonychosaur diversity in North America. *Canadian Journal of Earth Sciences* 54(9): 919–935.]

Stenopelix valdensis

> *Stenopelix valdensis* von Meyer, 1857 [Berriasian; Obernkirchen Member, Bückeberg Formation, Niedersachsen, Germany]

Generic name: *Stenopelix* ← {(Gr.) στενός/stenos: narrow + (Gr.) πέλυξ/pelyx: pelvis}

Specific name: *valdensis* ← {(stratal name) Wealden + -ensis}

'German Wealden' was superseded by the Bückeberg Formation (Casey *et al.* 1975) (*1).

Etymology: **narrow pelvis** from Wealden

Taxonomy: Ornithischia: Marginocephalia: Ceratopsia

cf. Ornithischia: Marginocephalia: Pachycephalosauria

[References: (*1) Hornung J, Böhme A, Reich M (2011). The 'German Wealden' and the Obernkirchen Sandstone – an introduction. In: Richter A, Reich M (eds), *Dinosaur Tracks 2011, an International Symposium, Obernkirchen, April 14–17, 2011*: 62–72.]

Stokesosaurus clevelandi

> *Stokesosaurus clevelandi* Madsen, 1974 [Kimmeridgian; Brushy Basin Member, Morrison Formation, Emery, Utah, USA]

Generic name: *Stokesosaurus* ← {(person's name) Stokes + (Gr.) σαῦρος/sauros: lizard}; in honor of "Professor William Lee Stokes, whose continued efforts made possible the development of the Cleveland-Lloyed Quarry" (*1).

Specific name: *clevelandi* ← {(place-name) Cleveland + -ī}; referring to the town of Cleveland.

Etymology: **Stokes' lizard** of Cleveland

Taxonomy: Theropoda: Tyrannosauroidea

Other species:

> *Stokesosaurus langhami* Benson, 2008 ⇒ *Juratyrant langhami* (Benson, 2008) Brusatte & Benson, 2013 [late Tithonian; upper member, Kimmeridge Clay Formation, Dorset, England, UK]

[References: (*1) Madsen JH (1974). A new theropod dinosaur from the Upper Jurassic of Utah. *Journal of Paleontology* 48: 27–31.]

Stormbergia dangershoeki

> *Lesothosaurus diagnosticus* Galton, 1978 [Hettangian; upper Elliot Formation, Lesotho]

= *Stormbergia dangershoeki* Butler, 2005 [Hettangian; upper Elliot Formation, South Africa]

Generic name: *Stormbergia* ← {(stratal name) Stormberg Group + -ia}; "referring to the Stormberg Group of South Africa and Lesotho, the rock sequence that has provided so much information on early dinosaurs" (*1).

Specific name: *dangershoeki* ← {(place-name/person's name) Dangershoek + -ī}; "referring to the locality from which the holotype (SAM-PK-K1105) was collected (*1). It was "collected from Dangershoek Farm, Herschel District, eastern Cape Province by C. E. Gow" (*1).

Etymology: Dangershoek's **one from Stormberg**

Taxonomy: Ornithischia: Genasauria

Notes: *Stormbergia dangershoeki* almost certainly represents the adult form of *Lesothosaurus* according to Knoll *et al.* (2009). *Stormbergia* is thought to be a junior subjective synonym of *Lesothosaurus*, according to Baron *et al.* (2016).

[References: (*1) Butler RJ (2005). The 'fabrosaurid' ornithischian dinosaurs of the Upper Elliot Formation (Lower Jurassic) of South Africa and Lesotho. *Zoological Journal of the Linnean Society* 145: 175–218.]

Streptospondylus altdorfensis

> *Streptospondylus altdorfensis* von Meyer, 1832 [late Callovian; Basse-Normandie, France]

= *Steneosaurus rostromajor* Saint-Hilaire, 1825 ⇒ *Streptospondylus rostromajor* (Saint-Hilaire, 1825) Owen, 1842 [late Callovian; Basse-Normandie, France]

= *Laelaps gallicus* Cope, 1867 [Bajocian; Inferior Oolite Group, Chipping Norton, UK]

Generic name: *Streptospondylus* ← {(Gr.) στρεπτος/streptos: reversed + σπονδυλος/spondylos: vertebra}

Specific name: *altdorfensis* ← {(place-name) Altdorf: a town in south-eastern Germany + -ensis}; referring to the place where Teleosauridae (Crocodylomorpha) remains had also been found. Cuvier (1808) considered the remains to be crocodilian (*1).

Etymology: **reversed vertebra** from Altdorf *

*Altdorf (not occurrence of this specimen).

Taxonomy: Theropoda: Megalosauria

[References: (*1) Cuvier G (1808). Sur les ossements fossiles de crocodiles et particulièrement sur ceux des environs du Havre et d'Honfleur, avec des remarques sur les squelettes de sauriens de la Thuringe. *Annales du Muséum d'Histoire naturelle de Paris* (12): 73–110.]

Struthiomimus altus

> *Ornithomimus altus* Lambe, 1902 ⇒ *Struthiomimus altus* (Lambe, 1902) Osborn,

1917 [Campanian; Oldman Formation, Alberta, Canada]

Generic name: *Struthiomimus* ← {(generic name) Strūthiō [(Gr.) στρουθός/struthos: the ostrich, Struthio] + (Gr.) μῖμος/mimos: mimic}; referring to "its mimicry of the ostrich (*Struthio*) in the skull, neck, and foot structure" (*1). *Struthiomimus* had "the extremely small head and slender jaws entirely without teeth which resemble those of the ostrich (Struthio)" (*1).

Specific name: *altus* ← {(Lat.) altus: tall, lofty}

Etymology: tall **ostrich mimic**

Taxonomy: Theropoda: Coelurosauria: Ornithomimosauria: Ornithomimidae

Other species:

> *Struthiomimus brevetertius* Parks, 1926 ⇒ *Dromiceiomimus brevetertius* Russell, 1972 [Campanian; Horseshoe Canyon Formation, Alberta, Canada]

= *Ornithomimus edmontonicus* Sternberg, 1933 [Campanian; Horseshoe Canyon Formation, Alberta, Canada]

= *Struthiomimus currellii* Parks, 1933 [Campanian; Horseshoe Canyon Formation, Alberta, Canada]

= *Struthiomimus ingens* Parks, 1933 [Maastrichtian; Horseshoe Canyon Formation, Alberta, Canada]

= *Struthiomimus samueli* Parks, 1928⇒ *Ornithomimus samueli* (Parks, 1928) Kuhn, 1965 ⇒ *Dromiceiomimus samueli* (Parks, 1928) Russell, 1972 [Campanian; Dinosaur Park Formation, Alberta, Canada]

[References: (*1) Osborn HF (1917). Skeletal adaptations of *Ornitholestes*, *Struthiomimus*, *Tyrannosaurus*. *Bulletin of the American Museum of Natural History* 35: 733–771.]

Struthiosaurus austriacus

> *Struthiosaurus austriacus* Bunzel, 1871 [early Campanian; Grünbach Formation, Niederosterreich, Austria]

= *Danubiosaurus anceps* Bunzel, 1871 (*partim*) [early Campanian; Grünbach Formation, Niederösterreich, Austria]

= *Crataeomus lepidophorus* Seeley, 1881 [early Campanian; Grünbach Formation, Niederösterreich, Austria]

= *Crataeomus pawlowitschii* Seeley, 1881 (*partim*) [early Campanian; Grünbach Formation, Niederösterreich, Austria]

= *Hoplosaurus ischyrus* Seeley, 1881 [early Campanian; Coal-Bearing Complex Formation, Austria]

= *Pleuropeltus suessi* Seeley, 1881 (*partim*) [early Campanian; Grünbach Formation, Niederösterreich, Austria]

= *Leipsanosaurus noricus* Nopcsa, 1918 [early Campanian; Coal-Bearing Complex Formation, Austria]

Generic name: *Struthiosaurus* ← {(Gr.) στρουθός/strouthos [strūthos]: the ostrich, Struthio + (Gr.) σαῦρος/sauros: lizard}

Specific name: *austriacus* ← {austriacus: of Austria}

Etymology: **ostrich lizard** of Austria

Taxonomy: Ornithischia: Thyreophora: Ankylosauria: Nodosauridae: Struthiosaurinae

Other species:

> *Struthiosaurus transylvanicus* Nopcsa, 1915 [late Campanian; Sânpetru Formation, Judetul Covurlui, Romania]

> *Struthiosaurus languedocensis* Garcia & Pereda-Suberbiola, 2003 [early Campanian; Languedoc-Roussillon, France]

Stygimoloch spinifer

> *Troodon wyomingensis* Gilmore, 1931 ⇒ *Pachycephalosaurus wyomingensis* (Gilmore, 1931) Brown & Schlaikjer, 1943 (*1) (*nomen conservandum*) [Maastrichtian; Lance Formation, Wyoming, USA]

= *Tylosteus ornatus* Leidy, 1872 (*nomen rejectum*) [unknown; Lance Formation?, Wyoming, USA]

= *Stygimoloch spinifer* Galton & Sues, 1983 [Maastrichtian; Hell Creek Formation, Montana, USA]

= *Stenotholus kohleri* Giffin *et al.*, 1988 [Maastrichtian; Hell Creek Formation, Montana, USA]

= *Dracorex hogwartsia* Bakker *et al.*, 2006 [Maastrichtian; Hell Creek Formation, South Dakota, USA]

Generic name: *Stygimoloch* ← {(Gr. myth.) Στύγ-/Styg- [Στύξ/Styx: the mythical underground river] + Moloch: in allusion to the bizarre appearance of the animal in life} (*2)

The squamosal of *Stygimoloch* was characterized by distinctive hypertrophied, spike-like nodes (*3).

Specific name: *spinifer* ← {(Lat.) spina: thorn + fero: to bear}; 'carring spines', in allusion to the squamosal horns (*2).

Etymology: **Styx (Hell Creek) devil** bearing thorns

Taxonomy: Ornithischia: Marginocephalia: Pachycephalosauria: Pachycephalosauridae

Notes: According to Horner & Goodwin (2009), Goodwin & Evans (2016), *Dracorex hogwartsia* is considered to be a juvenile *Pachycephalosaurus wyomingensis* and *Stygimoloch spinifer* to be a subadult *P. wyomingensis*.

[References: (*1) Galton PM, Sues H-D (1983). New data on pachycephalosaurid dinosaurs (Reptilia: Ornithischia) from North America. *Canadian Journal of Earth Sciences* 20: 462–472.]; (*2) Sues HD, Galton PM (1987). Anatomy and classification of the North American Pachycephalosauria (Dinosauria: Ornithischia). *Palaeontographica Abteilung A: Paleozoologie-Stratigraphie* 198(1–3): 1–40.; (*3) Sullivan RM (2006). A taxonomic review of the pachycephalosauridae (Dinosauria: Ornithischia). In: Lucas SG, Sullivan RM, (eds.) *Late Cretaceous Vertebrates from the Western Interior. New Mexico Museum of Natural History and Science Bulletin* 35: 347–365.]

Styracosaurus albertensis

> *Styracosaurus albertensis* Lambe, 1913 ⇒ *Centrosaurus albertensis* (Lambe, 1913) Paul, 2010 [late Campanian; Dinosaur Park Formation, Alberta, Canada]

= *Styracosaurus parksi* Brown & Schlaikjer, 1937 [late Campanian; Dinosaur Park Formation, Alberta, Canada]

Generic name: *Styracosaurus* ← {(Gr.) στυρακο- [στύραξ/styrax: spike on the end of spear] + σαῦρος/sauros: lizard}; referring to the spikes on its frill.

The skull of this species is remarkable for the large nasal horn-core, and for the spike-shaped processes on the posterior margin of the coalesced parietals [*1].

Specific name: *albertensis* ← {(place-name) Alberta + -ensis}

Etymology: **spiked lizard** from Alberta

Taxonomy: Ornithischia: Marginocephalia: Ceratopsia: Ceratopsidae: Centrosaurinae: Centrosaurini

Other species:

> *Styracosaurus ovatus* Gilmore, 1930 ⇒ *Centrosaurus ovatus* (Gilmore, 1930) Paul, 2010 ⇒ *Rubeosaurus ovatus* (Gilmore, 1930) McDonald & Horner, 2010 [Campanian; Two Medicine Formation, Montana, USA]

Notes: McDonald & Horner placed *Styracosaurus ovatus* in its own genus *Rubeosaurus* in 2010.

[References: (*1) Lambe LM (1913). A new genus and species of Ceratopsia from the Belly River Formation of Alberta. *The Ottawa Naturalist* 27(9): 109–116.; Ryan MJ, Holmes R, Russell AP (2007). A revision of the late Campanian centrosaurine ceratopsid genus *Styracosaurus* from the Western Interior of North America. *Journal of Vertebrate Paleontology* 27(4): 944–962.]

Suchomimus tenerensis

> *Suchomimus tenerensis* Sereno *et al.*, 1998 ⇒ *Baryonyx tenerensis* (Sereno *et al.*, 1998) Sues *et al.*, 2002 [Aptian; Elrhaz Formation, Niger]

= *Cristatusaurus lapparenti* Taquet & Russell, 1998 [Aptian–Albian; Elrhaz Formation, Niger]

Generic name: *Suchomimus* ← {(Gr.) Σοῦχος/Souchos [Sūchos]*: crocodile + μῖμος/mimos: mimic}; "named for the low elongate snout and piscivorous adaptations of the jaws" [*1].

* Egyptian crocodile god Sobek

Specific name: *tenerensis* ← {(place-name) Ténéré Desert + -ensis}; "for the region of the Sahara in which it was discovered" [*1].

Etymology: **crocodile mimic** from Ténéré Desert

Taxonomy: Theropoda: Megalosauroidea (= Spinosauroidea): Spinosauridae

[References: (*1) Sereno PC, Beck AL, Dutheil DB, Gado B, Larsson HCE, Lyon GH, Marcot JD, Rauhut OWM, Sadleir RW, Sidor CA, Varricchio DD, Wilson GP, Wilson JA (1998). A long-snouted predatory dinosaur from Africa and the evolution of spinosaurids. *Science* 282 (5392): 1298–1302.]

Suchosaurus cultridens

> *Suchosaurus cultridens* Owen, 1841 (*nomen dubium*) [Early Cretaceous; Wadhurst Clay, East Sussex, UK]

Generic name: *Suchosaurus* ← {(Gr.) Σοῦχος/Souchos [Sūchos]: the Egyptian crocodile god Sobek + σαῦρος/sauros: lizard}

Specific name: *cultridens* ← {(Lat.) cultri- [culter: dagger, knife] + dēns: tooth}

Etymology: **crocodile lizard** with knife-like teeth

Taxonomy: Theropoda: Spinosauridae

Other species:

> *Suchosaurus girardi* Sauvage, 1897 [Barremian; Papo Seco Formation, Portugal]

Notes: *Suchosaurus cultridens* and *Suchosaurus girardi* are considered *nomina dubia* [*1].

[References: (*1) Mateus O, Araújo R, Natário C, Castanhinha R (2011). A new specimen of the theropod dinosaur *Baryonyx* from the early Cretaceous of Portugal and taxonomic validity of *Suchosaurus*. *Zootaxa* 2827: 54–68.]

Sulaimanisaurus gingerichi

> *Sulaimanisaurus gingerichi* Malkani, 2004 *vide* Malkani, 2006 [Maastrichtian; Vitakri Member, Pab Formation, Balochistan, Pakistan]

Generic name: *Sulaimanisaurus* ← {(place-name) Sulaiman Fold Belt: the host mountain range for latest Cretaceous dinosaurs + (Gr.) σαῦρος/sauros: reptiles} [*1]

Specific name: *gingerichi* ← {(person's name) Gingerich + -ī}; "in honor of Dr. Phillip D. Gingerich, Museum of Paleontology, University of Michigan, USA for verifying the ideas of

author regarding first dinosaurs from Pakistan" (*1).
Etymology: Gingerich's **Sulaiman reptiles**
Taxonomy: Saurischia: Sauropodomorpha:
Sauropoda: Titanosauria: Pakisauridae (= Titanosauridae) (*1)

[References: (*1) Malkani MS (2006). Biodiversity of saurischian dinosaurs from the Latest Cretaceous Park of Pakistan. *Journal of Applied and Emerging Sciences* 1(3): 108–140.]

Sulcavis geeorum [Avialae]

> *Sulcavis geeorum* O'Connor *et al.*, 2013 [Aptian; Jiufotang Formation, Liaoning, China]

Generic name: *Sulcavis* ← {(Lat.) sulcus: groove, furrow + avis: bird}; referring to the grooves on the teeth (*1).
Specific name: *geeorum* ← {(person's name) Gee family + -ōrum}
Etymology: Gee family's **groove bird**
Taxonomy: Theropoda: Avialae: Enantiornithes:
Bohaiornithidae
Notes: *Sulcavis geeorum* had robust teeth. (*1)

[References: (*1) O'Connor JK, Zhang Y, Chiappe LM, Meng Q, Quanguo L, Di L (2013). A new enantiornithine from the Yixian Formation with the first recognized avian enamel specialization. *Journal of Vertebrate Paleontology* 33(1): 1–12.]

Supersaurus vivianae

> *Supersaurus vivianae* Jensen, 1985 [Kimmeridgian; Brushy Basin Member, Morrison Formation, Colorado, USA]
= *Ultrasaurus macintoshi* Jensen, 1985 (*partim*) ⇒ *Ultrasauros macintoshi* (Jensen, 1985) Olshevsky, 1991 [Kimmeridgian; Brushy Basin Member, Morrison Formation, Colorado, USA]
= *Dystylosaurus edwini* Jensen, 1985 [Kimmeridgian; Brushy Basin Member, Morrison Formation, Colorado, USA]

Generic name: *Supersaurus* ← {super*: above + (Gr.) σαῦρος/sauros: lizard}
*internationally published vernacular name (*1).
Specific name: *vivianae* ← {(person's name) Vivian + -ae}; in honor of Vivian Jones, co-discoverer of all the important Late Jurassic fossil localities on the Uncompahgre Upwarp (*1).
Etymology: Vivian's **super lizard**
Taxonomy: Saurischia: Sauropodomorpha:
Sauropoda: Diplodocidae: Diplodocinae
Other species:

> *Dinheirosaurus lourinhanensis* Bonaparte & Mateus, 1999 ⇒ *Supersaurus lourinhanensis* (Bonaparte & Mateus, 1999) Tschopp *et al.*, 2015 [Kimmeridgian; Lourinhã Formation, Portugal]

Notes: *Supersaurus vivianae* is estimated 34 m in total length, according to Holtz (2007). It had a long neck and a long tail.

[References: (*1) Jensen JA (1985). Three new sauropod dinosaurs from the Upper Jurassic of Colorado. *Great Basin Naturalist* 45: 697–709.; (*2) Curtice B, Stadtman K (2001). The demise of *Dystylosaurus edwini* and a revision of *Supersaurus vivianae*. In: McCord RD, Boaz D (eds), *Western Association of Vertebrate Paleontologists and Southwest Paleontological Symposium – Proceedings* 2001. *Mesa Southwest Museum Bulletin* 8: 33–40.]

Suskityrannus hazelae

> *Suskityrannus hazelae* Nesbitt *et al.*, 2019 [Turonian; Moreno Hill Formation, New Mexico, USA]

Generic name: *Suskityrannus* ← {(Zuni) suski: coyote + (Lat.) tyrannus = (Gr.) τύραννος/tyrannos: an absolute sovereign, tyrant}
Specific name: *hazelae* ← {(person's name) Hazel + -ae}; in honor of Hazel Wolfe, who supported fossil expeditions in the Zuni Basin.
Etymology: Hazel's **coyote tyrant**
Taxonomy: Theropoda: Tyrannosauroidea:
Pantyrannosauria

Suuwassea emilieae

> *Suuwassea emilieae* Harris & Dodson, 2004 [?Tithonian (*1) or Kimmeridgian–Tithonian; Brushy Basin Member, Morrison Formation, Montana, USA]

Generic name: *Suuwassea* [pron.: Soo-oo-WAH-see-uh] ← {(Crow) suu: thunder + wassa: ancient (*1)}; "an homage to the traditional appellation "thunder lizard" often applied to sauropods (following *Brontosaurus* Marsh, 1879). The use of a Crow term further reflects the position of the type locality in ancestral Crow territory as well as its proximity to the present Crow Reservation. The spelling of the name follows the best current orthography for the Crow language, which does not use Latin characters; the pronunciation is approximate and simplified" (*1).
Specific name: *emilieae* ← {(person's name) Emilie + -ae}; "in honor of the late Emilie de Hellebranth, paleontology advocate who generously funded the expeditions in 1999–2000 that recovered the specimen" (*1).
Etymology: Emilie's, **first thunder heard in spring**
Taxonomy: Saurischia: Sauropodomorpha:
Sauropoda: Diplodocoidea: Flagellicaudata:
Dicraeosauridae

[References: (*1) Harris JD, Dodson P (2004). A new diplodocoid Sauropod dinosaur from the Upper Jurassic Morrison Formation of Montana, USA. *Acta*

Palaeontologica Polonica 49(2): 197–210.]

Suzhousaurus megatherioides

> *Suzhousaurus megatherioides* Li *et al*., 2007 [Aptian–Albian; Xinminpu Group, Gansu, China]

Generic name: *Suzhousaurus* ← {(Chin. place-name) Sùzhōu肅州: ancient name of the Jiuquan area + (Gr.) σαῦρος/sauros: lizard} (*1)

Specific name: *megatherioides* ← {(genus) *Megatherium** + -oides: -like}

**Megatherium* [(Gr.) μέγας/megas: big, great + θηρίον/therion: beast]: the giant ground sloth.

Etymology: *Megatherium*-like, **Suzhou lizard**

Taxonomy: Theropoda: Coelurosauria: Therizinosauroidea

[References: (*1) Li D, Peng C, You H, Lamanna MC, Harris JD, Lacovara KJ, Zhang J (2007). A large therizinosauroid (Dinosauria: Theropoda) from the Early Cretaceous of Northwestern China. *Acta Geologica Sinica* 81(4): 539–549.]

Syngonosaurus macrocercus

> *Acanthopholis macrocercus* Seeley, 1869 (*partim*) (*nomen dubium*) ⇒ *Syngonosaurus macrocercus* (Seeley, 1869) Seeley, 1879 (*nomen dubium*) ⇒ *Anoplosaurus macrocercus* (Seeley, 1869) Kuhn, 1964 (*nomen dubium*) [Cenomanian; Cambridge Greensand, UK]

Generic name: *Syngonosaurus* ← {syngono- [(Gr.) σύν/syn: along with + γόνος/gonos: offspring] + σαῦρος/sauros: lizard} of the same parent, kindred

Specific name: *macrocercus* ← {(Gr.) μακρός/macros: large + κέρκος/kerkos: the tail of a beast}

Etymology: large-tailed **kindred lizard**

Taxonomy: Ornithischia: Ornithopoda: Iguanodontia

Notes: According to *The Dinosauria 2nd edition*, *Syngonosaurus* was seen as an ankylosaur, but Barrett *et al.* (2020) reinterpreted it as a basal iguanodontian.

Syntarsus rhodesiensis

> *Syntarsus rhodesiensis* Raath, 1969 (preoccupied) ⇒ *Megapnosaurus rhodesiensis* (Raath, 1969) Ivie *et al*., 2001 ⇒ *Coelophysis rhodesiensis* (Raath, 1969) Bristowe & Raath, 2004 [Hettangian; Forest Sandstone Formation, Matabeleland North, Zimbabwe]

Generic name: *Syntarsus* ← {(Gr.)σύν/syn: with, together + ταρσός/tarsos: tarsus}; for the coossified construction of its tarsal foot bones.

Specific name: *rhodesiensis* ← {(place-name) Rhodesia: the old name for Zimbabwe + -ensis}

Etymology: **syn- tarsus** from Rhodesia

Taxonomy: Theropoda: Ceratosauria: Coelophysoidea

Other species:

> *Syntarsus kayentakatae* Rowe, 1989 ⇒ *Megapnosaurus kayentakatae* (Rowe, 1989) Ivie *et al*., 2001 ⇒ *Coelophysis kayentakatae* (Rowe, 1989) Bristowe & Raath, 2004 [Sinemurian–Pliensbachian; Kayenta Formation, Arizona, USA]

Syrmosaurus disparoserratus

> *Syrmosaurus disparoserratus* Maleev, 1952 (*nomen dubium*) ⇒ *Talarurus disparoserratus* (Maleev, 1952) Maryańska, 1977 (*nomen dubium*) ⇒ *Maleevus disparoserratus* (Maleev, 1952) Tumanova, 1987 (*nomen dubium*) [Cenomanian–Santonian, Baynshire Formation, Ömnögovi, Mongolia]

Generic name: *Syrmosaurus* ← {(Gr.)συρμός/syrmos: any lengthened sweeping motion + σαῦρος/sauros: lizard}

Specific name: *disparoserratus* ← {(Lat.) dispār: unequal + serrātus: serrated}; referring to the unequal serrations on the teeth.

Etymology: unequal-serrated, **dragging lizard**

Taxonomy: Ornithischia: Ankylosauria: Ankylosauridae

Other species:

>*Pinacosaurus grangeri* Gilmore, 1933 [Campanian; Djadokhta Formation, Ömnögovi, Mongolia]

= *Pinacosaurus ninghsiensis* Young, 1935 [Late Cretaceous, Ningxia, China]

= *Syrmosaurus viminocaudus* Maleev, 1952 [Campanian; Djadokhta Formation, Ömnögovi, Mongolia]

Szechuanosaurus campi

> *Szechuanosaurus campi* Young, 1942 (*nomen dubium*) [Tithonian; Kuangyuan Formation, Sichuan, China]

Generic name: *Szechuanosaurus* ← {(place-name) Szechuan四川 [Pinyin: Sìchuān] + (Gr.) σαῦρος/sauros: lizard}

Specific name: *campi* ← {(person's name) Camp + -ī}; in honor of Charles Lewis Camp.

Etymology: Camp's **Szechuan (Sichuan) lizard**

Taxonomy: Theropoda: Neotheropoda

Other species:

> *Szechuanosaurus yandonensis* Dong *et al.*, 1978 (*nomen nudum*) ⇒*Yangchuanosaurus shangyouensis* Dong *et al*., 1978 [Kimmeridgian–Tithonian; Shaximiao Formation, Zigong, Sichuan, China]

> *Szechuanosaurus zigongensis* Gao, 1993 ⇒ *Yangchuanosaurus zigongensis* (Gao, 1993)

Carrano *et al.*, 2012 [Oxfordian; Shaximiao Formation, Zigong, Sichuan, China]

Notes: According to *The Dinosauria 2nd edition*, *Szechuanosaurus campi* and *S. yandonensis* are considered to be *nomina dubia*.

T

Tachiraptor admirabilis

> *Tachiraptor admirabilis* Langer *et al.*, 2014 [Hettangian; La Quinta Formation, Táchira, Venezuela]

Generic name: *Tachiraptor* ← {(place-name) Táchira: the Venezuelan state + (Lat.) raptor: thief}; "in reference to the probable predatory habits of the animal" (*1).

Specific name: *admirabilis* ← {(Lat.) admīrābilis: admirable}; "in honor of Simon Bolivar's 'Admirable Campaign' in which La Grita, the town where the type locality is located, played a strategic role" (*1).

Etymology: **Tachira thief** of Admirable Campaign

Taxonomy: Theropoda: Neotheropoda

[References: (*1) Langer MC, Rincón AD, Ramezani J, Solórzano A, Rauhut OWM (2014). New dinosaur (Theropoda, *stem*-Averostra) from the earliest Jurassic of the La Quinta Formation, Venezuelan Andes. *Royal Society Open Science* 1: 140184.]

Talarurus plicatospineus

> *Talarurus plicatospineus* Maleev, 1952 [Cenomanian–Turonian; Bayan/Bayn Shireh Formation, Dornogovi, Mongolia]

Generic name: *Talarurus* ← {(Gr.) τάλαρος/talaros: basket, wicker + οὐρά/oura [ūrā]: tail}; referring to a club on its tail.

Specific name: *plicatospineus* ← {(Lat.) plicātus: folded [plico: to fold, wind, coil] + spīneus: thorny}; referring to the corrugated "spines forming the external ornament of the armor" (*1).

Etymology: **basket tail** with plicated spines

Taxonomy: Ornithischia: Thyreophora: Ankylosauria: Ankylosauridae

Other species:

> *Syrmosaurus disparoserratus* Maleev, 1952 ⇒ *Talarurus disparoserratus* (Maleev, 1952) Maryańska, 1977 ⇒ *Maleevus disparoserratus* (Maleevus, 1952) Tumanova, 1987 (*nomen dubium*) [Cenomanian; Bayan Shireh Formation, Ömnögovi, Mongolia]

[References: (*1) Maleev EA (1952). Noviy ankilosavr is verchnego mela Mongolii [A new ankylosaur from the Upper Cretaceous of Mongolia]. *Doklady Akademii Nauk Soyúz Sovétskikh Sotsialistícheskikh Respúblik* 87: 273–276.]

Talenkauen santacrucensis

> *Talenkauen santacrucensis* Novas *et al.*, 2004 [Campanian–Maastrichtian; Cerro Fortaleza Formation* (= Pari Aike Formation), Santa Cruz, Argentina] *acc. Sickman *et al.* (2018).

Generic name: *Talenkauen* ← {(Tehuelche / Aónikenk) talenk: small + kauen: skull}; "in reference to the proportionally small head of the animal" (*1).

Specific name: *santacrucensis* ← {(place-name) Santa Cruz: the southern Argentine province + -ensis} (*1)

Etymology: **small skull** from Santa Cruz

Taxonomy: Ornithischia: Neornithichia: Elasmaria
cf. Ornithischia: Ornithopoda: Iguanodontia: Euiguanodontia (*1)

[References: (*1) Novas FE, Cambiaso AV, Ambrosio A (2004). A new basal iguanodontian (Dinosauria, Ornithischia) from the Upper Cretaceous of Patagonia. *Ameghiniana* 41(1): 75–82.]

Talos sampsoni

> *Talos sampsoni* Zano et al., 2011 [Campanian; Kaiparowits Formation, Utah, USA]

Generic name: *Talos* ← {(Gr. myth.) Τάλως/Talōs}; "referring to the mythological, fleet-footed protector of Crete, often depicted as winged, who succumbed to a wound on the ankle. The name is also a play on the English word 'talon' meaning a sharply hooked claw" (*1).

Specific name: *sampsoni* ← {(person's name) Sampson + -ī}; "in honor of Scott D. Sampson, architect of the Kaiparowits Basin Project" (*1).

Etymology: Sampson's **Talos**

Taxonomy: Theropoda: Coelurosauria: Troodontidae

[References: (*1) Zanno LE, Varricchio DJ, O'Connor PM, Titus AL, Knell MJ (2011). A new troodontid theropod, *Talos sampsoni* gen. *et* sp. nov., from the Upper Cretaceous Western Interior Basin of North America. *PLoS ONE* 6(9): e24487.]

Tambatitanis amicitiae

> *Tambatitanis amicitiae* Saegusa & Ikeda, 2014 [Albian; Sasayama Group, Hyogo, Japan]

Generic name: *Tambatitanis* ← {(Jpn. place-name) Tamba丹波 + (Gr. myth.) Τῑτᾱνίς/Tītānis (feminine form of 'Τῑτάν/Tītan'): giants}; "in reference to the Tamba, the northwestern region of Kansai Area, SW Japan, where the type specimen was collected" and titanis, "symbolic of the great size of the specimen" (*1).

Specific name: *amicitiae* ← {(Lat.) amicitia: friendship}; "referring to the friendship between Messrs. Shigeru Murakami and Kiyoshi Adachi who found the holotype skeleton of this new species, in August, 2006" (*1).

Etymology: **Tamba giant** of friendship

Taxonomy: Saurischia: Sauropodomorpha: Sauropoda: Titanosauriformes: Somphospondyli

Notes: The nickname is Tambaryū (丹波竜) meaning 'Tamba dragon'.

[References: Saegusa H, Ikeda T (2014). A new titanosauriform sauropod (Dinosauria: Saurischia) from the Lower Cretaceous of Hyogo, Japan. *Zootaxa* 3848(1): 1–66.]

Tangvayosaurus hoffeti

> *Tangvayosaurus hoffeti* Allain *et al.*, 1999 [Aptian–Albian; Grès supérieurs Formation, Tang Vay, Savannakhet, Laos]

Generic name: *Tangvayosaurus* ← {(place-name) Tang Vay + (Gr.) σαῦρος/sauros: lizard}

Specific name: *hoffeti* ← {(person's name) Hoffet + -ī}; "in honor of Josué-Heilmann Hoffet, member of the *Service Géologique de l'Indochine*, who discovered the first Australasian dinosaurs" (*1). He is a French paleontologist.

Etymology: Hoffet's **Tang Vay lizard**

Taxonomy: Saurischia: Sauropoda: Titanosauriformes: Titanosauria (*1)

[References: (*1) Allain R, Taquet P, Battail B, Dejax J, Richir P, Veran M, Limon-Duparcmeur F, Vacant R, Mateus O, Sayarath P, Khenthavong B, Phouyavong S (1999). Un nouveau genre de dinosaure sauropode de la formation des Grès supérieurs (Aptien–Albien) du Laos. *Comptes Rendus de l'Académie des Sciences à Paris, Sciences de la Terre et des Planètes* 329(8): 609–616.]

Tanius sinensis

> *Tanius sinensis* Wiman, 1929 [Campanian; Jiangjunding Formation, Wangshi Group, Shandong, China]

Generic name: *Tanius* ← {(person's name) Tan譚 + -ius}; in honor of Chinese paleontologist Tán Xīchóu譚錫疇.

Specific name: *sinensis* ← {(place-name) Sin-*: China + -ensis}

*Sin(o)- : (Med. Latin) Sina ← (Gr.) Sinai ← ? | (Arab) Sin ← (Chin.) Ch'in秦 (*1)

Etymology: **Tanius** from China [中国譚氏龍]

Taxonomy: Ornithischia: Ornithopoda: Hadrosauroidea: Hadrosauromorpha

Other species:

> *Bactrosaurus prynadai* Riabinin, 1939 (*nomen dubium*) ⇒ *Tanius prynadai* (Riabinin, 1939) Young, 1958 (*nomen dubium*) [Late Cretaceous; Dabrazhin Formation, Kyrkkuduk, Asht, Tajikistan]

> *Tsintaosaurus spinorhinus* Young, 1958 [Campanian; Jingangkou Formation, Wangshi Group, Laiyang, Shandong, China]

?= *Tanius chingkankouensis* Young, 1958 (*nomen dubium*) [Campanian; Jingangkou Formation, Shandong, China]

= *Tanius laiyangensis* Zhen, 1976 (*nomen dubium*) [Campanian; Jingangkou Formation, Shandong, China]

Notes: According to *The Dinosauria 2nd edition*, *Tanius chingkankouensis* was considered to be synonymous with *T. sinensis*, and *T. laiyangensis* was regarded as a synonym of *Tsintaosaurus spinorhinus*. Zhang et al. (2019) considered *Tanius laiyangensis* to be a member of Kritosaurini within Hadrosaurinae.

[References: (*1) Ooga M *et al.* (eds). (2007) *Shogakukan Robert Grand Dictionnaire Français-Japonais*. Shougakukan, Tokyo. (the first edition:1888).; Wiman C (1929). Die Kreide-Dinosaurier aus Shantung. *Palaeontologia Sinica Series C* 6(1): 1–67.]

Tanycolagreus topwilsoni

> *Tanycolagreus topwilsoni* Carpenter *et al.*, 2005 [Oxfordian–Tithonian; Morrison Formation, Wyoming, USA]

Generic name: *Tanycolagreus* ← {(Gr.) τανυ-/tany-: long, stretched out + κῶλον/kōlon: limb + ἀγρεύς/agreus: hunter} (*1)

Specific name: *topwilsoni* ← {(person's name)"Top"Wilson + -ī}; "named for George " Top" Wilson, retired, United States Marine Corps" (*1).

Etymology: "Top" Wilson's **long-limbed hunter**

Taxonomy: Theropoda: Coelurosauria

Notes: Some specimens that have been referred to *Ornitholestes hermanni* (in part) and *Stokesosaurus clevelandi* (in part) were reassigned to *Tanycolagreus topwilsoni*. (*1)

[References: (*1) Carpenter K, Miles C, Cloward K (2005). New small theropod from the Upper Jurassic Morrison Formation of Wyoming. In: Carpenter K (ed), *The Carnivorous Dinosaurs*. Indiana University Press, Bloomington: 23–48.]

Taohelong jinchengensis

> *Taohelong jinchengensis* Yang *et al.*, 2013 [Early Cretaceous; Unit 5 Formation, Hekou Group, Gansu, China]

Generic name: *Taohelong* ← {(Chin. place-name) Táohé洮河: Tao River + lóng龍: dragon}

Specific name: *jinchengensis* ← {(Chin. place-name) Jīnchéng金城 + -ensis}

Etymology: **Taohe dragon** from Jincheng

Taxonomy: Ornithischia: Thyreophora: Ankylosauria: Nodosauridae: Polacanthinae

[References: Yang J-T, You H-L, Li D-Q, Kong DL

(2013). First discovery of polacanthine ankylosaur dinosaur in Asia. *Vertebrata PalAsiatica* 51(4): 265–277.]

Tapejara wellnhoferi [Pterosauria]

> *Tapejara wellnhoferi* Kellner, 1989 [Albian; Romualdo Member, Santana Formation, Ceará, Brazil]

Generic name: *Tapejara* ← {(Tupi) Tapejara: the old being}

Specific name: *wellnhoferi* ← {(person's name) Wellnhofer + -ī}; in honor of German paleontologist Peter Wellnhofer.

Etymology: Wellnhofer's **old being**

Taxonomy: Pterosauria: Pterodactyloidea: Tapejaridae: Tapejarini

Other species:

> *Tapejara imperator* Campos & Kellner, 1997 ⇒ *Tupandactylus imperator* (Campos & Kellner, 1997) Kellner & Campos, 2007 ⇒ *Ingridia imperator* (Campos & Kellner, 1997) Unwin & Martin, 2007 [Aptian; Nova Olinda Member, Crato Formation, Ceará, Brazil]

> *Tapejara navigans* Frey *et al.*, 2003 ⇒ *Tupandactylus navigans* (Frey *et al.*, 2003) Kellner & Campos, 2007 ⇒ *Ingridia navigans* (Frey *et al.*, 2003) Unwin & Martin, 2007 [Aptian; Nova Olinda Member, Crato Formation, Ceará, Brazil]

Notes: *Tupandactylus* retained priority over the name *Ingridia*. *Ingridia* was named in memory of Ingrid Wellnhofer, wife of Peter Wellnhofer.

Tapuiasaurus macedoi

> *Tapuiasaurus macedoi* Zaher *et al.*, 2011 [Aptian; Quiricó Formation, Areado Group, Minas Gerais, Brazil]

Generic name: *Tapuiasaurus* ← {(demonym) Tapuia + (Gr.) σαῦρος/sauros: lizard}; in reference to Tapuia, "from the Jês indigenous language family used to designate tribes that inhabited in inner regions of Brazil"(*1).

Specific name: *macedoi* ← {(person's name) Macedo + -ī}; "in honor of Ubirajara Alves Macedo, who first discovered the deposits near Coração de Jesus"(*1).

Etymology: Macedo's **Tapuia lizard**

Taxonomy: Saurischia: Sauropodomorpha: Sauropoda: Titanosauria: Nemegtosauridae?

Notes: An almost complete skull was discovered (*1).

[References: (*1) Zaher H, Pol D, Carvalho AB, Nascimento PM, Riccomini C, Larson P, Juarez-Valìeri R, Pires-Domingues R, da Silva Jr NJ, Campos DA (2011). A complete skull of an Early Cretaceous sauropod and the evolution of advanced titanosaurians. *PLoS ONE* 6(2): e16663.]

Tarascosaurus salluvicus

> *Tarascosaurus salluvicus* Le Loeuff & Buffetaut, 1991 [Campanian; unnamed unit (Var), Provence-Alpes-Côte d'Azur, France]

Generic name: *Tarascosaurus* ← {(France, legend) Tarasque + (Gr.) σαῦρος/sauros: lizard}; referring to "the Tarasque, fabled animal, a type of dragon in Provençal legend" (*1).

Specific name: *salluvicus* ← {(Lat.) Sallūviī: a Gaulish tribe}; referring to "the Salluvians, a Gallic people from the environs of Marseilles"(*1).

Etymology: Gallic people's **Tarasque lizard**

Taxonomy: Theropoda: Ceratosauria: Abelisauridae

[References: (*1) Le Loeuff J, Buffetaut E (1991). *Tarascosaurus salluvicus* nov. gen., nov. sp., a theropod dinosaur from the Upper Cretaceous of southern France. *Geobios* 24(5): 585–594.]

Tarbosaurus bataar

> *Tyrannosaurus bataar* Maleev, 1955 ⇒ *Tarbosaurus bataar* (Maleev, 1955) Rozhdestvensky, 1965 ⇒ *Jenghizkhan bataar* (Maleev, 1955) Olshevsky & Ford, 1995 [late Campanian; Nemegt Formation, Ömnögovi, Mongolia]

= *Tarbosaurus efremovi* Maleev, 1955 [late Campanian; Nemegt Formation, Ömnögovi, Mongolia]

= *Gorgosaurus novojilovi* Maleev, 1955 ⇒ *Deinodon novojilovi* (Maleev, 1955) Maleev, 1964 ⇒ *Aublysodon novojilovi* (Maleev, 1955) Charig, 1967 ⇒ *Deinodon novojilovi* (Maleev, 1955) Kuhn, 1965 ⇒ *Maleevosaurus novojilovi* (Maleev, 1955) Pickering, 1984 ⇒ *Albertosaurus novojilovi* (Maleev, 1955) Mader & Bradley, 1989 ⇒ *Tyrannosaurus novojilovi* (Maleev, 1955) Glut, 1997 [Campanian; Nemegt Formation, Ömnögovi, Mongolia]

= ? *Shanshanosaurus huoyanshanensis* Dong, 1977 ⇒ *Aublysodon huoyanshanensis* (Dong, 1977) Paul, 1988 [Campanian; Subashi Formation, Xinjiang, China]

= *Albertosaurus periculosus* Riabinin, 1930 ⇒ *Deinodon periculosus* (Riabinin, 1930) Kuhn, 1965 [late Maastrichtian; Yuliangze Formation, Heilongjiang, China]

= *Gorgosaurus lancinator* Maleev, 1955 ⇒ *Deinodon lancinator* (Maleev, 1955) Kuhn, 1965 ⇒ *Aublysodon lancinator* (Maleev, 1955) Charig, 1967 [Maastrichtian; Nemegt Formation, Ömnögovi, Mongolia]

= ? *Chingkankousaurus fragilis* Young, 1958 [Campanian; Jingangkou Formation, Laiyang,

Shandong, China]
= ? *Tyrannosaurus luanchuanensis* Dong, 1979 ⇒ *Jenghizkhan luanchuanensis* (Dong, 1979) Olshevsky *vide* Olshevsky *et al.*, 1995 [Campanian; Qiupa Formation, Henan, China]
= ? *Tyrannosaurus turpanensis* Zhai *et al.*, 1978 [Campanian–Maastrichtian; Subashi Formation, Shanshan, Xinjiang, China]

Generic name: *Tarbosaurus* ← {(Gr.) τάρβος/tarbos: fear, terror, alarm, dread + σαῦρος/sauros: lizard}
Specific name: *bataar* ← {(Mong.) баатар/baatar: hero}
Etymology: heroic, **terrible lizard**
Taxonomy: Theropoda: Coelurosauria: Tyrannosauroidea: Tyrannosauridae: Tyrannosaurinae

[References: Maleev EA (1955). Giant carnivorous dinosaurs of Mongolia. *Doklady, Academy of Sciences USSR* 104(4): 634–637.]

Tarchia gigantea

> *Dyoplosaurus giganteus* Maleev, 1956 ⇒ *Tarchia gigantea* (Maleev, 1956) Maryańska, 1977 (*nomen dubium**) [Campanian–Maastrichtian; Nemegt Formation, Ömnögovi, Mongolia] *according to Arbour *et al.* (2014).

Generic name: *Tarchia* ← {(Mong.) Тархи: brain + -ia}; "because of a relatively large brain case" (*1).
Specific name: *gigantea* ← {(Lat.) giganteus = (Gr.) γιγάντειος/giganteios: large}.
Etymology: large, **brainy one**
Taxonomy: Ornithischia: Thyreophora: Ankylosauria: Ankylosauridae
Other species:
> *Tarchia kielanae* Maryańska, 1977 [Campanian; Barun Goyot Formation, Khulsan, Ömnögovi, Mongolia]
= ? *Minotaurasaurus ramachandrani* Miles & Liles, 2009 (*2) ("*M. ramachandrani* was purchased from the Tucson Gem, Mineral and Fossil Showcase in Arizona, USA without provenance data." (*2)).
> *Tarchia teresae* Penkalski & Tumanova, 2016 [Maastrichtian; Nemegt Formation, Ömnögovi, Mongolia]

Notes: According to *The Dinosauria 2nd edition*, *Tarchia gigantea* was valid, but it was considered to be dubious by Arbour (2014). *Tarchia kielanae* had a larger brain case, almost twice the height of *Saichania chulsanensis* which has a skull almost the same size as *T. kielanae* (*1). Penkalski & Tumanova (2016) consider *Minotaurasaurus* to be a distinct taxon from *Tarchia*.

[References: (*1) Maryañska T (1977). Ankylosauridae (Dinosauria) from Mongolia. *Palaeontologia Polonica* 37: 85–151.]; (*2) Arbour VM, Currie PJ, Badamgarav D (2014). The ankylosaurid dinosaurs of the Upper Cretaceous Baruungoyot and Nemegt formations of Mongolia. *Zoological Journal of the Linnean Society* 172(3): 631–652.]

Tastavinsaurus sanzi

> *Tastavinsaurus sanzi* Canudo *et al.*, 2008 [early Aptian; Xert Formation, Maestrazgo Basin, Teruel, Aragón, Spain]

Generic name: *Tastavinsaurus* ← {(Catalan place-name) Tastavins + (Gr.) σαῦρος/sauros: lizard}; "Tastavin, meaning 'wine taster', a word that also lends its name to the Tastavins River, which in turn gives its name to the village, Peñarroya de Tastavins, where the fossils were found" (*1).
Specific name: *sanzi* ← {(person's name) Sanz + -ī}; "in honor of Professor José Luis Sanz, Madrid, for his studies of Spanish dinosaurs" (*1).
Etymology: Sanz's **Tastavins lizard**
Taxonomy: Saurischia: Sauropodomorpha: Camarasauromorpha: Laurasiformes (*2)
cf. Saurischia: Sauropoda: Neosauropoda: Titanosauriformes: Somphospondyli (*1)

[References: (*1) Canudo JI, Royo-Torres R, Cuenca-Bescós G (2008). A new sauropod: *Tastavinsaurus sanzi* gen. *et* sp. nov. from the Early Cretaceous (Aptian) of Spain. *Journal of Vertebrate Paleontology* 28(3): 712–731.; (*2) Royo-Torres R, Alcalá L, Cobos A (2012). A new specimen of the Cretaceous sauropod *Tastavinsaurus sanzi* from El Castellar (Teruel, Spain), and a phylogenetic analysis of the Laurasiformes. *Cretaceous Research* 34: 61–83.]

Tatankacephalus cooneyorum

> *Tatankacephalus cooneyorum* Parsons & Parsons, 2009 [Albian; Himes Member, Cloverly Formation, Montana, USA]

Generic name: *Tatankacephalus* ← {(Lakota) tatanka: bison + cephalus [(Gr.) κεφᾰλή/kephalē: head]}
Specific name: *cooneyorum* ← {(person's name) Cooney + -ōrum}; in honor of the family of John Patrick Cooney.
Etymology: Cooney family's **bison head**
Taxonomy: Ornithischia: Thyreophora: Ankylosauria: Nodosauridae (*1)

[References: Parsons WL, Parsons KM (2009). A new ankylosaur (Dinosauria: Ankylosauria) from the Lower Cretaceous Cloverly Formation of central Montana. *Canadian Journal of Earth Sciences* 46(10): 721–738.; (*1) Thompson RS, Parish JC, Maidment SCR, Barrett PM (2012). Phylogeny of the ankylosaurian dinosaurs (Ornithischia: Thyreophora). *Journal of Systematic Palaeontology* 10(2): 301–312.]

Tatankaceratops sacrisonorum

> *Tatankaceratops sacrisonorum* Ott & Larson, 2010 [Maastrichtian; Hell Creek Formation, South Dakota, USA]

Generic name: *Tatankaceratops* ← {(Lakota) Tatanka: American Bison + ceratops [(Gr.) κέρας/keras: horn + ὤψ/ōps: face]}; "in reference to this specimen being roughly the size of an American Bison and its locality being very close to the town of Buffalo, South Dakota, and to recognize and honor the Lakota Sioux Tribe, who are the prior inhabitants of the area"(*1).

Specific name: *sacrisonorum* ← {(person's name) Sacrison + -ōrum}; "in honor of Stan and Steven the Sacrison, the twin brothers from Buffalo, South Dakota, who discovered and collected the specimen" (*1).

Etymology: Sacrison brothers' **Bison horned face (ceratopsian)**

Taxonomy: Ornithischia: Ceratopsia: Ceratopsidae: Chasmosaurinae

[References: (*1) Ott CJ, Larson PL (2010). A new, small ceratopsian dinosaur from the latest Cretaceous Hell Creek Formation, northwest South Dakota, United States: a preliminary description. In: Ryan MJ, Chinnery-Allgeier BJ, Eberth DA (eds), *New Perspectives on Horned Dinosaurs. The Royal Tyrrell Museum Ceratopsian Symposium,* Indiana University Press, Bloomington: 656 pp.]

Tataouinea hannibalis

> *Tataouinea hannibalis* Fanti *et al*., 2013 [early Albian; Oum ed Diab Member, Ain el Guettar Formation, Tataouine, Tunisia]

Generic name: *Tataouinea* ← {(place-name) Tataouine Governorate + -a} (*1).

Specific name: *hannibalis* ← {(Lat. person's name) "Hannibal Barca (247–183 BC): Carthaginian military commander who marched an army including war elephants across Southern Europe" (*1)}

Etymology: Hannibal's **one from Tataouine**

Taxonomy: Saurischia: Sauropodomorpha: Sauropoda: Rebbachisauridae

[References: (*1) Fanti F, Cau A, Hassine M, Contessi M (2013). A new sauropod dinosaur from the Early Cretaceous of Tunisia with extreme avian-like pneumatization. *Nature Communications* 4 (2080): 1–7.]

Tatisaurus oehleri

> *Tatisaurus oehleri* Simmons, 1965 ⇒ *Scelidosaurus oehleri* (Simmons, 1965) Lucas, 1996 [Sinemurian; Dark Red Beds of the Lower Lufeng Series, Yunnan, China]

Generic name: *Tatisaurus* ← {(Chin. place-name) Tati大地 [Pinyin: Dàdì]: village | Ta Ti (Locality No.8): the Dark Red Beds of the Lower Lufeng Series (*1) + (Gr.) σαῦρος/sauros: lizard}

Specific name: *oehleri* ← {(person's name) Oehler + -ī}; in honor of Edgar Oehler.

Etymology: Oehler's **Tati lizard**

Taxonomy: Ornithischia: Thyreophora

Notes: Lucas (1996) reclassified *Tatisaurus oehleri* as a species of *Scelidosaurus*. According to *The Dinosauria 2nd edition*, *Scelidosaurus oehleri* is a synonym of *Tatisaurus oehleri*.

[References: (*1) Simmons DJ (1965). The non-therapsid reptiles of the Lufeng Basin, Yunnan, China. *Fieldiana Geology, Chicago Natural History Museum* 15: 1–93.]

Taurovenator violantei

> *Taurovenator violantei* Motta *et al*., 2016 [middle Cenomanian; Huincul Formation, Río Negro, Argentina]

Generic name: *Taurovenator* ← {(Lat.) taurus: bull + vēnātor: hunter} (*1)

Specific name: *violantei* ← {(person's name) Violante + -ī}; "in honor of Enzo Violante, owner of the farm where the specimen was discovered"(*1).

Etymology: Violante's **bull hunter**

Taxonomy: Theropoda: Tetanurae: Allosauroidea: Carcharodontosauridae

[References: (*1) Motta MJ, Aranciaga Roland AM, Rozadilla S, Agnolín FE, Chimento NR, Egli FB, Novas FE (2016). New theropod fauna from the Upper Cretaceous (Huincul Formation) of northwestern Patagonia, Argentina. *Cretaceous Period: Biotic Diversity and Biogeography. New Nexico Museum of Natural History and Science Bulletin* 71: 231–253.]

Taveirosaurus costai

> *Taveirosaurus costai* Antunes & Sigogneau-Russell, 1991 [Maastrichtian; Argilas de Aveiro Formation, Coimbra, Portugal]

Generic name: *Taveirosaurus* ← {(place-name) Taveiro: the village + (Gr.) σαῦρος/sauros: lizard}

Specific name: *costai* ← {(person's name) Costa + -ī}; in honor of Portuguese geologist João Carrington da Costa.

Etymology: Costa's **Taveiro lizard**

Taxonomy: Ornithischia

Notes: According to *The Dinosauria 2nd edition*, *Taveirosaurus costai* is considered as a *nomen dubium*.

Tawa hallae

> *Tawa hallae* Nesbitt *et al*., 2009 [Norian; Petrified Forest Member, Chinle Formation, New Mexico, USA]

Generic name: *Tawa* ← {(Hopi, myth.) Tawa; "the Puebloan sun god" (*1)}.
Specific name: *hallae* ← {(person's name) Hall + -ae}; "after Ruth Hall, who collected many of the specimens that formed the genesis of the Ghost Ranch Ruth Hall Museum of Paleontology collections" (*1).
Etymology: Hall's **sun god Tawa**
Taxonomy: Theropoda

[References: (*1) Nesbitt SJ, Smith ND, Irmis RB, Turner AH, Downs A, Norell MA (2009). A complete skeleton of a Late Triassic saurischian and the early evolution of dinosaurs. *Science* 326(5959): 1530–1533.]

Tazoudasaurus naimi

> *Tazoudasaurus naimi* Allain *et al.*, 2004 [Pliensbachian–Toarchian; Toundoute Continental Series, Ouarzazate, Morocco]

Generic name: *Tazoudasaurus* ← {(place-name) Tazouda + (Gr.) σαῦρος/sauros: lizard} (*1)
Specific name: *naimi* ← {(Arabic) naïmi: slender}; referring to the small size of the holotype (*1).
Etymology: slender, **Tazouda lizard**
Taxonomy: Saurischia: Sauropodomorpha: Sauropoda: Vulcanodontidae

[References: (*1) Allain R, Aquesbi N, Dejax J, Meyer C, Monbaron M, Montenat C, Richir P, Rochdy M, Russell D, Taquet P (2004). A basal sauropod dinosaur from the Early Jurassic of Morocco. *Comptes Rendus Palevol* 3(3): 199–208.]

Technosaurus smalli [Silesauridae]

> *Technosaurus smalli* Chatterjee, 1984 [Norian; Cooper Canyon Formation, Texas, USA]

Generic name: *Technosaurus* ← {(Gr.) τεχνο-/techno- [τέχνη/technē: skill, art] + (Gr.) σαῦρος/sauros: lizard}; in honor of Texas Tech University which sponsored the dig.
Specific name: *smalli* ← {(person's name) Small + -ī}; in honor of Bryan J. Small for preparation of the material (*1).
Etymology: Small's **Texas Tech University lizard**
Taxonomy: Dinosauriformes: Silesauridae
Notes: Chatterjee first described the fossil as an ornithischian (*1).

[References: (*1) Chatterjee S (1984). A new ornithischian dinosaur from the Triassic of North America. *The Science of Nature* 71(12): 630–631.]

Tehuelchesaurus benitezii

> *Tehuelchesaurus benitezii* Rich *et al.*, 1999 [Kimmeridgian; Cañadón Asfalto Formation, Chubut, Argentina] (*1)

Generic name: *Tehuelchesaurus* ← {(Tehuelche) Tehuelche: a native American people of Argentina + (Gr.) σαῦρος/sauros: lizard}
Specific name: *benitezii* ← {(person's name) Benitez + -iī}; in honor of Aldo Benitez who discovered the holotype.
Etymology: Benitez' **Tehuelche lizard**
Taxonomy: Saurischia: Sauropodomorpha: Sauropoda: Eusauropoda: Camarasauromorpha: Camarasauridae

[References: (*1) Rich TH, VickersRich P, Gimenez O, Cúneo R, Puerta P, Vacca R (1999). A new sauropod dinosaur from Chubut province, Argentina. In: Tomida Y, Rich TH, Vickers-Rich (eds), *Proceedings of the Second Gondwanan Dinosaur Symposium, National Science Museum Monographs* 15: 61–84.]

Teihivenator macropus

> *Laelaps macropus* Cope, 1868 (preoccupied) ⇒ *Dryptosaurus macropus* (Cope, 1868) Hay, 1902 ⇒ *Teihivenator macropus* (Cope, 1868) Yun, 2017 [early Maastrichtian; Navesink Formation, Monmouth Group, New Jersey, USA]

Generic name: *Teihivenator* ← {(Arapaho) Teihiihan: strong + (Lat.) vēnātor: hunter} (*1)
Specific name: *macropus* ← {(Gr.) μακρός/makros: long, large + πούς/pous [pūs]: foot}.
Etymology: **strong hunter** with long feet
Taxonomy: Theropoda: Tyrannosauroidea (*1)
Notes: Brownstein (2017) concluded that *Teihivenator macropus* is a chimera and a *nomen dubium*. (*2)

[References: (*1) Yun C (2017). *Teihivenator* gen. nov., a new generic name for the tyrannosauroid dinosaur *"Laelaps" macropus* (Cope, 1868; preoccupied by Koch, 1836). *Journal of Zoological and Bioscience Research* 4(2): 7–13.; (*2) Brownstein CD (2017). Theropod specimens from the Navesink Formation and their implications for the diversity and biogeography of ornithomimosaurs and tyrannosauroids on Appalachia. *PeerJ Preprints* 5: e3105v1.]

Telmatosaurus transsylvanicus

> *Limnosaurus transsylvanicus* Nopcsa, 1899 (preoccupied) (*1) ⇒ *Telmatosaurus transsylvanicus* (Nopcsa, 1899) Nopcsa, 1903 ⇒ *Hecatasaurus transsylvanicus* (Nopcsa, 1899) Brown, 1910 ⇒ *Orthomerus transsylvanicus* (Nopcsa, 1899) Nopcsa, 1915 [Campanian–Maastrichtian; Sâmpetru Formation, Hunedoara, Romania]

Generic name: *Telmatosaurus* ← {(Gr.) τελματο-/telmato-* [τέλμα/telma: swamp, marsh] + σαῦρος/sauros: lizard}
**cf.* limno- [λίμνη/limnē: a marshy lake, mere]
Specific name: *transsylvanicus* ← {(place-name) Transsylvania [*sic*]: Transylvania* + -icus}

*a historical region which today is located in central Romania.

Etymology: **swamp lizard** of Transsylvania

Taxonomy: Ornithischia: Ornithopoda: Hadrosauroidea: Hadrosauromorpha

cf. Ornithischia: Ornithopoda: Hadrosauridae (according to *The Dinosauria 2nd edition*)

Notes: *Limnosaurus* had already been used for a crocodilian (later reclassified *Pristichampsus*) by Marsh (1871), so Nopcsa renamed. (*1)

Other species:

> *Trachodon cantabrigiensis* Lydekker, 1888 (*nomen dubium*) ⇒ *Hadrosaurus cantabrigiensis* (Lydekker, 1888) Newton, 1892 (*nomen dubium*) ⇒ *Telmatosaurus cantabrigiensis* (Lydekker, 1888) Olshevsky, 1978 (*nomen dubium*) [early Cenomanian; Cambridge Greensand Member, West Melbury Chalk Formation, Cambridgeshire, England, UK]

[References: (*1) Nopcsa F (1903). *Telmatosaurus*, new name for the dinosaur *Limnosaurus*. *Geological Magazine, decade 4* 10: 94–95.]

Tendaguria tanzaniensis

> *Tendaguria tanzaniensis* Bonaparte *et al*., 2000 [Kimmeridgian; Tendaguru Formation, Mtwara, Tanzania]

Generic name: *Tendaguria* ← {(place-name) Tendaguru + -ia}; referring to Tendaguru in Tanzania. "The remains were collected at Nambango, near Tendaguru Hill" (*1).

Specific name: *tanzaniensis* ← {(place-name) Tanzania + -ensis}

Etymology: **Tendaguru's one** from Tanzania

Taxonomy: Saurischia: Sauropodomorpha: Sauropoda

Notes: The finds, which were discovered by Bornhardt in 1911, were described by Janensch in 1929, but not named.

[References: (*1) Bonaparte JF, Heinrich W-D, Wild R (2000). Review of *Janenschia* Wild with the description of a new sauropod from the Tendaguru beds of Tanzania and a discussion on the systematic value of procoelous caudal vertebrae in the Sauropoda. *Palaeontographica Abteilung A* 256: 25–76.]

Tengrisaurus starkovi

> *Tengrisaurus starkovi* Averianov & Skutschas, 2017 [Barremian–Aptian; Mogoito Member, Murtoi Formation, Buryatia, Russia]

Generic name: *Tengrisaurus* ← {(myth.) Tengri: the primary chief deity in Mongolian-Turkish mythology + (Gr.) σαῦρος/sauros: lizard} (*1)

Specific name: *starkovi* ← {(person's name) Starkov + -ī}; "after Alexey Starkov for his generous assistance and contribution to the study of Early Cretaceous vertebrates of Transbaikalia" (*1).

Etymology: Starkov's **Tengri lizard**

Taxonomy: Saurischia: Sauropoda: Titanosauriformes: Titanosauria: Lithostrotia

[References: (*1) Averianov A, Skutschas P (2017). A new lithostrotian titanosaur (Dinosauria, Sauropoda) from the Early Cretaceous of Transbaikalia, Russia. *Biological Communications* 62(1): 6–18.]

Tenontosaurus tilletti

> *Tenontosaurus tilletti* Ostrom, 1970 [Albian; Himes Member, Cloverly Formation, Montana, USA]

Generic name: *Tenontosaurus* ← {(Gr.) τενοντο-/tenonto- [τένων/tenon, *m.*: sinew] + σαῦρος/sauros, *m.*: lizard}; "in reference to ossified tendons along the vertebral column" (*1).

Specific name: *tilletti* ← {(person's name) Tillett + -i}; in honor of Lloyd Tillett Family of Lovell, Wyoming, for their assistance and hospitality extended to the field parties. (*1)

Etymology: Tillett's **sinew lizard**

Taxonomy: Ornithischia: Ornithopoda: Iguanodontia

Other species:

> *Tenontosaurus dossi* Winkler *et al*., 1997 [late Aptian; Twin Mountain Formation, Texas, USA]

[References: (*1) Ostrom JH (1970). Stratigraphy and paleontology of the Cloverly Formation (Lower Cretaceous) of the Bighorn Basin area, Wyoming and Montana. *Bulletin of the Peabody Museum of Natural History* (35): 1–234.]

Teratophoneus curriei

> *Teratophoneus curriei* Carr *et al*., 2011 [Campanian; Kaiparowits Formation, Utah, USA]

Generic name: *Teratophoneus* ← {(Gr.) τερατο-/terato- [τέρας/teras: monster] + φονεύς/phoneus: murderer}

Specific name: *curriei* ← {(person's name) Currie + -ī}; in honor of Dr. Philip J. Currie.

Etymology: Currie's **monstrous murderer**

Taxonomy: Theropoda: Coelurosauria: Tyrannosauroidea: Tyrannosauridae: Tyrannosaurinae

[References: Carr TD, Williamson TE, Britt BB, Stadtman, K (2011). Evidence for high taxonomic and morphologic tyrannosauroid diversity in the Late Cretaceous (late Campanian) of the American southwest and a new short-skulled tyrannosaurid from Kaiparowits Formation of Utah. *Naturwissenschaften* 98(3): 241–246.]

Teratosaurus suevicus [Loricata]

> *Teratosaurus suevicus* Meyer, 1861 [Norian;

Löwenstein Formation, Baden-Württemberg, Bayern, Germany]

Generic name: *Teratosaurus* ← {(Gr.) τέρας/teras: monster + σαῦρος/sauros: lizard}

Specific name: *suevicus* ← {(place-name) Suevia: Swabia, a region in southwestern Germany + -icus}

Etymology: **monster lizard** of Swabia

Taxonomy: Reptilia: Rauisuchidae

Other species:

> *Teratosaurus bengalensis* Das-Gupta, 1928 [Induan; Panchet Formation, West Bengal, India]
> *Teratosaurus lloydi* Owen, 1841 [Anisian; Bromsgrove Sandstone Formation, England, UK]
> *Teratosaurus trossingensis* von Huene, 1907–1908 (*nomen dubium*)
> *Teratosaurus minor* von Huene, 1907–1908 (*nomen dubium*) ⇒ *Efraasia minor* (von Huene, 1907–1908) Yates, 2003 [Norian, Löwenstein Formation, Baden-Württemberg, Germany]
> *Teratosaurus silesiacus* Sulej, 2005 ⇒ *Polonosuchus silesiacus* (Sulej, 2005) Brussatte *et al.*, 2009 [Carnian; Poland]

Terminocavus sealeyi

> *Terminocavus sealeyi* Fowler & Freedman Fowler, 2020 [Campanian; Kirtland Formation, New Mexico, USA]

Generic name: *Terminocavus* ← {termono-: coming to the end of [(Lat.) terminus: bound, end] + cavus: hollow, a hole; after the nearly-closed parietal embayment} (*1)

Specific name: *sealeyi* ← {(person's name) Sealey + -ī}; "after Paul Sealey who discovered the holotype species" (*1).

Etymology: Sealey's **closing cavity**

Taxonomy: Ornithischia: Ceratopsidae: Chasmosaurinae

[References: (*1) Fowler DW, Freedman Fowler EA (2020). Transitional evolutionary forms in chasmosaurine ceratopsid dinosaurs: evidence from the Campanian of New Mexico. *PeerJ* 8: e9251.]

Tethyshadros insularis

> *Tethyshadros insularis* Dalla Vecchia, 2009 [late Campanian; Liburnian Formation, Trieste, Friuli-Venezia Giulia, Italy]

Generic name *Tethyshadros* ← {(Gr.) Τηθύς/Tēthys: "an ocean that occupied the general position of the Alpine-Himalayan orogenic belt" + ἁδρός/hadros: "hadrosauroid"} (*1)

Specific name: *insularis* ← {(Lat.) insular: "island dweller" (*1) [īnsula: island]}

Etymology: **Tethyan hadrosauroid** of Island dweller

Taxonomy: Ornithischa: Ornithopoda: Iguanodontia: Hadrosauroidea

Notes: The small size of the specimens suggests that *Tethyshadros insularis* may be an insular dwarf. (*1)

[References: (*1) Dalla Vecchia FM (2009). *Tethyshadros insularis*, a new hadrosauroid dinosaur (Ornithischia) from the Upper Cretaceos of Italy. *Journal of Vertebrate Paleontology* 29(4): 1100–1116.]

Texacephale langstoni

> *Texacephale langstoni* Longrich *et al.*, 2010 [late Campanian; upper shale member, Aguja Formation, Texas, USA]

Generic name: *Texacephale* ← {(place-name) Texas + (Gr.) κεφαλή/kephale: head}

Specific name: *langstoni* ← {(person's name) Langston + -ī}; "in honor of Wann Langston, for his contributions to the vertebrate palaeontology of the Big Bend region" (*1).

Etymology: Langston's **Texas head (pachycephalosaurid)**

Taxonomy: Ornithischia: Marginocephalia: Pachycephalosauria: Pachycephalosauridae

[References: (*1) Longrich NR, Sankey J, Tanke D (2010). *Texacephale langstoni*, a new genus of pachycephalosaurid (Dinosauria: Ornithischia) from the upper Campanian Aguja Formation, southern Texas, USA. *Cretaceous Research* 31(2): 274–284.]

Texasetes pleurohalio

> *Texasetes pleurohalio* Coombs, 1995 [late Albian; Paw Paw Formation, Texas, USA]

Generic name: *Texasetes* ← {(place-name) Texas + (Gr.) ἔτης/etēs: dweller} (*1)

Specific name: *pleurohalio* ← {pleuro-: by the side or adjacent to [(Gr.) πλευρά/pleura: the rib, the side] + ἅλιος/halios: of the sea} (*1)

Etymology: **dweller in Texas** adjacent to the sea (*1)

Taxonomy: Ornithischia: Thyreophora: Ankylosauria: Nodosauridae

[References: (*1) Coombs WP (1995). A nodosaurid ankylosaur (Dinosauria: Ornithischia) from the Lower Cretaceous of Texas. *Journal of Vertebrate Paleontology* 15(2): 298–312.]

Teyuwasu barberenai

> *Staurikosaurus pricei* Colbert, 1970 [Carnian; Santa Maria Formation, Rio Grande do Sul, Brazil] (*1)

= *Teyuwasu barberenai* Kischlat, 1999 [Carnian; Santa Maria Formation, Rio Grande do Sul, Brazil]

Generic name: *Teyuwasu* ← {(Tupi) *teyuwasu*: big lizard}

Specific name: *barberenai* ← {(person's name)

Barberena + -ī}; in honor of paleontologist Dr. M. C. Barberena.[*1]

Etymology: Barberena's **big lizard**

Taxonomy: Saurischia

Notes: According to *The Dinosauria 2nd edition*, *Teyuwasu barberenai* was considered *a nomen dubium.* But it is now referred as a second species of *Staurikosaurus pricei* [*1].

[References: (*1) Garcia MS, Müller RT, Dias-Da-Silva S (2019). On the taxonomic status of *Teyuwasu barberenai* Kischlat, 1999 (Archosauria: Dinosauriformes), a challenging taxon from the Upper Triassic of southern Brazil. *Zootaxa* 4629(1): 146–150.; Kischlat E-E. (1999). A new dinosaurian "rescued" from the Brazilian Triassic: *Teyuwasu barbarenai*, new taxon. *Paleontologia em Destaque, Boletim Informativo da Sociedade Brasileira de Paleontologia* 14(26): 58.]

Thanatotheristes degrootorum

> *Thanatotheristes degrootorum* Voris *et al.*, 2020 [Campanian; Foremost Formation, Alberta, Canada]

Generic name: *Thanatotheristes* ← {(Gr.) θάνᾰτος/thanathos: death | (Gr. myth.) Θάνᾰτος/Thanatos: Death + θεριστής /theristes: harvester, reaper}

Specific name: *degrootorum* ← {(person's name) De Groot + -ōrum}; in honor of John and Sandra De Groot, who discovered the type specimen.

Etymology: John and Sandra De Groot's **harvester of death**

Taxonomy: Saurischia: Theropoda: Tyrannosauridae: Tyrannosaurinae: Daspletosaurini

Notes: According to Voris *et al.* (2020), *Thanatotheristes* is found to be the sister taxon to *Daspletosaurus*.

[References: Voris JT, Therrien F, Zelenitzky DK, Brown CM (2020). A new tyrannosaurine (Theropoda: Tyrannosauridae) from the Campanian Foremost Formation of Alberta, Canada, provides insight into the evolution and biogeography of tyrannosaurids. *Cretaceous Research* 110: 104388.]

Thanos simonattoi

> *Thanos simonattoi* Delcourt & Iori, 2018 [Santonian; São José do Rio Preto Formation, São Paulo, Brazil]

Generic name: *Thanos* ← {(Gr.) θάνατος/thanatos: death | (Marbel Comic character) Thanos: a supervillain, created by Jim Starlin}

Specific name: *simonattoi* ← {(person's name) Simonatto + -ī}; in honor of Sérgio Simonatto, the discoverer of the specimen.

Etymology: Simonatto's **Thanos**

Taxonomy: Saurischia: Theropoda: Abelisauridae: Brachyrostra

Thecocoelurus daviesi

> *Thecospondylus daviesi* Seeley, 1888 (*nomen dubium*) ⇒ *Coelurus daviesi* (Seeley, 1888) Nopcsa, 1901 (*nomen dubium*) ⇒ *Thecocoelurus daviesi* (Seeley, 1888) von Huene, 1923 (*nomen dubium*) [Barremian; Wessex Formation, Isle of Wight, UK]

Generic name: *Thecocoelurus* ← {(Gr.) θήκη/theke: socket, sheath + *Coelurus*}; a contraction of *Thecospondylus* and *Coelurus*.

Specific name: *daviesi* ← {(person's name) Davies + -ī}; in honor of William Davies, a British palaeontologist.

Etymology: Davies' ***Thecospondylus* plus *Coelurus***

Taxonomy: Theropoda: Ornithomimosauria

Notes: According to *The Dinosauria 2nd edition*, *Thecospondylus daviesi* Seeley, 1888 (type of *Thecocoelurus* von Huene, 1923) was considered a *nomen dubium*. Seeley first described the fossil in 1888 and named *Thecospondylus daviesi.* Nopcsa renamed it *Coelurus daviesi* in 1901, and von Huene gave them their generic name *Thecocoelurus* in 1923. According to Allain *et al.* (2014), *Valdoraptor oweni* is a possible junior synonym of *Thecocoelurus daviesi*, but they are considered to be *nomina dubia*.

Thecodontosaurus antiquus

> *Thecodontosaurus antiquus* Morris, 1843 [Rhaetian; Magnesian Conglomerate Formation, Avon, UK]

= ?*Agrosaurus macgillivrayi* Seeley, 1891 (*nomen dubium*) ⇒ *Thecodontosaurus macgillivrayi* (Seeley, 1891) von Huene, 1906 (*nomen dubium*) [Rhaetian; Magnesian Conglomerate Formation, Bristol, UK]

= *Thecodontosaurus platyodon* Marsh, 1892 [Rhaetian; Magnesian Conglomerate Formation, Bristol, UK]

= *Thecodontosaurus cylindrodon* von Huene, 1908 (*partim*) [Rhaetian; Magnesian Conglomerate Formation, Bristol, UK]

= ?*Thecodontosaurus caducus* Yates, 2003 ⇒ *Pantydraco caducus* (Yates, 2003) Galton *et al.*, 2007 [Rhaetian; Pant-y-ffynnon Quarry, Bonvilston, South Wales, UK]

The generic name *Thecodontosaurus* was named by Riley & Stutchbury, *vide* Owen (1842), and the type species *T. antiquus* was provided by Morris (1843).

Generic name: *Thecodontosaurus* Riley & Stutchbury, 1836 ← {(Gr.) θήκη/thēkē: socket + οδοντο-[ὀδών/odōn: tooth] + σαῦρος/sauros: lizard}

Specific name: *antiquus* ← {(Lat.) antīquus: ancient}

Etymology: ancient, **socket-toothed lizard**

Taxonomy: Saurischia: Sauropodomorpha: Prosauropoda

Other species:

> *Thecodontosaurus latespinatus* von Huene, 1905 (*nomen dubium*) [Ladinian; Muschelkalk Group, Lorraine, France]

> *Thecodontosaurus primus* von Huene, 1907–1908 (*nomen dubium*) [Anisian; Chorzow Formation, Opole, Poland]

> *Thecodontosaurus subcylindrodon* von Huene, 1905 (*nomen dubium*) [Carnian; Schilfsandstein Member, Stuttgart Formation, Baden-Württemberg, Germany]

> *Thecodontosaurus minor* Haughton, 1918 (*nomen dubium*) [Hettangian; upper member, Elliot Formation, Eastern Cape, South Africa]

> *Thecodontosaurus dubius* Haughton, 1924 (*nomen dubium*) [Stormberg Series, South Africa]

> *Thecodontosaurus minimus* Ellenberger, 1970 (*nomen dubium*) [Hettangian; upper member, Elliot Formation, Mafeteng, Lesotho]

> *Thecodontosaurus gibbidens* Cope, 1878 (*nomen dubium*) ⇒ *Galtonia gibbidens* (Cope, 1878) Hunt & Lucas, 1994* [Late Triassic; Pennsylvania, USA]
*non-dinosaurian archosaur

> *Thecodontosaurus elisae* Sauvage, 1907 (*nomen dubium*) ⇒ *Plateosaurus elizae* (Sauvage, 1907) von Huene 1908 [*sic*] ⇒ *Plateosaurus elisae* (Sauvage, 1907) von Huene, 1914 (*nomen dubium*) [Rhaetian; Grès de Infralias Formation, Champagne-Ardennes, France]

> *Megadactylus polyzelus* Hitchcock, 1865 ⇒ *Amphisaurus polyzelus* (Hitchcock, 1865) Marsh, 1882 ⇒ *Anchisaurus polyzelus* (Hitchcock, 1865) Marsh, 1885 ⇒ *Thecodontosaurus polyzelus* (Hitchcock, 1865) von Huene, 1914 [Hettangian; Portland Formation, Massachusetts, USA]

> *Sellosaurus gracilis* Meyer; von Huene, 1905 ⇒ *Plateosaurus gracilis* (von Huene, 1905) von Huene, 1926 [Norian; Löwenstein Formation, Baden-Wurttemberg, Germany]

= *Thecodontosaurus hermannianus* von Huene, 1905 (*nomen dubium*) ⇒ *Sellosaurus hermannianus* (von Huene, 1905) von Huene, 1915 [Norian; Löwenstein Formation, Baden-Wurttemberg, Germany]

> *Teratosaurus minor* von Huene, 1908 ⇒ *Efraasia minor* (von Huene, 1908) Yates, 2003 [Alaunian; Löwenstein Formation, Baden-Württemberg, Germany]

= *Sellosaurus fraasi* von Huene, 1908 [Norian; Löwenstein Formation, Baden-Württemberg, Germany]

= *Thecodontosaurus diagnosticus* Fraas, 1913 (*nomen dubium*) ⇒ *Palaeosaurus diagnosticus* von Huene, 1932 ⇒ *Palaeosauriscus diagnosticus* (von Huene, 1932) Charig, 1967 ⇒ *Efraasia diagnostica* (von Huene, 1932) Galton, 1973 [Norian; Löwenstein Formation, Baden-Württemberg, Germany]

> *Massospondylus browni* Seeley, 1895 (*nomen dubium*) ⇒ *Thecodontosaurus browni* (Seeley, 1895) von Huene, 1932 [Hettangian; Elliot Formation, Eastern Cape, South Africa]

> *Thecodontosaurus alophos* Haughton, 1932 (*nomen dubium*) ⇒ *Teleocrater alophos* (Haughton, 1932) Charig, 1856 (*nomen nudum*) [Anisian; Manda Formation, Ruvuma, Tanzania]

= *Nyasasaurus parringtoni* Nesbitt *et al.*, 2013 [Anisian; Manda Formation, Tanzania]

Notes: According to Ballell *et al.* (2020), *Thecodontosaurus caducus* is considered to be a juvenile of *Thecodontosaurus antiquus*.

[References: (*1) Galton P (2007). Notes on the remains of archosaurian reptiles, mostly basal sauropodomorph dinosaurs, from the 1834 fissure fill (Rhaetian, Upper Triassic) at Clifton in Bristol, southwest England. *Revue de Paleobiologie* 26(2): 505–591.]

Thecospondylus horneri

> *Thecospondylus horneri* Seeley, 1882 (*nomen dubium*) [Berriasian; Wealden Group, England, UK]

Generic name: *Thecospondylus* ← {(Gr.) θήκη/thēkē: sheath + σπόνδῠλος/spondylos: vertebrae}

Specific name: *horneri* ← {(person's name) Horner + -ī}

Etymology: Horner's **sheath vertebra**

Taxonomy: Saurischia

Other species:

> *Thecospondylus daviesi* Seeley, 1888 (*nomen dubium*) ⇒ *Coelurus daviesi* (Seeley, 1888) Nopcsa, 1901 (*nomen dubium*) ⇒ *Thecocoelurus daviesi* (Seeley, 1888) von Huene, 1923 (*nomen dubium*) [Barremian; Wessex Formation, Isle of Wight, UK]

Theiophytalia kerri

> *Theiophytalia kerri* Brill & Carpenter, 2007 [Aptian; Lytle Member, Purgatoire Formation, Colorado, USA]

Generic name: *Theiophytalia* ← {(Gr.) θεῖος/theios: belonging to the gods + φυταλία/phytalia: garden}; "referring to 'Garden of the Gods Park' where the specimen was found"(*1).

Specific name: *kerri* ← {(person's name) Kerr [pron.: "care"] + -ī}; "after James Hutchison Kerr, who discovered the specimen" (*1).
Etymology: Kerr's **Garden of the Gods**
Taxonomy: Ornithischia: Ornithopoda: Styracosterna
[References: (*1) Brill K, Carpenter K (2007). A description of a new ornithopod from the Lytle Member of the Purgatoire Formation (Lower Cretaceous) and a reassessment of the skull of *Camptosaurus*. In: Carpenter K (ed), *Horns and Beaks: Ceratopsian and Ornithopod Dinosaurs*. Indiana University Press, Bloomington: 49–67.]

Therizinosaurus cheloniformis

> *Therizinosaurus cheloniformis* Maleev, 1954 [Maastrichtian; Nemegt Formation, Ömnögovi, Mongolia]

Generic name: *Therizinosaurus* ← {(Gr.) therizino- [θερίζω/therizō: to reap, to cut off] + σαῦρος/sauros: lizard}; referring to its scythe-like claws.
Specific name: *cheloniformis* ← {(Gr.) χελώνη/chelōnē: turtle + (Lat.) forma: form}; turtle-like, turtle-formed.
Etymology: turtle-like **scythe lizard** (*1)
Taxonomy: Theropoda: Coelurosauria: Therizinosauridae
Notes: Maleev assumed that *Therizinosaurus* was a gigantic turtle-like reptile (*1).
[References: (*1) Maleyev YA (1954). Noviy chyeryepoobrazniy yashschyer Mongolii [A new turtle-like reptile from Mongolia]. *Priroda* 1954 (3): 106–108.]

Thescelosaurus neglectus

> *Thescelosaurus neglectus* Gilmore, 1913 [Maastrichtian; Lance Formation, Niobrara, Wyoming, USA]

Generic name: *Thescelosaurus* ← {(Gr.) θέσκελος/theskelos: marvellous, wondrous [θέσκελος = θεο-είκελος/theo-eikelos: godlike] + σαῦρος/sauros: lizard}
Specific name: *neglectus* ← {(Lat.) neglēctus: neglected}
"This specimen had remained in the original packing boxes" for many years (*1).
Etymology: neglected, **wonderful lizard**
Taxonomy: Ornithischia: Neornithischia: Parksosauridae / Thescelosauridae: Thescelosaurinae
Other species:

> *Thescelosaurus edmontonensis* Sternberg, 1940 [late Maastrichtian; Scollard Formation, Alberta, Canada]

> *Thescelosaurus garbanii* Morris, 1976 ⇒ *Bugenasaura garbanii* (Morris, 1976) Galton, 1995 (*2) [Maastrichtian; Hell Creek Formation, Montana, USA]

> *Bugenasaura infernalis* Galton, 1995 (*2) [Maastrichtian; Hell Creek Formation, South Dakota, USA]

> *Thescelosaurus assiniboiensis* Brown *et al.*, 2011 [Maastrichtian; Frenchman Formation, Saskatchewan, Canada]

> *Thescelosaurus warreni* Parks, 1926 ⇒ *Parksosaurus warreni* (Parks, 1926) Sternberg, 1937 (*2) [Maastrichtian; Horseshoe Canyon Formation, Alberta, Canada]

Notes: According to *The Dinosauria 2nd edition*, *Thescelosaurus garbanii* was considered as a possibly synonym of *Bugenasaura infernalis*.
[References: (*1) Gilmore CW (1913). A new dinosaur from the Lance Formation of Wyoming. *Smithsonian Miscellaneous Collections* 61(5): 1–5.; (*2) Boyd CA, Brown CM, Scheetz RD, Clarke JA (2009). Taxonomic revision of the basal neornithischian taxa *Thescelosaurus* and *Bugenasaura*. *Jounal of Vertebrate Paleontology* 29(3): 758–770.]

Thespesius occidentalis

> *Thespesius occidentalis* Leidy, 1856 ⇒ *Hadrosaurus occidentalis* (Leidy, 1856) Cope, 1871 ⇒ *Trachodon occidentalis* (Leidy, 1856) Kuhn, 1936 [Maastrichtian; Lance Formation, South Dakota, USA]

= *Agathaumas milo* Cope, 1874 ⇒ *Hadrosaurus milo* (Cope, 1874) Hay, 1901 [Maastrichtian; Denver Formation, Colorado, USA]

Generic name: *Thespesius* ← {(Gr.) θεσπέσιος/thespesios: wondrous}
Specific name: *occidentalis* ← {(Lat.) occidentālis: western}
Etymology: western, **wondrous one**
Taxonomy: Ornithischia: Ornithopoda: Hadrosauridae: Saurolophinae
Other species:

> *Claosaurus annectens* Marsh, 1892 ⇒ *Trachodon annectens* (Marsh, 1892) Gilmore, 1915 ⇒ *Thespesius annectens* (Marsh, 1892) Gilmore, 1915 ⇒ *Anatosaurus annectens* (Marsh, 1892) Lull & Wright, 1942 ⇒ *Edmontosaurus annectens* (Marsh, 1892) Brett-Surman *vide* Chapman & Brett-Surman, 1990 [late Maastrichtian; Lance Formation, Wyoming, USA]

> *Trachodon altidens* Lambe, 1902 (*nomen dubium*) ⇒ *Didanodon altidens* (Lambe, 1902) Osborn, 1902 ⇒ *Pteropelyx altidens* (Lambe, 1902) Lambe, 1902 ⇒ *Procheneosaurus altidens* (Lambe, 1902) Lull & Wright, 1942 ⇒ *Thespesius altidens* (Lambe, 1902) Steel,

1969 [middle Campanian; Dinosaur Park Formation, Alberta, Canada]

> *Thespesius saskatchewanensis* Sternberg, 1926 ⇒ *Anatosaurus saskatchewanensis* (Sternberg, 1926) Lull & Wright, 1942 ⇒ *Edmontosaurus saskatchewanensis* (Sternberg, 1926) Brett-Surman, 1990 [late Maastrichtian; Frenchman Formation, Saskatchewan, Canada]

> *Edmontosaurus regalis* Lambe, 1917 [Maastrichtian; Horseshoe Canyon Formation, Alberta, Canada]

= *Thespecius edmontonensis* Gilmore, 1924 ⇒ *Thespecius edmontoni* (Gilmore, 1924) Gilmore, 1924 ⇒ *Anatosaurus edmontonensis* (Gilmore, 1924) Lull & Wright, 1942 ⇒ *Anatosaurus edmontoni* (Gilmore, 1924) Lull & Wright, 1942 [Maastrichtian; Edmonton Group, Alberta, Canada]

Notes: According to *The Dinosauria 2nd edition*, *Thespesius occidentalis* was considered as a *nomen dubium*.

Tianchisaurus nedegoapeferima

> "Jurassosaurus" *nedegoapeferima* ⇒ *Tianchisaurus nedegoapeferima* Dong, 1993 [Middle Jurassic; Toutunhe Formation, Xinjiang, China]

The generic name "Jurassosaurus" *nedegoapeferima*, which was proposed by Spielberg in 1993, is now a *nomen nudum*.

Generic name: *Tianchisaurus* ← {(Chin. place-name) Tianchi + (Gr.) σαῦρος/sauros: lizard}; Tianchi "meaning the Heavenly Pool (Tiān天: 'heaven', Chí池: 'pool or lake') and being a famous lake in the Tian Shan Mountains".[*1]

Specific name: *nedegoapeferima* ← {(acronym) Ne, De, Go, A, Pe, Fe, Ri, Ma}; for the surnames of the main stars of the 1993 film "Jurassic Park", **Ne**ill, **De**rn, **Go**ldblum, **A**ttenborough, **Pe**ck, **Fe**rrero, **Ri**chards and **Ma**zzello. [*1]

Etymology: **heavenly lake lizard** of stars of the film 'Jurassic Park'

Taxonomy: Ornithischia: Ankylosauria

Notes: According to *The Dinosauria 2nd edition* (2004), *T. nedegoapeferima* was considered as a *nomen dubium*. It lacked a bony club at the tip of its tail which other ankylosaurids had.

[References: (*1) Dong Z (1993). An ankylosaur (ornithischian dinosaur) from the Middle Jurassic of the Junggar Basin, China. *Vertebrata PalAsiatica* 31: 258–264.]

Tianyulong confuciusi

> *Tianyulong confuciusi* Zheng *et al*., 2009 [late Bathonian–Oxfordian; Tiaojishan Formation, Jianchang, Liaoning, China]

Generic name: *Tianyulong* ← {(Chin.) Tiānyǔ天宇 + lóng龍: dragon}; "referring to the Shandong Tianyu Museum of Nature, where the specimen is housed"[*1].

Specific name: *confuciusi* ← {(Lat.) Cōnfūcius + -ī}; "dedicated to Confucius, the founder of Confucianism"[*1].

Etymology: **Tianyu dragon** of Confucius

Taxonomy: Ornithischia: Ornithopoda: Heterodontosauridae

[References: (*1) Zheng X-T, You H-L, Xu X, Dong Z-M (2009). An Early Cretaceous heterodontosaurid dinosaur with filamentous integumentary structures. *Nature* 458(7236): 333–336.]

Tianyuraptor ostromi

> *Tianyuraptor ostromi* Zheng *et al*., 2010 [Barremian; Yixian Formation, Lingyuan, Liaoning, China]

Generic name: *Tianyuraptor* ← {(place-name) Tiānyǔ天宇 + (Lat.) raptor: robber}; "derived from the name of the museum that has the holotype"[*1].

Specific name: *ostromi* ← {(person's name) Ostrom + -ī}; "in honor of John Ostrom, who contributed greatly to the study of dromaeosaurid fossils"[*1].

Etymology: Ostrom's **Tianyu robber**

Taxonomy: Theropoda: Maniraptora: Dromaeosauridae

[References: (*1) Zheng X, Xu X, You H, Zhao Q, Dong Z (2009). A short-armed dromaeosaurid from the Jehol Group of China with implications for early dromaeosaurid evolution. *Proceedings of the Royal Society B* 277(1679): 211–217.]

Tianzhenosaurus youngi

> *Tianzhenosaurus youngi* Pang & Cheng, 1998 [Cenomanian; upper member, Huiquanpu Formation, Tianzhen, Shanxi, China]

Generic name: *Tianzhenosaurus* ← {(Chin. place-name) Tiānzhèn天鎮: a county in Shanxi Province + (Gr.) σαῦρος/sauros: lizard}

Specific name: *youngi* ← {(person's name) Young + -ī}; in honor of Chinese paleontologist C. C. Young楊 [Pinyin: Yáng].

Etymology: Young's **Tianzhen lizard**

Taxonomy: Ornithischia: Ankylosauria: Ankylosauridae: Ankylosaurinae

[References: Pang Q, Cheng Z (1998). A new ankylosaur of the late Cretaceous from Tianzhen, Shanxi. *Progress in Natural Science* 8(3): 326–334.]

Tienshanosaurus chitaiensis

> *Tienshanosaurus chitaiensis* Yang, 1937 [Oxfordian; Shishugou Formation, Xinjiang,

China]

Generic name: *Tienshanosaurus* ← {(Chin. place-name) Tienshan天山 [Pinyin: Tiānshān] + (Gr.) σαῦρος/sauros: lizard}; referring to the Tian Shan.

Specific name: *chitaiensis* ← {(Chin. place-name) Chitai奇台 [Pinyin: Qītái] + -ensis}

Etymology: **Tienshan lizard** from Chitai [奇台天山龍]

Taxonomy: Saurischia: Sauropodomorpha: Sauropoda: Mamenchisauridae

Notes: A fossilized egg was also found. According to *The Dinosauria 2nd edition* (2004), *Tienshanosaurus chitaiensis* is considered as a *nomen dubium*.

Timimus hermani

> *Timimus hermani* Rich & Vickers-Rich, 1994 [late Aptian; Eumeralla Formation, Victoria, Southern Australia, Australia]

Generic name: *Timimus* ← {(person's name) Tim: Timothy Flannery and Timothy Rich (son of authors) + (Gr.) μῖμος/mīmos: mimic}

Specific name: *hermani* ← {(person's name) Herman + -ī}; in honor of John Herman.

Etymology: Herman's **Tim mimic**

Taxonomy: Theropoda: Coelurosauria: Tyrannosauroidea

Notes: *Timimus* was thought to be basal coelurosaurs (including tyrannosauroids and possibly ornithomimosaurs) (1994) [*1], and, according to *The Dinosauria 2nd edition* (2004), it was considered to be a *nomen dubium*. But it was later reclassified as tyrannosauroids (2012) [*2].

[References: (*1) Rich TH, Vickers-Rich P (1994). Neoceratopsians and ornithomimosaurs: dinosaurs of Gondwana origin? *National Geographic Research and Exploration* 10(1): 129–131.; (*2) Benson RBJ, Rich TH, Vickers-Rich P, Hall M (2012). Theropod fauna from southern Australia indicates high polar diversity and climate-driven dinosaur provinciality. *PLoS ONE* 7(5): e37122.]

Timurlengia euotica

> *Timurlengia euotica* Brusatte *et al*., 2016 [middle Turonian; Bissekty Formation, Kyzylkum Desert, Navoi Viloyat, Uzbekistan]

Generic name: *Timurlengia* ← {Timurleng [(Persian) Temur-(i) Lang: Timur the Lame] + -ia}; "in reference to the fourteenth-century Central Asian ruler Timurleng (English: Tamerlane)" [*1].

Specific name: *euotica* ← {(Gr.) εὖ/eu: well + ὠτο-/ōto-[οὖς: ear, hearing] + (suffix) -ikos}; "'well-eared', in reference to the large inner ear of the holotype" [*1].

Etymology: well-eared **Tamerlane's one**

Taxonomy: Theropoda: Coelurosauria: Tyrannosauroidea

[References: (*1) Brusatte SL, Averianov A, Sues H-D, Muir A, Butler IB (2016). New tyrannosaur from the mid-Cretaceous of Uzbekistan clarifies evolution of giant body sizes and advanced senses in tyrant dinosaurs. *Proceedings of the National Academy of Sciences of the United States of America.* 113(13): 3447–3452.]

Titanoceratops ouranos

> *Titanoceratops ouranos* Longrich, 2011 [Campanian; Fruitland Formation or Kirtland Formation, New Mexico, USA]

Generic name: *Titanoceratops* ← {(Gr. myth.) Τιτάν/Titan: mythical race of ancient giants + κέρας/keras (= ceras): horn + ὤψ/ōps: face} [*1]

Specific name: *ouranos* ← {(Gr. myth.) Οὐρανός/Ouranos: the father of the Titans in Greek mythology} [*1]

Etymology: Ouranos' **giant horned-face (ceratopsian)**

Taxonomy: Ornithischia: Ceratopsia: Ceratopsidae: Chasmosaurinae: Triceratopsini

Notes: According to Wick and Lehman (2013), *Titanoceratops ouranos* is a subjective synonym of *Pentaceratops sternbergi*.

[References: (*1) Longrich NR (2011). *Titanoceratops ouranos*, a giant horned dinosaur from the late Campanian of New Mexico. *Cretaceous Research* 32(3): 264–276.]

Titanosaurus indicus

> *Titanosaurus indicus* Lydekker, 1877 [Maastrichtian; Main/Lower Limestone Member, Lameta Formation, Madhya Pradesh, India]

Generic name: *Titanosaurus* ← {(Gr.) Τιτάν/Tītan: a mythial giant + σαῦρος/sauros: lizard}

Specific name: *indicus* ← {(Lat.) Indicus: Indian, of India}

Etymology: **Titan lizard** of India

Taxonomy: Saurischia: Sauropodomorpha: Sauropoda: Camarasauropodomorpha: Titanosauriformes: Somphospondyli: Titanosauria

Other species:

> *Titanosaurus blanfordi* Lydekker, 1879 [Maastrichtian; Lameta Formation, Maharashtra, India]

> *Titanosaurus falloti* Hoffet, 1942 (*nomen dubium*) [Aptian–Albian; Grès supérieurs Formation, Savannakhet, Laos]

> *Titanosaurus madagascariensis* Depéret, 1896 (*nomen dubium*) ⇒ *Laplatasaurus madagascariensis* (Depéret, 1896) von Huene & Matley, 1933 (*nomen dubium*) [Coniacian; Ankazomihaboka Formation, Mahajanga Basin, Madagascar]

> *Titanosaurus nanus* Lydekker, 1893 (*nomen dubium*) [Santonian; Bajo de la Carpa Formation, Sierra Roca, Río Neuquén, Argentina]
> *Titanosaurus rahioliensis* Mathur & Srivastava, 1987 [Maastrichtian; Lameta Formation, Gujarat, India]
> *Titanosaurus valdensis* von Huene, 1929 (*nomen dubium*) ⇒ *Iuticosaurus valdensis* (von Huene, 1929) le Loeuff *et al.*, 1993 [Berriasian–Barremian; Wessex Formation, Isle of Wight, UK]
> *Titanosaurus lydekkeri* von Huene, 1929 (*nomen dubium*) ⇒ *Iuticosaurus lydekkeri* (von Huene, 1929) le Loeuff *et al.*, 1993 [Albian; upper Greensand Formation, Isle of Wight, UK]
> *Titanosaurus australis* Lydekker, 1893 ⇒ *Saltasaurus australis* (Lydekker, 1893) McIntosh, 1990 ⇒ *Neuquensaurus australis* (Lydekker, 1893) Powell, 1992 [Santonian; Bajo de la Carpa Formation, Neuquén, Argentina]
> *Titanosaurus robustus* von Huene, 1929 ⇒ *Saltasaurus robustus* (von Huene, 1929) McIntosh, 1990 ⇒ *Neuquensaurus robustus* (von Huene, 1929) Powell, 1992 (*nomen dubium*) [Campanian–Maastrichtian; Allen Formation, Cinco Saltos, Río Negro, Argentina]
> *Titanosaurus colberti* Jain & Bandyopadhyay, 1997 ⇒ *Isisaurus colberti* (Jain & Bandyopadhyay, 1997) Wilson & Upchurch, 2003 [Maastrichtian; Lameta Formation, Maharashtra, India]
> *Titanosaurus montanus* Marsh, 1877 ⇒ *Atlantosaurus montanus* (Marsh, 1877) Marsh, 1877 (*nomen dubium*) [Kimmeridgian; Morrison Formation, Colorado, USA]
> *Titanosaurus dacus* Nopcsa, 1915 ⇒ *Magyarosaurus dacus* (Nopcsa, 1915) von Huene, 1932 [Maastrichtian; Sânpetru Formation, Szentpéterfalva, Hateg Basin, Romania]

Notes: According to *The Dinosauria 2nd edition*, *Titanosaurus indicus* and *Titanosaurus blandfordi* were considered to be valid.

[References: Lydekker R (1877). Notices of new and other vertebrata from Indian Tertiary and secondary rocks. *Records of the Geological Survey of India* 10(1): 30–43.; Wilson JA, Upchurch P (2003). A revision of *Titanosaurus* Lydekker (Dinosauria – Sauropoda), the first dinosaur genus with a 'Gondwanan' distribution. *Journal of Systematic Palaeontology* 1(3): 125–160.]

Tochisaurus nemegtensis

> *Tochisaurus nemegtensis* Kurzanov & Osmólska, 1991 [late Campanian; Nemegt Formation, Ömnögovi, Monglia]

Generic name: *Tochisaurus* ← {(Mong.) toch': ostrich + (Gr.) σαῦρος/sauros: lizard}; "because of the functionally didactylous foot" (*1).

Specific name: *nemegtensis* ← {(place-name) Nemegt + -ensis}

Etymology: **ostrich lizard** from Nemegt

Taxonomy: Theropoda: Coelurosauria: Deinonychosauria: Troodontidae

[References: (*1) Kurzanov SM, Osmólska H (1991). *Tochisaurus nemegtensis* gen. *et* sp. n., a new troodontid (Dinosauria, Theropoda) from Mongolia. *Acta Palaeontologica Polonica* 36(1): 69–76.]

Tonganosaurus hei

> *Tonganosaurus hei* Li *et al.*, 2010 [Early Jurassic; Yimen Formation, Tong'an, Huili, Sichuan, China]

Generic name: *Tonganosaurus* ← {(Chin. place-name) Tōng'ān通安 + (Gr.) σαῦρος/sauros: lizard} (*1)

Specific name: *hei* ← {(Chin. person's name) He + -ī}; in honor of Hé Xìnlù何信禄, "who spent a lifetime in dinosaur research" (*1).

Etymology: He's Tong'an lizard [何氏通安龍]

Taxonomy: Saurischia: Sauropodomorpha: Eusauropoda: Mamenchisauridae

[References: (*1) Li K, Yang C-Y, Liu J, Wang Z-X (2010). A new sauropod from the lower Jurassic of Huili, Sichuan, China. *Vertebrata PalAsiatica* 48(3): 185–202.]

Tongtianlong limosus

> *Tongtianlong limosus* Lü *et al.*, 2016 [Maastrichtian; Nanxiong Formation, Jiangxi, China]

Generic name: *Tongtianlong* ← {(Chin. place-name) Tōngtiān通天 + lóng龍: dragon}; referring to "Tongtianyan of Ganzhou, the first grotto south of the Yangtze River", and also meaning "the road to heaven, a fitting epitaph for a deceased dinosaur preserved with outstretched arms" (*1).

Specific name: *limosus* ← {(Lat.) limōsus: muddy}; "referring to the holotype specimen being found in an unusual posture in a mudstone" (*1).

Etymology: muddy **Tongtian dragon**

Taxonomy: Theropoda: Maniraptora: Oviraptorosauria: Oviraptoridae

[References: (*1) Lü J, Chen R, Brusatte SL, Zhu Y, Shen C (2016). A Late Cretaceous diversification of Asian oviraptorid dinosaurs: evidence from a new species preserved in an unusual posture. *Scientific Reports* 6(35780).]

Tornieria africana

> *Gigantosaurus africanus* Fraas, 1908 ⇒ *Tornieria africana* (Fraas, 1908) Sternfeld, 1911 ⇒ *Barosaurus africanus* (Fraas, 1908)

Janensch, 1929 [Tithonian; upper Dinosaur Member, Tendaguru Formation, Lindi, Tanzania]

Generic name: *Tornieria* ← {(person's name) Tornier + -ia}; in honor of German paleontologist Gustav Tornier.

Specific name: *africana* ← {(Lat.) Āfricāna (feminine form of Āfricānus): African}

Etymology: **for Tornier** from Africa

Taxonomy: Saurischia: Sauropodomorpha: Sauropoda: Diplodocidae: Diplodocinae

Other species:

> *Barosaurus gracilis* Russell *et al.*, 1980 (*nomen nudum*) ⇒ *Tornieria gracilis* (Russell *et al.*, 1980) Olshevsky, 1991 (*nomen nudum*) [Tendaguru, Tanzania]

> *Gigantosaurus dixeyi* Haughton, 1928 (preoccupied) ⇒ *Tornieria dixeyi* (Haughton, 1928) von Huene, 1932 ⇒ *Malawisaurus dixeyi* (Haughton, 1928) Jacobs *et al.*, 1993 [Aptian; Dinosaur Beds Formation, Northern, Malawi]

> *Gigantosaurus robustus* Fraas, 1908 ⇒ *Tornieria robusta* (Fraas, 1908) Sternfeld, 1911 ⇒ *Janenschia robusta* (Fraas, 1908) Wild, 1991 [Tithonian; Tendaguru Formation, Lindi, Tanzania]

Notes: Remes (2006) concluded *Tornieria* was a valid genus. (*1)

[References: (*1) Remes K (2006). Revision of the Tendaguru sauropod dinosaur *Tornieria africana* (Fraas) and its relevance for sauropod paleobiogeography. *Journal of Vertebrate Paleontology* 26(3): 651–669.]

Torosaurus latus

> *Torosaurus latus* Marsh, 1891 [Maastrichtian; Lance Formation, Wyoming, USA]

= *Torosaurus gladius* Marsh, 1891 [Maastrichtian; Lance Formation, Wyoming, USA]

Generic name: *Torosaurus* ← {(Gr.) toro-: perforated [τορέω/toreō: pierce, perforate] + σαῦρος/sauros: lizard}; referring to a pair of large openings in its frill (*1).

Specific name: *latus* ← {(Lat.) lātus: broad, wide}

Etymology: wide, **perforated lizard**

Taxonomy: Ornithischia: Marginocephalia: Ceratopsia: Ceratopsidae: Chasmosaurinae: Triceratopsini (= Torosaurini)

Other species:

> *Arrhinoceratops utahensis* Gilmore, 1946 ⇒ *Torosaurus utahensis* (Gilmore, 1946) Lawson, 1976 ⇒ *Triceratops utahensis* (Gilmore 1946) Longrich, 2014 [Maastrichtian; North Horn Formation, Utah, USA]

[References: (*1) Marsh OC (1891). Notice of new vertebrate fossils. *The American Journal of Science, series 3* 42: 265–269.]

Torvosaurus tanneri

> *Torvosaurus tanneri* Galton & Jensen, 1979 ⇒ *Megalosaurus tanneri* (Galton & Jensen, 1979) Paul, 1988 [Kimmeridgian; Brushy Basin Member, Morrison Formation, Colorado, USA]

= *Edmarka rex* Bakker *et al.*, 1992 [Kimmeridgian; Talking Rocks Member, Morrison Formation, Wyoming, USA]

Generic name: *Torvosaurus* ← {(Lat.) torvus: savage, cruel, wild + (Gr.) σαῦρος/sauros: lizard} (*1)

Specific name: *tanneri* ← {(person's name) Tanner + -ī}; "in honor of N. Eldon Tanner, first counselor in the First Presidency of The Church of Jesus Christ of Latter-day Saints" (*1).

Etymology: Tanner's **savage lizard**

Taxonomy: Theropoda: Megalosauroidea (= Spinosauroidea): Megalosauridae (= Torvosauridae): Megalosaurinae

Other species:

> *Torvosaurus gurneyi* Hendrickx & Mateus, 2014 [late Kimmeridgian; Amoreira-Porto Novo Member, Lourinhã Formation, Leiria, Portugal]

[References: (*1) Galton PM, Jensen JA (1979). A new large theropod dinosaur from the Upper Jurassic of Colorado. *Brigham Young University Geology Studies* 26(1): 1–12.]

Tototlmimus packardensis

> *Tototlmimus packardensis* Serrano-Brañas *et al.*, 2016 [Campanian; Packard Shale Formation, Sonora, Mexico]

Generic name: *Tototlmimus* ← {(Nahuatl) tototl: bird + (Lat.) mīmus = (Gr.) μῖμος/mīmos: mimic}

Specific name: *packardensis* ← {(stratal name) Packard Shale Formation + -ensis}

Etymology: **bird mimic** from Packard

Taxonomy: Theropoda: Ornithomimosauria: Ornithomimidae

Trachodon mirabilis

> *Trachodon mirabilis* Leidy, 1856 (*nomen dubium*) ⇒ *Hadrosaurus mirabilis* (Leidy, 1856) Leidy, 1868 ⇒ *Diclonius mirabilis* (Leidy, 1856) Cope, 1883 [middle Campanian; Judith River Formation, Montana, USA]

Generic name: *Trachodon* ← {(Gr.) τραχύς/trachys: rough + ὀδών (= ὀδούς/odous [odūs]: tooth)}

Specific name: *mirabilis* ← {(Lat.) mīrābilis: marvelous}

Etymology: marvelous, **rough tooth**

Taxonomy: Ornithischia: Ornithopoda: Hadrosauridae
Other species:
> *Trachodon cantabrigiensis* Lydekker, 1888 (*nomen dubium*) ⇒ *Hadrosaurus cantabrigiensis* (Lydekker, 1888) Newton, 1892 (*nomen dubium*) ⇒ *Telmatosaurus cantabrigiensis* (Lydekker, 1888) Olshevsky, 1978 (*nomen dubium*) [Cenomanian; Cambridge Greensand Member, West Melbury Chalk Formation, UK]
> *Trachodon altidens* Lambe, 1902 (*nomen dubium*) ⇒ *Didanodon altidens* (Lambe, 1902) Osborn, 1902 ⇒ *Pteropelyx altidens* (Lambe, 1902) Lambe, 1902 ⇒ *Procheneosaurus altidens* (Lambe, 1902) Lull & Wright, 1942 ⇒ *Thespesius altidens* (Lambe, 1902) Steel, 1969 [middle Campanian; Dinosaur Park Formation, Alberta, Canada]
> *Trachodon marginatus* Lambe, 1902 (*nomen dubium*) ⇒ *Stephanosurus marginatus* (Lambe, 1902) Lambe, 1914 [late Campanian; Dinosaur Park Formation, Alberta, Canada]
> *Trachodon selwyni* Lambe, 1902 (*nomen dubium*) [mid-Cretaceous; Berry River Series]
> *Trachodon amurensis* Riabinin, 1925 (*nomen dubium*) ⇒ *Mandschurosaurus amurensis* (Riabinin, 1925) Riabinin, 1930 [Maastrichtian; Yuliangze Formation, China]
> *Sanpasaurus imperfectus* Young, 1944 (*nomen dubium*) ⇒ *Trachodon imperfectus* (Young, 1944) Kuhn, 1964 (*nomen dubium*) [Middle Jurassic; Kuangyuan Series, Weiyuan, Sichuan, China]
> *Edmontosaurus regalis* Lambe, 1917 [Campanian; Horseshoe Canyon Formation, Alberta, Canada]
= *Trachodon atavus* Cope, 1871
> *Claosaurus annectens* Marsh, 1892 ⇒ *Anatosaurus annectens* (Marsh, 1892) Lull & Wright, 1942 ⇒ *Edmontosaurus annectens* (Marsh, 1892) Brett-Surman *vide* Chapman & Brett-Surman, 1990 [Maastrichtian; Lance Formation, Wyoming, USA]
= *Trachodon longiceps* Marsh, 1890 ⇒ *Anatotitan longiceps* (Marsh, 1890) Olshevsky, 1991 [Maastrichtian; Lance Formation, Wyoming, USA]
> *Hadrosaurus cavatus* Cope 1871 ⇒ *Trachodon cavatus* (Cope, 1871) Hay, 1902 (*nomen dubium*) [Maastrichtian; Navesink Formation, New Jersey, USA]

Tralkasaurus cuyi
> *Tralkasaurus cuyi* Cerroni *et al.*, 2020 [Cenomanian–Turonian; Huincul Formation, Río Negro, Argentina]
Generic name: *Tralkasaurus* ← {(Mapudungun) tralka: thunder + (Gr.) σαῦρος/sauros: lizard}
Specific name: *cuyi* ← {(place-name) El Cuy: village + -ī}
Etymology: **thunder lizard** of El Cuy
Taxonomy: Saurischia: Theropoda: Abelisauridae

Tratayenia rosalesi
> *Tratayenia rosalesi* Porfiri *et al.*, 2018 [Santonian; Bajo de la Carpa Formation, Neuquén, Argentina]
Generic name: *Tratayenia* ← {(place-name) Tratayen: the locality + -ia} (*1)
Specific name: *rosalesi* ← {(person's name) Rosales + -ī}; "in honor of Diego Rosales, the discoverer of the specimen" (*1).
Etymology: Rosales' **one from Tratayen**
Taxonomy: Theropoda: Megaraptora: Megaraptoridae
Notes: A medium-sized megaraptoran.
[References: (*1) Porfiri JD, Juárez Valieri RD, Santos DDD, Lamanna MC (2018). A new megaraptoran theropod dinosaur from the Upper Cretaceous Bajo de la Carpa Formation of northwestern Patagonia. *Cretaceous Research* 89: 302–319.]

Traukutitan eocaudata
> *Traukutitan eocaudata* Juárez Valieri & Calvo, 2011 [Santonian; Bajo de la Carpa Formation, Neuquén, Argentina]
Generic name: *Traukutitan* ← {(Araucanian legend) Trauku: the Araucanian mountain spirit, usually represented like a giant + (Gr. myth.) τιτάν/titan: giants} (*1)
Specific name: *eocaudata* ← {(Gr.) ἠώς/ēōs: dawn + (Lat.) cauda: tail}; "in reference to the basal morphology displayed in the middle caudal vertebrae present in this form" (*1).
Etymology: **Trauku giants** with a basal tail
Taxonomy: Saurischia: Sauropodomorpha: Sauropoda: Titanosauria
[References: (*1) Juárez Valieri RD, Calvo JO (2011). Revision of MUCPv 204, a Senonian basal titanosaur from northern Patagonia. In: Calvo, González, Riga, Porfiri, Dos Santos (eds), *Paleontología y Dinosarios des de América Latina*: 143–152.]

Triceratops horridus
> *Ceratops horridus* Marsh, 1889 ⇒ *Triceratops horridus* (Marsh, 1889) Marsh, 1889 [Maastrichtian; Lance Formation, Niobrara, Wyoming, USA]
= *Triceratops flabellatus* Marsh, 1889 ⇒ *Sterrholophus flabellatus* (Marsh, 1889) Marsh, 1891 ⇒ *Agathaumas flabellatus* (Marsh, 1889) Burkhardt, 1892 [Maastrichtian;

Lance Formation, Wyoming, USA]
= *Triceratops serratus* Marsh, 1890 [Maastrichtian; Laramie Formation, Wyoming, USA]
= *Triceratops elatus* Marsh, 1891 [Maastrichtian; Lance Formation; Wyoming, USA]
= *Triceratops calicornis* Marsh, 1898 [Maastrichtian; Lance Formation; Wyoming, USA]
= *Triceratops obtusus* Marsh, 1898 [Maastrichtian; Lance Formation, Wyoming, USA]
= *Triceratops eurycephalus* Schlaikjar, 1935 [Maastrichtian; Lance Formation, Wyoming, USA]
= *Triceratops albertensis* Sternberg, 1949 [Maastrichtian; Scollard Formation, Alberta, Canada]

Generic name: *Triceratops* ← {(Gr.) τρι-/ tri-: 3 + κερατ-/ kerat- [κέρας/ keras: horn] + ὤψ/ōps: face}. "In addition to the pair of massive horn-cores on the top of the skull, there is a third horn-core on the nose" (*1).

Specific name: *horridus* ← {(Lat.) horridus: rough}

Etymology: rough, three horned face

Taxonomy: Ornithischia: Marginocephalia: Ceratopsia: Ceratopsidae

Other species:

> *Agathaumas sylvestris* Cope, 1872 (*nomen dubium*) ⇒ *Triceratops sylvestris* (Cope, 1872) Kuhn, 1936 (*nomen dubium*) [Maastrichtian; Lance Formation, Wyoming, USA]
> *Bison alticornis* Marsh, 1887 ⇒ *Ceratops alticornis* (Marsh, 1887) Marsh, 1889 ⇒ *Triceratops alticornis* (Marsh, 1887) Hatcher *et al.*, 1907 (*nomen dubium*) [Maastrichtian; Denver Formation, Colorado, USA]
> *Triceratops galeus* Marsh, 1889 (*nomen dubium*) [Maastrichtian; Denver Formation, Colorado, USA]
> *Triceratops prorsus* Marsh, 1890 ⇒ *Agathaumas prorsus* (Marsh, 1890) Lydekker, 1893 [Maastrichtian; Laramie Formation, Wyoming, USA]
= *Triceratops brevicornis* Hatcher, 1905 [Maastrichtian; Lance Formation, Wyoming, USA]
= *Polyonax mortuarius* Cope, 1874 ⇒ *Agathaumas mortuarius* (Cope, 1874) Hay, 1901 [Maastrichtian; Laramie Formation, Colorado, USA]
> *Triceratops sulcatus* Marsh, 1890 [Maastrichtian; Laramie Formation, Wyoming, USA]
> *Diceratops hatcheri* Lull *vide* Hatcher, 1905 (preoccupied) ⇒ *Triceratops hatcheri* (Lull *vide* Hatcher, 1905) Lull, 1933 ⇒ *Nedoceratops hatcheri* (Lull *vide* Hatcher, 1905) Ukrainsky, 2007 ⇒ *Diceratus hatcheri* (Lull *vide* Hatcher, 1905) Mateus, 2008 [Maastrichtian; Laramie Formation, Wyoming, USA]
> *Triceratops ingens* Lull, 1915 (*nomen dubium*) [Maastrichtian; Lance Formation, Wyoming, USA]
> *Triceratops maximus* Brown, 1933 (*nomen dubium*) [Maastrichtian; Hell Creek Formation, Montana, USA]
> *Arrhinoceratops utahensis* Gilmore, 1946 ⇒ *Torosaurus utahensis* (Gilmore, 1946) Lawson, 1976 ⇒ *Triceratops utahensis* (Gilmore, 1946) Longrich, 2014 [Maastrichtian; North Horn Formation, Utah, USA]

[References: (*1) Marsh OC (1889). Notice of gigantic horned Dinosauria, from the Cretaceous. *American Journal of Science* 3(38): 173–176.]

Trierarchuncus prairiensis

> *Trierarchuncus prairiensis* Fowler *et al.*, 2020 [Maastrichtian; Hell Creek Formation, Montana, USA]

Generic name: *Trierarchuncus* ← {trierarch [(Lat.) trierarchus = (Gr.) τριήραχος/triērarchos: "specifically seafaring ship's captain (trireme ships of ancient Greece)"] + (Lat.) uncus: hook} "Captain Hook"; referring to the hook-handed pirate of *Peter Pan*.

Specific name: *prairiensis* ← {(place-name) the prairie + -ensis}; "referring to the gentle plains of eastern Montana (in particular the American Prairie Reserve) where the new material was discovered".

Etymology: **Captain Hook** of the prairie

Taxonomy: Theropoda: Alvarezsauridae: Parvicursorinae

cf. Theropoda: Maniraptora: Alvarezsauridae (according to Fowler *et al.*, 2020)

Trigonosaurus pricei

> *Trigonosaurus pricei* Campos *et al.*, 2005 [Maastrichtian; Marília Formation, Minas Gerais, Brazil]

Generic name: *Trigonosaurus* ← {(Gr.) τρίγωνος/ trigónos [(place name) "Triângulo Mineiro" from the Minas Gerais State] + σαῦρος/sauros: reptile}

Specific name: *pricei* ← {(person's name) Price + -ī}; "in honor of Llewellyn Ivor Price, a very important vertebrate paleontologist, whose birth day centenary is celebrated in 2005. L. I. Price collected this and several other specimens and inspired the authors of this paper, some of which had the pleasure to work with him (D. A. Campos and R. J. Bertini)" [*1].

Etymology: Price's **reptile from Triângulo Mineiro**

Taxonomy: Saurischia: Sauropodomorpha: Sauropoda: Titanosauriformes: Titanosauria: Titanosauridae

[References: (*1) Campos DA., Kellner AWA, Bertini RJ, Santucci RM (2005). On a titanosaurid (Dinosauria, Sauropoda) vertebral column from the Bauru Group, Late Cretaceous of Brazil. *Arquivos do Museu Nacional, Rio de Janeiro* 63(3): 565–593.]

Trinisaura santamartaensis

> *Trinisaura santamartaensis* Coria *et al.*, 2013 [late Campanian; Snow Hill Island Formation, Santa Marta Cove, James Ross Island, Antarctica]

Generic name: *Trinisaura* ← {(person's name) Trinidad Diaz + (Gr.) σαύρα/saura (feminine form of "σαῦρος/sauros": lizard)}

Specific name: *santamartaensis* ← {(place-name) Santa Marta Cove + -ensis}

"The materials were found on the surface enclosed in a hard sandstone concretion collected near Santa Marta Cove, James Ross Island, from the lower levels of the Snow Hill Island Formation (Campanian)" [*1]

cf. Santa Marta Formation (Santonian–Campanian) underlies Snow Hill Island Formation.

Etymology: **Trinidad Diaz's lizard** from Santa Marta Cove

Taxonomy: Ornithischia: Ornithopoda

[References: (*1) Coria RA, Moly JJ, Reguero M, Santillana S, Marenssi S (2013). A new ornithopod (Dinosauria; Ornithischia) from Antarctica. *Cretaceous Research* 41: 186–193.]

Triunfosaurus leonardii

> *Triunfosaurus leonardii* Carvalho *et al.*, 2017 [Berriasian–lower Hauterivian; Rio Piranhas Formation, Triunfo Basin, Paraíba, Brazil]

Generic name: *Triunfosaurus* ← {(place-name) Triunfo Basin + (Gr.) σαῦρος/sauros: lizard, reptile} [*1]

Specific name: *leonardii* ← {(person's name) "in honor of the paleontologist Giuseppe Leonardi, who dedicated greater part of his life to the study of the reptile ichnofauna from the northeastern Brazil" [*1].

Etymology: Leonardi's **Triunfo reptile**

Taxonomy: Saurischia: Sauropoda: Titanosauriformes: Titanosauria

[References: (*1) Carvalho IS, Salgado L, Lindoso RM, de Araújo-Júnior HI, Costa Nogueira FC, Soares JA (2017). A new basal titanosaur (Dinosauria, Sauropoda) from the Lower Cretaceous of Brazil. *Journal of South American Earth Sciences* 75: 74–84.]

Troodon formosus

> *Troodon formosus* Leidy, 1856 [Campanian; Judith River Formation, Montana, USA]

= *Polyodontosaurus grandis* Gilmore, 1932 (*nomen dubium*) [Campanian; Dinosaur Park Formation, Montana, USA]

= *Pectinodon bakkeri* Carpenter, 1982 ⇒ *Troodon bakkeri* (Carpenter, 1982) Olshevsky, 1991 (*nomen dubium*) [Lancian; Lance Formation, Wyoming, USA]

= *Stenonychosaurus inequalis* Sternberg, 1932 ⇒ *Saurornithoides inequalis* (Sternberg, 1932) Carpenter, 1982 ⇒ *Troodon inequalis* (Sternberg, 1932) Currie, 2005 [Campanian; Belly River Group, Alberta, Canada]

Generic name: *Troodon* ← {tro-: wounding [(Gr.) τρώγω/trōgō: to gnaw, nibble] + ὀδών/odōn: tooth}

Specific name: *formosus* ← {(Lat.) fōrmōsus; finely formed, beautiful}.

Etymology: finely formed, **wounding tooth**

Taxonomy: Theropoda: Maniraptora: Troodontidae

Other species:

> *Saurornithoides mongoliensis* Osborn, 1924 ⇒ *Troodon mongoliensis* (Osborn, 1924) Paul, 1988 [Campanian; Djadokhta Formation, Ömnögovi, Mongolia]

Notes: According to *The Dinosauria 2nd edition*, *Polyodontosaurus grandis* Gilmore, 1932, *Stenonychosaurus inequalis* Sternberg, 1932 and *Pectinodon bakkeri* Carpenter, 1982 were considered to be synonyms of *Troodon formosus* Leidy, 1856. *Stenonychosaurus* was thought a separate genus from the possibly dubious *Troodon* and reverted (Evans *et al.* 2017, van der Reest & Currie, 2017).

[References: Holtz TR, Brinkman DL, Chandler CL (1998). Denticle morphometrics and a possibly omnivorous feeding habit for the theropod dinosaur *Troodon*. *Gaia* 15: 159–166.]

Tsaagan mangas

> *Tsaagan mangas* Norell *et al.*, 2006 [Campanian; Djadokhta Formation, Ömnögovi, Mongolia] [*1] [*2]

Generic name: *Tsaagan* ← {(Mong.) цагаан: white} [*2]

Specific name: *mangas* ← {(Mong.) мангас: monster} [*2]

Etymology: **white** monster

Taxonomy: Theropoda: Coelurosauria: Maniraptora: Dromaeosauridae: Velociraptorinae

[References: (*1) Dingus L, Loope DB, Dashzeveg D, Swisher III CC (2008). The geology of Ukhaa Tolgod (Djadokhta Formation, Upper Cretaceous, Nemegt Basin, Mongolia). *American Museum Novitates*

3616: 1–40.; (*2) Norell MA, Clark JM, Turner AH, Makovicky PJ, Barsbold R, Rowe T (2006). A new dromaeosaurid theropod from Ukhaa Tolgod (Ömnögov, Mogolia). *American Museum Novitates* 3545: 1–51.]

Tsagantegia longicranialis

> *Tsagantegia longicranialis* Tumanova, 1993 [Cenomanian–Santonian; Bayan Shireh Formation, Burkhant, Gobi Desert, Mongolia]

Generic name: *Tsagantegia* ← {(place-name) Tsagaan-Teg/Tsagaan Teeg*: locality + -ia}
 *Tsagaan-Teeg [tsagaan: white+ teeg/teg: landform] (*1)

Specific name: *longicranialis* ← {(Lat.) longus: long + (Gr.) κρανίον/kranion: skull}

Etymology: **Tsagaan-Teg's one** with a long skull

Taxonomy: Ornithischia: Ankylosauria: Ankylosauridae

[References: (*1) Mongolian place names and stratigraphic terms. 〈https://artscimedia.case.edu/wp-content/uploads/sites/108/2017/05/17213028/Benton-et-al-2000-Dinosaurs-of-Russia-00c3-mongolian-place-stratigraphic-names.pdf〉]

Tsintaosaurus spinorhinus

> *Tsintaosaurus spinorhinus* Young, 1958 [Campanian; Jingangkou Formation, Laiyang, Shandong, China]

= *Tanius laiyangensis* Zhen, 1976 [Campanian; Jingangkou Formation, Laiyang, Shandong, China]

Generic name: *Tsintaosaurus* ← {(place-name) Tsingtao青島 [Pinyin: Qīngdǎo] + (Gr.) σαῦρος/sauros: lizard}

Specific name: *spinorhinus* ← {(Lat.) spina: thorn, spine + (Gr.) ῥινός/rhinos [ῥίς/rhis: nose]}; referring to the distinctive spine-like crest on the snout.

Etymology: **Qingdao lizard** with a nose spine

Taxonomy: Ornithischia: Ornithopoda: Hadrosauridae: Lambeosaurinae: Tsintaosaurini

Notes: *Tsintaosaurus* had a unicorn-like crest on its skull. According to *The Dinosauria 2nd edition*, *Tanius laiyangensis* is regarded as a junior synonym of *Tsintaosaurus spinorhinus*.

Tugulusaurus faciles

> *Tugulusaurus faciles* Dong, 1973 [late Aptian; Lianmuqin/Lianmugin Formation, Tugulu Group, Xinjiang, China]

Generic name: *Tugulusaurus* ← {(Chin. place-name) Tugulu吐谷魯: Tuguru Group + (Gr.) σαῦρος/sauros: lizard}

Specific name: *faciles* ← {(Lat.) facilis: easily moving}

Etymology: nimble, **lizard from Tugulu**

Taxonomy: Theropoda: Coelurosauria: Alvarezsauria
 cf. Theropoda: Tetanurae: Coelurosauria (*1)

[References: (*1) Rauhut OWM, Xu X (2005). The small theropod dinosaurs *Tugulusaurus* and *Phaedrolosaurus* from the Early Cretaceous of Xinjiang, China. *Journal of Vertebrate Paleontology* 25(1): 107–118.]

Tuojiangosaurus multispinus

> *Tuojiangosaurus multispinus* Dong *et al.*, 1977 [Oxfordian; upper Shaximiao Formation, Zigong, Sichuan, China]

Generic name: *Tuojiangosaurus* ← {(Chin. place-name) Tuójiāng沱江 + (Gr.) σαῦρος/sauros}; referring to "the Tuojiang river valley, which is a tributary to the Sidajianghe River" (*1) (*2).

Specific name: *multispinus* ←{(Lat.) multispinus: many-spined}; referring to "17 pairs of plated armor, which constitutes the most heavily armored genus in the subfamily" (*2).

Etymology: many-spined **Tuojiang lizard**

Taxonomy: Ornithischia: Thyreophora: Stegosauria: Stegosauridae

[References: (*1) Dong Z, Li X, Zhou S, Zhang Y (1977). On the stegosaurian remains from Zigong (Tzekung), Szechuan province. *Vertebrata PalAsiatica* 15(4): 307–312.; (*2) Dong Z, Zhou S, Zhang Y (1983). Dinosaurs from the Jurassic of Sichuan. *Palaeontologica Sinica* 162 C 23: 1–136.]

Tupandactylus imperator [Pterosauria]

> *Tapejara imperator* Campos & Kellner, 1997 ⇒ *Tupandactylus imperator* (Campos & Kellner, 1997) Kellner & Campos, 2007 ⇒ *Ingridia imperator* (Campos & Kellner, 1997) Unwin & Martin, 2007 [Aptian–Albian; Nova Olinda Member, Crato Formation, Ceará, Brazil]

Generic name: *Tupandactylus* ← {(Tupian culture) Tupa or Tupan: god of the thunder + (Gr.) δάκτυλος/dactylos: finger}

Specific name: *imperator* ← {(Lat.) imperātōr: emperor}

Etymology: **Tupan's finger**, emperor

Taxonomy: Pterosauria: Pterodactyloidea: Tapejaromorpha: Tapejaridae: Tapejarinae

Other species:

> *Tapejara navigans* Frey *et al.*, 2003 ⇒ *Tupandactylus navigans* (Frey *et al.*, 2003) Kellner & Campos, 2007 ⇒ *Ingridia navigans* (Frey *et al.*, 2003) Unwin & Martin, 2007 [Aptian; Nova Olinda Member, Crato Formation, Ceará, Brazil]

Turanoceratops tardabilis

> *Turanoceratops tardabilis* Nessov *et al.*, 1989 [middle Turonian; Bissekty Formation, Dzharakuduk, Navoi, Uzbekistan]

Generic name: *Turanoceratops* ← {(Persian, place-name) Turan: Turkestan region + ceratops [(Gr.) κέρας/keras: horn + ὤψ/ōps: face]}

Specific name: *tardabilis* ← {tardabilis [(Lat.) tardus: slow, tardy] retarding; referring to the protracted research.

Etymology: protracted, **Turan horned face (ceratopsian)**

Taxonomy: Ornithischia: Neoceratopsia: Ceratopsoidea

Notes: According to Sues & Averianov (2009), *Turanoceratops tardabilis* represents the first definite ceratopsid dinosaur from Asia. [*1]

[References: (*1) Sues H-D, Averianov A (2009). *Turanoceratops tardabilis*—the first ceratopsid dinosaur from Asia. Naturwissenschaften 96(5): 645–652.]

Turiasaurus riodevensis

> *Turiasaurus riodevensis* Royo-Torres *et al.*, 2006 [Kimmeridgian–Valanginian; Villar del Arzobispo Formation, Teruel, Aragón, Spain]

Generic name: *Turiasaurus* ← {(Lat. place-name) Turia: Teruel + (Gr.) σαῦρος/sauros: lizard}

Specific name: *riodevensis* ← {(place-name) Riodeva + -ensis}

Etymology: **Turia (Teruel) lizard** from Riodeva

Taxonomy: Saurischia: Sauropodomorpha: Sauropoda: Turiasauria

[References: Royo-Torres R, Cobos A, Alcalá L (2006). A giant European Dinosaur and a new sauropod clade. *Science* 314(5807): 1925–1927.]

Tylocephale gilmorei

> *Tylocephale gilmorei* Maryańska & Osmólska, 1974 [Campanian; Barun Goyot Formation, Ömnögovi, Mongolia]

Generic name: *Tylocephale* ← {(Gr.) tylo- [τύλος/tylos = τύλη/tylē: swelling on the skin] + κεφᾰλή/kephale: head}; "because of the thickening of the skull roof" [*1].

Specific name: *gilmorei* ← {(person's name) Gilmore + -ī}; "in honor of the late Ch. W. Gilmore, who first gave the detailed description of a pachycephalosaurid species" [*1].

Etymology: Gilmore's **swelling head**

Taxonomy: Ornithischia: Ornithopoda: Marginocephalia: Pachycephalosauria: Pachycephalosauridae

[References: (*1) Maryańska T, Osmólska H (1974). Pachycephalosauria, a new suborder of ornithischian dinosaurs. *Palaeontologia Polonica* 30: 45–102.]

Tyrannosaurus rex

> *Tyrannosaurus rex* Osborn, 1905 [Maastrichtian; Hell Creek Formation, Montana, USA]

= *Dynamosaurus imperiosus* Osborn, 1905 ⇒ *Tyrannosaurus imperiosus* (Osborn, 1905) Osborn, 1906 [Maastrichtian; Lance Formation, Wyoming, USA]

= *Manospondylus gigas* Cope, 1892 (*nomen dubium*) [Maastrichtian; Hell Creek Formation, South Dakota, USA]

= *Albertosaurus megagracilis* Paul, 1988 ⇒ *Dinotyrannus megagracilis* (Paul, 1988) Olshevsky *et al.*, 1995 [Maastrichtian; Hell Creek Formation, Montana, USA]

= *Aublysodon molnari* Paul, 1988 ⇒ *Stygivenator molnari* (Paul, 1988) Olshevsky, 1995 [Maastrichtian; Hell Creek Formation, Montana, USA]

= *Gorgosaurus lancensis* Gilmore, 1946 ⇒ *Deinodon lancensis* (Gilmore, 1946) Kuhn, 1965 ⇒ *Albertosaurus lancensis* (Gilmore, 1946) Russell, 1970 ⇒ *Nanotyrannus lancensis*? (Gilmore, 1946) Bakker *et al.*, 1988 [Maastrichtian; Hell Creek Formation, Montana, USA]

Generic name: *Tyrannosaurus* ← {(Gr.) τύραννος/tyrannos: tyrant + σαῦρος/sauros: lizard}

Specific name: *rex* ← {(Lat.) rēx: king}.

Etymology: **tyrant lizard** king

Taxonomy: Theropoda: Coelurosauria: Tyrannosauroidea: Tyrannosauridae: Tyrannosaurinae

Other species:

> *Tyrannosaurus bataar* Maleev, 1955 ⇒ *Tarbosaurus bataar* (Maleev, 1955) Rozhdestvensky, 1965 ⇒ *Jenghizkhan bataar* (Maleev, 1955) Olshevsky *et al.*, 1995 [late Campanian; Nemegt Formation, Ömnögovi, Mongolia]

= *Tarbosaurus efremovi* Maleev, 1955 ⇒ *Tyrannosaurus efremovi* (Maleev, 1955) Rozhdestvensky, 1977 [late Campanian; Nemegt Formation, Ömnögovi, Mongolia]

= *Gorgosaurus novojilovi* Maleev, 1955 ⇒ *Deinodon novojilovi* (Mallev, 1955) Maleev, 1964 ⇒ *Deinodon novojilovi* (Maleev, 1955) Kuhn, 1965 ⇒ *Aublysodon novojilovi* (Maleev, 1955) Charig, 1967 ⇒ *Tarbosaurus novojilovi* (Maleev, 1955) Olshevsky, 1978 ⇒ *Albertosaurus novojilovi* (Maleev, 1955) Mader & Bradley, 1989 ⇒ *Maleevosaurus novojilovi* (Maleev, 1955) Carpenter, 1992 ⇒ *Tyrannosaurus novojilovi* (Maleev, 1955) Glut, 1997 [Campanian; Nemegt Formation,

Ömnögivi, Mongolia]
= ?*Tyrannosaurus luanchuanensis* Dong, 1979 ⇒ *Jenghizkhan luanchuanensis* (Dong, 1979) Olshevsky & Ford, 1995 [Campanian; Qiupa Formation, Henan, China]
= ?*Tyrannosaurus turpanensis* Zhai *et al.*, 1978 (*nomen dubium*) [Campanian–Maastrichtian; Subashi Formation, Shanshan, Xinjiang, China]
> *Tyrannosaurus lanpingensis* Yeh, 1975 (*nomen dubium*) [Late Cretaceous; Jingxing Formation, Yunnan, China]

Notes: According to *The Dinosauria 2nd edition*, *Aublysodon amplus* Marsh, 1892, *Aublysodon cristatus* Marsh, 1892, *Aublysodon lateralis* Cope, 1876 and *Aublysodon mirandus* Leidy, 1868 are considered to be *nomina dubia*.

[References: Osborn HF (1905). *Tyrannosaurus* and other Cretaceous carnivorous dinosaurs. *Bulletin of the American Museum of Natural History* 21(14): 259–265.]

Tyrannotitan chubutensis

> *Tyrannotitan chubutensis* Novas *et al.*, 2005 [Albian; Cerro Castaño Member, Cerro Barcino Formation, Chubut, Argentina]

Generic name: *Tyrannotitan* ← {(Lat.) tyrannus = (Gr.) τύραννος/tyrannos: tyrant + Τῑτάν/Tītan: giant}

Specific name: *chubutensis* ← {(place-name) Chubut Province + -ensis}

Etymology: **tyrant giant** from Chubut

Taxonomy: Theropoda: Carnosauria: Allosauroidea: Carcharodontosauridae

[References: Novas FE, de Valais S, Vickers-Rich P, Rich TM (2005). A large Cretaceous theropod from Patagonia, Argentina, and the evolution of carcharodontosaurids. *Naturwissenschaften* 92(5): 226–230.]

U

Uberabatitan ribeiroi

> *Uberabatitan ribeiroi* Salgado & Carvalho, 2008 [Maastrichtian; Serra da Galga Member, Marília Formation, Uberaba, Minas Gerais, Brazil]

Generic name: *Uberabatitan* ← {(place-name) Uberaba: the city of Uberaba + (Gr. myth.) τιτάν/titan: giant}

Specific name: *ribeiroi* ← {(person's name) Ribeiro + -ī}; "in honor of Luiz Carlos Borges Ribeiro, director of the Centro de Pesquisas Paleontológicas Lewellyn Price, for his consistent support of palaeontological research in Minas Gerais State"(*1).

Etymology: Ribeiro's **Uberaba giant**

Taxonomy: Saurischia: Sauropodomorpha: Sauropoda: Titanosauria

[References: (*1) Salgado L, Carvalho IdS (2008). *Uberabatitan ribeiroi*, a new titanosaur from the Marília Formation (Bauru Group, Upper Cretaceous), Minas Gerais, Brazil. *Palaeontology* 51(4): 881–901.]

Ubirajara jubatus

> *Ubirajara jubatus* Smyth *et al.*, 2020 [Aptian; Crato Formation, Brazil]

Generic name: *Ubirajara* ← {(Tupi) Ubira: spear + jara: lord} lord of the spear; referring to "the stiffened, elongate integumentary structures associated with the specimen".

Specific name: *jubatus* ← {(Lat.) jubatus: maned}

Etymology: maned **Lord of the Spear**

Taxonomy: Saurischia: Theropoda: Compsognathidae

Notes: *Ubirajara* preserves filamentous integument.

Udanoceratops tschizhovi

> *Udanoceratops tschizhovi* Kurzanov, 1992 [Campanian; Djadokhta Formation, Ömnögovi, Mongolia]

Generic name: *Udanoceratops* ← {(place-name) Udan-Sayr: the locality where the holotype was found + (Gr.) κερατ- [κέρας/keras: horn] + ὤψ/ōps: face}

Specific name: *tschizhovi* ← {(person's name) Tschizhov + -ī}; in honor of D. O. Tschizhov.

Etymology: Tschizhov's **ceratops (horn face) from Udan Sayr**

Taxonomy: Ornithischia: Marginocephalia: Ceratopsia: Leptoceratopsidae

[References: Kurzanov SM (1992). A giant protoceratopsid from the Upper Cretaceous of Mongolia. *Paleontological Journal* 26: 103–116.]

Ugrosaurus olsoni

> *Ugrosaurus olsoni* CoBabe & Fastovsky, 1987 [Maastrichtian; Hell Creek Formation, Montana, USA]

Generic name: *Ugrosaurus* ← {(Scandinavian) ugro: ugly + (Gr.) σαῦρος/sauros: lizard}

Specific name: *olsoni* ← {(person's name) Norman Olson + -ī}; in honor of Norman Olson, the owner of the ranch where the specimen was found.

Etymology: Olson's **ugly lizard**

Taxonomy: Ornithischia: Marginocephalia: Ceratopsia: Ceratopsidae: Chasmosaurinae

Notes: According to *The Dinosauria 2nd edition*, *Ugrosaurus olsoni* was regarded as a *nomen dubium*.

[References: CoBabe EA, Fastovsky DE (1987). *Ugrosaurus olsoni,* a new ceratopsian (Reptilia: Ornithischia) from the Hell Creek Formation of eastern Montana. *Journal of Paleontology* 61(1): 148–154.]

Ugrunaaluk kuukpikensis

> *Edmontosaurus regalis* Lambe, 1917 [late Campanian; Horseshoe Canyon Formation, Alberta, Canada]

= *Ugrunaaluk kuukpikensis* Mori *et al*., 2015 [Maastrichtian; Prince Creek Formation, Alaska, USA]

Generic name: *Ugrunaaluk* ← {(Alaskan Iñupiaq) "ugrunnaq, referring to a grazing animal with a long set of grinding teeth" + "-aluk: old"}; "in honor of the Alaskan Native Iñupiaq culture from the area where the type material was discovered" (*1).

Specific name: *kuukpikensis* ← {(Iñupiaq) kuukpik: the Colville River, Alaska + -ensis} (*1)

Etymology: **ancient grazer** from Colville River

Taxonomy: Ornithischia: Iguanodontia: Hadrosauridae: Saurolophinae: Edmontosaurini

Notes: *Ugrunaaluk* is generally regarded as a synonym of *Edmontosaurus*.

[References: (*1) Mori H, Druckenmiller PS, Erickson GM (2016). A new Arctic hadrosaurid from the Prince Creek Formation (lower Maastrichtian) of northern Alaska. *Acta Palaeontologica Polonica* 61(1):15–32.; Takasaki R, Fiorillo AR, Tykoshi RS, Kobayashi Y (2020). Re-examination of the cranial osteology of the Arctic Alaskan hadrosaurine with implications for its taxonomic status. *PLoS ONE* 15(5): e0232410.]

Ultrasauros macintoshi

> *Supersaurus vivianae* Jensen, 1985 [late Kimmeridgian; Morrison Formation, Colorado, USA]

= *Dystylosaurus edwini* Jensen, 1985 [late Kimmeridgian; Morrison Formation, Colorado, USA]

= *Ultrasaurus macintoshi* Jensen, 1985 (*partim*) ⇒ *Ultrasauros macintoshi* (Jensen, 1985) Olshevsky, 1991 [late Kimmeridgian; Morrison Formation, Colorado, USA]

Generic name: *Ultrasauros* ← {(Lat.) ultra + (Gr.) σαῦρος/sauros: lizard}

Specific name: *macintoshi* ← {(person's name) MacIntosh + -ī}

Etymology: MacIntosh's **ultra lizard**

Taxonomy: Saurischia: Sauropodomorpha: Sauropoda: Diplodocoidea: Diplodocidae

Notes: According to *The Dinosauria 2nd edition*, *Ultrasauros macintoshi* Jensen, 1985 (*partim*) is regarded to be a synonym of *Supersaurus vivianae,* and *Dystylosaurus edwini* was valid.

Ultrasaurus tabriensis

> *Ultrasaurus tabriensis* Kim, 1983 [Aptian–Albian; Gugyedong Formation, Gyeongsangbuk-do, South Korea]

Generic name: *Ultrasaurus* ← {(Lat.) ultra + (Gr.) σαῦρος/sauros: lizard}

Specific name: *tabriensis* ← {(place-name) Tabri + -ensis}

Etymology: **ultra lizard** from Tabri

Taxonomy: Saurischia: Sauropodomorpha: Sauropoda

Notes: According to *The Dinosauria 2nd edition*, *Ultrasaurus tabriensis* is regarded a *nomen dubium*.

Umoonasaurus demoscyllus [Plesiosauria]

> *Umoonasaurus demoscyllus* Kear *et al*., 2006 [Aptian; Bulldog Shale Formation, South Australia, Australia]

Generic name: *Umoonasaurus* ← {(Antakirinja*) Umoona: indigenous name for the Coober Pedy** area + (Gr.) σαῦρος/sauros: lizard}

*Australian Aboliginal language.

** (local Aboriginal term) kupa-piti: 'whitefella hole'.

Specific name: *demoscyllus* ← {(Gr.) δῆμος/dēmos: of the people + (Gr. myth.) Σκύλλα/Skylla: a sea monster}; referring to the type locality and purchase of the holotype specimen by public donations (*1).

Etymology: **Umoona (Coober Pedy) lizard**, Scylla of the people

Taxonomy: Plesiosauria: Leptocleididae

Notes: Opalized skeleton of *Umoonasaurus* was discovered from the Andamooka opal fields, west of Coober Pedy. (*1).

[References: (*1) Kear BP, Schroeder NI, Lee MSY (2006). An archaic crested plesiosaur in opal from the Lower Cretaceous high-latitude deposits of Australia. *Biology Letters* 2(4): 615–619.]

Unaysaurus tolentinoi

> *Unaysaurus tolentinoi* Leal *et al*., 2004 [Norian; Caturrita Formation, Paraná Basin, Rio Grande do Sul, Brazil]

Generic name: *Unaysaurus* ← {(Tupi / Tupy) unay [pron.: u-na-hee]: Água Negra, meaning "black water" + (Gr.) σαῦρος/sauros: lizard} (*1)

Specific name: *tolentinoi* ← {(person's name) Tolentino + -ī}; in honor of Mr. Tolentino Flores Marafiga, who discovered the specimen (*1).

Etymology: Tolentino's **Agua Negra (black water) lizard**

Taxonomy: Saurischia: Sauropodomorpha: Unaysauridae

[References: (*1) Leal LA, Azevedo SAK, Kellner AWA, Da Rosa ÁAS (2004). A new early dinosaur (Sauropodomorpha) from the Caturrita Formation (Late Triassic), Paraná Basin, Brazil. *Zootaxa* 690: 1–24.]

Unenlagia comahuensis

> *Unenlagia comahuensis* Novas & Puerta, 1997 [late Turonian; Portezuelo Formation, Neuquén, Argentina]

Generic name: *Unenlagia* ← {(Mapuche) uñen: half + lag: bird} [*1]

Specific name: *comahuensis* ← {(place-name) Comahue: a Mapuche name referring to North-West Patagonia + -ensis} [*1]

Etymology: **half bird** from Comahue

Taxonomy: Theropoda: Dromaeosauridae: Unenlagiinae

cf. Theropoda: Coelurosauria: Maniraptora [*1]

Other species:

> *Unenlagia paynemili* Calvo *et al*., 2004 [late Turonian; Portezuelo Formation, Neuquén, Argentina]

Notes: *Unenlagia* is a medium-sized maniraptoran dinosaur, nearly 2 m long [*1].

Makovicky *et al*. (2005) and Turner *et al*. (2007) consider *Neuquenraptor* Novas & Pol, 2005 to be a junior subjective synonym of *Unenlagia* Novas & Puerta, 1997.

[References: (*1) Novas FE, Puerta PF (1997). New evidence concerning avian origins from the late Cretaceous of Patagonia. *Nature* 387: 390–392.]

Unescoceratops koppelhusae

> *Unescoceratops koppelhusae* Ryan *et al*., 2012 [Campanian; Dinosaur Park Formation, Alberta, Canada]

Generic name: *Unescoceratops* ← {(acronym) UNESCO + ceratops [(Gr.) κέρας/keras: horn + ὤψ/ōps: face]}; "referring to the World Heritage Site designation conferred upon the holotype locality (Dinosaur Provincial Park, Alberta) by the United Nations Educational, Scientific and Cultural Organization" [*1]. The genus name was proposed to be *Dinosaurprovincialparkaceratops*, but it was thought too long.

Specific name: *koppelhusae* ← {(person's name) Koppelhus + -ae}; "in honor of Eva B. Koppelhus, in order to recognize her contributions to vertebrate paleontology and palynology" [*1].

Etymology: Koppelhus' **UNESCO horned face (ceratopsian)**

Taxonomy: Ornithischia: Ceratopsia: Neoceratopsia: Leptoceratopsidae [*1]

[References: (*1) Ryan MJ, Evans DC, Currie PJ, Brown CM, Brinkman D (2012). New leptoceratopsids from the Upper Cretaceous of Alberta, Canada. *Cretaceous Research* 35: 69–80.]

Unquillosaurus ceibalii

> *Unquillosaurus ceibalii* Powell, 1979 [Campanian; Los Blanquitos Formation, Salta, Argentina]

Generic name: *Unquillosaurus* ← {(place name) Unquillo + (Gr.) σαῦρος/sauros: lizard}

Specific name: *ceibalii* ← {(place name) Ceibal + -ī}; referring to El Ceibal, the place closest to the locality.

Etymology: El Ceibal's **Unquillo lizard**

Taxonomy: Theropoda: Coelurosauria: Maniraptora: Dromaeosauridae: Unenlagiinae

Urbacodon itemirensis

> *Urbacodon itemirensis* Averianov & Sues, 2007 [Cenomanian; Dzharakuduk Formation, Itemir, Navoi, Uzbekistan]

Generic name: *Urbacodon* ← {(acronym) URBAC: Uzbekistan, Russia, Britain, America and Canada + (Gr.) ὀδών: tooth}; in reference to "the international joint expeditions to the Kyzylkum Desert" [*1].

Specific name: *itemirensis* ← {(place-name) Itemir: the type locality + -ensis} [*1].

Etymology: **URBAC's tooth** from Itemir

Taxonomy: Theropoda: Troodontidae

[References: (*1) Averianov AO, Sues H-D (2007). A new troodontid (Dinosauria: Theropoda) from the Cenomanian of Uzbekistan, with a review of troodontid records from the territories of the former Soviet Union. *Journal of Vertebrate Paleontology* 27(1): 87–98.]

Utahceratops gettyi

> *Utahceratops gettyi* Sampson *et al*., 2010 [late Campanian; Kaiparowits Formation, Utah, USA]

Generic name: *Utahceratops* ← {(place-name) Utah + ceratops [(Gr.) κέρας/keras: horn + ὤψ/ōps: face]} [*1]

Specific name: *gettyi* ← {(person's name) Getty + -ī}; "in honor of Mike Getty, who discovered the holotype and who has played a pivotal role in the recovery of fossils from GSENM (Grand Staircase-Escalante National Monument)" [*1].

Etymology: Getty's **Utah horned face (ceratopsian)**

Taxonomy: Ornithischia: Marginocephalia: Ceratopsia: Ceratopsidae: Chasmosaurinae [*1]

[References: (*1) Sampson SD, Loewen MA, Farke AA, Roberts EM, Forster CA, Smith JA, Titus AL

(2010). New horned dinosaurs from Utah provide evidence for intracontinental dinosaur endemism. *PLoS ONE* 5(9): e12292.]

Utahraptor ostrommaysorum

> *Utahraptor ostrommaysorum* Kirkland *et al.*, 1993 [Barremian; Yellow Cat Member, Cedar Mountain Formation, Utah, USA]

Generic name: *Utahraptor* ← {(place-name) Utah + (Lat.) raptor: thief, plunderer}; "referring to the occurrence of this formidable predatory dinosaur in Utah, 'Utah's predator'" (*1).

Specific name: *ostrommaysorum* ← {Ostrom & Mays + -ōrum}; "in honor of Dr. John Ostrom of Yale University for his ground breaking research on *Deinonychus* and its relationship to birds: and in honor of Chris Mays, president of Dinamation International Corporation, who in founding the Dinamation International Society set the stage for the research presented herein" (*1).

Etymology: Ostrom & Mays' **Utah's predator**

Taxonomy: Theropoda: Coelurosauria: Maniraptora: Dromaeosauridae

Notes: *Utahraptor* had large curved claws on the second toes.

[References: (*1) Kirkland JI, Gaston R (1993). A large dromaeosaur (Theropoda) from the Lower Cretaceous of Utah. *Hunteria* 2(10): 1–16.]

Utatsusaurus hataii [Ichthyopterygia]

> *Utatsusaurus hataii* Shikama *et al.*, 1978 [Olenekian; Osawa Formation, Minamisanriku-cho, Miyagi, Japan]

Generic name: *Utatsusaurus* ← {(place-name) Utatsu-chō 歌津町: now Minamisanriku-chō, Miyagi Prefecture + (Gr.) σαῦρος/sauros: lizard}

Specific name: *hataii* ← {(person's name) Hatai畑井 + -ī}; in honor of paleontologist, Kotora Hatai (畑井小虎).

Etymology: Hatai's **Utatsu lizard**

Taxonomy: Ichthyopterygia: Utatsusauridae

Notes: *Utatsusaurus* had no dorsal fin. It is considered one of the most primitive types of ichthyosaur. In Japanese it is called "Utatsugyoryū歌津魚竜 or Utatsuryū歌津竜".

[References: Shikama T, Kamei T, Murata M (1977). Early Triassic Ichthyosaurus, *Utatsusaurus hataii* gen. *et* sp. nov. from the Kitakami Massif, Northeast Japan. *Science Reports of the Tohoku University Second Series (Geology)* 48(1-2): 77–97.]

Uteodon aphanoecetes

> *Camptosaurus aphanoecetes* Carpenter & Wilson, 2008 ⇒ *Uteodon aphanoecetes* (Carpenter & Wilson, 2008) McDonald, 2011 [early Tithonian; Morrison Formation, Utah, USA]

Generic name: *Uteodon* ← {(demonym) Ute ["yewt"] (*1): the Native American people who inhabit northeastern Utah + (Gr.) ὀδών/odōn: tooth}

Specific name: *aphanoecetes* ← {*cf.* [(Gr.) ἀφανής/aphanēs: unseen]}; "in reference to the new species having been "hidden in plain sight" because it was on exhibit for over 75 years" (*1).

Etymology: **Ute tooth** hidden in plain sight

Taxonomy: Ornithischia: Ornithopoda: Iguanodontia

[References: (*1) Carpenter K, Wilson Y (2008). A new species of *Camptosaurus* (Ornithopoda: Dinosauria) from the Morrison Formation (Upper Jurassic) of Dinosaur National Monument, Utah, and a biomechanical analysis of its forelimb. *Annals of Carnegie Museum* 76(4): 227–263.]

V

Vagaceratops irvinensis

> *Chasmosaurus irvinensis* Holmes *et al.*, 2001 ⇒ *Vagaceratops irvinensis* (Holmes *et al.*, 2001) Sampson *et al.*, 2010 [late Campanian; Dinosaur Park Formation, Irvine, Alberta, Canada]

Generic name: *Vagaceratops* ← {(Lat.) vagus: wanderer, wandering + ceratops [(Gr.) κέρας/keras: horn + ὤψ/ōps: face]}; "in reference to the occurrence of this clade in the north (Alberta) and south (Utah) of Laramidia during the late Campanian" (*1).

Specific name: *irvinensis* ← {(place-name) Irvine + -ensis}

Etymology: **wandering horned-face (ceratopsian)** from Irvine

Taxonomy: Ornithischia: Marginocephalia: Ceratopsidae: Chasmosaurinae

[References: (*1) Sampson SD, Loewen MA, Farke AA, Roberts EM, Forster CA, Smith JA, Titus AL (2010). New horned dinosaurs from Utah provide evidence for intracontinental dinosaur endemism. *PLoS ONE* 5(9): e12292.]

Vahiny depereti

> *Vahiny depereti* Curry Rogers *et al.*, 2014 [Maastrichtian; Anembalemba Member, Maevarano Formation, Mahajanga, Madagascar]

Generic name: *Vahiny* ← {(Malagasy) vahiny: traveller, visitor, foreigner}; "reflecting the rarity of this taxon in the Mahajanga Basin" (*1).

Specific name: *depereti* ← {(person's name) Depéret + -ī}; "in honor of Charles Depéret, who described the original dinosaur material from Madagascar and presciently recognized the sauropod nature of the Malagasy osteoderms" (*1).

Etymology: Depéret's **traveller**

Taxonomy: Saurischia: Sauropodomorpha: Sauropoda: Neosauropoda: Titanosauria

[References: (*1) Curry Rogers K, Wilson JA (2014). *Vahiny depereti*, gen, *et* sp. nov., a new titanosaur (Dinosauria, Sauropoda) from the Upper Cretaceous Maevarano Formation, Madagascar. *Journal of Vertebrate Paleontology* 34(3): 606–617.]

Valdoraptor oweni

> *Megalosaurus oweni* Lydekker, 1889 ⇒*Altispinax oweni* (Lydekker, 1889) von Huene, 1923 ⇒ *Valdoraptor oweni* (Lydekker, 1889) Olshevsky, 1991 [late Valanginian; Tunbridge Wells Sand Formation, Weald Clay Group, West Sussex, UK]

Generic name: *Valdoraptor* ← {(Lat. place-name) Valdus: Wealden + (Lat.) raptor: plunderer, robber}; referring to the Upper Wealden of Cuckfield, West Sussex, England

Specific name: *oweni* ← {(person's name) Owen + -ī}; in honor of Sir Richard Owen.

Etymology: Owen's **Wealden robber**

Taxonomy: Theropoda: Tetanurae: Ornithomimosauria

Notes: According to Allain *et al.* (2014), *Valdoraptor* was considered to be likely one of the ornithomimosaurs, and possible junior synonym of *Thecocoelurus*.

[References: Olshevsky G (1991). A revision of the parainfraclass Archosauria Cope, 1869, excluding the advanced Crocodylia. *Mesozoic Meanderings* 2 pp. 1–196.]

Valdosaurus canaliculatus

> *Dryosaurus canaliculatus* Galton, 1975 ⇒ *Valdosaurus canaliculatus* (Galton, 1975) Galton, 1977 [late Barremian; Wessex Formation, Wealden Group, Isle of Wight, UK]

= *Camptosaurus valdensis* Lydekker, 1888 [late Barremian; Isle of Wight, UK]

Generic name: *Valdosaurus* ← {(Lat.place-name) Valdus: Wealden + (Gr.) σαῦρος/sauros: lizard}; referring to the Early Cretaceous Wealden deposits.

Specific name: *canaliculatus* ← {(Lat.) canāliculātus; with a small channel}; with reference to the deep anterior intercondylar groove (*1).

Etymology: **Wealden lizard** with a small channel

Taxonomy: Ornithischa: Ornithopoda: Iguanodontia: Dryosauridae

Other species:

> *Valdosaurus nigeriensis* Galton & Taquet, 1982 ⇒ *Elrhazosaurus nigeriensis* (Galton & Taquet, 1982) Galton, 2009 [Aptian; Elrhaz Formation, Agadez, Niger]

Notes: In *The Dinosauria 2nd edition*, *Camptosaurus valdensis* was considered to be a synonym of *Valdosaurus canaliculatus*. *Valdosaurus nigeriensis* was given its own genus, *Elrhazosaurus* Galton, 2009.

[References: (*1) Galton PM (1975). English hypsilophodontid dinosaurs (Reptilia: Ornithischia). *Palaeontology* 18(4): 741–752.]

Vallibonavenatrix cani

> *Vallibonavenatrix cani* Malafaia *et al.*, 2019 [Barremian; Arcillas de Morella Formation, Castellón, Spain]

Generic name: *Vallibonavenatrix* ← {(place-name) Vallibona: the town where the holotype was found + vēnātrīx: huntress}

Specific name: *cani* ← {(person's name) Cano + -ī}; in honor of Juan Cano Forner, who found the holotype specimen.

Etymology: Cano's **huntress from Vallibona**

Taxonomy: Saurischia: Theropoda: Spinosauridae: Spinosaurinae

Variraptor mechinorum

> *Variraptor mechinorum* Le Loeuff & Buffetaut, 1998 [late Campanian–early Maastrichtian; Grès à Reptiles Formation, La Bastide Neuve, Fox-Amphoux, Var, France] (*1)

Generic name: *Variraptor* ← {(place-name) Var: a river and an administrative department + (Lat.) raptor: thief} (*1)

Specific name: *mechinorum* ← {(person's name) Méchin + -ōrum}; "after Patrick and Annie Méchin, who collected the material and kindly presented the holotype to the Musée des Dinosaures, Espéraza" (*1).

Etymology: Mechin couple's **Var thief**

Taxonomy: Theropoda: Maniraptora: Dromaeosauridae

[References: (*1) Le Loeuff J, Buffetaut E (1998). A new dromaeosaurid theropod from the Upper Cretaceous of southern France. *Oryctos* 1: 105–112.]

Vayuraptor nongbualamphuensis

> *Vayuraptor nongbualamphuensis* Samathi *et al.*, 2019 [Barremian; Sao Khua Formation, Nong Bua Lamphu, Thailand]

Generic name: *Vayuraptor* ← {(Sanskrit) Vayu: God of Wind + (Lat.) raptor: thief}; "in reference to its long and slender tibia, which suggest a fast running animal" (*1).

Specific name: *nongbualamphuensis* ← {(place-name) Nong Bua Lamphu Province + -ensis}

Etymology: **Wind thief (raptor)** from Nong Bua Lamphu

Taxonomy: Theropoda: Coelurosauria

[References: (*1) Samathi A, Chanthasit P, Martin

Sander P (2019). Two new basal coelurosaurian theropod dinosaurs from the Lower Cretaceous Sao Khua Formation of Thailand. *Acta Palaeontologica Polonica* 64(2): 239–260.]

Vectaerovenator inopinatus

> *Vectaerovenator inopinatus* Barker *et al.*, 2020 [late Aptian; Lower Greensand, Ferruginous Sands Formation, Isle of Wight, UK]

Generic name: *Vectaerovenator* ← {(Lat. place-name) Vectis: Isle of Wight + (Gr.) ἀήρ/aēr: air + (Lat.) venator: hunter}; "referring to its high degree of skeletal pneumaticity".

Specific name: *inopinatus* ← {(Lat.) inopīnātus: unexpected}; referring to its surprise discovery in the notably dinosaur-poor Lower Greensand strata of the Isle of Wight.

Etymology: unexpected **Isle of Wight air-filled hunter**

Taxonomy: Theropoda: Tetanurae

Vectidraco daisymorrisae [Pterosauria]

> *Vectidraco daisymorrisae* Naish *et al.*, 2013 [late Barremian; Vectis Formation, Wealden Group, Isle of Wight, UK]

Generic name: *Vectidraco* ← {(Lat. place-name) Vectis: Isle of Wight + (Lat.) draco: dragon}

Specific name: *daisymorrisae* ← {(person's name) Daisy Morris [*1]: finder of the holotype + -ae}

Etymology: Daisy Morris' **dragon from the Isle of Wight**

Taxonomy: Pterosauria: Pterodactyloidea: Lophocratia: Azhdarchoidea [*2]

Notes: Authors described that *Vectidraco* was a small toothless pterosaur which featured a crest on its snout.

[References: (*1) Isle of Wight girl Daisy Morris has flying prehistoric beast named after her. BBC. 20 March 2013.; (*2) Naish D, Simpson M, Dyke G (2013). A new small-bodied azhdarchoid pterosaur from the Lower Cretaceous of England and its implications for pterosaur anatomy, diversity and phylogeny. *PLoS ONE* 8(3): e58451.]

Velafrons coahuilensis

> *Velafrons coahuilensis* Gates *et al.*, 2007 [Campanian; Cerro del Pueblo Formation, Coahuila, Mexico]

Generic name: *Velafrons* ← {(Spanish) vela: sail + (Lat.) frons: forehead}; "in reference to the sale-like crest on the forehead of this taxon" [*1].

Specific name: *coahuilensis* ← {(place-name) Coahuila: the Mexican state + -ensis} [*1]

Etymology: **sailed forehead** from Coahuila

Taxonomy: Ornithischia: Ornithopoda: Hadrosauridae: Lambeosaurinae: Lambeosaurini

[References: (*1) Gates TA, Sampson SD, Delgado De Jesús CR, Zanno LE, Eberth D, Hernandez-Rivera R, Aguillón Martínez MC, Kirkland JI (2007). *Velafrons coahuilensis*, a new lambeosaurine hadrosaurid (Dinosauria: Ornithopoda) from the Late Campanian Cerro del Pueblo Formation, Coahuila, Mexico. *Journal of Vertebrate Paleontology* 27(4): 917–930.]

Velocipes guerichi

> *Velocipes guerichi* von Huene, 1932 (*nomen dubium*) [Norian; Lissauer Breccia Formation, Opole, Poland]

Generic name: *Velocipes* ← {(Lat.) veloci- [vēlōx: swift] + pēs: foot}

Specific name: *guerichi* ← {(person's name) Guerich + -ī}; in honor of German geologist, paleontologist, and botanist Georg Julius Ernst Gürich.

Etymology: Gürich's **swift foot**

Taxonomy: Theropoda?

Notes: According to *The Dinosauria 2nd edition*, *V. guerichi* is considered to be a *nomen dubium*.

Velociraptor mongoliensis

> *Velociraptor mongoliensis* Osborn, 1924 [Campanian; Djadokhta Formation, Ömnögovi, Mongolia]

Generic name: *Velociraptor* ← {(Lat.) veloci- [vēlōx: high-speed, swift] + raptor: plunderer, robber} Generic name was applied that it seemed to have been an alert, swift-moving carnivorous dinosaur [*1].

Specific name: *mongoliensis* ← {(place-name) Mongolia + -ensis}

Etymology: **high-speed plunderer** from Mongolia

Taxonomy: Theropoda: Coelurosauria: Dromaeosauridae: Velociraptorinae

Other species:

> *Velociraptor osmolskae* Godefroit *et al.*, 2008 [Campanian; Bayan Mandahu Formation, Inner Mongolia, China]

> *Deinonychus antirrhopus* Ostrom, 1969 ⇒ *Velociraptor antirrhopus* (Ostrom, 1969) Paul, 1988 [Aptian; Cloverly Formation, Montana, USA]

Notes: *Velociraptor* had very large, strongly recurved ungual phalages [*1]. "Locked in combat" specimen is famous. A collapsing sand dune may have buried the two combatants, a carnivorous velociraptor and a plant-eating protoceratops, according to the New York Museum [*2].

[References: (*1) Osborn HF (1924). Three new Theropoda, *Protoceratops* zone, central Mongolia. *American Museum Novitates* 144: 1–12.; (*2) Museum displays fossil of dinosaurs locked in combat. CNN. com. nature. May 17, 2000. By Stenger, R.]

Velocisaurus unicus

> *Velocisaurus unicus* Bonaparte, 1991 [Santonian; Bajo de la Carpa Formation, Neuquén, Argentina]

Generic name: *Velocisaurus* ← {(Lat.) veloci- [vēlōx: swift] + (Gr.) σαῦρος/sauros: lizard}

Specific name: *unicus* ← {(Lat.) ūnicus: unique}; referring to its foot.

Etymology: unique, **swift lizard**

Taxonomy: Theropoda: Abelisauroidea: Noasauridae

Venenosaurus dicrocei

> *Venenosaurus dicrocei* Tidwell *et al.*, 2001 [Barremian; Poison Strip Member, Cedar Mountain Formation, Utah, USA]

Generic name: *Venenosaurus* ← {(Lat.) venēnum: poison + (Gr.) σαῦρος/sauros: reptile}; referring to "the Poison Strip Sandstone Member, Cedar Mountain Formation, from which the type specimen was collected" (*1).

Specific name: *dicrocei* ← {(person's name) DiCroce + -ī}; in honor of "Anthony DiCroce, who discovered the specimen" (*1).

Etymology: DiCroce's **lizard from Poison Strip Member**

Taxonomy: Sauropoda: Titanosauriformes: Brachiosauridae

[References: (*1) Tidwell V, Carpenter K, Meyer S (2001). New titanosauriform (Sauropoda) from the Poison Strip Member of the Cedar Mountain Formation (Lower Cretaceous), Utah. In: Tanke DH, Carpenter K (eds), *Mesozoic Vertebrate Life*. Indiana University Press, Bloomington: 139–165.]

Vescornis hebeiensis [Avialae]

> *Vescornis hebeiensis* Zhang *et al.*, 2004 [Valanginian–Aptian; Huajiying Formation*, Hebei, China] *according to Jin *et al.* (2008)

Generic name: *Vescornis* ← {(Lat.) vescus: small + (Gr.) ὄρνις/ornis: bird}; "indicating the short and undeveloped alular digit and other ungula phalanges of the manus" (*1).

Specific name: *hebeiensis* ← {(Chin. place-name) Hébĕi河北 + -ensis}; "derived from the locality of this bird, the Hebei Province" (*1).

Etymology: **small bird** from Hebei

Taxonomy: Theropoda: Avialae: Enantiornithes: Gobipterygidae

cf. Aves: Enantiornites: Euenantiornithes (*1)

Notes: *Jibeinia* Hou, 1997 and *Vescornis* Zhang *et al.*, 2004 are possibly synonymous, but it is impossible to conclude that they are the same species. (*1) The holotype of *Jibeinia luanhera* is lost.

[References: (*1) Zhang F, Ericson Per GP, Zhou Z (2004). Description of a new enantiornithine bird from the Early Cretaceous of Hebei, northern China. *Canadian Journal of Earth Science* 41: 1097–1107.]

Vespersaurus paranaensis

> *Vespersaurus paranaensis* Langer *et al.*, 2019 [early Late Cretaceous; Rio Paraná Formation, Paraná, Brazil]

Generic name: *Vespersaurus* ← {(Lat.) vesper: evening, west + (Gr.) σαῦρος/sauros: lizard, saurian (*1)}; "in reference to the name of the town, *i.e.* Cruzeiro do Oeste (= "Western Cross"), where the fossils were found" (*1).

Specific name: *paranaensis* ← {(place-name) Paraná + -ensis}; "referring to the Paraná State, of which *Vespersaurus paranaensis* represents the first non-avian dinosaur record" (*1).

Etymology: **western lizard** from Paraná

Taxonomy: Theropoda: Ceratosauria: Abelisauroidea: Noasauridae: Noasaurinae

[References: (*1) Langer MC, de Oliveira Martins N, Manzig PC, de Souza Ferreira G, de Almeida Marsola JC, Fortes E, Lima R, Sant'ana LCF, da Silva Vidal L, da Silva Lorençato RH, Ezcurra MD (2019). A new desert-dwelling dinosaur (Theropoda, Noasaurinae) from the Cretaceous of south Brazil. *Scientific Reports* 9: 9379.]

Veterupristisaurus milneri

> *Veterupristisaurus milneri* Rauhut, 2011 [Kimmeridgian–earliest Tithonian; Middle Dinosaur Member, Tendaguru Formation, Lindi, Tanzania] (*1)

Generic name: *Veterupristisaurus* ← {(Lat.) veterus [vetus: old] + (Gr.) πρίστις/pristis: a large sea monster, often used for sharks + (Gr.) σαῦρος/sauros: lizard}; "in reference to the status of the new taxon as the currently oldest known representative of the shark-toothed lizards, the carcharodontosaurids" (*1).

Specific name: *milneri* ← {(person's name) Milner + -ī}; "in honor of Angela C. Milner, for her many contributions to vertebrate palaeontology, including numerous works on the theropod dinosaurs" (*1).

Etymology: Milner's **old shark lizard**

Taxonomy: Theropoda: Carnosauria: Carcharodontosauridae

[References: (*1) Rauhut OWM (2011). Theropod dinosaurs from the Late Jurassic of Tendaguru (Tanzania). *Palaeontology* 86: 195–239.]

Viavenator exxoni

> *Viavenator exxoni* Filippi *et al.*, 2016 [Santonian;

Bajo de la Carpa Formation, Neuquén, Argentina]

Generic name: *Viavenator* ← {(Lat.) via: road + vēnātor: hunter}

Specific name: *exxoni* ← {(company's name) Exxon Mobil: the oil and gas company + -ī}; "in recognition of Exxonmobil's commitment to the preservation of paleontological heritage of the La Invernada area, Rincón de los Sauces, Neuquén, Patagonia, Argentina" (*1).

Etymology: Exxonmobil's **hunter of the road**

Taxonomy: Theropoda: Ceratosauria: Abelisauridae: Furileusauria (*1)

[References: (*1) Filippi LS, Méndez AH, Juárez Valieri RD, Garrido AC (2016). A new brachyrostran with hypertrophied axial structures reveals an unexpected radiation of latest Cretaceous abelisaurids. *Cretaceous Research* 61: 209–219.]

Vitakridrinda sulaimani

> *Vitakridrinda sulaimani* Malkani, 2006 [Maastrichtian; Vitakri Member, Pab Formation, Sulaiman Foldbelt, Barkhan, Balochistan and Punjab provinces, Pakistan]

Generic name: *Vitakridrinda* ← {(place-name) Vitakri*: the dinosaurs' type locality / Vitakri village + (Urdu / Seraiki) drinda: beast} (*1).

*Vitakri member of upper part of Pab Formation in Alam Kali Kakor locality of Vitakri region.

Specific name: *sulaimani* ← {(place-name) Sulaiman + -ī}; "after the name of Sulaiman Foldbelt which acts as a Cretaceous park for terrestrial ecosystem" (*1).

Etymology: **Vitakri beast** from the Sulaiman Fold Belt

Taxonomy: Theropoda: Ceratosauria: Abelisauridae

Notes: Formally described by M. S. Malkani, *Vitakridrinda* is now regarded as a possible *nomen nudum*. The rostrum, which was thought originally to be referred to *Vitakridrinda*, was reclassified as a new genus *Induszalim* Malkani, 2006.

[References: (*1) Malkani MS (2006). Biodiversity of saurischian dinosaurs from the Latest Cretaceous Park of Pakistan. *Journal of Applied and Emerging Sciences* 1(3):108–140.]

Vitakrisaurus saraiki

> *Vitakrisaurus saraiki* Malkani, 2010 (*nomen dubium*) [Maastrichtian; Vitakri Member, Pab Formation, Pakistan]

Generic name: *Vitakrisaurus* ← {(place-name) Vitakri: host locality / the village Vitakri + (Gr.) σαῦρος/sauros: lizard}

Specific name: *saraiki* ← {(language) Saraiki + -ī}; "in honor of the Saraiki language of host Sulaiman Range and Daman area."

Etymology: Saraiki's **Vitakri lizard**

Taxonomy: Theropoda: ?Noasauridae

Volgatitan simbirskiensis

> *Volgatitan simbirskiensis* Averianov & Efimov, 2018 [Hauterivian; Speetoniceras versicolor ammonoid zone, Volga, Ulyanovsk, Russia]

Generic name: *Volgatitan* ← {(place-name) Volga River + (Gr. myth.) Τιτάν/titan*}

*a member of the second order of divine beings, descended from the primordial deities and preceding the Olympian deities in Greek mythology (*1)

Specific name: *simbirskiensis* ← {(place-name) Simbirsk: the former name of Ulyanovsk city + -ensis} (*1)

Etymology: **Volga titan** from Simbirsk

Taxonomy: Saurischia: Sauropoda: Titanosauriformes: Titanosauria: Lithostrotia

[References: (*1) Averianov A, Efimov V (2018). The oldest titanosaurian sauropod of the Northern Hemisphere. *Biological Communications* 63(3): 145–162.]

Volkheimeria chubutensis

> *Volkheimeria chubutensis* Bonaparte, 1979 [late Toarcian; Cañadón Asfalto Formation, Chubut, Argentina]

Generic name: *Volkheimeria* ← {(person's name) Volkheimer + -ia}; in honor of Argentine paleontologist Wolfgang Volkheimer.

Specific name: *chubutensis* ← {(place-name) Chubut + -ensis}

Etymology: **for Volkheimer** from Chubut

Taxonomy: Saurischia: Sauropodomorpha: Sauropoda: Gravisauria

[References: Bonaparte JF (1979). Dinosaurs: a Jurassic assemblage from Patagonia. *Science* 205(4413): 1377–1379.]

Vorona berivotrensis [Avialae]

> *Vorona berivotrensis* Forster *et al.*, 1996 [Maastrichtian; Maevarano Formation, Mahajanga Basin, Madagascar]

Generic name: *Vorona* ← {(Malagasy) vorona [pron.: voo-roo-na]: bird} (*1).

Specific name: *berivotrensis* ← {(place-name) Berivotra: a village in Mahajanga + -ensis} (*1).

Etymology: **bird** from Berivotra

Taxonomy: Theropoda: Avialae: Ornithothoraces: Euornithes

Notes: Vorona had a sickle claw on each foot.

[References: (*1) Forster CA, Chiappe LM, Krause DW, Sampson SD (1996). The first Cretaceous bird from Madagascar. *Nature* 382: 532–534.]

Vouivria damparisensis

> *Vouivria damparisensis* Mannion *et al*., 2017 [middle Oxfordian; Calcaires de Clerval Formation, Damparis, Jura, Franche-Comté, France]

Generic name: *Vouivria* ← {(Old French) vouivre: the wyvern, a legendary winged reptile [← (Lat.) vīpera: viper, adder] + -ia}. "In the homonym novel by Marcel Aymé, 'La Vouivre' is a beautiful woman who lives in the swamps in the neighbourhood of Dôle (Franche-Compté) and protects a spectacular ruby" (*1).

Specific name: *damparisensis* ← {(place-name) Damparis: the type locality + -ensis} (*1).

Etymology: **wyvern (winged reptile)** from Damparis

Taxonomy: Saurischia: Sauropodomorpha: Titanosauriformes: Brachiosauridae

[References: (*1) Mannion PD, Allain R, Moine O (2017). The earliest known titanosauriform sauropod dinosaur and the evolution of Brachiosauridae. *PeerJ* 5: e3217.]

Vulcanodon karibaensis

> *Vulcanodon karibaensis* Raath, 1972 [Sinemurian–Pliensbachian*(*1); Forest Sandstone Formation, Mashonaland North, Rhodesia (now Zimbabwe)]

*Viglitti *et al*. (2018) described that *Vulcanodon karibaensis* is from the uppermost Forest Sandstone. The stratigraphic position is older than previously thought. (*1)

Generic name: *Vulcanodon* ← {(Lat. myth.) Vulcānus: the Roman god of fire + (Gr.) ὀδών/odōn [ὀδούς/odūs]: tooth}; referring to the fact that the partial skeleton was thought to have been interbedded sandstones of the Batoka Basalt Formation. (*1) (*2) According to Cooper (1984), the teeth belonged to an unidentified theropod.

Specific name: *karibaensis* ← {(place-name) Kariba: Lake Kariba + -ensis}; referring to the place of discovery on a small island (Sibilobilo area) in Lake Kariba.

Etymology: **Vulcanus tooth** from Lake Kariba

Taxonomy: Saurischia: Sauropodomorpha: Sauropoda: Gravisauria: Vulcanodontidae
cf. Saurischia: Sauropodomorpha: Sauropoda (according to *The Dinosauria 2nd edition*)

Notes: *Vulcanodon* is the most primitive known sauropod.

[References: (*1) Viglietti PA, Barrett PM, Broderick TJ, Munyikwa D, MacNiven R, Broderick L, Chapelle K, Glynn D, Edwards S, Zondo M, Broderick P, Choiniere JN (2018). Stratigraphy of the *Vulcanodon* type locality and its implications for regional correlations within the Karoo Supergroup. *Journal of African Earth Sciences* 137: 149–156.; (*2) Bond G, Wilson JF, Raath MA (1970). Upper Karroo pillow lava and a new sauropod horizon in Rhodesia. *Nature* 227: 1339.]

W

Wakinosaurus satoi

> *Wakinosaurus satoi* Okazaki, 1992 [Hauterivian; Sengoku Formation, Kwanmon Group, Fukuoka, Japan]

Generic name: *Wakinosaurus* ← {(Jpn. place-name) Wakino脇野: the sub-group + σαῦρος/sauros: lizard}

Specific name: *satoi* ← {(Jpn. person's name) Satō 佐藤 + -ī}; in honor of Masahiro Sato (佐藤正弘) who found the tooth of theropod in1990.

Etymology: Sato's **Wakino lizard**

Taxonomy: Theropoda

Notes: In 1990, a single damaged tooth (57.4mm) with serrations was found (*1). According to *The Dinosauria 2nd edition*, *W. satoi* is considered to be a *nomen dubium*. The nickname of *Wakinosaurus* is Wakinosatoiryū.

[References: (*1) Okazaki Y (1992). A new genus and species of carnivorous dinosaur from the Lower Cretaceous Kwanmon Group, northern Kyushu. *Bulletin of the Kitakyushu Museum of Natural History* 11: 87–90.]

Walgettosuchus woodwardi

> *Walgettosuchus woodwardi* von Huene, 1932 (*nomen dubium*) [Cenomanian; Griman Creek Formation, Walgett, New South Wales, Australia]

Generic name: *Walgettosuchus* ← {(place-name) Walgett + (Gr.) Σοῦχος/Soukhos: Egyptian crocodile god Sobek}

Specific name: *woodwardi* ← {(person's name Woodward + -ī)}; in honor of English paleontologist Sir Arthur Smith Woodward.

Etymology: Woodward's **Walgett crocodile god Sobek**

Taxonomy: Saurischia: Theropoda: Tetanurae

Notes: An opalised vertebra was discovered in 1905 at Lightning Ridge near Walgett.

Wamweracaudia keranjei

> *Wamweracaudia keranjei* Mannion *et al*., 2019 [Tithonian; Tendaguru Formation, Lindi, Tanzania]

Generic name: *Wamweracaudia* ← {(demonym) Wamwera: a tribe in the Lindi-region, Tanzania + (Lat.) cauda: tail}

Specific name: *keranjei* ← {(person's name) Keranje + -ī}; in honor of Mohammadi Keranje, the chief excavator of this specimen.

Etymology: Keranje and **Wamwera's tail**

Taxonomy: Saurischia: Sauropodomorpha: Sauropoda: Mamenchisauridae

[References: Mannion PD, Upchurch P, Schwarz D, Wings O (2019). Taxonomic affinities of the putative titanosaurs from the Late Jurassic Tendaguru Formation of Tanzania: phylogenetic and biogeographic implications for eusauropod dinosaur evolution. *Zoological Journal of the Linnean Society* 185(3): 784–909.]

Wannanosaurus yansiensis

> *Wannanosaurus yansiensis* Hou, 1977 [early Maastrichtian; Xiaoyan Formation, Yansi, Shexian, Anhui, China]

Generic name: *Wannanosaurus* ← {(Chin. place-name) Wǎnnán皖南* + (Gr.) σαῦρος/sauros: lizard}

*皖 [Pinyin: wǎn]: the abbreviation for 安徽 [Pinyin: Ānhuī]

Specific name: *yansiensis* ← {(Chin. place name) Yánsì巌寺 + -ensis}

Etymology: **Wannan lizard** from Yansi [巌寺皖南龍]

Taxonomy: Ornithischia: Pachycephalosauria: Pachycephalosauridae (*1)

Notes: Estimated to be 60 cm in total length, the individual was an adult at death.

[References: (*1) Hou L (1977). A primitive pachycephalosaurid from the Cretaceous of Anhui, China, *Wannanosaurus yansiensis* gen. *et* sp. nov. *Vertebrata PalAsiatica* 15(3): 198–202.]

Weewarrasaurus pobeni

> *Weewarrasaurus pobeni* Bell *et al.*, 2018 [Cenomanian; Griman Creek Formation, New South Wales, Australia]

Generic name: *Weewarrasaurus* ← {(place-name) Wee Warra: the locality + (Gr.) σαῦρος/sauros: lizard}

Specific name: *pobeni* ← {(person's name) Poben + -ī}; "in recognition of Mike Poben who acquired and donated the holotype" (*1).

Etymology: Poben's **Wee Warra lizard**

Taxonomy: Ornithischia: Ornithopoda

Notes: The type specimen was found at the Wee Warra opal mine near Lightning Ridge, New South Wales. The jawbone was preserved in opal.

[References: (*1) Bell PR, Herne MC, Brougham T, Smith ET (2018). Ornithopod diversity in the Griman Creek Formation (Cenomanian), New South Wales, Australia. *PeerJ* 6: e6008.]

Wellnhoferia grandis [Avialae]

> *Wellnhoferia grandis* Elżanowski, 2001 [early Tithonian; upper Solnhofen Lithographic Limestone, Bayern, Germany]

Generic name: *Wellnhoferia* ← {(person's name) Wellnhofer + -ia}; "in honor of Dr. Peter Wellnhofer, Chief Curator Emeritus, Bayerische Staatssammlung für Paläontologie und historische Geologie, Munich" (*1).

Specific name: *grandis* ← {(Lat.) grandis: big, great}

Etymology: big, **Wellnhofer's one**

Taxonomy: Theropoda: Avialae: Archaeopterygidae *cf.* Aves: Archaeopterygidae (*1)

[References: (*1) Elżanowski A (2001). A new genus and species for the largest specimen of *Archaeopteryx*. *Acta Palaeontologica Polonica* 46(4): 519–532.]

Wendiceratops pinhornensis

> *Wendiceratops pinhornensis* Evans & Ryan, 2015 [middle Campanian; Oldman Formation, Alberta, Canada]

Generic name: *Wendiceratops* ← {(person's name) Wendy + ceratops [(Gr.) κέρας/ceras: horn + ὤψ/ōps: face]}; "in honor of Wendy Sloboda, who discovered the type locality, combined with ceratops (horned-face) from the Greek, a common suffix for horned dinosaur generic names" (*1).

Specific name: *pinhornensis* ← {(place-name) Pinhorn + -ensis}; "referring to the Pinhorn Provincial Grazing Reserve in Alberta, Canada, where the type locality is located" (*1).

Etymology: **Wendy's horned-face (ceratopsian)** from Pinhorn Reserve

Taxonomy: Ornithischia: Ceratopsidae: Centrosaurinae

[References: (*1) Evans DC, Ryan MJ (2015). Cranial anatomy of *Wendiceratops pinhornensis* gen. *et* sp. nov., a centrosaurine ceratopsid (Dinosauria: Ornithischia) from the Oldman Formation (Campanian), Alberta, Canada, and the evolution of ceratopsid nasal ornamentation. *PLoS ONE* 10(7): e0130007.]

Wiehenvenator albati

> *Wiehenvenator albati* Rauhut *et al.*, 2016 [Callovian; Ornatenton Formation, Minden, Northrhine-Westphalia, Germany]

Generic name: *Wiehenvenator* ← {(Ger. place-name) the Wiehengebirge: a chain of hills south of Minden + (Lat.) vēnātor: hunter}. The gender of this genus is masculine. (*1)

Specific name: *albati* ← {(person's name) Albat + -ī}; "in honor of Friedrich Albat, who found the holotype specimen" (*1).

Etymology: Albat's **Wiehengebirge hunter**

Taxonomy: Theropoda: Megalosauroidea: Megalosauridae

[References: (*1) Rauhut OWM, Hübner TR, Lanser K-P (2016). A new megalosaurid theropod dinosaur from the late Middle Jurassic (Callovian) of north-western Germany: Implications for theropod evolution and faunal turnover in the Jurassic. *Palaeontologia Electronica* 19(2)26A: 1–165.]

Willinakaqe salitralensis

> *Willinakaqe salitralensis* Juárez Valieri *et al.*, 2010 [late Campanian; lower member, Allen Formation, Río Negro, Argentina]

Generic name: *Willinakaqe* ← {(Mapuche) willi: South + iná: mimic + kaqe: duck};"the duck-mimic of the South" (*1).

Specific name: *salitralensis* ← {(place-name) Salitral Moreno locality + -ensis} (*1).

Etymology: **duck-mimic of the South** from Salitral Moreno

Taxonomy: Ornithischa: Ornithopoda:
  Hadrosauridae: Saurolophinae
   cf. Ornithopoda: Ankylopollexia:
    Hadrosauroidea: Saurolophidae (*1)

Notes: According to Caballero & Coria (2016), *Willinakaqe salitralensis* is considered to be a *nomen vanum* (empty name). The holotype is too weathered.

[References: (*1) Juárez Valieri RD, Haro JA, Fiorelli LE, Calvo JO (2010). A new hadrosauroid (Dinosauria: Ornithopoda) from the Allen Formation (Late Cretaceous) of Patagonia, Argentina. *Revista del Museo Argentino de Ciencias Naturales neuvo serie* 12 (2): 217–231.]

Wintonotitan wattsi

> *Wintonotitan wattsi* Hocknull *et al.*, 2009 [latest Albian; Winton Formation, Queensland, Australia]

Generic name: *Wintonotitan* ← {(place-name) Winton: the town of Winton + (Gr. myth) Τῑτάν/Tītan: Giant} (*1)

Wintonotitan is known from the Winton Formation of central-western Queensland, Australia (*2). It was discovered from "Triangle Paddock", Elderslie Station (a pastoral lease), approximately 60 km north-west of Winton.

Specific name: *wattsi* ← {(person's name) Watts + -ī}; in honor of "Keith Watts who discovered the type specimen and donated it to the Queensland Museum in 1974" (*1).

Etymology: Watts' **giant from Winton**

Taxonomy: Saurischia: Sauropoda:
  Titanosauriformes

Notes: Estimated to be about 15 m in total length.

[References: (*1) Hocknull SA, White MA, Tischler TR, Cook AG, Calleja ND, Sloan T, Elliott DA (2009). New mid-Cretaceous (latest Albian) dinosaurs from Winton, Queensland, Australia. *PLoS ONE* 4(7): e6190.; (*2) *Wintonotitan wattsi* 〈https://australian.museum/learn/dinosaurs/fact-sheets/wintonotitan-wattsi/〉]

Wuerhosaurus homheni

> *Wuerhosaurus homheni* Dong, 1973 ⇒ *Stegosaurus homheni* (Dong, 1973) Maidment *et al.*, 2008 [late Aptian; Lianmugin Formation, Tugulu Group, Xinjiang, China]

Generic name: *Wuerhosaurus* ← {(place-name) Wuerho烏爾禾 + (Gr.) σαῦρος/sauros: lizard}

Specific name: *homheni* ← {homheni "flat and wide"}, in reference to the sacral region of this stegosaur.

Etymology: flat and wide, **lizard from Wuerho** [平坦烏爾禾龍]

Taxonomy: Ornithischia: Thyreophora: Stegosauria:
  Stegosauridae

Other species:

> *Wuerhosaurus ordosensis* Dong, 1993 (*nomen dubium*) [Valanginian; Ejinhoro Formation, Inner Mongolia, China]

> *Wuerhosaurus mongoliensis* Ulansky, 2014 *vide* Galton & Carpenter, 2016 (*nomen nudum*) ⇒ *Mongolostegus exspectabilis* Tumanova & Alifanov, 2018 [Aptian–Albian; Dzunbain Formation, Dornogovi, Mongolia]

Notes: Tumanova & Alifanov (2018) formally named *Wuerhosaurus* "mongoliensis" *Mongolostegus exspectabilis*.

[References: Maidment SCR, Norman DB, Barrett PM, Upchurch P (2008). Systematics and phylogeny of Stegosauria (Dinosauria: Ornithischia). *Journal of Systematic Palaeontology* 6(4): 367–407.]

Wulagasaurus dongi

> *Wulagasaurus dongi* Godefroit *et al.*, 2008 [late Maastrichtian; Yuliangze Formation, Wulaga, Heilongjiang, China]

Generic name: *Wulagasaurus* ← {(place-name) Wulaga: the type locality + (Gr.) σαῦρος/sauros: lizard} (*1)

Specific name: *dongi* ← {(Chin.person's name) Dǒng + -ī}; "in honor of Dong Zhi-Ming, one of the most famous dinosaur specialists, for his fundamental contribution to the knowledge of dinosaurs in China" (*1).

Etymology: Dong's **Wulaga lizard**

Taxonomy: Ornithischia: Ornithopoda:
  Hadrosauridae: Saurolophinae
   cf. Ornithischia: Ornithopoda:
    Hadrosauridae: Hadrosaurinae (*1)

[References: (*1) Godefroit P, Hai S, Yu T, Lauters P (2008). New hadrosaurid dinosaurs from the uppermost Cretaceous of northeastern China. *Acta Palaeontologica Polonica* 53(1): 47–74.]

Wulatelong gobiensis

> *Wulatelong gobiensis* Xu *et al.*, 2013 [Campanian; Wulansuhai Formation (or Bayan Mandahu Formation), Linhe, Inner Mongolia, China]

Generic name: *Wulatelong* ← {(place-name) Wulate 烏拉特 [pron.: Woo-la-tuh] (*1) = Urad + (Chin.) lóng龍: dragon} (*1)

Specific name: *gobiensis* ← {(place-name) the Gobi desert + -ensis} (*1)

Etymology: **Wulate dragon** from Gobi

Taxonomy: Theropoda: Maniraptora: Oviraptorosauria: Oviraptoridae

[References: (*1) Xu X, Tan Q-W, Wang S, Sullivan C, Hone DWE, Han F-L, Ma Q-Y, Tan L, Xiao D (2013). A new oviraptorid from the Upper Cretaceous of Nei Mongol, China, and its stratigraphic implications. *Vertebrata PalAsiatica* 51(2): 85–101.]

Wulong bohaiensis

> *Wulong bohaiensis* Poust *et al.*, 2020 [Aptian; Jiufotang Formation, Liaoning, China]

Generic name: *Wulong* ← {(Chin.) wǔ舞: dance + lóng龍: dragon}; "for the individual's sprightly pose and inferred nimble habits" (*1).

Specific name: *bohaiensis* ← {(place-name) Bóhǎi 渤海 + -ensis}; "in honor of its accession in the collections of the Dalian Natural History Museum (DNHM) situated on the shore of the Bohai strait" (*1).

Etymology: **dancing dragon** from Bohai

Taxonomy: Saurischia: Theropoda: Paraves: Dromaeosauridae (*1)

[References: (*1) Poust AW, Gao C, Varricchio DJ, Wu J, Zhang F (2020). A new microraptorine theropod from the Jehol Biota and growth in early dromaeosaurids. *The Anatomical Record. American Association for Anatomy* 303(4): 963–987.]

X

Xenoceratops foremostensis

> *Xenoceratops foremostensis* Ryan *et al.*, 2012 [middle Campanian; Foremost Formaton, Foremost, Alberta, Canada]

Generic name: *Xenoceratops* ← {(Gr.) ξένος/xenos: foreign, alien + ceratops [κέρας/ keras: horn + ὤψ/ōps: face}; "referring to the lack of ceratopsian material known from the Foremost Formation" (*1).

Specific name: *foremostensis* ← {(place-name) the Village of Foremost, Alberta + -ensis}

Etymology: **foreign horned face (ceratopsian)** from Foremost

Taxonomy: Ornithischia: Ceratopsia: Ceratopsidae: Centrosaurinae

cf. Ornithischia: Ceratopsia: Neoceratopsia: Ceratopsidae: Chasmosaurinae (*1)

[References: (*1) Ryan MJ, Evans DC, Shepherd KM (2012). A new ceratopsid from the Foremost Formation (middle Campanian) of Alberta. *Canadian Journal of Earth Sciences,* 49: 1251–1262.]

Xenoposeidon proneneukos

> *Xenoposeidon proneneukos* Taylor & Naish, 2007 [late Berriasian; Ashdown Formation, Wealden Group, East Sussex, UK]

Generic name: *Xenoposeidon* ← {(Gr.) ξένος/xenos: strange, alien + (Gr. myth.) Ποσειδῶν/Poseidōn: the god of earthquakes and the sea}; "in reference to the sauropod *Sauroposeidon* Wedel, Cifelli & Sanders, 2000" (*1).

Specific name: *proneneukos* ← {(Lat.) pronus: forward sloping}; "describing the characteristic morphology of the neural arch" (*1). "Neural arch slopes anteriorly 35 degrees relative to the vertical" (*1).

Etymology: **strange Poseidon** with forward-sloping neural arch

Taxonomy: Saurischia: Sauropodomorpha: Sauropoda: Neosauropoda: Diplodocoidea: Rebbachisauridae

[References: (*1) Taylor MP, Naish D (2007). An unusual new neosauropod dinosaur from the Lower Cretaceous Hastings beds group of East Sussex, England. *Palaeontology*, 50(6): 1547–1564.]

Xenotarsosaurus bonapartei

> *Xenotarsosaurus bonapartei* Martínez *et al.*, 1986 [late Cenomanian; lower Bajo Barreal Formation, Chubut, Argentina]

Generic name: *Xenotarsosaurus* ← {(Gr.) ξένος/xenos: strange, peculiar + (Lat.) tarsus = (Gr.) ταρσος/tarsos: ankle and heel + (Gr.) σαῦρος/sauros: lizard}; "allude a la peculiar conformación del astrágalo-calcáneo y su relación con la tibia" (*1).

Specific name: *bonapartei* ← {(person's name) Bonaparte + -ī}; "en homenaje al Dr. José Fernando Bonaparte, eminente estudiosa de los vertebrados mes ozoicos" (*1).

Etymology: Bonaparte's **strange-ankled lizard**

Taxonomy: Theropoda: Ceratosauria: Abelisauridae (*1)

[References: (*1) Martínez RD, Giménez O, Rodríguez J, Bochatey G (1986). *Xenotarsosaurus bonapartei* nov. gen. *et* sp. (Carnosauria, Abelisauridae), un nuevo Theropoda de la Formación Bajo Barreal, Chubut, Argentina. IV *Congreso Argentino de Paleontologia y Bioestratigrafia*: 23–31.]

Xiangornis shenmi [Avialae]

> *Xiangornis shenmi* Hu *et al.*, 2012 [Aptian;

Jiufotang Formation, Liaoning, China]

Generic name: *Xiangornis* ← {(Chin.) xiáng翔: free flight + (Gr.) ὄρνις/ornis: bird} (*1)

Specific name: *shenmi* ← {(Chin.) shénmì神秘: mysterious} (*1)

Etymology: mysterious, **free flight bird** [神秘翔鳥]

Taxonomy: Aves: Enantiornithes (*1)

[References: (*1) Hu D, Xu X, Hou L, Sullivan C (2012). A new enantiornithine bird from the Lower Cretaceous of western Liaoning, China, and its implications for early avian evolution. *Journal of Vertebrate Paleontology* 32(3): 639–645.]

Xianshanosaurus shijiagouensis

> *Xianshanosaurus shijiagouensis* Lü *et al.*, 2009 [Cenomanian; Mangchuan Formation, Henan, China]

Generic name: *Xianshanosaurus* ← {(Chin. place name) Xiànshān峴山 + (Gr.) σαῦρος/sauros: lizard}

Specific name: *shijiagouensis* ← {(Chin. place name) Shǐjiāgōu 史家溝 + -ensis}

Etymology: **Xianshan lizard** from Shijiangou

Taxonomy: Saurischia: Sauropodomorpha: Sauropoda: Titanosauria

Notes: The Haoling Formation (originally part of the Mangchuan Formation) was initially thought to be Cenomanian in age (*1), but is now considered Aptian–Albian (*2).

[References: (*1) Lü J, Xu L, Jiang X, Jia S, Li M, Yuan C, Zhang X, Ji Q (2009). A preliminary report on the new dinosaurian fauna from the Cretaceous of the Ruyang Basin, Henan Province of central China. *Journal of the Paleontological Society of Korea* 25(1): 43–56.]; (*2) Xu L, Pan ZC, Wang ZH, Zhang XL, Jia SH, Lü JC, Jiang BL (2012). Discovery and significance of the Cretaceous system in Ruyang Basin, Henan Province. *Geological Review* 58: 601–613.]

Xiaosaurus dashanpensis

> *Xiaosaurus dashanpensis* Dong & Tang, 1983 [Bajocian; lower Shaximiao Formation, Sichuan, China]

Generic name: *Xiaosaurus* ← {(Chin.) xiǎo曉: dawn + (Gr.) σαῦρος/sauros: lizard}; referring to the age of the fossil.

Specific name: *dashanpensis* ← {(Chin. place-name) Dàshānpū大山鋪 + -ensis}

Etymology: **dawn lizard** from Dashanpu [大山鋪曉龍]

Taxonomy: Ornithischia: Neornithischia

Notes: According to the *Dinosauria 2nd edition*, *Xiaosaurus dashanpensis* is considered a *nomen dubium*. However, Barrett *et al.* concluded it to be provisionally valid (*1).

[References: (*1) Barrett PM, Butler RJ, Knoll F (2005). Small-bodied ornithischian dinosaurs from the Middle Jurassic of Sichuan, China. *Journal of Vertebrate Paleontology* 25(4): 823–834.]

Xiaotingia zhengi [Anchiornithidae]

> *Xiaotingia zhengi* Xu *et al.*, 2011 [late Bathonian; Tiaojishan Formation, Liaoning, China]

Generic name: *Xiaotingia* ← {(Chin. person's name) Xiǎotíng曉廷 + -ia}; "in honor of Zheng Xiaoting for his efforts in establishing the Shandong Tianyu Museum of Nature as a repository for vertebrate fossils from China" (*1).

Specific name: *zhengi* ← {(person's name) Zhèng鄭 + -ī}

Etymology: **for** Zheng **Xiaoting** [鄭氏曉廷龍]

Taxonomy: Theropoda: Avialae (?): Anchiornithidae
cf. Theropoda: Coelurosauria: Archaeopterygidae (*1)

[References: (*1) Xu X, You H, Du K, Han F (2011). An archaeopteryx-like theropod from China and the origin of Avialae. *Nature* 475 (7357): 465–470.]

Xingtianosaurus ganqi

> *Xingtianosaurus ganqi* Qiu *et al.*, 2019 [Early Cretaceous; Yixian Formation, Liaoning, China]

Generic name: *Xingtianosaurus* ← {(Chin.) Xíngtiān 刑天: "Chinese deity recorded in *Shānhǎijīng* 山海経 who continued to fight even after his head had been cut off" (*1) + (Gr.) σαῦρος/sauros: lizard}; "referring to the skull-less holotype" (*1).

Specific name: *ganqi* ← {(Chin.) Gānqī干戚: "the weapon of Xingtian recorded in Shanhaijing"}

Etymology: weaponed **Xingtian lizard** [干戚刑天龍]

Taxonomy: Theropoda: Oviraptorosauria: Caudipteridae

[References: (*1) Qiu R, Wang X, Wang Q, Li N, Zhang J, Ma Y (2019). A new caudipterid from the Lower Cretaceous of China with information on the evolution of the manus of Oviraptorosauria. *Scientific Reports* 9(6431): 1–10.]

Xingxiulong chengi

> *Xingxiulong chengi* Wang *et al.*, 2017 [Hettangian; Shawan Member, Lufeng Formation, Yunnan, China]

Generic name: *Xingxiulong* ← {(Chin.) Xīngxiù星宿: 'constellation' + lóng龍: dragon}; "derived from the name of the ancient 'Xingxiu Bridge' in Lufeng County, which was built during the Ming Dynasty (1368–1644)" (*1).

Specific name: *chengi* ← {(Chin. person's name) Chéng程 + -i}; "dedicated to Prof. Zheng-Wu Cheng (1931–2015), for his lifetime contribution to Chinese terrestrial biostratigraphy, including the Lufeng Basin" (*1).

Etymology: Cheng's **Xingxiu dragon**

Taxonomy: Saurischia: Sauropodomorpha: Sauropodiformes [*1]

[References: (*1) Wang Y-M, You H-L, Wang T (2017). A new basal sauropodiform dinosaur from the Lower Jurassic of Yunnan Province, China. *Scientific Reports* 7: 41881.]

Xinjiangovenator parvus

> *Xinjiangovenator parvus* Rauhut & Xu, 2005 [late Aptian; Lianmugin Formation, Tugulu Group, Xinjiang, China]

Generic name: *Xinjiangovenator* ← {(Chin. place-name) Xīnjiāng新疆: Xinjiang Uygur Autonomous Region + (Lat.) vēnātor: hunter} [*1]

Specific name: *parvus* ← {(Lat.) parvus: small}; "referring to the small size of the specimen" [*1].

Etymology: small **hunter from Xinjiang**

Taxonomy: Theropoda: Coelurosauria: Maniraptora

[References: (*1) Rauhut OWM, Xu X (2005). The small theropod dinosaurs *Tugulusaurus* and *Phaedrolosaurus* from the early Cretaceous of Xinjiang, China. *Journal of Vertebrate Paleontology* 25(1): 107–118.]

Xinjiangtitan shanshanesis

> *Xinjiangtitan shanshanesis* Wu *et al.*, 2013 [Middle Jurassic; Qiketai Formation, Shanshan, Xinjiang, China]

Generic name: *Xinjiangtitan* ← {(Chin. place-name) Xīnjiāng新疆 + (Gr. myth.) Τιτάν/Tītan: giant} [*1].

Specific name: *shanshanesis* [*sic*] ← {(Chin. place-name) Shànshàn鄯善 + -e(n)sis} [*1].

Etymology: **Xinjiang Titan** from Shanshan

Taxonomy: Saurischia: Sauropodomorpha: Sauropoda: Mamenchisauridae

[References: (*1) Wu W-H, Zhou C-F, Wings O, Sekiya T, Dong Z-M (2013). A new gigantic sauropod dinosaur from the Middle Jurassic of Shanshan, Xinjiang. *Global Geology* 32(3): 437–446.]

Xiongguanlong baimoensis

> *Xiongguanlong baimoensis* Li *et al.*, 2010 [Aptian; Xinminbao Group, White Ghost Castle area, Yujingzi Basin, Gansu, China]

Generic name: *Xiongguanlong* ← {(Chin, place-name) Xióngguān雄關: Grand Pass, a historic name for the nearby city of Jiayuguan嘉峪關 + (Chin.) lóng龍: dragon} [*1].

Specific name: *baimoensis* ← {(Chin.) báimó白魔: White Ghost + -ensis}; "in reference to a prominent topographic feature in the field area dubbed the White Ghost Castle" [*1], a rock formation near the fossil site.

Etymology: **Xiongguan (Grand Pass) dragon** from White Ghost Castle [白魔雄關龍]

Taxonomy: Theropoda: Coelurosauria: Tyrannosauroidea

[References: (*1) Li D, Norell MA, Gao K-Q, Smith ND, Makovicky PJ (2009). A longirostrine tyrannosauroid from the Early Cretaceous of China. *Proceedings of the Royal Society B: Biological Sciences* 277 (1679): 183–190.]

Xixianykus zhangi

> *Xixianykus zhangi* Xu *et al.*, 2010 [late Coniacian; Majiacun Formation, Xixia, Henan, China]

Generic name: *Xixianykus* ← {(Chin. place-name) Xīxiá西峡 + (Gr.) ὄνυξ/onyx: claw} [*1].

Specific name: *zhangi* ← {(Chin. person's name) Zhāng張 + -ī}; "in honor of Prof. Zhang Wentang, who has contributed greatly to the study of paleontology in Henan Province" [*1].

Etymology: Zhang's **Xixia claw**

Taxonomy: Theropoda: Alvarezsauridae: Parvicursorinae [*1]

Notes: Estimated to be 0.5 m in total length, *Xixianykus* is one of smallest dinosaurs.

[References: (*1) Xu X, Wang D-Y, Sullivan C, Hone DWE, Han F-L, Yan R-H, Du F-M (2010). A basal parvicursorine (Theropoda: Alvarezsauridae) from the Upper Cretaceous of China. *Zootaxa* 2413: 1–19.]

Xixiasaurus henanensis

> *Xixiasaurus henanensis* Lü, *et al.*, 2010 [Coniacian; Majiacun Formation, Henan, China] [*1]

Generic name: *Xixiasaurus* ← {(Chin. place-name) Xīxiá西峡 + (Gr.) σαῦρος/sauros: lizard}; "referring to the theropod dinosaur found in the Chinese administrative unit Xixia County of Henan Province" [*1].

Specific name: *henanensis* ← {(Chin, place-name) Hénán河南 + -ensis}; "referring to Henan Province, in which the holotype site in Xixia County is found" [*1].

Etymology: **Xixia lizard** from Henan Province

Taxonomy: Theropoda: Tetanurae: Maniraptora: Troodontidae [*1]

[References: (*1) Lü J, Xu L, Liu Y, Zhang X, Jia S, Ji Q (2010). A new troodontid theropod from the Late Cretaceous of central China, and the radiation of Asian troodontids. *Acta Palaeontologica Polonica* 55(3): 381–388.]

Xixiposaurus suni

> *Xixiposaurus suni* Sekiya, 2010 [Hettangian; Lufeng Formation, Yunnan, China]

Generic name: *Xixiposaurus* ← {(Chin. place-name)

Xìxìpō細細坡 + (Gr.) σαῦρος/sauros: lizard}
Specific name: *suni* ← {(Chin. person's name) Sūn 孫 + -ī}; in honor of Professor Sūn Gé孫革 of Jiling University.
Etymology: Sun's **lizard from Xixipo Village** [孫氏細細坡龍]
Taxonomy: Saurischia: Sauropodomorpha: Prosauropoda
[References: Sekiya T (2010). A new prosauropod dinosaur from Lower Jurassic in Lufeng of Yunnan. *Global Geology* 29(1): 6–15.]

Xiyunykus pengi
> *Xiyunykus pengi* Xu *et al.*, 2018 [Barremian–Aptian?; upper part of Tugulu Group, Junggar Basin, Xinjiang, China]
Generic name: *Xiyunykus* ← {(Chin.) Xīyù西域: 'western regions'; referring to Central Asia including Xinjiang + (Gr.) ὄνυξ/onyx: claw} (*1)
Specific name: *pengi* ← {(person's name) Péng彭 + -ī};"in honor of Professor Peng Xiling, who has contributed greatly to the study of geology in Xinjiang" (*1)
Etymology: Peng's **claws from Central Asia**
Taxonomy: Theropoda: Alvarezsauria
[References: (*1) Xu X, Choiniere J, Tan Q, *et al.* (2018). Two Early Cretaceous fossils document transitional stages in alvarezsaurian dinosaur evolution. *Current Biology* 28(17): 2853–2860.e3.]

Xuanhanosaurus qilixiaensis
> *Xuanhanosaurus qilixiaensis* Dong, 1984 [Bathonian–Callovian; lower Shaximiao Formation, Qilixia, Xuanhan, Sichuan, China]
Generic name: *Xuanhanosaurus* ← {(Chin. place-name) Xuānhàn 宣漢 + (Gr.) σαῦρος/sauros: lizard}
Specific name: *qilixiaensis* ← {(Chin. place-name) Qīlǐxiá七里峡 + -ensis}
Etymology: **Xuanhan lizard** from Qilixia
Taxonomy: Theropoda: Tetanurae: Allosauroidea: Metriacanthosauridae
Notes: Estimated to be 4.5 m in total length. (according to Paul, 2010)
[References: Dong Z (1984). A new theropod dinosaur from the Middle Jurassic of Sichuan Basin. *Vertebrata PalAsiatica* 22(3): 213–218.]

Xuanhuaceratops niei
> "Xuanhuasaurus" Zhao, 1985 (*nomen nudum*) ⇒ *Xuanhuaceratops niei* Zhao *et al.*, 2006 [Late Jurassic; Houcheng Formation, Xuanhua, Hebei, China]
Generic name: *Xuanhuaceratops* ← {(Chin. place-name) Xuānhuà 宣化: the geographic region that includes the type locality + ceratops [(Gr.) κέρας/keras: horn + ὤψ/ōps: face]}
Specific name: *niei* ← {(person's name) Nie + -ī}; in honor of Nie Rongzhen who kindly provided the authors with the specimen (*1).
Etymology: Nie's **horned-face (ceratopsian) from Xuanhua District**
Taxonomy: Ornithischia: Ceratopsia: Chaoyangsauridae
[References: (*1) Zhao X, Cheng Z, Xu X, Makovicky PJ (2006). A new ceratopsian from the Upper Jurassic Houcheng Formation of Hebei, China. *Acta Geologica Sinica* 80(4): 467–473.]

Xunmenglong yingliangis
> *Xunmenglong yingliangis* Xing *et al.*, 2019 [Hauterivian; Huajiying Formation, Hebei, China]
Generic name: *Xunmenglong* ← {(Chin.) xùnměng 迅猛: swift + lóng龍: dragon}
Specific name: *yingliangis* ← {(Chin.) Yīngliáng英良}; after Yingliang Group, China.
Etymology: **swift dragon** of Yingliang [英良迅猛龍]
Taxonomy: Saurischia: Theropoda: Compsognathidae

Xuwulong yueluni
> *Xuwulong yueluni* You *et al.*, 2011 [Aptian; Xinminpu Group, Gansu, China]
Generic name: *Xuwulong* ← {(Chin. Courtesy name) Xùwǔ敘五 + (Chin.) lóng龍: dragon}; referring to Professor Wang Yue-lun.
Specific name: *yueluni* ← {(person's name) Yue-lun 曰倫 + -ī}; in honor of Wang Yue-lun (王曰倫), the precursor of the Gansu Geological Museum.
Etymology: Wang Yue-lun **(Xuwu)'s dragon** [曰倫敘五龍]
Taxonomy: Ornithischia: Ornithopoda: Hadrosauroidea
[References: You H, Li D, Liu W (2011). A new hadrosauriform dinosaur from the Early Cretaceous of Gansu Province, China. *Acta Geologica Sinica* 85(1): 51–57.]

Y

Yamaceratops dorngobiensis
> *Yamaceratops dorngobiensis* Makovicky & Norell, 2006 (*1) [Late Cretaceous (*2)*; Javkhlant Formation, Khugenetslavkant, Dorngovi, Gobi Desert, Mongolia]
*The age, previously had been assessed as late Early Cretaceous (Makovicky & Norell, 2006)
Generic name: *Yamaceratops* ← {(Buddhism) Yama

+ ceratops [(Gr.) κέρας/ keras: horn + ωψ/ōps: face]}; "Yama, a Tibetan tantric Buddhist deity, who is the Lord of Death and one of the eight Dharmapalas, or protectors, of Buddhist teaching. Yama has the head of a water buffalo and bears horns, a trait from which ceratopsians derive their name" (*1).

Specific name: *dorngobiensis* ← {(place-name) Dorngobi + -ensis}; "referring to the Eastern Gobi provenance of this taxon" (*1).

Etymology: **Yama horned face (ceratopsian)** from Dorngobi

Taxonomy: Ornithischia: Marginocephalia: Ceratopsia: Neoceratopsia

[References: (*1) Makovicky PJ, Norell MA (2006). *Yamaceratops dorngobiensis*, a new primitive ceratopsian (Dinosauria: Ornithischia) from the Cretaceous of Mongolia. *American Museum Novitates* 3530: 1–42.]; (*2) Eberth DA, Kobayashi Y, Lee Y-N, Mateus O, Therrien F, Zelenitsky DK, Norell M A (2009). Assignment of *Yamaceratops dorngobiensis* and associated redbeds at Shine Us Khudag (eastern Gobi, Dorngobi Province, Mongolia) to the redescribed Javkhlant Formation (Upper Cretaceous). *Journal of Vertebrate Paleontology* 29(1): 295–302.]

Yamanasaurus lojaensis

> *Yamanasaurus lojaensis* Apesteguía *et al.*, 2019 [Maastrichtian; Río Playas Formation, Loja, Ecuador]

Generic name: *Yamanasaurus* ← {(place-name) Yamana region + (Gr.) σαῦρος/sauros: lizard}

Specific name: *lojaensis* ← {(place-name) Loja Province + -ensis}

Etymology: **Yamana lizard** from Loja

Taxonomy: Saurischia: Sauropoda: Saltasaurinae

[References: Apesteguía S, Soto Luzuriaga JE, Gallina PA, Tamay Granda J, Guamán Jaramillo GA (2019). The first dinosaur remains from the Cretaceous of Ecuador. *Cretaceous Research* 108: 104345.]

Yandusaurus hongheensis

> *Yandusaurus hongheensis* He, 1979 [Oxfordian; lower Shaximiao Formation, Sichuan, China]

Generic name: *Yandusaurus* ← {(Chin. place-name) Yándū塩都*: the ancient name for Zìgòng自貢 + (Gr.) σαῦρος/sauros: lizard}

*Yandu [yan塩: salt +du都: capital]: referring to salt historically being the major economic product of the region (*1).

Specific name: *hongheensis* ← {(Chin. place-name) Hónghè鴻鶴 + -ensis}. "The skeleton was discovered near Hongheba Dam, southeast of Zigong municipality, Sichuan" (*1).

Etymology: **Yandu lizard** from Honghe

Taxonomy: Ornithischia: Ornithopoda

cf. Ornithischia: Ornithopoda: Hypsilophodontidae (*1)

Other species:

> *Yandusaurus multidens* He & Cai, 1983 ⇒ *Agilisaurus multidens* (He & Cai, 1983) Peng, 1992 ⇒ *Hexinlusaurus multidens* (He & Cai, 1983) Barrett *et al.*, 2005 [Bajocian; lower Shaximiao Formation, Sichuan, China]

[References: (*1) He X, Cai K (1984). *The Middle Jurassic Dinosaurian Fauna from Dashanpu, Zigong, Sichuan Vol. 1. The Ornithopod Dinosaurs.* Sichuan Scientific and Technological Publishing House, Chengdu: 1–71.]

Yangavis confucii [Avialae]

> *Yangavis confucii* Wang & Zhou, 2018 [Aptian; Yixian Formation, Liaoning, China]

Generic name: *Yangavis* ← {(person's name) Yáng + (Lat.) avis: bird}; in honor of the late distinguished Chinese palaeontologist Zhongjian Yang.

Specific name: *confucii* ← {(person's name) + -ī}; in honor of Confucius, and referring to the confuciusornithids.

Etymology: Confucius and **Yang's bird**

Taxonomy: Theropoda: Confuciusornithidae

Yangchuanosaurus shangyouensis

> *Yangchuanosaurus shangyouensis* Dong *et al.*, 1978 [Oxfordian; upper Shaximiao Formation, Sichuan, China]

= *Yangchuanosaurus magnus* Dong *et al.*, 1983 [Oxfordian; upper Shaximiao Formation, Sichuan, China]

= *Szechuanosaurus yandonensis* Dong *et al.*, 1978 (*nomen dubium*) [Oxfordian (or Kimmeridgian–Thitonian); upper Shaximiao Formation, Sichuan, China]

= "*Szechuanoraptor dongi*" Chure, 2001 (*nomen ex dissertatione*) [Oxfordian; upper Shaximiao Formation, Sichuan, China]

Generic name: *Yangchuanosaurus* ← {(Chin. place-name) Yangchuan 永川 [Pinyin: Yǒngchuān] + (Gr.) σαῦρος/sauros: lizard}

Specific name: *shangyouensis* ← {(Chin. place-name) Shàngyóu上游 + -ensis}

Etymology: **Yangchuan (Yongchuan) lizard** from Shangyou

Taxonomy: Theropoda: Neotheropoda

cf. Theropoda: Allosauroidea: Sinraptoridae (according to *The Dinosauria 2nd edition*)

cf. Theropoda: Carnosauria: Megalosauridae (*1)

Other species:

> *Szechuanosaurus zigongensis* Gao, 1993 ⇒

Yangchuanosaurus zigongensis (Gao, 1993) Carrano *et al*., 2012 [Bajocian; lower Shaximiao Formation, Sichuan, China]

> *Yangchuanosaurus hepingensis* Gao, 1992 ⇒ *Sinraptor hepingensis* (Gao, 1992) Currie & Zhao, 1994 [Oxfordian (or Bathonian–Callovian); upper Shaximiao Formation, Sichuan, China]

[References: (*1) Dong Z, Zhang Y, Li X, Zhou S (1975). A new carnosaur from Yongchuan County, Sichuan Province. *Ke Xue Tong Bao [Science Newsletter]* 23(5): 302–304.]

Yanornis martini [Avialae]

> *Yanornis martini* Zhou & Zhang, 2001 [Aptian; Jiufotang Formation, Liaoning, China]

= *Archaeovolans repatriatus* Czerkas & Xu, 2002 (*1) [Aptian; Jiufotang Formation, Liaoning, China]

= *Aberratiodontus wui* Gong *et al*., 2004 [Aptian; Jiufotang Formation, Liaoning, China]

Generic name: *Yanornis* ← {(Chin.) Yān燕: the ancient Chinese Yan Dynasty + (Gr.) ὄρνις/ornis: bird}

Specific name: *martini* ← {(person's name) Martin + -ī}; in honor of Larry D. Martin for his contribution to the study of Mesozoic birds.

Etymology: Martin's **Yan bird**

Taxonomy: Avialae: Ornithuromorpha: Yanornithiformes: Songlingornithidae / Yanornithidae

cf. Aves: Ornithurae: Yanornithiformes: Yanornithidae (*2)

Other species:

> *Yanornis guozhangi* Wang *et al*., 2013 [Barremian; Yixian Formation, Liaoning, China]

[References: (*1) Zhou Z, Clarke JA, Zhang F (2002). *Archaeoraptor*'s better half. *Nature* 420: 285.; (*2) Zhou Z, Zhang F (2001). Two new ornithurine birds from the Early Cretaceous of western Liaoning, China. *Chinese Science Bulletin* 46(15): 1258–1264.]

Yaverlandia bitholus

> *Yaverlandia bitholus* Galton, 1971 [late Hauterivian; Wessex Formation, Yaverland Battery, Sandown, Isle of Wight, UK]

Generic name: *Yaverlandia* ← {(place-name) Yaverland + -ia} (*1)

Specific name: *bitholus* ← {(Lat.) bi- : two + tholus: dome}; "with reference to the double dome" (*1). It had a skull cap thickened with two small domes.

Etymology: **Yaverland's one** with double-dome

Taxonomy: Theropoda: Maniraptora

cf. Ornithischia: Ornithopoda: Pachycephalosauridae (*1)

Notes: *Yaverlandia* was thought to be a member of Pachycephalosauridae (*1), but according to Naish & Martill (2008), it is actually a maniraptoran.

[References: (*1) Galton PM (1971). A primitive dome-headed dinosaur (Ornithischia: Pachycephalosauridae) from the Lower Cretaceous of England and the function of the dome of pachycephalosaurids. *Journal of Paleontology* 45(1): 40–47.]

Yehuecauhceratops mudei

> *Yehuecauhceratops mudei* Rivera-Sylva *et al*., 2017 [late Campanian; Aguja Formation, Coahuila, Mexico]

Generic name: *Yehuecauhceratops* ← {(Nahuatl*) yehuecauh: ancient + ceratops [(Gr.) κέρας/kera: horn + ὤψ/ōps: face]} (*1).

*a language or group of languages of the Uto-Aztecan language family. Nahuatl has been spoken in central Mexico.

Specific name: *mudei* ← {(acronym) MUDE: Museo del Desierto + -ī}; "in honor of the Museo del Desierto, at Saltillo, Coahuila, Mexico" (*1).

Etymology: **ancient horned face (ceratopsian)** of Museo del Desierto

Taxonomy: Ornithischia: Ceratopsidae: Centrosaurinae

[References: (*1) Rivera-Sylva HE, Frey E, Stinnesbeck W, Guzmán-Gutiérrez JR, González-González AH (2017). Mexican ceratopsids: Considerations on their diversity and biogeography. *Journal of South American Earth Sciences* 75: 66–73.]

Yi qi [Scansoriopterygidae]

> *Yi qi* Xu *et al*., 2015 [Callovian; Tiaojishan Formation, Qinglong, Hebei, China]

Generic name: *Yi* ← {(Chin.) yì翼: wing}; referring to the bizarre wings of this animal (*1)

Specific name: *qi* ← {(Chin.) qí奇: strange}

Etymology: strange **wing**

Taxonomy: Theropoda: Eumaniraptora: Avialae (?): Scansoriopterygidae

Notes: *Yi* had membranous wings (*1).

[References: (*1) Xu X, Zheng X, Sullivan C, Wang X, Xing L, Wang Y, Zhang X, O'Connor JK, Zhang F, Pan Y (2015). A bizarre Jurassic maniraptoran theropod with preserved evidence of membranous wings. *Nature* 521: 70–73.]

Yimenosaurus youngi

> *Yimenosaurus youngi* Bai *et al*., 1990 [Hettangian; Fengjiahe Formation, Jiaojiadian, Yimen, Yunnan, China]

Generic name: *Yimenosaurus* ← {(Chin. place-name) Yìmén易門 + (Gr.) σαῦρος/sauros: reptile} (*1)

Specific name: *youngi* ← {(person's name) Young

[Pinyin:Yáng 楊] + -ī}; in honor of "Professor Yang Zhongjian (C. C. Young), the founder of vertebrate paleontology in China and a provider of outstanding contributions to prosauropod research in Yunnan Province" (*1).

Etymology: Young's **Yimen reptile**

Taxonomy: Saurischia: Sauropodomorpha: Prosauropoda: Plateosauridae

[References: (*1) Bai Z, Yang J, Wang G (1990). *Yimenosaurus*, a new genus of Prosauropoda from Yimen County, Yunnan Province. *Yuxiwenbo (Yuxi Culture and Scholarship)* 1: 14–23.]

Yingshanosaurus jichuanensis

> *Yingshanosaurus jichuanensis* Zhu, 1994 [Oxfordian; upper Shaximiao Formation, Yingshan, Sichuan, China]

Generic name: *Yingshanosaurus* ← {(Chin. place-name) Yíngshān + (Gr.) σαῦρος/sauros: lizard}

Specific name: *jichuanensis* ← {(Chin. place-name) Jǐchuān: the location of the site + -ensis}

Etymology: **Yingshan lizard** from Jichuan [濟川營山龍]

Taxonomy: Ornithischia: Thyreophora: Stegosauria: Stegosauridae: Stegosaurinae

Notes: This taxon had been considered invalid due to an insufficient description in 1984. However, in 1994 Zhu fully described the animal.

[References: Zhu S (1994). 記四川盆地營山県一剣龍化石 [Record of a fossil stegosaur from Yingshan in the Sichuan Basin]. *Sichuan Cultural Relics* (S1): 8–14.]

Yinlong downsi

> *Yinlong downsi* Xu *et al.*, 2006 [Oxfordian; upper Shishugou Formation, Wucaiwan, Junggar Basin, Xinjiang, China]

Generic name: *Yinlong* ← {(Chin.) yǐn隠: hiding + lóng龍: dragon}; "derived from the movie *Crouching Tiger, Hidden Dragon* which was filmed in the locality where the holotype was found" (*1).

Specific name: *downsi* ← {(person's name) Downs + -ī}; "in memory of Mr. Will Downs, who joined many palaeontological expeditions in China including the one with us in 2003, shortly before his death" (*1).

Etymology: Downs' **hidden dragon**

Taxonomy: Ornithischia: Marginocephalia: Ceratopsia

[References: (*1) Xu X, Forster CA, Clark JM, Mo J (2006). A basal ceratopsian with transitional features from the Late Jurassic of northwestern China. *Proceedings of the Royal Society B: Biological Sciences* 273(1598): 2135–2140.]

Yixianornis grabaui [Avialae]

> *Yixianornis grabaui* Zhou & Zhang, 2001 [Aptian; Jiufotang Formation, Yixian, Liaoning, China]

Generic name: *Yixianornis* ← {(Chin. place-name) Yixian + (Gr.) ὄρνις/ornis: bird}

Its remains have been found in the Jiufotang Formation which overlies Yixian Formation.

Specific name: *grabaui* ← {(person's name) Grabau + -ī}; "dedicated to late American geologist Amadeus William Grabau, a pioneering geologist in Liaoning, for his contribution to the study of the Jehol Biota" (*1).

Etymology: Grabau's **Yixian bird**

Taxonomy: Theropoda: Avialae: Ornithuromorpha: Yanornithiformes: Songlingornithidae

[References: (*1) Zhou Z, Zhang F (2001). Two new ornithurine birds from the Early Cretaceous of western Liaoning, China. *Chinese Science Bulletin* 46: 1258–1264.]

Yixianosaurus longimanus [Anchiornithidae]

> *Yixianosaurus longimanus* Xu & Wang, 2003 [Barremian–Aptian; Yixian Formation, Liaoning, China]

Generic name: *Yixianosaurus* ← {(Chin. place-name) Yìxiàn義縣: Yixian County + (Gr.) σαῦρος/sauros: reptile} (*1)

Specific name: *longimanus* ← {(Lat.) longus: long + manus: hand}; referring to "the significantly elongated manus" (1).

Etymology: long-handed **Yixian reptile**

Taxonomy: Theropoda: Maniraptora: Avialae (?): Anchiornithidae

cf. Theropoda: Maniraptora (*1)

[References: (*1) Xu X, Wang X-L (2003). A new maniraptoran dinosaur from the Early Cretaceous Yixian Formation of western Liaoning. *Vertebrata PalAsiatica* 41(3): 195–202.]

Yizhousaurus sunae

> *Yizhousaurus sunae* Zhang *et al.*, 2018 [Hettangian; Zhangjiaao Member, Lufeng Formation, Yunnan, China]

Generic name: *Yizhousaurus* ← {(Chin. place-name) Yìzhōu益州 + (Gr.) σαῦρος/sauros: lizard}; "referring to the Chuxiong Yi Autonomous Prefecture of Yunnan Province" (*1).

Specific name: *sunae* ← {(Chin. person-name) Sūn 孫 +-ae}; "in honor of Professor Ai-Ling Sun, for her great contribution to Chinese vertebrate fossils, including those from Lufeng" (*1).

Etymology: Sun's **Yizhou lizard** [孫氏益州郡龍]

Taxonomy: Saurischia: Sauropodomorpha: Sauropoda

[References: (*1) Zhang Q-N, You H-L, Wang T, Chatterjee S (2018). A new sauropodiform dinosaur with a 'sauropodan' skull from the Lower Jurassic Lufeng Formation of Yunnan Province, China. *Scientific Reports* 8: 13464.]

Yongjinglong datangi

> *Yongjinglong datangi* Li *et al.*, 2014 [Early Cretaceous; Hekou Group, Lanzhou-Minhe Basin, Gansu, China]

Generic name: *Yongjinglong* ← {(Chin. place-name) Yǒngjìng永靖 + lóng龍: dragon}; "referring to Yongjing County, which is close to the fossil location of the new sauropod and also yields numerous dinosaur track fossils" (*1).

Specific name: *datangi* ← {(Chin.) Dà Táng*大唐 | (person's name) Táng + -ī}; referring to "the Táng dynasty and also in honor of Mr. Zhi-Lu Tang from the Institute of Vertebrate Paleontology and Paleoanthropology, Beijing for his numerous contributions to the research of dinosaurs" (*1).

*poetic name for Tang dynasty

Etymology: Tang's **Yongjing dragon**

Taxonomy: Saurischia: Sauropoda: Titanosauriformes: Titanosauria

[References: (*1) Li L-G, Li D-Q, You H-L, Dodson P (2014). A new titanosaurian sauropod from the Hekou Group (Lower Cretaceous) of the Lanzhou-Minhe Basin, Gansu Province, China. *PLoS ONE* 9(1): e85979.]

Yuanmousaurus jiangyiensis

> *Yuanmousaurus jiangyiensis* Lü *et al.*, 2006 [Middle Jurassic; Zhanghe Formation, Jiangyi, Yuanmou, Yunnan, China]

Generic name: *Yuanmousaurus* ← {(Chin. place-name) Yuánmóu元謀 + (Gr.) σαῦρος/sauros: lizard}

Specific name: *jiangyiensis* ← {(Chin. place-name) Jiāngyì 姜驛 + -ensis}

Etymology: **Yuanmou lizard** from Jiangyi

Taxonomy: Saurischia: Sauropodomorpha: Sauropoda: Eusauropoda: Mamenchisauridae

[References: Lü J, Li S, Ji Q, Wang G, Zhang J, Dong Z (2006). New eusauropod dinosaur from Yuanmou of Yunnan Province, China. *Acta Geologica Sinica-English Edition* 80(1): 1–10.]

Yueosaurus tiantaiensis

> *Yueosaurus tiantaiensis* Zheng *et al.*, 2012 [Aptian–Cenomanian; Liangtoutang Formation, Tiantai, Zhejiang, China]

Generic name: *Yueosaurus* ← {(Chin. place-name) Yuè越: ancient name for Zhejiang Province + (Gr.) σαῦρος/sauros: lizard}

Specific name: *tiantaiensis* ← {(Chin. place-name) Tiāntái 天台 + -ensis}

Etymology: **Yue (Zhejiang) lizard** from Tiantai

Taxonomy: Ornithischia: Ornithopoda

[References: Zheng W, Zin X, Shibata M, Azuma Y, Yu F (2012). A new ornithischian dinosaur from the Cretaceous Liangtoutang Formation of Tiantai, Zhejiang Province, China. *Cretaceous Research* 34: 208–219.]

Yulong mini

> *Yulong mini* Lü *et al.*, 2013 [Cenomanian–Maastrichtian; Qiupa Formation, Henan, China]

Generic name: *Yulong* ← {(Chin. place-name) Yù 豫: the abbreviated name of Henan Province河南省 + (Chin.) lóng 龍: dragon} (*1)

Specific name: *mini* ← {mini: small}; referring to the small size of known specimens (*1).

Etymology: small **Yu (Henan) dragon**

Taxonomy: Theropoda: Coelurosauria: Oviraptoridae

[References: (*1) Lü J, Currie PJ, Xu L, Zhang X, Pu H, Jia S (2013). Chicken-sized oviraptorid dinosaurs from central China and their ontogenetic implications. *Naturwissenschaften* 100(2): 165–175.]

Yumenornis huangi [Avialae]

> *Yumenornis huangi* Wang *et al.*, 2013 [Aptian; Xiagou Formation, Changma Basin, Gansu, China]

Generic name: *Yumenornis* ← {(Chin. place-name) Yùmén玉門: the name of the city + (Gr.) ὄρνις/ornis: bird}

Specific name: *huangi* ← {(Chin. person's name) Huáng黄 + -ī}; "dedicated to Mr. Zhao-Chu Huang of the Institute of Vertebrate Paleontology and Paleoanthropology (IVPP) for his long-lasting support to several generations of IVPP staff" (*1).

Etymology: Huang's **bird from Yumen** [黄氏玉門鳥]

Taxonomy: Avialae: Ornithuromorpha
cf. Aves: Pygostylia: Ornithothoraces: Ornithuromorpha (*1)

[References: (*1) Wang Y-M, O'Connor JK, Li D-Q, You H-L (2013). Previously unrecognized ornithuromorph bird diversity in the Early Cretaceous Changma Basin, Gansu Province, Northwestern China. *PLoS ONE* 8(10): e77693.]

Yunganglong datongensis

> *Yunganglong datongensis* Wang *et al.*, 2013 [Cenomanian; Zhumapu Formation, Datong, Shanxi, China]

Generic name: *Yunganglong* ← {(Chin. place-name) Yúngāng雲崗 + lóng 龍: dragon}; "after 'Yungang Grottoes' 雲崗石窟, a UNESCO World Heritage built in the 5th and 6th centuries about 50 km east of the fossil locality" (*1).

Specific name: *datongensis* ← {(Chin. place-name) Dàtóng大同: the city in which the locality is situated + -ensis} [*1].
Etymology: **Yungang dragon** from Datong City [大同雲崗石窟龍]
Taxonomy: Ornithischia: Ornithopoda: Hadrosauroidea: Hadrosauromorpha
[References: (*1) Wang R-F, You H-L, Xu S-C, Wang S-Z, Yi J, Xie L-J, Jia L, Li Y-X (2013). A new hadrosauroid dinosaur from the early Late Cretaceous of Shanxi Province, China. *PLoS ONE* 8(10): e77058.]

Yungavolucris brevipedalis [Avialae]

> *Yungavolucris brevipedalis* Chiappe, 1993 [Maastrichtian; Lecho Formation, Salta, Argentina]

Generic name: *Yungavolucris* ← {(place-name) Yunga (or Yungas): the phytogeographic region in which El Brete is located + (Lat.) volucris: bird} [*1]
Specific name: *brevipedalis* ← {(Lat.) brevis: short + pedalis [pēs: foot]} [*1]
Etymology: short-footed **Yungas bird**
Taxonomy: Theropoda: Avialae: Enatiornithes
cf. Aves: Enantiornithes [*1]
[References: (*1) Chiappe LM (1993). Enantiornithine (Aves) tarsometatarsi from the Cretaceous Lecho Formation of Northwestern Argentina. *American Museum Novitates* 3083: 1–27.]

Yunmenglong ruyangensis

> *Yunmenglong ruyangensis* Lü *et al.*, 2013 [Aptian; Haoling Formation, Ruyang Basin, Henan, China]

Generic name: *Yunmenglong* ← {(Chin. place-name) Yúnmèng雲夢: Yunmengshan area + lóng龍: dragon}
Specific name: *ruyangensis* ← {(Chin. place-name) Rŭyáng Basin汝陽盆地 + -ensis}
Etymology: **Yunmeng dragon** from Ruyang [汝陽雲夢龍]
Taxonomy: Saurischia: Sauropodomorpha: Sauropoda: Somphospondyli
[References: Lü J, Xu L, Pu H, Zhang X, Zhang Y, Jia S, Chang H, Zhang J, Wei X (2013). A new sauropod dinosaur (Dinosauria, Sauropoda) from the late Early Cretaceous of the Ruyang Basin (central China). *Cretaceous Research* 44: 202–213.]

Yunnanosaurus huangi

> *Yunnanosaurus huangi* Young, 1942 [Sinemurian; Zhangjiawa Member, Lufeng Formation, Yunnan, China]

Generic name: *Yunnanosaurus* ← {(Chin. place-name) Yúnnán Province雲南省 + (Gr.) σαῦρος/sauros: lizard}
Specific name: *huangi* ← {(place-name) village of Huangchiatien*[*1] + -ī}
*Huang-chiatien: about 1 kilometer north of Shawan, Lufeng, Yunnan Province, China (acc. Young, C-C. (1944))
Etymology: **Yunnan lizard** of Huangchiatien
Taxonomy: Saurischia: Sauropodomorpha: Plateosauria: Sauropodiformes
cf. Saurischia: Sauropodomorpha: Prosauropoda (according to *The Dinosauria 2nd edition*)
Other species:

> *Yunnanosaurus youngi* Lu *et al.*, 2007 [Middle Jurassic; Zhanghe Formation, Yunnan, China]
> *Yunnanosaurus robustus* Young, 1951 [*2] [Sinemurian; Zhangjiawa Member, Lufeng Formation, Yunnan, China]

[References: (*1) Barrett PM, Upchurch P, Zhou X-D, Wang X-L (2007). The skull of *Yunnanosaurus huangi* Young, 1942 (Dinosauria: Prosauropoda) from the Lower Lufeng Formation (Lower Jurassic) of Yunnan, China. *Zoological Journal of the Linnean Society* 150(2): 319–341.]; (*2) Sekiya T, Jin X, Zheng W, Shibata M, Azuma Y (2014). A new juvenile specimen of *Yunnanosaurus robustus* (Dinosauria: Sauropodomorpha) from Early to Middle Jurassic of Chuxiong Autonomous Prefecture, Yunnan Province, China. *Historical Biology, An International Journal of Paleobiology* 26(2): 252–277.]

Yunyangosaurus puanensis

> *Yunyangosaurus puanensis* Dai *et al.*, 2010 [Middle Jurassic; Xintiangou Formation; Chongqing, China]

Generic name: *Yunyangosaurus* ← {(place-name) Yúnyáng County雲陽縣 + (Gr.) σαῦρος/sauros: reptile}
Specific name: *puanensis* ← {(place-name) Pŭān普安 + -ensis}
Etymology: **Yunyang reptile** from Puan
Taxonomy: Theropoda: Tetanurae

Yurgovuchia doellingi

> *Yurgovuchia doellingi* Senter *et al.*, 2012 [Barremian?–Aptian; lower Yellow Cat Member, Cedar Mountain Formation, Utah, USA] [*1]

Generic name: *Yurgovuchia* ← {(Ute) yurgovuch: coyote + -ia}; "in honor of the Ute Tribe of northeastern Utah and derived from the Ute word *yurgovuch*, meaning "coyote", a predator of similar size to *Y. doellingi* that currently inhabits the same region" [*1].
Specific name: *doellingi* ← {(person's name) Doelling + -ī}; "in honor of Helmut Doelling in recognition of his 50-plus years of geological research and

mapping of Utah for the Utah Geological Survey" (*1).

Etymology: Doelling's **coyote**

Taxonomy: Theropoda: Coelurosauria: Dromaeosauridae

[References: (*1) Senter P, Kirkland JI, DeBlieux DD, Madsen S, Toth N (2012). New dromaeosaurids (Dinosauria: Theropoda) from the Lower Cretaceous of Utah, and the evolution of the dromaeosaurid tail. *PLoS ONE* 7(5): e36790.]

Yutyrannus huali

> *Yutyrannus huali* Xu *et al*., 2012 [Barremian–Aptian; Yixian Formation, Liaoning, China]

Generic name: *Yutyrannus* ← {(Chin.) yǔ 羽: feathers + (Lat.) tyrannus = (Gr.) τύραννος/tyrannos: king or tyrant} (*1)

Specific name: *huali* ← {(Chin.) huálì華麗: beautiful}; "referring to the beauty of the plumage of this animal" (*1)

Etymology: beautiful, **feathered tyrant**

Taxonomy: Theropoda: Coelurosauria: Tyrannosauroidea

Notes: *Yutyrannus huali* bears long filament feathers. *Y. huali* is the largest-known species of dinosaur with direct evidence of feathers (*1).

[References: (*1) Xu X, Wang K, Zhang K, Ma Q, Xing L, Sullivan C, Hu D, Cheng S, Wang S (2012). A gigantic feathered dinosaur from the Lower Cretaceous of China. *Nature* 484(7392): 92–95.]

Z

Zalmoxes robustus

> *Mochlodon robustum* Nopcsa, 1899 ⇒ *Rhabdodon robustum* (Nopcsa, 1899) Nopcsa, 1915 ⇒ *Zalmoxes robustum* (Nopcsa, 1899) Weishampel *et al*., 2003 [late Maastrichtian; Sânpetru Formation, Hunedoara, Romania]

= *Onychosaurus hungaricus* Nopcsa, 1902 [Late Cretaceous, Transylvania, Romania]

= *Camptosaurus inkeyi* Nopcsa, 1900 [late Campanian; Sânpetru Formation, Județul Covurlui, Romania]

Generic name: *Zalmoxes* ← {(legend) Zalmoxes: the Dacian deity Zalmoxis}

"Zalmoxes (alternative spelling Zalmoxis) is said to have been a freed slave of Pythagoras who, upon being freed, travelled to Dacia (ancient Romania) and became a teacher, healer, vegetarian and high priest. He was later deified by the Dacians as a god of The Mystery, ecstasy, the underworld and immortality (Eliade 1972). The relevance of the name for this dinosaur is its Dacian orign and the fact that the subterranean crypt' of this herbivore was originally opened and that (taxonomic) immortality then came from Nopcsa's original work" (*1).

Specific name: *robustus* ← {(Lat.) rōbustus: robust}

Etymology: robust **Zalmoxes**

Taxonomy: Ornithischia: Ornithopoda: Iguanodontia: Rhabdodontidae

Other species:

> *Zalmoxes shqiperorum* Weishampel *et al*., 2003 [late Maastrichtian; unnamed unit, Alba, Romania]

Notes: According to *The Dinosauria 2nd edition*, *Onychosaurus hungaricus* and *Camptosaurus inkeyi* were regarded as synonyms of *Zalmoxes robustus*.

[References: (*1) Weishampel DB, Jianu C.-M, Csiki Z, Norman DB (2003). Osteology and phylogeny of *Zalmoxes* (n. g.), an unusual euornithopod dinosaur from the latest Cretaceous of Romania. *Journal of Systematic Palaentology* 1(2): 65–123.]

Zanabazar junior

> *Saurornithoides junior* Barsbold, 1974 ⇒ *Zanabazar junior* (Barsbold, 1974) Norell *et al*., 2009 [Maastrichtian; Nemegt Formation, Ömnögovi, Mongolia]

Generic name: *Zanabazar* ← {(Tibetan Buddhism) Zanabazar}; "in honor of Zanabazar (1635–1723), the first Bogd Gegen of Mongolia" (*1).

Specific name: *junior* ← {(Lat.) junior (*comp*. of 'juvenis'): younger}; in reference to it being younger than *Saurornithoides mongoliensis*, Campanian.

Etymology: younger **Zanabazar**

Taxonomy: Theropoda: Troodontidae: Troodontinae

Notes: The holotype of *Zanabazar junior* is larger than most other troodontid specimen.

[References: (*1) Norell MA, Makovicky PJ, Bever GS, Balanoff AM, Clark JM, Barsbold R, Rowe T (2009). A review of the Mongolian Cretaceous dinosaur *Saurornithoides* (Troodontidae: Theropoda). *American Museum Novitates* 3654: 1–63.]

Zapalasaurus bonapartei

> *Zapalasaurus bonapartei* Salgado *et al*., 2006 [Barremian–Aptian; Piedra Parada Member, La Amarga Formation, Neuquén, Argentina]

Generic name: *Zapalasaurus* ← {(place-name) Zapala + (Gr.) σαῦρος/sauros: lizard} ; "in reference to Zapala, a city of the Neuquén Province, Argentina, located some 80 km to the north of the holotype locality" (*1).

Specific name: *bonapartei* ← {(person's name) Bonaparte + -ī}; "in homage to Dr. J. Bonaparte, who collected the holotype material, and in recognition of his professional career, and to his important work in understanding Mesozoic vertebrates" (*1).

Etymology: Bonaparte's **Zapala lizard**
Taxonomy: Saurischia: Sauropodomorpha: Sauropoda: Diplodocoidea: Rebbachisauridae
cf. Saurischia: Saurpodomorpha: Sauropoda: Diplodocoidea (*1)

[References: (*1) Salgado L, Carvalho IS, Garrido AC (2006) *Zapalasaurus bonapartei*, un nuevo dinosaurio saurópodo La Formación La Amarga (Cretácico Inferior) noroeste de Patagonia, Provincia de Neuquén, Argentina. [*Zapalasaurus bonapartei*, a new sauropod dinosaur from La Amarga Formation (Lower Cretaceous), northwestern Patagonia, Neuquén Province, Argentina. *Geobios* 39(5): 695–707.]

Zapsalis abradens

> *Zapsalis abradens* Cope, 1876 [Campanian; Judith River Formation, Montana, USA]

Generic name: *Zapsalis* ← {(Gr.) ζά-/za-: thorough + ψᾰλίς/psalis: a pair of scissors}
Cope named *Zapsalis* based on a tooth found in Montana (*1).
Specific name: *abradens* ← {(Lat.) abradens: abrading}; referring to the preserved tooth which displays considerable attrition, especially on the flat side (*1).
Etymology: abrading, **thorough scissor**
Taxonomy: Theropoda: Dromaeosauridae: Dromaeosaurinae
Notes: According to *The Dinosauria 2nd edition*, *Zapsalis abradens* was regarded as a *nomen dubium*, but, in 2013 Larson and Currie recognized *Zapsalis* as a valid taxon.

[References: (*1) Cope ED (1876). On some extinct reptiles and Batrachia from the Judith River and Fox Hills beds of Montana. *Proceedings of the Academy of Natural Sciences of Philadelphia* 28: 340–359.]

Zaraapelta nomadis

> *Zaraapelta nomadis* Arbour *et al.*, 2014 [Campanian; Barun Goyot Formation, Gobi Desert, Mongolia] (*1)

Generic name: *Zaraapelta* ← {(Mongolian) зараа/zaraa: hedgehog + (Lat.) pelta= (Gr.) πέλτη/peltē: a light shield}; "in reference to the spiky appearance of its skull, and the osteoderms found on all ankylosaurs" (*1).
Specific name: *nomadis* ← {(Lat.) Nomas: nomad}; "in reference to Mongolian travel company Nomadic Expeditions, which has facilitated many years of palaeontological fieldwork in the Gobi Desert" (*1).
Etymology: Nomadic Expeditions' **hedgehog shield**
Taxonomy: Ornithischia: Thyreophora: Ankylosauria: Ankylosauridae: Ankylosaurinae

[References: (*1) Arbour VM, Currie PJ, Badamgarav D (2014). The ankylosaurid dinosaurs of the Upper Cretaceous Baruungoyot and Nemegt formations of Mongolia. *Zoological Journal of the Linnean Society* 172(3): 631–652.]

Zby atlanticus

> *Zby atlanticus* Mateus *et al.*, 2014 [Kimmeridgian; Amoreira–Porto Novo Member, Lourinhã Formation, Lourinhã, Portugal]

Generic name: *Zby* ← {(person's name) Zbyszewski}; "after the Russian-French paleontologist Georges Zbyszewski (1909–1999), who devoted his career to the geology and paleontology of Portugal" (*1).
Specific name: *atlanticus* ← {(Lat.) atlanticus: Atlantic, of the Atlantic Ocean}; "because the specimen was found in a scenic bay over the Atlantic Ocean" (*1).
Etymology: for **Zbyszewski** of the Atlantic
Taxonomy: Saurischia: Sauropoda: Eusauropoda: Turiasauria

[References: (*1) Mateus O, Mannion PD, Upchurch P (2014). *Zby atlanticus*, a new turiasaurian sauropod (Dinosauria, Eusauropoda) from the Late Jurassic of Portugal. *Journal of Vertebrate Paleontology* 34(3): 618–634.]

Zephyrosaurus schaffi

> *Zephyrosaurus schaffi* Sues, 1980 [Aptian–Albian; Himes Member, Cloverly Formation, Montana, USA]

Generic name: *Zephyrosaurus* ← {(Lat.) Zephyrus = (Gr.myth.) Ζέφυρος/Zephyros: a Greek god of west wind + σαῦρος/sauros: lizard}
Specific name: *schaffi* ← {(person's name) Schaff + -i}; in honor of Charles R. Schaff, who found the specimen.
Etymology: Schaff's **Zephyrus (west-wind) lizard**
Taxonomy: Ornithischia: Ornithopoda: Parksosauridae / Thescelosauridae: Orodrominae
cf. Ornithischia: Ornithopoda: Heterodontosauridae: Euornithopoda (according to *The Dinosauria 2nd edition*)

Zhanghenglong yangchengensis

> *Zhanghenglong yangchengensis* Xing *et al.*, 2014 [Santonian; Majiacun Formation, Xixia, Henan, China]

Generic name: *Zhanghenglong* ← {(Chin. person's name) Zhang Heng + lóng龍: dragon}; "derived from the full name of Mr. Zhang Heng, a famous Chinese astronomer, mathematician, inventor, poet, and statesman who lived during the Eastern Han Dynasty (AD 25–220) of China. The figure was born in the outskirts of Nanyang in southwestern Henan Province, quite close to the Xixia Basin" (*1).

Specific name: *yangchengensis* ← {(Chin. place-name) Yangcheng + -ensis}; "derived from a large administrative region called Yangcheng that was established in the Spring and Autumn period (BC 770–403) of the Eastern Zhou Dynasty of China. This ancient administrative region included what is now southwestern Henan Province" (*1).

Etymology: **Zhang Heng's dragon** from Yangcheng

Taxonomy: Ornithischia: Ornithopoda: Iguanodontia: Hadrosauriformes: Hadrosauroidea

[References: (*1) Xing H, Wang D, Han F, Sullivan C, Ma Q, He Y, Hone DWE, Yan R, Du F, Xu X (2014). A new basal hadrosauroid dinosaur (Dinosauria: Ornithopoda) with transitional features from the Late Cretaceous of Henan Province, China. *PLoS ONE* 9(6): e98821.]

Zhejiangosaurus lishuiensis

> *Zhejiangosaurus lishuiensis* Lü *et al.*, 2007 (*nomen dubium**) [Cenomanian; Chaochuan Formation, Lishui, Zhejiang, China]
*according to Arbour & Currie (2015).

Generic name: *Zhejiangosaurus* ← {(Chin. place-name) Zhèjiāng浙江 + (Gr.) σαῦρος/sauros: lizard}

Specific name: *lishuiensis* ← {(Chin. place-name) Lìshuǐ 麗水 + -ensis}

Etymology: **Zhejiang lizard** from Lishui

Taxonomy: Ornithischia: Thyreophora: Ankylosauria: Nodosauridae

[References: Lü J, Jin X, Sheng Y, Li Y, Wang G, Azuma Y (2007). New nodosaurid dinosaur from the Late Cretaceous of Lishui, Zhejiang Province, China. *Acta Geologica Sinica* 81(3): 344–350.]

Zhenyuanlong suni

> *Zhenyuanlong suni* Lü & Brusatte, 2015 [Barremian; Yixian Formation, Jehol Group, Liaoning, China]

Generic name: *Zhenyuanlong* ← {(Chin. person's name) Zhènyuán振元 + lóng龍: dragon}; "in honor of Mr. Zhenyuan Sun, who secured the specimen for study" (*1).

Specific name: *suni* ← {(Chin. person's name) Sun 孫 + -ī}

Etymology: Sun's **Zhenyuan dragon** [孫氏振元龍]

Taxonomy: Theropoda: Coelurosauria: Maniraptora: Dromaeosauridae

[References: (*1) Lü J, Brusatte SL (2015). A large, short-armed, winged dromaeosaurid (Dinosauria: Theropoda) from the Early Cretaceous of China and its implications for feather evolution. *Scientific Reports* 5: 11775.]

Zhongjianornis yangi [Avialae]

> *Zhongjianornis yangi* Zhou & Li, 2010 [Early Cretaceous; Jiufotang Formation, Jianchang, Liaoning, China]

Generic name: *Zhongjianornis* ← {(Chin. person's name) Zhōngjiàn鍾健 + (Gr.) ὄρνις/ornis: bird}; "in honor of the late Professor Zhongjian Yang (Chung-Chien Young), father of Chinese vertebrate paleontology and founder of the Institute of Vertebrate Paleontology and Paleoanthropology" (*1).

Specific name: *yangi* ← {(person's name) Yáng楊 + -ī} (*1).

Etymology: Yang's **Zhongjian bird**

Taxonomy: Theropoda: Avialae: Ornithothoraces: Euornithes *cf.* Aves (*1)

[References: (*1) Zhou Z, Zhang F, Li Z (2010). A new Lower Cretaceous bird from China and tooth reduction in early avian evolution. *Proceedings of the Royal Society B: Biological Sciences* 277(1679): 219–227.]

Zhongjianosaurus yangi

> *Zhongjianosaurus yangi* Xu & Qin, 2017 [Aptian; Yixian Formation, Lingyuan, Liaoning, China]

Generic name: *Zhongjianosaurus* ← {(Chin. person's name) Zhōngjiàn鍾健 + (Gr.) σαῦρος/sauros: lizard}; "in honor of Yang Zhongjian (C. C. Young), who is the founder of vertebrate paleontology in China" (*1).

Specific name: *yangi* ← {(person's name) Yáng楊 + -ī} (*1)

Etymology: Yang's **Zhongjian lizard** [楊氏鍾健龍]

Taxonomy: Theropoda: Tetanurae: Dromaeosauridae

[References: (*1) Xu X, Qin Z-C (2017). A new tiny dromaeosaurid dinosaur from the Lower Cretaceous Jehol Group of western Liaoning and niche differentiation among the Jehol dromaeosaurids. *Vertebrata PalAsiatica* 55(2): 129–144.]

Zhongornis haoae [Avialae]

> *Zhongornis haoae* Gao *et al.*, 2008 [Early Cretaceous; upper Yixian Formation, Dawangzhangzi Bed, Lingyuan, Liaoning, China]

Generic name: *Zhongornis* ← {(Chin.) zhōng 中: middle or intermediate + (Gr.) ὄρνις/ornis: bird} (*1)

Specific name: *haoae* ← {(Chin. person's name) Hǎo + -ae}; "in honor of Ms. Hao, who kindly donated the specimen to the Dalian Nature History Museum" (*1).

Etymology: Hao's **intermediate bird**

Taxonomy: Theropoda: Avialae: Euavialae: Avebrevicauda *cf.* Theropoda: Aves (*1)

[References: (*1) Gao C, Chiappe LM, Meng Q, O'Connor JK, Wang X, Cheng X, Liu J (2008). A new basal lineage of Early Cretaceous birds from China and its implications on the evolution of the avian tail. *Palaeontology* 51(4): 775–791.]

Zhongyuansaurus luoyangensis

> *Gobisaurus domoculus* Vickaryous *et al.*, 2001 [Turonian; Ulansuhai Formation, Inner Mongolia, China]

= *Zhongyuansaurus luoyangensis* Xu *et al.*, 2007 [Cenomanian; Mangchuan Formation, Sichuan Group, Ruoyang, Henan, China]

Generic name: *Zhongyuansaurus* ← {Chin. place-name) Zhōngyuán中原 + (Gr.) σαῦρος/sauros: lizard}; "for the area south of Yellow River area, Henan Province" (*1).

Specific name: *luoyangensis* ← {(Chin. place-name) Luòyáng洛陽 + -ensis}; "referring to the fossil site, Ruyang County, the administrative regionalization of the Luoyang area" (*1).

Etymology: **Zhongyuan lizard** from Luoyang

Taxonomy: Ornithischia: Thyreophora: Ankylosauria: Ankylosauridae

cf. Thyreophora: Eurypoda: Ankylosauria: Nodosauridae (*1)

Notes: According to Arbour (2014), *Zhongyuansaurus* is concluded to be a probable junior synonym of *Gobisaurus*.

[References: (*1) Xu L, Lu J, Zhang X, Jia S, Hu W, Zhang J, Wu Y, Ji Q (2007). New nodosaurid ankylosaur from the Cretaceous of Ruyang, Henan Province. *Acta Geologica Sinica* 81(4): 433–438.]

Zhouornis hani [Avialae]

> *Zhouornis hani* Zhang *et al.*, 2013 [Aptian; Jiufotang Formation, Jehol Group, Liaoning, China]

Generic name: *Zhouornis* ← {(Chin. person's name) Zhou + (Gr.) ὄρνις/ornis: bird}; "in homage to Dr. Zhou Zhonghe, who has contributed greatly to the study of the early evolution of birds" (*1).

Specific name: *hani* ← {(Chin. person's name) Han + -ī}; "in honor of the holotype's collector, Mr. Lizhuo Han" (*1).

Etymology: Han and **Zhou's bird**

Taxonomy: Theropoda: Avialae: Enantiornithes: Bohaiornithes

cf. Aves: Pygostylia: Ornithothoraces: Enantiornithes (*1)

[References: (*1) Zhang Z, Chiappe LM, Han G, Chinsamy A (2013). A large bird from the Early Cretaceous of China: new information on the skull of Enantiornithines. *Journal of Vertebrate Paleontology* 33(5): 1176–1189.]

Zhuchengceratops inexpectus

> *Zhuchengceratops inexpectus* Xu *et al.*, 2010 [Campanian; Xingezhuang Formation, Wangshi Group, Zhucheng, Shandong, China]

Generic name: *Zhuchengceratops* ← {(Chin. place-name) Zhūchéng諸城 + (Latinized Gr.) ceratops [(Gr.) κέρας/keras: horn + ὤψ/ōps: face]} (*1)

Specific name: *inexpectus* ← {(Lat.) inexspectātus: unexpected}; "referring to the unexpected discovery of an articulated skeleton in the Zhucheng bone-beds" (*1).

Etymology: unexpected **horned face (ceratopsian) from Zhucheng**

Taxonomy: Ornithischia: Ceratopsia: Leptoceratopsidae

[References: (*1) Xu X, Wang K, Zhao X, Sullivan C, Chen S (2010). A new leptoceratopsid (Ornithischia: Ceratopsia) from the Upper Cretaceous of Shandong, China and its implications for neoceratopsian evolution. *PLoS ONE* 5(11): e13835.]

Zhuchengosaurus maximus

> *Shantungosaurus giganteus* Hu, 1973 (*1) [Campanian; Xingezhuang Formation, Shandong, China]

= *Zhuchengosaurus maximus* Zhao *et al.*, 2007 [Campanian; Xingezhuang Formation, Shandong, China]

= *Huaxiaosaurus aigahtens* Zhao *et al.*, 2011 [Campanian; Xingezhuang Formation, Shandong, China]

Generic name: *Zhuchengosaurus* ← {(Chin. place-name) Zhūchéng諸城 + (Gr.) σαῦρος/sauros: lizard}

Specific name: *maximus* (Lat.) māximus (= sup. of "māgnus": large, great)

Etymology: largest **Zhucheng lizard**

Taxonomy: Ornithischia: Ornithopoda: Iguanodontia: Hadrosauridae: Saurolophinae

Notes: *Zhuchengosaurus maximus* and *Huaxiaosaurus aigahtens* are considered to be synonymic with *Shantungosaurus giganteus* Hu, 1973.

[References: (*1) Ji Y, Wang X, Liu Y, Ji Q (2011). Systematics, behavior and living environment of *Shantungosaurus giganteus* (Dinosauria: Hadrosauridae). *Acta Geologica Sinica* 85(1): 58–65.]

Zhuchengtitan zangjiazhuangensis

> *Zhuchengtitan zangjiazhuangensis* Mo *et al.*, 2017 [Campanian; Xingezhuang Formation, Zhucheng, Shandong, China]

Generic name: *Zhuchengtitan* ← {(Chin. place-name) Zhūchéng諸城 + (Gr.) Τῑτάν/Tītan}

Specific name: *zangjiazhuangensis* ← {(Chin. place-

name) Zāngjiāzhuāng臧家莊 + -ensis}
Etymology: **Zhucheng titan** from Zangjiazhuang [臧家莊諸城巨龍]
Taxonomy: Saurischia: Sauropodomorpha:
Sauropoda: Titanosauria: Saltasauridae
[References: Mo J, Wang K, Chen S, Wang P, Xu X (2017). A new titanosaurian sauropod from the Late Cretaceous strata of Shan-dong Province. *Geological Bulletin of China* 36(9): 1501–1505.]

Zhuchengtyrannus magnus

> *Zhuchengtyrannus magnus* Hone *et al.*, 2011 [Campanian; Xingezhuang Formation, Shandong, China]
= *Tyrannosaurus zhuchengensis*? Hu *et al.*, 2001 [Campanian; Xingezhuang Formation, Shandong, China]

Generic name: *Zhuchengtyrannus* ← {(Chin. place-name) Zhūchéng諸城 + (Lat.) tyrannus = (Gr.) τύραννος/tyrannos: king, tyrant}
Specific name: *magnus* ← {(Lat.) māgnus: great}; in reference to the size of the animal. *Zhuchengtyrannus* was a "large theropod, comparable in size to both *Tarbosaurus* and *Tyrannosaurus*" (*1).
Etymology: great **tyrant from Zhucheng**
Taxonomy: Theropoda: Coelurosauria:
Tyrannosauridae: Tyrannosaurinae
[References: (*1) Hone DWE, Wang K, Sullivan C, Zhao X, Chen S, Li D, Ji S, Ji Q, Xu X (2011). A new, large tyrannosaurine theropod from the Upper Cretaceous of China. *Cretaceous Research* 32(4): 495–503.]

Ziapelta sanjuanensis

> *Ziapelta sanjuanensis* Arbour *et al.*, 2014 [late Campanian; De-na-zin Member, Kirtland Formation, San Juan Basin, New Mexico, USA]

Generic name: *Ziapelta* ← {(demonym) the Zia + (Lat.) pelta: a small shield}; "Zia, referring to the Zia sun symbol, a stylized sun with four groups of rays, having religious significance to the Zia people of New Mexico and the iconic symbol on the state flag of New Mexico; pelta, in reference to the osteoderms found on all ankylosaurids" (*1).
Specific name: *sanjuanensis* ← {(place-name) San Juan + -ensis}; "in reference to San Juan County and the structural basin from which the specimen was derived" (*1).
Etymology: **Zia shield** from San Juan
Taxonomy: Ornithischia: Thyreophora:
Ankylosauria: Ankylosauridae
Notes: The body of *Ziapelta* was protected by osteoderms. A phylogenetic analysis allies *Ziapeta* to the northern North American ankylosaurids *Ankylosaurus, Anodontosaurus, Euoplocephalus, Dyoplosaurus,* and *Scolosaurus* (*1).
[References: (*1) Arbour VM, Burns ME, Sullivan RM, Lucas SG, Cantrell AK, Fry J, Suazo TL (2014). A new ankylosaurid dinosaur from the Upper Cretaceous (Kirtlandian) of New Mexico with implications for ankylosaurid diversity in the Upper Cretaceous of western North America. *PLoS ONE* 9(9): e108804.]

Zigongosaurus fuxiensis

> *Zigongosaurus fuxiensis* Hou *et al.*, 1976 ⇒ *Mamenchisaurus fuxiensis* (Hou *et al.*, 1976) Zhang & Chen, 1996 [Bathonian–Callovian; lower Shaximiao Formation, Sichuan, China]

Generic name: *Zigongosaurus* ← {(Chin. place-name) Zìgòng自貢 + (Gr.) σαῦρος/sauros: lizard}
Specific name: *fuxiensis* ← {(Chin. place-name) Fŭxī釜溪 + -ensis}
Etymology: **Zigong lizard** from Fuxi [釜溪自貢龍]
Taxonomy: Saurischia: Sauropodomorpha:
Eusauropoda: Mamenchisauridae
Notes: According to *The Dinosauria 2nd edition* (2004), *Zigongosaurus fuxiensis* is a synonym of *Mamenchisaurus fuxiensis.* However, some sources assigned it to *Omeisaurus*, and some to own genus.

Zizhongosaurus chuanchengensis

> *Zizhongosaurus chuanchengensis* Dong *et al.*, 1983 [Aalenian; Daanzhai Member, Ziliujing Formation, Zizhong, Sichuan, China]

Generic name: *Zizhongosaurus* ← {(Chin. place-name) Zīzhōng 資中 + σαῦρος/sauros: reptile} (*1)
Specific name: *chuanchengensis* ← {(Chin. place-name) Chuánchéng船城: a local town + -ensis} ; "in reference to Chuancheng (translated as "Boat City") which is on a small mountain named Yuezhongloushan that resembles a boat, and as such the people of the municipality of Zizhong refer to it as Chuancheng" (*1).
Etymology: **Zizhong reptile** from Chuancheng
Taxonomy: Saurischia: Sauropodomorpha:
Sauropoda: Vulcanodontidae
Notes: According to *The Dinosauria 2nd edition*, *Zizhongosaurus* is considered to be a *nomen dubium*.
[References: (*1) Dong Z, Zhou S, Zhang Y (1983). Dinosaurs from the Jurassic of Sichuan. *Palaeontologica Sinica, New Series C* 162(23): 1–136.]

Zuniceratops christopheri

> *Zuniceratops christopheri* Wolfe & Kirkland, 1998 [Turonian; lower Moreno Hill Formation, New Mexico, USA]

Generic name: *Zuniceratops* ← {((demonym) the

Zuni + ceratops [(Gr.) κέρας/keras: horn + ὤψ/ ōps: face]}; "in honor of the Zuni people, whose ancestral homelands include the region where the specimens were discovered" (*1).

Specific name: *christopheri* ← {(person's name) Christopher + -ī}; "referring to Christopher James Wolfe, co-discoverer of the holotype" (*1).

Etymology: Christopher's **Zuni horned-face (ceratopsian)**

Taxonomy: Ornithischia: Ceratopsia: Neoceratopsia

[References: (*1) Wolfe DG, Kirkland JI (1998). *Zuniceratops christopheri* n. gen. & n. sp., a ceratopsian dinosaur from the Moreno Hill Formation (Cretaceous, Turonian) of west-central New Mexico. *Lower and Middle Cretaceous Terrestrial Ecosystems, New Mexico Museum of Natural History and Science Bulletin* 24: 307–317.]

Zuolong salleei

> *Zuolong salleei* Choiniere *et al.*, 2010 [Oxfordian; upper member, Shishugou Formation, Xinjiang, China]

Generic name: *Zuolong* ← {(Chin. person's name) Zuǒ Zōngtáng左宗棠 + (Chin.) lóng龍: dragon}; "referring to General Zuǒ Zōngtáng (also known as General Tso), who conquered portions of Xinjiang during the Qing dynasty" (*1).

Specific name: *salleei* ← {(person's name) Sallee + -ī}; in honor of "Hilmar Sallee, whose bequest partially funded excavations at Wucaiwan" (*1).

Etymology: Sallee and **Zuo's dragon**

Taxonomy: Theropoda: Tetanurae: Coelurosauria (*1)

[References: (*1) Choiniere JN, Clark JM, Forster CA, Xu X (2010). A basal coelurosaur (Dinosauria: Theropoda) from the Late Jurassic (Oxfordian) of the Shishugou Formation in Wucaiwan, People's Republic of China. *Journal of Vertebrate Paleontology* 30(6): 1773–1796.]

Zuoyunlong huangi

> *Zuoyunlong huangi* Wang *et al.*, 2017 [Cenomanian; Zhumapu Formation, Shanxi, China]

Generic name: *Zuoyunlong* ← {(Chin. place-name) Zuǒyún左雲: Zuoyun County + (Chin.) lóng龍: dragon} (*1).

Specific name: *huangi* ← {(Chin. person's name) Huáng 黄+ -ī}; "in honor of Mr Huang, Wei-Long, who excavated the first dinosaurs in Zuoyun County and even Shanxi Province in 1957. These dinosaurs have been studied by Young (1958)" (*1).

Etymology: Huang's **dragon from Zuoyun** [黄氏左雲龍]

Taxonomy: Ornithischia: Ornithopoda: Ankylopollexia: Styracosterna: Hadrosauriformes: Hadrosauroidea

[References: (*1) Wang R-F, You H-L, Wang S-Z, Xu S-C, Yi J, Xie L-J, Jia L, Xing H (2017). A second hadrosauroid dinosaur from the early Late Cretaceous of Zuoyun, Shanxi Province, China. *Historical Biology: An International Journal of Paleobiology* 29 (1): 17–24.]

Zupaysaurus rougieri

> *Zupaysaurus rougieri* Arcucci & Coria, 2003 [Norian; Ros Colorados Formation, La Rioja, Argentina]

Generic name: *Zupaysaurus* ← {(Quechua) zupay: devil* + (Gr.) σαῦρος/sauros: lizard} (*1).
*the Incan Death God

Specific name: *rougieri* ← {(person's name) Rougier + -ī}; "in acknowledgment of Dr. Guillermo Rougier, under whose direction the specimen was found and collected" (*1).

Etymology: Rougier's **devil lizard**

Taxonomy: Theropoda: Neotheropoda
cf. Theropoda: Tetanurae (*1)

[References: (*1) Arcucci AB, Rodolfo AC (2003). A new Triassic carnivorous dinosaur from Argentina. *Ameghiniana* 40(2): 217–228.]

Zuul crurivastator

> *Zuul crurivastator* Arbour & Evans, 2017 [Campanian; Coal Ridge Member, Judith River Formation, Montana, USA]

Generic name: *Zuul* ← {(movie character) Zuul: "referring to Zuul the Gatekeeper of Gozer, a fictional monster from the 1984 film *Ghostbusters*" (*1)}.

Specific name: *crurivastator* ← {(Lat.) crus: shin, shank + vāstātor: destroyer}; in reference to the sledgehammer-like tail club" (*1).

Etymology: **Zuul**, the shin destroyer

Taxonomy: Ornithischia: Thyreophora: Ankylosauria: Ankylosauridae: Ankylosaurinae: Ankylosaurini (*1)

[References: (*1) Arbour VM, Evans DC (2017). A new ankylosaurine dinosaur from the Judith River Formation of Montana, USA, based on an exceptional skeleton with soft tissue preservation. *Royal Society Open Science* 4: 161086.]

Writer:

Mayumi Matsuda was born in Tokyo, in 1952, and grew up in Chiba, Japan. She graduated from Keio University Faculty of Letters in 1999. She is the author of the *Scientific Names of Dinosaurs and Their Etymologies* published in 2017, Hokuryukan, Tokyo. Her favorite scientific name is *Yi qi*.

Superviser:

Shin-ichi Fujiwara was born in Chiba, in 1979, and grew up in Saitama, Japan. He got a Doctor of Science in 2008 from the University of Tokyo, and works at Nagoya University Museum since 2012. He studies functional morphology on locomotor apperatus of extant and extinct tetrapods.

The Dictionary of the Etymology of Dinosaur Names

2021 年 12 月 20 日　初版発行

監修　藤原慎一
著者　松田眞由美

発行者　福田久子

発行所　株式会社 北隆館
〒153-0051 東京都目黒区上目黒3-17-8
電話03(5720)1161 振替00140-3-750
http://www.hokuryukan-ns.co.jp/
e-mail : hk-ns2@hokuryukan-ns.co.jp

印刷所　株式会社 東邦

ISBN978-4-8326-1011-8 C3545